Lecture Notes in Computer Science 8582

Commenced Publication in 1973
Founding and Former Series Editors:
Gerhard Goos, Juris Hartmanis, and Jan van Leeuwen

Editorial Board

Beniamino Murgante   Sanjay Misra
Ana Maria A.C. Rocha   Carmelo Torre
Jorge Gustavo Rocha   Maria Irene Falcão
David Taniar   Bernady O. Apduhan
Osvaldo Gervasi (Eds.)

Computational Science and Its Applications – ICCSA 2014

14th International Conference
Guimarães, Portugal, June 30 – July 3, 2014
Proceedings, Part IV

 Springer

Volume Editors

Beniamino Murgante, University of Basilicata, Potenza, Italy
E-mail: beniamino.murgante@unibas.it

Sanjay Misra, Covenant University, Ota, Nigeria
E-mail: sanjay.misra@covenantuniversity.edu.ng

Ana Maria A.C. Rocha, University of Minho, Braga, Portugal
E-mail: arocha@dps.uminho.pt

Carmelo Torre, Politecnico di Bari, Bari, Italy
E-mail: torre@poliba.it

Jorge Gustavo Rocha, University of Minho, Braga, Portugal
E-mail: jgr@di.uminho.pt

Maria Irene Falcão, University of Minho, Braga, Portugal
E-mail: mif@math.uminho.pt

David Taniar, Monash University, Clayton, VIC, Australia
E-mail: david.taniar@infotech.monash.edu.au

Bernady O. Apduhan, Kyushu Sangyo University, Fukuoka, Japan
E-mail: bob@is.kyusan-u.ac.jp

Osvaldo Gervasi, University of Perugia, Perugia, Italy
E-mail: osvaldo.gervasi@unipg.it

ISSN 0302-9743   e-ISSN 1611-3349
ISBN 978-3-319-09146-4   e-ISBN 978-3-319-09147-1
DOI 10.1007/978-3-319-09147-1
Springer Cham Heidelberg New York Dordrecht London

Library of Congress Control Number: 2014942987

LNCS Sublibrary: SL 1 – Theoretical Computer Science and General Issues

Typesetting: Camera-ready by author, data conversion by Scientific Publishing Services, Chennai, India

Printed on acid-free paper

Springer is part of Springer Science+Business Media (www.springer.com)

Welcome Message

On behalf of the Local Organizing Committee of ICCSA 2014, it is a pleasure to welcome you to the 14th International Conference on Computational Science and Its Applications, held during June 30 – July 3, 2014. We are very proud and grateful to the ICCSA general chairs for having entrusted us with the task of organizing another event of this series of very successful conferences.

ICCSA will take place in the School of Engineering of University of Minho, which is located in close vicinity to the medieval city centre of Guimarães, a UNESCO World Heritage Site, in Northern Portugal. The historical city of Guimarães is recognized for its beauty and historical monuments. The dynamic and colorful Minho Region is famous for its landscape, gastronomy and vineyards where the unique *Vinho Verde* wine is produced.

The University of Minho is currently among the most prestigious institutions of higher education in Portugal and offers an excellent setting for the conference. Founded in 1973, the University has two major poles: the campus of Gualtar in Braga, and the campus of Azurém in Guimarães.

Plenary lectures by leading scientists and several workshops will provide a real opportunity to discuss new issues and find advanced solutions able to shape new trends in computational science.

Apart from the scientific program, a stimulant and diverse social program will be available. There will be a welcome drink at Instituto de Design, located in an old Tannery, that is an open knowledge centre and a privileged communication platform between industry and academia. Guided visits to the city of Guimarães and Porto are planned, both with beautiful and historical monuments. A guided tour and tasting in Porto wine cellars, is also planned. There will be a gala dinner at the Pousada de Santa Marinha, which is an old Augustinian convent of the 12th century refurbished, where ICCSA participants can enjoy delicious dishes and enjoy a wonderful view over the city of Guimarães.

The conference could not have happened without the dedicated work of many volunteers, recognized by the coloured shirts. We would like to thank all the collaborators, who worked hard to produce a successful ICCSA 2014, namely Irene Falcão and Maribel Santos above all, our fellow members of the local organization.

On behalf of the Local Organizing Committee of ICCSA 2014, it is our honor to cordially welcome all of you to the beautiful city of Guimarães for this unique event. Your participation and contribution to this conference will make it much more productive and successful.

We are looking forward to see you in Guimarães.

Sincerely yours,

Ana Maria A.C. Rocha
Jorge Gustavo Rocha

Preface

These 6 volumes (LNCS volumes 8579-8584) consist of the peer-reviewed papers from the 2014 International Conference on Computational Science and Its Applications (ICCSA 2014) held in Guimarães, Portugal during 30 June – 3 July 2014.

ICCSA 2014 was a successful event in the International Conferences on Computational Science and Its Applications (ICCSA) conference series, previously held in Ho Chi Minh City, Vietnam (2013), Salvador da Bahia, Brazil (2012), Santander, Spain (2011), Fukuoka, Japan (2010), Suwon, South Korea (2009), Perugia, Italy (2008), Kuala Lumpur, Malaysia (2007), Glasgow, UK (2006), Singapore (2005), Assisi, Italy (2004), Montreal, Canada (2003), and (as ICCS) Amsterdam, The Netherlands (2002) and San Francisco, USA (2001).

Computational science is a main pillar of most of the present research, industrial and commercial activities and plays a unique role in exploiting ICT innovative technologies, and the ICCSA conference series has been providing a venue for researchers and industry practitioners to discuss new ideas, to share complex problems and their solutions, and to shape new trends in computational science.

Apart from the general track, ICCSA 2014 also included 30 workshops, in various areas of computational sciences, ranging from computational science technologies, to specific areas of computational sciences, such as computational geometry and security. We accepted 58 papers for the general track, and 289 in workshops. We would like to show our appreciation to the workshops chairs and co-chairs.

The success of the ICCSA conference series, in general, and ICCSA 2014, in particular, was due to the support of many people: authors, presenters, participants, keynote speakers, workshop chairs, Organizing Committee members, student volunteers, Program Committee members, Advisory Committee members, international liaison chairs, and people in other various roles. We would like to thank them all.

We also thank our publisher, Springer–Verlag, for their acceptance to publish the proceedings and for their kind assistance and cooperation during the editing process.

We cordially invite you to visit the ICCSA website http://www.iccsa.org where you can find all relevant information about this interesting and exciting event.

June 2014

Osvaldo Gervasi
Jorge Gustavo Rocha
Bernady O. Apduhan

Organization

ICCSA 2014 was organized by University of Minho, (Portugal) University of Perugia (Italy), University of Basilicata (Italy), Monash University (Australia), Kyushu Sangyo University (Japan).

Honorary General Chairs

Antonio M. Cunha	Rector of the University of Minho, Portugal
Antonio Laganà	University of Perugia, Italy
Norio Shiratori	Tohoku University, Japan
Kenneth C. J. Tan	Qontix, UK

General Chairs

Beniamino Murgante	University of Basilicata, Italy
Ana Maria A.C. Rocha	University of Minho, Portugal
David Taniar	Monash University, Australia

Program Committee Chairs

Osvaldo Gervasi	University of Perugia, Italy
Bernady O. Apduhan	Kyushu Sangyo University, Japan
Jorge Gustavo Rocha	University of Minho, Portugal

International Advisory Committee

Jemal Abawajy	Daekin University, Australia
Dharma P. Agrawal	University of Cincinnati, USA
Claudia Bauzer Medeiros	University of Campinas, Brazil
Manfred M. Fisher	Vienna University of Economics and Business, Austria
Yee Leung	Chinese University of Hong Kong, China

International Liaison Chairs

Ana Carla P. Bitencourt	Universidade Federal do Reconcavo da Bahia, Brazil
Claudia Bauzer Medeiros	University of Campinas, Brazil
Alfredo Cuzzocrea	ICAR-CNR and University of Calabria, Italy

Marina L. Gavrilova   University of Calgary, Canada
Robert C. H. Hsu   Chung Hua University, Taiwan
Andrés Iglesias   University of Cantabria, Spain
Tai-Hoon Kim   Hannam University, Korea
Sanjay Misra   University of Minna, Nigeria
Takashi Naka   Kyushu Sangyo University, Japan
Rafael D.C. Santos   National Institute for Space Research, Brazil

Workshop and Session Organizing Chairs

Beniamino Murgante   University of Basilicata, Italy

Local Organizing Committee

Ana Maria A.C. Rocha   University of Minho, Portugal (Chair)
Jorge Gustavo Rocha   University of Minho, Portugal
Maria Irene Falcão   University of Minho, Portugal
Maribel Yasmina Santos   University of Minho, Portugal

Workshop Organizers

Advances in Complex Systems: Modeling and Parallel Implementation (ACSModPar 2014)

Georgius Sirakoulis   Democritus University of Thrace, Greece
Wiliam Spataro   University of Calabria, Italy
Giuseppe A. Trunfio   University of Sassari, Italy

Agricultural and Environment Information and Decision Support Systems (AEIDSS 2014)

Sandro Bimonte   IRSTEA France
Florence Le Ber   ENGES, France
André Miralles   IRSTEA France
François Pinet   IRSTEA France

Advances in Web Based Learning (AWBL 2014)

Mustafa Murat Inceoglu   Ege University, Turkey

Bio-inspired Computing and Applications (BIOCA 2014)

Nadia Nedjah   State University of Rio de Janeiro, Brazil
Luiza de Macedo Mourell   State University of Rio de Janeiro, Brazil

Computational and Applied Mathematics (CAM 2014)

Maria Irene Falcao   University of Minho, Portugal
Fernando Miranda   University of Minho, Portugal

Computer Aided Modeling, Simulation, and Analysis (CAMSA 2014)

Jie Shen   University of Michigan, USA

Computational and Applied Statistics (CAS 2014)

Ana Cristina Braga   University of Minho, Portugal
Ana Paula Costa Conceicao
   Amorim   University of Minho, Portugal

Computational Geometry and Security Applications (CGSA 2014)

Marina L. Gavrilova   University of Calgary, Canada
Han Ming Huang   Guangxi Normal University, China

Computational Algorithms and Sustainable Assessment (CLASS 2014)

Antonino Marvuglia   Public Research Centre Henri Tudor,
                                         Luxembourg
Beniamino Murgante   University of Basilicata, Italy

Chemistry and Materials Sciences and Technologies (CMST 2014)

Antonio Laganà   University of Perugia, Italy

Computational Optimization and Applications (COA 2014)

Ana Maria A.C. Rocha   University of Minho, Portugal
Humberto Rocha   University of Coimbra, Portugal

Cities, Technologies and Planning (CTP 2014)

Giuseppe Borruso   University of Trieste, Italy
Beniamino Murgante   University of Basilicata, Italy

Computational Tools and Techniques for Citizen Science and Scientific Outreach (CTTCS 2014)

Rafael Santos   National Institute for Space Research, Brazil
Jordan Raddickand   Johns Hopkins University, USA
Ani Thakar   Johns Hopkins University, USA

Econometrics and Multidimensional Evaluation in the Urban Environment (EMEUE 2014)

Carmelo M. Torre   Polytechnic of Bari, Italy
Maria Cerreta   University of Naples Federico II, Italy
Paola Perchinunno   University of Bari, Italy
Simona Panaro   University of Naples Federico II, Italy
Raffaele Attardi   University of Naples Federico II, Italy

Future Computing Systems, Technologies, and Applications (FISTA 2014)

Bernady O. Apduhan   Kyushu Sangyo University, Japan
Rafael Santos   National Institute for Space Research, Brazil
Jianhua Ma   Hosei University, Japan
Qun Jin   Waseda University, Japan

Formal Methods, Computational Intelligence and Constraint Programming for Software Assurance (FMCICA 2014)

Valdivino Santiago Junior   National Institute for Space Research
                             (INPE), Brazil

Geographical Analysis, Urban Modeling, Spatial Statistics (GEOG-AN-MOD 2014)

Giuseppe Borruso   University of Trieste, Italy
Beniamino Murgante   University of Basilicata, Italy
Hartmut Asche   University of Potsdam, Germany

High Performance Computing in Engineering and Science (HPCES 2014)

Alberto Proenca   University of Minho, Portugal
Pedro Alberto   University of Coimbra, Portugal

Mobile Communications (MC 2014)

Hyunseung Choo   Sungkyunkwan University, Korea

Mobile Computing, Sensing, and Actuation for Cyber Physical Systems (MSA4CPS 2014)

Saad Qaisar   NUST School of Electrical Engineering and
                            Computer Science, Pakistan
Moonseong Kim   Korean Intellectual Property Office, Korea

New Trends on Trust Computational Models (NTTCM 2014)

Rui Costa Cardoso   Universidade da Beira Interior, Portugal
Abel Gomez   Universidade da Beira Interior, Portugal

Quantum Mechanics: Computational Strategies and Applications (QMCSA 2014)

Mirco Ragni   Universidad Federal de Bahia, Brazil
Vincenzo Aquilanti   University of Perugia, Italy
Ana Carla Peixoto Bitencourt Universidade Estadual de Feira de Santana
                            Brazil
Roger Anderson   University of California, USA
Frederico Vasconcellos Prudente Universidade Federal de Bahia, Brazil

Remote Sensing Data Analysis, Modeling, Interpretation and Applications: From a Global View to a Local Analysis (RS2014)

Rosa Lasaponara   Institute of Methodologies for Environmental
                                Analysis National Research Council, Italy
Nicola Masini   Archaeological and Monumental Heritage
                                Institute, National Research Council, Italy

Software Engineering Processes and Applications (SEPA 2014)

Sanjay Misra   Covenant University, Nigeria

Software Quality (SQ 2014)

Sanjay Misra   Covenant University, Nigeria

Advances in Spatio-Temporal Analytics (ST-Analytics 2014)

Joao Moura Pires   New University of Lisbon, Portugal
Maribel Yasmina Santos   New University of Lisbon, Portugal

Tools and Techniques in Software Development Processes (TTSDP 2014)

Sanjay Misra   Covenant University, Nigeria

Virtual Reality and its Applications (VRA 2014)

Osvaldo Gervasi   University of Perugia, Italy
Lucio Depaolis   University of Salento, Italy

Workshop of Agile Software Development Techniques (WAGILE 2014)

Eduardo Guerra   National Institute for Space Research, Brazil

Big Data:, Analytics and Management (WBDAM 2014)

Wenny Rahayu   La Trobe University, Australia

Program Committee

Jemal Abawajy	Daekin University, Australia
Kenny Adamson	University of Ulster, UK
Filipe Alvelos	University of Minho, Portugal
Paula Amaral	Universidade Nova de Lisboa, Portugal
Hartmut Asche	University of Potsdam, Germany
Md. Abul Kalam Azad	University of Minho, Portugal
Michela Bertolotto	University College Dublin, Ireland
Sandro Bimonte	CEMAGREF, TSCF, France
Rod Blais	University of Calgary, Canada
Ivan Blecic	University of Sassari, Italy
Giuseppe Borruso	University of Trieste, Italy
Yves Caniou	Lyon University, France
José A. Cardoso e Cunha	Universidade Nova de Lisboa, Portugal
Leocadio G. Casado	University of Almeria, Spain
Carlo Cattani	University of Salerno, Italy
Mete Celik	Erciyes University, Turkey
Alexander Chemeris	National Technical University of Ukraine "KPI", Ukraine
Min Young Chung	Sungkyunkwan University, Korea
Gilberto Corso Pereira	Federal University of Bahia, Brazil
M. Fernanda Costa	University of Minho, Portugal
Gaspar Cunha	University of Minho, Portugal
Alfredo Cuzzocrea	ICAR-CNR and University of Calabria, Italy
Carla Dal Sasso Freitas	Universidade Federal do Rio Grande do Sul, Brazil
Pradesh Debba	The Council for Scientific and Industrial Research (CSIR), South Africa
Hendrik Decker	Instituto Tecnológico de Informática, Spain
Frank Devai	London South Bank University, UK
Rodolphe Devillers	Memorial University of Newfoundland, Canada
Prabu Dorairaj	NetApp, India/USA
M. Irene Falcao	University of Minho, Portugal
Cherry Liu Fang	U.S. DOE Ames Laboratory, USA
Edite M.G.P. Fernandes	University of Minho, Portugal
Jose-Jesus Fernandez	National Centre for Biotechnology, CSIS, Spain
Maria Antonia Forjaz	University of Minho, Portugal
Maria Celia Furtado Rocha	PRODEB and Universidade Federal da Bahia, Brazil
Akemi Galvez	University of Cantabria, Spain
Paulino Jose Garcia Nieto	University of Oviedo, Spain
Marina Gavrilova	University of Calgary, Canada
Jerome Gensel	LSR-IMAG, France

Maria Giaoutzi	National Technical University, Athens, Greece
Andrzej M. Goscinski	Deakin University, Australia
Alex Hagen-Zanker	University of Cambridge, UK
Malgorzata Hanzl	Technical University of Lodz, Poland
Shanmugasundaram Hariharan	B.S. Abdur Rahman University, India
Eligius M.T. Hendrix	University of Malaga/Wageningen University, Spain/Netherlands
Hisamoto Hiyoshi	Gunma University, Japan
Fermin Huarte	University of Barcelona, Spain
Andres Iglesias	University of Cantabria, Spain
Mustafa Inceoglu	EGE University, Turkey
Peter Jimack	University of Leeds, UK
Qun Jin	Waseda University, Japan
Farid Karimipour	Vienna University of Technology, Austria
Baris Kazar	Oracle Corp., USA
DongSeong Kim	University of Canterbury, New Zealand
Taihoon Kim	Hannam University, Korea
Ivana Kolingerova	University of West Bohemia, Czech Republic
Dieter Kranzlmueller	LMU and LRZ Munich, Germany
Antonio Laganà	University of Perugia, Italy
Rosa Lasaponara	National Research Council, Italy
Maurizio Lazzari	National Research Council, Italy
Cheng Siong Lee	Monash University, Australia
Sangyoun Lee	Yonsei University, Korea
Jongchan Lee	Kunsan National University, Korea
Clement Leung	Hong Kong Baptist University, Hong Kong
Chendong Li	University of Connecticut, USA
Gang Li	Deakin University, Australia
Ming Li	East China Normal University, China
Fang Liu	AMES Laboratories, USA
Xin Liu	University of Calgary, Canada
Savino Longo	University of Bari, Italy
Tinghuai Ma	NanJing University of Information Science and Technology, China
Sergio Maffioletti	University of Zurich, Switzerland
Ernesto Marcheggiani	Katholieke Universiteit Leuven, Belgium
Antonino Marvuglia	Research Centre Henri Tudor, Luxembourg
Nicola Masini	National Research Council, Italy
Nirvana Meratnia	University of Twente, The Netherlands
Alfredo Milani	University of Perugia, Italy
Sanjay Misra	Federal University of Technology Minna, Nigeria
Giuseppe Modica	University of Reggio Calabria, Italy

Mario Valle   Swiss National Supercomputing Centre,
                                 Switzerland
Pablo Vanegas   University of Cuenca, Equador
Piero Giorgio Verdini   INFN Pisa and CERN, Italy
Marco Vizzari   University of Perugia, Italy
Koichi Wada   University of Tsukuba, Japan
Krzysztof Walkowiak   Wroclaw University of Technology, Poland
Robert Weibel   University of Zurich, Switzerland
Roland Wismüller   Universität Siegen, Germany
Mudasser Wyne   SOET National University, USA
Chung-Huang Yang   National Kaohsiung Normal University, Taiwan
Xin-She Yang   National Physical Laboratory, UK
Salim Zabir   France Telecom Japan Co., Japan
Haifeng Zhao   University of California at Davis, USA
Kewen Zhao   University of Qiongzhou, China
Albert Y. Zomaya   University of Sydney, Australia

Reviewers

Abdi Samane   University College Cork, Ireland
Aceto Lidia   University of Pisa, Italy
Afonso Ana Paula   University of Lisbon, Portugal
Afreixo Vera   University of Aveiro, Portugal
Aguilar Antonio   University of Barcelona, Spain
Aguilar José Alfonso   Universidad Autónoma de Sinaloa, Mexico
Ahmad Waseem   Federal University of Technology Minna,
                                 Nigeria
Aktas Mehmet   Yildiz Technical University, Turkey
Alarcon Vladimir   Universidad Diego Portales, Chile
Alberti Margarita   University of Barcelona, Spain
Ali Salman   NUST, Pakistan
Alvanides Seraphim   Northumbria University, UK
Álvarez Jacobo de Uña   University of Vigo, Spain
Alvelos Filipe   University of Minho, Portugal
Alves Cláudio   University of Minho, Portugal
Alves José Luis   University of Minho, Portugal
Amorim Ana Paula   University of Minho, Portugal
Amorim Paulo   Federal University of Rio de Janeiro, Brazil
Anderson Roger   University of California, USA
Andrade Wilkerson   Federal University of Campina Grande, Brazil
Andrienko Gennady   Fraunhofer Institute for Intelligent Analysis
                                 and Informations Systems, Germany
Apduhan Bernady   Kyushu Sangyo University, Japan
Aquilanti Vincenzo   University of Perugia, Italy
Argiolas Michele   University of Cagliari, Italy

Athayde Maria Emília Feijão Queiroz	University of Minho, Portugal
Attardi Raffaele	University of Napoli Federico II, Italy
Azad Md Abdul	Indian Institute of Technology Kanpur, India
Badard Thierry	Laval University, Canada
Bae Ihn-Han	Catholic University of Daegu, South Korea
Baioletti Marco	University of Perugia, Italy
Balena Pasquale	Polytechnic of Bari, Italy
Balucani Nadia	University of Perugia, Italy
Barbosa Jorge	University of Porto, Portugal
Barrientos Pablo Andres	Universidad Nacional de La Plata, Australia
Bartoli Daniele	University of Perugia, Italy
Bação Fernando	New University of Lisbon, Portugal
Belanzoni Paola	University of Perugia, Italy
Bencardino Massimiliano	University of Salerno, Italy
Benigni Gladys	University of Oriente, Venezuela
Bertolotto Michela	University College Dublin, Ireland
Bimonte Sandro	IRSTEA, France
Blanquer Ignacio	Universitat Politècnica de València, Spain
Bollini Letizia	University of Milano, Italy
Bonifazi Alessandro	Polytechnic of Bari, Italy
Borruso Giuseppe	University of Trieste, Italy
Bostenaru Maria	"Ion Mincu" University of Architecture and Urbanism, Romania
Boucelma Omar	University Marseille, France
Braga Ana Cristina	University of Minho, Portugal
Brás Carmo	Universidade Nova de Lisboa, Portugal
Cacao Isabel	University of Aveiro, Portugal
Cadarso-Suárez Carmen	University of Santiago de Compostela, Spain
Caiaffa Emanuela	ENEA, Italy
Calamita Giuseppe	National Research Council, Italy
Campagna Michele	University of Cagliari, Italy
Campobasso Francesco	University of Bari, Italy
Campos José	University of Minho, Portugal
Cannatella Daniele	University of Napoli Federico II, Italy
Canora Filomena	University of Basilicata, Italy
Cardoso Rui	Institute of Telecommunications, Portugal
Caschili Simone	University College London, UK
Ceppi Claudia	Polytechnic of Bari, Italy
Cerreta Maria	University Federico II of Naples, Italy
Chanet Jean-Pierre	IRSTEA, France
Chao Wang	University of Science and Technology of China, China
Choi Joonsoo	Kookmin University, South Korea

Choo Hyunseung	Sungkyunkwan University, South Korea
Chung Min Young	Sungkyunkwan University, South Korea
Chung Myoungbeom	Sungkyunkwan University, South Korea
Clementini Eliseo	University of L'Aquila, Italy
Coelho Leandro dos Santos	PUC-PR, Brazil
Colado Anibal Zaldivar	Universidad Autónoma de Sinaloa, Mexico
Coletti Cecilia	University of Chieti, Italy
Condori Nelly	VU University Amsterdam, The Netherlands
Correia Elisete	University of Trás-Os-Montes e Alto Douro, Portugal
Correia Filipe	FEUP, Portugal
Correia Florbela Maria da Cruz Domingues	Instituto Politécnico de Viana do Castelo, Portugal
Correia Ramos Carlos	University of Evora, Portugal
Corso Pereira Gilberto	UFPA, Brazil
Cortés Ana	Universitat Autònoma de Barcelona, Spain
Costa Fernanda	University of Minho, Portugal
Costantini Alessandro	INFN, Italy
Crasso Marco	National Scientific and Technical Research Council, Argentina
Crawford Broderick	Universidad Catolica de Valparaiso, Chile
Cristia Maximiliano	CIFASIS and UNR, Argentina
Cunha Gaspar	University of Minho, Portugal
Cunha Jácome	University of Minho, Portugal
Cutini Valerio	University of Pisa, Italy
Danese Maria	IBAM, CNR, Italy
Da Silva B. Carlos	University of Lisboa, Portugal
De Almeida Regina	University of Trás-os-Montes e Alto Douro, Portugal
Debroy Vidroha	Hudson Alley Software Inc., USA
De Fino Mariella	Polytechnic of Bari, Italy
De Lotto Roberto	University of Pavia, Italy
De Paolis Lucio Tommaso	University of Salento, Italy
De Rosa Fortuna	University of Napoli Federico II, Italy
De Toro Pasquale	University of Napoli Federico II, Italy
Decker Hendrik	Instituto Tecnológico de Informática, Spain
Delamé Thomas	CNRS, France
Demyanov Vasily	Heriot-Watt University, UK
Desjardin Eric	University of Reims, France
Dwivedi Sanjay Kumar	Babasaheb Bhimrao Ambedkar University, India
Di Gangi Massimo	University of Messina, Italy
Di Leo Margherita	JRC, European Commission, Belgium

Di Trani Francesco	University of Basilicata, Italy
Dias Joana	University of Coimbra, Portugal
Dias d'Almeida Filomena	University of Porto, Portugal
Dilo Arta	University of Twente, The Netherlands
Dixit Veersain	Delhi University, India
Doan Anh Vu	Université Libre de Bruxelles, Belgium
Dorazio Laurent	ISIMA, France
Dutra Inês	University of Porto, Portugal
Eichelberger Hanno	University of Tuebingen, Germany
El-Zawawy Mohamed A.	Cairo University, Egypt
Escalona Maria-Jose	University of Seville, Spain
Falcão M. Irene	University of Minho, Portugal
Farantos Stavros	University of Crete and FORTH, Greece
Faria Susana	University of Minho, Portugal
Faruq Fatma	Carnegie Melon University,, USA
Fernandes Edite	University of Minho, Portugal
Fernandes Rosário	University of Minho, Portugal
Fernandez Joao P	Universidade da Beira Interior, Portugal
Ferreira Fátima	University of Trás-Os-Montes e Alto Douro, Portugal
Ferrão Maria	University of Beira Interior and CEMAPRE, Portugal
Figueiredo Manuel Carlos	University of Minho, Portugal
Filipe Ana	University of Minho, Portugal
Flouvat Frederic	University New Caledonia, New Caledonia
Forjaz Maria Antónia	University of Minho, Portugal
Formosa Saviour	University of Malta, Malta
Fort Marta	University of Girona, Spain
Franciosa Alfredo	University of Napoli Federico II, Italy
Freitas Adelaide de Fátima Baptista Valente	University of Aveiro, Portugal
Frydman Claudia	Laboratoire des Sciences de l'Information et des Systèmes, France
Fusco Giovanni	CNRS - UMR ESPACE, France
Fussel Donald	University of Texas at Austin, USA
Gao Shang	Zhongnan University of Economics and Law, China
Garcia Ernesto	University of the Basque Country, Spain
Garcia Tobio Javier	Centro de Supercomputación de Galicia (CESGA), Spain
Gavrilova Marina	University of Calgary, Canada
Gensel Jerome	IMAG, France
Geraldi Edoardo	National Research Council, Italy
Gervasi Osvaldo	University of Perugia, Italy

Giaoutzi Maria	National Technical University Athens, Greece
Gizzi Fabrizio	National Research Council, Italy
Gomes Maria Cecilia	Universidade Nova de Lisboa, Portugal
Gomes dos Anjos Eudisley	Federal University of ParaÃba, Brazil
Gomez Andres	Centro de Supercomputación de Galicia, CESGA (Spain)
Gonçalves Arminda Manuela	University of Minho, Portugal
Gravagnuolo Antonia	University of Napoli Federico II, Italy
Gregori M. M. H. Rodrigo	Universidade Tecnológica Federal do Paraná, Brazil
Guerlebeck Klaus	Bauhaus University Weimar, Germany
Guerra Eduardo	National Institute for Space Research, Brazil
Hagen-Zanker Alex	University of Surrey, UK
Hajou Ali	Utrecht University, The Netherlands
Hanzl Malgorzata	University of Lodz, Poland
Heijungs Reinout	VU University Amsterdam, The Netherlands
Henriques Carla	Escola Superior de Tecnologia e Gestão, Portugal
Herawan Tutut	University of Malaya, Malaysia
Iglesias Andres	University of Cantabria, Spain
Jamal Amna	National University of Singapore, Singapore
Jank Gerhard	Aachen University, Germany
Jiang Bin	University of Gävle, Sweden
Kalogirou Stamatis	Harokopio University of Athens, Greece
Kanevski Mikhail	University of Lausanne, Switzerland
Kartsaklis Christos	Oak Ridge National Laboratory, USA
Kavouras Marinos	National Technical University of Athens, Greece
Khan Murtaza	NUST, Pakistan
Khurshid Khawar	NUST, Pakistan
Kim Deok-Soo	Hanyang University, South Korea
Kim Moonseong	KIPO, South Korea
Kolingerova Ivana	University of West Bohemia, Czech Republic
Kotzinos Dimitrios	Université de Cergy-Pontoise, France
Lazzari Maurizio	CNR IBAM, Italy
Laganà Antonio	Department of Chemistry, Biology and Biotechnology, Italy
Lai Sabrina	University of Cagliari, Italy
Lanorte Antonio	CNR-IMAA, Italy
Lanza Viviana	Lombardy Regional Institute for Research, Italy
Le Duc Tai	Sungkyunkwan University, South Korea
Le Duc Thang	Sungkyunkwan University, South Korea
Lee Junghoon	Jeju National University, South Korea

Lee KangWoo	Sungkyunkwan University, South Korea
Legatiuk Dmitrii	Bauhaus University Weimar, Germany
Leonard Kathryn	California State University, USA
Lin Calvin	University of Texas at Austin, USA
Loconte Pierangela	Technical University of Bari, Italy
Lombardi Andrea	University of Perugia, Italy
Lopez Cabido Ignacio	Centro de Supercomputación de Galicia, CESGA
Lourenço Vanda Marisa	University Nova de Lisboa, Portugal
Luaces Miguel	University of A Coruña, Spain
Lucertini Giulia	IUAV, Italy
Luna Esteban Robles	Universidad Nacional de la Plata, Argentina
Machado Gaspar	University of Minho, Portugal
Magni Riccardo	Pragma Engineering SrL, Italy, Italy
Malonek Helmuth	University of Aveiro, Portugal
Manfreda Salvatore	University of Basilicata, Italy
Manso Callejo Miguel Angel	Universidad Politécnica de Madrid, Spain
Marcheggiani Ernesto	KU Lueven, Belgium
Marechal Bernard	Universidade Federal de Rio de Janeiro, Brazil
Margalef Tomas	Universitat Autònoma de Barcelona, Spain
Martellozzo Federico	University of Rome, Italy
Marvuglia Antonino	Public Research Centre Henri Tudor, Luxembourg
Matos Jose	Instituto Politecnico do Porto, Portugal
Mauro Giovanni	University of Trieste, Italy
Mauw Sjouke	University of Luxembourg, Luxembourg
Medeiros Pedro	Universidade Nova de Lisboa, Portugal
Melle Franco Manuel	University of Minho, Portugal
Melo Ana	Universidade de São Paulo, Brazil
Millo Giovanni	Generali Assicurazioni, Italy
Min-Woo Park	Sungkyunkwan University, South Korea
Miranda Fernando	University of Minho, Portugal
Misra Sanjay	Covenant University, Nigeria
Modica Giuseppe	Università Mediterranea di Reggio Calabria, Italy
Morais João	University of Aveiro, Portugal
Moreira Adriano	University of Minho, Portugal
Mota Alexandre	Universidade Federal de Pernambuco, Brazil
Moura Pires João	Universidade Nova de Lisboa - FCT, Portugal
Mourelle Luiza de Macedo	UERJ, Brazil
Mourão Maria	Polytechnic Institute of Viana do Castelo, Portugal
Murgante Beniamino	University of Basilicata, Italy
NM Tuan	Ho Chi Minh City University of Technology, Vietnam

Nagy Csaba	University of Szeged, Hungary
Nash Andrew	Vienna Transport Strategies, Austria
Natário Isabel Cristina Maciel	University Nova de Lisboa, Portugal
Nedjah Nadia	State University of Rio de Janeiro, Brazil
Nogueira Fernando	University of Coimbra, Portugal
Oliveira Irene	University of Trás-Os-Montes e Alto Douro, Portugal
Oliveira José A.	University of Minho, Portugal
Oliveira e Silva Luis	University of Lisboa, Portugal
Osaragi Toshihiro	Tokyo Institute of Technology, Japan
Ottomanelli Michele	Polytechnic of Bari, Italy
Ozturk Savas	TUBITAK, Turkey
Pacifici Leonardo	University of Perugia, Italy
Pages Carmen	Universidad de Alcala, Spain
Painho Marco	New University of Lisbon, Portugal
Pantazis Dimos	Technological Educational Institute of Athens, Greece
Paolotti Luisa	University of Perugia, Italy
Papa Enrica	University of Amsterdam, The Netherlands
Papathanasiou Jason	University of Macedonia, Greece
Pardede Eric	La Trobe University, Australia
Parissis Ioannis	Grenoble INP - LCIS, France
Park Gyung-Leen	Jeju National University, South Korea
Park Sooyeon	Korea Polytechnic University, South Korea
Pascale Stefania	University of Basilicata, Italy
Passaro Pierluigi	University of Bari Aldo Moro, Italy
Peixoto Bitencourt Ana Carla	Universidade Estadual de Feira de Santana, Brazil
Perchinunno Paola	University of Bari, Italy
Pereira Ana	Polytechnic Institute of Bragança, Portugal
Pereira Francisco	Instituto Superior de Engenharia, Portugal
Pereira Paulo	University of Minho, Portugal
Pereira Ricardo	Portugal Telecom Inovacao, Portugal
Pietrantuono Roberto	University of Napoli "Federico II", Italy
Pimentel Carina	University of Aveiro, Portugal
Pina Antonio	University of Minho, Portugal
Pinet Francois	IRSTEA, France
Piscitelli Claudia	Polytechnic University of Bari, Italy
Piñar Miguel	Universidad de Granada, Spain
Pollino Maurizio	ENEA, Italy
Potena Pasqualina	University of Bergamo, Italy
Prata Paula	University of Beira Interior, Portugal
Prosperi David	Florida Atlantic University, USA
Qaisar Saad	NURST, Pakistan

Quan Tho	Ho Chi Minh City University of Technology, Vietnam
Raffaeta Alessandra	University of Venice, Italy
Ragni Mirco	Universidade Estadual de Feira de Santana, Brazil
Rautenberg Carlos	University of Graz, Austria
Ravat Franck	IRIT, France
Raza Syed Muhammad	Sungkyunkwan University, South Korea
Ribeiro Isabel	University of Porto, Portugal
Ribeiro Ligia	University of Porto, Portugal
Rinzivillo Salvatore	University of Pisa, Italy
Rocha Ana Maria	University of Minho, Portugal
Rocha Humberto	University of Coimbra, Portugal
Rocha Jorge	University of Minho, Portugal
Rocha Maria Clara	ESTES Coimbra, Portugal
Rocha Maria	PRODEB, San Salvador, Brazil
Rodrigues Armanda	Universidade Nova de Lisboa, Portugal
Rodrigues Cristina	DPS, University of Minho, Portugal
Rodriguez Daniel	University of Alcala, Spain
Roh Yongwan	Korean IP, South Korea
Roncaratti Luiz	Instituto de Fisica, University of Brasilia, Brazil
Rosi Marzio	University of Perugia, Italy
Rossi Gianfranco	University of Parma, Italy
Rotondo Francesco	Polytechnic of Bari, Italy
Sannicandro Valentina	Polytechnic of Bari, Italy
Santos Maribel Yasmina	University of Minho, Portugal
Santos Rafael	INPE, Brazil
Santos Viviane	Universidade de São Paulo, Brazil
Santucci Valentino	University of Perugia, Italy
Saracino Gloria	University of Milano-Bicocca, Italy
Sarafian Haiduke	Pennsylvania State University, USA
Saraiva João	University of Minho, Portugal
Sarrazin Renaud	Université Libre de Bruxelles, Belgium
Schirone Dario Antonio	University of Bari, Italy
Schneider Michel	ISIMA, France
Schoier Gabriella	University of Trieste, Italy
Schutz Georges	CRP Henri Tudor, Luxembourg
Scorza Francesco	University of Basilicata, Italy
Selmaoui Nazha	University of New Caledonia, New Caledonia
Severino Ricardo Jose	University of Minho, Portugal
Shakhov Vladimir	Russian Academy of Sciences, Russia
Shen Jie	University of Michigan, USA
Shon Minhan	Sungkyunkwan University, South Korea

Wachowicz Monica	University of New Brunswick, Canada
Walkowiak Krzysztof	Wroclav University of Technology, Poland
Xin Liu	Ecole Polytechnique Fédérale Lausanne, Switzerland
Yadav Nikita	Delhi Universty, India
Yatskevich Mikalai	Assioma, Italy
Yeoum Sanggil	Sungkyunkwan University, South Korea
Zalyubovskiy Vyacheslav	Russian Academy of Sciences, Russia
Zunino Alejandro	Universidad Nacional del Centro, Argentina

Sponsoring Organizations

ICCSA 2014 would not have been possible without the tremendous support of many organizations and institutions, for which all organizers and participants of ICCSA 2014 express their sincere gratitude:

Universidade do Minho
Escola de Engenharia
Universidade do Minho
(http://www.uminho.pt)

UNIVERSITÀ DEGLI STUDI
DI PERUGIA

University of Perugia, Italy
(http://www.unipg.it)

University of Basilicata, Italy (http://www.unibas.it)

 Monash University, Australia
(http://monash.edu)

K*U
九州産業大学
KYUSHU SANGYO UNIVERSITY

Kyushu Sangyo University, Japan
(www.kyusan-u.ac.jp)

Associação Portuguesa de Investigação Operacional
(apdio.pt)

Table of Contents

Workshop on Geographical Analysis, Urban Modeling, Spatial Statistics (GEOG-and-MOD 2014)

Workshop on High Performance Computing in Engineering and Science (HPCES 2014)

Workshop on Mobile Communications (MC 2014)

Workshop on Mobile-Computing, Sensing, and Actuation - Cyber Physical Systems (MSA4CPS 2014)

UAV-Based Orthophoto Generation in Urban Area: The Basilica of Santa Maria di Collemaggio in L'Aquila

Luigi Barazzetti, Raffaella Brumana, Daniela Oreni,
Mattia Previtali, and Fabio Roncoroni

Politecnico di Milano, p.za Leonardo da Vinci 32
20133 - Milan, Italy
{luigi.barazzetti,raffaella.brumana,daniela.oreni,
mattia.previtali,fabio.roncoroni}@polimi.it

Abstract. This paper presents the photogrammetric pipeline behind the generation of the UAV-based orthophoto of the Basilica of Santa Maria di Collemaggio (L'Aquila, Italy). The 2009 L'Aquila earthquake caused serious damage to the basilica and a restoration work is currently in progress. A part of the research carried out by the authors was the investigation of UAV technology in urban context for supporting the surveying phase carried out with modern techniques that include total station data, laser scans, and close-range photogrammetry. The image acquisition phase by means of an UAV platform is illustrated and discussed along with the implemented algorithms used to generate an orthophoto with a flight over the whole basilica. We would like to prove that UAV technology has reached a significant level of maturity and image acquisition can be carried out in fully automated way. On the other hand, image processing software today available on the commercial market could be insufficient for accurate and detailed reconstructions. The implementation of ad-hoc algorithms is therefore mandatory to exploit the full potential and automate the photogrammetric processing workflow.

Keywords: 3D texturing, image matching, image orientation, UAV, orthophoto production, texture-mapping.

1 Introduction

The Basilica of Santa Maria di Collemaggio (Fig. 1) is the principal monument of Abruzzo architecture. Begun in 1287 on the orders of Pope Celestino V, the basilica has a splendid facade, adorned with pink and white decorations, three rose windows and three portals. On the left of the basilica is the Porta Santa, which is opened on occasion of the Festa della Perdonanza (the feast of forgiveness) in late August.

The basilica is currently being restored: Eni (www.eni.com) and the City Council embarked on the preliminary phase of technical investigations and historical research. These investigations are aimed at setting the framework required to address subsequent design and planning actions. With the signing of this agreement, Eni is committing to allocating the economic resources necessary for the realisation of the

B. Murgante et al. (Eds.): ICCSA 2014, Part IV, LNCS 8582, pp. 1–13, 2014.

project, as well as providing the technical and project management expertise necessary to complete the restoration and redevelopment. The works will focus primarily on safety, seismic improvements and the restoration of the Basilica, and will also include the redevelopment of the nearby Parco del Sole. Indeed, parts of the structure were significantly damaged in the 2009 earthquake in L'Aquila. While the church's front is intact, its cupola, transept vaults and the triumphal arches have collapsed.

One of the surveying techniques (laser scanning and photogrammetry) used by the authors to create the BIM [1-2] of the structure was UAV technology, that proved to be a powerful emerging methodology to handle the digital reconstruction in the urban area here presented.

Fig. 1. The basilica of Collemaggio from the Falcon (the metallic provision roof is visible)

Nowadays, the growing interest for these flying platforms is motivated by several new applications. Aerial surveys allow acquisition of high resolution images thanks to the possibility of flying at very low altitudes. Manned aircrafts or helicopters rarely reach these low altitudes. The possibility of mounting moving cameras on UAVs can be useful for capturing images of some kinds of objects that are difficult to acquire with standard aerial images. Typical examples are building façades, dams, and vertical rock faces [3]. In addition, some categories of objects cannot be reconstructed with images taken from the ground (e.g. roofs). Therefore, UAV information is an intermediate product between aerial and close-range applications. SLR and low-cost digital cameras and video-cameras are mounted on UAVs for photogrammetric applications. In some cases, even multispectral sensors were employed [4]. Their use is increasing, and the results obtained thus far are quite promising.

The use of an autonomous navigation/positioning system enables most UAVs to follow a predefined flight-plan and to record orientation parameters. These sensors are normally based on GNSS/INS technology, but alternative solutions integrating other kinds of data (e.g., scans and images) have been explored [5-6].

UAVs can also fly without an ad-hoc navigation system. This is an advantage in case of emergency [7] because the system can provide a set of images. This preliminary investigation can be extremely useful in ensuring the safety of human operators.

Nowadays, researchers are working to improve UAV capabilities, starting from the navigation system and continuing to the acquisition of the images. There exist numerous examples in technical literature in which UAVs have become powerful platforms for mapping applications, with limited costs and a quick production workflow [8].

The new scenario that UAVs are opening is so relevant that Colomina et al. [9]state that a new paradigm for photogrammetry has arrived, after close-range, aerial, and satellite photogrammetry. UAV technology may solve many problems that currently cannot be overcome using terrestrial or aerial surveys. In several close-range applications (e.g., cultural heritage documentation and conservation), UAVs have been successfully used (e.g., to map archaeological sites), whereas other surveying techniques are not suitable (see, for example [10]). Agriculture represents another potential field for the extensive use of UAVs [11].

UAV image blocks can be processed following the standard photogrammetric pipeline. According to Eisenbeiss [12], the workflow for a generic block of images (with the traditional aerial network geometry) includes these steps:

- flight planning - image acquisition;
- measurement of tie points;
- measurement of control points;
- bundle block adjustment;
- DSM generation;
- orthophoto production.

These steps are fully automated, except for the identification of ground control points (manual). Tie points for image blocks can be collected in an automated way when the standard feature-based (FB) procedures can be employed. Scale differences, illumination changes, occlusions, and convergent imagery result in failure of standard AAT (automatic aerial triangulation) with commercial software, since these applications are much closer to those usually solved in close-range applications. On the other hand, this complex category of images can be processed without user's interaction after the implementation of algorithms like SIFT [13], obtaining the (relative) orientation parameters of the images.

For this reason, it is not clear how to classify UAV projects. Sometimes they are aerial projects, but in some cases, images are more similar to those of close-range applications. The block acquired in this work has a geometry similar to that of aerial project, but baselines and angles are slightly irregular because of the limited accuracy of the navigation systems, that is surely worse than standard sensors used in aerial photogrammetry.

In the next sections the procedure for image collection, processing and orthophoto production are described. Most algorithms were implemented by the authors in order to overcome some limitations of existing commercial packages.

2   Data Acquisition and Processing

The generation of the orthophoto was carried out by combining images with total station data. This section describes the different data collected and their registration into a unique and stable (physically materialized) reference system.

2.1   The UAV Platform and Image Acquisition

The photogrammetric survey of the site was carried out with the UAV platform AscTec Falcon 8. The system is equipped with a RGB camera Sony NEX-5N photogrammetrically calibrated [14].

The Falcon 8 (70 cm x 60 cm, weight 2 kg) is equipped with 8 motors and is able to fly up to 15-20 minutes with a single battery. The electronic equipment includes a GPS antenna and a system of accelerometers determining the system roll, pitch and yaw. The communication system allows the ground station to receive telemetry data and video signals from the on-board sensors.

The average flying height was 60 m, obtaining a pixel size (on the roof) of about 13.5 mm, i.e. more than sufficient to obtain an orthophoto with scale factor 1:100.

Fig. 2. The UAV system during the flight over the Basilica of Collemaggio

The whole block is made up of 52 images acquired with the software AscTec AutoPilot Control. The software allows the operator to import a georeferenced image and then waypoints can be added manually or in an automated way by defining the overlap. This data will be then transferred to the Falcon, whereas the user has to take off and land.

Fig. 3. The trajectory of the Falcon reprojected on georeferenced cartography

2.2   The Geodetic Network

The geometrical survey started with the creation of a geodetic network made up of several stations around the basilica. This allowed the creation of a stable reference system where all data (images and laser scans) were registered.

The instrument is a robotic total station Leica TS30 (distance precision ±0.6 mm and angular precision ±0.15 mgon) and the adopted scheme is shown in Fig. 4. Several laser targets were also directly measured in this phase in order to avoid repositioning errors and obtain more precise results. These targets include retro-reflective tapes mounted on stable objects (e.g. metal pillars), laser targets for the following scan registration phase, and photogrammetric targets placed on the ground and then used as GCPs for UAV image triangulation.

In all, the network consists in 2774 observations (922 distances, 923 directions, 923 zenith angles, and 6 inner constraints) and 7014 unknowns; global redundancy is therefore 2060 degrees of freedom.

It is important to notice that several points were measured from multiple stations and all tripods were fixed during the whole survey. Least Squares adjustment provided a sub-millimetre precision for stations and points with multiple intersections, whereas the remaining points have an average precision of about ±1.5 mm.

Some ground control points (GCPs) were then measured in order to fix the reference system of the photogrammetric projects. In this case some targets (black circular mark on a white background) where placed around the basilica. The measurement of their centre coordinates was carried out with a prism on a pole, obtaining a precision of about ±2 mm. In addition, some natural points on the roof were collimated by using the reflector-less function of the total station. These points were inserted in order to include objects of the basilica and to obtain a more stable datum for photogrammetric purposes.

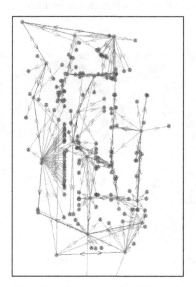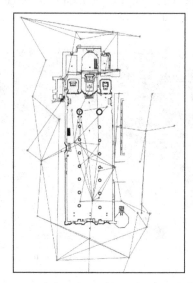

Fig. 4. Global geodetic network with also laser targets and the same scheme (stations only and their connections) with the map of the basilica

2.3   Automated Image Orientation

The automatic orientation procedure was carried out with the ATiPE software [15] in order to handle complex image configurations [16]. ATiPE can be considered as a multi-step process in which several parameters are estimated or refined. The procedure can be properly tuned to deal more effectively with different image sets, thanks to the optional choice between techniques and input parameters.

The input elements for the automated tie point extraction and transfer are the images and the calibration parameters of the camera used. A generic block of n images can be considered composed of $(n^2-n)/2$ combinations of stereo-pairs, which are firstly analysed independently for the identification of the correspondences, and then progressively combined.

The automatically extracted pixel coordinates of homologous image points can be imported and used for image orientation and sparse geometry reconstruction in most commercial photogrammetric packages (with, for example, PhotoModeler, Leica Photogrammetry Suite, and Australis – iWitness – iWitnessPro).
The innovative aspects of the developed method are:

- effectiveness on a large variety of unorganised image datasets;
- capability of working with high-resolution images;
- accurate image measurements based on Least Squares Matching;

- combination of feature-based (FB) and area-based (AB) matching algorithms;
- procedures for image point reduction and regularization;

ATiPE was designed for photogrammetric and detailed 3D modelling applications, therefore a real-time data processing is not an issue of primary importance: there is a particular attention to the final accuracy and uniform distribution of the extracted tie points. The implemented FBM strategies, coupled with an initial network geometry analysis (called visibility map), allow a CPU time of few hours for image blocks composed of several tenth of high resolution images used at their original size. The robust detection of correspondences with the proposed procedure is achieved by combining the accuracy of traditional photogrammetric methods with the automation of computer vision approaches.

The input elements of ATiPE are the images, the full set of interior orientation parameters (optional) and a visibility map between the images (optional). All images are normally used with their calibration parameters in order to avoid self-calibration which is generally not appropriate and reliable in practical 3D modelling projects.

The sparse image block acquired, similar to a classical aerial one, is shown in Fig. 5. The block has several strips with 52 images, which produce 1326 possible image combinations. For each image pair, the features were compared in less than 5 seconds with a kd-tree approach.

18 ground control points (targets and points on the roof measured by means of a total station) were included in the adjustment to fix the datum. Sigma-naught after Least Squares adjustment (collinearity equations are the mathematical model employed) is 0.674 pixels, whereas the RMS of image coordinates was 0.67 pixels. The RMSE values on 5 check point coordinates were 5.3 mm (X), 6.1 mm (Y), and 11.2 mm (Z), respectively. This provides the exterior orientation parameters of the images along with the 3D coordinates (i.e. a preliminary sparse reconstruction) used to run the dense matching phase for orthophoto generation.

Fig. 5. A 3D view of image orientation results

3   Orthophoto Generation via Dense Matching and Texture-Mapping

The generation of the orthophoto is a two-step procedure where a dense model of the basilica is created with multi-image matching techniques and a true-orthophoto is derived by reprojecting the images. In this second phase occlusions are taken into consideration in order to correctly represent the different objects.

3.1   The Dense Reconstruction via Multi-image Correlation

The extraction of the dense point cloud was carried with an in-house software called MGCM+ [17]. This approach is an area-based matching algorithm based on the Least Squares matching technique extended to multiple images. The MGCM+ procedure is based on the concept of multi-image matching guided from the object space, therefore a variable number of images can be matched with the integrated epipolar constraint. This allows one to reduce multiple candidates caused by surface discontinuities, occlusions, and repetitive patterns.

On the other hand, the nonlinearity of the problem requires an approximate 3D position of the object points and a preliminary location of the homologous point positions in the images. These problems are overcome in the current implementation by using a rough object model (called "seed model") derived from the previous adjustment of 3D coordinates. Once the approximate position is defined in the 3D space an elevation interval along the visual ray is set. The back-projection of the point onto the search images gives the bounds of the epipolar segments along with the solution can be found. Moreover, the back-projection process provides the approximate positions of homologous points in the search images.

Another aspect is the selection criterion of images to be matched. This aspect is even more critical because of the distinction between reference and "slave" images. The selection of the reference image for each facet of the "seed model" needs the estimation of the local normal vectors. Then, the reference image is selected according to the normal vector closer to the surface normal direction. This means that the choice of the reference image can progressively change depending on the considered triangle.

The second problem concerns the choice of the "slave" images. This optimization is still performed by exploiting the "seed" model. In the first phase only the images whose normal vector forms an angle lower than 90° with the local surface normal are used. Then, an additional selection is performed in order to discard images with large perspective deformations with respect to the reference image (where the assumption of an affine transformation as geometric model can become inadequate). The candidate images are further analysed in a second selection phase and only those for which MGCM can provide reliable results are effectively processed. In particular each triangle in the "seed" model is back-projected onto the reference and potential "slave" images. Two shape parameters are evaluated: the area of the triangle and its angular deformations. If the area of a back-projected triangle on a potential "slave" image is less than 50% of the area of the triangle (in the reference) and the angular variation exceed 40%, we assume that perspective deformations or scale variations are too large for a correct application of the affine model.

In the case of the basilica, the MGCM+ algorithm was run with a coarser point cloud model built by using as approximate model the triangulation of the 3D tie point coordinates derived from bundle adjustment. The size of the grid was set to 7cm×7cm, obtaining a point cloud of about 2 million points. In the second step (next iteration) the point cloud was triangulated obtaining a new approximate model for a denser reconstruction, where the new grid size is 2cm×2cm and is made up of 8 million points (Fig. 6).

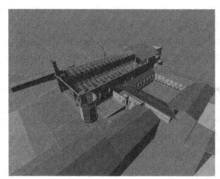

Fig. 6. The BIM model of the basilica and a visualization with the point cloud extracted from dense image matching (MGCM+)

3.2   Orthophoto Generation

The implemented procedure for texture-mapping of 3D objects can be split into two almost independent steps: a (i) geometric part, which includes the visibility analysis and the texture assignment, and a (ii) radiometric one, where the radiometric adjustment correction of the images is carried out.

First of all, a visibility analysis is run to detect occluded areas in the images. The basic idea of this analysis is a check of a generic triangle T_j. If the triangle is not visible from the viewpoint I_i there will be at least another triangle T_k that gives an occlusion. Starting from an object surface described by a triangulated structure, the texture-mapping algorithm is based on the automatic back-projection of all triangle vertices onto the image space by using the pinhole camera model.

The occlusion in the image space is given by the intersection between two back-projected triangles. The reciprocal distance between the vertices of the triangles and the image projection centres are calculated in order to understand the position of the occluded triangle. The farther triangle is occluded, whereas the closer is the occluding one. The CPU cost of this operation depends on $O(m^2)$ and some strategies were implemented to speed up data processing: (i) view frustum culling, (ii) back-facing culling; and (iii) triangle distance culling.

The view frustum culling is based a progressive labelling of the triangles outside the camera view frustum. This can be done by back-projecting the vertices of the triangles. If the vertices of the triangle fall outside the image boundaries the triangle can be labelled as hidden.

A further reduction (for an oriented mesh) of triangles can be obtained by using the back-facing culling. If the mesh is oriented the normal vector of each triangle can be calculated by fitting a plane to the triangle vertex. Given the triangle normal and the camera position it is possible to define the vector normal to the image plane. If the angle between the normal and the viewing direction of the camera is larger than $\pi/2$ the triangle is geometrically not visible.

These two checks may allow a significant reduction of the number of the triangles to be inspected in the visibility analysis. However, the check phase may result in lot of tries for triangles that back-projected in the image space are far from the reference, obtaining a series of not occluded results. On the other hand, the probability of intersection is higher for triangles that back-projected in the image are close to the reference one. For this reason, the intersection test is limited to a certain amount of close triangles. The number of triangles to be considered depends on the quality of the geometrical model and on the image scale.

After the assignment of the visible triangles from each camera position, the best texture for each triangle can be determined. Two quality parameters are considered for this choice: the resolution of the image in object space and the camera viewing direction. The image whose quality parameters reach the maximum is used as texture. Finally, the texture coordinates for the triangle are calculated by back-projecting the triangle coordinates in the object space onto the selected image by using collinearity equations.

As the described algorithm works independently for each triangle, the brightness and colour levels between adjacent triangles can be significantly different, causing sharp differences in the textured model. This inhomogeneity can be reduced with a colour/brightness correction that binds all the images into a unique texture.

The radiometric correction is not run in the RGB space, but in the L*a*b* space (ISO 1164A). Indeed, the L*a*b* colour space is designed to approximate human vision. In particular, the L*channel represents the lightness of the colour (L*=0=black, L*=100=diffuse white), whereas the a* and b* channels define the colour plane. The a* axis is positioned along the red/magenta and greed direction (a*<0=green, a*>0=magenta), the b*axis along the yellow and blue direction (b*<0=blue, b*>0=yellow). This channel distinction can be used to perform accurate colour balance corrections by modifying the a* and b* components.

The proposed correction exploits the overlapping areas between consecutive images. The common points detected using the ATiPE algorithm are employed to initialize this phase: differences can be estimated and interpreted as samples. Starting from these samples, differences functions can be interpolated over the whole image.

Three key factors have a significant influence on the results: (i) the density of common points; (ii) their position; and (iii) the interpolation function used.

The point density has a fundamental role for an accurate correction. It is necessary to have a high number of common points to obtain a good approximation for the colour/brightness difference functions (sparse and bad distributed common points can cause some artefacts). For this reason manual measurement of common points between images, with some tenth of points, is not sufficient for an effective correction. The ATiPE implementation is useful to overcome these two problems.

Finally, starting from the textured digital model of the basilica the orthophoto is obtained by projecting the model texture onto the defined projection plane. In this case the orthographic projection was carried out in the XY plane defined by the local reference system of geodetic network. The final map is shown in Fig. 7.

Fig. 7. The final orthophoto of the basilica derived from UAV images and photogrammetric data processing

4   Conclusion

The paper described the methodology used to generate an accurate and detailed orthophoto from images acquired with an UAV over the Basilica of Collemaggio (L'Aquila, Italy). This orthophoto is only one of the metric product used to support the restoration works of the structure: the whole project consists not only in 2D representation (CAD drawings), but also in a BIM model. It deserve to be mentioned that the required output is more than a 3D model, as a BIM becomes a complete and exhaustive description of the object by means of a database, where the orthophoto was integrated.

The generation of the final orthophoto was carried out by combining standard surveying techniques (total station) and images. Some specific algorithms (ATiPE, MGCM+) were implemented by the authors and adapted to the case of UAV images in order to automate the photogrammetric processing pipeline. The acquisition of the images was not a complex task because UAV technology has reached a high level of maturity. In particular, it was sufficient to setup a proper flight plan and the Falcon 8 was able to cover to whole longitudinal profile of the object.

On the other hand, the image processing software today available on the commercial market could be insufficient to handle different practical situations, and the implementation and use of ad-hoc algorithms can overcome most limitations. The final orthophoto obtained with the automated (i) image matching, (ii) image orientation, (iii) dense surface matching, and (iv) texture-mapping algorithms has a geometric resolution of 1.35 cm (GSD) and remains the first photorealistic map of the roof of the basilica with this high level of detail.

References

1. Oreni, D., Brumana, R., Cuca, B.: Towards a Methodology for 3D Content Models. The Reconstruction of Ancient Vaults for Maintenance and Structural Behaviour in the logic of BIM management. In: Virtual Systems in the Information Society, NJ, USA, Milan, Italy, pp. 475–482 (2012)
2. Oreni, D., Cuca, B., Brumana, R.: Three-dimensional virtual models for better comprehension of architectural heritage construction techniques and its maintenance over time. In: Ioannides, M., Fritsch, D., Leissner, J., Davies, R., Remondino, F., Caffo, R. (eds.) EuroMed 2012. LNCS, vol. 7616, pp. 533–542. Springer, Heidelberg (2012)
3. Eisenbeiss, H.: The autonomous mini helicopter: a powerful platform for mobile mapping. International Archives of Photogrammetry, Remote Sensing and Spatial Information Sciences 37(Part B1), 977–983 (2008)
4. Nebiker, S., Annen, A., Scherrer, M., Oesch, D.: A light-weight multispectral sensor for micro UAV - opportunities for very high resolution airborne remote sensing. International Archives of Photogrammetry, Remote Sensing and Spatial Information Sciences 37(Part B1), 1193–1200 (2008)
5. Eugster, H., Nebiker, S.: UAV-based augmented monitoring – real-time georeferencing and integration of video imagery with virtual globes. International Archives of Photogrammetry, Remote Sensing and Spatial Information Sciences 37(Part B1), 1229–1235 (2008)
6. Wang, Z., Brenner, C.: Point based registration of terrestrial laser data using intensity and geometry features. International Archives of Photogrammetry, Remote Sensing and Spatial Information Sciences 37(Part B5), 583–590 (2008)
7. Kerle, N., Heuel, S., Pfeifer, N.: Real-time data collection and information generation using airborne sensors. In: Zlatanova, S., Li, J. (eds.) Geospatial Information for Emergency Response, pp. 43–74. Taylor & Francis, London (2008)
8. Zongjian, L.: UAV for mapping – low altitude photogrammetric survey. International Archives of Photogrammetry, Remote Sensing and Spatial Information Sciences 37(Part B1), 1183–1186 (2008)

9. Colomina, I., Blázquez, M., Molina, P., Parés, M.E., Wis, M.: Towards a new paradigm for high-resolution low-cost photogrammetry and remote sensing. International Archives of Photogrammetry, Remote Sensing and Spatial Information Sciences 37(Part B1), 1201–1206 (2008)

10. Eisenbeiss, H., Sauerbier, M.: Investigation of UAV systems and flight modes for photogrammetric applications. The Photogrammetric Record 26(136), 400–421 (2011)

11. Grenzdörffer, G.J., Engel, A., Teichert, B.: The photogrammetric potential of low-cost UAVs in forestry and agriculture. International Archives of Photogrammetry, Remote Sensing and Spatial Information Sciences 37(Part B1), 1207–1213 (2008)

12. Eisenbeiss, H.: Applications of photogrammetric processing using an autonomous model helicopter. Revue Francaise de Photogrammetrie et de Teledetection. In: Symposium ISPRS Commission Technique I "Des capteurs a l'Imagerie", Saint-Mande Cedex, France, vol. 185 (2007-1), 6 pages (2007)

13. Lowe, D.G.: Distinctive image features from scale-invariant keypoints. International Journal of Computer Vision 60(2), 91–110 (2004)

14. Previtali, M., Barazzetti, L., Roncoroni, F.: Spatial data management for energy efficient envelope retrofitting. In: Murgante, B., Misra, S., Carlini, M., Torre, C.M., Nguyen, H.-Q., Taniar, D., Apduhan, B.O., Gervasi, O. (eds.) ICCSA 2013, Part I. LNCS, vol. 7971, pp. 608–621. Springer, Heidelberg (2013)

15. Barazzetti, L., Remondino, F., Scaioni, M.: Orientation and 3D modelling from markerless terrestrial images: Combining accuracy with automation. The Photogrammetric Record 25(132), 356–381 (2010)

16. Barazzetti, L., Forlani, G., Roncella, R., Remondino, F., Scaioni, M.: Experiences and achievements in automated image sequence orientation for close-range photogrammetric projects. In: Proc. of SPIE Optics+Photonics, Munich, Germany, May 23-26, vol. 8085 (2011)

17. Previtali, M., Barazzetti, L., Scaioni, M., Yixiang, T.: An automatic multi-image procedure for accurate 3D object reconstruction. In: 4th International Congress on Image and Signal Processing, IEEE Conference Record Number 18205, Shanghai, October 15-17, 5 pages (2011)

A Cartographic and GIS Perspective of Geodiversity Analysis: The Iberian Peninsula/ Spain Case Study

Rufino Pérez-Gómez[1], Juan-José Ibáñez[2], and Antonio Vázquez Hoehne[1]

[1] Universidad Politécnica de Madrid (UPM), Madrid, Spain
[2] Centro de Investigaciones sobre Desertificación, CIDE (CSIC-UV), Km 405,
Apdo. Oficial 46113, Valencia, Spain
{rufino.perez,antonio.vazquez.hoehne}@upm.es

Abstract. This paper describes some research that has been conducted on a topic of growing interest among earth scientists like geodiversity from a cartographic and GIS perspective. Many soil and earth scientists consider that geodiversity measurement and quantification should play a key role when evaluating the ecological, environmental and social value of a territory. This geodiversity project is based on the Spain and the Iberian Peninsula case study. Many earth scientists use multiple mathematical techniques, such as diversity statistics and models, to end up with summary statistics which best describe the general properties of the phenomena. In our study, we will focus in the use of cartography and GIS concepts and tools in order to conduct the modeling and analysis stages. The geodiversity study is based in drainage basins as basic spatial units for the analysis. Additionally, some diversity analysis has been conducted in relation to the 3D components relief zones.

Keywords: Cartography, GIS, spatial analysis, 3D models, geodiversity, soil diversity, lithodiversity, Iberian Peninsula.

1 Introduction

It is noteworthy the strong technology push that has taken place in many Cartography and GI fields in the last years. These range from data capture and new mapping technologies, which offer both fast and accurate acquisition of topographic data, to Internet technologies. For this reason, one of the main commitments of the International Cartographic Association (ICA) is the continuous research about improvements in the use of geospatial information in the benefit of science and society, as stated in its Strategic Plan and its research agenda [17]. In line with this strategy we selected the research topic of geodiversity to evaluate the capabilities of modern cartography and GIS concepts as applied in a particular study case.

The literature review reveals how biologists have been using for many years different diversity statistics in order to conduct ecological studies. In doing so many mathematical indices and models of diversity have been devised and used to try to understand and explain the structure and organization of ecosystems. Because there were some periods of time in which the extinction of species was observed, as a result

B. Murgante et al. (Eds.): ICCSA 2014, Part IV, LNCS 8582, pp. 14–30, 2014.

of human activities, that caused a growing interest in estimating diversity and implementing biodiversity conservation policies. However, it was not observed a similar interest to compare and quantify the diversity of other environmental factors like soils, land forms or lithology. The interest in these abiotic factors is much more recent what is a bit surprising. In relation to this subject, some such as Ibáñez et al. [8, 9] express the relevance of this matter by pointing out that "the characterization and quantification of diversity of landform, rock and soil, as non-renewable natural resources, should be taken into account when estimating a territory's ecological value".

In spite that many indices and models have been devised and used to measure and quantify diversity, these may be grouped into three classes [13]: indices of richness, abundance distribution models, and indices based on proportional abundance of objects. In this project, we decided to apply indices and diversity statistics belonging to the third group, that means those related with the relative or proportional abundance of objects. A detailed description of the mathematical indices will be given later during the explanation of the methodology. In a similar strategy to the biologists' approach, when conducting landscape analysis, some soil scientists decided to use the diversity indices and models to explore, compare, evaluate and quantify the complexity of soil patterns in different areas and environments. Some authors such as Ibáñez et al. [6,7,8] analyzed the potentialities of biodiversity estimation methodologies in the soil survey and in the development of Soil Information Systems. They showed how the richness of soil types increases as the area sampled increases. But the area is not the unique factor in diversity and the habitat heterogeneity concept is often mentioned in the literature. The habitat heterogeneity hypothesis suggests that increasing areas can support more species because of increasing habitat heterogeneity, such as pedodiversity and relief diversity [9,10]. Nowadays, geodiversity is a research topic that interest to more and more soil and earth scientists and, as a result of this, there is a growing number of papers on this subject. In these published papers, some terms like geoparks, geoheritage, environmental services and environmental legislation appear as valuable applications of geodiversity projects and studies.

Though cartographers have been involved for many centuries in the design and production of all types of maps, it was thanks to the work of the French cartographer Jacques Bertin what produced a very important theoretical development by proposing basic cartographic principles for map design [2,3,4]. Nowadays, The Semiology of Graphics is considered the "grammar of the cartographic language". Bertin's legacy has provided an overall theoretical framework of making and using cartographic symbols as a sign system [15,16]. Nowadays, GIS environment has increased a lot its performance levels and allows to obtain many different products: collection of thematic maps, graphs, reports, 3D models and surfaces, 3D perspectives, profiles, animations, online virtual globes representations, etc., that will help to identify, understand and communicate the most relevant elements of the phenomenon studied in the project [5]. This project will help us to evaluate how well cartography and GI products can supplement and improve the traditional mathematical analysis and graphs published by earth scientists when working in geodiversity studies of different ecosystems or study areas.

2   Materials and Methods

First of all, an extensive literature review on the subject matter was done in order to identify the key elements. Afterwards, a specific case study was proposed by some experts in soil science and geodiversity. Finally, all the available data related to the project area, reference and thematic data, was gathered and structured into a spatial geodatabase to conduct the works described in this paper following the "conceptual modeling" of the geodiversity experts.

2.1   Iberian Peninsula/Spain Case Study

The Iberian Peninsula was selected as an interesting case study because is a well-defined territorial and geographical unit. Besides that, the Peninsula constitutes a microplate situated in the convergence zone between the Eurasian and African tectonic plates, which collided and joined with the Mesomediterranean Plate [1]. It is believed that contains a great diversity in soils, lithology, tectonics and landforms. Unfortunately, only the soil database covers the whole peninsula while the datasets available for the other variables (lithology, quaternary deposits and vegetation series) are related to the peninsular part of Spain because the lack of available data of Portugal area. So, Ceuta, Melilla and the Balearic and Canary Islands are not considered in this study. The working scale for this project is 1:1.000.000 and the spatial data used for the study was partially captured in a previous project [6] and further updated with the following sources:

- Soil Map: an existing 1:1.000.000 soil map based in the FAO classification was used. Because is a small scale is rather generalized and only the soil associations or Soil Mapping Units (SMUs) are graphically represented. The soil components or Soil Typological Units (STUs) are only recorded as attributes in the database.
- Vegetation series Map: It was obtained from the 1:400.000 vegetation series map of Spain as developed by Salvador Rivas Martinez. It can be downloaded from the Ministry of Agriculture, Food and Environment website.
- Lithology: It was obtained by digitizing the 1:1.000.000 hydrogeologic map of Spain produced by the Spanish Geological Institute (IGME).
- Quaternary Deposits: It was obtained by digitizing the 1:1.000.000 map of Spain of Quaternary Deposits produced by the Spanish Geological Institute (IGME).
- Reference data: Administrative units, hydrography, DEMs, etc. were downloaded from the National Mapping Agency (IGN/CNIG) website.

All this reference and thematic data was integrated into a spatial geodatabase. For this purpose the ArcGIS 10.1 Software was used. Different conversion and management processes such as format and projection conversion were needed, and afterwards, some editing and updating tasks were conducted to homogenize and prepare the data for the analysis. At this stage, it´s worth mentioning some modifications of the soil map database structure for its particularities. Soil maps, following the FAO or WRB classifications, have similar structure. Because it's a small scale map it is rather generalized and only the soil associations or Soil Mapping Units (SMUs) are graphically represented. The soil components or Soil Typological Units (STUs) are only recorded as attributes in the database with its percentage within that particular soil polygon (SMU). So, we needed to modify the database structure of the soil table by adding attributes for soil components area (STUs) in order to carry out the intended geodiversity analysis.

Once all the available data has been integrated in the project geodatabase, we designed and executed specialized geoprocessing workflows in order to obtain the required mathematical elements for the geodiversity analysis. These elements will be later copy into a spreadsheet for the implementation of the chosen indices and models.

2.2   Geoprocessing Workflows

When geoprocessing operations are used for analysis, they create new data that can be used to answer geographic questions. In fact, the great number of geoprocessing tools is what makes GIS technology so powerful. So, once we have our original database, with the required reference and thematic data, we need to design a clear and efficient workflow aimed to achieve the specific analysis that, in our case, is the geodiversity study of the project area we defined. During the literature review we could check that many indices and models have been used to estimate and quantify diversity. We decided to use the proportional abundance of objects because is the most frequent way of estimating diversity. In these methods two main components are identified. The first is named "richness" and relates to the number of different object classes present in the area (e.g. the number of different landforms, lithology types or soil types according to a certain classification). The second element is called "evenness" or "equitability" and refers to the way in which the individuals are distributed among the existing object classes. So, this means the relative o proportional abundance of each object type (i.e. the relative area occupied by each type of soil in a drainage basin). To clarify these components, we should observe that for two portions of land of the same area and richness, the one with greater diversity will be the one which have the different types with similar area or relative abundance. So, this means that the greater the homogeneity in the proportional abundance of the object classes the greater the diversity of that area (e.g. the diversity of a drainage basin). Though there are several indices used in the literature, most ecologists prefer Shannon's Index for its ease in calculation. In calculating Shannon's Index, any logarithmic base can be adopted. The natural logarithm will be used in this article. Its mathematical expression is as follows:

$$H' = \sum_{i=1}^{n} p_i \cdot \ln p_i, \tag{1}$$

where H' is the negative entropy (negentropy) or diversity of the population, and p_i is the proportion of individuals found in this ith object. In our study we calculate p_i as the percentage of surface area occupied by a particular object (soil type, lithology type, etc.) within the sample or statistical area considered that, in our case, is each individual drainage basin of the project area. The index is maximum for any S (richness) if all objects have equal numbers of individuals and minimum if the individuals are maximally concentrated in one object.

In order to conduct the proposed geodiversity analysis a geoprocessing workflow diagram was outlined with the necessary steps to obtain the various elements of the diversity statistics. Figure 1 shows details of this workflow.

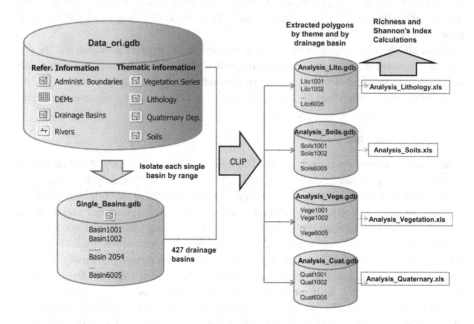

Fig. 1. Geoprocessing workflow for diversity statistics components estimation

Once we obtained the original database we proceed to isolate each individual drainage basin in the project area. Each drainage basin was assigned with a rank by the Horton-Strahler method which represents the stream order or its branching complexity. These single drainage basins will behave as the sample units for diversity statistics estimation in any of the considered environment variables (lithology, soils, vegetation series and quaternary deposits). So, the first processing step was to create a new database in which we stored each single drainage basin as a new isolated feature class table. In the Iberian Peninsula project we identified 427 drainage basins. Whenever we use Spain, as project area, the number of drainage basins is reduced to 363.

Every single drainage basin will be used to clip or extract the polygons, of a particular theme. For this purpose we created 4 geodatabases, each one stores all the thematic polygons extracted by the individual basins (see figure 1). After the clipping operations, the new soil tables need further updating. Though the areas of the new polygons are automatically updated by the geodatabase, the 60 fields with the surface area of the soil components are static and require to be modified accordingly. To update many attribute fields in many tables, what may means several thousands of time–consuming manual editing operations, is solved through a model that execute the geoprocessing tasks automatically and speed up the process.

2.3   Mathematical Analysis

During the literature search several indices and models for diversity statistics were reviewed. Among these, the proportional abundance of objects methods are the most often manner of estimating diversity. Though there are several indices of this type, most ecologists and earth scientists prefer Shannon's Index for its ease in calculation. This is also the manner we used to estimate geodiversity in this project.

After this process, we obtained richness and Shannon's Index values for all the drainage basins and the four environmental variables (lithology, soils, vegetation series and quaternary deposits). So, any individual basin has now eight new individual numeric attributes to work with and analyze. At this stage most ecologists and earth scientists use different mathematical techniques to estimate various diversity statistics and to end up with some summary statistics. Later, they interpret and connect these statistics with some particular properties of the phenomenon under study (e.g. spatial structure and degree of complexity, forming factors, evolution processes, etc).

One of the ways to process diversity statistics, widely used in biodiversity studies, is through dispersion graphs which show the richness-area relationships. In these graphs the neperian logarithm of the richness (*Y-axis*) is drawn against the neperian logarithm of the sample area (*X-axis*). Soil experts can interpret these graphs and reach conclusions about the structure and complexity of the different landscapes. At this point, it's worth mentioning some findings obtained in previous studies. Ibáñez et al. [7],[9] were the first to show how the evolution of fluvial systems induces an increase in complexity of pedogeomorphological landscapes. In order to do these studies, many ecologists have used archipelagos as natural laboratories to research the mechanisms that produce biodiversity patterns and, at the same time, to test diversity estimation techniques. Finally, some authors such as Ibáñez and Efland [11] are considering a "Theory of Island Pedogeography" in a similar approach of the "Theory of Island Biogeography" of MacArthur and Wilson [12].

In our study the richness-area relationship was analyzed for the four environmental variables. In all cases a power law function were fitted to the diversity data. Because the slopes of the fitted lines are similar in the four cases, the increase of richness with the area follows similar patterns. Some soil and earth scientists suggest that there might exist some general diversity laws which affect both biotic and abiotic factors.

At the moment there is an ongoing research, conducted by soil scientists and mathematicians, on the Iberian Peninsula/Spain diversity statistics that, hopefully will provide some results and findings. These, in turn, will be compared and analyzed in relation to previous studies. However, the approach in this paper is not an in depth-analysis of the diversity statistics with mathematical techniques, but rather to focus in cartography and GIS methodologies to analyze and visualize spatial diversity patterns.

3   Geomatic Analysis

The richness-area relationship and other diversity analysis are well suited to be treated with cartography and GIS concepts and tools. GIS technology provides a powerful environment for modeling, analysis and visualization. The GIS project integrates accurate and updated maps of the study area, together with their attribute databases, in order to obtain detailed 2D and 3D map-like representations of the phenomena. The geomatic analysis, which follows, encompasses three types of works: thematic cartography, 3D visualization techniques and zonal statistical analysis with raster techniques.

3.1   Thematic Cartography

Maps have always been effective means of communication when conveying spatial distributions, patterns and relationships. Traditional cartographers have been involved for many years not only in the design and production of all types of maps, but also in the reading and understanding of the represented phenomenon. In this project all drainage basins were assigned a rank by the Horton-Strahler method, and during the mathematical calculations richness and Shannon's Index values for all the drainage basins were obtained. These values were estimated for the four environmental variables. All this information, generated in a spreadsheet, is integrated back in the spatial database as numeric attributes to enable the map production workflows. In any project we can produce a collection of thematic maps which best represent the main aspects of the phenomenon under study base on the database contents. To illustrate this capability we include and comment some maps related to our case study.

Figure 2 shows a soil diversity map of the Iberian Peninsula. The map portrays two layers: the soil richness (i.e. the number of soil types present in a certain basin) and the Shannon's index value as a measure of the soil diversity or pedodiversity. The richness is depicted my means of proportional symbols while the soil diversity is shown by assigning appropriate colors of a "value scale" to the Shannon's index

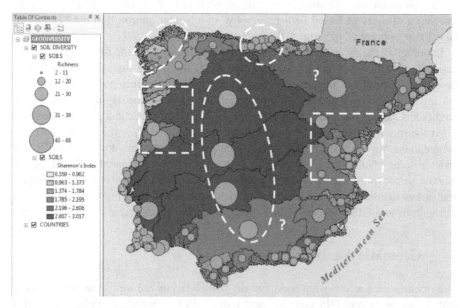

Fig. 2. GIS map representing soil diversity in the Iberian Peninsula

val-ues (choropleth technique). By doing so, the map contains two spatial distributions and as a result of that any of the 429 drainage basins are encoded with 2 diversity properties. By reading the map we can draw up a few features of the mapped area:

- There is a clear "global pattern" in which both, the richness and Shannon's index, increase with the rank and area of the basin. This is more or less a radial pattern which goes from the coastline (low values) to the big river valleys in the central part of the Peninsula (high values).
- The map user can also easily identify some "homogeneous zones" like the areas of high values, middle values or low values (see ellipses on the map). This is achieved through the selective property of visual variables like color and size.
- Likewise some "heterogeneous zonal patterns" can be located in the map in areas where the diversity properties varies more quickly as we move throughout the space (see rectangles on the map).
- There are "particular cases" like Ebro and Guadalquivir rivers (rank 6) that show lower diversity values than expected.

Similar maps have been produced to analyze the lithology, vegetation series and quaternary deposits by drainage basin. However, these maps only show the peninsular Spain area because of the lack of information from Portugal area. For this reason, some big river basins, in rank and size, which go to the Atlantic watershed, appear uncompleted. This could produce an underestimation of richness and diversity in those basins, though geodiversity experts think, and have tested with the available soil layer, that the power law functions which relate richness and area are hardly affected. All in all, the geodiversity study in Spain is a worthy an interesting study case.

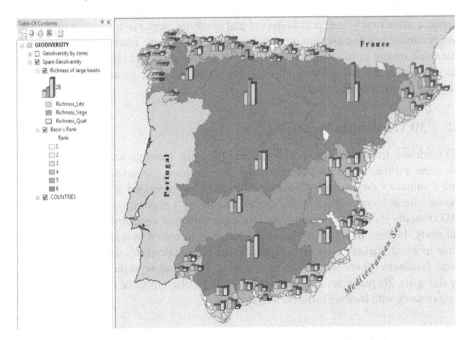

Fig. 3. GIS map representing richness values by basin rank in Spain

Besides the described mapping processes in which we study and represent one environmental variable at a time (analytical maps), we can also compare the diversity values (e.g. richness values) of several environmental variables at the same time and in relation with the rank of the basin with a polyanalytic map. This strategy is used in the figure 3 where the map shows the three richness values, in different colors, with their associated proportional symbols. At the same time, the basin rank layer provides a suitable background to analyze the diversity statistics variation. The reading and interpretation of this map allows us to make some comments:

- As expected, there is a clear general trend in which the richness of the three variables grows with the rank/area of the basin. Low rank basins have smaller areas because they are related to small rivers which have their origin near the coastline. The map has hidden the symbols of the smallest basins (rank 1 and 2) to prevent a too cluttered representation and to make the map clearer and more legible. This way we have a more readable map to identify the main patterns through visual interpretation. We also complemented this general map with some zonal maps which represent the same themes in a more detailed manner.
- There are zones and basins with "similar richness" values of the 3 variables, others with "significant differences" and a third group with "large differences" in their values. The variation of these spatial properties is of especial relevance when studying the richness-area mathematical relationships in the diversity analysis projects.
- A big vegetation series decrease is observed in the North-South direction (Ebro-Tajo- Guadiana river basins)

Because the current GIS technology capabilities we can produce a collection of thematic maps (analytical, polyanalytical and synthetic maps) in a user friendly manner. A careful reading and interpretation of each map, as we made in the 2 previous examples, enable us to identify general trends, zonal patterns and particular cases that the earth scientist and geodiversity experts should be able to attach or link with adequate "geographical meaning".

3.2   3D Visualization Techniques

3D modeling techniques allow us to encode the same contents we include in a map but using a different representation strategy. For instance, the richness and Shannon's Index values of the basins, that we represented in the separated analytical maps, are shown here as components of a 3D model. The figure 4 illustrates this technique with "3D diversity models" applied to the four environmental variables we considered in our study. In these models the Shannon` index is represented by a color ramp or value scale in which darker colors are assigned to the highest diversity values. On the other hand, "richness values" are shown by means of extrusions or vertical extensions and, by doing so, 2D polygons turn into solids or 3D blocks. Once the models are created we can work with them in different ways:

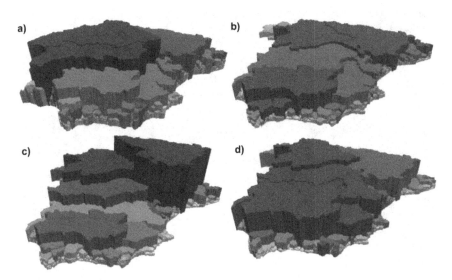

Fig. 4. Diversity models: a) soils, b) lithology, c) vegetation series and d) quaternary deposits

- Analyze each model separately to identify global and zonal patterns and local or particular cases as we made in the map reading and interpretation phase
- Compare several models simultaneously to identify different spatial properties of the basins.
- Create and store perspective or 3D views that best convey the previous findings.
- Create animations that show the results and most relevant findings in a "dynamic" manner. Alternatively these animations can be exported to video to exchange with other users.

In order to illustrate the previous statements we mention here some examples of 3D models interpretations:

- All 3D models show clearly the increase of the richness and Shannon's index values as the rank/area of the basin increase. So, as we move from the coastline to the center of the study area we get higher and darker blocks in the four models.
- Looking at the center of Spain, it seems clear that lithology and quaternary deposits richness values are rather similar, the soil richness values varies slightly while the vegetation series richness change drastically.
- Looking at a specific basin in the four models we can appreciate multiple spatial properties simultaneously. For instance, the Ebro river shows a very high vegetation series richness, while it has lower soil and quaternary deposits richness values in relation with its neighbors of the same rank (e.g. Duero and Tajo rivers).
- If we observe one model at a time we can reach specific findings and interpretations. For example, looking at the 3D vegetation series diversity model we can clearly appreciate a North- South decrease of richness (see the sequence of the Ebro-Duero-Tajo and Guadiana river basins).

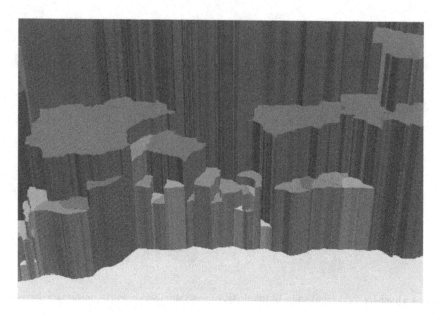

Fig. 5. Details of soil diversity in the South of the Iberian Peninsula

Apart from the previous general findings and interpretations, we can focus in specific zones of any of the models. As an example, figure 5 shows some details of the soil diversity model in a small area of the South of the Iberian Peninsula. We can note various aspects and make some comments:

- Several basins of the same rank have different richness, what is shown by differences in the height of the extruded polygon. From a geographical point of view this means that some basins may have a higher degree of complexity as the soil landscapes are made up of a larger number of soil types present in the area.
- We can identify basins of the same rank and richness, but different Shannon' index (soil diversity or pedodiversity). The darker the 3D block the higher the diversity and more homogeneous is the distribution or relative abundance of the soil types within the basin. On the contrary, the lighter the colors the smaller is the diversity and more heterogeneous (less equitable) is the soil distribution in the basin.

An additional way to analyze and represent a spatial distribution is to compute the differences of all the attribute values with the mean value of the statistical series. This strategy produces a double list of positive and negative values that can be extruded upwards and downwards and represented by green and red colors respectively. As an example, to illustrate this concept, it has been applied to the soil richness values in the Iberian Peninsula shown in the figure 6. Because there are many small area basins (rank 1 and 2) along the coastline, the average is low. On the other hand, the basins in the central part with higher richness values, because of their rank/area properties, appear very dominant replicating the radial pattern we described in the map reading process.

Finally, besides the 3D modeling techniques described previously, we can also display, explore and analyze models in on-line virtual globes such as ArcGlobe and Google Earth. These globes use spherical surfaces as reference surfaces.

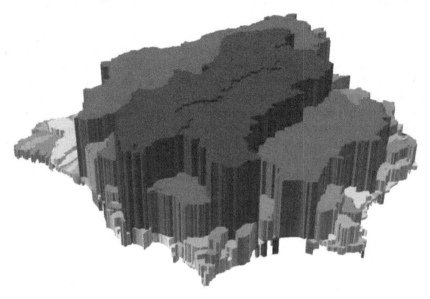

Fig. 6. Model with soil richness differences to mean value in the Iberian Peninsula

In short, 2D mapping and 3D modeling techniques complement each other when analyzing overall, zonal and detail aspects of a geographical phenomenon.

3.3   Zonal Statistical Analysis with Raster Techniques

The previously described mathematical and geomatic studies about the geodiversity in the Iberian Peninsula and Spain have been conducted using the drainage basins as sample unit for the analysis. However, we can consider other ways to gain supplemental information about the diversity of the environmental variables. For this reason, the approach we are going to follow now is to observe and compute the diversity statistics variation with the components of the 3D environment: elevation, slope and orientation or aspect. In order to achieve this objective, a 3D model of the project area has to be built. GIS programs allow us to model reality with surfaces (TIN, raster, terrains) and features (2D, 3D) in order to create virtual landscapes of the study area.

Within GIS technologies, the raster data structure provides the richest modelling environment and operators for spatial analysis. Raster analysis techniques include a set of statistical functions which makes descriptive statistics a part of our geographic analysis. We can use statistical functions to identify trends and spatial patterns or particularities of the geographical distributions. They also allow us to detect and evaluate changes in the landscape. The statistical functions are usually classified as follows: cell statistics, neighborhood or focal statistics and zonal statistics. After the previous 3D visual exploration, the get familiar with the study area, we analyze landscape characteristics using zonal statistical analysis with raster techniques. Figure 7 gives an overview of the process. Any of the 3D components of the landscape can be represented by a separate 3D surface which can be adequately reclassified. This process change the continuous surfaces into stepped surfaces. By doing so, we obtain three raster surfaces with elevation zones, slope zones and aspect zones respectively. On the other hand, we have four vector layers that describe the environment variables (soils, lithology, vegetation series, and quaternary deposits) in the area of interest (AOI). These layers usually have thousands of thematic polygons and are converting into a raster format (VRC) for the analysis. Once we have the zone layers and the thematic layers in raster format (also called "value layers") we can conduct the zonal statistical analysis. Instead of a new raster, the zonal statistic function produces a table of descriptive statistics and an optional graph or chart. The statistics are based on the value field (e.g. soil type) of the value raster (e.g. elevation zone). The right part of figure 7 shows an example of this type to illustrate the workflow.

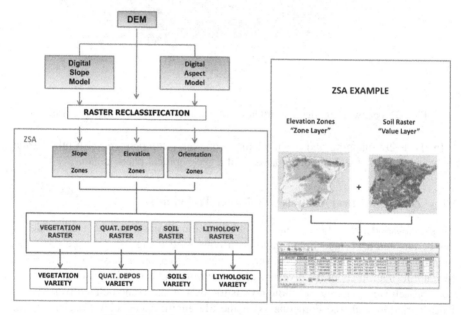

Fig. 7. General schema of the zonal statistical analysis

Because we consider four environment variables and three groups of zones (elevation zones, slope zones and aspect zones), we carried out the raster analysis workflow 12 times. In the first study we analyzed the behavior of the soil types with the elevation zones. In this case, because of the small scale, we used soil associations that are represented in the map as Soil Mapping Units (SMUs) or legend units. In any of the processes we obtain, as output, a table of descriptive statistics that we can also visualize by means of graphs. These statistics will be meaningful or meaningless depending whether the cell values are measurements or codes. Because the value field is a code (soil type or SMU code), not all the statistics are meaningful and only "Variety", "Majority" and "Minority" are useful in this case. The "Majority" represents the most often soil type in each zone. Similarly the "Minority" field stores the least often SMU. Though this information may be useful for some studies, we rather focus in the "Variety" statistics because reflects the number of different soil types in each zone, so this means that coincides with the "richness" concept in biodiversity. For this reason we chose this variable to be displayed in the associated graph. The graph displays in the X-axis the five elevation zones and in the Y-axis the variety values in each zone. So, the image reflects a clear and steady decrease of the number of soil types or SMUs ("richness") as the elevation increase.

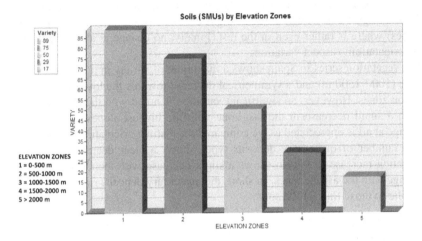

Fig. 8. Change of soil (SMUs) variety with the elevation zones in Iberian Peninsula

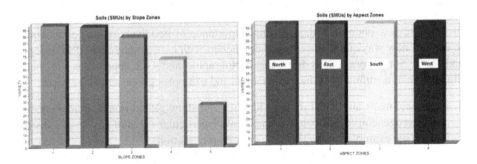

Fig. 9. Change of soil (SMUs) variety with the slope zones (left) and aspect zones (right) in the Iberian Peninsula

The next step was to conduct a similar study to analyze the variation of soil types with the slope zones and orientation or aspect zones. For the particular case of slope zones, their class limits were obtained after a first tentative approach with the "natural breaks" (Jenks) classification method, and then, rounding off the numbers for an easier map reading. Figure 9 records these results and shows a significant decrease of the number of soil types as slope increase (left image), but not as strong as with elevation, and no change is observed with the orientation zones (right image).

Similar procedures have been applied to analyze the other variables (lithology, vegetation series and quaternary deposits) in relation to the three described zone layers in order to study their particular spatial properties. We just need to recall at this point that these variables are analyzed in the peninsular Spain, as study area, because Portugal data was not available at the moment of the study. The main findings are mentioned here:

- The number of lithology types is similar in the first zones, up to 1500 m of elevation, and then decrease quite significantly in the upper zones. On the other hand, the lithology variety only shows a slight decrease in the last slope zone (highest slopes) where is rather even in the rest. Finally, no change at all is observed with the orientation o aspect variation.
- The vegetation series in Spain shows a maximum of variety in the middle elevation zone (1000-1500 m) and a symmetrical decrease towards the lower and higher elevation zones. However, no significant variation is noted with the slope zones apart from a small decrease in the last zone with the highest slopes. Besides that, no change at all is appreciated as we compared the distinct orientation zones.
- The number of quaternary deposits types shows a clear decrease whenever the elevation increase. It is also shows a slight decrease with the slope zones and no change with the aspect zones. In short, the quaternary deposits richness patterns are similar to those find for the soils.

4   Conclusions

Geodiversity is described by many soil and earth scientists as an important factor to estimate the ecological, environmental and social value of a territory. Because the literature review underlines the growing importance of geodiversity studies is worth to explore new ways to analyze this phenomenon with cartography and GIS concepts and tools. Some conclusions from this cartographic and GIS approaches are mentioned here as a summary of the explanations and examples given across the paper:

1. Analytical maps are useful tools to identify and show global trends, homogeneous and heterogeneous zonal patterns and local or particular cases of any spatial distribution (see examples in the thematic cartography section).
2. Polyanalytic maps enable the simultaneous representation the different variables at the same time. This allows the analysis and comparison of several spatial properties to identify relationships or recognize areas of special characteristics.
3. 3D models are also a very visual and attractive manner to reveal the structure and main patterns of a geographic distribution. Once the models are created, we can analyze each model separately or compare several models simultaneously (see details and examples in the 3D visualization section).

4. The zonal statistical analysis offers some interesting features such as the study of richness variation with elevation, slope and orientation zones. These studies supplement the geodiversity analysis based on drainage basins and allow us to gain new findings and conclusions as described in the paper.
5. The results of the geodiversity studies might be used in environmental management and assessment activities, the design of natural reserves or geoparks or the elaboration of environmental legislation.

In short, we have tried to illustrate how maps and Geographic Information products contribute to the process of knowledge acquisition and are very important elements supporting scientific investigations.

Acknowledgments. We are grateful to our Master Student Laura Bernardino Velasco for her cooperation, hard work and motivated participation in the Iberian Peninsula/Spain Project.

References

1. Benito-Calvo, A., Pérez-González, A., Magri, O.Y., Meza, P.: Assessing regional geodiversity: Thei berianpeninsula. Earth Surface Processes Landforms 34, 1433–1445 (2009)
2. Bertin, J.: Graphics and Graphic Information Processing. de Gruyter, Berlin (1981)
3. Bertin, J.: Semiology of Graphics: Diagrams, Networks, Maps. University of Wisconsin Press, Madison (1983)
4. Bertin, J.: Semiology of Graphics: Diagrams Networks Maps. ESRI Press, Redlands (2010)
5. Heer, J., Bostock, M., Ogievetsky, V.: A Tour Through the Visualization Zoo. Communications of the ACM 53(6), 59–67 (2010)
6. Carrera, C.: Implantación de un GIS en el Centro de CC. Medioambientales. Technical Report. Technical University of Valencia (UPV), 155 p. and annexes and maps (1998)
7. Ibáñez, J.J., Jiménez-Ballesta, R., García-Álvarez, A.: Soil landscapes and drainage basins in Mediterranean mountain areas. Catena 17, 573–583 (1990)
8. Ibáñez, J.J., De-Alba, S., Bermúdez, F.F., García-Álvarez, A.: Pedodiversity: Concepts and measures. Catena 24, 215–232 (1995)
9. Ibáñez, J.J., Caniego, F.J., San-José, F., Carrera, C.: Pedodiversity–area relationships for islands. Ecol. Model. 182, 257–269 (2005a)
10. Ibáñez, J.J., Caniego, F.J., García-Álvarez, A.: Nested subset analysis and taxarange size distributions of pedological assemblages: implications for biodiversity studies. Ecol. Model. 182, 239–256 (2005b)
11. Ibáñez, J.J., Effland, W.R.: Toward a Theory of Island Pedogeography: Testing the driving forces for pedological assemblages in archipelagos of different origins. Geomorphology 135, 215–223 (2011)
12. MacArthur, R.H., Wilson, E.O.: The Theory of Island Biogeography, p. 218. Princeton Univ. Press, Princeton (1967)
13. Magurran, A.E.: Ecological Diversity and Its Measurement, p. 179. Croom Helm, London (1988)
14. McIntosh, R.P.: An index of diversity and the relation of certain concepts to diversity. Ecology 48, 392–403 (1967)

15. Morita, T.: Reflections on the Works of Jacques Bertin: From Sign Theory to Cartographic Discourse. The Cartographic Journal 48(2), 86–91 (2011)
16. Ormeling, F.: Cartography as a Tool for Supporting Geospatial Decisions. GIM International 27(8) (2013)
17. Virrantaus, K., Fairbairn, D., Kraak, M.J.: ICA Research Agenda on Cartography and GI Science. The Cartographic Journal 46(2), 63-75(13) (2009)

The Agent-Based Spatial Simulation
to the Burglary in Beijing

Chen Peng[1] and Justin Kurland[2]

[1] Policing Information Engineering Institute,
People's Public Security University of China, Beijing, People's Republic of China, 102600
uctzpch@gmail.com
[2] Department of Security and Crime Science, University College London,
35 Tavistock Square, London, England, WC1H 9EZ
uctzjku@live.ucl.ac.uk

Abstract. Since the Agent-based simulation tool was introduced into criminology research, most work concentrated on crime theory validation or hypothesis testing, little was contributed to crime spatial pattern replication. In this paper, using street network and subway network as the landscape and proposing a statistic-based instead of predefined human mobility pattern as the individual's routine activity, the spatial distribution of burglary in Beijing is simulated and valid by the actual pattern. The result indicates that the Agent-based modeling method partly detects the crime hotspots and the spatial pattern of crime, and specifically the crime level on the nodes with different accessibility is proved to be identical to the actual one. The study made in this work demonstrates that Agent-based modeling is a potential tool to predict or explain crime pattern in space, and also some further work which aims to improve its validation is discussed in the end of this paper.

Keywords: Agent, Crime hotspot, spatial pattern, simulation.

1   Introduction

During the past few years, as the development of Complex Adaptive System (CAS) theory, the Agent-based simulation tool was introduced into criminology research. Traditionally, the criminology was viewed as social science and numerous relevant studies were concentrated on social investigation as well as data analysis to valid crime theories. But overall, it is accepted broadly that the crimes being distributed around the corners in cities are generated from the interactions between people and people as well as people and environment, so using traditional methods to study crime meet some difficulties in disclosing the dynamics behind the crime activities. However, CAS does.

The first work about Agent-based simulation about crime was proposed by Brantingham and Brantingham in 2005 [1]. In their work, they suggested a framework how to simulate individual's activities in space using Agent. Following their work, Malleson, borrowing from Brantinghams' ideas, constructed an artificial city environment

B. Murgante et al. (Eds.): ICCSA 2014, Part IV, LNCS 8582, pp. 31–43, 2014.
© Springer International Publishing Switzerland 2014

and simulated how burglary occurs within the backcloth of human routine activity [2]. After that, Groff contributed crime simulation by proposing a work in which the street network in Seattle was used to carry on individuals activities, and her work indicated that people taking the risk of robbery was influenced by the intervals people spending outside of their house [3][4][5]. Further, Malleson advanced Agent-based simulation of crime by integrating population into the model [6].

Since much work has been done on crime simulation with Agent-based model, some disadvantages still exist. One of the highlighted problems is that the purposes of Agent-based simulation work are testing crime theory or relevant hypotheses, but replicating of crime pattern in space was seldom taken into account. Thus, in this paper, the spatial pattern of burglary in Beijing is replicated using Agent-based modeling method, and its validation is tested with actual data.

2   Landscape Definition

The environment in which offenders, victims and passers-by lives include various elements, such as road, street, square, home, entertainment sites, school, office, factory, government, et al. However, in backcloth of individual movement all these can be simplified into two types of structure: node and path [7]. The nodes are where people spend their time living, working, entertaining, and path connects the nodes so that people could reach any place they want. Currently, as modern traffic becomes more and more convenient, many people prefer to take public traffic to shorten the distance of traveling, thus, the landscape where individuals perform routine activities could be represented by integration of street network and traffic network (see Figure 1).

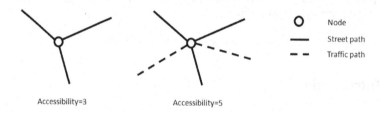

Fig. 1. Topology structure of landscape

In Figure 2 the street network and subway network in Beijing is demonstrated. The subway network is selected as the traffic network because it is the busiest line across the city, each day millions of people take the line to travel. The topology of street network and subway network are analyzed using software package ArcEngine, which is a development tool in ArcGIS. The outcomes show that the streets are intersected into 13008 nodes and 20975 paths, namely each node has 3.22 connections on average. While the subway lines are composed of 119 nodes and 131 edges, each node has 1.1 connections on average.

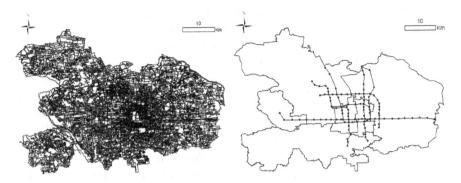

Fig. 2. Street network and subway network in Beijing

3   Mobility Behavior Definition

The individual's mobility pattern was well described by crime pattern theory [8]. Brantingham and Brantingham thought people performed their routine activities in the places around the nodes and paths where they feel safe and comfortable, and they called places awareness space. Thus, people's routine activities are described as the movements in the awareness spaces.

In previous work about crime simulation, the individual's movement in space was predefined [4][5][6], which means the agent moves along the fixed routes following the shortest path between any pair of nodes. Additionally other researchers let agents move totally randomly in space. Actually, both treatments neglect the aspects that people's movement is the integration of deterministic and stochastic. People indeed travel between their homes, schools, companies, entertainment sites in lowest cost, but meanwhile they also occasionally walk aside to take a drink on the way home or work, or go to see a friend on holiday. So another way of simulating agent movement behavior is proposed. Based on Gonzalez's work in which he detected the mobile phone's signal and discovered that people's staying time obeys heavy-tail distribution [9], a statistic-based mobility behavior rule is suggested by Ni [10]. However, his model failed to consider people have to travel via different networks. So after making corrections to the model, a new mobility behavior definition is described below:

1) *Initialization*: each individual agent h is initially located at node i, which can be thought of as their home location.
2) *Traveling*: At each time step, individuals leave the node they are currently located on with a probability of p_{move} and choose either to move along the paths (street or subway network) with probability p_{move}. At each time step, a uniform random number generator is used to determine which choice is made. For example, in the case of deciding which network (street or subway) to move along, if the random number generated (μ) is less than p_{select}, individual h will walk on the street network; otherwise individual h will move using the subway. If the agent decides to move, it randomly selects one of the nodes directly connected to their

current location from the relevant network and moves to one of the adjacent nodes; otherwise, the agent remains at its current location.

3) *Waiting*: when agent h moves to a node j, it is assigned a waiting time interval t_w drawn from a heavy-tail distribution $p(t_w) \sim t_w^{-(1+\lambda)}$, where $\lambda > 0$ and $1 \leq t_w \leq T_w$. The agent will wait for t_w time steps at node j before it moves again.

4) *Returning*: after waiting t_w time steps at node j, agent h returns back to node i (i.e. the home node) with probability p_{back}, and then proceeds from step 2); otherwise, the agent will travel to another node k randomly selected from the neighboring nodes of node j, then followed by step 3).

After introducing the mobility rule, the individual's movement behavior is dominated by statistical parameter p_{back}, p_{move}, p_{select}, t_w and λ instead of unchanged routes.

4   Agent Rules

According to routine activity theory [11], three types of agent are defined in the model: police, offender, and civilian. For the civilian agent, their homes are the potential targets of offenders and they perform their routine activities by using the mobility behavior model. But for the other two agents, the police is the crime reduction and prevention force while the offender is the crime opportunity searching agent.

Police
The task of police in the model is to detecting offender and preventing burglary occurring. So police includes basically three functions, where are learning, decision-making and mobility.

Learning. The police in the model need to learn how many crimes have occurred on the node where he is staying and directed neighbored ones so that he can decide where to move. So the police learning function is to compute the count of burglary occurred on the nodes in past intervals.

Decision-making. As the main task of police is to prevent or stop the burglary occurring, the decision-making process is to select the targeted node which owns the most frequent burglary. So in this step the police will reorder the computed burglary table and choose the highest crime-level-node as the target.

Mobility. When the targeted node on which most burglaries occurs is selected, police will directly move to that node. Otherwise if the node meeting the condition does not exist, police will randomly select a neighboring node and move.

It could be inferred that the three functions have to be iterated consecutively in each time step, thus the police will keep moving in space all the time.

Offender
The offender's functions are relatively complicated compared to police because in addition to the basic routine activities like civilian he has several stages to change from a civilian into formal criminal. The first is motivation emergence. Many previous studies support the view that economy's rise or reduction will lead to the

change of crime rate [12][13]. The reason for this phenomenon is that the unemploy-ment produced from economy's fluctuations has deeply impact on crime rate, which means the people with lower incomes would choose to commit property offence to maintain their lives. Thus the personal owning property level directly decides whether the offender would be risked to commit crime. The second stage is decision-making process. The rational choice theory indicated that offender have to balance his risk and income from committing crimes [14]15]. The potential risk comes from the being discovered by passers-by or witnesses, being arrested by police or stopped by guar-dians. While their incomes are consist of not only being the money or property from offence, but also the satisfactory and accomplishment feelings. In following parts two functions are described.

Motivation emergence. A variable named *personal property* was defined to describe offender's identity. When offender's personal property is zero, its offending motiva-tion emerges and the agent becomes a potential offender who is ready to commit crime [4]. Once he has changed from civilian into a potential offender he will keep searching for suitable targets for maintaining his lives all the time. If the offender's personal property is non-zero, the agent will keep maintaining in civilian identity and perform the routine activities, but at meantime he has to consume 1 unit of property each day.

Decision-making. When the agent is motivated to become a potential offender, he would become into the predatory criminals who keeping searching for suitable targets [16]. During this process one factor that influences it's succeeding or failing to finish the illegal activities is the guardians. The guardians could be the police or passers-by, they play the roles of preventing or stopping burglary on one site, however, their roles are different. The ones who located in their awareness space are called formal guar-dians because they would stop the property crimes occurring in strong wills. As a contrary, the ones located in their non-awareness space are called informal guardians who would prevent property crimes occurring in lower will. An investigation made by Reynald indicated that 23% of passers-by would intervene when they witness abnor-mal actions [17]. Therefore, in this work 20% is measured as the guardianship to the civilian who locate in their non-awareness spaces, but to the ones who locate in, their crime prevention level is defined as 100%. In this sense, on one node 5 guardians will totally stop offender from committing crimes. However, it doesn't mean offender have no chance in this case, the offender still could succeed in committing crime when guardians number is lower than 5, but some risks have to be paid by them. Thus, another variable – *offending risk* – is defined to describe if offender could make a successful offence on the node. The variable is generated from a random number generator, if the number is lower than the existed guardianship the offender would drop off the offending action, but if the generated number is higher than the guardian-ship the offender would commit crimes in risks.

In Figure 3 the motivation and decision-making processes of offender is demon-strated. It could be referred from the flow chart that offender has to experience several stages and make multi- decisions to commit a successful crime, but before he makes it he would keep iterating this process until he does an offence and gets benefits from it.

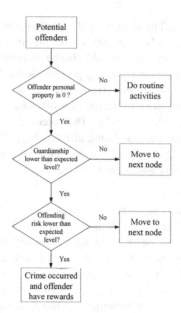

Fig. 3. Motivation and decision-making process of the offender agent

5   Initial Conditions and Parameter

Before simulation is started one work has to be made is to populate the civilians, police and offender agents on the nodes. Based on the principle that each node has to be
covered by at least one civilian agent, the number of civilian agents deployed on the
node is determined according to the following rules: if the population density on node
A is four times to the one on node B, there will be one civilian agent placed on node
B while four on node A. The police agents are initially placed randomly on the nodes
and their proportion to the civilian is kept in lower level, namely 1%. The number of
offender agent is also controlled at 1% of the total civilian agent because the population living in lower standard in Beijing is estimated to be 1.1%. As there is no data
available about where these people live, they are initially distributed in space in proportion to the street node density by output areas (There are eight output areas in
Beijing, see Appendix A).

   Another condition has to be predefined is the time step. It is assumed that each individual agent takes 4 minutes on average to finish the distance between any pair of
node however connected, so each day is consist of 360 time steps.

   In Table 1 the parameters and other initial conditions are listed. The parameters
are referenced from Beijing Annual Statistic Book (2007), which includes both
demographic and traveling pattern data.

Table 1. Parameters to the simulation work

Parameter	Rationale
$N_{civilian}$=17716	Total number of civilian agents
$N_{offender}$=177	Total number of potential offenders
N_{police}=177	Total number of police agent
p_{jump}=1.0	The probability of civilian and offender agent leaving their home node
p_{select}=0.17	The probability of civilian and offender agent traveling in tube network
p_{back}=0.5	The probability of civilian and offender agent returning to their home node
T_{max}=360	Time step in each day
λ=0.6	The parameter describing how people spend their time across a day
personal property=(1,5)	The variable determine the identity of offender agent, generated from a uniform random number generator
offending risk=(0,1)	The variable of offender agent risked to offence, generated from a uniform random number generator
$E_{guardianship}$=0.2	The guardianship non-owner of the node prefer to intervene property crime
iterations=20	The times of the simulation running to smooth randomness

6   Result

The model is configured and simulated using a personal computer. It takes 10 hours to run for 20 times to smooth the randomness and the final result is the mean level of all variables. The result is compared to the actual data for validation from perspective of visualization and crime level on nodes in different accessibility.

Firstly, the simulated crime points and actual data are displayed in Figure 4, and the mean center as well as standard deviational ellipse are used to describe the central as well as directional tendency of a point pattern. Results indicate that the distance between the actual and simulation distribution mean center is 3,550.34 meters, but is remarkably small when considering that these eight output areas in Beijing are 1,368.35 km^2 in size. The resulting standard deviational ellipses indicate that the real residential burglary pattern is a little different from the simulated pattern. From the picture it could be seen that the pattern generated from simulation distribution is more dispersed than the actual distribution of residential burglary events but their orientations are similar.

The hotspots of both crime patterns are studied with kernel density estimation method (KDE). The cell sizes of the two analyses are treated equally but the searching radius are varied so that crime hotspots in different levels could be identified and comparable. It could be inferred from the Figure 5 and 6 that simulated crime patterns demonstrate a similar spatial distribution with actual data. The hotspots identified in both analyses mainly concentrate in the zones around the center areas, but in simulated pattern the hotspots on the edge areas are not valid by the actual result.

In further, the count of burglary on each node is computed and treated as the attributes of the polygon where the node sitting in. After that the local Moran's I which is a useful statistic tool in Local Indicator of Spatial Analysis is used to identify the hotspots that are in significant statistic level [18]. The street blocks where crimes are located in are treated as the basic unit to make the statistic, and the result is demonstrated in Figure 7. It could be seen from the graph that some significant hotspots are detected by the model in the zones around the center areas, but it could also be found that some hotspots are failed to be simulated by the model. For example, the hotspots simulated by the model concentrated on North-West, South-West and North-East corner of Beijing are proved to be invalid by the actual pattern.

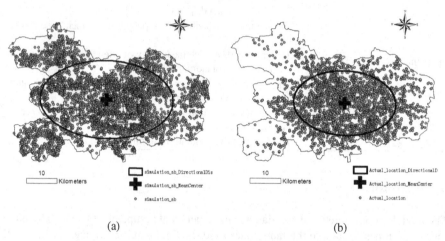

Fig. 4. Crime distribution in space, (a) simulated pattern, (b) actual pattern

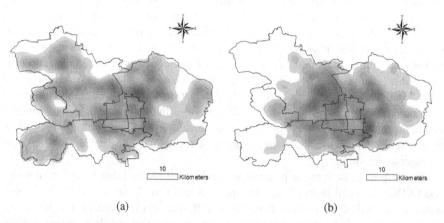

Fig. 5. Crime distribution identified by KDE in space, (a) simulated pattern, (b) actual pattern, searching radius is 3.0Km, cell size is 250 meters

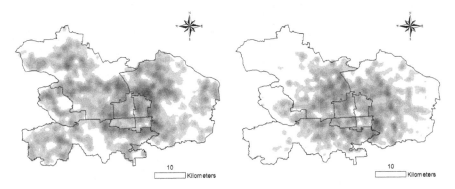

Fig. 6. Crime distribution identified by KDE in space, (a) simulated pattern, (b) actual pattern, searching radius is 1.5Km, cell size is 250 meters

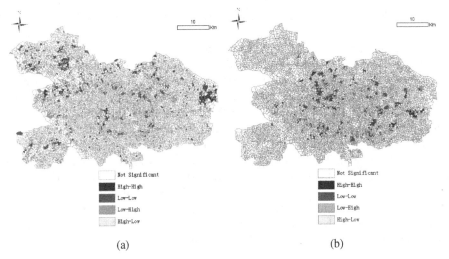

(a)   (b)

Fig. 7. Crime distribution identified by Local Moran's I in space, (a) simulated pattern, (b) actual pattern

From descriptive analysis in visualization it can be seen that the Agent-based modeling study partly identifies the hotspots and also some clues are inferred from the result. First, the original location of offender might have impact on crime distribution in space. As indicated by theory of journey to crime (JTC) that offenders are often searching for crime opportunity in least effort principle [19], therefore the initial locations of criminals would decide where crimes occur and concentrated. So, the locating initial position of offender agent by output area might cause the improper distribution of hotspots. Second, the unreported offence is probably another reason to cause the difference between simulated and actual crime pattern. Based on previous research made by Zhang who reported that it was estimated that only 77% residential burglary were recorded by the police [20], so in this work it is believed that more than 20% unrecorded offence is not known and brings negative influence to outcomes.

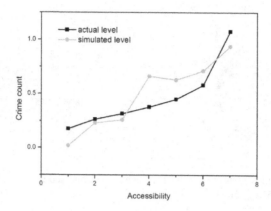

Fig. 8. Simulated crime count against actual count by node accessibility

Finally, the influence of node accessibility to the crime count is tested. As described by previous literatures, the accessibility to the node dominates how the place being reached by civilian, offender and police, then the crime pattern is influenced [21][22][23]. So the simulated crime count against actual crime count by varied accessibility is displayed in Figure 8. It could be inferred that both simulated and actual crime pattern demonstrate higher crime count concentrated on the node with higher accessibility, or in another words, the higher crime risk occurs on the places that are more easily accessed. This is identical to the previous work made by Newman and Johnson who stated that the crime risk is lowest on cul-de-sac because strangers often go unchallenged for ambiguity in these places [24][25]. To date though how crime distributed by street segment accessibility is still a controversy, the simulation shapes similar distribution with actual one proves the model works well in replicating crime risks in the places by different connection in Beijing.

7   Conclusion

This paper sought to replicate crime distribution pattern in Beijing with Agent-based modeling method. In order to accomplish this target the street network and subway network are treated as the landscape, a statistic-based human mobility behavior rule is proposed as agent routine activities and motivation driving decision-making process to the offender agent is constructed. In study, the spatial distribution of burglary is simulated and compared with the actual crime pattern to valid the model. The visualization descriptive analysis results demonstrate though the hotspots on the edge areas are failed to be detected the locations in center area of Beijing are identified by the model. Finally, the influence of the node accessibility to the crime level is tested, and the findings indicate that the model well replicate the crime level by different connected node.

The simulation result in this paper is proved to be a potential tool to replicate and study crime distribution in space, but more future work have to be made in order to achieve a better effect of crime pattern replication. The first thing is to include offender home locations into the model. As JTC theory has indicated that offenders usually search for targets and commit crimes with least effort, thus the fine-scale information on offenders should be collected and added into the initial condition of simulation. The second issue is that more data about land use and environment should be included. The lands with different functions are often having different attractions to the offenders and civilians. The important sites as shopping center, bar and station usually play the roles of crime generators or crime attractors in shaping crime distribution [26]. Therefore introducing more basic infrastructure data would improve the validation of Agent-based model. Another important problem have to be emphasized is the crime data collection. As more than 20% crime data is in short the replication effect of the model is difficult to be proved, so more accurate crime information will benefit to the model validation.

Acknowledgement. This work was sponsored by National Science Foundation (71203229), and Professor Du Shihong was greatly appreciated for his suggestions in improving this paper.

References

1. Brantingham, P.L., Glasser, U., Kinney, B., et al.: A computational model for simulating spatial aspects of crime in urban environments. In: IEEE International Conference on Systems, Man and Cybernetics, vol. 4, pp. 3667–3674 (2005)
2. Malleson, N., Heppenstall, A., See, L.: Crime reduction through simulation: An Agent-based model of burglary. Computers, Environment and Urban Systems 34(3), 236–250 (2009)
3. Groff, E.R.: Simulation for theory testing and experimentation: An example using routine activity theory and street robbery. Journal of Quantitative Criminology 23, 75–103 (2007a)
4. Groff, E.R.: 'Situating' simulation to model human spatial-temporal interactions: An example using crime events. Transactions in GIS 11(4), 507–530 (2007b)
5. Groff, E.R.: Adding the Temporal and Spatial Aspects of Routine activities: A further test of routine activity theory. Security Journal 21, 95–116 (2008)
6. Malleson, N., Birkin, M.: Analysis of crime pattern through the integration of an agent-based model and a population microsimulation. Computer, Environment and Urban system 36(6), 551–561
7. Brantingham, P.L., Brantingham, P.J.: Nodes, paths and edges: considerations on the complexity of crime and the physical environment. Journal of Environmental Psychology 13, 3–28 (1993)
8. Brantingham, P.L., Brantingham, P.J.: Mobility, notoriety, and crime: A study of crime patterns in urban nodal points. Journal of Environmental Systems 11, 89–99 (1982)
9. Gonzálaz, M., Hidalgo, C., Barbasi, A.: Understanding individual human mobility patterns. Nature 453(06958) (2008)
10. Ni, S.J., Weng, W.G.: Impact of travel patterns on epidemic dynamics in Heterogeneous Spatial Metapopulation networks. Physical Review E 79, 016111 (2009)

11. Cohen, L.E., Felson, M.: Social change and crime rate trends: A routine activity approach. American Sociological Review 44(4), 588–608 (1979)
12. Pyle, D.J., Deadman, D.: Crime and the business cycle in post-war Britain. British Journal of Criminology 34, 339–357 (1994)
13. Deadman, D., Pyle, D.J.: Forecasting Recorded Property Crime Using a Time-Series Econometric Model. British Journal of Criminology 37, 437–445 (1997)
14. Clarke, R.V., Cornish, D.B.: Rational choice. In: Paternoster, R., Bachman, R. (eds.) Explaining Criminals and Crime, pp. 23–42. Roxbury Publishing Co., Los Angeles (2001)
15. Cornish, D., Clarke, R.: The Reasoning criminal: Rational choice perspectives on offending. Springer, New York (1986)
16. Clarke, R.V., Cornish, D.B.: Modeling offender's decisions: A framework for research and policy. In: Tonry, M., Morris, N. (eds.) Crime and Justice: An Annual Review of Research, vol. 6. University of Chicago Press, Chicago (1985)
17. Reynald, D.M.: Guardians on Guardianship: Factors affecting the willingness to supervise, the ability to detect potential offenders, and the willingness to intervene. Journal of Research in Crime and Delinquency 47(3), 358–390 (2010)
18. Anselin, L., Syabri, I., Kho, Y.: GeoDa: An Introduction to Spatial Data Analysis. Geographical Analysis 38(1), 5–22 (2006)
19. Zipf, G.K.: Human Behavior and the Principle of Least Effort. Addison-Wesley Press, Oxford (1949)
20. Zhang, L., Messner, S., Liu, J.: An Exploration of the Determinants of Reporting Crime to the Police in the City of Tianjin. China Criminology 45, 959–983 (2007)
21. Beavon, D.J.K., Brantingham, P.L., Brantingham, P.J.: The influence of street networkson the patterning of property offenses. In: Clarke, R.V. (ed.) Crime Prevention Studies, vol. 2. Criminal Justice Press, New York (1994)
22. Bevis, C., Nutter, J.B.: Changing Street Layouts to Reduce esidential Crimes. Paper presented at the Annual Meeting of the American Society of Criminology, Atlanta, GA (1977)
23. Hiller, B.: Can streets be made safe? Urban Design 9, 31–45 (2004)
24. Newman, O.: Defensible Space: Crime Prevention Through Urban Design. Macmillan, New York (1972)
25. Johnson, S.D., Bowers, K.J.: Permeability and crimes risk: Are cul-de-sacs safer? Journal of Quantitative Criminology 26(1), 89–111 (2010)
26. Brantingham, P.J., Brantingham, P.L.: Criminality of place: Crime generators and crime attractors. European Journal of Criminal Policy and Research 3, 5–26 (1995)

Appendix A: Map of Output Areas in Beijing

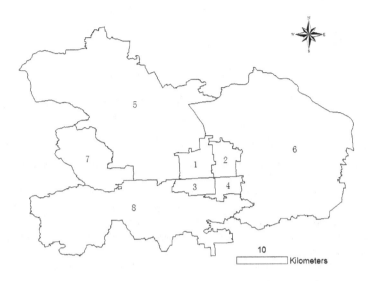

Fig. A.1. Beijing's output areas

Historical Map Registration via Independent Model Adjustment with Affine Transformations

Luigi Barazzetti, Raffaella Brumana, Daniela Oreni, and Mattia Previtali

Politecnico di Milano, p.za Leonardo da Vinci 32
20133 - Milan, Italy
{luigi.barazzetti,raffaella.brumana,
daniela.oreni,mattia.previtali}@polimi.it

Abstract. This paper presents a methodology for historical map registration with a multi-image affine transformation estimated with a simultaneous Least Squares independent model adjustment. The method is based on two set of equations for (i) image-to-ground and (ii) image-to-image points and provides the rectification parameters with a global approach where all the unknowns are simultaneously estimated. It avoids error accumulation of traditional progressive registration approaches implemented in commercial and open-source software packages. Different case studies along with pro and cons are illustrated and discussed.

Keywords: accuracy, adjustment, historical atlas, historical map, Least Squares, precision, registration.

1   Introduction

The integrated use of geo-referenced historical maps and spatial data generated at similar comparable scale is an important tool to understand the different transformations occurred in a territory over the centuries [1-2]. Nowadays, land planning is usually based on an as-built situation analysis along with social and economic analysis integrated with the surrounding region and historical documents. On the other hand, the use of historical maps with new data sources can be intended as an additional tool from different points of view, e.g. the (i) study and preservation of the traditional characteristics of the area (agricultural, residential, touristic, religious), the (ii) identification of the ancient water courses, canal courses and networks, and (iii) the recognition of old settlements and archaeological sites.

The opportunity to overlay historical maps, modern maps (raster and vector), and aerial and satellite images, makes it possible to run processing algorithms and to identify the changes occurred overtime.

The chance to compare and overlay cartographic data requires the co-registration of the data in a common reference system. For this reason, in the case of data processing in GIS environments a preliminary fundamental step is the creation of a structure where the inspection of a "vertical profile" concerns the analysis of the same pixel in terms of spatial coordinates.

B. Murgante et al. (Eds.): ICCSA 2014, Part IV, LNCS 8582, pp. 44–56, 2014.

Historical maps are scanned (turned into digital images [3]) and then georeferenced in a cartographic reference system. This last step is not trivial as many aspects should be taken into consideration: the choice of the reference dataset, the geometric model used to "adapt" the scanned image, the precision and accuracy obtainable, etc. [4-7].

Most GIS software (e.g. ESRI ArcMap, Quantum GIS) or packages for Remote Sensing (e.g. ENVI, ERDAS Autosynch) have special tools for map georeferencing. Their functioning is relatively simple: an existing map is chosen as reference and the measurement of some image-to-ground correspondences (usually point coordinates) allows the estimation of a geometric transformation between scanned image and reference. The operator has (i) to interactively extract a set of corresponding points and (ii) to select a geometric model for image georeferencing, then the scanned image will be resampled and georeferenced.

On the other hand, this procedure is often independently carried out for the different images, meaning that the method is a progressive registration where errors can be accumulated. The solution proposed in this paper tries to avoid this limit by considering the images as a 2D block (like a photogrammetric dataset, see [8-15]) that can be simultaneously registered by Least Squares independent model adjustment [16-18]. The implemented algorithm provides a simultaneous estimate of all the unknowns (transformation parameters and ground coordinates of image-to-image points) by creating two set of equations integrating traditional (i) image-to-ground measurements, (ii) image-to-image-to-ground collimations, and additional (iii) image-to-image points (usually border points). The final result is a global estimation that avoids error accumulation, whereas image-to-image correspondences plays as tie points and strengthen block geometry (Fig. 1), like in photogrammetric projects where tie points and ground control points are combined.

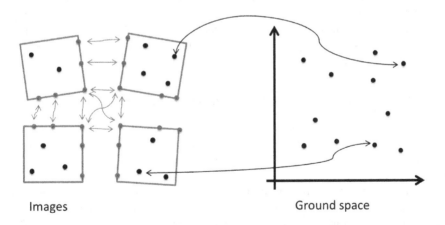

Images   Ground space

Fig. 1. The idea behind the implemented independent model adjustment. Image-to-ground (black) and image-to-image (red) points are adjusted simultaneously with an multi-affine transformation as mathematical model.

The dataset used in this work is already available in the platform 'Atl@nte dei Catasti Storici e delle Carte Topografiche della Lombardia' (Atlas of Historical Cadastres and Topographic maps of Lombardy), that is a project funded by Fondazione Cariplo (http://www.atlantestoricolombardia.it/). It aims to implement a geo-portal for management of historical geo-referenced maps at territorial-regional and cadastral-local scale [19]. The main objective of Atl@nte is to fully combine historical data within geospatial information into decision making processes, addressing the government of territory by Public Authorities (PA) and supporting professional related activities.

2   Mathematical Model for Least Squares Adjustment

This section provides a synthetic description of algorithms and procedures implemented in most GIS packages (section 2.1) and a more complete explanation of the implemented multi-map procedure (section 2.2). The problem of outlier rejection is also addressed (section 2.3).

2.1   Registration with Common Georeferencing Tools

As mentioned in the previous statement, this section briefly presents the standard image-to-ground registration approach implemented in commercial and open source packages. More information can be simply found in the manual. The main limit of this methodology is the independent use of the images, that are processed by using a progressive registration where errors could accumulated. Starting from a set of image pixel-to-ground (e.g. an existing georeferenced map) points, the software calculates a geometric transformation between the images and produces a registered image. Different software have different transformations, e.g. polynomial, homography, similarity, or splines, among the others. One of the most common transformation is the affine, as it preserves straight lines [20].

2.2   Multi-image Affine-Based Registration via Independent Models

The mathematic model across different images includes standard image-to-ground points and image-to-image points. Data need to be stored in an ordered structure including image coordinates (x_{ij}, y_{ij}) and ground coordinates (X_i, Y_i), where i is point index and j image index.

Given an image-to-ground correspondence expressed with homogeneous coordinates (such point can be visible in a single image < image-to-ground > or in a generic number of images < image-to-image-to-ground >), the chosen planar affine model can be expressed as:

$$\mathbf{X}_{ij} = \begin{bmatrix} X_{ij} \\ Y_{ij} \\ 1 \end{bmatrix} = \begin{bmatrix} \mathbf{L}_j & \mathbf{t}_j \\ \mathbf{0}^T & 1 \end{bmatrix} \begin{bmatrix} x_{ij} \\ y_{ij} \\ 1 \end{bmatrix} = \mathbf{H}_j \mathbf{x}_{ij} \tag{1}$$

where:

$$\mathbf{L}_j = \begin{bmatrix} a_j & b_j \\ d_j & e_j \end{bmatrix} \tag{2}$$

is a non-singular matrix, and $\mathbf{t}_j = [c_j, f_j]^T$ a translation vector. An affine transformation includes a non-isotropic scaling , therefore lengths and angles are not preserved under an affinity. On the other hand, this transformation preserves parallel lines, i.e. a very important parameter during historical map registration. Other important invariants are the ratio of lengths of parallel line segments, the ratio of areas, and the line at infinity.

One of the advantage of the proposed methodology is the simultaneous use of both image-to-ground and image-to-image points. This means that two different sets of equations can be formulated depending on the matched point. Equations are then included in the same system that provides a simultaneous estimate of all the transformation parameters (transformation parameters and ground coordinates of tie points) via independent model adjustment with a planar affine transformation instead of a standard 2D affinity.

The equations for an image-to-ground point (i.e. ground coordinates (X_i, Y_i) are known values) can be written as:

$$\begin{cases} \underline{a}_j x_{ij} + \underline{b}_j y_{ij} + \underline{c}_j - X_i = v_{x,ij} \\ \underline{d}_j x_{ij} + \underline{e}_j y_{ij} + \underline{f}_j - Y_i = v_{y,ij} \end{cases} \tag{3}$$

where the underlined terms are the unknown and v are residuals. In the case of image-to-image points (e.g. border points measured in adjacent maps, Fig. 2) the corresponding ground coordinates become additional unknowns. Obviously, the same pint must be visible in at least two images in order to (i) strengthen block geometry (points on the border have a better geometric distribution) and (ii) increase the redundancy of Least Squares adjustment. If such a point is visible in k images, it will provide $2k$ equations and 2 unknowns.

Fig. 2. Some examples of border points, i.e. image-to-image points represented with a "notch" visible in the blue circle with transparency

The observation equations for image-to-image matches are similar but ground coordinates (X_i, Y_i) are now unknown values:

$$\begin{cases} \underline{a}_j x_{ij} + \underline{b}_j y_{ij} + \underline{c}_j - \underline{X}_i = v_{x,ij} \\ \underline{d}_j x_{ij} + \underline{e}_j y_{ij} + \underline{f}_j - \underline{Y}_i = v_{y,ij} \end{cases} \tag{4}$$

The linear functional (including both points – see Fig. 3) and stochastic models have the form:

$$\mathbf{Ax} - \mathbf{b} = \mathbf{v}$$
$$\mathbf{C_{bb}} = \mathbf{I} \tag{5}$$

where $\mathbf{A}$ is the design matrix, $\mathbf{x}$ the unknown vector (including transformation parameters and ground coordinates of image-to-image points), $\mathbf{b}$ the observation vector, $\mathbf{v}$ the residual vector, and $\mathbf{C_{bb}}$ the variance matrix of observations (the identity matrix is used). The rank defect of $\mathbf{A}$ (i.e. no inverse for the singular normal equations) can be removed if at least three non-collinear ground points are included in the adjustment.

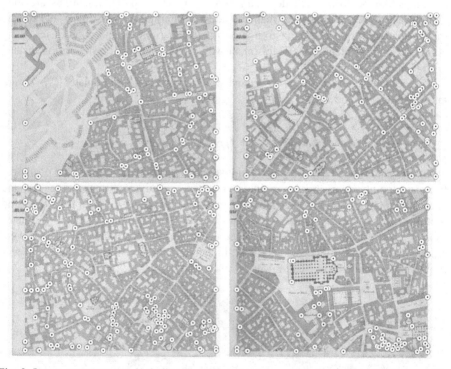

Fig. 3. Image measurements including the different category of point for 4 adjacent scanned maps. More details about the dataset used in this work (urban map of Milan – 1807) are given in section 3.

The linear system of normal equations has the form:

$$Nx = \begin{bmatrix} A_1 & \bar{N} \\ \bar{N}^T & A_2 \end{bmatrix} \begin{pmatrix} x_1 \\ x_2 \end{pmatrix} = \begin{pmatrix} n_1 \\ 0 \end{pmatrix} = n \tag{6}$$

where x_1 contains the unknown transformation parameters (a_j, b_j, c_j, d_j, e_j, f_j) of and x_2 the image coordinates of the image-to-image point projected onto the ground reference system.

The redundancy depends on the number of images (m) i.e., six geometric parameter per every image, the number of image points (n) and the number of 2D points (u) used as image-to-image points. The redundancy (r) of this system becomes $2(n-u-3m)$.

The normal matrix N has a particular banded form, where A_1 is a hyper-diagonal matrix with sub-matrices 6×6 corresponding to individual images, whereas A_2 is made up of a sequence of 2×2 matrices.

The solution (and its precision C_{xx}) is obtained with standard Least Squares techniques:

$$\hat{x} = N^{-1}b$$

$$\sigma_0^2 = \frac{(N\hat{x} - b)^T(N\hat{x} - b)}{r} \tag{7}$$

$$C_{xx} = \sigma_0^2 N^{-1}$$

Data snooping was also included in the adjustment process to discard wrong matches by means of the standardized residuals ($u = v \, / \, \sigma_v$) computed from the covariance matrix of residuals [21].

On the other hand, a more robust methodology was added to the algorithm in order to remove wrong matches before running the LS adjustment algorithm. More details are given in the next section.

2.3 Outlier Detection

In the case of datasets made up of several maps, the interactive measurement of many image points can lead to some gross errors (mismatches). Although these errors can be found during the adjustment phase (data snooping), a more robust procedure was implemented to identify and discard possible mismatches during the matching step.

The RANdom SAmple Consensus (RANSAC, [22]) algorithm between image pairs is run during the manual collimation of both image-to-ground (x_{ij}, y_{ij}) $\leftrightarrow$ (X_i, Y_i) and image-to-image (x_{ij}, y_{ij}) $\leftrightarrow$ (x'_{ij}, y'_{ij}) points. RANSAC estimates the number of trials N and then extracts a minimum number of correspondences to calculate the affine transformation between the considered point. Then, the transfer error is evaluated, and given a threshold T the extraction of inliers is carried out.

The number of trials can be estimated as follows:

$$N = \frac{log\,(1-P)}{log\big(1-(1-e)\big)^{s}} \tag{8}$$

where P is the probability (usually 0.99), s the size of sample (3 in this case), e the expected percentage of outliers. Outliers can be identified and removed with an iterative process carried out with points randomly extracted during which several affine transformations $\mathbf{H}$ are estimated (for the same dataset and any subsample k of correspondences).

The transfer error, i.e. the error for a correspondence from an affine matrix $\mathbf{H}$, is given by the symmetric error:

$$d_r^2 = d(\mathbf{x}, \mathbf{H}^{-1}\mathbf{x}')^2 + d(\mathbf{x}', \mathbf{H}\mathbf{x})^2 \tag{9}$$

The proposed pairwise RANSAC approach is intended as a powerful additional tool to reject outliers before running the independent model adjustment. The operator can use the algorithm when the dataset includes many point matches and several images, whereas for small datasets outliers can be manually removed with a visual check.

3   Case Study

The dataset used in this paper is the urban map of Milan (Italy - scale factor 1:1000) derived from a survey in the years 1807-1808 carried out by Astronomi dell'Osservatorio Milanese di Brera. It is made up of 39 homogeneous plates plus an addition plate showing the position of the single maps (Fig. 4). The digital georeferenced version (global mosaic) can be interactively explored on the website http://www.atlantestoricolombardia.it/.

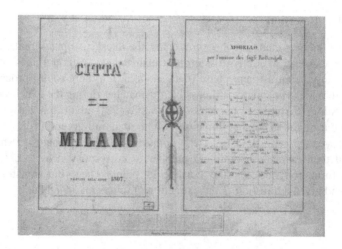

Fig. 4. A detail of the map including the grid index

Data processing with the multi-image affine-based algorithm was carried out with 24 plates. Some images were only registered by means of image-to-image points (border points) in order to check the feasibility of the proposed methodology. Ground points (X_i, Y_i) were measured by means of modern maps in the Gauss-Boaga projection, whereas image points (x_{ij}, y_{ij}) were expressed in pixels. An expert operator measured 1,671 image points, obtaining 3,342 equations and 754 unknowns (several points are visible in more than an image). Global redundancy was 2,588 and sigma-naught after independent model adjustment was ±2.4 m, that is a value higher than expected.

For this reason, the robust RANSAC check was run on all image-to-image and image-to-ground pair combinations, removing 31 matches. The new adjustment (data redundancy is 2,557) gave a sigma-naught of about ±1.2 m and provided the affine transformation parameters of the images and ground coordinates of image-to-image matches.

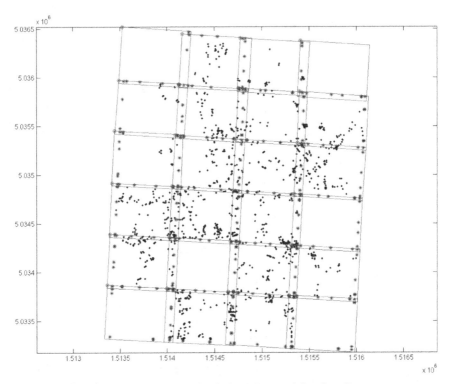

Fig. 5. Map scheme after independent model registration

A scheme of registration results, showing image-to-ground points ("black") and both image-to-image and image-to-image-to-ground (points visible in 1 or more images – "blue") is shown in Fig. 5. As can be seen, 5 maps were registered with border points, whereas the remaining ones have a combination of ground and border points.

Shown in Fig. 6 is the normal matrix **N** of Least Squares along with (left) a zoom on the hyper-diagonal matrix with sub-matrices 6×6 corresponding to affine parameters. This part of the matrix ($\mathbf{A_1}$) can be considered composed of 3×3 matrices for x and y, respectively.

The covariance matrix $\mathbf{C_{xx}} = \mathbf{N^{-1}b}$ provided the precision of unknown parameters. The average precisions of translation parameters $(\bar{\sigma}_{cj}, \bar{\sigma}_{fj})$ of all images were ±1.2 m and ±1.0 m (for X and Y), whereas the average precisions of ground coordinates computed from image-to-image correspondences $(\bar{\sigma}_X, \bar{\sigma}_Y)$ were both ±1.1 m.

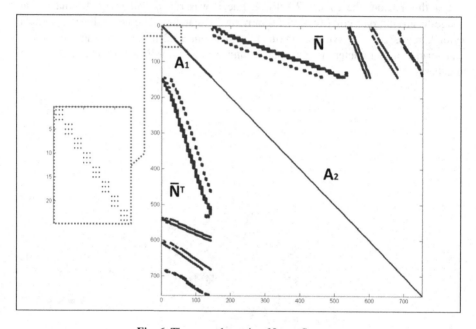

Fig. 6. The normal matrix of Least Squares

A simple inspection (accuracy evaluation) between the registered maps and some ground points from modern cartography confirmed the previous values. The registered maps were generated with the estimated affine parameters along with bilinear interpolation for the radiometric values. Shown in Fig. 7 are some details along with a scale bar. The chosen format is GeoTiff.

A global visualization of registration results including the images resampled and mosaicked starting from the estimated parameters is shown in Fig. 8. The border of the images were manually edited with seamless lines in order to obtain a global mosaic.

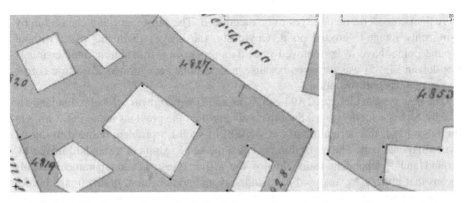

Fig. 7. Points from modern cartography (blue) overlapped on the georeferenced maps

Fig. 8. The final mosaic (GeoTiff format) derived with the proposed affine-based independent model adjustment

A final consideration deserves to be mentioned. The same algorithm can be run by using only image-to-ground point. Obviously, the images (5 in this case) without ground points have to be removed from data processing. In addition, another image (top left in Fig. 5) was removed because ground points are located close to a corner and have a worse distribution.

The new system includes 2,194 equations and 108 unknowns (global redundancy is 2,086). Sigma-naught was ±0.8 m, that is better than the previous value (±1.2 m). On the other hand, the normal matrix (128×128) of such LS problem reduced to the A_1 hyper-diagonal part. The algorithm provided the same results of a standard image-to-ground and independent estimation of the affine transformation parameters. This means that images are analysed independently and no constraint on the image border is provided with the possibility of larger discontinuities between consecutive maps. The previous solution obtained by including additional matches in a combined adjustment is more reliable notwithstanding a worse precision after Least Squares adjustment.

4   Conclusion

This paper presented a procedure for historical map registration where multiple images are simultaneously processed starting from a set of (i) image-to-ground, (ii) image-to-image-to-ground, and (iii) image-to-image correspondences. The method replicates the effect of independent model adjustment with an affine transformation as mathematical model.

The main advantages include the use of additional tie points (like in photogrammetric projects) that improve block geometry (especially for areas close to the borders of the images) and can reduce the number of ground point measurements.

The implemented package performs a rigorous Least Squares estimate of unknowns (affine transformation parameters and ground coordinates of image-to-image points) and their precision, obtaining a rigorous solution in a functional and stochastic sense.

Moreover, as maps are processed by considering a block of images, the algorithm avoids standard registration techniques where a single map is independently registered with some image-to-ground measurements. Obviously, the opportunity to include image-to-image matches requires measurements in different images, i.e. points should be stored in an ordered structure along with their multiplicity. This means that the method is more reliable and robust than existing approaches implemented in software packages, notwithstanding the more time consuming multi-map matching phase.

Acknowledgments. This work was partially supported by Fondazione Cariplo, Bando 2008 "Creare e divulgare cultura attraverso gli archivi storici" (rif.n.2009-2006 Divulgare Cultura).

References

1. Cuca, B., Oreni, D., Brumana, R.: Digital cartographic heritage in service to the society: Landscape analysis for informed decision making. In: 18th International Conference on Virtual Systems and Multimedia (VSMM), pp. 499–506 (2012)
2. Brumana, R., Oreni, D., Cuca, B., Rampini, A., Pepe, M.: Open access to historical Atlas: Sources of information and services for landscape analysis in an SDI framework. In: Murgante, B., Gervasi, O., Misra, S., Nedjah, N., Rocha, A.M.A.C., Taniar, D., Apduhan, B.O. (eds.) ICCSA 2012, Part II. LNCS, vol. 7334, pp. 397–413. Springer, Heidelberg (2012)
3. Fleet, C.: The ABC of map digitization, Map Library. National Library of Scotland (2009)
4. Boutoura, C., Livieratos, E.: Some fundamentals for the study of the geometry of early maps by comparative methods. e-Perimetron 1(1), 60–70 (2006)
5. Bitelli, G., Cremonini, S., Gatta, G.: Ancient map comparisons and georeferencing techniques: A case study from the Po River Delta (Italy). e-Perimetron 4(4), 221–233 (2009)
6. Jenny, B., Hurni, L.: Studying cartographic heritage: Analysis and visualization of geometric distorsions. Computers & Graphics 35(2), 402–411 (2011)
7. Brovelli, M.A., Minghini, M., Giori, G., Beretta, M.: Web Geoservices and Ancient Cadastral Maps: The Web C.A.R.T.E. Project. Transactions in GIS 16(2), 125–142 (2012)
8. Granshaw, S.I.: Bundle Adjustment Methods in Engineering Photogrammetry. The Photogrammetric Record 10(56), 181–207 (1980)
9. Fraser, C.S.: Photogrammetric measurement to one part in a million. Photogrammetric Engineering & Remote Sensing 58, 305–310 (1992)
10. Kraus, K.: Photogrammetry: Geometry from Images and Laser Scans, 2nd edn., p. 459. Walter de Gruyter (2008)
11. Mikhail, E.M., Bethel, J., McGlone, J.C.: Introduction to Modern Photogrammetry, p. 479. John Wiley & Sons (2001)
12. Barazzetti, L., Remondino, F., Scaioni, M.: Combined use of photogrammetric and computer vision techniques for fully automated and accurate 3D modeling of terrestrial objects. In: Videometrics, Range Imaging and Applications X, Proc. of SPIE Optics+Photonics, 2–3, vol. 7447 (2009)
13. Furukawa, Y., Curless, B., Seitz, S.M., Szeliski, R.: Towards Internet-scale Multi-view Stereo. In: IEEE Conference on Computer Vision and Pattern Recognition CVPR, San Francisco (2010)
14. Roncella, R., Re, C., Forlani, G.: Performance evaluation of a structure and motion strategy in architecture and cultural heritage. IAPRS&SIS, 38(5/W16) on CD-ROM (2011)
15. Barazzetti, L., Forlani, G., Roncella, R., Remondino, F., Scaioni, M.: Experiences and achievements in automated image sequence orientation for close-range photogrammetric projects. In: Proc. of SPIE Optics+Photonics, Munich, Germany, May 23-26, vol. 8085 (2011)
16. Williams, V.A., Brazier, H.H.: The method of adjustment of independent models. Huddersfield test strip. The Photogrammetric Record 5(26), 123–130 (1965)
17. Williams, V.A., Brazier, H.H.: Aerotriangulation by independent models: A comparison with other methods. Photogrammetria 21(3), 95–99 (1966)
18. Crosilla, F., Beinat, A.: Use of generalised Procrustes analysis for the photogrammetric block adjustment by independent models. ISPRS Journal of Photogrammetry and Remote Sensing 56(3), 195–209 (2002)

19. Cuca, B., Brumana, R., Scaioni, M., Oreni, D.: Spatial Data Management of Temporal Map Series for Cultural and Environmental Heritage. International Journal of Spatial Data Infrastructures Research 6 (2011)
20. Hartley, R.I., Zisserman, A.: Multiple View Geometry in Computer Vision, p. 672. Cambridge University Press (2004)
21. Baarda, W.: Statistical Concepts in Geodesy, Netherland Geodetic Commision. New Series 2(4) (1967)
22. Fischler, M.A., Bolles, R.C.: Random sample consensus: A paradigm for model fitting with applications to image analysis and automated cartography. Communications of the ACM 24(6), 381–395 (1981)

Method of Wellbeing Estimation
in Territory Management

Tatiana Penkova

Institute of Computational Modelling SB RAS, Krasnoyarsk, Russia
penkova_t@icm.krasn.ru

Abstract. This paper presents the new technique of territory wellbeing estimation based on creation of a geographically-oriented wellbeing standard and estimation of the territory wellbeing level. The author proposes a method of wellbeing level estimation that provides the calculation of territory wellbeing index taking into account various aspects of vital activity and semantic interpretation of index values using fuzzy logic. The practical result of the work is an implementation of suggested solutions for social monitoring of the territory.

Keywords: Wellbeing estimation, Decision making support, Territory management.

1   Introduction

Automation of territory management requires the development of methods and algorithms that provide the decision making support. The preparation of reasonable decisions requires a comprehensive data analysis about the current state of the territory. The information generalization of about processes in various sectors of socio-economic vital activity is a critical condition for effective territory management.

The challenges of analytical decision making support of the territory management have been investigated by many scientists [1,2,3,4]. Generally, the estimation of the control object condition is a key stage in the decision making process [4,5,6,7,8]. The quality of decisions depends largely on a system of applied indicators and methods of their identification. Many existing methods [1], [4], [9], [10] use set of primary indicators to calculate wellbeing index and do not enable to have a comprehensive estimate. Some authors [3], [7], [8] suggest a hierarchical model of indicators that gives main index and several intermediate indexes for various fields of territory activity but the ways of calculating indexes do not enable to have estimates relative to territorial normative. In additional, most of the methods [5], [6], [10] are based on comparison with actual maximum and minimum values for totality of territories. Unfortunately, these approaches allow us to measure territories relative to each other only and do not give us an opportunity to assess a territory independently by analyzing its characteristics.

B. Murgante et al. (Eds.): ICCSA 2014, Part IV, LNCS 8582, pp. 57–68, 2014.

Taking into consideration the advantages and disadvantages of the existing approaches, the development of territory wellbeing estimation method is a topical problem. First of all, the method should allow an analyst to detect the risk factors and identify the priority directions of territory evolution for achieving the target level of wellbeing. Moreover, to estimate the state of social and economic objects the method should provide the analysis of heterogeneous indicators and assessment of their changes according to geographically-oriented standard.

This paper presents a technique of territory wellbeing estimation based on creation of a geographically-oriented wellbeing standard and estimation of the territory wellbeing level. The author proposes a method of wellbeing level estimation that provides the calculation of territory wellbeing index taking into account the various aspects of vital activity and semantic interpretation of index values using fuzzy logic. The practical result of work is an implementation of suggested solutions for social monitoring of the territory and decision making support.

The outline of paper is as follows. Section 2 presents a new technique of territory wellbeing estimation. Section 3 describes the author's method of territory wellbeing level estimation. Section 4 considers the implementation of proposed solutions for social monitoring. Section 5 comprises the conclusion and issues for further research.

2   Technique of the Territory Wellbeing Estimation

The new technique of territory wellbeing estimation is based on creation of a geographically-oriented wellbeing standard and estimation of the territory wellbeing level. Figure 1 presents context IDEF0 diagram of the territory wellbeing estimation process.

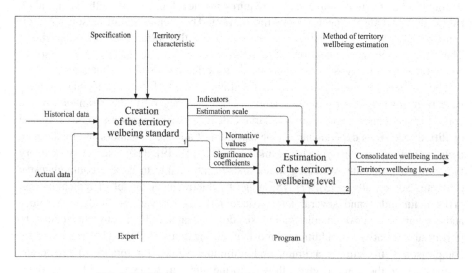

Fig. 1. IDEF0 diagram of the territory wellbeing estimation

IDEF0 methodology provides a function modelling and graphical notation for formalization of business processes using language of Structured Analysis and Design Technique (SADT) [11, 12]. IDEF0 model is a hierarchy of diagrams (i.e. single descriptions of the system). Each diagram describes the functions (e.g. activities, actions, processes or operations); inputs and outputs as the data needed to perform the function and the data that is produced as a result of the function respectively; controls which constrain or govern the function and mechanisms which can be thought of as a person or device which performs the function.

The territory wellbeing estimation process consists of two basic stages:

— Creation of the territory wellbeing standard
— Estimation of the territory wellbeing level

Creation of the territory wellbeing standard is a process of developing the geographically-oriented wellbeing standard according to Chechenin's approach [13]. Wellbeing standard is a target level of wellbeing and is required for correct estimation of current state of the territory. Creation of wellbeing standard is a process performed by experts using historical and actual data based on territory characteristic and specification. Figure 2 presents the detailed IDEF0 diagram of the territory wellbeing standard creation process.

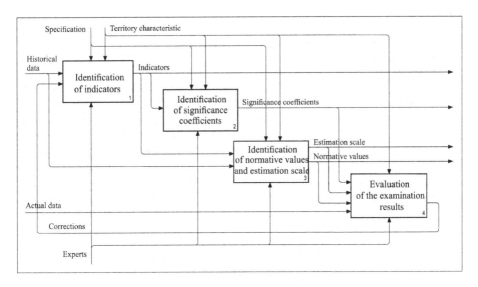

Fig. 2. IDEF0 diagram of the territory wellbeing standard creation

Estimation of the territory wellbeing level is a process of calculating wellbeing index and the territory wellbeing level identification proposed by author. This process is performed by program using actual data and created wellbeing standard. Figure 3 presents the detailed IDEF0 diagram of the territory wellbeing level estimation process.

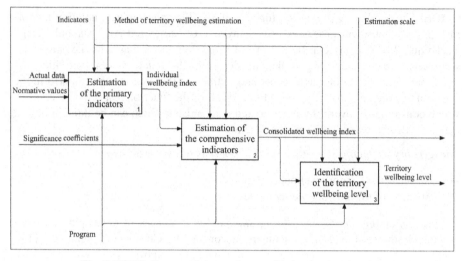

Fig. 3. IDEF0 diagram of the territory wellbeing level estimation

Territory wellbeing standard creation process (figure 2) consists of the following stages [13]:

— Identification of indicators
— Identification of significance coefficients
— Identification of normative values and estimation scale
— Evaluation of the examination results

Identification of indicators is a process of forming the hierarchy of indicators (i.e. the set of primary indicators and levels of their aggregation) which characterize the positive and negative aspects of territory wellbeing and reflect the current state of a socio-economic entity. The set of primary indicators contains the heterogeneous indicators: indicators which should be increased to improve wellbeing level (e.g. birth rate, family income and housing) and indicators which should be decreased to improve wellbeing level (e.g. death rate, morbidity and crime rate). Hierarchical structure allows the analyst to investigate the various fields in detail and to detect the risk factors. Figure 4 shows the example of aggregation levels for social monitoring [14].

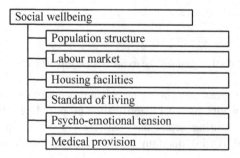

Fig. 4. Aggregation levels for social monitoring

Identification of significance coefficients is a process of calculating the significance coefficients of indicators using method of direct estimation. Firstly, experts give estimates for each indicator of hierarchy level on the scale from one to ten. Table 1 presents a fragment of expert points.

Table 1. Fragment of expert points

Indicators	Exp.1	Exp.2	Exp.3	Exp.4	Exp.5
Population structure	9.00	8.00	8.00	7.00	8.00
Labour market	8.00	9.00	7.00	8.00	8.00
Housing facilities	10.00	10.00	9.00	10.00	10.00
Standard of living	10.00	10.00	10.00	9.00	10.00
Psycho-emotional tension	10.00	9.00	9.00	8.00	9.00
– Children's drug addiction (per 100000)	10.00	10.00	10.00	10.00	10.00
– Teenage drug addiction (per 100000)	10.00	9.00	10.00	10.00	9.00
– Mortality from suicide (per 100000)	8.00	8.00	9.00	8.00	7.00
– Drug addiction (per 100000)	8.00	7.00	7.00	7.00	6.00
– Alcoholism (per 100000)	9.00	7.00	7.00	7.00	5.00
...					
Medical provision	7.00	7.00	10.00	8.00	9.00

Then, the significance coefficients are calculated by:

$$u_k = \frac{R_{k\Sigma}}{\sum_{k=1}^{n} R_{k\Sigma}}, \tag{1}$$

where u_k – is a significance coefficient of k-th indicator, $R_{k\Sigma}$ – is a sum of points of k-th indicator, n – is a number of indicators of hierarchy level. For example, according to Formula 1 the significance coefficient of "Children's drug addiction" indicator is $u = 50/238 = 0.21$, where 50 – is a sum of points of "Children's drug addiction" indicator; 238 – is a sum of points for all indicators of "Psycho-emotional tension" aggregation level. The consistency of expert opinions is identified using Kendall's concordance coefficient [15].

Identification of normative values and estimation scale is a basic process of creating the geographically-oriented wellbeing standard. On this stage, experts identify the normative values of indicators by investigating historical data and taking account the peculiarity of the territory. The normative values of indicators are the typical values that characterize normal state of the territory in expert opinion. This approach is unique in the way that the normative values of indicators are produced as the ranges instead of single value. In this case, experts identify lower and upper limits of normative values for each indicator. For example, for social monitoring of a Siberian industrial city the range of "Children's drug addiction" is 4.20-7.20, the range of "Mortality from suicide" is 32.00-35.00.

The estimation scale is created based on fuzzy logic theory [15]. The linguistic variable "Wellbeing level" is identified by the set of values and membership functions $\mu_j(x)$, where $x-$ is a value of estimate, $x \in \{I\}$, $j = \overline{1,7}-$ is a number of wellbeing level. Table 2 presents the estimation scale developed for social monitoring of a Siberian industrial city.

Table 2. Example of social wellbeing estimation scale

Values of linguistic variable "Wellbeing level"	Membership functions
Critical	$\mu_1 = \begin{cases} 1, & x < 0 \\ 0, & x \geq 0 \end{cases}$
Very low	$\mu_2 = \begin{cases} 1, & 0 \leq x \leq 0,18 \\ 25(0,22 - x), & 0,18 < x < 0,22 \\ 0, & 0,22 \leq x \end{cases}$
Low	$\mu_3 = \begin{cases} 0, & 0 \leq x < 0,18 \\ 25(x - 0,18), & 0,18 < x < 0,22 \\ 1, & 0,22 \leq x \leq 0,38 \\ 25(0,42 - x), & 0,38 < x < 0,42 \\ 0, & 0,42 \leq x \end{cases}$
Middle	$\mu_4 = \begin{cases} 0, & 0 \leq x < 0,38 \\ 25(x - 0,38), & 0,38 < x < 0,42 \\ 1, & 0,42 \leq x \leq 0,58 \\ 25(0,58 - x), & 0,58 < x < 0,62 \\ 0, & 0,62 \leq x \end{cases}$
Satisfactory	$\mu_5 = \begin{cases} 0, & 0 \leq x < 0,58 \\ 25(x - 0,58), & 0,58 < x < 0,62 \\ 1, & 0,62 \leq x \leq 0,78 \\ 25(0,82 - x), & 0,78 < x < 0,82 \\ 0, & 0,82 \leq x \end{cases}$
Acceptable	$\mu_6 = \begin{cases} 0, & 0 \leq x < 0,78 \\ 25(x - 0,78), & 0,78 < x < 0,82 \\ 1, & 0,82 \leq x \leq 1 \end{cases}$
Perfect	$\mu_7 = \begin{cases} 1, & x > 1 \\ 0, & x \leq 1 \end{cases}$

Evaluation of the examination results is a final process where experts can check the created standard by applying it to actual data and make the necessary corrections.

Territory wellbeing level estimation process (figure 3) consists of the following stages:

— Estimation of the primary indicator
— Estimation of the comprehensive indicator
— Identification of the territory wellbeing level

Estimation of the primary indicator is a process of calculating the individual wellbeing index using actual data and normative values of the primary indicators.

Estimation of comprehensive indicator is a process of calculating the consolidated wellbeing index using individual indexes and significance coefficients of indicators.

Identification of wellbeing level is a process of semantic interpretation of index values based on estimation scale.

All these stages are performed according to author's method of the territory wellbeing level estimation that is described below.

3   Method of the Territory Wellbeing Level Estimation

The method of the territory wellbeing level estimation proposed by author in this paper is an improvement to the approach to estimation of complex socio-economic objects [8]. This method provides the calculation of individual and consolidated wellbeing indexes based on integration of heterogeneous indicators. In contrast to existing methods the author suggests use the ranges of the normative values as a wellbeing standard instead of the single value and estimate the significance of indicator change relative to standard range.

The individual wellbeing index is calculated by:

$$i_k = 1 + \Delta P_k \cdot S_k , \tag{2}$$

where i_k – individual index of k-th indicator; ΔP_k – compliance coefficient of actual values of k-th indicator with standard; $S_k = \pm 1$ – coefficient which characterizes the "polarity" of k-th indicator, $S_k = 1$ when the change of indicator is proportional to index and $S_k = -1$ when the change of indicator is inversely proportional to index.

Compliance coefficient is calculated by:

$$\Delta P_k = \begin{cases} 0, \text{if } P_k \in [N_k, Z_k] \\ \dfrac{P_k - Z_k}{Z_k - N_k}, \text{if } P_k > Z_k \\ \dfrac{P_k - N_k}{Z_k - N_k}, \text{if } P_k < N_k \end{cases} \tag{3}$$

where ΔP_k – is a compliance coefficient of actual values of k-th indicator with standard; P_k – actual value of k-th indicator; $[N_k, Z_k]$ – is a range of normative value of k-th indicator, N_k – is a lower limit of range, Z_k – is an upper limit of range.

In the case when actual value of indicator falls within the range, the compliance coefficient $\Delta P_k = 0$; In the case when actual value of indicator is above the upper limit of range, the compliance coefficient has a positive value $\Delta P_k > 0$; In case when actual value of indicator is below the lower limit of range the compliance coefficient has a negative value $\Delta P_k < 0$.

As a result, the value of individual index can be identified as $i_k > 1$ that demonstrates significant improvement of indicator or as $i_k < 1$ that demonstrates significant degradation of indicator. This feature of index is extremely important for estimation of wellbeing and understanding the current state of the territory.

Consolidated wellbeing index is calculated by:

$$I = \sum_{k=1}^{n} u_k \cdot i_k , \qquad (4)$$

where I – is a consolidated wellbeing index; u_k – is a significance coefficient of k-th indicator, $u_k > 0$ and $\sum u_k = 1$, n – number of indicators. The value of consolidated index can also be identified as $I > 1$ or $I < 1$ that demonstrates significant improvement or significant degradation of the territory wellbeing respectively.

In addition, the proposed method provides the semantic interpretation of index values based on estimation scale. Each index value corresponds to a value of linguistic variable "Wellbeing level" with expert confidence level.

Consider the example of wellbeing level estimation for indicators of "Psycho-emotional tension" level. The data for social monitoring are shown in Table 3.

Table 3. Example of indicators for social wellbeing level estimation

Indicators	u_k	$[N_k, Z_k]$		P_k	Index
Population structure	0.15	-	-	-	0.61
Labour market	0.15	-	-	-	0.80
Housing facilities	0.19	-	-	-	0.75
Standard of living	0.19	-	-	-	0.60
Psycho-emotional tension	0.17	-	-	-	0.53
Children's drug addiction	0.21	4.20	7.20	7.16	1.00
Teenage drug addiction	0.20	120.00	150.00	180.56	-0.02
Mortality from suicide	0.17	32.00	35.00	35.00	1.00
Drug addiction	0.15	200.00	250.00	265.38	0.69
Alcoholism	0.15	1100.00	1300.00	1333.63	0.83
...	...	...	...	...	...
Medical provision	0.16	-	-	-	0.85

At the first stage, for primary indicators we calculate the compliance coefficient according to Formula 3. As can be seen from Table 3, the actual values of "Children's drug addiction" and "Mortality from suicide" indicators fall within the normative range, therefore the coefficient of compliance is identified as: $\Delta P = 0$. The actual values of other indicators above the upper limit of the range, therefore, the compliance coefficient is calculated according to the second condition in Formula 3. For "Teenage drug addiction" indicator the compliance coefficient is identified as: $\Delta P = (180.00 - 150.00)/(150.00 - 120.00) = 1.02$. For "Drug addiction" and "Alcoholism" indicators the compliance coefficient equals 0.31 and 0.17 respectively.

Then, we calculate the individual wellbeing index according to Formula 2. Taking into account the negative "polarity" for "Teenage drug addiction" indicator the individual index is identified as: $i = 1 + 1.02 \cdot (-1) = -0.02$. The individual indexes for other indicators are following: "Children's drug addiction" – 1.00; "Mortality from suicide" – 1.00; "Drug addiction" – 0.69 and "Alcoholism" – 0.83.

At the second stage, for comprehensive indicators we calculate the consolidated wellbeing indexes according to Formula 4. The consolidated index for "Psycho-emotional tension" level is identified as: $I = 0.21 \cdot 1.00 + 0.20 \cdot (-0.02) + 0.17 \cdot 1.00 + ... = 0.53$. The consolidated indexes for other levels are following: "Population structure" – 0.61; "Labour market" – 0.80; "Housing facilities" – 0.75; "Standard of living" – 0.60; "Medical provision" – 0.85. The general consolidated index for social wellbeing is 0.69.

At the third stage, we interpret the index values based on estimation scale (Table 2). Expert confidence level is identified by value of membership functions. The result of interpretation of index values for comprehensive indicators is represented in Table 4. As can be seen, the wellbeing level of "Medical provision" is identified as "Acceptable", "Housing facilities" – "Satisfactory" and "Psycho-emotional tension" – "Middle" with 100%. The wellbeing levels of "Labour market", "Standard of living" are identified as value between "Acceptable" and "Satisfactory" that informs us about the equal possibility of improvement or degradation. The wellbeing level of "Population structure" tend to "Satisfactory". The general social wellbeing level of region is identified as "Satisfactory".

Table 4. Example of interpretation of index values

Comprehensive indicators	Values of membership functions						
	μ_1	μ_2	μ_3	μ_4	μ_5	μ_6	μ_7
Social wellbeing	0	0	0	0	1	0	0
Population structure	0	0	0	0,25	0,75	0	0
Labour market	0	0	0	0	0,5	0,5	0
Housing facilities	0	0	0	0	1	0	0
Standard of living	0	0	0	0,5	0,5	0	0
Psycho-emotional tension	0	0	0	1	0	0	0
Medical provision	0	0	0	0	0	1	0

Thus, the method allows the analyst to estimate the current wellbeing level of the territory by analyzing the heterogeneous indicators and assessing their changes according to geographically-oriented standard. The results of estimation depend largely on correctness of standard that requires the constant update of normative values.

4   Social Monitoring of the Territory Based on Wellbeing Indicators

The proposed method of the territory wellbeing level estimation is used for analytical decision making support in the system of social-hygienic monitoring [17]. Figure 5 illustrates the wellbeing estimation results as a cross-table. OLAP tools enable the user to analyze data interactively from multiple perspectives (e.g. territory, date, indicator and wellbeing level) using basic analytical operations: consolidation (roll-up), drill-down, and slicing and dicing. In Figure 5 the left section of cross-table there are names of comprehensive indicators and names of territories (city districts), in the right section of cross-table the values of consolidated social wellbeing indexes for 2005, 2006 and 2007 years are represented. Figure 6 demonstrates the wellbeing estimation results as a cartogram. GIS tools provide the data visualization by color of map objects according to estimation scale.

Social wellbeing index				
		Measure ▲ ▼ Year ▲ ▼		
	🔳	Wellbeing index		
Indicator ▲ ▼	**Territory** ▲ ▼	2005	2006	2007
Housing facilities	Zavodskoy	0,71	0,74	0,75
	Kuznetskiy	0,52	0,66	0,69
	Kujbushevskiy	0,73	0,77	0,79
	Novoilynskiy	0,95	0,98	0,98
	Ordzhonikidzevskiy	0,70	0,78	0,78
	Centralniy	0,87	0,88	0,88
Medical proposition	Zavodskoy	0,76	0,84	0,97
	Kuznetskiy	0,62	0,76	0,86
	Kujbushevskiy	0,53	0,64	0,71
	Novoilynskiy	0,78	0,85	1,00
	Ordzhonikidzevskiy	0,64	0,76	0,84
	Centralniy	0,86	0,90	1,00
Psycho-emotional tension	Zavodskoy	0,21	0,34	0,09
	Kuznetskiy	0,52	0,52	0,45
	Kujbushevskiy	0,43	0,34	0,21
	Novoilynskiy	0,72	0,72	0,64
	Ordzhonikidzevskiy	0,64	0,46	0,38
	Centralniy	0,83	0,75	0,55

Fig. 5. Representation of wellbeing estimation results as a cross-table

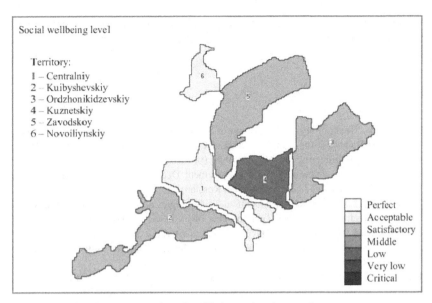

Fig. 6. Representation of wellbeing estimation results as a map

The implementation of proposed method allows the decision maker to detect the risk factors and make reasonable control decisions.

5   Conclusion

This paper presents the technique of the territory wellbeing estimation that includes the creation of the geographically-oriented wellbeing standard and estimation of the territory wellbeing level. The wellbeing standard is a target level of wellbeing which provides the correct estimation of current state of the territory. In order to estimate the territory wellbeing level the author has proposed a method which provides the calculation of individual and consolidated wellbeing indexes based on integration of heterogeneous indicators and estimation of their changes relative to ranges of normative values. Also, the suggested method provides semantic interpretation of index values. The method allows the decision maker to detect the risk factors, identify the priority directions of territory evolution for achieving the target level of wellbeing and make reasonable control decisions.

The practical result is an implementation of suggested solutions for social monitoring of the territory. Operational control of social state of the territory by means of getting the comprehensive estimates and realisation of appropriate administrative actions have led to improvement of basic factors and increase of social wellbeing index (e.g. 2005 – 0.68; 2006 – 0.74; 2007 – 0.73 ... 2011 – 0.79; 2012 – 0.80) that reflect the improvement of life quality and effective management.

Suggested solutions can be used successfully in other fields. For instance, for territory vulnerabilities estimation and early prevention of natural and anthropogenic emergencies; for estimation and planning of organization activity and other problems.

The future research will be connected with developing methods for generation of control recommendations based on results of comprehensive wellbeing estimation and expert knowledge.

References

1. Tsybatov, V.: Technology of Forecasting of Socio-Economic Activity of Region Based on Methods of Balances. In: 13th International Conference on the Application of Artificial Intelligence in Engineering, pp. 69–72 (1998) (in Russian)
2. Trahtengerts, E.A.: Computer-Aided management Decision-Making Support Systems. J. Problemy Upravleniya 1, 13–28 (2003) (in Russian)
3. Nozhenkova, L.F.: Information-Analytical Technologies and Systems of Region Management Support. J. Computational Technologies 14(6), 71–81 (2009) (in Russian)
4. Makarov, A.M., Galkina, N.V., Savenkov, B.V.: Criteria of Social and Economic Efficiency in the Estimation of Territory Viability. J. of Chelyabinsk State University 1(2), 57–62 (2003) (in Russian)
5. Kalinina, V.V.: Modern Approaches to the Estimation of the Industrial Complex of Region. J. Volglgrad National University 2(19), 62–69 (2011) (in Russian)
6. Murias, P., Martinez, F., Miguel, C.: An Economic Wellbeing Index for the Spanish Provinces: A Data Envelopment Analysis Approach. Social Indicator Research 77, 395–417 (2006)
7. Becht, M.: The Theory and Estimation of Individual and Social Welfare Measures. Journal of Economic Surveys 9(1), 53–87 (1995)
8. Borisova, E.V.: Index Method of Comprehensive Quantitative Estimation of the Quality of Complex Objects. In: XII International Conference "Mathematics, Computer, Education", Izhevsk, vol. (1), pp. 249–159 (2005)
9. Helliwell, J.F., Putnam, R.D.: The Social Context of Well-Being. J. Philosophical Transactions of the Royal Society 359(1449), 1435–1446
10. Diamantopoulos, A., Winklhofer, H.M.: Index Construction With Formative Indicators: An Alternative to Scale Development. J. Marketing Research 38(2), 269–277 (2001)
11. Marca, D.A., McGowan, C.L.: SADT: Structured Analysis and Design Technique, p. 392. McGraw-Hill Book Co., New York (1987)
12. Davis, W.S.: Business systems analysis and design. Business & Economics, 534 (1994)
13. Chechenin, G.I., Saprykina, T.V., Zhilina, N.M.: Approach of Group Examination for Developing the Automated System of Socio-Hygienic Monitoring. In: Regional Conference "Informational Bowels of Kuzbass", pp. 183–185 (2004) (in Russian)
14. Penkova, T.G., Shaldybina, K.V.: Structuring Data for Monitoring of Region Socio-economical Evolution In IV Siberian Congress Women-Mathematics. J. "Open Education", 171–174 (2006) (in Russian)
15. Kendall, M.G., Smith, B.B.: The problem of m rankings. Annals of Mathematical Statistics 10(3), 275–287 (1939)
16. Dubois, D.J.: Fuzzy Sets and Systems: Theory and Applications, p. 393. Academic Press (1980)
17. Zhilina, N.M., Chechenin, G.I., Penkova, T.G.: Information and Algorithmic Models of Automated System of Socio-Hygienic Monitoring. In: XXXIX International Conference "Public Health. Occupational Health. Ecology", pp. 87–90 (2004) (in Russian)

From Volunteered Geographic Information to Volunteered Geographic OLAP: A VGI Data Quality-Based Approach

Sandro Bimonte[1], Omar Boucelma[2], Olivier Machabert[2], and Sana Sellami[2]

[1] Irstea, TSCF
9 avenue Blaise Pascal CS20085, 63178 Aubière, France
sandro.bimonte@irstea.fr
[2] LSIS, Aix Marseille University, CNRS
Av. Escadrille Normandie-Niemen, 13397 Marseille, France
{omar.boucelma,olivier.machabert,sana.sellami}@lsis.org

Abstract. Volunteered Geographic Information (VGI) represents a valuable source of information. At the same time this data presents new data quality issues such as credibility. In this paper, we investigate the integration of VGI in Spatial OLAP (SOLAP) systems, which allow analyzing huge volume of geographic datasets. By means of a real case study, we highlight the similarities and differences of these two kinds of systems. We propose a methodology to handle VGI data quality issues during the warehousing and OLAPing phases. In particular, we define a new aggregation ETL operator based on the VGI credibility. We also present a spatio-multidimensional model that provides decision-makers with a general description of the correctness of the SOLAP aggregations.

Keywords: Spatial OLAP, Volunteered Geographic Information, Data Quality.

1 Introduction

Spatial Data Warehouse (SDW) and Spatial OLAP (SOLAP) tools [12] have been developed for exploring and synthesizing huge volume of geo-referenced data. These systems rely on data that are extracted from external official data sources, cleaned, transformed and loaded in the spatial data warehouses by means of ETL tools (Extract-Transform-Load) accordingly with a spatio-multidimensional model. SOLAP has been effectively used in several different applications domains (such as health, urban, marketing, environment, etc.) [3].

New technologies for geospatial positioning, web mapping, wiki tools have led to collaborative web-based maps systems that allow experts and amateurs (volunteers) to create and share themed oriented geographic information in using tools such as OpenStreetMap. This has led to the concept of Volunteered Geographic Information (VGI) that has been defined as "*the harnessing of tools to create, assemble, and disseminate geographic data provided voluntarily by individuals*" [9]. Along this line,

B. Murgante et al. (Eds.): ICCSA 2014, Part IV, LNCS 8582, pp. 69–80, 2014.

VGI systems have already proved useful for risk monitoring situations [10], or more generally for geographical applications where recent data, knowledge and participation are necessary in the decision-making process [16]. However VGI data have some drawbacks: data quality (data precision and data credibility) [8][1], metadata are scarce, and data do not always comply with "official data" [11]. Moreover, VGI systems do not provide any technique to handle efficient analysis of historical data [16], which could be mandatory for understanding and explaining current geographic phenomena [3]. Consequently, there is a need for integrating VGI into SOLAP. (Un)Fortunately warehousing and analytics methodologies provided by existing SOLAP systems are not able to handle VGI data since they are conceived for well-known official data [17] [5]. The work described in this paper aims to fulfil this need.

The contributions of this paper are as follows:

1. We provide a comparative analysis of SOLAP and VGI systems according to some important criteria such as data, analysis, architectural issues. This work is mandatory for a better comprehension of crucial theoretical and implementation issues that are raised by the integration process.
2. We detail issues related to the integration of VGI data in the SDW development process. Thus we propose a new methodology that takes into account the credibility of VGI data in the warehousing and on-line multidimensional analysis phases. In particular, we propose a novel ETL operator based on the credibility of the VGI data.
3. We propose some new SOLAP measures that allow decision-makers to explore VGI and official warehoused data according to the degree of awareness they tolerate the bad quality of VGI data.
4. Finally, in order to show the feasibility of new ETL operators and measures, we present an implementation in a classical Relational SOLAP architecture.

This paper is organized as follows. Section 2 highlights differences between SOLAP and VGI, followed by a description of an application scenario in Section 3. In Section 4, we discuss our quality-based approach, while, in Section 5 we describe our implementation.

2   SOLAP and VGI: A Comparative Analysis

In this section, we provide a comparative study between SOLAP and VGI along the line of data and architecture (Table 1).

SDW are populated using conventional and spatial data, i.e., SDW contain data that could be natively stored in a Spatial Database Management System (DBMS). VGI is defined as a set of geographic information (spatial and alphanumeric data) coupled with contextual multidimensional and textual information (i.e. complex data such as photo, video, comments, etc.). This means that SOLAP data can be considered as a subset of possible data types provided in VGI. Despite of this apparently important divergence, (S)OLAP literature covers work that aims to integrate complex data as contextual multimedia information for warehoused spatial data and OLAP analysis. For example, [14] integrate textual documents in data warehouse to enhance and extend OLAP analysis.

Table 1. SOLAP and VGI Comparison

	Spatial OLAP	*VGI*
Data	High quality	Average quality
	Huge volume	Small data sets
	Spatial, conventional and historical data	Semi-structured and unstructured data
		Spatial data
Architecture	Scalability, security, spatio-multidimensional analysis	Collaborative and simple spatial analysis
	ETL-based	

Data historicity is another difference between SOLAP and VGI systems. Indeed, as for classical transactional systems, VGI systems have been designed for sharing and publishing data in a real-time fashion [16]. Warehoused spatial data is loaded in the SOLAP systems with a very low frequency because (i) during the processing of data sources, data must be cleaned and transformed in order to comply with the spatial data warehouses schema while getting rid of quality problems and (ii) setting up techniques that grant good performances (indexes, materialized views, etc.) may be a very time consuming task.

Different criteria have been proposed to assess spatial data quality: lineages, positional precision, attribute precision, temporal precision, data completeness and logical consistency [6]. It is difficult to assess all these criteria for VGI data. In particular, completeness depicts the omission and completeness in information. Several works have been proposed to address completeness in data warehouse storage and querying [7]. In SOLAP systems, evaluation of data completeness is easily achieved by means of the experts' knowledge since the data sets that are mandatory in the decision-making process are known at the SDW design stage. On the other hand, in a VGI project, assessing the completeness is a critical issue: as stated in [13] the interesting area for volunteers can be as huge as the number of volunteers and it is not possible to force volunteers to provide needed data.

In [1] authors define VGI precision: for spatial precision a GPS or a reference map can be used; for attribute precision authors advocate the need for automatically collecting information and using the density of similar messages to define data reliability. Precision issues can also be raised in SOLAP systems: precision can be defined as a schema compliance ratio, i.e., how much the data sources schemas comply with the spatio-multidimensional schema, or simply as precision of the datasets. Schema precision is simply handled during the conceptual SDW design phase, where according to the data-driven design approach the spatio-multidimensional schema is created using only sources schemas that match decision-makers needs [12]. Data precision is usually handled by means of integrity constraints that are rules, defined in the ETL process, allowing filtering bad precision data [4]. However, in a VGI flexible data context, it seems very difficult to define a set of predefined logical rules and schema mapping as in SDW.

Credibility is another quality criteria. Assessing users' credibility (experts or citizens) is important for evaluating the overall reliability of their contribution: according to [8], it is difficult to give an accurate definition of credibility. Thus, it can be considered as a believability degree of information based on notions of its trustworthiness and expertise. On the other hand, in SDW systems credibility is not taken into account since it is believed that all data sources are credible [17] or it is handled only in the visual reporting tool [5]. Thus, this means that a new kind of metadata, such as credibility, should be associated to the warehoused data and used in further SOLAP analysis.

Finally, VGI data can be considered as versioning data in the sense that for each geographical object several users can upload/modify its spatial and thematic content at different time [15]. The same geographical phenomena modelled using the VGI approach can have different points of view since several different users can independently upload/modify related geographic information. This feature/issue is not supported in existing SOLAP tools, since factual information is uniquely identified by dimensional data. However, some research works do address dimensions and facts versioning by means of ad-hoc multidimensional logical and physical schema [2].

To achieve scalability, security and maintainability, SOLAP systems usually rely on Relational OLAP architectures [12]. Here data sources are integrated into the SDW tier by means of ETL tools (e.g., Spatial Data Integrator). Spatial DBMS are used to store spatial data (SDW tier) (e.g. PostGis). The concepts of measures, facts and dimensions are modelled in the Spatial OLAP server (e.g., GeoMondrian), which also implements SOLAP operators. The SOLAP client tier (e.g., Map4Decision) provides interactive cartographic and tabular displays to visualize SOLAP queries results and allowing decision-makers to trigger SOLAP operators by the simple interaction with the user interface.

The spatial data warehouse populating task is typically handled by ETL tools allowing transformation (e.g. aggregation, join, etc.) cleaning (e.g. filter) and uniformity of data sources according to the SDW model before being loaded. ETL tools typically access data sources using classical interfaces and protocols. Thus, ETL tools should integrate new functionalities to handle data quality and schema issues of VGI data. Moreover, SOLAP tools should be able to allow data insertion by means of alphanumeric and cartographic displays in an easy and effective way, preempting classical ETL protocols.

From the above discussion about similarities and difference of SOLAP and VGI approaches some mandatory issues emerge that we consider as mandatory for achieving an effective integration.

- Data quality issues impact data analysis. Indeed, SOLAP users are used to explore warehouse data without providing attention to (aggregated) data quality since they consider all data official and correct. Thus, when integrating (poor quality) data, final decision-makers should be supported in interpreting the quality of data used in order to adapt their final decisions. Then, it is necessary to manage credibility, precision and versioning issues when VGI data are integrated in SOLAP systems since in SOLAP systems all data sources are considered credible and good.

- From an architectural point of view, it is necessary to adopt a VGI solution that provides data in a way and a format that are compatible with classical ETL and SOLAP tools.

3   Motivating Example

In this section, we present a real scenario where volunteers can report information about flooding in using Wikimapia, an open-content collaborative mapping project that allows the description and annotation of various geographical objects. Fig. 1 below illustrates such situation where volunteers post information and possibly meta-information such as water levels, damages, etc. Because volunteers have different levels of expertise (amateurs vs. professionals) an important issue to consider is their credibility. In our approach credibility is derived from Wikimapia by means of a voting mechanism for trustworthiness and the number of alerts (inputs from users) for expertise, according to [8].

Fig. 1. Flooding Information on Wikimapia

A flooding alert or simply an alert is characterized by a location (geometry), a date (at the hour granularity), a user (with an associated credibility) and some metadata supplied by the user. A simple application scenario related to this example could be expressed as follows: "What is the average impact of riverbanks of floods per city in year 2012?". This query may help define priorities for urban planning operations for reducing the risks of floods impacts on riverbanks and urban neighborhoods elements (roads, shops, etc.).

In order to answer this query, we could develop an SOLAP application shown on Figure 2 using the UML profile defined in [4]. The profile defines a stereotype or a tagged value for each spatio-multidimensional element; for example "Fact" for the fact, "SpatialAggLevel" for the spatial dimension levels whose geometries are represented with geometric attributes stereotyped "LevelGeometry", and

"AggRelationship" to model the aggregation relationships that link dimension levels. Note that the model contains a spatial dimensions ("Location") and a temporal dimension ("Time"). The measure is a value (e.g. "Low") describing the event according to its type (e.g. "Flood", "Fire", etc.).

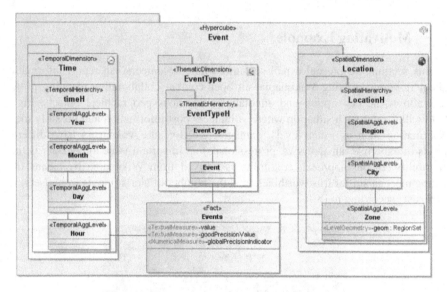

Fig. 2. VGOLAP spatio-multidimensional model

Using this SOLAP model decision-makers can analyze events by space and time in order to have a global vision on the behavior of the flooding hazard.

In order to allow decision-makers to trigger SOLAP queries for analyzing the evolution of a climatic phenomenon and define urban priorities, new data about climatic events are necessary. It is important to underline that these data are volatile: it is difficult to retrieve flooding descriptions when the event is finished. To this aim, VGI systems represent a useful data source for feeding the SDW and allowing the spatial decision-making tool to proceed. In integrating VGI data with the SDW, we maintain the execution of SOLAP queries on incoming climatic events.

4   Quality-Based Approach

We describe in this section our methodology for (1) integrating VGI data with SOLAP system, and (2) performing analysis.

First, VGI data are aggregated using an ad-hoc ETL method (credibility-based aggregation) (Sec 4.1), based on VGI users credibility. This allows compliance of VGI data with the SDW schema, and avoids versioning and credibility related problems. Then, VGI data are loaded in the SDW and in the SOLAP server to impact the SOLAP analysis process with their quality (Sec. 4.2). We propose some metadata rules, defined as spatio-multidimensional measures, representing (aggregated) data quality. This SOLAP metadata provides decision-makers with a useful meta-information about the quality of the results of SOLAP queries, avoiding misleading data interpretation.

4.1   Quality-Based ETL

In our approach is necessary to enrich the ETL phase with credibility information carried out by VGI data. In our case study, volunteers with different credibility (Table 2) may differently evaluate the impact on the riverbanks (Table 3). These different values need to be aggregated leading to one fact and measure in the SDW presented in Figure 2. In other terms, a set of geographic objects provided by volunteers has to be aggregated to define one fact.

Table 2. Users' credibility

User	*Credibility*
Jean-Pierre	3
Florence	2
Bob	1

Table 3. VGI data (VGIobjects)

VGIObjects	*User*	*Answer*	*Category*	*Location*	*Timestamp*
VE$_1$	Jean-Pierre	1 - low	impact on riverbanks	zone3-1	12h-1-11-2012
VE$_2$	Jean-Pierre	1 - low	impact on riverbanks	zone3-2	12h-1-11-2012
VE$_3$	Florence	2 - avg	impact on riverbanks	zone3-1	12h-1-11-2012
VE$_4$	Florence	3 - high	impact on riverbanks	zone3-1	12h-1-11-2012
VE$_5$	Bob	3 - high	impact on riverbanks	zone3-1	10h-1-11-2012

Then, a *credibility-based aggregation operator* Δ *that* takes a set of geographic objects ($VE_1 ...., VE_m$) as input and returns an event (E) is defined (see Definition 1). A numerical aggregation function (*AVGVE*) weighted by credibility is applied to calculate the numerical value of the event (*Em*). Spatial (e.g., bounding box) (*SpatialAgg*) and temporal (*TemporalAgg*) aggregation functions are applied to calculate the location and the timestamp of the event, respectively (*Ezone and Etimestamp*).

Thus, our ETL operator allows several geographic objects to be merged in considering their users' credibility for a unique fact (event). In this way, the users' information is not represented in the spatio-multidimensional model (c.f. Fig. 2).

Definition 1 – Credibility-based Aggregation.

The credibility-based aggregation operator Δ *is defined as follows:*

- *Let SpatialAgg and TemporalAgg be the spatial and temporal aggregation functions, respectively.*
- *Let VE_1 ...., VE_m be VGI geographic objects.*

$E = \Delta(VE_1$ ...., VE_m, *SpatialAgg, TemporalAgg) where:*

- *The value of measure of E is $\sum_{i=1}^{m} VO_i.value / m$ where VO_i are the VGI geographic objects provided by users with the highest credibility*
- *Ezone = SpatialAgg ($VE_1.VElocation$, ..., $VE_m.VElocation$)*
- *Etimestamp = TemporalAgg ($VE_1. VEtimestamp$, ..., $VE_m. VEtimestamp$)*

When applying the aggregation operator using a convex hull for the spatial aggregation (zone3-1 and zone3-2 are geographic regions included in zone3), the alerts in Table 2 are aggregated into one VGI event with a *low value* to represent the impact to the riverbanks provided by Jean-Pierre, who is the most credible user. Then, this event that is loaded into the spatial data warehouse, and analyzed using the measures described in the next section.

4.2   Quality-Based Analysis

In order to feed decision-makers with data credibility, we extend the spatio-multidimensional schema of the official SDW with some new measures representing the quality value obtained in the previous phase. In particular, we propose two measures to allow decision-makers to analyze data according to their tolerance to the usage risk caused by (users and multidimensional) quality of VGI data. The first measure is used when decision-makers are sensible to "not good" data, and they want to "see" only good data. We call that measure: *GoodPrecisionValue* (see Definition 2). The second measure is used when decision-makers are risk aware and they aim to "see" all data but with a feedback of the quality of data. We call that measure: *GlobalPrecisionIndicator*.

Let us formalize those measures.

Definition 2 – GoodPrecisionValue

Let VGIFacts the set of VGI data integrated into the data warehouse and NoVGIFacts the historical one. Let m the measure of the facts and its value denoted as $F_i.mvalue$ and Agg the aggregation function applied to it (e.g. SUM, AVG, etc.).

Then: ***GoodPrecisionValue*** *is $AGG_{i=1}^{n} F_i.mvalue$ where $F_i \in NoVGIFacts$*

GoodPrecisionValue allows aggregating only facts coming from official sources and consequently having not problems of credibility, which represents a "safe" aggregated value for decision-makers.

Let's consider the subset of the SDW data related to the impact of riverbanks as represented in Table 4, where the first 4 tuples are historical data (NoVGIFacts) and the other ones are VGI data (VGIFacts).

Table 4. Facts of the case study SOLAP application

Time	*Location*	*Value*	*EventType*
12h-19-10-2011	Zone1	**Low**	Impact on riverbank
12h-20-10-2011	Zone2	**Low**	Impact on riverbank
12h-2-11-2011	Zone3	**Low**	Impact on riverbank
12h-2-11-2011	Zone4	**High**	Impact on riverbank
12h-1-11-2012	Zone3	**Low**	Impact on riverbank
11h-1-11-2012	Zone4	**High**	Impact on riverbank

Then, the result of aggregation on all zones, all years and impact on riverbank is shown on Table 5. It can be noticed that it is different from the aggregation (i.e. avg) with all values meaning that the VGI data affect the result of the aggregation.

Table 5. Quality-based measures

Time	*Location*	Value	*Goodprecisi onValue*	*GlobalPrecisi onIndicator*	*EventType*
All years	All zones	**Avg**	**Low**	**0.6**	Impact on riverbank

We now define the GlobalPreciseIndicator (see Definition 3), which is used to represent how much the aggregated measure is calculated using VGI data.

Definition 3 – Global Precision Indicator

GlobalPrecisionIndicator is *VGIfactNb/ FactNb* where *FactNb is the number of facts, and VGIfactNb is the number of VGI facts used in the aggregation.*

In our example the global precision indicator is 0.6 meaning that 2/3 of measures values used in the aggregation are issued from VGI data. Then the measure value avg should be not safely considered in the decision-making process.

5   Implementation

In this section we present the VGOLAP system and architecture we deployed to implement our quality-based approach (Fig. 3).

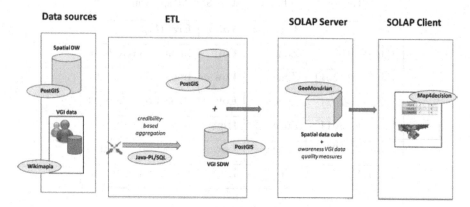

Fig. 3. VGOLAP Architecture

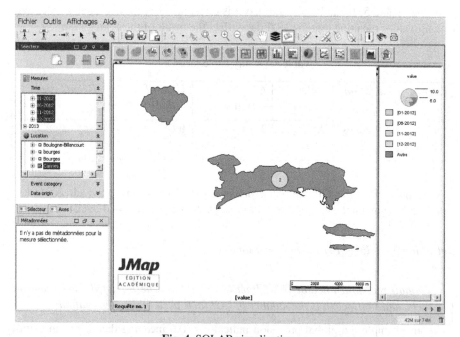

Fig. 4. SOLAP visualization

ETL methods described in Section 4 have been implemented in java and PL/SQL scripts. The SDW tier is implemented in the Spatial DBMS PostGIS that provides a native support for spatial data. The SOLAP Server relies on GeoMondrian, a full-featured SOLAP server. Measures and aggregation described in Section 4 are defined as well known MDX expressions. Tabular and cartographic visualization of SOLAP queries are provided by means of the SOLAP system Map4Decision. Fig. 4 shows the results of the SOLAP query retrieving the average impact riverbanks for one city at different months.

6   Conclusion and Future Work

Volunteered Geographic Information systems (VGI) have proven useful in allowing Internet contributors to create spatial datasets, or to provide valuable information during or after natural hazards. Nevertheless, VGI raises data quality issues such as credibility and precision. In this paper, we investigate the integration of VGI in Spatial OLAP (SOLAP) systems to ensure a better decision-making process. In using an environmental case study, we highlighted the similarities and differences of these systems, and came up with a methodology, a multidimensional model and a decision-support prototype that integrate both concepts and systems. In particular, we proposed a new ETL aggregation operator based on VGI credibility and some new measures.

We are currently working on the evaluation of our proposal on the complete real dataset of our case study. For future work, we are planning to investigate the usage of linguistic measures (such as good, bad, etc.), which are more familiar to decision-makers as already investigated in [18] by means of fuzzy theory.

References

1. Galindo, A., Díaz, P.: A quality approach to volunteer geographic information. In: Proc. 7th International Symposium on Spatial Data Quality, pp. 109–114 (2011)
2. Arigon, A., Tchounikine, A., Miquel, M.: Handling multiple points of view in a multimedia data warehouse. TOMCCAP 2(3), 199–218 (2006)
3. Bernier, E., Gosselin, P., Badard, T., Bédard, Y.: Easier surveillance of climate-related health vulnerabilities through a web-based spatial olap application. International Journal of Health Geographics 8, 8–18 (2009)
4. Boulil, K., Bimonte, S., Pinet, F.: A UML & spatial OCL based approach for handling quality issues in SOLAP systems. In: Proc. 14th International Conference on Enterprise Information Systems, pp. 99–104 (2012)
5. Daniel, F., Casati, F., Palpanas, T., Chayka, O.: Managing Data Quality in Business Intelligence Applications. In: Proc. QDB/MUD, pp. 133–143 (2008)
6. Devillers, R., Jeansoulin, R.: Fundamentals of Spatial Data Quality. ISTE, London (2006)
7. Dyreson, C., Pedersen, T., Jensen, C.: Incomplete information in multidimensional databases. Multidimensional Databases, 282–309 (2003)
8. Flanagin, A., Metzger, M.: The credibility of volunteered geographic information. GeoJournal 72, 137–148 (2008)
9. Goodchild, M.: Citizens as sensors: The world of volunteered geography. GeoJournal 69, 211–221 (2007)
10. Goodchild, M., Glennon, J.: Crowdsourcing geographic information for disaster response: A research frontier. Int. J. Digital Earth 3, 231–241 (2010)
11. Gouveia, C., Fonseca, C.: New approaches to environmental monitoring: the use of ICT to explore volunteered geographic information. GeoJournal 72, 185–197 (2008)

12. Malinowski, E., Zimányi, E.: Advanced Data Warehouse Design: From Conventional to Spatial and Temporal Applications. Springer (2008)
13. Maué, P., Schade, S.: Quality Of Geographic Information Patchworks. In: Proc. 11th AGILE International Conference on Geographic Information Science, pp. 1–8 (2008)
14. Pérez-Martínez, J.M., Llavori, R.B., Cabo, M.J.A., Pedersen, T.B.: Contextualizing data warehouses with documents. Decision Support Systems 45, 77–94 (2008)
15. Rinner, C., KeSSler, C., Andrulis, S.: The use of web 2.0 concepts to support deliberation in spatial decision-making. Computers, Environment and Urban Systems 32, 386–395 (2008)
16. Roche, S., Propeck-Zimmermann, E., Mericskay, B.: Geoweb and crisis management: issues and perspectives of volunteered geographic information. GeoJournal, 1–20 (2011)
17. Vassiliadis, P.: A Survey of Extract-Transform-Load Technology. IJDWM 5(3), 1–27 (2009)
18. Feng, L., Dillon, S.: Using Fuzzy Linguistic Representations to Provide Explanatory Semantics for Data Warehouses. IEEE Trans. Knowl. Data Eng. 15(1), 86–102 (2003)

Arranging Advertisement as a Communication Media of Society towards the Enhancement of Urban Visual Quality

Endy Marlina[1], Punto Wijayanto[1], Dita Ayu Rani[1],
Desrina Ratriningsih[1], and Tutut Herawan[2,3]

[1] Department of Architecture
Faculty of Science and Technology
Universitas Teknologi Yogyakarta
Jl. Ring Road Utara, Jombor, Sleman, Yogyakarta, Indonesia
[2] Department of Information System
Universiti Malaya
50603 Pantai Valley, Kuala Lumpur, Malaysia
[3] AMCS Research Center, Yogyakarta, Indonesia
{endy_marlina,punto.wijayanto,dita_arch3,
desrina.128}@yahoo.com, tutut@um.edu.my

Abstract. Advertisement plays a role to create an effective communication among societies and urban environment. In the very dynamic era of Yogyakarta society today, advertisement undeniably has been becoming one of the essential components in modern urban landscape. However, inaccuracy of advertisement design conversely can emerge inconvenience and interruption for society. Again it can also lead to the decline of the urban visual quality. In this paper, a study on advertisement regulation in Yogyakarta is conducted using a descriptive-qualitative approach based on the analysis on the communication aspect and spatial aspect. The results are used in recommending for the advertisement regulation based on three criteria: *aesthetics*, *ethics*, and *safety*. Aesthetics is in consideration of cultural values suitable in Yogyakarta and visual consideration. Meanwhile, ethics and safety are in accordance with Government Regulation.

Keywords: Aesthetics, Ethics, Safety, Arrangement of Advertisement.

1 Introduction

In Yogyakarta, the emergence of mushrooming advertisement as one of commercial communication devices is remarked by how complex the issue is in installing an advertising device or sign in certain area for involving certain interests, they are the interest of the advertisement installers, the interest of the authority of the area where the advertising sign is placed, and the interest of society seeing the advertisement. On the other side, the existence of the advertising signs in an urban space frequently blocks the sight to the urban elements that, in fact, are interesting such as the building

B. Murgante et al. (Eds.): ICCSA 2014, Part IV, LNCS 8582, pp. 81–94, 2014.

architecture and the urban landscape elements. Hence, an advertisement regulation is deemed essential. Inaccuracy in advertisement designs can emerge the inconvenience and interruption for society. Even, it can lead to the decline of the urban visual quality [1].

Based on the dynamics of development of installing the advertising signage in Yogyakarta, a number of issues can be identified as follows:

a. Discrepancy between the existing master plan for the advertisement regulation and the development of the recent condition of the advertising signage layout.
b. Unavailability of evaluation data for the existing condition of advertisement regulation.
c. Unidentified issues and potentials for the development of advertisement regulation.
d. Unavailability of a comprehensive plan for advertisement regulation complete with a draft appropriate with urban development and environment condition.
e. Unavailability of recommendation for the policy of advertisement regulation.

Therefore, this paper presents a study on advertisement regulation in Yogyakarta which is conducted using a descriptive-qualitative approach based on the analysis on the communication aspect and spatial aspect.

2   Related Works

Advertisement is known as *Street Graphics* [2]. Advertisement also defined as a thing, tool, action or media in which the form and motif of them are designed for a commercial purpose, introducing, suggesting, promoting or for attract the public attention to the product, service, people or body that can be seen, listened, felt, or enjoyed by public [3]. An advertisement is functioned to create an effective communication between society and environment in urban area. It cannot be denied that the advertisement, also called as *outdoor publicity*, is an important component in a landscape of modern city. It is suggested that an advertisement is harmoniously integrated with the built environment design [4]. An ideal principle of advertisement should be [5]: a) Capable of reflecting the visual characteristics of a region, b) Capable of guaranteeing a viewpoint that can be clearly seen, c) Having a form that is appropriate with the architecture of the building where the *signage* is posted, d) As the elements harmoniously integrated with the building – not as an additional element that can distract the visual, and e) Capable of uniting the communication directly or indirectly.

Advertisement is categorized as building that can be considered as a physical form of a construction work integrated with the space of its position either partly or entirely on/under land underground and/or water [6]. The requirements of the building architecture account for appearance, balance, compatibility, and harmony with its environment and the consideration towards the existing balance between the local social-cultural values and the implementation of any development of architectures and engineering. The appearance of the building located in a cultural site must be designed by considering the conservational principles [6], [7]. The appearance of the building built bordering on the conserved buildings must be designed by considering the principles of building aesthetics. Infrastructure and facilities of a building refer to the interior and exterior facilities of a building that can support the fulfillment of building function integrally or independently [8]. The infrastructure and facilities of the building also function as advertisement/signage including billboards, or advertising boards separately or in the form of wall.

The government does a control to the advertisement of tobacco products [9]. This control is conducted through printed media, broadcasting media, media of Information Technology and/or outdoor media. The outdoor media must fulfill some conditions as follows: a) Not being placed at non-smoking area, b) Not being placed at the main or protocol streets c) Must be placed parallel with the roadside and cannot cross the road or be transverse, and d) Not allowed to be more than 72 m2 in size.

Road Usage Space (*Ruang Manfaat Jalan*) and Right On Way (*Ruang Milik Jalan*) can be used for the advertising construction and information media. This utilization, however, must fulfill the following requirements [10]:

a. Not distracting security and safety of the road users
b. Not distracting the sight and the concentration of the motorists
c. Not distracting the function and the construction of the roads and its complementary buildings
d. Not distracting and reducing the function of the signs and other traffic control facilities
e. In accordance with the local regulation and/or the regulation of relevant authorities
f. Form of the advertisement and information media cannot be equal or resemble to the traffic signs

The building construction of the advertisement and information media must be designed safer so that when the buildings are damaged or collapsed, it will not endanger the road users, road construction and road building complementary. To ensure security and safety of the road users, the construction of the advertisement and information media and electrical installation in the advertisement and information media must comply with the following technical requirements: a) Regulation on building load, b) Regulation on steel building plan, c) Regulation on building materials, d) Regulation on concrete building plan, and e) Regulation on electrical installation.

Related to the spatial plan, advertisement should have been a regulatory element at the level Detail Spatial Plan. Meanwhile its designated area (zoning) is at the level of Spatial Land Settlement Plan. In more detail, advertisement comes to be a part considered in Urban Design Pattern and Urban Structure which commonly are issued in Plan for Building and Environment. Based on those regulations, the advertisement regulation should concern with the following points:

a. *Environment Aesthetics Standard* in order to support the aesthetic of environment.
b. *Advertisement Saturation* related to the number of the advertising signage installed in a particular area, thus needed to be limited.

c. The significance of the decision regarding the ban on the advertising signage installation in free zone and restricted zone for any advertisements.
d. The significance of control system from authority in monitoring the illegal advertisements
e. The significance of precision in design and size of installed advertising signage based on the proposed design and size.

3   Method

This study was conducted using a qualitative-descriptive approach based on the analysis on the aspects of communication and space. Communication aspect here is used as the base for the advertisement regulation by considering that advertisement is one of social or commercial communication media. Meanwhile, spatial aspect is used as the base of the advertisement regulation by considering that advertisement is one of the urban design components. Survey on data collection is a stage of identifying the existing condition of advertisement order that includes the order of the advertising signs, buildings, and space nearby. Data collection was conducted either directly to obtain primary data or indirectly to obtain secondary one. In collecting the primary data, two methods were used they are observation and field survey and interview.

Analysis is a phase of analyzing the existing condition about the advertisement structure by referring to the existing policies, literatures and direction for the spatial development [11-17] and trend of society dynamics in future. The results of this analysis were then used as the bases in formulating recommendations for advertisement order. The data and information obtained were systematically organized to obtain any matters that must be analyzed such as policy review and literature review.

Formulating the concept of recommendations for advertisement regulation is done using a participatory planning strategy in a limited scope through representatives based on their expertise. Some competent parties in this case were involved such as government, private sector, and stakeholders in advertising world. Meanwhile, participation of community was accommodated through a number of interviews.

4   Results and Discussion

4.1   Form of Advertisement

To accommodate the growth and dynamics of public life today and in future, advertisement can be classified into the following types: a) advertising boards/ billboards, b) videotron, c) cloth billboard, d) vinyl/plastic billboard, e) attached advertisement/stickers, f) leaflet advertisement, g) walking advertisement (including in vehicle), h) aerial advertisement, i) floating advertisement, j) voice advertisement, k) show advertisement, and l) light/movie/slide advertisement. The form of advertisement can be categorized as shown in table 1:

Table 1. Recommendation of Advertisement Form

No	Types of Advertisement	Based on the position to the road		Based on the position of length and width		Based on the Angle of View			Based on size		
		Longitudinal/ unidirectional	Transverse	Vertical	Horizontal	One-sided	Two-sided	> two-sided	Max (25 m2–32 m2)	Med (12 m2–24 m2)	Mini (< 12 m2)
1	Advertisement Board / Billboard	v	v	v	v	v	v	v	For the spot of advertisement installed on the artery roadside	For the spot of advertisement installed on the collector roadside	v
2	Videotron/ Megatron	v	v	v	v	v	v		For the spot of advertisement installed on the artery roadside	For the spot of advertisement installed on the collector roadside	v
3	Fabric	Incidental, maximal 1 month for installation on the determined road									
4	vinyl/ plastic	As a form of compensation of contribution to public facilities with the knowledge and approval of authorized government									
5	Attached/ Sticker	As a form of compensation of contribution to public facilities with the knowledge and approval of authorized government									
6	Leaflets	Incidental, not emerging any disturbance to environment, fulfilling the consideration in aesthetics, ethics, and safety									
7	Walking	Incidental, not emerging any disturbance to environment, fulfilling the consideration in aesthetics, ethics, and safety									
8	Aerial	Adjusted with the width of media used									
9	Floating	Adjusted with the width of media used									
10	Voice	Adjusted with the width of media used									
11	Show	Incidental, not emerging any disturbance to environment, fulfilling the consideration in aesthetics, ethics, and safety									
12	Light/film/ slide	Adjusted with the width of media used									

a. *Based on its position on the road,* the advertisement is divided into the longitudinal one or unidirectional to the road and the transverse one or crossing the road.

b. *Based on its length and width,* the advertisement is categorized into vertical one and horizontal one.

c. *Based on angle of view,* advertisement is categorized into one-sided advertisement, two-sided advertisement and multi-sided one.

d. *Based on the size*, the advertisement is divided into *big-sized* one if it is 25 m2 - 32 m2 in size, medium-sized one if in the range of 12 m2 - < 24 m2 in size, and small-sized one if sized < 12 m2.

Of the existing types of advertisement, particularly for the size of light advertisement, the form will be adapted based on the width of media used. Table 1 presents the recommendation of the form of advertisements in detail.

The image of Yogyakarta city is addressed to reflect the aspects of education, struggle, tourism, and cultural-based service. This is in line with the vision of Yogyakarta: *"Yogyakarta as Qualified Education City, Cultural-Based Tourism and Centre for Environmentally Sound Service"* [18]. To support this vision, the aesthetics aspect must be well-considered in the implementation of advertisement. This consideration then is becoming the base of the following recommendations for the advertisement implementation related to the facade of building and historical building as follows:

a. The installation of the advertising signage attached on building must be maximum 40% (forty per cent) of the facade width. Such installation commonly is used to the billboard of the shop name and recommended to be the one that must be attached on the facade without any additional constructions. The installation of the advertising signs attached on the facade can be done by using the space on the facade or by vertically attaching the advertisement on the facade as shown in Figure 1. The types of the advertisements placed on the facade can include Billboard, Videotron or three-dimensional one.

   Other installations for the Advertisement of Company Name or Profession Identification are as follows:

   1) The advertising signs are installed horizontally or unidirectional and vertically or crossing the road. The advertising signs should not block the building ornaments including windows and detail of architecture. For the installation of advertising signs in the vertical position to the road, it is recommended as follows:

      a) The maximum distance for the installation is 1 meter or half of pavement (depending on which one firstly taken) with the height distance of minimal 3 meter from pavement.
      b) The design of advertisement form can be one-sided and two-sided with 15 cm in maximum for the thickness of the advertising board.
      c) The position of the length and width of advertisement form might be vertical or horizontal.

   2) The advertisement in horizontal position (unidirectional to the road) must be really visible at distance to make the information delivered much easier to be transferred to the users.

b. A cultural-site building is not allowed to be used as an advertising media with the exceptions:

   1) Business/Profession Advertisement with the provision of 10% at maximum form the facade width and 1.5 m at maximum height and/or
   2) Light advertisement

Fig. 1. Recommendation for the types of Advertisement Installation on Facade
Source: Construction of Researcher, 2013

The detailed recommendations for the installation of advertising construction on the building facade are presented in Table 2.

Table 2. Recommendation for Advertising Construction Installation on the Building Façade

No	Types of Advertisement	Installation by Attaching on Facade	
		Horizontally/unidirectional to the road	**Vertically/ Crossing the road**
1	Advertising Board /billboard	1. Not allowed to block the building ornaments including windows and architectural detail	1. Not allowed to block the building ornaments including windows and architectural detail
		2. Advertisement design must also be integrated to the building	2. Advertisement design must also be integrated to the building
		3. The dimension of the advertisement is maximally 40% (forty per cent) of the facade width	3. The maximum distance to build is 1 meter or half of the pavement width (depending on which one firstly taken) with the height distance of minimal 3 meters from pavement
		4. For the advertisement to 2) the cultural site building, it is only allowed for the business/professional advertisement with the provisions at maximal 10% (ten per cent) of the width of facade and the highest level of height at 1.5 m (one point five meter)	4. The design of the advertisement can be in a one-sided or two-sided form with maximal 15 cm in thickness of advertising board.
			1. The length and width position of the advertisement form might be vertical and horizontal.
2	videotron/me gatron	1.Not allowed to block the building ornaments including windows and architectural detail	1. Not allowed to block the building ornaments including windows and architectural detail
		2. Advertisement design must also be integrated to the building	2. Advertisement design must also be integrated to the building

No	Types of Advertisement	Installation by Attaching on Facade	
		Horizontally/unidirectional to the road	Vertically/ Crossing the road
		3. The dimension of the advertisement is maximally 40% (forty per cent) of the facade width	3. The maximum distance to build is 1 meter or half of the pavement width (depending on which one firstly taken) with the height distance of minimal 3 meters from pavement
		4. For the advertisement to 2) the cultural site building, it is only allowed for the business/professional advertisement with the provisions at maximal 10% (ten per cent) of the width of facade and the highest level of height at 1.5 m (one point five meter)	4. The design of the advertisement can be in a one-sided or two-sided form with maximal 15 cm in thickness of advertising board.
			4. The length and width position of the advertisement form might be vertical and horizontal.
3	Fabric	Not recommended	Not recommended
4	vinyl/plastic	1. Not allowed to block the building ornaments including windows and architectural detail	1. Not allowed to block the building ornaments including windows and architectural detail
		2. Advertisement design must also be integrated to the building	2. Advertisement design must also be integrated to the building
		3. The dimension of the advertisement is maximally 40% (forty per cent) of the facade width	3. The maximum distance to build is 1 meter or half of the pavement width (depending on which one firstly taken) with the height distance of minimal 3 meters from pavement
		4. For the advertisement to 2) the cultural site building, it is only allowed for the business/professional advertisement with the provisions at maximal 10% (ten per cent) of the width of facade and the highest level of height at 1.5 m (one point five meter)	4. The design of the advertisement can be in a one-sided or two-sided form with maximal 15 cm in thickness of advertising board.
			5. The length and width position of the advertisement form might be vertical and horizontal.
5	attached/sticker	1. Not allowed to block the building ornaments including windows and architectural detail	1. Not allowed to block the building ornaments including windows and architectural detail
		2. Advertisement design must also be integrated to the building	2. Advertisement design must also be integrated to the building
		3. The dimension of the advertisement is maximally 40% (forty per cent) of the facade width	3. The maximum distance to build is 1 meter or half of the pavement width (depending on which one firstly taken) with the height distance of minimal 3 meters from pavement
		4. For the advertisement to 2) the cultural site building, it is only allowed for the business/professional advertisement with the provisions at	4. The design of the advertisement can be in a one-sided or two-sided form with maximal 15 cm in thickness of advertising board.

No	Types of Advertisement	Installation by Attaching on Facade	
		Horizontally/unidirectional to the road	**Vertically/ Crossing the road**
		maximal 10% (ten per cent) of the width of facade and the highest level of height at 1.5 m (one point five meter)	5. The length and width position of the advertisement form might be vertical and horizontal.

4.2 Advertising Placement

The strategic value of the advertisement spots is highly determined by a number of factors such as spatial usage or the potential of the region in achieving the target of installing the advertising sign, size of the advertisement, angle of view of advertisement, road class, and the price of spot of the advertising sign installation [19]. This is influenced by the consideration of the advertisers in selecting the location of the advertising sign installation: spatial usage, road class, and road cut-offs [20]. In this way, the advertisement settling can be managed as follows:

a. The advertising placement on the bulge land of personal individual or institution can be done in hall (see Table 1), by attaching it on the frontage and on the side of building or inside the building (see Table 2).

    The advertising placement on the bulge land of personal individual follows certain rules regarding form, placement, and size of the advertising signage adjusted to the road class in front. At the primary road class, the advertisement settling on the bulge land of individual is allowed until a maximal size with the distance among the $\geq$ 9m2 advertisements at least at 200 meters and its maximum number for 7 advertisements in each road space. The number of beamed advertisement in one space of bulge land is for 1 advertising construction at maximum.

b. The advertising placement on the bulge land of government and/or public facilities can be done on the pavement, bus halt, market/terminal/*Smart Park* /parking Park, schools or government institution.

c. The advertising placement on the bulge land of government and/or public facilities is *not recommended* to be done on the lamp posts, traffic lamp post, pergola, overpass, portal, clock pillar, police station and signpost except as a compensation for the contribution to a public facility given with the knowledge and approval of authorized government with provision that the compensational advertisement sized at 5400 cm2 or 30% of facade width (depending on which one firstly taken).

d. The advertising placement in an environmental park area or in Green Space Area is *not recommended* except as a compensation for the contribution to public facilities consigned with the knowledge and approval of authorized government provided that the maximum size of the compensation advertisement is at 2400 cm2 or 20% of facade width, depending on which is firstly achieved.

One of essential aspects needed to be well considered in the practice of advertisement implementation is the aspect of safety [21]. The policy has been a base of the following recommendations in the advertising placement in Yogyakarta:

a. Advertisement is not allowed to be conducted on divider roads and Green Space Area
b. Advertisement is not allowed to be conducted in the form of *wall painting* in considering that it can distract the concentration of the road users – particularly for motorists.
c. Advertisement is not allowed to be conducted in the form of portal or in any other constructing types that cross the road - particularly the one intended to an advertisement implementation.
d. Advertisement is not allowed to be conducted in the form of fabric except for banners and flags. One of the methods of installing an advertising sign mostly found is by crossing on the roads. This absolutely is not recommended as such advertisement is easy to be worn and will distract the sight of the road users.
e. Advertisement is not allowed to be conducted in the form of large *front light* billboard. In this case, the front light can cause a light reflection that can distract the visual comfort of the road users.

In addition to safety, aesthetics also becomes a base for the recommendation for not implementing the advertisement using a media of fabric/vinyl/plastics in a space of main roads in Yogyakarta including at the streets of *Laksda Adi Sucipto, Urip Sumoharjo, Jend Sudirman, P. Mangkubumi, Malioboro* and *Ahmad Yani*.

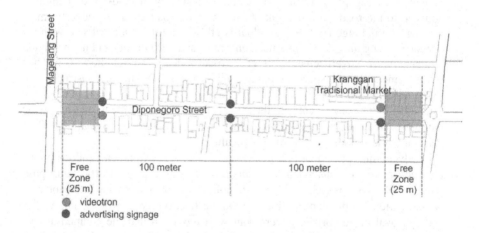

Fig. 2. Recommendation of Advertisement Spots (e.g. on the cut-off road at Diponegoro Street con) Source: Construction of Researcher, 2013

Similarly, the aspects of ethics and aesthetics are also highly needed to be considered in making the implementation of advertisement can support the policy on spatial planning in order to make Yogyakarta beautiful and harmonious. To prevent the stacks of advertising boards on the cross road, a free-advertisement area on the cut-off road in the radius of 25 meters from the cross road is highly recommended then. To keep maintaining the urban appearance, it is recommended that the installation of advertising signage on a road cut-off is placed by 200 meters inter-spots of advertisement. Other recommendation to minimize the number of advertising media is by giving priority towards the use of Videotron. Two spots of videotron are placed at the corner of the road cut-off. If the road is one-way direction, the videotron will only be placed in one of the corners. Meanwhile, in two-way direction, the videotron will be placed in each corner of the roads cut-off in accordance with the direction of the vehicles as shown in figure 2.

Referring to the Regulation of Yogyakarta Mayor Number 25 Year 2013, a number of special zones that are recommended are including:

a. Area of Tugu (pillar) with the radius of 50 m (fifty meters) from *Tugu Pal Putih*;
b. *Pangeran Mangkubumi* Street
c. *Malioboro* Street
d. *Ahmad Yani* Street
e. Area of 0 (zero) km with the radius of 50 m (fifty meters) from the middle of intersection
f. *Trikora* Street
g. *Alun -alun utara*
h. *Alun-alun selatan*
i. *Alun-alun Sewandanan Pakualaman*
j. Buildings of *Plengkung Gading* and *Plengkung Wijilan*
k. *Pojok Beteng* Area

4.3   Advertisement Content

In general, based on its content, an advertisement can be categorized into commercial and Visual Aids advertisement. Commercial advertisement includes the ones for professional business, products, and business and products. Meanwhile, what is meant by visual aid advertisement is the type of advertisement regarding non-commercial advertisement that might contain social and political information. Visual aid can be categorized into visual aids and visual aids and advertisement. The latter is due to the existence of advertisements found to contain both commercial and non-commercial information at all once.

The essential point needed to be concerned here is to make the content of advertisement in line with the local cultural values. It is then highly recommended if the content of an advertisement can emphasize on the main local cultures. This is to support the vision of Yogyakarta as a qualified education city, cultural-based tourism and centre for environmental service as mandated in the policy of spatial city plan of Yogyakarta in 2010. The more detail regulation about the content of cultural values needs to be further explained to avoid the emergence of ambiguity among society.

In this study, any advertisements containing tobacco product are found. The recommendation about such advertisement is based on the Government Regulation Number 109/2012 about the security on the additive substances in the form of

Tobacco Products. In this policy, the government conducts a control on the tobacco product advertisement. This policy is becoming a base for the following recommendations on the settling of an advertisement containing information regarding the tobacco products or any alcoholic liquor.

a. Not to place any advertisements containing cigarette and/or alcoholic liquor products in or outside school areas in the radius of 50 meters (fifty meters). However, the existence of public facilities in Yogyakarta, some of which are near schools makes it possible to be an exception for the regulation that is for an incidental advertisement containing cigarette products at the area of Mandala Krida Stadium and Kridosono Stadium in relation to the events in that areas.

b. Not to place any cigarette product advertisement in any non-smoking area at the main or protocol streets, either in the crosswise or in the crossing position on the road and the size not exceeding 72 m^2 in size (seventy-two square meters).

5   Conclusion

To support the vision of Yogyakarta, the aesthetics aspect must be well-considered in implementation of advertisement related to the facade of building and historical building. The advertisement implementation in Yogyakarta region can be classified into three zones [22]:

a. Special Zone – It is a zone free from any advertising administration –except for the types of advertisement with the following provisions:

1) Advertisement for business/professional signage attached on the building.

2) For the service on public information and product advertisement, government or in cooperation with other parties determines and provides the spots of the advertisement.

3) The incidental advertisement for any event calendar is allowed only in the area of Alun-Alun Utara, Alun - Alun Selatan and Alun - Alun Sewandanan Pakualaman.

A number of special zones that are recommended are including: Area of Tugu (pillar) with the radius of 50 m (fifty meters) from *Tugu Pal Putih, Pangeran Mangkubumi* Street, *Malioboro* Street, *Ahmad Yani* Street, Area of 0 (zero) km with the radius of 50 m (fifty meters) from the middle of intersection, *Trikora* Street, *Alun -alun utara, Alun-alun selatan, Alun-alun Sewandanan Pakualaman,* Buildings of *Plengkung Gading* and *Plengkung Wijilan,* and *Pojok Beteng* Area.

b. *Tight Control Zone* refers to a zone where the advertisement administration is allowed by considering the cultural site zone.

c. Medium Control Zone refers to a zone besides special zone and tight control zone.

The design of advertising board must possess the uniqueness of certain ornaments derived from the society culture. To support the achievement of vision of Yogyakarta, design is recommended to be developed using the various ornaments of Yogyakarta.

To support the ethics aspect, the government does a control to the advertisement of tobacco products [9], which is conducted through printed media, broadcasting media, media of Information Technology and/or outdoor media.

In the practice of advertisement implementation on the aspect of security and safety of the road users, the construction of the advertisement and information media and electrical installation in the advertisement and information media must comply with technical requirements. Hence, like a building structure, advertisement is also given certain requirements of building structure including: a) Architectural requirements comprising appearance, balance, compatibility and harmony of advertisement to environment, and b) Appearance Requirement – an advertisement must concern with the form and characteristic of the architecture and environment nearby.

Acknowledgement. The authors thank to Universitas Teknologi Yogyakarta for supporting this research. The work of Tutut Herawan is supported by University of Malaya High Impact Research Grant no vote UM.C/625/HIR/MOHE/SC/13/2 from Ministry of Education Malaysia.

References

1. Sunarimahingsih. Signifikansi Papan Reklame Di Pusat Kota Semarang. Jurnal Seri Kajian Ilmiah, 15(1) (2013) (in Indonesia)
2. Ewald, Mendelker: Street graphics: A concept and a system. American Society of Landscape Architects Foundation, USA (1971)
3. Major of Yogyakarta. Regulation about the Master plan of Advertisement and Visual Aids in Yogyakarta City (26) (2010)
4. Cullen, Gordon. Townscape. The Architectural Press, London (1962)
5. Shirvani, Hamid: The Urban Design Process. Van Nostrand Reinhold, New York (1985)
6. Indonesian government. Regulation about Building Construction (28) (2002)
7. Indonesian government. Regulation about the Implementation of Law (36) (2005)
8. Major of Yogyakarta. Regulation about Building Construction (2) (2002)
9. Indonesian government. Regulation about the Controlling Additives Substances in the form of Tobacco Products (109) (2012)
10. Ministry of Public Works. Regulation about Guidelines in Utilization and Usage of Road Parts (20) (2010)
11. Silva, R., Pires, J.M., Santos, M.Y.: Spatial Clustering in SOLAP Systems to Enhance Map Visualization. International Journal of Data Warehousing and Mining 8(2), 23–43 (2012)
12. Schoier, G., Borruso, G.: Spatial Data Mining for Highlighting Hotspots in Personal Navigation Routes. International Journal of Data Warehousing and Mining 8(3), 45–61 (2012)
13. Zhao, G., Xuan, K., Rahayu, W., Taniar, D., Safar, M., Gavrilova, M., Srinivasan, B.: Voronoi-Based Continuous k Nearest Neighbor Search in Mobile Navigation. IEEE Trans. on Industrial Electronics 58(6), 2247–2257 (2011)
14. Xuan, K., Zhao, G., Taniar, D., Srinivasan, B.: Continuous Range Search Query Processing in Mobile Navigation. In: IEEE Proceedings of the 14th International Conference on Parallel and Distributed Systems (ICPADS 2008), pp. 361–368 (2008)

15. Xuan, K., Zhao, G., Taniar, D., Safar, M., Srinivasan, B.: Voronoi-based multi-level range search in mobile navigation. Multimedia Tools Appl. 53(2), 459–479 (2011)
16. Al-Khalidi, H., Taniar, D., Safar, M.: Approximate algorithms for static and continuous range queries in mobile navigation. Computing 95(10-11), 949–976 (2013)
17. Bimonte, S., Bertolotto, M., Gensel, J., Boussaid, O.: Spatial OLAP and Map Generalization: Model and Algebra. International Journal of Data Warehousing and Mining 8(1), 24–51 (2012)
18. Yogyakarta government. District Spatial Plan of Yogyakarta 2010–2013 (2012)
19. The Decree of Ministry for Internal Affair No 15 year 1999 states that
20. Kasali, Rhenal: ManajemenPeriklanan. PT Citra Aditya Bakti, Bandung (1993)
21. Minister of Public Works. Regulation about Guideline of the Utilization and Usage of Road Parts (20) (2010)
22. Indonesian government. Regulation about the security on the additive substances in the form of Tobacco Products (109) (2012)

Spatial Control of Post-earthquake Market
Based on *Paseduluran*

Endy Marlina[1], Arya Ronald[2], Sudaryono[2], Rozaida Ghazali[3], and Tutut Herawan[4,5]

[1] Department of Architecture
Faculty of Science and Technology
Universitas Teknologi Yogyakarta
Jl. Ring Road Utara, Jombor, Sleman, Yogyakarta, Indonesia
[2] Department of Architecture
Faculty of Engineering
Universitas Gadjah Mada
Jl. Grafika 2, Yogyakarta, Indonesia, 55281
[3] Faculty of Science Computer and Information Technology
Universiti Tun Hussein Onn Malaysia
86400 Parit Raja, Batu Pahat, Johor, Malaysia
[4] Department of Information System
University of Malaya
50603 Pantai Valley, Kuala Lumpur, Malaysia
[5] AMCS Research Center, Yogyakarta, Indonesia
{endy_marlina,aryaronald,sudaryono_satrosasmito}@yahoo.com,
rozaida@uthm.edu.my, tutut@um.edu.my

Abstract. This research finds a theory which is very likely related to local wisdom during the post-disaster emergency period. The concept of control of space of post-earthquake market is the realization of the community's vigilance arising due to the earthquake and was influenced by the culture of the society who has been there and continues to grow to the present. Deep exploration will reveal the rationale of the society's strategy in arising vigilance in emergency period. The concepts which were found then were further explored with transcendental depth and managed to find the *paseduluran* as the basis of the control of the post-quake market space. The value of *paseduluran* underlying the consensus of control of space of post-earthquake market includes consensus on spatial distance, space boundary, and order of space, and space control, which is proved to accelerate the recovery of the community to normal life.

Keywords: Boundary of space, Space arrangement, Spatial control, Spatial distance, *Paseduluran*.

1   Introduction

After created, a space will grow and develop along with the human who dwells in it [1]. The understanding of space includes the physical and psychological aspects [2]. Humans will give different responses to their physical environment, depending on the

B. Murgante et al. (Eds.): ICCSA 2014, Part IV, LNCS 8582, pp. 95–108, 2014.

understanding, ideas and thought perception, which is closely related to their cultural background.

Eearthquake with the magnitude 6.9 on the Richter scale paralyzing DIY (Yogyakarta Special Region) on May 27, 2006 causing many deaths, injuries, physical damage, infrastructure damage, and economic damage. This disaster then raises public awareness. The society with low economic level was the group with the most social economic proneness to the disaster [3,4]. Nevertheless, in the earthquake of DIY in 2006, this group proved to be able to manage to stand on its feet again. This was indicated by the rise of the economic activities of the societies as their life support. The response of the society to the disaster would be likely to bring on the ideas, habits, and beliefs that are preserved in the society memory and motivate them to carry out particular actions to solve it [5]. In the periods of disaster preparedness until the rehabilitation and reconstruction after-earthquake DIY 2006, the activities were mostly in forms of attitude and behavior of the market society to the surrounding. These attitude and behavior also influenced the physical form of space where they did their activities.

Basically, a market can be seen as the psychological, social, and cultural reflection of the society that uses it [6,7]. Therefore the development process and response or reaction to market spatial after the disaster of DIY 2006 can be studied to reveal the basis of thinking, value, and culture as the background. This is interesting to study further concerning the possibility to discover spatial theory which very likely related to the local wisdom in the post-disaster emergency period that is evident to be able to guide the society to recover from the disaster in relatively short period. The culture and values can be the builder of local defense of the community in the disaster. In the disaster management cycle, this can be likely to contribute specifically to the response, disaster preparedness level, rescue, rehabilitation and construction, and disaster mitigation. In this cycle, local culture and values increase the resilience which is the ability of a community to face the disaster and get recovered to the normal condition [8].

Therefore, this research finds a theory which is very likely related to local wisdom during the post-disaster emergency period. The concept of control of space of post-earthquake market is the realization of the community's vigilance arising due to the earthquake and was influenced by the culture of the society who has been there and continues to grow to the present.

2   Related Works

The process of development and the responses toward the market spatial in the period of disaster preparedness until rehabilitation and reconstruction after the earthquake of DIY 2006 can be dug up to reveal the basis of thinking, values and culture playing as the background [9]. This specific character can form an identity to a place in general which is called sense of place [3]. The understanding of space involves the physical form and the activities of the human inside of it. In this relation, human is the change agent with the power to organize and control the space as what they want [6].

Culture and local values of the society can enhance the resilience and guide them in facing the disaster to be recovered to the normal conditions, which are a form of local wisdom [9,10]. In the disaster management cycle, this local wisdom has made a significant contribution, especially on the stages of response/disaster preparedness, rescue, rehabilitation and reconstruction, and disaster mitigation [9,11,12].

The tightness between space spatially and human doing the activities gives the basis in understanding that setting includes three aspects, namely: human as actor, type of activity, value system, and culture formed the interaction process between human and the space [13]. The cultural component includes language, knowledge system, social organization, life tools and technology system, economic system, religion system, and art. The form of culture includes idea, activity, and physical work [14]. Every human being has his own uniqueness. The uniqueness will influence his surrounding environment and his behavior [15]. The surrounding environment is not only a place for human to do activities, but it is also an integral part of human behavior pattern.

The relation between space and its occupant (human) is a record of the human life which if it is investigated, it gives a picture of his behavior and its change during the occupancy period [13]. In this relation, there is an emphasis on the human cultural background, such as life perspective, belief, values and norms which will determine the individual behavior reflecting in his way of life and role chosen in the society [16]. Moreover, this cultural and social contest will determine the human activity system [7]. Different society will result in different concepts and forms of space. Behavior setting is the interaction between an activity and more specific space [15].

Market is a place used to serve the community economic activities. Space can be observed from two different views, those are: a) space as a product – space is seen as a result/product of an activity. This view sees space as something formed and influenced by the activities of the user inside of it, which can take place in a short or long period of time [1,2,17]. In this case, space is created due to setting demand of an activity; and b) space as work – a space is seen as thing that can form or direct an activity. In this view, a space can be categorized as subject, which is active and influences or forms the activities of the users.

The connection between space and human needs to be explored using phenomenology paradigm in line with the adjustment between the architectural phenomenon and intentionality aspect as one of the stages of thinking. Human in this scenario is viewed as the origin of intentionality genesis [18]. Another important phenomenology stage is the reduction including eidetic reduction, phenomenological reduction, and transcendental reduction, which emphasize on the whole reality to get the awareness that gives transcendental meaning to what is truly an integral part of our awareness [19].

3   Method

The study on the connection between the spatial and the activities in the market is explored in the period of disaster preparedness until rehabilitation and reconstruction period after disaster with the entry point the earthquake event in May 2006. Related to that historical event, we will find the philosophical theoretical and conceptual formula that probably influences the control of market spatial. The control of market spatial is viewed as a phenomenon that needs to be understood as what it is. Consequently, we need to focus our attention on the phenomenon without any prejudice and without imposing theories. The attention needs to be concerned on the system to reveal the principles [20].

The research question 1 is: "What kind of market spatial has developed after the earthquake (from the disaster preparedness period until the rehabilitation and reconstruction period) of DIY 2006?" To answer this question, the connection between space as physical order, humans (with their activities and culture), and environment as context is explored according to the natural condition. In the beginning the research was conducted in 8 (eight) markets in Bantul Regency (Piyungan Market, Jodog Market, Pundong Market, Ngangkruk Market, Turi Market, Gatak Market, Niten Market, and Bantul Market) and 3 (three) markets in Gunungkidul Regency (Wonosari Market, Merdeka Market, and Playen Market) to find the research focus. Exploration conducted by phenomenological depth found empirical themes. Furthermore, the exploration was focused on the 6 (six) out of 11 (eleven) markets, those are Piyungan Market, Jodog Market, Pundong Market, Ngangkruk Market, Turi Market, and Gatak Market. This is based on the empirical themes repetition during the research process.

Research question 2 is: "What responses influence the development of the market spatial after earthquake (from the disaster preparedness period until the rehabilitation and reconstruction period) of DIY 2006?" To answer this question, the search on realities and information on field was conducted with the eidetic depth. Empirical themes found in 6 (six) markets was deeply analyzed to find the inductive concepts of control of market spatial. In the end, the themes were specified into 5 (five) markets, namely Pundong Market, Ngangkruk Market, Turi Market, Gatak Market, and Jodog Market. Of the five markets explored, Pundong Market was the most intensively explored market because the most empirical themes were found here.

The research question 3 is: "What matters causes the responses of the users toward the market spatial developed after earthquake (from the disaster preparedness period until the rehabilitation and reconstruction period) of DIY 2006?" To answer this question, information investigation was carried out until transcendental depth, which is the understanding of objects through continuous and intensive acquaintance [21]. In this process, the information was fortified and induced into the themes, which then constructed into the concepts. The intensive closeness between the researchers and the objects was maintained for the clarity of the phenomena studied and the success in creating and finding of the principles of control of market spatial. As a result, the essence/principle of control of market spatial is near the subjective knowledge instead of objective knowledge. This subjectivity can be subsided and changed into objectivity through inter subjective dialog process. To avoid the researcher subjectivity, the triangulation was conducted – finding other similar data to test the information correctness [22]. The research was not conducted once only but in iterative observation.

4   Spatial Control of Post-earthquake Market Spatial in Bantul Regency

The concepts found in this research were specifically built in the emergency condition after the earthquake. The research that was conducted during the period of disaster preparedness until the rehabilitation and reconstruction period after earthquake found the empirical themes related to market spatial. The careful consideration on the themes successfully revealed the response of the market society that further influenced the development of market spatial.

The emergency context after the earthquake has built specific understanding of the market society of Bantul Regency toward market spatial as a place for the activities to soothe the mind and economic activities that were performed based on the brotherhood value. In this case, the market spatial developed to fulfill the tranquility needs of the users. The need to calm them appeared due to the earthquake of May 2006. The strategy used by the market society was by looking for the feeling of safe and comfortable by increasing the intensity to be together among them. The market spatial also developed to fulfill the economic needs of the users. The strategies taken by the society to response the earthquake of May 27, 2006 were oriented on the interaction management inter-individuals. Those strategies are: 1) strengthening the old brotherhood; 2) building new brotherhood; 3) minimizing the conflict potentials; and 4) attempting common goodness.

Generally, the concepts of space found in the related markets shows that in general life practices of the society in the research location were related to the efforts to manage the interaction between one person and another person. This strategy was oriented to create a harmonious relationship between each other. This is based on the brotherhood (*paseduluran*) spirit/value as shown in Figure 1.

The earthquake of May 2006 caused awareness to the market society. This awareness resulted in the disorder to the human life process. Based on deep exploration, the reflection of the strategy reveals the faith of the market users which serves as the basis of all of their thoughts, attitudes, and behaviors which is also reflected in the physical form of market spatial. The faith is *paseduluran* (brotherhood) value.

In this research can be captured visually distinctive character of post-earthquake market spatial in Bantul Regency. Development of the market space in the emergency response to the rehabilitation and reconstruction after the earthquake is related to the distance between spaces, the space boundary, space arrangement, and space control. Physical realization of control of market spatial during the emergency period was based on the value *paseduluran* and shows some consensuses on the control of post-earthquake market spatial.

4.1   Paseduluran in the Context of Consensus of Distance between Spaces

Damaged and collapse of *bakulan* spaces due to the earthquake in 2006 was responded with the improvement and development of new *bakulan* spaces.

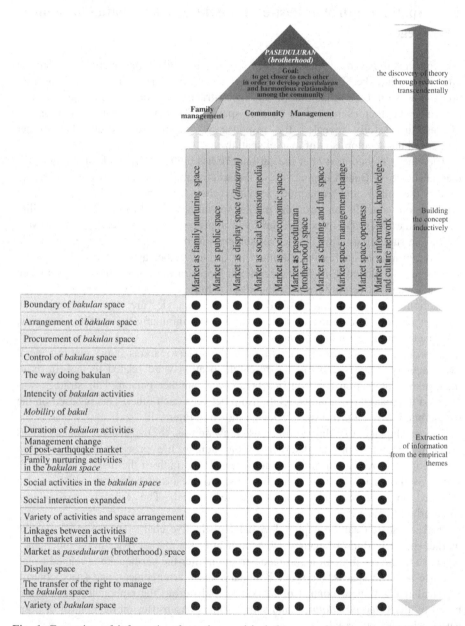

Fig. 1. Extraction of information from the empirical themes to build concept and find the brotherhood value (*paseduluran*)
Source: Researcher construction, 2012

Paseduluran is a local value which is underlying these responses which was implemented in developing new *bakulan* spaces are clustered adjacent to the distance between the spaces one another. This is the realization of the needs of market users to get closer to each other in order to develop *paseduluran* and harmonious relationship among the community as shown in Figure 2.

In larger areas, the mobility of *bakulan* activity in various markets in the system of *pekenan* also establish a specific atmosphere on control of post-earthquake market spatial. The system build many cycles of pekenan users in different market cycles, each consisting of a series of five market according to the Javanese calendar. Rotation of pekenan users which is occurs in a variety of market cycles are not always equal.

The system is build different slices of the market society. The motivation to develop a harmonious brotherhood as the response of post-earthquake emergency conditions realized by increasing the intensity of activity in the market. This enlarges the slices of market society in rotation system of *bakulan* activities (pekenan). Although physically the market spaces is away from each other, but in non-physically the users of the market is close from each other. The distance between market users are not understood as physical distance, but rather non-physical distance that is the brotherhood / *paseduluran*. This means that the distance among *bakulan* spaces is determined by the proximity between the user each other. This consensus is building people's understanding of the market space after the earthquake in Bantul Regency that the distance between spaces is social distance / proximity between users of the market.

Fig. 2. Consensus of distance between spaces in control of post-quake market spatial in Bantul Regency developed based on the *Paseduluran*
Source: Construction researcher, 2012

4.2   Paseduluran in the Context of Consensus of Boundary of Space

Considering the safety problems was underlying the development of market space by minimizing the use of space boundary. *Bakulan* space with minimal space boundary provides an opportunity for the intensivelly social contact between the consumer of markets. This market space arrangement in accordance with the market community needs to increase the intensity of social interactions among them. Social contact not only done visually, but the minimal space boundary also provides an opportunity for the occurrence of direct physical contact as shown in Figure 3. Based on deep exploration, post-quake market space conceived as a communal space. The space is expanded to accommodate many *bakulan* activities. *Bakulan* activities developed as an economic activity condensed by social values. *Paseduluran* which underlying these activities developing closer ties one another. Physically, this built a consensus of boundary of space.

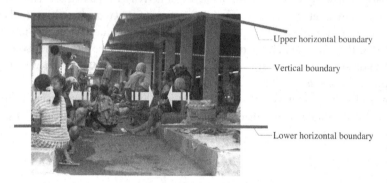

Fig. 3. Consensus of boundary of space in post-quake market spatial in Bantul Regency developed based on the *Paseduluran*
Source: Construction researcher, 2012

4.3   Paseduluran in the Context of Consensus of Space Arrangement

Awareness that characterizes the post-earthquake emergency condition is also seen visually with the phenomenon of changing the arrangement of market spaces. The *bakul* reduce *Bakulan* equipment used on *Bakulan* activities or changing the way they organize and clean up his wares. This is intended to enable them to settle their *Bakulan* activity when an emergency occurs aftershocks. In addition, the reduction of *Bakulan* equipment is also aimed at getting a more spacious room that allows market users to accept more social interaction in their *bakulan* space. This strategy encourages market users to accept for other people into the circle of their brotherhood as shown in Figure 4.

During the response to the rehabilitation and reconstruction after the earthquake, the market is a medium to intensify the development of the social relation of society. This is the strategy chosen in response to earthquake.

space arrangement
is simple and easily adjusted
in order to accomodate
the social relationship

Fig. 4. The consensus in the space arrangement of market spatial after the earthquake in Bantul Regency are developed based on the value of *Paseduluran*
Sources: Construction researcher, 2012

4.4 Paseduluran in the Context of Consensus of Space Control

Bakulan activities was understanding as an non formal activities. It is used to manage the activities of the market flexibly adapted to their needs. Precautions were awakened in response to the earthquake motivate people to develop a harmonious relationship between one another. Mixing of several activities in this *bakulan* space effect on the control of space. Each space users lose control of the movement area. In this case, each user space allow others to acces into their area. Within the scope of the market in general, pe*seduluran* also affect space control after the earthquake. This implemented in a way to loser their control of area. The proximity between the *bakul* (seller) and the wong tuku (buyer) based on the value of *paseduluran*. Development of market spatial during the response to the rehabilitation and reconstruction after the earthquake is based on the value of *paseduluran* build public awareness in terms of controlling space as shown in Figure 5.

Space control consider
the appropriateness of social

Fig. 5. Consensus of space control of market after the earthquake in Bantul Regency are developed based on the value *Paseduluran*
Sources: Construction researcher, 2012

Orientation to the development of harmonious social relationships was build a consensus of 'freedom' access to market space. In this case, market space conceived as a medium to develop social relationships, meaning 'to be' accessible to their brothers. Definition of 'should' be accessed contains an agreement that the free access of market space was carried out within the limits of social propriety. It is not written, the public agreed that the boundaries of social decency attitudes and behavior towards people who are not harming others and cause conflict between them. That is, the consensus of space control on the market after the earthquake in Bantul regency is based on the social propriety, that is *paseduluran*.

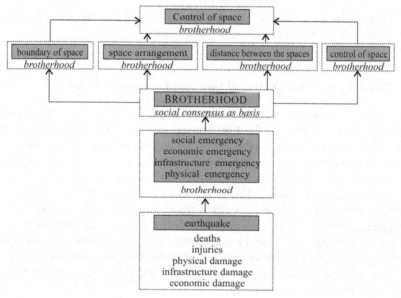

Fig. 6. Consensus of space control of market spatial post-earthquake in Bantul Regency based on the *paseduluran*
Sources: Construction researcher, 2012

Results of this study revealed that phenomenon of post-earthquake market space in Bantul is due to the space consensus that associated with the distance between spaces, the boundary of space, fabric of space, and space control. Consensus was a reflection of people's minds embodied in attitudes, actions, and behaviors, which was implemented in the control of market space as a response to the vigilance since May 2006 earthquake.

Paseduluran value then mobilize the community to take action in response to the vigilance. Physical implementation of these responses can be visually seen in the control of market space during the post-earthquake period. Control of post-earthquake market space has been build consensus of space as shown in Figure 6.

Market spatial post-earthquake build consensus of space as follows:

1. distance between the spaces on the market spatial after the earthquake in Bantul Regency based on *paseduluran* is the social distance / proximity between market users.
2. boundary of space on the market spatial after the earthquake in Bantul Regency which based on pa*seduluran* is social closeness.
3. space arrangement on the market spatial after the earthquake in Bantul Regency based on *paseduluran* is social order.
4. control of space on the market spatial after the earthquake in Bantul Regency which based on pa*seduluran* is social propriety.

Consensus about the distance among spaces, boundary of space, and the space arrangement was basically related to the control of space. Consensus control of space in market spatial after the earthquake in Bantul Regency based on *paseduluran* is the social propriety. Implementation of this consensus is forming the distinctive character of the market after the earthquake in Bantul Regency.

5   Conclusion

The control of post-earthquake market space specifically awoke in a state of emergency. This condition builds people's understanding of market space after the earthquake as follows:

1) Shaped as semi-open communal space which is controlled by considering the distance between spaces, the boundary of space, arrangement of space, and space control.
2) Shaped as semi-open personal space which is controlled by considering the distance between spaces, the boundary of space, arrangement of space, and space control.
3) Associated with human activity as a user, economic activity, physical environment, and the social order
4) Developed specifically in the context of post-earthquake emergency

In the context of earthquake, the development of market spatial during the period of disaster preparedness until the period of rehabilitation and reconstruction after the earthquake is influenced by: 1) tranquility needs, 2) safety needs, and socio-economic needs. The analysis until the transcendental depth in this research found brotherhood value as the value which gave basis the physical development of market spatial during the period of disaster preparedness until the period of rehabilitation and reconstruction after the earthquake. It means that brotherhood is a society cultural value that was used as the strategic basis to response the earthquake.

The implementation of the spatial strategies [23-27] during the period of disaster preparedness until the period of rehabilitation and reconstruction after the earthquake produced various activities and market spatial orders that in this research was found to have high social value content. The orientation of the fulfillment of the tranquility and safety needs had born orientation of social togetherness and motivated the market society to develop some activities influencing the market spatial order. Therefore, the spatial order formed can be referred to as *socio-cultural spatial*.

The implementation of the brotherhood theory in control of market spatial after earthquake is not limited only to one market, but it is also implemented in the wider scope to the market rotation system in Bantul Regency. The implementation of the strategies in brotherhood value in the wider scope in Bantul Regency through *pekenan* system was admitted by the society to be able to strengthen the society defense and accelerate the society recovery from the damage after the earthquake.

Consider that *paseduluran* value can help the society of Bantul Regency to recover from the wound due to the earthquake of May 2006; the researcher recommends further similar studies in other regions in Indonesia. The fact that more disasters occur in Indonesia gives the foundation of recommendation to conduct similar studies in other regions of Indonesia. This is aimed at finding the local strength of the community, especially related to the disaster management activity. The development of the local society strength is expected enhance the resilience of the society to deal with the disaster and recover to the normal condition.

5.1   Contribution for Disaster Management

The recovery of Bantul society can be categorized as fast one, so it can be a reference for disaster mitigation in other regions. In the research of the market spatial conducted during the period of disaster preparedness until the period of rehabilitation and reconstruction after the earthquake, the brotherhood value is the value that developed in the society and has been proven as one of resilience shaping components during the disaster.

More specifically in the development of a market in the district of Bantul in particular, or within the scope of the Java community in general, to consider the provision of open spaces to accommodate the development aspirations of the people who may need to be implemented in the physical order of the market space. The existence of an open area in the market which is a common space has the potential to be used as a space to develop emergency facility in the event of a disaster. In the normal condition, the existence of this open area provides opportunities social cultural development of the community who can provide a positive influence on the social development of society in general equilibrium.

Acknowledgement. This work is supported by University of Malaya High Impact Research Grant no vote UM.C/625/HIR/MOHE/SC/13/2 from Ministry of Education Malaysia.

References

1. Habraken: The Structure of the Ordinary: Form and Control in the Built Environment. Graphic Composition Inc., USA (1998)
2. Schulz, Christian-Norberg: The Intention in Architecture. Rizolli, New York (1977)

3. Marlina, Endy: "Rukun": Kearifan Lokal Jawa Yang Membimbing Pemulihan Masyarakat Pasca Gempa 2006. Jurnal Sosiohumaniora 12(3) (2010) (in Indonesia) ISSN 1411-0911
4. Marlina, Endy: Community Empowerment of Wonosari: Supporting Society Economic Post Earthquake 2006. In: Prosiding Regional Workshop "Action For Effective Management of Post-Disaster Recovery". Gadjah Mada University, Yogyakarta (2009) (in Indonesia)
5. Alwisol: Psikologi Kepribadian. UPT Penerbitan Universitas Muhammadiyah, Malang (2008) (in Indonesia)
6. Lawson, Bryan: The Language of Space. Architectural Press, London (2001)
7. Koentjaraningrat: Pengantar Ilmu Antropologi. PT Rineka Cipta, Jakarta (1990) (in Indonesia)
8. Thywissen, Katharina: Component of Risk. A Comparative Glossary. Institute For Environment And Human Security, United Nations University (2006)
9. Carter, W.N.: Disaster Management: A Disaster Manager's Handbook. ADB, Manila (1991)
10. Sutikno: Bahan Sosialisasi Gempa di Bantul dan Sekitarnya. Refresher Course Geo-Information for Hazard and Disaster Management, Gadjah Mada University, Yogyakarta (2003) (in Indonesia)
11. Faizal, L.: A View Strategy of Indonesia on Disaster Mitigation. Research and Development Centre for Human Settlements, Ministry of Public Works Clark, Jakarta (2007) (in Indonesia)
12. Sukandarrumidi: Gempa Tektonik dan Akibatnya. Seminar Penyusunan Arahan Pemanfaatan Ruang Kawasan Bencana Alam Di Kawasan Pantai Selatan Pulau Jawa, Badan Perencanaan dan Pembangunan Daerah Propinsi DIY, Yogyakarta (2006) (in Indonesia)
13. Rappoport, A.: System of Activities and System of Setting. Cambridge University Press, Cambridge (1990)
14. Rappoport, A.: Some Perspectives on Human Use Organization of Space. Melbourne (1972)
15. Haryadi, Setiawan, B.: Arsitektur Lingkungan dan Perilaku. Gadjah Mada University Press, Yogyakarta (1995) (in Indonesia)
16. Geertz, C.: The Interpretation of Cultures. Harper Collins Publisher, USA (1973)
17. Van de Ven, C.: Space in Architecture. Van Gorcum & Comp., Eindhoven (1987)
18. Lincourt, M.: In Seacrh of Elegance. Toward an Architecture Satisfaction. McGill-Queen's University Press, Monreal (1999)
19. Welton, D.: The Essential Husserl. Indiana Unversity Press. Bloomington and Indianapolis
20. Zubaedi (2007) Filsafat Barat: Dari Logika Baru Rene Descartes Hingga Revolusi Sains ala Thomas Kuhn. Ar-Ruzz Media, Yogyakarta (1999) (in Indonesia)
21. Peursen, C.A.: Orientasi di Alam Filsafat. Dick Hartoko, Translater (1988)
22. Muhajir, N.: Metodologi Penelitian Kualitatif. Rake Sarasin, Jakarta (2004) (in Indonesia)
23. Mahboubi, H., Bimonte, S., Deffuant, G., Chanet, J.-P., Pinet, F.: Semi-Automatic Design of Spatial Data Cubes from Simulation Model Results. International Journal of Data Warehousing and Mining 9(1), 70–95 (2013)
24. Bimonte, S., Bertolotto, M., Gensel, J., Boussaid, O.: Spatial OLAP and Map Generalization: Model and Algebra. International Journal of Data Warehousing and Mining 8(1), 24–51 (2012)
25. Silva, R., Moura-Pires, J., Santos, M.Y.: Spatial Clustering in SOLAP Systems to Enhance Map Visualization. International Journal of Data Warehousing and Mining 8(2), 23–43 (2012)

26. Schoier, G., Borruso, G.: Spatial Data Mining for Highlighting Hotspots in Personal Navigation Routes. International Journal of Data Warehousing and Mining 8(3), 45–61 (2012)
27. Taniar, D., Safar, M., Tran, Q.T., Rahayu, W., Park, J.H.: Spatial Network RNN Queries in GIS. The Computer Journal 54(4), 617–627 (2011)

Spatial Analysis of School Network Applying Configurational Models

Bárbara M. Giaccom Ribeiro[*], Laís Corteletti, Leonardo Lima,
and Clarice Maraschin

Graduate Program on Urban and Regional Planning - PROPUR, Federal University of Rio
Grande do Sul - UFRGS, Porto Alegre, Brazil
{bgiaccom,laiscorteletti}@gmail.com, leonardolima_@hotmail.com,
clarice.maraschin@ufrgs.br

Abstract. Contemporary urban planning is demanding new tools for analyzing and evaluating impacts of continuous changes occurring in urban systems. The systemic and quantitative character of urban models makes them important tools in the construction of indicators of urban performance. This work aims to: a) present a methodology for spatial analysis based on configurational approach; b) develop an empirical application for the analysis of spatial distribution of a specific urban facility, the schools network in Novo Hamburgo, Brazil. This methodology systemically articulates variables related to urban space (morphological description), to the population (socioeconomic data) and to school facility itself (e.g., type, size and standard). We apply *Space Centrality* models [1, 2], supported in a GIS (Geographic Information System) environment. The application demonstrates the great potential of the methodology for generating performance indicators, which can support decision-making in planning the school network.

Keywords: Configurational models, GIS (Geographic Information System), performance indicators, schools network.

1 Introduction

The complexity of urban systems and the nature of their transformation processes have challenged urban research to seek new approaches that can manage these phenomena adequately. In the field of urban modeling, several methodologies have been developed, addressing the city as a complex system, consisting of many elements and relationships in a state out of equilibrium [3, 4, 5]. In this approach, the urban system is formed by a large number of agents taking location decisions, generating non-linear dynamics, and therefore a vast range of possible futures for urban form.

These new insights also defy the nature and role of urban planning. There is a growing awareness of the uselessness of attempting to define and impose one given form on the city. New knowledge about the dynamics of the form of the city points to the need to

---

[*] Corresponding author.

B. Murgante et al. (Eds.): ICCSA 2014, Part IV, LNCS 8582, pp. 109–124, 2014.

provide tools to government and society to analyze different possible trajectories of the urban system and their socio-spatial implications [4]. The evaluation process earns a central role in urban planning system, demanding the development of systematic methods for monitoring the evolution of the urban system in order to anticipate the effects of changes proposed by the agents. In this scenario, urban modeling is a potential resource that can act as measurement and evaluation instrument. Urban models can be understood as simplified representations of reality, thus, allowing the choice of particular aspects of reality and its quantitative representation. As a result of these common roots, indicators and urban models have strong logical and operational relationships as: a) complementarity relations, which justify the use of both tools; and b) similarity relations, which imposes conditions of logic and operational consistency [6]. According to these authors, data and model outputs together constitute an expanded geographical information system which can provide the basis for performance indicator calculations.

The objectives of this paper are: 1) to present a spatial analysis methodology based in urban configurational models; and 2) develop an empirical application to demonstrate the analysis of the spatial distribution of a specific urban facility – the network of schools – in the city of Novo Hamburgo, Brazil.

It is known that the population's access to education facilities is an important indicator of quality of life in cities. At the same time, investments in schools, both public and private, demand efficient criteria in the allocation of resources, both at the time of the location of new schools, as in investing in pre-existing institutions. Dealing with this complex problem is an opportunity to explore the spatial analysis through urban models, which are able to describe the state of the urban system in terms of access to school facilities as well as their effectiveness. Specifically, this analysis aims to answer the following questions: 1) how is schools accessibility distributed within the study area?; 2) what is the relationship between the location of schools and the residents income?; 3) what are the best residence locations in relation to schools, considering every level of education?; and 4) which schools tend to capture more population? (considering the population has the same displacement conditions).

The proposed methodology is based on configurational models, that allow joining, systemically, variables related to space (distances), to population (socioeconomic data) and to school facility itself (size, pattern, etc.). We apply Accessibility, Centrality, Convergence and Spatial Opportunity models, supported in a GIS (Geographic Information System) environment. We consider that the outputs of these models have the potential to act as basis for relevant performance indicators to the problem at hand.

The paper is organized into four sections, besides this introduction. In the following item, performance indicators and configurational models are briefly contextualized. The methodology of the study is presented in section 3, followed by the empirical application and its main results. The paper concludes with a discussion about the strengths and limitations of the methodology presented.

2   Indicators and Urban Models

Contemporary cities reflect their socioeconomic inequalities through large contrasts in terms of quality of housing and environment and also through the population's access to goods and (public and private) urban services. In Brazil, for example, despite having income level above the world average, the cities stand out for the high inter-personal income inequality. In 2008, even with its recent fall, the country was still the world's fifth most unequal [7]. There is a great demand for methods to quantify and analyze the variations in the quality of life of cities aiming to support the decisions of urban planning and public investment.

Indicators have been used since the 1960's to produce information on the cities' quality of life [8]. An indicator can be defined as a variable, a measure or value that conveys relevant information on the state of a particular phenomenon. Examples of best-known urban indicators are rates of green area *per capita*, or number of hospital beds *per capita*. Such indicators have limitations, since both can hide large differences in the distribution of facilities within the city, as well as discrepancies in the distribution of real users.

Performance indicators [6] are more complex measures, able to reflect the interdependencies among the components of the urban system, and they are, therefore, based on urban models. Their goals are to provide a concise overview on the state of the urban system at a given time (description, measure), as well as assess likely impacts of considered or proposed urban transformation actions. There are basically two types of spatial performance indicators [6]. The former are effectiveness or equity indicators and report the quality of urban life. They are related to individuals and residents, based on the residential location and how it is served by the organizations. The second type of indicator, which relates to provider organizations (public or private), are the efficiency indicators, and refer to the economy as a whole system, i.e., the rational use of resources and the general costs of living and of economic production. These indicators differ from traditional social indicators as they consider the spatial interaction between people and organizations and are built from urban models, therefore, the selected variables are related in a systemic way. What differentiates them from traditional indicators is precisely its ability to globally evaluate the urban system, with reference to different service sectors (e.g., fundamental level schools), different social strata (e.g., middle classes) and specific urban areas (e.g., a street, a neighborhood).

As simplified representations of reality, urban models operate with a selection of variables and significant relationships for the purpose of the model [9, 3]. It is important to emphasize again that the models presented here are not performance indicators *per se*, but a step in their elaboration. An appraisal of urban performance needs to face the task of judging the results, according to standards, norms or planning objectives [8]. Several models driven to capture and description of urban spatial differentiation have been suggested, and are better known as the various forms of accessibility measuring. Arentze et al. [10] cite three groups of commonly used accessibility measures, ranging from the simplest ones, based in travel costs faced by consumers in the process of satisfying their demands, to more complex measures, which express accessibility in terms of surplus value, benefit or utility consumers gain from facilities. These models, associated to spatial interaction theory (gravity), have little concern with specific spatial situations [11]. Searching for a more comprehensive description of the space, Hillier and Hanson [12] proposed a space syntax measure, Relative Asymmetry, taking accessibility as a topological mean distance from each space to all others in the same spatial system. Axial lines are used to describe connectivity; they describe space more efficiently than traditional zones or links used in spatial interaction models.

Krafta [2, p. 37] ranks these accessibility models according to the complexity of their inner structure, as follows: (i) measures that do not distinguish point hierarchy, that is, do not differentiate origins (demand points) from   destinations (supply points), for example, relative asymmetry; (ii) measures that do not distinguish different destinations in terms of their quality, such as the models in which the distance to the nearest supply point is considered; (iii) measures which do hierarchize supply point and do not take into account increasing or decreasing probabilities of choice derived from chain destinations, such as measures of consumer welfare; and (iv) measures which handle supply hierarchies, as well as plural destinations, for example, the multistop travel model [10]".

Space Centrality models [1, 2] can be seen as a bridge between these different families of models, to the extent that they bring together spatial and functional particularities of urban systems. Configurational models assume the city has a hierarchical pattern of spatial differentiation (configuration) which characteristics influence other aspects, as pedestrian circulation and land use. Such models apply methodologies of disaggregating the city into components (basic units of space, spatial attributes) and their relationships (topological descriptions, adjacencies, centrality). In these cases, graph theory provides the analytical basis for the calculation of different measures and properties of the urban network. The models assume the shortest path hypothesis, i.e., the connections between cells of the network will always be made by the shortest paths. Thus, any city would exhibit a spatial differentiation, i.e., a hierarchy in which some cells (spaces) are distinguished by their relative position and/or the number of connections with others. Having in Haggett and Chorley [13] its main pathfinders, this research field lately gained special relevance because of its ability to represent complex systems and thus be part of numerous investigations that explore the limits of scientific knowledge. This effort has been shared, among others, between some recent studies on social networks [14]; on the analysis of intra-urban networks [15-17]; on cognition and spatial dynamics [18]; on random networks [19, 20]; and, especially, on networks structure and dynamics [21].

Many of these studies have used some kind of centrality measures aiming to perform structural analysis. There are different ways to calculate centrality in networks; the main ones are those based on connectivity, straightness, closeness, betweenness and information [22]. The simplest spatial network is one in which nodes represent portions of space (zones, axial lines, segments, corners, etc.) and lines are adjacency or distances. More elaborate networks may consider weights for nodes and for lines, loading the first, for instance, with quantities of land uses, activities, residents, or jobs, and the latest, with metric distances, travel times, travel costs, etc. Weighted spatial networks have proven extremely useful to explore urban systems [1, 2, 22, 23].

In this context of configurational models, we selected four centrality measures [1, 2], which are distinguished by admitting weighting from attributes (areas, activities, attractiveness, etc.) that can be allocated to each spatial cell. Thus, these models work with a systemic and weighted description, and allow various levels of spatial disaggregation, overcoming the limitations of descriptions aggregated by sectors or zones. The models are able to capture nuances of the spatial distribution of urban facilities in a degree of spatial accuracy as thorough as the spatial description is adopted in the study.

The first model chosen for this spatial analysis is *Accessibility*, which is a measure of relative distance of a space in the system. It is related to the facilities and difficulties to reach a certain point or place in the urban system, and can be defined as the property of a particular component to be closest to all other network elements, considering the shortest (or preferred) paths between them [13, 23].

Broadly speaking, Centrality is the property of a cell (space) being along the path that connects two other cells, and their hierarchy is given by the total number of times this one cell appears in the paths connecting all pairs of cells of a system [24]. Krafta [1] proposes a Weighted Betweenness Centrality (*Freeman-Krafta Centrality*), introducing the notions of tension and distance: the tension reflects the relationship between two points expressed by the product of its contents; the distance refers to the extension of the shortest path between each pair of points, and this increases as the centrality of each cell interposed in the path decreases. Krafta *Planar Centrality* model considers only the spatial differentiation originated from the system of spaces and their connections, while the model of *Centrality* (or Weighted Centrality) considers the presence and the unequal distribution of built forms and also the activities in these spaces [1]. The model allows the parameterization of the attributes by stating the activities associated with the built forms, to which one can assign different weights. The model considers that the tension generated by each pair of cells (product of loads) will suffer dissipation, i.e., it will be distributed among all cells which are part of the shortest paths, thereby considering the influence of distance.

Applying directed graphs, Krafta [2] also formulates other urban spatial analysis models, derived from the original Weighted Betweenness Centrality model. The use of a directed graph allows one to associate different attributes to the origins and destinations of the directed vectors. This study uses two of these measures: *Convergence* and *Spatial Opportunity*. The Convergence measure [2] can be defined as the prime location of supply points of a particular service, depending on the distribution of potential consumers and other supply points of the same service, i.e., the convergence portrays the efficiency or ability to attract users to different supply points of a given service. The Spatial Opportunity, in turn, is a measure of privilege on the home location considering a service system [2]. This is a measure that instruments the analysis of effectiveness, for it accurately describes the ease of access of each demand point to a supply system [25]. In planning, equitable distribution involves locating resources or facilities for the benefit of the largest possible number of different spatially defined social groups [26]. Table 1 synthesizes the configurational models adopted in this work and their main outputs.

Table 1. Configurational models selected for application and their main outputs

Models	Measure (output)
Accessibility	Relative distance of the spaces in the system.
Freeman-Krafta Centrality (planar)	Relative importance of the spaces in the system interconnection (betweenness).
Convergence	Ranking of the most central supply locations in relation to the demand spaces (polarization power or gravitation). Related to the notion of service efficiency.
Spatial Opportunity	Ranking of the demand spaces in relation to their relative accessibility to supply locations. Related to the notion of location effectiveness or equity.

In the next two sections, we focus on the methodology of empirical application of these selected models to spatial analysis of the school network.

3   Material and Methods

3.1   Study Area

The empirical study is developed using data from the city of Novo Hamburgo, localized in the Metropolitan Region of Porto Alegre (RMPA), state of Rio Grande do Sul, in the south of Brazil. It's distant 40 km. to the state capital (Porto Alegre). The city comprises an area of 223.60 km², and almost 70% are rural areas (156.31 km²) [27]. Being the eighth most populous city in the state, with 247,781 inhabitants [28], Novo Hamburgo has high quality of life indicators: the HDI - Human Development Index is 0.809 [29]; it has the seventh largest GDP - Gross Domestic Product of the state, and the GDP *per capita* is R$ 23,009.67 [28]. Figure 1 presents the study area location.

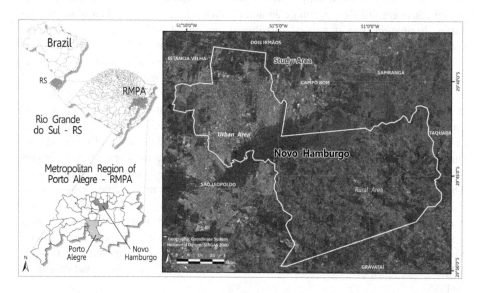

Fig. 1. The study area comprises the urban area of Novo Hamburgo, a city localized in the north portion of the Metropolitan Region of Porto Alegre (RMPA). Base image source: available in ArcGIS Desktop v. 10.1 [30] - ESRI / DigitalGlobe, 2010-2011.

3.2   Input Data

The data used in this analysis include: (i) census data: population by age group and by average monthly *per capita* income[1] groups [31]; (ii) geographic location of the

---

[1]   The average monthly *per capita* income is presented as fractions and multiples of the minimum wage, which, in 2010, by the time of the IBGE Population Census was R$510.00 or US$274.19.

schools, which were classified according to the education network (public or private) and the level of education [32]; and (iii) vector representation of the road system in the form of "street segments" [33].

The census data is divided into *demand* and *supply* data. The *demand* corresponds to possible school students, which are divided by age groups. The *supply* consists in school enrollments, classified according to the education network (public and private) and to the level of education.

The Brazilian Educational System is divided into three levels: fundamental (7 to 14 years of age), intermediate (15 to 17 years of age) and higher education, the latter comprising two different levels: undergraduate and graduate. Preschool or infant education is added to this hierarchical structure, for the purpose of providing assistance to children less than 7 years of age. Any youth or adult who did not follow or finish regular schooling at the appropriate age has the possibility of making up for the delay by attending courses and supplementary examinations customizing the mode of education to this special type of student.

In this scenario, the schools are classified according to the education network, namely municipal and state schools and private schools; and according to the education level: L1 - preschool or infant education and fundamental education; L2 - intermediate; and L3 - education for adults and young people and vocational education. In the study area, 125 education facilities were took into account for this study, and the educational services they offer are as follows: 86 schools offer preschool or infant education, 85 the fundamental level, 18 the intermediate level; 10 institutions have adults and young people education, and 9 facilities offer vocational courses [32].

To characterize the equivalent demand groups, population data, provided by the 2010 IBGE Population Census [31] are classified according to age groups: A1 - population aged between 0 and 14 years old; A2 - between 15 and 17 years; and finally A3 - literate population aged between 18 and 30 years added to the illiterate population aged between 18 and 40 years old. This data grouping enables a more appropriate exam of the relationship between schools supply and students demand, considering the actual target audience for each grade level.

Allied to the point location of every school, coverage areas are applied [34]. There are a number of criteria to be met by educational institutions, such as location, minimum sizing, specific neighborhoods, tree planting, among others [35]. The distribution of such facility in the urban area must meet the maximum distance between student's homes and schools, considering the travel time. The schools are divided into three categories according to the level of education: a) kindergarten and fundamental schools: it is recommended a time of displacement on foot no greater than 10 minutes and covering a radius of 400 meters; b) intermediate level: maximum 30 minutes-walk and coverage radius of 800 meters; and c) higher education, vocational education and education for adults and young people: the travel time on foot is undetermined, but the coverage radius suggested is 1,600 meters [34].

These data are inserted into each street segment as a load. The present methodology demands the system under study to be represented in the form of networks [9]. This representation allows defining the discrete unit of urban space to be used, according to the study objectives. The suggested approach is the use of the spatial unit

of the "street segment" (illustrated in Figure 2), which is the segment of a street localized between two street corners, a portion of space between the intersections of
routes, discontinuities or changes in direction (i.e., the face of a block) [33, 36]. The
choice of this spatial unit with high level of detail considers two factors: the flow
characteristics and distribution of local practices in the city.

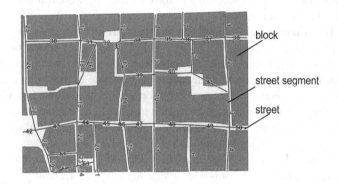

Fig. 2. A system of public spaces in which the spatial unit is the street segment.
(Source: adapted from [25], p.146).

3.3   Methodology

As previously mentioned, the proposed methodology is based on configurational
models, that allow joining, systemically, variables related to space (distances), to
population (socioeconomic data) and to school facility itself (size, pattern, etc.).

The methodological steps are summarized in the block diagram in Figure 3. The
database was built on a GIS (Geographic Information System) environment (ArcGIS
v. 10.1 [20]), where the road network and demand and supply data were integrated.
Demand and supply data were added to the vectors of the street segments using geoprocessing techniques.

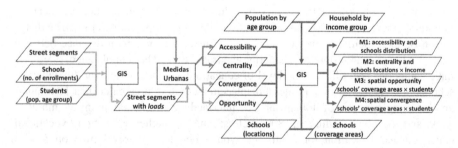

Fig. 3. Methodological procedures developed in this study

The spatial basis consists of a map of the study area, i.e., the urban road network,
represented by street segments. This vector database, consisting of 7,716 street segments, was imported into *Medidas Urbanas* software [37], which allows linking a

database to the spatial basis, enabling the insertion of loads into the spatial units to represent the spaces different attributes. Thus, the amount of schools and residents, as well as their characteristics (e.g., income, age, education level, etc.), in each street segment is calculated as loads. The software *Medidas Urbanas* performed the configurational analysis, applying the selected models presented in the previous section: Accessibility, Planar Centrality, Convergence and Spatial Opportunity.

The results obtained by applying these models were re-introduced into the geospatial database, and the relationship between the spatial system and the socioeconomic data were analyzed and evaluated. For Convergence and Spatial Opportunity analysis, the coverage areas of each educational facility are also applied. Eight thematic maps were produced, which are presented in section 4.

4   Results and Discussion

The final results are presented as four map sets, which seek to answer the proposed research questions.

1. **Schools spatial distribution in the study area:** the map shown in Figure 4 relates the system's Acessibility with the population density of the enumeration areas and the location of public and private schools. We observe the direct correlation between the location of private schools along the routes of greater accessibility of road system and in areas of higher population density (inhabitants per hectare). An interesting aspect is that the methodology also allows establishing a classification (ranking) of schools according to the results of the accessibility measure. Thus, one can identify schools that have prime locations with respect to the urban system as a whole.
2. **Relationship between schools location and population income:** the map shown in Figure 5 shows the relationship between the system's Planar Centrality and the households distribution according to average monthly per capita income groups and the location of public and private schools. The analysis points the direct correlation between the distribution of private schools along with the street segments of the road network with highest Planar Centrality values and in the areas of highest income *per capita.*
3. **Locational privilege of schools:** the maps presented in Figure 6 show the relationship between the system's convergence, the schools location, as well as their coverage area, and the population distribution according to age groups. The convergence indicates where schools with a higher likelihood of capturing users are, considering they have the same displacement opportunities. The analysis of relative convergence is able to reveal hierarchy patterns present at different school levels, which results can assist in planning concerning these issues. The relative convergence represents the ability of schools to polarize the target population.
4. **Privileged residential location towards schools location:** the maps presented in Figure 7 show the relationship between the spatial opportunity of the road network, the location of schools, with their respective coverage areas, and the population distribution according to age groups. The spatial opportunity shows where the residential locations, with the highest privilege as to the schools ranges, are, considering that all users have the same displacement conditions.

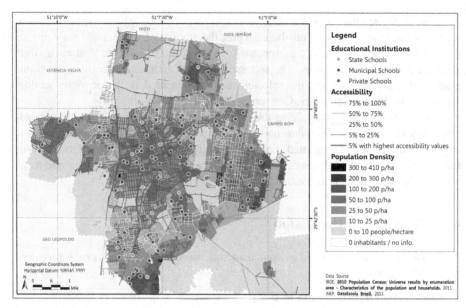

Fig. 4. The results of the Accessibility measure applied to the spatial system (road network represented by street segments) and distribution of public and private schools; base map: population density (by enumeration area)

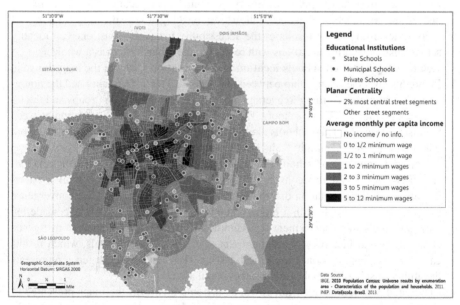

Fig. 5. The spatial system's Planar Centrality and schools location; base map: households' distribution by income ranges

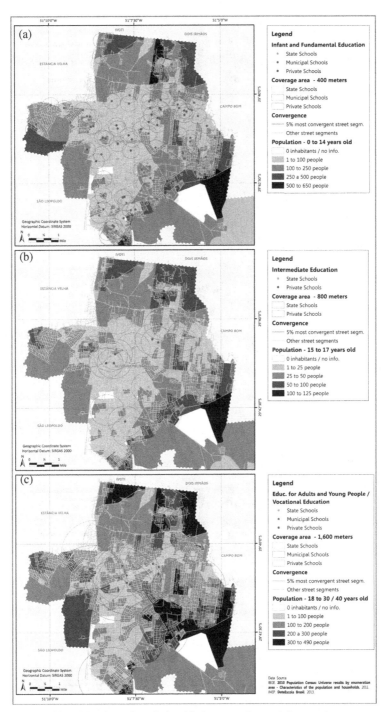

Fig. 6. Maps (a) (b) and (c) show the relationship between the spatial system's convergence and: a) schools L1 and students A1, b) schools L2 and students A2, and c) schools L3 and students A3

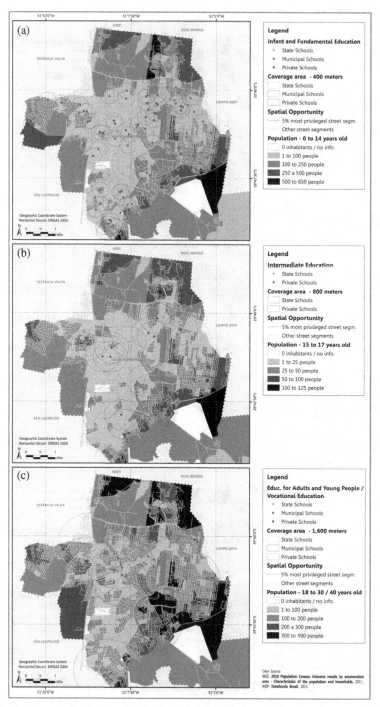

Fig. 7. Maps (a) (b) and (c) show the relationship between the spatial system's spatial opportunity and: a) schools L1 and students A1, b) schools L2 and students A2, and c) schools L3 and students A3

The analysis performed for each of the three educational levels evidences the lower level institutions (i.e., L1 - infant and fundamental schools) have a more even distribution of their relative convergence values. In the opposite side, the L3 level (i.e., education for adults and young people and vocational education) has the biggest discrepancies on the distribution of the relative convergence values between the schools: the first-placed institution captures 22.51% of the target population, while the second place captures only 7.18%. The same scenario is observed with the intermediate education data. The graph shown on Figure 8 presents the distribution of relative convergence and the schools of the three educational levels analyzed.

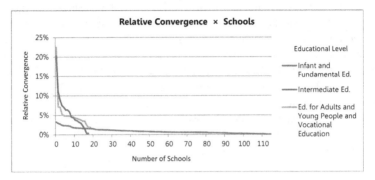

Fig. 8. Relative convergence values for every school of Novo Hamburgo education system

5   Conclusion

This paper presents an application of configurational models aiming to produce a spatial analysis of school facilities. We analyze Accessibility and Centrality of these facilities, considering the different socioeconomic strata of the population. The measure of Convergence can be taken as an efficiency measure, i.e., the ability of schools to polarize the target population. This is an important goal for schools, related both to the economic viability of private institutions, as well as to the proper allocation of resources in public institutions. The measure of Spatial Opportunity model represents a measure of equity, revealing the distribution of access opportunities for different social groups. Into a global urban perspective, it is of public interest that schools generate a suitable composition of types, education levels and occurrence locations, distributing opportunities in the extension of urban territory, efficiently and equitably.

An advantage of the models used in this work is to consider a more complex and comprehensive measure of accessibility to the facilities than traditional spatial interaction models. As previously mentioned, an advantage of weighted spatial networks based models is to bring together spatial and functional particularities of urban structure. Space Centrality models allowed to identify the locational strategic advantages of each school in the urban system and also to rank the spatial performance of these facilities in relation to the uneven distribution of the population and income over the territory. These models allow to compute geometric or topological distances, and also to consider impedance effects to generate shortest paths in the street grid (not used in this work).

The presented methodology allows performing various types of analysis and also proposes many measures. This flexibility allows one to create different analysis and to cover various spatial scales. Different measures are calculated, considering different attributes of the supply service, as well as the population demanding the service. Another positive aspect is to enable a systemic analysis, i.e., to consider the impacts of local changes over the entire spatial system. The models can be used to evaluate locational strategies for new school facilities, helping to build "what-if" scenarios to explore the consequences of possible location decisions.

A limit of this application is the development stage of the technology, given that the software used for the calculation of configurational measures still needs programming improvement. Considering the case of the measures of Convergence and Spatial Opportunity, their calculation requires a large processing capacity and computational memory, preventing the application on very large systems (in terms of number of spaces). The software also has limitations regarding the graphical presentation of results, hence the need for association with more developed GIS systems, which offer a variety of graphical and analytical capabilities.

This study demonstrates that the application of urban models is a promising alternative as a basis for the construction of such indicators of the efficiency and fairness associated with particular spatial arrangements, supporting decision processes.

Acknowledgements. The authors would like to acknowledge CAPES (Coordination for the Improvement of Higher Education Personnel) for financial support during the development of this study, through scholarships for master's and doctoral degrees; and Romulo Celso Krafta, for permission to use, for purposes of academic research, the Medidas Urbanas v.1.5 software.

References

1. Krafta, R.: Modelling intraurban configurational development. Environment & Planning B 21, 67–82 (1994), doi:10.1068/b210067
2. Krafta, R.: Urban convergence: morphology and attraction. Environment & Planning B 23(1), 37–48 (1996), doi:10.1068/b230037
3. Batty, M.: Cities and Complexity. Understanding Cities with Cellular Automata, Agent-Based Models and Fractals. MIT Press, Cambridge (2005)
4. Portugali, J.: Self-Organization and the City. Springer, Germany (2000)
5. Allen, P.M.: Cities and Regions as Self-Organizing Systems. Models of Complexity. OPA, Amsterdam (1997)
6. Bertuglia, C.S., Rabino, G.A.: Performance Indicators and evaluation in contemporary urban modelling. In: Bertuglia, C.S., Clarke, G.P., Wilson, A.G. (eds.) Modelling the City: Performance, Policy and Planning. Routledge, London (1994)
7. IPEA – Institute of Applied Economic Research. In: Boueri, R., Costa, M.A. (eds.): Brasil em Desenvolvimento 2013. Estado, Planejamento e Políticas Públicas, vol. 2. Brasília, Brazil (2013),
   http://www.ipea.gov.br/portal/
   index.php?option=com_content&view=article&id=20730

8. Clarke, G.P., Wilson, A.G.: Performance indicators in urban planning: The historical context. In: Bertuglia, C.S., Clarke, G.P., Wilson, A.G. (eds.) Modelling the City: Performance, Policy and Planning. Routledge, London (1994)

9. Echenique, M.: Modelos Matemáticos de la Estructura Espacial Urbana, Aplicaciones en América Latina. Ediciones SIAP/Ediciones Nueva Visión, Buenos Aires, Argentina (1975)

10. Arentze, T.A., Borges, A.W.J., Timmermans, H.J.P.: Multistop-based Measurement of Accessibility in a GIS Environment. International Journal of Geographical Information Systems 8(4), 343–356 (1994), doi:10.1080/02693799408902005

11. Timmermans, H.: Decision Support Systems in Urban Planning. E & FN SPON. University of Eindhoven (1997)

12. Hillier, B., Hanson, J.: The social logic of space. Cambridge University Press, Cambridge (1984)

13. Haggett, P., Chorley, R.J.: Network analysis in geography. Edward Arnold, London (1969)

14. Rosvall, M., Sneppen, K.: Modeling self-organization of communication and topology in social networks. Physical Review E 74(1), 016108 (2006), doi:10.1103/PhysRevE.74.016108

15. Buhl, J., Gautrais, J., Reeves, N., Solé, R.V., Valverde, S., Kuntz, P., Theraulaz, G.: Topological patterns in street networks of self-organized urban settlements. The European Physical Journal B - Condensed Matter and Complex Systems 49(4), 513–522 (2006), doi:10.1140/epjb/e2006-00085-1

16. Porta, S., Crucitti, P., Latora, V.: The network analysis of urban streets: A dual approach. Physica A: Statistical Mechanics and its Applications 369(2), 853–866 (2006), doi: http://dx.doi.org/10.1016/j.physa.2005.12.063

17. Lämmer, S., Gehlsen, B., Helbing, D.: Scaling laws in the spatial structure of urban road networks. Physica A: Statistical Mechanics and its Applications 363(1), 89–95 (2006), doi:10.1016/j.physa.2006.01.051

18. Portugali, J.: Toward a cognitive approach to urban dynamics. Environment and Planning B: Planning and Design 31(4), 589–613 (2004), doi:10.1068/b3033

19. Kim, B.J., Jun, T., Kim, J.-Y., Choi, M.Y.: Network marketing on a small-world network. Physica A: Statistical Mechanics and its Applications 360, 493–504 (2006), doi:10.1016/j.physa.2005.06.059

20. Minnhagen, P., Rosvall, M., Sneppen, K., Trusina, A.: Self-organization of structures and networks from merging and small-scale fluctuations. Physica A: Statistical Mechanics and its Applications 340(4), 725–732 (2004), doi:http://dx.doi.org/10.1016/j.physa.2004.05.019

21. Newman, M., Barabási, A.-L., Wats, D.J.: The Structure and Dynamics of Networks. Princeton University Press, Princeton (2006)

22. Crucitti, P., Latora, V., Porta, S.: Centrality measures in spatial networks of urban streets. Physical Review E 73(3), 036125 (2006), doi:10.1103/PhysRevE.73.036125

23. Ingram, D.R.: The concept of accessibility: A search for an operational form. Regional Studies 5(2), 101–107 (1971), doi:10.1080/09595237100185131

24. Freeman, L.C.: A set of measures of centrality based on betweenness. Sociometry 40(1), 35–41 (1977)

25. Krafta, R.: Notas de Aula de Morfologia Urbana. Ed. UFRGS, Porto Alegre, Brazil (2014)

26. Talen, E.: Visualizing fairness: Equity maps for planners. Journal of the American Planning Association 64(1), 22–38 (1998), doi:10.1080/01944369808975954

27. FEE – Siegfried Emanuel Heuser Economics and Statistics Foundation: Novo Hamburgo (2014), http://www.fee.tche.br/sitefee/pt/content/resumo/pg_municipios_detalhe.php?municipio=Novo+Hamburgo

28. IBGE – Brazilian Institute of Geography and Statistics: Cidades. IBGE, Rio de Janeiro (2014), http://cidades.ibge.gov.br/xtras/home.php

29. UNDP – United Nations Development Programme: Human Development Report 2013. The Rise of the South: Human Progress in a Diverse World. UNDP, New York (2013), http://hdr.undp.org/sites/default/files/reports/14/hdr2013_en_complete.pdf

30. ESRI – Environmental Systems Research Institute, Inc.: ArcGIS. Professional GIS for the desktop, v. 10.1. Redlands, USA (2012)

31. IBGE – Brazilian Institute of Geography and Statistics: 2010 Population Census: Universe results by enumeration area - Characteristics of the population and households. Rev. 02/22/2013. IBGE, Rio de Janeiro (2011),
ftp://ftp.ibge.gov.br/Censos/Censo_Demografico_2010/
Resultados_do_Universo/Agregados_por_Setores_Censitarios/

32. INEP – NationalInstitute for Educational Studies and Research Anísio Teixeira: DataEscola Brasil. INEP, Brasília, Brazil (2013),
http://www.dataescolabrasil.inep.gov.br/

33. Batty, M.: A new theory on Space Syntax. CASA, UCL, working paper n. 75 (2004)

34. Castello, I.R.: Bairros, Loteamentos e Condomínios - Módulos Didáticos. CD-Rom. Editora da UFRGS, Porto Alegre, Brazil (2008)

35. Santos, C.N.F.: A cidade como um jogo de cartas. EDUFF, Niterói, Brazil (1988)

36. Zechlinski, A.P.P.: Configuration and Practices in Urban Space: An analysis of urban spatial structure. PHD Thesis. UFRGS, Porto Alegre, Brazil (2013)

37. Polidori, M.C., Granero, J., Krafta, R.: Medidas Urbanas. v 1.5. software. FAUrb-UFPel, Pelotas, Brazil (2001)

Governing the Historical City: Transformation is Necessary to Counteract the Further Waste of Extra-Urban Land

Pier Luigi Paolillo[1], Massimo Rossati, and Mattia Andrea Rudini

[1] Department of Architecture and Urban Studies, Politecnico di Milano, Italy
pierluigi.paolillo@polimi.it
http://paolillo.professor.polimi.it/

Abstract. The safeguarding of building heritage identity is motivated by three important attributes that are: the complexity of its physical-architectural characteristics; the richness of its articulation; and the socio-economic, environmental, and landscape interdependencies inherent to it. However, the ancient built environment must also measure itself against an increasingly pressing necessity for adjustment to the needs of work and living environments. Furthermore, it is challenged to do so in such a way that reduces the wasteful use of extra-urban land by redirecting urban transformations back within the existing city [15]. The map of *intervenibility* on heritage, proposed in this work, thus represents a technical response, assisted by multivariate geostatistics and Geographic Information Systems. It responds to the necessity to reach a synthesis that – after examination of the effects of urban history on contemporary places – identifies its virtues and contradictions. Thus, the map suggests the degree and modes of building interventions through a multi-dimensional classification of real estate by cultural value, visual field, urban morphology/typology, structural quality, and form of open spaces.

Keywords: Building Heritage, Map of Intervenibility, Geographical Information Systems, Multivariate Geostatistical Analysis.

1 Estimating the Degree of Intervenibility on Building Heritage

The safeguarding of the architectural quality of historical real estate is an important aspect of the preservation of overall Italian cultural heritage. However, restriction on allowing the provision of additional agricultural land to the urbanization process requires a different attitude toward the preexistent built environment. This attitude involves amplifying settlement capacity where opportunities – in terms of market and structural physiognomy as well as harmonization with the context – present themselves. In order to evaluate the pertinent qualitative and problematic factors, that is whether to allow building transformation interventions, the judgment must derive from the multidimensional estimate of the descriptive components of an area. Therefore, an analytical process with the ultimate end of estimating the interactions of the economic, environmental, and landscape subsystems should be followed [15]. In this sense, the capacity of Geographical Information Systems to allow applications,

B. Murgante et al. (Eds.): ICCSA 2014, Part IV, LNCS 8582, pp. 125–139, 2014.
© Springer International Publishing Switzerland 2014

evaluations, and classifications that were impossible up until some years ago is enormous [12]. Treating the indicators by multivariate analysis [3] proves to be preferable – with respect to other applications, such as multi-criteria analysis [9] – for the possibilities: i) to synthesize in non hierarchical terms an often heterogeneous database, maintaining the complexity of its characterizing factors unaltered, and ii) to redirect factors to a limited number of variables whose significant latencies are then able to emerge [1]; [3]; [4]; [7]. The construction of the map of building heritage intervenibility acts within this context [12]; [14] and represents a technical elaboration with an elevated efficiency to yield the complex characteristics of each real estate property. These characteristics derive from the convergence of the in-depth analyses of landscape, morpho/settlement and socio/economic factors into one single unit. The plan for the historical city is extracted from the interdependencies of these factors to respond to the demands of housing pressures.

2   Integrating the Archives with the Urban Census

The assumption of an investigation routed in quantitative explorations [4]; [6]; [7]; [10] resides in the construction of an appropriate database, such as to develop analytical systems that force the complex characteristics of the studied space to emerge. So the examination of ancient building heritage will first involve the verification of the completeness of available archives and their degree of exhaustiveness, update, usability, quality, and treatability in the Geographical Information System domain.

Considering the elevated bearing of the variables – whose fundamental construction almost never initially permits the generation of punctual and updated data – it will be necessary to conduct an urban census to collect the data through direct investigations. This presents the opportunity to collect information, which would otherwise not be obtainable and to organize the database:

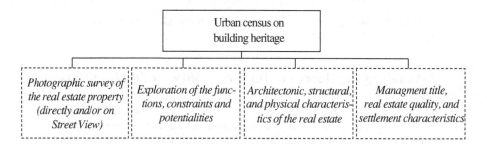

Fig. 1. Characteristics extractable through the urban census of heritage

The archive of census data, to be then spilled into the attribute tables of the georeferenced shape files in GIS [13], will include: *i*) photographs of the real estate (taken either directly or by utilizing the platform Street View, which is today sufficiently representative of the state of existing real estate); *ii*) functions included within the real estate (distinguished by floors, property units and prevalence, etc.); *iii*) characteristics of the real estate (such as dating, state of conservation, floors above ground, modality of access, number of rooms, presence of empty spaces, superfetation and/or elements of quality, etc.); and *iv*) characteristics of the property (including management regime, presence and type of appurtenances and parking, etc.).

3   The Redirection of Building Heritage into Spatial Units of Intervention: the Identification of Appurtenances

In the examination of the building heritage of the historical city, its valuable fabric, its settlement layers, and its stylistic value, the choice of admissible interventions on artifacts and on physical space has two intentions. On the one hand, the choice individuates real estate characteristics, and on the other hand, it identifies the spatial units most likely to spill into the results of the analyses. This last argument is not to be taken lightly, especially if one considers that the information gathered in the censure phase does not exhaust itself in the investigation of only building heritage but extends to the factors of vitality or marginalization, rebirth or regression, stagnation or change that imprint the existence of the centers [15]. The matter is not, in fact, limited to the mere recuperation of existing buildings and/or the coactive repopulation of historic places. On the contrary, beyond the restoration of their traditional poly-functional asset, there is a clear exigency to not only intervene upon the built environment but to also guarantee the persistence of the relational fabric of the appurtenances (spatial units of reference for the analyses).

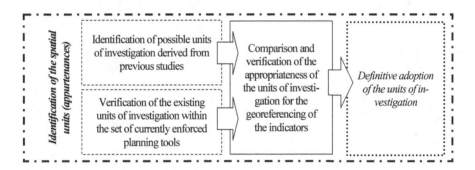

Fig. 2. Diagram of the identification of the units of investigation

4   The Variables Adopted to Construct the Map of Intervenibility

The census and georeferencing of the collected variables allowed for the detailed formatting of a series of operations to follow in order to identify strategic guidelines for intervention on the historical center. The following analysis concerns the construction of the map of intervenibility on building heritage in the historical center, of which the logical steps are demonstrated in the following diagram (Fig. 4).

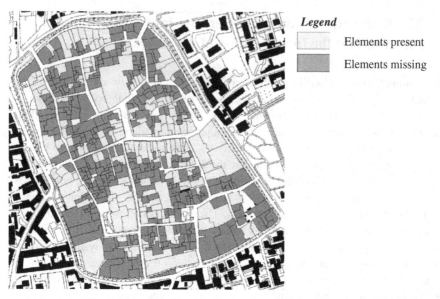

Legend

Elements present

Elements missing

Fig. 3. Map of quality elements, derived from the urban census

The preceding diagram highlights the way in which the different descriptors were grouped into three macro-indicators that express the quality of the built space. These macro-indicators were subsequently grouped in classes of homogeneous character through geostatistical multivariate treatment in order to ultimately allow the classification of the intervenibility on the appurtenances identified in the initial phase.

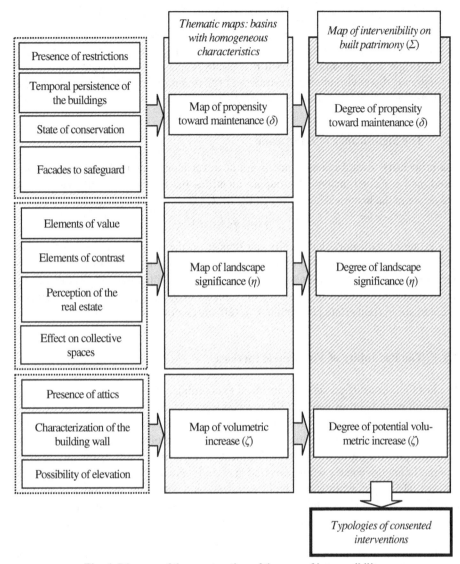

Fig. 4. Diagram of the construction of the map of intervenibility

4.1   The Propensity Toward Maintenance of the Constructions

It is thus necessary to recognize the factors of effective quality in the built physiognomy of a historical fabric. This consideration allows for both the maintenance of the intrinsic value of the housing patrimony through conservative interventions and the identification of the artifacts of least prestige and most apt to requalification interventions, restructuring, or demolition. Consequently, given the function:

$$f(\delta 1, \delta 2, \delta 3, \delta 4, \delta 5, \delta 6)$$

This first analytical step estimates the (δ) buildings' propensity toward maintenance on the basis of a series of synthetic indicators that directly impinge upon the nature, origin, and condition of each real estate property: δ_1 (historical-monumental restrictions), δ_2 (environmental restrictions), δ_3 (temporal persistence of the buildings), δ_4 (conservation), δ_5 (facades to safeguard), δ_6 (superfetation).

4.2   The Significance of Landscape

The propensity toward maintenance is not however alone among themes to be considered, and a second analytical grouping identifies the landscape significance of the real estate in the historical center (η) on the basis of:

$$f(\eta_1, \eta_2, \eta_3, \eta_4)$$

as a group of indicators that signify the impact on the recipient subject of the built space, which can be redirected to the identification of qualifying factors or to the impact connected with the appurtenances of the properties with f, therefore, explained by the independent variables η_1 (presence of elements of value), η_2 (presence of elements of contrast), η_3 (real estate perceptions), η_4 (effects on collective spaces).

4.3   The Possibility of Volumetric Increase

The last topic considered supports the need to coordinate the valuable traits of the heritage with more onerous needs of today's residency through the possible modification of shapes and volumes. It is therefore necessary to establish which real estate properties result as having the potential for elevation, addition of a new floor, or recuperation of a preexisting attic. The establishment of these potentials: i) permits ulterior volume in appurtenances whose covered surface is already based on compatible values, but where intervention is to be incentivized on a patrimony that today is not particularly desired; and ii) admits the alignment of building curtain walls (through the reorganization of the gutter quotas of the roofs that face public spaces) in the direction of a general policy of physical-architectural improvement.

The pursuit of such objectives has made it possible to identify the permissibility of incremental volumetric increase (ζ) on the basis of the third analytical grouping:

$$f(\zeta_1, \zeta_2, \zeta_3, \zeta_4, \zeta_5)$$

founded upon four indicators: ζ_1 (presence of building curtain walls expressive of at least 60% of the buildings with aligned gutter quotas), ζ_2 (presence of specialized buildings, not considerable for elevation), ζ_3 (presence of buildings of reference for elevation of the curtain wall), ζ_4 (presence of buildings not considerable for elevation).

5   Applications of Multivariate Analyses to Estimate Intervenibility

Taking advantage of the great computational capacity of territorial information systems and the use of the multivariate statistical software AddaWin [5], the three macro-indicators were grouped into clusters of homogeneous traits through multivariate statistical treatment given by the following steps:

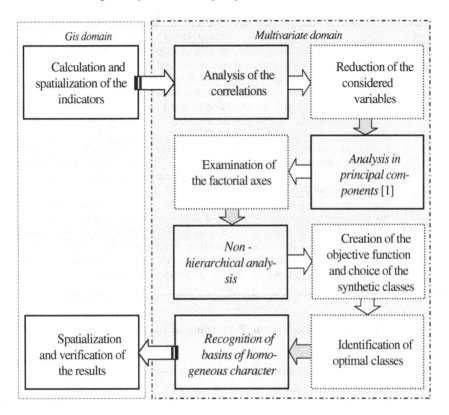

Fig. 5. Operative, multivariate steps in Geographical Information System domain

The result of such analyses is the identification of an optimal number of classes with respect to which the behavior of the units of investigation adopted must be described. The homogeneous clusters here circumscribed express the possibility to add, for determined characteristics, the statistical units studied. However, it is still necessary to highlight, precisely because the results of non-hierarchical analyses do not offer perfect correspondences between the classes and attributes of the real estate, the necessity to introduce algorithms. These algorithms consider the effective characteristics of each unit of real estate, thereby taking advantage of the entire set of variables [16].

5.1   The Real Estate's Propensity Toward Maintenance

In correspondence to the previously discussed landscapes, one first result can produce (in order to demonstrate an example) 13 classes, opportunely redirected into smaller clusters. In this case, they are 5, each respectively expressive of a propensity toward high, medium/high, medium, medium/low and low, and of the characteristics hereby reassumed:

Table 1. Table of the description of the classes of propensity toward maintenance

Classes	Description
Class 1	This class includes those buildings subject to restriction, to elevated temporal persistence, in a good state of conservation, and with characteristics of notable historical-architectural quality (approximately 4% of the total)
Class 2	Excluding real estate with specified restrictions, this class includes 52% of the total appurtenances and is constituted of buildings characterized by maximum temporal persistence, in a good state of conservation, and with facades to be safe-guarded
Class 3	Class composed of two appurtenance clusters: the first describes those properties with high temporal persistence but in mediocre or terrible conservation state, the second case refers buildings with a limited persistence and involved in modifications throughout the course of the years
Class 4	The fourth class corresponds to appurtenances characterized by medium/low propensity to maintenance and in a mediocre state of conservation, absence of elements of quality and subject to interventions of restructuring or reconstruction (9% in total)
Class 5	Appurtenances that present an intermediate state of temporal persistence but with mediocre or terrible maintenance states and construction works underway. A characterizing factor in this class is also the presence of superfetation in appurtenance spaces

5.2   The Potential Defined by the Perception of Landscape Significance of the Buildings

Considering how the perception/cognition of built space represents a factor of elevated significance with regard to the urban elements of landscape meaningfulness, especially in contexts of high visibility [8], the carrier of such values is the second macro indicator. Even in this case, the application of multivariate statistics allows for the identification of at least 11 classes that can be reduced to 5 groups of landscape significance, each of which can be described as follows:

5.3   The Latent Potential Derived from the Volumetric Increase

The last macro indicator considered requires the coordination of the characteristics of the value of building heritage with their real estate potentials through the possible modification of shapes and volumes. Thus, it tends toward the recognition of that real estate which is potentially includable by volumetric elevation: i) to allow new volume in areas where the covered surface area is already based upon high values, incentivizing the intervention on a patrimony which today is almost entirely not appraised; ii)

Table 2. Description of the classes of landscape value

Classes	Description
Class 1	Class characterized by an elevated index of perception, by factors of value (facades or physical/ architectural elements of rank or of the presence of historical/monumental restrictions (19% of the total appurtenances).
Class 2	Class which extends across 15% of the total, marked by real estate with characteristics of value, linguistic coherence with the surrounding, absence of superfetation, however still in conditions of not optimal visual perception.
Class 3	Class (11% of the total) constituted of two different clusters, the first composed of elements lacking in factors of value and of contrast and localized in central positions (easily perceptible from collective spaces), the second, on the other hand, characterized by the presence of elements just as negative as they are positive, implicating the intervention on elements of contrast.
Class 4	Class that most impinges on the total (48%), composed of the aggregation of three groups characterized by appurtenances with limited perception, absence of qualifying elements and, in some cases, factors of contrast to be eliminated in order to favor perception.
Class 5	Class containing two clusters with limited perceptive levels, distinguished for the absence of quality elements and for the relevant factors of contrast, that impinge negatively on the perception of public spaces.

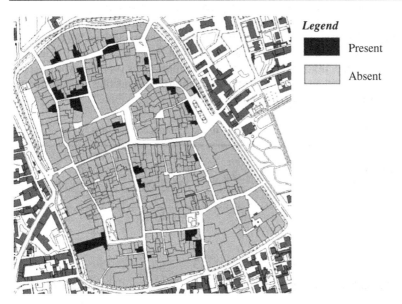

Fig. 6. Map of the possibilities of volumetric increase for the realignment of the building curtain walls

to allow an alignment of buildings' curtain walls in the direction of a general policy of physical/architectural betterment. A significant entity of elevation can emerge from the statistical treatment of indicators. Simultaneously, the characteristic historical/architectural language can be respected. It is localized on secondary axes and in portions of the fabric that are less central, less involved, and consequently routed in the determinant visual prospective and in the empty space/ filled space rapport by the imaginary collective. The result, exhibited by the subsequent map, prescribes the presence of buildings with the possibility of volumetric increase to elevate the existing real estate and realign the curtain walls facing the public space.

6   The Generation of Algorithms Dedicated to the Scenarios of Intervention

After the estimate of the partial indicators, to construct the map of intervenibility, alternative scenarios must be constructed through:

$$f = U_A = \{[(U_B + U_C) \times U_D] + (U_E \times U_F)\}$$

with:

U_A = real estate classification pertaining to goods of historical-monumental value
U_B = synthesis of the inherent variables and physical characteristics of the real estate
U_C = visibility of the real estate with respect to the public space
U_D = potential volumetric increase
U_E = specific indicator for the existing real estate of communal mark
U_F = specific indicator for the real estate of historical/ monumental mark

The application of the entire algorithm attributes a parameter to each unit of investigation that results in different scenarios. Each scenario is characterized by a different number of classes, utilizing the disaggregation of all of the real estate properties per incidence % of buildings involved in significant transformations (in other words, classifiable units of high and medium/high transformation). The profiles of 5 different identifiable scenarios are reported in the following table.

Table 3. Degrees of intervenibility of the 5 produced scenarios

Scenario (n.)	Degree of inter-venibility	Units of high transformation		Units of me-dium/high trans-formation		Units affected by significant trans-formations	
		Tot.	%	Tot.	%	Tot.	%
1	High	88	16.9	210	40.2	298	57.1
2	Medium	98	18.8	141	27.0	239	45.8
3	Low	89	17.0	65	12.5	154	29.5
4	Medium/Low	96	18.4	63	12.1	159	30.5
5	Medium/High	102	19.5	182	34.9	84	54.4

To test the degree of accuracy of the obtained scenarios, these last scenarios will be paralleled by two other concerting analyses. The first analysis is the imprint of the relational space; more specifically, the configuration of the physical space of a historical center with respect to the different uses to which its built spaces lend themselves in favor of the residents and the habitual or occasional users. The second analysis is the latent potentialities in a historical center, quantified by the study of the residential, extra-residential, perceptive, and configurative assets. The latter is significant in the immediate recognition of the appurtenances on which the necessity for intervention is priority, with respect to the objectives of renewal and requalification of the historic nucleus.

6.1   The Quality of Relational Space for the Choice of Scenarios of Intervenibility

Though conscious of the difficulty in connecting data on vital statistics to the dimension of the single appurtenances, it will be necessary to identify 3 indicators useful in the examination of the vivacity of the relations in the realm of ancient building heritage. The 3 indicators are as follows: i) the residential stability (peculiar demographic aspects such as fragile strata, foreign population, family economics, etc.); ii) the extra-residential vitality (presence of activities, heterogeneity, commercial exchanges); and iii) the proximity to structures of community use (citizens services, public structures). The subsequent multivariate analysis will thus permit the attainment of basins of equal phenomenal intensity. The m classes (result of a non hierarchical application) will be reduced to the n new, reduced classes, expressive of the quality of the relational space and the social and economic dynamism.

6.2   The Identification of Latent Potentials and Intervention Priorities

In order to identify the latent potentialities within the appurtenances of the historical center, it will be necessary to apply the informative results produced up to this point in the identification of an index of intervention priority. The benefit that the single-building intervention could have on the entire urban fabric can be associated to this index.

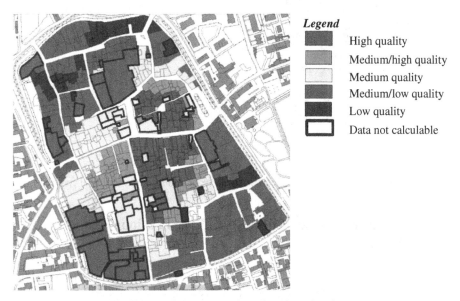

Fig. 7. Map of the degree of quality of relational space

The calculation of this index should involve 4 indicators: i) potentiality derived from residential uses (possibility of elevating, level of utilization of the real estate, quantity of empty apartments); ii) potentiality derived from commercial uses (presence of empty extra-residential spaces at the ground floor, presence of equipped spaces for the settlement of extra-residential activities); iii) perceptive potentiality (landscape relevance, state of conservation, presence of elements of contrast); and iv) current centrality (index of global and local centrality). The determination of the 3 degrees of potentiality of intervention will quantify the benefit associated with the prospective of recuperating, or managing more effectively, the existing building heritage. On the other hand, the index of current centrality – whose interaction between *Global* e *Local Closeness* is recognized by [2], [18] and [16] –, will allow the comparison of the potential levels with the current condition of the real estate. After having subjected the result to multivariate analysis and to the spatialization of the outcomes [17], it is possible to distinguish the different degrees of intervention associated with the units of investigation (otherwise defined real estate appurtenances). Thus, the possible actions with respect to the corresponding building capacities can be established. In turn, the degrees of priority with respect to the established settlement need can be identified.

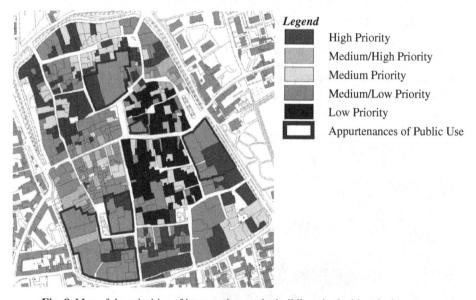

Fig. 8. Map of the priorities of intervention on the buildings in the historical center

7   The Degrees of Intervenibility and the Previous Urban Legislation

To arrive at the calculation of the effective propensity toward transformation (and classify the degrees of building intervenibility on historical real estate), it will be necessary to compare the 5 scenarios produced. They must be compared with respect to their admissible function determined by the previous urban legislation and the classes recoded with respect to the consented actions.

An analysis of the correlations will be applied to the recoded classes of intervention. In the transformability profiles, it will be necessary (even before reading the results) to consider if it is worth accepting that with the level of intervenibility closest to the highest correlation between the new scenario and the previous urban legislation classification. Usually, such an examination reveals the excessive progressive restrictiveness of such legislation. Such regulation is the fruit of failing strategies that have been too liberally experimented in the past five hundred years. These strategies have been contradistinguished more by obsessive conservative paroxysm than by precise solutions based on local specificities. The results of this are the unmotivated sequences of constraints and restrictions affixed by many urban municipal provisions. These provisions are almost always impotent in the increasingly infinitesimal regulation on the subject of historical buildings. The regulatory outcomes are truculent but uncertain; therefore, they actually favor an even more distant solution to the problem.

Table 4. Criteria of recoding of the classes of intervenibility

Classes of previous urban legislation	Recode	Classes of building intervenibility
Restoration/renewal	1	Historical/monumental buildings to be preserved Historical/monumental buildings to be upgraded
Maintenance with restrictions	2	Low degree of intervenibility Medium/low degree of intervenibility
Reconstruction	3	Medium/high degree of intervenibility High degree of intervenibility

It is thus necessary to introduce technical discontinuity with respect to the past urban consuetude. According to how much is expected of an analysis of the correlations targeted at loosening the apposition of "just because" constraints, the optimal scenario where the most refinements of the model converge seems to be the following:

Table 5. Degree of correlation between the scenarios of produced intervenibility, the previous urban legislatorial instruments, and the new opportunities for intervention

	Intervenibility		Distance from the classification of previous urban legislation (b)	Total distance from desirable values (a, b)
	Intensity	Distance from median value (a)		
Scenario 1	(+ +)	(– – – – –)	(–)	(– – –)
Scenario 2	(–)	(·)	(– – – – –)	(– – –)
Scenario 3	(– –)	(– – – – –)	(– – –)	(– – – –)
Scenario 4	(–)	(– – –)	(– – – –)	(– – –)
Scenario 5	(+)	(– – –)	(– –)	(– –)

Thus, the following building categories, recognized in the analysis of the group of possibilities to modify the existent built environment, will be attributed to the attitudinal classes of transformation, identified within the conclusive map of intervenibility (derived from the scenario deemed optimal). These building categories include: *i*) *Extremely limited intervenibility*, with interventions allowed within the limits of the categories of conservation and of restoration of the restricted real estate of high historical and monumental quality; *ii*) *Limited intervenibility*, with interventions allowed up to the renewal of the real estate, without demolition and reconstruction; *iii*) *Medium/Low intervenibility*, with interventions allowed up to the partial substitution of the real estate; *iv*) *Medium/High intervenibility*, with interventions allowed up to the total substitution of the real estate (exception made for those elements to be safeguarded, such as valuable facades); *v*) *Elevated intervenibility*, with interventions allowed up to the built restructuring with total substitution of the real estate; *vi*) *Maximum intervenibility*, with interventions allowed up to urban restructuring; and *vii*) *Exceptional intervenibility*, with intervention allowed up to demolition without reconstruction.

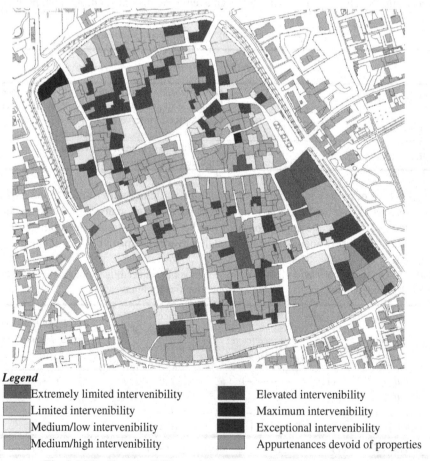

Legend

Extremely limited intervenibility	Elevated intervenibility
Limited intervenibility	Maximum intervenibility
Medium/low intervenibility	Exceptional intervenibility
Medium/high intervenibility	Appurtenances devoid of properties

Fig. 9. Map of intervenibility on the patrimony of the historical center

The constructed framework suggests modes of intervention for situations in which the building heritage is of ancient nature, like in Italian historical centers. From here on, there is a need to change the operative logic in the construction sector: from the postwar period, dedicated to the realization of new buildings, as opposed to the requalification of the form and the structure of the existing buildings. Unfortunately, the construction sector has prescribed, taking advantage of the "hunger for houses" of the last ten years, residential styles to which demand has had to adjust itself until the recent real estate market crisis (which first trampled the United States and then the rest of the world, including Italy). This is thus the moment to radically change course and to shift demand toward existing building heritage, particularly its historical units.

References

1. Brunoro, G.: Analisi delle componenti principali. Franco Angeli, Milano (1988)
2. Crucitti, P., Latora, V., Porta, S.: The Network Analysis of Urban Streets: A Dual Approach. In: Physica A, Statistical Mechanics and its Applications, vol. 369(2) (2006)
3. Fabbris, L.: Statistica multivariata. Analisi esplorativa dei dati. McGraw Hill, Milano (1997)
4. Fraire, M., Rizzi, A.: Statistica. Metodi esplorativi e inferenziali, Carocci, Roma (2005)
5. Griguolo, S.: Addati. Un pacchetto per l'analisi esplorativa dei dati – Guida all'uso. Istituto Universitario di Architettura di Venezia, Venezia (2008)
6. Matthews, J.A.: Metodologia statistica per la ricerca geografica. Franco Angeli, Milano (1981)
7. Morrison, D.F.: Metodi di analisi statistica multivariata. Ambrosiana, Milano (1976)
8. Murgante, B.: Il contributo dell'analisi visiva alla definizione della qualità di un territorio. Università degli Studi della Basilicata, Potenza (2008)
9. Nijkamp, P., Rietveld, P., Voogd, H.: Multicriteria Evaluation in Physical Planning. North Holland, Amsterdam (1990)
10. Palermo, P.C., Griguolo, S. (eds.): Nuovi problemi e nuovi metodi di analisi territoriale. Franco Angeli, Milano (1984)
11. Paolillo, P.L.: New survey instruments: studies for the environmental assessment report of the general plan in a case in Lombardy. In: Rabino, G., Caglioni, M. (eds.) Planning, Complexity and New Ict, pp. 215–224. Alinea, Firenze (2009)
12. Paolillo, P.L.: Sistemi informativi e costruzione del piano. Metodi e tecniche per il trattamento dei dati ambientali. Maggioli, Rimini (2010a)
13. Paolillo, P.L.: L'innovazione della carta dell'intervenibilità nell'esperienza dei capoluoghi lombardi di Como e Sondrio. In: Atti XXXI Conferenza scientifica annuale AISRe Associazione Italiana di Scienze Regionali, Aosta, settembre 20-22, pp. 1–10 (2010b)
14. Paolillo, P.L. (ed.): Il piano di governo del territorio di Como. Materiali di ricerca. Maggioli, Rimini (2011)
15. Paolillo, P.L.: L'urbanistica tecnica. Costruire il piano comunale. Maggioli, Rimini (2012)
16. Paolillo, P.L.: Limbiate, dalla condizione indifferenziata di «corea» alla scoperta delle nuove centralità. In: Territorio, vol. 66 (2013)
17. Paolillo, P.L., Rossati, M.Y., Baresi, U.: Una nueva disciplina del patrimonio inmobiliario para revitalizar los centros históricos italianos: la evaluación del grado de «intervenibilidad». In: VIII Simposio Internacional Desafíos en el Manejo y Gestión de la Ciudad, Camaguey (República de Cuba), Febrero 1-4 (2014)
18. Porta, S., Latora, V.: Multiple Centrality Assessment. Centralità e ordine complesso nell'analisi spaziale e nel progetto urbano. In: Territorio, vol. 39 (2006)

Analyzing Urban Extensions and Its Effects over the Commercial Activity of an Urban Network

Taras Agryzkov[1], José Luis Oliver[2], Leandro Tortosa[1], and José Vicent[1]

[1] Departamento de Ciencia de la Computación e Inteligencia Artificial
Universidad de Alicante
Ap. Correos 99, E–03080, Alicante, Spain
[2] Departamento de Expresión Grafica y Cartografia
Universidad de Alicante
Ap. Correos 99, E–03080, Alicante, Spain

Abstract. In this paper, we present a way for analyzing and visualizing extensions or enhancements of an existing urban network, as well as the effects that cause those extensions on commercial activity taking place in the network itself. This analysis is based on an algorithm for classifying the nodes of the network, depending on the type and number of facilities allocated to each node. Using this classification, it is possible to visualize the network according to a gradient color scale, allowing us to identify the most important nodes in the network and the part of the network more influenced by the introduction of new facilities. With this classification algorithm we can simulate and evaluate the effect that produces any reform plan over an existing urban street network.

To understand the process, a detailed example of an urban extension is proposed and studied. Different land uses are assigned and some endowments and facilities are allocated in the new area. A comparison between the existing urban network and the new urban area created from the extension is performed, with the aim to determine the influence of the new commercial activity introduced respect to the activity taking place in the existing one.

Keywords: Urban networks, Urban development, PageRank algorithm, Graph visualization.

1 Introduction

A complex network provides a framework for modeling many real-work phenomena in the form of a network. Many complex systems can be described by networks, in which the constituent component are represented by vertices and the connections between the components are represented by edges between the corresponding vertices. We can consider urban networks as a specific type of complex networks.

The most common approach to convert physical streets' network into an abstract mathematical structure (a complex network or a graph), is mapping street

B. Murgante et al. (Eds.): ICCSA 2014, Part IV, LNCS 8582, pp. 140–152, 2014.

intersections as nodes while mapping street intersections as links. This is the well-known primal approach. This approach is based on the metrics of the geographical networks and this is the natural way to represent a network in Geographic Information Systems (GIS).

In the other hand, space syntax [6],[7] is a well-known analysis theory and tool about spatial form based on graph theory and GIS. It provides a new rational design and research method of urban space. Since the space syntax has been created, it receives extensive attention in the field of city research and applications. There are a large number of examples to prove the correctness and importance of the syntactic theory on urban and architectural spaces understanding and simulation, particularly in the transformation of some historical sites and the renewal program of the design of history urban. Many studies have shown that the space syntax theory contributes to the understanding of urban space and the interpretation of simulation, will help indepth understanding of the nature and functions of urban space [10]. At present, space syntax as a tool for understanding spatial structure is applied to many aspects of urban research, including road network analysis [5], traffic flow [9], land grading [19], urban spatial morphology [11,13,18], geometric accessibility [4] and some other fields [14], but most are lack of a great insight into the syntactic parameters.

In this paper we follow a primal approach to represent any urban network. We can see [17] to understand the benefits of the primal approach for the network analysis of geographic systems as those of urban streets. The study of complex networks presents new methods for detecting the important or relevant nodes in a given network and to distinguish core nodes from peripheral network elements.

Several techniques are proposed, ranging from structure preserving model reduction, shortest path trees, network motifs, as well as variations of Google PageRank algorithm (see [2],[3],[12] and [15]). The PageRank model uses the structure of the Web to build a Markov chain considering a nonnegative and irreducible matrix M for transition probability called also a primitive matrix. This property guarantees that, according to Frobenius test for primitivity, a stationary vector exists as a solution for the PageRank problem. Classification algorithms based on PageRank concept have been also used in spatial data analysis. In [8] the authors use a weighted PageRank algorithm as an indicator for predicting aggregate flow in large street networks.

In this paper a PageRank classification algorithm is used to simulate and assess new extensions of urban street networks. To achieve this, we firstly transform the urban street network into a primal graph. Secondly, a ranking of the nodes is performed in order to sort the nodes according to their importance in the network. This classification of the nodes allows us to visualize graphically the network. Finally, we use this model to perform simulations of urban expansions and to evaluate the impact of these new actions in the immediate surroundings, as well as in the global network.

2   An Algorithm for Ranking the Nodes in an Urban Network

The PageRank method [15] was proposed to compute a ranking for every Web page based on the graph of the Web. Therefore, PageRank constitutes a global ranking of all Web pages, regardless of their content, based solely on their location in the Web's graph structure. The purpose of the method is obtaining a vector, called PageRank vector, which gives the relative importance of the pages. Since this vector is calculated based on the structure of the Web connections, it is said to be independent of the request of the person performing the search. For a more detailed description of the PageRank algorithm see [16].

In [2], Agryzkov et al. propose an adaptation of the PageRank model to establish a ranking of nodes in an urban network, considering the influence of external activities. In the following, we refer to this algorithm as the Adapted Pagerank Algorithm (APA algorithm). Although the APA algorithm is applied to urban networks, it is perfectly applicable to any network, whenever we want to analyze or represent additional information from the network itself, through a numerical assignment to the different nodes on the network.

The central idea behind the APA algorithm for ranking the nodes is the construction of a data matrix D, which allows us to represent numerically the information of the network that we are going to analyze and measure. The algorithm proposed in [2] is:

APA Algorithm. Let us assume that we have a graph representing an urban network with n nodes, representing squares or intersections. We proceed with the following steps.

1. Obtain the transition matrix A from the graph of the network.
2. Consider the different characteristics k_i associated to each of the nodes for the problem studied; evaluate them in each node. With these data, we construct the matrix D (data matrix).
3. Construct a vector v_0, according to the importance of each of the characteristics evaluated. This vector represents a multiplicative factor.
4. Obtain a vector v by multiplying $Dv_0 = v$.
5. Normalize $v \to v^*$, using the standar method.
6. Construct the matrix V, from v^*.
7. Construct the matrix $M' = (1 - \alpha)A + \alpha V$, from A and V.
8. Compute the eigenvector associated to the eigenvalue 1 for the matrix M'. That is our ranking vector.

A comment on the parameter α must be introduced. Consider a random surfer that starts from a random page, and at every time chooses the next page by clicking on one of the links in the current page (selected uniformly at random among the links present in the page). Clearly, important pages will be visited more often. However, in our case, because of the characteristics of the vector v^* used to construct V, a surfer starts now at nodes with higher values of the corresponding entry in vector v.

PageRank is formally defined as the stationary distribution of a stochastic process whose states are the nodes of the web graph. The process itself is obtained by combining the normalized adjacency matrix of the web graph with a trivial uniform process that is needed to make the combination irreducible and aperiodic, so that the stationary distribution is well defined. The combination depends on the damping factor α. When $\alpha = 0$, it means that $M' = A$, which is a matrix that describes a random walk on a graph. On the other hand, when $\alpha = 1$, then $M' = V$ which does not depend on the graph. In PageRank algorithms, the parameter α usually takes the value $\alpha = 0.15$. The output of applying this algorithm to a network is a ranking vector $r = r_1, r_2, \ldots, r_n$ with n components, where the i-th component represents the ranking of i-th node within the overall network.

Summarizing, we can say that the mean feature of this algorithm is the construction of the matrix D and the vector v_0. Firstly, the matrix D allows us to represent numerically the information we want to study; secondly, the vector v_0 allows us to establish the importance of each of the factors or characteristics that have been measured by means of D. In other words, we could say that the algorithm constitutes a model to establish a ranking of nodes in a network, with the objective to assigns a numerical value to each node according to its significance.

3   Visualizing the Network According to the Ranking of the Nodes

In [1], a model to visualize and analyze information obtained from a street network is proposed. The process of displaying information in a graphical way can be summarized in the following steps

1. Collect data.
2. Numerical assigment of the data to each of the nodes of the network.
3. Run the APA algorithm for ranking the nodes.
4. Visualize the network according to the value of the ranking of each node.

The first step is to construct a primal graph representing the urban street (nodes represent the streets intersections and the edges are the streets). In this model the nodes constitutes the key point in the process. This is because to the fact that the nodes store the numerical values of the information we are measuring or analyzing. For instance, if we are analyzing the commercial activity in an urban network, we have to assign numerical values to each node depending on the type of business activity that takes place in their environment.

The process to allocate the information to the nodes is not as easy as it seems. It is necessary to convert a qualitative information to quantitative values. The criteria applied is to assign to each node only the information that is allocated in the edges that concur in it. Therefore, following the example of commercial activity, we should assign to each node only those endowments or facilities that pertain to the edges of it. Once assigned numerical values to each of the nodes of the street network, the APA algorithm must be run to obtain a classification of the nodes according to their importance in the network. Afterwards, we only have to represent graphically the nodes of the network following a model who allows us to clearly distinguish the nodes according to their importance. To visualize the result of the process in whole network, It will be used a gradient scale color that range from red (most important nodes) to blue (least important nodes).

4   Simulating Urban Network Extensions by Ranking Their Nodes

4.1   General Considerations

As it was already mentioned in the previous sections, the APA algorithm allows us to establish a classification of the nodes of the network according to their importance. Moreover, other relevant feature is that we can graphically represent some kind of information related to urban networks. The visualization and classification of this information constitutes a powerful tool to better understand the characteristics of a city.

In our model, it is possible to simulate and visualize extensions of any urban network where we can allocate the endowments and facilities that we consider appropriate. We can also evaluate the impact that the extensions of the network cause on the neighboring of it. In other words, it is possible to introduce modifications or extensions in the topology of the urban network with the aim to evaluate the effect of such modifications over the whole network.

The objective of the simulation developed in this section is to detect and measure the alteration of the commercial sector in the consolidated urban land, once proposed the new extension of the city. This type of evaluation may be very useful when we analyze the consequences of future development plans.

4.2   An Example of Simulation

We present with detail an example of an extension of an urban network and its impact over the surrounding area. In Figure 1 we can see a fragment of an urban area, which is going to be the objective of our study. We have divided the urban zone into two areas: the east part (light gray color) represents the map of the currently urban network with the streets and buildings, while the west part of the figure represents the area that constitutes the expansion of the urban network.

Currently, the expansion area is classified as rural land and because of this, we can design an extension of the city and evaluate its impact on the neighboring, paying special attention to the effect that the expansion causes on the commercial activity. In Figure 2 we have a map of the current state of the area of the urban network studied in this example. We construct the primal graph of this fragment of the city and consider the commercial activity as the information that will be represented and analyzed using the APA algorithm.

Fig. 1. Map of the urban network chosen for the example with the existing urban area (east part in light gray color) and the extension of the network (west part)

The commercial activity can be divided in the following types: bars and restaurants (type I), shops (type II), bank and offices (type III), and shopping centers or supermarkets (type IV). So, taking into account this division we introduce in the urban network some endowments and facilities like restaurants, offices, shops and department stores. The purpose of choosing a small number of facilities is to verify how an area with low commercial activity is strongly influenced by a new urban expansion area with a large presence of commercial activity.

With the collection of the data, we run the APA algorithm to establish a ranking of the nodes, according to their importance in the commercial activity of the city (represented now by means of the primal graph).

In Figure 3 we can see a graphical representation of the ranking of the nodes when the vector v_0 used in the APA algorithm is $[1, 1, 1, 1]$. This means that we are giving the same importance to the four types of commercial activities. Therefore, we obtain a general visualization without particularizing in any of the four sectors of the commercial activities).

We must remark again that we introduced a small quantity of endowments and facilities in the urban network in order to observe the differences when we perform an extension. So, we can appreciate that many areas of the urban network are displayed in blue color; this is a consequence of a low level of commercial activity.

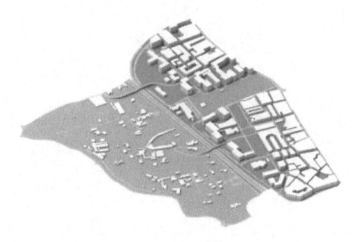

Fig. 2. Current state of the rural area that constitutes the expansion of the urban network

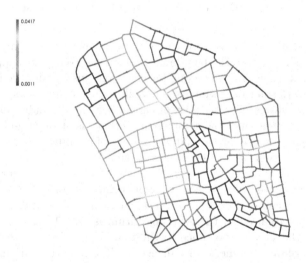

Fig. 3. Visualization of the ranking of the nodes given by the APA algorithm, applied to the existing urban network

With this visualiation process, we have developed the first part of the simulation: the evaluation of the current urban network before performing any extension on it. At this time, to continue the simulation, we need to design and represent the extension of the network, with new facilities and endowments that lead us to a new urban design.

The new partial plan that we set includes two types of land use: one is linked to the residential sector and the other is ready to host large public and private facilities. The residential area, located in the central part, concentrate the majority of commercial activity that takes place in this new urban space. The western part of the extension has been designed to accommodate public facilities, while the eastern zone is destined to the location of large department stores. The details of the urban design and the exact composition of the urban extension is shown in Figure 4. In the left image, we have represented the expansion area, while in the right one we represent the different endowments and facilities proposed in the design.

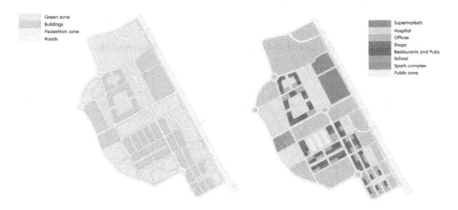

Fig. 4. Distribution of land uses and facilities in the extension proposed for the urban network studied

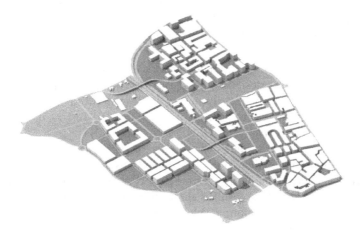

Fig. 5. View of the integration between the existing urban network and the extension zone proposed

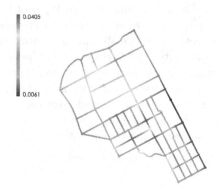

Fig. 6. Applying the APA algorithm to the area proposed to expand the urban network

Fig. 7. Visualizing the commercial activity of the whole area with $v_0 = [1, 1, 1, 1]$

In Figure 5 we can see how the new network expansion integrates with the existing urban fabric. Once designed the features of the new partial plan, we can make their assessment independently of the existing network, with the aim of studying the commercial activity taking place (Figure 6). Just as in the Figure 3 we have used the vector $v_0 = [1, 1, 1, 1]$.

Finally, the simulation evaluates the impact of the new endowments of the proposed extension on the allocations that already existed. To do this we need to reclassify the nodes. Therefore, we run the APA algorithm considering the whole urban area studied. The result is shown in Figure 7 and if we compare the results with those obtained before developing the extension, distinct differences that should be analyzed carefully can be observed.

If we focus on the central part of the existing urban area, without considering the extention (see Figure 3), it is observed a red color area just at the center. This area does not exist if we consider the expansion of the network (see Figure 7).

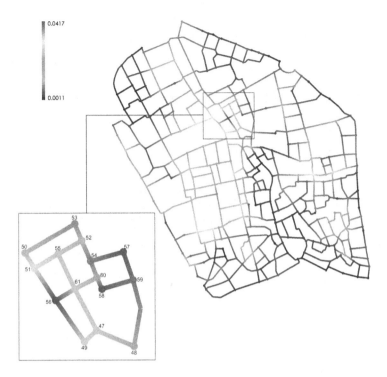

Fig. 8. Visualization of the ranking of the nodes of the area of interes before applying the expansion

This difference is remarkable and it is produced when we measure the global commercial activity, without particularizing in any commercial sector. Therefore, the image of Figure 7 shows the decrease in the values of the ranking of the nodes in this area. This is due to the influence of the commercial activity located in the new enlarged area. We can conclude that the new commercial area designed in the extension, is like an attractor with respect to the commercial activity located in the existing urban network.

In order to clarify and better understand the influence of the urban expansion in an existing area, we are going to compare the value of the ranking of the nodes, located in the area we are analyzing, before and after the expansion. This comparison is shown in Figures 8 and 9. In Figure 8, we have a graphical representation of the ranking of the nodes regardless of the expansion, while in Figure 9 it is shown the ranking of the nodes after applying the expansion. In both figures we highlight the area of interest where are located the nodes that will be analyzed in more detail.

Figure 10 shows the exact values of the ranking of the nodes given by the APA algorithm before and after performing the urban expansion, while Figure 11 shows the variation (absolute value) for the ranking of the nodes analyzed.

Fig. 9. Visualization of the ranking of the nodes of the area of interes after applying the expansion

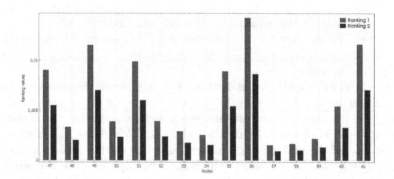

Fig. 10. Values of the ranking before and after applying the expansion

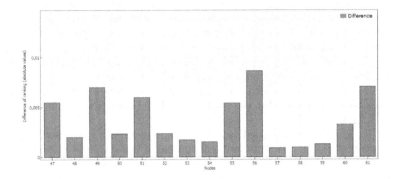

Fig. 11. Difference between the ranking of the nodes before and after of applying the expansion

The numerical study carried out with these nodes, we observe that those nodes that heve a higher ranking values are suffering a greater decrease in their value, once considered the extension of the network.

5   Conclusions

As a conclusion, we can say that it is useful to use a PageRank algorithm for classifying the nodes of a network to simulate and evaluate new expansions of any urban network. We can design the specific characteristics of a new urban area to determine the impact that the new facilities and endowments introduced produce over the existing urban street network. This computation and visualization process allows us to develop some extensions of the urban network with the objective to improve some characteristics of it. It was demostrated with an example how the introduction of a new commercial area in an extension of an urban network may affect the commercial activity of the existing urban network.

References

1. Agryzkov, T., Oliver, J.L., Tortosa, L., Vicent, J.F.: A model to visualize information in a complex streets' network. In: Omatu, S., Neves, J., Rodriguez, J.M.C., Paz Santana, J.F., Gonzalez, S.R. (eds.) Distrib. Computing & Artificial Intelligence. AISC, vol. 217, pp. 129–136. Springer, Heidelberg (2013)
2. Agryzkov, T., Oliver, J.L., Tortosa, L., Vicent, J.F.: An algorithm for ranking the nodes of an urban network based on concept of PageRank vector. Applied Mathematics and Computation 219, 2186–2193 (2012)
3. Berkhin, P.: A survey on PageRank computing. Internet Mathematics 2, 73–120 (2005)
4. Cheng, C., Zhang, W., Chen, J., Cai, J.: Evaluating the ac-cessibility about Beijing's subways in 2008 based on spatial syntax. GeoInformation Science 9(6), 31–35 (2007)

5. Duan, Z., Wang, Q.: Road network analysis and evaluation of Huizhou City based on space syntax. In: International Conference on Measuring Technology and Mechatronics Automation (ICMTMA), Hunan, China, vol. 3, pp. 579–582 (2009)
6. Hillier, B.: Space is the machine. Cambridge University Press, London (1996)
7. Hillier, B.: Centrality as a process: accounting for attraction inequalities in deformed grids. Urban Design International 4, 107–127 (2000)
8. Jiang, B., Jia, T.: Agent-based simulation of human movement shaped by the underlying street structure. International Journal of Geographical Information Science 25, 51–64 (2011)
9. Jiang, B., Liu, C.: Street-based topological representations and analyses for predicting traffic flow in GIS. International Journal of Geographical Information Science 23(9), 1119–1137 (2009)
10. Jiang, P., Peponis, J.: Historic and emerging urban centres in the metropolitan Atlanta region: spatial dynamics and morphogenesis. In: Proceedings of the 5th International Symposium on Space Syntax, Delft University of Technology, vol. 1, pp. 283–294 (2005)
11. Landre, M.: Analyzing yachting patterns in the Biesbosch National Park using GIS technology. Technovation 29(9), 576–642 (2009)
12. Langville, A.M., Mayer, C.D.: Deeper inside PageRank. Internet Mathematics 1, 335–380 (2005)
13. Li, J., Duan, J.: Multi-scale representation of urban spatial morphology based on GIS and spatial syntax. Journal of Central China Normal University (Natural Sciences) 38(2), 383–387 (2004)
14. McCahill, C., Garrick, N.W.: The applicability of space syntax to bicycle facility planning. Transportation Research Record 2074, 46–51 (2008)
15. Page, L., Brin, S., Motwani, R., Winogrand, T.: The pagerank citation ranking: Bringing order to the web. Technical report 1999-66, Stanford InfoLab (1999)
16. Pedroche, F.: Metodos de cálculo del vector PageRank. Boletin Sociedad Espanola Matematica Aplicada 39, 7–30 (2007)
17. Porta, S., Crucitti, P., Latora, V.: The network analysis of urban streets: a primal approach. Environment and Planning B: Planning and Design 33, 705–725 (2006)
18. Yi, Z., Li, B., Xiao, G., Yu., W.: The role of space syntax in urban formation analysis. Bulletin of Surveying and Mapping (2), 44–47 (2008)
19. Zhao, H., Li, L., Zhu, H.: Application of extended space syntax in urban land grading. Journal of Geomatics 32(2), 9–11 (2007)

Using Urban Modelling and Geographical Analyses to Tackle Emergent Urban Logistics Issues

Developing a Decision-Making Tool for Urban Goods Distribution

Raphaëlle Ducret

Mines ParisTech, PhDstudent, Paris, France
raphaelle.ducret@mines-paristech.fr

Abstract. For the most part, urban goods transportation is neglected by urban studies and urban planning, as opposed to urban passenger transportation. This article will present outcomes of a research project whose goal is to develop an efficient decision-making tool for urban logistics from a spatial and territorial perspective, based on urban modelling and geographical analyses. The hypothesis of this research project is that understanding urban freight through spatial structures will contribute to improving territorial diagnosis, which is essential to the understanding of urban goods distribution issues before any decision can be made and will result in specific and efficient last-mile delivery solutions. This article will describe the methodology of the decision-making tool, and discuss its limitations, inputs and perspectives.

Keywords: urban logistics, geographical analysis, urban modelling, decision-making tools, urban parcel distribution.

1 Introduction

Over the past fifteen years urban logistics has become a shared issue in several big cities around the world, even if it remainsan under-addressed problem [22], [13], [11]. For the most part, goods transportation is neglected by the field of urban planning; this could be said to contrast with the centrality of the study ofurban passenger transportation. Among disciplines that have tried to understand logistics organisations and the distribution of urban goods, geography and spatial studies have always taken a backseat compared to economy, management and especially transportation engineering sciences and modelling approach[15, 16, 17, 18], [26]. As a result, diagnosis and decision-making tools for urban freight based on geographical analyses and urban modelling are scarce in the urban planning field. Modelling tools for urban freight have been implemented by researchers since the 1980s, in order to estimate demand, optimize distribution routes and logistics organizations, reduce negative environmental externalities and help public authorities in decision-making[3], [14].With the exception of LUTI models (land use transportation interactions models), most of them

B. Murgante et al. (Eds.): ICCSA 2014, Part IV, LNCS 8582, pp. 153–168, 2014.
© Springer International Publishing Switzerland 2014

have no connection to the area's characteristics. Nevertheless, from our point of view, spatial approaches could complement technical approaches because city characteristics influence urban goods transportation and distribution. By city characteristics we mean a city's location and layout, morphology and shape, urban land use, urban sprawl or density, street design and transportation network, among others. French cites share common features that have an impact on urban freight and that could be studied and generalised into a diagnosis tool thanks to urban modelling. This would help choose the most suitable organisation and efficient policy for urban logistics. Yet so far, urban logistics organisations as well as city logistics measures have not taken this fact into account.

In this paperwe would like to present the outcomes of a research project[1] conducted by the French Post Operator, La Poste, the aim of which is to trial a decision-making tool for urban logistics in cities of all sizes, drawing on a spatial and territorial perspective and using urban modelling and geographical analyses. The research project is based on the hypothesis that a deeper understanding of urban freight premised on spatial characteristics, through the analytical tool of modelling, will contribute to improving territorial diagnosis, which is essential to the comprehension of issues of urban goods distribution before any decision can be made. We also think that understanding urban freight from a spatial point of view will lead to specific and thus efficient last-mile parcel delivery solutions, whilst avoiding unexpected and negative effects. Furthermore, despite real improvements in data-collecting methods, one of the most important issues for city logistics is the poor availability of freight data and its difficult matching to the purpose of both researchers and practitioners [10], [7], and [21]. In order to overcome this difficulty we have decided to combine operational data from the French Post Operator with classic spatial and urban datain an innovative way. These data have been analysed and reworked from an urban logistics and geographical point of view. This method translates our hypothesis of a real value of geographical analysis in tackling urban freight issues in a practical way and will allow us to verify it. It will also enable us to examine the use of various types of data for urban freight modelling.

The article will be structured as follows. First, we will provide a short international review of previous research in urban logistics, before focusing on those aspects which, according to our research hypothesis, plead for the use of geographical analysis and urban modelling to address existing issues. Secondly, we will describe and explain the methodology of the decision-making tool for urban logistics based on urban modelling, geographical analyses and operational data which has been developed throughout the research project.We will present an operational case study based on the comparison of the cities of Angers and Avignon. Finally, some limitations of this work, inputs of the modelling trial and research perspectives for urban studies and geography in the field of urban logistics will be discussed.

---

[1] On-going PhD in the Ecole des Mines financed by the French Post operator - La Poste.

2   Urban Logistics from a Geographical Perspective: A Predominantly Unexplored Field

2.1   Urban Logistics: Challenges for Researchers and Practitioners

The Urban Logistics Field: A Review. For fifteen years urban logistics has become a significant subject for both researchers and practitioners, though for the moment an under-tackled field for the latter [26], [21]. Since the 1970s urban logistics has appeared as an essential function for cities to develop and citizens to live in those cities. Pioneering research was conducted at that time in the Anglo-Saxon world and in bigger cities in the developed world, and this research focused on truck traffic management in particular. During the eighties interest in urban freight declined. Since the late 1980s and especially the 1990s there has been improvements in the understanding of urban logistics systems, especially in Europe, the United States and Japan, thanks to various programmes, large surveys and conferences [2], [11], [36]. Urban goods distribution is covered by both operational and planning studies. Nowadays,even if urban logistics has become a research field on its own, as Odgen stated in 1992 in the first book entirely devoted to urban freight, it remainsan under-addressed issue, as opposed to urban passenger transport[28], [1], [7]. Local authorities' interest in urban goods distribution has also grown over the past few years [7], [11], [21].

Definitions and Challenges. Urban freight transport or urban goods distribution can be defined as « the movement of things (as distinct from people) to, from, within, and through urban areas » [28: 14], movements to and from business units, households, administrations, stores, etc. The urban freight system is extremely complex. In fact, hundreds of supply chains can be observed within a city with at least one for each sector [20], [11]. An example of this can be found in cities' main logistics chains: the parcel sector, the convenience store and retail sectors, home deliveries, individual purchase trips, waste collection, industrial supply, construction supply, food market logistics organization, and so on[20],[11], [30]. These supply chains present different scales and organisations, for example, micro local deliveries, hub and spoke organisation, single run delivery, and so forth; transport vehicles and motorization; size, packaging and shipment from fresh produce to waste and parcels; delivery times, among others. Moreover,the urban freight system is composed of a wide range of stakeholders and participants, such as suppliers, logistics providers, shippers, consumers, city-makers and planners, real estate agents, eachwith their own interest, opinion and concern regarding the urban area[16]. Even if freight activity is essential for a city's economic and social life, urban freight also has a negative impacts of various natures. The most frequently mentioned issue is environmental. In fact, urban freight is partially responsible for both global emissions of greenhouse gases and local polluting emissions [6], [11]. Urban goods vehicles are partially responsible for accessibility issues when they stop in unauthorised areas for loading or unloading in the street. Economic competitiveness of cities and their attractiveness can suffer owing to its high level of road congestion. It is also an economic issue for freight providers because of consequences on fuel consumption and the time it takes to run a delivery round. Another linked issue is road safety. Similarly, quality of life can be affected by urban freight noise and disturbance. Finally, the fact that freight takes place in an urban environment increases the level of complexity one has to deal with. Urban logistics has to face problems of public space sharing with other economic activities, passenger transport and citizens; and economics and urban planning trade-off. Trade-offs often disadvantage logistics, as shown by the unexpected effects of various regulation measures, economic and urban planning arbitrations resulting in logistics sprawl and increasing logistics and environmental costs, etc.

As a result of the process of metropolisation and urbanisation, cities have to deal with an increasing number of collections and deliveries of large amounts of divided and fragmented freight. Besides, large urban areas have to be as attractive and competitive as sustainable. As a consequence, urban freight appears both as a positive lever which needs to be protected, and a negative externality which needs to be regulated and optimized. Over the past fifteen years, awareness of the issue of urban logistics has grown, improvements regarding data collection and global knowledge of urban goods distribution have been made, and various measures and initiatives from both private and public sectors have been aimed at improving urban freight [11], [26]. But in spite of those encouraging changes, local authorities still lack full understanding of the urban logistics issue and above all a view of global transport planning which would enable them to properly manage both freight and passenger transport in an urban and complex context [21].

2.2 Geographical Analyses and Urban Modelling to Tackle Urban Logistics Issues

Urban Freight, Geographical Analyses and Urban Studies: A Gap. Geography and spatial studies have not sufficiently investigated urban freight as a research field and tried to understand the role of spatial organisations in schemes of distribution [26], [15, 16, 17, 18], [36]. In fact, among disciplines that have tried to understand the organisation of urban logistics and the distribution of urban goods, geography and spatial studies have always taken a backseat as compared to the economy, management, political science and transportation engineeringsciences.Contrary to urban passenger transport studies and the interactions between passenger transport and urban form, which have been largely explored in urban studies and more broadly in geography, freight transport has been neglected until now [1]. So far, when spatial studies have looked at issues of freight, it was on a small scale (global, national or regional scale) [9], [17] rather than large scale (urban areas or cities, spatial morphology and organisation, neighbourhoods, and even street design). Geography has for example provided studies of warehouse, port and heavy freight facility locations, transportation in a logistics context, and urban logistics sprawl phenomena[4], [9], [32], [35]. However, when exploring the link between urban characteristics and urban logistics, some researchers have established that urban freight is more influenced by the nature of economic activities in cities than by features of spatial organization (with the exception of the size of the city) [20], [29], [10].

Urban Freight from a Spatial Point of View. The intuition that urban freight studies should be reinforced by urban studies and geographical analyses is shared by researchers in the urban logistics community. Dablanc has claimed that urban freight "varies according to city's geographical area" [11: 11]. Even if the suggested scheme is a classic radio-concentric one, this is a clear attempt to think urban logistics from a spatial point of view. More recently some researchers have argued that urban transport activities are also affected by spatial and geographical factors and certain "urban form prerequisites", like the city's size and density, layout and urban form, street design, urban morphology, the location of activities, land use, and the position of the city in the supply chain [1], [22]. Allen et al. note that the interactions between vehicle trip, urban form and land use have been under-researched. Thanks to a precise study of urban areas in the United Kingdom, they subsequently demonstrated that several geographical, spatial and land use factors such as the facility's location, the city's size and location in the city network, street design, settlement size and density, city layout, and commercial and industrial land-use patterns are likely to influence the efficiency and intensity of freight journeys [1].

Even if over the past few years, geographers as well as researchers in urban freight have become aware that the impact of geographical features on logistics organisations has until recently been largely forgotten in analyses [36], [23], so far very few researchers have tried to apply and verify this hypothesis. Macario's concept of "logistics profile", which identifies "for some well-defined areas inside a city, reasonably homogenous groups of logistics needs" in order to provide the most adapted solution, is based on three key points,among which "the urban characteristics of the area", which takes into account commercial density and homogeneity, conditions of accessibility as well as time restrictions [25]. Despite its spatial approach, the logistics profile appears to be incomplete, especially regarding spatial features, and does not exactly correspond to what could be expected when referring to a geographical analysis

Research Hypothesis. We believe spatial analysis of urban characteristics and features to be relevant for a better understanding of urban freight and an efficient organisation of the activity by practitioners. Our hypothesis is that urban areas, their spatial and geographical organisation and layout at different scales, their morphology at a very large scale, have been forgotten in the majority of last-mile delivery solutions currently implemented in urban logistics. In practice, even if some territorial-based innovations or initiatives can be observed, for example in the urban parcel delivery sector, very few urban freight providers show sufficient interest inprecise territorial diagnosis. On the other hand, local governments, with a few exceptions, underestimate the use of urban studies and diagnosis prior to making decisions regarding urban distribution. In fact, when implementing measures and initiatives in urban goods movements, local authorities most of the time neglect territorial diagnosis, whilst conducting traffic diagnostics, cost analyses and surveys mapping stakeholder'needs and requirements, which are also pivotal to the understanding of urban freight and its efficient management. However, the lack of a geographical comprehension of urban freight, as a part of a global diagnosis, can lead to unpredicted and sometimes negative long- or short-term effects [21], [12]. In this article, we would like to demonstrate the significance of spatial analysis prior to urban logistics decisions, arguing that a better understanding of urban areas' specificities iscentral to the definition of suitable last-mile solutions which minimize unwanted effects and maximize acceptability by all stakeholders.

From our point of view, the city's and urban area's size, settlement density and urban sprawl, urban morphology and layout, landuse, street design, type of housing, as well as the location of the city within the city network and the logistics network are likely to influence last-mile delivery efficiency. Besides, the economic and functional profile of an area and the level of urban freight policy and governance are factors to be taken into account. Even if each city has different characteristics and "contextual differences" [21], it is possible to build a general framework mapping cities' spatial and morphological features in order to provide specific geographical analysis in a global approach which will prove central to thetackling of urban logistics issues.

2.3   Modelling Tools in Urban Logistics: "Trends and Gaps"

Urban Freight Modelling: A Review. Among the urban logistics tools for city planners and transport and logistics stakeholders, freight models have been developed since the mid-seventies. First developments have been slow and mainly designed by and for research purposes. Since the 1990s and especially the first decade of our century, when urban goods movements became a significant issue for researchers and practitioners, urban freight modelling can be said to havedeveloped [3], [31], [28]. Basically, urban logistics modelling tools built over the past twenty years provide information and data about urban goods movements, modal share, vehicle trips and tours, traffic flow, loading rates, emissions levels, consumers' behaviour, thanks to data recordingvehicle type, routes, location of the shipper and hauler, traffic volume, loading rates, location, consumers' and stakeholders' movements, etc. [14], [2,][3]. Different typologies have been proposed to classify the great variety of models. We can distinguish models by the object studied: commodity-based models [8], truck-based models [19], [34], and behavioural-models [27], by the models' perspective (planning perspective, policy and technology perspectives, behavioural perspective, etc.), or by their objectives (economic efficiency, environment, road safety, etc.) [3]. We can also identify differences by looking at their functions (operational models, evaluating models, estimation models and simulation models). As such, urban freight modelling trends have followed urban freight changes toward operational research and the evaluation of practical measures and initiatives, namely urban distribution centres, time windows delivery, environmental regulations, etc. There are also multiple model building approaches. First models were multi-step models or gravitational models used for passengers but adapted to urban freight to provide estimation of generation trips or for distribution. There were also input-output models among others. Following from this, more specific and urban freight oriented models have been implemented in recent times [31].

Urban Freight Modelling from a Spatial Point of View. Contrasting with the fact that relations between urban transport and urban planning have been studied since the

1970s and modelled in various ways since then through successful land-use transportation interaction models [5], links between spatial organisation and goods distribution in urban areas have been overlooked. To the best of our knowledge, no urban goods model puts spatial analysis and geographical features at the core of its calculation, trying to match land use and urban freight transport. Even in the French model Freturb, which is a land-use transportation based model, spatial characteristics of cities are blurred because the main descriptors are the economic and functional features of the city [33]. Anand et al. have analysed the descriptors of a great number of city logistics models and have shown that the most commonly used descriptors are traffic flow and commodity flow, freight and trip generation, loading rate, pollution level and transportation cost. Location and land use, which are more spatial oriented descriptors, are used comparatively little [3]. Even planning models are not always based on spatial oriented descriptors. In that context, we would like to open urban freight modelling to spatial data and generalise spatial study for urban freight, providing a decision-making tool based on spatial data and territorial analysis.

Urban Freight: A Data Issue. Though relevant improvements have been madeover the past few years in the collection of data and more broadly in the expansion of knowledge of urban freight around the world [10], urban freight modelling is limited by the availability of data. Regarding public data, the frequent absence of an urban freight authority and the lack of interest of local authorities in urban goods movements could explain it. Moreover, data collection is an extremely complex process when it comes to aspects such as truck loading, truck origin and destination, GG emission levels but also consumer behaviours or stakeholder responses to different measures. Long, heavy and expensive surveys have to be carried out. Besides, in the competitive context of freight transport and the last-mile market, logistics and operational data are sensitive and confidential. Scarce cooperation and lack of dialogue between private stakeholders, local authorities and researchers also prevent the use of these highly significant and revealing data. In fact, operational data are stable, most of the time on a long time step, precise, provided at different scales and above all available, but not always urban freight-oriented. In this study we will make us ofoperational data collected by the French Post Operator. This facilitates the carrying out of an effective analysis of urban logistics issues without necessitating a heavy and complex data-collecting campaign.This will also allow us to provide an example of the way in which various types of data can be used in a process of urban freight modelling and also contribute to geographical analysis.

3   Methodology of the Decision-Making Tool: Urban Modelling and Geographical Analyses to Address Urban Logistics Issues

3.1   Presentation of the Decision-Making Tool for Urban Freight

Structure of the Urban Freight Decision-Making Tool. The goal of the research project is to develop a first draft of a decision-making tool, thanks to a territorial diagnosis and a precise understanding of the influence of cities' specificities on urban freight organizations, in order to formulate specific and efficient last-mile delivery solutions. The decision-making tool is composed of two modules: one module of territorial diagnosis; and one recommendations module for urban freight delivery solutions for the last-mile. In this article, our intention is to focus on the first module, which is partially based on a modelling approach. It comprises two levels of analysis in order to provide a diagnosis of urban areaswhich takesinto account both general features and specificities, immobility and urban dynamics. At the first level, a segmentation of the city into urban sub-groups identifies homogenous spatial sub-groups whose spatial organisation and demand features influence last-mile delivery and the nature of the delivery services provided.A modelling approach is employed at that stage. Thanks to the secondlevel of Module 1, the decision-making tool could also take into account citycharacteristics in terms of urban freight policy (i.e. regulation, urban freight planning and dialogue between stakeholders, etc.), and other spatial and economic features (i.e. transportation network, specific activities, and so on) which significantly impact on urban distribution. In this paper, we will focus on level 1, at which we test a modelling and cartographic approach, combined with a geographical analysis.

3.2   The Modelling Approach of the Decision-Making Tool for Urban Freight

Function and Perspectives of the Modelling Approach. The need to build an operational model for urban distribution providers became apparent. But it could also be useful for local authorities to understand how the organisation of their city and their choices with respect to urban planning could have an impact on urban freight. For that reason, we chose to map the results with ArcGis in order to make the tool comprehensible and to strengthen its geographical component. The decision was made to work on a 200x200 meters squares grid based on a European projection system in order to provide a precise representation of spatial features and analyse the very last mile of delivery (this means around 700 000 squares for the sixteen cities studied in this research project). The INSEE (France's National Institute for Statistics and Economic Studies) has been providing the grid for free for a few years, covering the whole of the French territory. In France, the Freturb Model provides maps at that scale but so far few urban freight studies or models have used such a high scale and grid. However, a square grid is far more stable on a long-term approach than a division into administrative zones and more neutral than other political limits because it is based on a pure geometric definition. Thus, it is easier to implement, analyse, and share; calculations of interactions are similarly facilitated. This scale also resolves heterogeneity issues [24]. Finally, it allows cross-political limits studies of urban freight phenomena. This offers a new approach and at the same time opens analysis to different scales, from a regional scale to a global one.

The Model Descriptors. Model criteria have been chosen based on the fact that spatial features influence urban freight. Criteria have been selected thanks to a literature review, the empirical database obtained by means of interviews with professional delivery providers and finally a multidimensional analysis based on a principal component analysis (PCA).

We have conducted twenty-two interviews with urban goods parcel distribution providers in nineteen differentFrench cities (where choice of cities was determined by the aim to studydifferentsizes, urban morphologies, layouts, street design networks, and functional andeconomic specialisations[2]). The aim was to understand what spatial and morphological features of a city influence freight delivery and how they do so. Basically, the conclusions of these interviews have revealed that parcel providers are true geographers and that their knowledge of spatial organisations is pivotal to their activity and its efficiency. But none of them are explicitly aware of that ability and they underestimate urban organizations in location strategies and new services implementation.

Interviews have confirmed that city size has an influence on urban freight. It appears that high or low density has a great influence on urban delivery productivity. The transport network influences final delivery because it is responsible for connectivity and accessibility to the recipient. Thus, congestion and low density networks are constraints providers have to adapt to. Clearly, city centres have appeared as the main bottleneck for last-mile delivery because core centres frequently concentrates all those issues. City layout and transport infrastructure at small and high scales are also likely to have an impact on delivery, including the fine street meshing of the city centre, street design and type of housing. Even urban planning projects like publicpassenger transportation network, pedestrian zones could challenge urban delivery.

Based on the results of the interviews, we have proposed an extended list of model descriptors describing the density and morphology of an urban area, the type of housing and construction and to characterise the street network and road transport network, its density and level of accessibility. In addition to space-related descriptors we have added demand-oriented descriptors (related toboth business and households) which are likely to influence the type of item delivered and the nature of the delivery service and thus eventually, the organisation.

Thanks to the PCA (which identified correlated parameters and redundant parameters) conducted on an extensive number of spatial and functional descriptors, we have been able to determine the model's descriptors. A list of six descriptors has been decided upon (see table 1). Each of them describes how spatial and functional characteristics could influence urban freight organisation.

Data Collection and Processing. The postal database, the INSEE database and France's National Institute of Geography (IGN) database have provided data to define the model's descriptors. Some descriptors, based on postal and spatial data, have been formulated thanks to the geo-processing tools of the geographical information system ArcGis. In the end, all data are quantitative. Even qualitative –spatial– information

---

[2] List of the urban areas studied : Strasbourg, Montpellier, Lyon, Toulouse, Angers, Marseille, Douai, Toulon, Bayonne, Avignon, Orléans, Besançon, Limoges, Aix en Provence, Calais, Lens, Mâcon, Tours, and Montbéliard. Selection is based on a segmentation of the French urban areas. Criteria of the segmentation are economic and spatial, and the level of governance regarding urban logistics.

has been processed into quantitative data. Half of the descriptors could be defined as spatial-oriented descriptors, whether the original data is postal, logistical or spatial, and half are functional descriptors (table 1).

Whereas INSEE data already fitted the 200x200meter grid, an important work of data collection has been made to adjust the geo-localised postal data to the grid. Projection and error correction of the postal data have been an important moment of the modelling process, but also a pivotal step because it was of paramount importance to be able to analyse the professional data from a spatial point of view.

Table 1. Model descriptors

Descriptor	Nature of the descriptor	Nature of the data
Population density	Spatial descriptor : density	INSEE data
Number of street sections	Spatial descriptor :features of the street network, meshing	IGN data processed by ArcGis
Number of delivery point of a distribution point	Spatial descriptor :type of housing, verticality of the housing	Postal data
Linear density of the distribution points	Spatial descriptor : density and morphology Functional descriptor : nature of the demand	Postal data
Median income	Functional descriptor : standard of living	INSEE data
Proportion of professional distribution points	Functional descriptor : nature of the demand	Postal data

The Modelling Approach. With respect to the modelling approach, we have completed a K-Means analysis on sixteen French urban areas[3] to provide spatial segmentation and classification of homogenous spatial sub-groups. The urban areas have been chosen for their different sizes, spatial organisations and functional profile so that a general conclusion on the city could be given[4]. This scale allows studying urban freight at a global level which is unfortunately rarely done by local authorities.

We have conducted a K-means analysis to provide a segmentation of all the squares of the sixteen urban areas studied into coherent urban sub-groups that is to say to group similar spatial and functional-featured squares homogeneously and in a coherent way. K-means, which is an automated classification algorithm, was applied

---

[3] An urban area is a French statistics scale created by the National Institute for Statistics and Economic Studies. It refers to a group of municipalities – a major urban centre and the municipalities surrounding it (in the urban periphery), in which at least 40% of the population works in the major centre or the urban area.

[4] List of the urban areas studied: Toulouse, Compiègne, Lille, Montbéliard, Avignon, Orléans, Besançon, Annecy, Royan, Calais, Rennes, Strasbourg, Sens, Dreux, Angers, and Montpellier. Selection is based on a segmentation of the French urban areas. Criteria of the segmentation are economic and spatial, and the level of governance regarding urban logistics.

to all the 700 000 squares and to all the sixteen urban areas studied in a single movement because we were searching for a model providing a general image of the city regarding urban freight. The hypothesis was that some general urban sub-groups will appear for all cities and some of them will be more specific to a precise group a city.We have decided not to introduce a notion of distance between the squares or an image processing algorithm, assuming that homogeneity and connectedness would appear between same-featured squares and give the segmentation coherence.

Table 2. Description of the eight urban sub-groups

	Urban sub-group	1	2	3	4	5	6	7	8
	Descriptors	Core centre area, dense and wealthy	Pericentral and central dense areas, medium to low incomes	Pericentral and central areas, medium incomes	Collective housing areas	Wealthy-low-density residential areas	Low density residential areas, medium incomes	Low density residential areas, medium to low	Professional activity area
SPATIAL FEATURES	Number of street sections	high	high	High	high	low density	low density	very low density	-
	Number of delivery points of a distribution point	medium	medium	medium	high	low	low	low to very low	-
	Population density	high	high	High	medium	low	low	low to very low	-
	Density of distribution points	very high and high	high	High	medium	medium to low	medium to low	low	medium to high
	Linear density of distribution points	high	high	High	medium	medium	medium	medium	-
DEMAND CHARACTERISTICS	Type of housing	medium to dense housing, no specificity	medium to dense housing, no specificity	medium to dense housing, no specificity	mainly collective housing and apartment buildings	low density single-family housing	low density single-family housing	low density single-family housing	-
	Proportion of professional distribution points	low-medium	low	Low	low	low-medium	low	low	high
	Median income	high	medium to low	medium	medium to low	high	medium	medium to low	/

However, after the first K-means analysis, we discovered that the processing of both spatial and functional descriptors together did not allow the identification of clear urban sub-groups with homogenous spatial and functional patterns. As a consequence, we decided to effect two different segmentations, one spatial and one functional, before combining them. Thus, we can achieve a segmentation of the urban area in several sub-group and areas which are coherent but also convincing when it comes to their spatial and functional features and the way those features are meant to influence urban freight distribution. The crossing of the two sub-segmentations provided twenty urban sub-groups, which is not efficient when trying to understand cities' influence on urban freight and cities' specificities and when attempting to read and analyse sub-groups on a map. Consequently, we conducted a statistical analysis and a geographic analysis of the segmentation of the sixteen urban areas in order to identify a smaller number of sub-groups. We eliminated sub-groups which represented low proportions of population but not those which are representative of a specific type city. Finally, we decided to select eight homogenous and coherent urban sub-groups. Table 2 describes each one regarding spatial and functional descriptors related to urban freight.

3.3   Case Studies: The Urban Logistics Profiles of the Cities of Avignon and Angers

Presentation of the Case Studies. Sixteen territorial diagnoses of French urban areas of different sizes, spatial organisations and socio-economic profiles have been carried out based on this methodology, and tests have been conducted, with parcel distribution providers and thanks to geographical analyses, to assess the efficiency of the tool. The capacityof the tool to provide well-differentiated, contiguous and homogenous spatial sub-groups or zones has been examined. Besides, thanks to a geographical analysis, we have verified that the identified zones, whosefeatures are considered to have an impact on urban freight, present truly specific characteristics regarding urban freight. Finally, the cartographic result of the modelling approach has been analysed in order to decide whether or not the model gives a coherent picture of the spatial organisation of the area, whilst logistical data have been used and mixed with geographical data. In this section we will focus on two case studies involving the cities of Angers and Avignon, two "large" urban areas.

Modelling Results and Interpretation in an Urban Freight Context. Figure 2 clearly show that without any image processing the typology provides relatively clear and homogenous zones with contiguous squares. From a geographical point of view, the areas in question do not provide a new image of city organisation but fit traditional and general urban analyses and city development. Yet at the same time, the identified zones could also provide specific information about cities. In fact, some urban sub-groups appear for all French cities and some of them appear or do not depending on the characteristics of the area studied. For example, the smaller city of Calais in the North of France is only described by four sub-groups while Angers and Avignon contains even sub-groups. As for Angers, the map does not present sub-group number 2 but sub-group 3 is strongly represented within the core city – this is specific to this city and corroborated by a socio-economic analysis of the area. Avignon shows a different profile with respect to sub-groups 2 and 3.Both urban areas present low population density levels and are experiencing the urban sprawl phenomenon. As a matter of fact, urban sub-groups 7 and 6 are well represented in both urban areas in the model. Even the city itself is sparsely dense, in both figures low density area can be observed in the city centre and also within the first ring of cities around the major urban centre contrary to, for example, large cities such as Toulouse.

Moreover, the model is able to translate differentiated morphological organizations at a high scale. The city of Avignon is a medieval city with a core historical centre surrounded by ramparts. Its historical centre presents important characteristics regarding the network, street design and accessibility, which influence the productivity of delivery tours and its logistics organisation. Even if the road network is dense, streets are narrower and entering the core city is difficult because of the ramparts and rare and narrow gateways. It is also a commercial heart and it remains a central residential area of Avignon. The area is precisely shaped in the map by the sub-group 2 and 1 with a high meshing density and a high distribution point density and linear density.

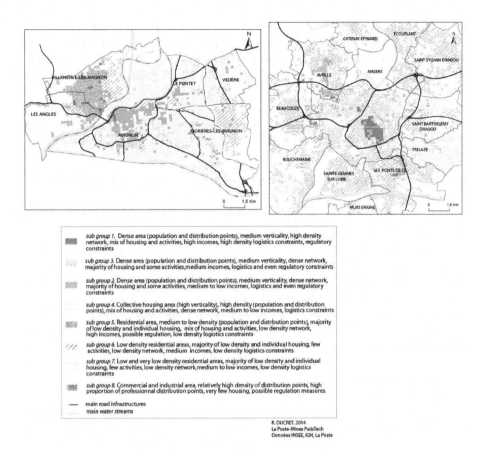

sub group 1. Dense area (population and distribution points), medium verticality, high density network, mix of housing and activities, high incomes, high density logistics constraints, regulatory constraints

sub group 3. Dense area (population and distribution points), medium verticality, dense network, majority of housing and some activities, medium incomes, logistics and even regulatory constraints

sub group 2. Dense area (population and distribution points), medium verticality, dense network, majority of housing and some activities, medium to low incomes, logistics and even regulatory constraints

sub group 4. Collective housing area (high verticality), high density (population and distribution points), mix of housing and activities, dense network, medium to low incomes, logistics constraints

sub group 5. Residential area, medium to low density (population and distribution points), majority of low density and individual housing, mix of housing and activities, low density network, high incomes, possible regulation, low density logistics constraints

sub group 6. Low density residential areas, majority of low density and individual housing, few activities, low density network, medium incomes, low density logistics constraints

sub group 7. Low and very low density residential areas, majority of low density and individual housing, few activities, low density network, medium to low incomes, low density logistics constraints

sub group 8. Commercial and industrial area, relatively high density of distribution points, high proportion of professionnal distribution points, very few housing, possible regulation measures

— main road infrastructures
··· main water streams

R. DUCRET. 2014
La Poste-Mines ParisTech
Données INSEE, IGN, La Poste

Fig. 1. Cartography of the urban sub-groups of the city of Avignon and Angers and cities of their first ring

In Angers, the city centre is less dense and constrained but nevertheless presents high population and distribution point density levels which are represented by sub-group 1. Thus, we can conclude that the eight urban sub-groups chosen for this study are able to precisely and coherently describe spatial organisations but also differentiate cities and give reliable geographical pictures. This means that, from a geographical perspective and despite its simplicity, the modelling tool is a relevant urban modelling tool thanks to a majority of logistical data combined with geographical data.

From an operational point of view, thanks to the tool a parcel provider is able to identify coherent zones with similar spatial and demand features, and could decide to apply specific services or organisation into or for the zone. The segmentation provides specific elements about spatial and organisational constraints on urban delivery related to the infrastructure network, the density of the distribution point and the verticality of the housing. For example, areas defined by the sub-group 8, which are industrial and commercial areas, do not need the same services as residential areas (sub-groups 1 to 7). Within the residential areas, the types of housing, the density of the distribution point as well as the level of income are equally relevant information. For example that the duration of delivery is not the same for apartment buildings and individual houses. Besides, the productivity of deliveries is not the same in dense central areas and low density residential areas. Taking into account the productivity levels of the areas, specific optimizations could be proposed. Moreover, the mapping of the sub-groups allows the grouping of spatially closed areas with close features to achieve sufficient demand and efficiency in delivery service. It could also help implement efficient routes, choose the right vehicle and even locate logistical facilities.

4   Conclusions, Limitations and Inputs for Future Research

The model described in this article is relatively simple and represents an attempt to model spatial features from the point of view of an urban freight. Even if this first test raises questions about urban modelling for urban freight (what type of data is more relevant, how can we formalise qualitative and spatial features into quantitative data, what type of mathematic formalisation, and so on), this first step, aimed at bringing urban analysis and urban freight closer, is encouraging in many ways.

First of all, this work proves that data that is not necessarily urban freight-oriented, such as spatial data or professional data, could provide relevant information for urban logistics practitioners. Logistic data have been used to build spatial descriptors and successfully used to analyse the way in which spatial organisations and logistics work together influencing the delivery. It may solve the issue of poor data availability for urban freight modelling.

From a geographical perspective, this modelling tool, based on a majority of data related to logistics mixed with geographical data, is a relevant urban modelling tool because it provides an accurate and relevant picture of urban organisation. Thus, we argue that spatial studies can contribute highly relevant information about urban goods movement and logistics.

The geographical analysis coupled with a logistical approach and logistic features provides a relevant diagnosis for urban logistics practitioners and local authorities. Urban modelling of urban freight and more broadly speaking spatial research could help local authorities to better understand logistics and could even provide operational responses to tackle urban freight organisations and services. These conclusions emphasize the necessity for urban logistics studies and planning to include in-depth spatial analyses and for urban freight modelling to enlarge the descriptors to spatial features.

Acknowledgments. The author would like to thank Mokhtar Beji for his contribution to modelling during his internship in La Poste in 2013, Jesus Gonzales Feliu for his help regarding the urban freight modelling review and Alicia Tromp for her proofreading.

References

1. Allen, J., Browne, M., Cherrett, T.: Investigating relationships between road freight transport, facility location, logistics management and urban form. Journal of Transport Geography 24, 45–57 (2012)
2. Ambrosini, C., Routhier, J.-L.: Objectives, Methods and Results of Surveys Carried out in the Field of Urban Freight Transport: An International Comparison. Transport Reviews 24(1), 57–77 (2004)
3. Anand, N., van Duin, R., Tavasszy, L., Quak, H.: City Logistics Modeling Efforts: Trends and Gaps - A Review. Procedia - Social and Behavioral Sciences 39, 101–115 (2012)
4. Andriankaja, D.: The location of parcel terminals: links with the locations of clients. Procedia - Social and Behavioral Sciences 38, 677–686 (2012)
5. Batty, M.: Urban modelling. International Encyclopedia of human geography, pp. 51–58 (2009)
6. CERTU, Transport de marchandises en ville: quels enjeux pour les collectivités? Editions du Certu, 12p. (2013)
7. Cherrett, T., et al.: Understanding urban freight activity – key issues for freight planning. Journal of Transport Geography 24, 22–32 (2012)
8. Comi, A., Delle Site, P., Filippi, F., Nuzzolo, A.: Urban Freight Transport Demand Modelling: A State of the Art. European Transport/TrasportiEuropei 51(7), 1–17 (2012)
9. Dablanc, L., Rakotonarivo, D.: The impacts of logistics sprawl: How does the location of parcel transport terminals affect the energy efficiency of goods' movements in Paris and what can we do about it? In: 6th International Conference on City Logistics, Puerto Vallarta, Mexico, June 20-July 2, 2009, pp. 251–264 (2010)
10. Dablanc, L.: City distribution, a key element of the urban economy: Guidelines for practitioners. City Distribution and Urban Freight Transport: Multiples Perspectives, 13–36 (2011)
11. Dablanc, L.: Freight transport for development toolkit: Urban Freight. The World Bank (2009)
12. Diziain, D., Gardrat, M., Routhier, J.-L.: Far from the Capitals: what are the relevant city logistics public policies? In: 13th World Conference on Transport Research, Rio de Janeiro, Brésil, Juillet 15-18, 16 p. (2013)
13. Giuliano, G., O'Brien, T., Dablanc, L., Holliday, K.: NCFRP Project 36(05). Synthesis of Freight Research in Urban Transportation Planning. Washington D.C., National Cooperative Freight Research Program (2013)
14. Gonzales-Feliu, J., Routhier, J.-L.: Modeling Urban Goods Movement: How to be Oriented with so Many Approaches? Procedia - Social and Behavioral Sciences 39, 89–100 (2012)
15. Hall, P.V., Hesse, M., Rodrigue, J.-P.: Guest editorial: Re-exploring the interface between economic and transport geography. Environment and Planning A 38, 1401–1408 (2006)
16. Hesse, M.: The city as a terminal: the urban context of logistics and freight transport. Ashgate Publishing, Grande Bretagne (2008)
17. Hesse, M., Rodrigue, J.P.: Introduction: dossier on freight transport and logistics. Tijdschriftvoor Economische en Sociale Geografie 95(2), 133–134 (2004)
18. Hesse, M., Rodrigue, J.P.: The transport geography of logistics and freight distribution. Journal of Transport Geography 12, 171–184 (2004b)
19. Holguin-Veras, J., Thorson, E., Zorilla, J.C.: Commercial Vehicle Empty Trip Models With Variable Zero Order Empty Trip Probabilities. Networks and Spatial Economics 10(2), 241–259 (2010)

20. LET, Diagnostic du transport de marchandises dans une agglomération, Paris (2000)
21. Lindholm, M.: Enabling sustainable development of urban freight from local authority perspective. In: 13th World Conference on Transport Research, working paper (2013)
22. Lindholm, M.: How Local Authority Decision Makers Address Freight Transport in the Urban Area. Procedia - Social and Behavioral Sciences 39, 134–145 (2012)
23. Lindholm, M.: A sustainable perspective on urban freight transport: factors affecting local authorities in the planning procedures. Procedia - Social and Behavioral Sciences (2), 6205–6216 (2010)
24. Lipatz, J.-L.: L'analyse spatiale a-t-elle encore besoin de maillages géographiques? Rencontres SIG (2013)
25. Macário, R., Galelo, A., Martins, P.M.: Business models in urban logistics. Ingenería & Desarollo (24), 77–96 (2008)
26. Macharis, C., Melo, S. (eds.): Introduction- city distribution: challenges for cities and researchers. City distribution and urban freight transport: multiples perspectives, pp. 1–9. Edward Elgar Publishing, Northampton (2011)
27. Marcucci, E., Gatta, V.: Behavioral modeling of urban freight transport. Testing non-linearpolicy effects for retailers. In: Gonzalez-Feliu, J., Semet, F., Routhier, J.L. (eds.) Sustainable Urbanlogistics: Concepts, Methods and Information Systems. Springer, Berlin (2013)
28. Ogden, K.: Urban goods movement: a guide to policy and planning, Brookfield, Ashgate Publishing Company (1992)
29. Patier, D., Routhier, J.L.: La logistique urbaine, acquis et perspectives. Introduction au dossier. Les Cahiers Scientifiques du Transport 55, 5–10 (2009)
30. Patier-Marque, D.: La logistique dans la ville, CELSE. Paris (2002)
31. Russo, F., Comi, A.: A modeling system to link end-consumers and distribution logistics. European Transport/ TransportiEuropei 28, 6–19 (2004)
32. Rodrigue, J.P., Comtois, C., Slack, B.: The Geography of Transport Systems. Routledge, Taylor & Francis Group (2008)
33. Routhier, J., Toilier, F.: Freturb: simuler la logistique urbaine. Modéliser la ville: Formes urbaines et politiques de transport, Méthodes et Approches, 438p. (2010)
34. Taniguchi, E., Thompson, R.G., Yamada, T., Van Duin, R.: City logistics: network modeling and intelligent transportation systems. Pergamon, Oxford (2001)
35. Taaffe, E.J., Gauthier, H.L., O'Kelly, M.E.: Geography of Transportation, 2nd edn. Prentice Hall (1996)
36. Woudsma, C.: Understanding the movement of goods, not people: issues, evidence and potential. Urban Studies 38(13), 2439–2455 (2001)

Patterns of Internal Migration of Mexican Highly Qualified Population through Network Analysis

Camilo Caudillo-Cos and Rodrigo Tapia-McClung[*]

Centro de Investigación en Geografía y Geomática "Ing. Jorge L. Tamayo", A.C.
Contoy #137 esq. Chemax, Col. Lomas de Padierna, Tlalpan, 14240, México, D.F.
{ccaudillo,rtapia}@centrogeo.org.mx

Abstract. Many real, social, technological, biological and information systems can be described as complex networks. Nonetheless, few studies treat migration from this standpoint. Some migration studies focus on people, and some others on places. The former require very detailed data, while the latter are based on aggregate data. This study is based on places and uses aggregate data, taking flows as an observable and their resulting patterns are the object of study. Mexican censual events take place every ten years, the most recent on 2010, and it has been only until recently that there are enough capabilities and tools available to visualize and model internal migration to the municipal level. Few studies have focused on analyzing migratory movements of such detail, opting instead for the state level. Network analysis allows the identification of communities with a certain degree of spatial structure, that is, the importance that geographical proximity plays in migration.

Keywords: Internal migration, Complex networks, Social network analysis, Highly qualified human capital.

1 Introduction

Many real, social, technological, biological and information systems can be described as complex networks. Nonetheless, few studies have been found that treat migration from this point of view. Notable exceptions are: one focusing on multiscale mobility in the United Kingdom [1] and another one dealing with internal state migration of the United States [2].

Since the decade of the 1990s the study of complex networks and some of its properties have gained importance. Such is the case of scale-free and small-world networks. We want to study internal migration through the analysis of complex networks based on graph theory, where an undirected graph $G = (V, E)$ is defined as a set of V vertices or nodes linked through a set of E edges. This simple construct turns out to be of great utility since it allows the representation of a broad range of real world systems such as: transportation, energy and ecology among others. In social networks, each individual is a node: such as a twitter account or the information flow between

---

[*] Corresponding author.

B. Murgante et al. (Eds.): ICCSA 2014, Part IV, LNCS 8582, pp. 169–184, 2014.

companies. In this case, as in movement models in general, nodes represent places (metropolitan areas) and the edges (that become arcs in a directed graph) represent flows of people.

Sociologists have studied the interaction between individuals and have created a series of parameters that allow the characterization of the roles each actor plays in terms of their relationships. This is known as social network analysis [3].

In this study focus is given to recent migration, understood as population that in 2005 lived someplace different to the place they lived during the censual moment of 2010, the general population (thus yielding a general population group of people ages 5 or more) and highly qualified population (identifying the population that during the last censual event reported their change of residence and are either studying or have finished college and/or graduate studies).

In 2012, Martínez-Viveros et al. [4] approached the identification of places susceptible to innovation through the characterization of stocks of human capital, measured through the classic indicator of educational achievement as well as the geography of ability (being understood as the contribution of creative occupations). In the study, a significant relationship was found between the indicators of human capital and the concentration of industrial activities with medium-high and high technological base. This implies that the stock of human capital is important for clustering. Nonetheless, the demographic mobility was left out and is presented in the current paper. A classical demographic analysis is carried out in which focus is given to the structure of the general and the migrating population, and differentiating in age and gender groups for these populations, through the construction of age pyramids and specialization indicators.

Furthermore, networks are characterized according to their properties of centrality and clustering into communities. That is, community detection from the interactions that will allow identifying the roles each metropolitan area plays according to their position in their group, and the interactions with the rest of the groups. In this case, the study is restricted to the analysis of 59 metropolitan areas defined by the National Population Council and the National Institute of Statistics and Geography [5].

This paper is structured as follows: the methodology of the study is presented in Section 2, along with a short description of the data used in Section 3. Section 4 presents a discussion on how to characterize human migration while Section 5 looks into the relationship between migration and development. Finally, Section 6 presents a collection of empirical results and conclusions. Throughout the paper, the reader will find that most of the figures can be explored interactively when clicked on the links provided in the figure captions.

2   Methodology

Two main groups of interest are identified for this study: the first, the recent population movements in general and the second, highly qualified migrants defined as those who are studying or have finished a university or graduate degree. Focus is also given to knowledge-intensive occupations presented in Martínez-Viveros et al. [4]. It is important to note that these flows are expected to be small since they can only be identified if they are part of the labor market. In this sense, a hypothesis is that these flows are subrepresented because the censual definition may not capture intermediate migratory movements, as it only cares if someone changed their place of residence five years before the census takes place.

First, work is carried out with migration matrices in absolute terms and the rate of migration matrix (relative values). These can easily be transformed into networks by creating an edge between metropolitan areas that have flows between each other. Once such a network is created, techniques of social network analysis can be applied to them.

As Bender-deMoll mentions in the network mapping report [6], a classic field of application for social network analysis is the characterization of flows of goods and people. Networks are used to represent flow patterns between sets of entities and constitute a useful analysis of movement structures. Results of some studies on trade flows have shown to provide more knowledge and have helped predict global resources flow between countries. By analyzing data on both forced and voluntary migrations, a strong correlation has been found between the geography and the relationships shown by aggregate flows. In the same way, these flows reflect the social links of migrants, as they may move to places where relatives and/or friends are located, or to places that information networks have detected to be viable for development [7]. Indeed, Massey [8] and Marconi [9] also refer to this behavior citing Hagerstrand's theories on migration.

2.1   Network Visualization and Characterization

An important characteristic of this study is network visualization. By geographically placing nodes, some network features can be highlighted according to node parameters. It also allows the identification of special structures in the flow patterns. A common practice in migration studies is to aggregate data according to the analysis unit. In this case, starting with the origin-destination matrix, networks are built and then characterized. Furthermore, the square matrix is transformed into an array of minimum information that avoids information redundancy and also allows for the dynamic exploration of flows between metropolitan areas.

Networks are distinguished from other datasets due to their emphasis in relationships. That is, a network has properties that emerge from the links between its elements. This allows the calculation of some parameters that summarize local aspects from the global structure of the network:

Degree Distribution. It is one of the most important parameters to characterize a graph. The degree of a node is the number of connections it has to other nodes, regardless of their direction, while degree distribution is the probability distribution of the degrees over the network. Other parameters derived from this one are the degree of entry and the degree of exit which, in a weighted network, correspond to the entry or exit flow in a node.

Node Centrality. It highlights aspects of the network structure capturing how close a node is to the rest (closeness) or the role each node plays in the subset of shortest paths of the network (betweenness) [10, 11, 12, 13].

Community Detection. Other fundamental elements are network communities. Using connectivity and flows, nodes that are strongly related can be detected. This can be done in two different ways: the first, by identifying cliques or complete graphs that share edges (as in Figure 2); the second, by means of the modularity algorithm [14]. In principle, community detection is possible if the graph is weighted and its weights have a heterogeneous distribution [15].

Brokerage Roles. According to Gould and Fernandez [16], there are five possible brokerage roles of nodes according to their position in the group and with respect to other groups: coordinators, itinerant brokers, representatives, gatekeepers and liaisons. These are exemplified in Figure 1.

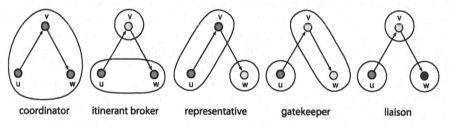

coordinator   itinerant broker   representative   gatekeeper   liaison

Fig. 1. Possible brokerage roles of node v (taken from de Nooy et al. [17])

In Figure 1, the role of node v is identified according to its position within the group and towards other groups. Of the possible roles, two of them imply the mediation between group members. Firstly, in the coordinator role, the mediator belongs to the group. Secondly, in the itinerant broker case, two members of the same group use an external mediator. The three remaining cases are brokerage roles between different groups described as follows: the mediator acts as a representative of its own group in another one because it regulates the flow of information or goods of its own group. The gatekeeper regulates the flow of information or goods towards its own group. Lastly, in the liaison role, there is a mediator between groups and it does not necessarily belong to any of the groups.

3   Data

Data for this study is taken from the 2010 Mexican General Population Census sample database containing recent migratory movements from 2005 to 2010 between Mexican municipalities. These are aggregated to capture the movements involving 59 metropolitan areas. It is necessary to point out that the censual definition leaves out of the view those movements that take place in a smaller temporal window and that could be relevant given the mobility of highly qualified population. It is also important to note that issues related to sample size are beyond the scope of this report, even though it is recognized that for some of the groups under study this would have a significant impact.

Mexico is divided into 32 states, constituting the largest aggregation unit of the country. At this scale, a migration matrix is easy to handle. Figure 2 shows the graph corresponding to such matrix. Node size varies with gross migration and line width varies with the amount of flows between entities. This is a uniform graph, since each node has *n-1* arcs. However, many processes are hidden at this scale. The 59 metropolitan areas correspond to the medium scale study in which the municipalities for each area are aggregated. Figure 3 shows the Mexican states and the metropolitan areas considered in this study.

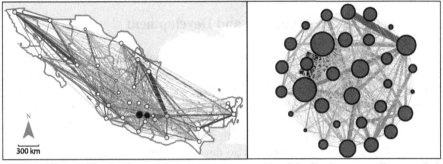

a) Geographic distribution of nodes   b) Distribution according to their weighted degree

Fig. 2. Graph of the migration between the 32 Mexican states (source: authors' construction with data from the 2010 General Population Census sample database using Gephi [18])

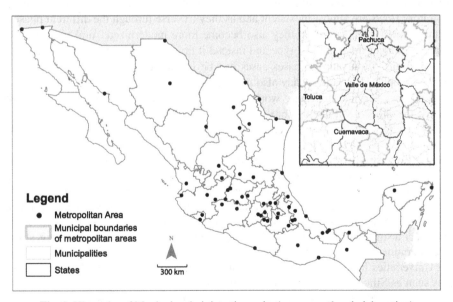

Fig. 3. Hierarchy of Mexico's administrative units (source: authors' elaboration)

4   Definitions to Characterize Human Migration

A *demographic current* is the *flow* that occurs between two territories. The *dominating current* is the most important one, numerically speaking, while the flow with the lower value is called *countercurrent*. Net migration balance is defined as the difference between a current and its countercurrent. The same components used for Elridge's definitions [19] are used in Mexico's censual instruments since 2000, with the restriction that it is valid at the state level. Given that, in this study we drift somewhat away from this definition to focus on recent migration, that is, the population that changed residence between the years 2005 and 2010.

5   Approaches on Migration and Development

Wilbur Zelinsky postulated a model denominated the mobility transition hypothesis, which pretended to complement the general model of demographic transition. He starts with the basic postulate that social scientists try to find patterns and regularities in human activities by means of theoretical constructs which, in the case of Geography and Demography, are rather scarce. That is why he uses the demographic transition (or vital transition) and the laws of migration postulated by Ravenstein in 1885 [20]. According to this model, societies go through different phases of development, depending on their levels of urbanization, industrialization, and scale of modernity. The hypothesis is based on the relationship between different types of mobility and the general development process. It starts from the conceptualization of a diffusion problem and identifies important transformations in the types of migration (change of usual place of residence) and circulation (work mobility). In traditional societies, migratory flows are almost inexistent and as they traverse through the different phases of the demographic transition they also become more modernized, rural-rural migration is no longer the dominant type and instead it favors metropolitan and international migration: "In advanced phases cases can be identified in which at an individual level —people's lifecycle— they also go through a series of movements from education, military service, marriage, work and its implications in different changes and finally, the choosing of a place for retirement" [21].

From economics it is recognized that economic activity is, in first instance, concentrated in the territory by the physical and geographical conditions imposed by such diverse accessibility niches and differential transportation costs. However, on a second instance, it is concentrated by the exploitation of clustering economies. Abel and Deitz [22] analyze if the graduate degrees and Research and Development activities of colleges and universities are related with the human capital in metropolitan areas where they are located. Results indicate that they have a positive relationship, but rather a small one in terms of human capital stock and production, thus suggesting that migration plays an important role in the geographical distribution of human capital. Universities have been regarded as motors to develop local economy, with examples such as Silicon Valley and the Boston route 128 corridor, but also recognized as a fundamental part in the current transition to knowledge-based economies.

There is also the strong belief that by retaining graduates in their ranks, colleges and universities in declining regions can help alleviate their economic sufferings. In fact, they find that the amount of human capital of a region is one of the most powerful predictors of economic vitality [23]. Empirical evidence is explained by the fact that human capital increases productivity and the generation of ideas at an individual level. Therefore, by extension, a higher level of human capital in a region boosts its regional productivity. Given the importance of human capital in economic performance of a region, it is surprising that so little research exists that analyzes the factors that conduct the differences of human capital throughout space.

6   Empirical Results

Two main empirical results of this study are presented: the comparison between features of general migration (population ages 5 and older) and highly educated population and the characterization of each population group's flow networks. All the figures in this section include hyperlinks to interactive online graphics and maps. Graphics were prepared using the Tableau software [24].

6.1   Mexico's Internal Migration and the Migration of Highly Qualified Human Capital

A first relevant element of internal migration is its evolution with respect to previous patterns. In Mexico, during the first half of the 20th century, the first migration attractor was Mexico City and it experienced a huge growth. From the decade of the 1980s, Mexico City's metropolitan area experienced a social decrease. However, flows were mainly directed to municipalities close to the main urban center, thus generating a vast metropolitan area, a pattern that remains up until the beginning of the 21st century [25, 26]. Less is known about the dynamics of other regions in the country, but border cities have also experienced an important growth due to mainly two processes: the establishment of urban-assembly plant corridors and the closeness with the main US border crossing points. In the mobility transition scheme, Mexico is in a recessive phase of the dominant city (the Mexico City basin that shows, back from the 1980 census data, a tendency to repel the central city towards peripheral municipalities and cities nearby).

As far as the general process of vital (or demographic) transition goes, there is a correlation between the population age structure and the change in population education. The distribution of these elements is unequal in the territory since metropolitan areas have a great variation in size and regional preeminence. Population pyramids for ages 20-75 and education level are shown in Figure 4 so it is possible to observe its evolution from 1990 to 2010. As a default, the total for the 59 metropolitan areas is shown but data for each metropolitan area considered in the study can be explored individually. The pyramid shows how population without formal education starts to disappear among the youngest (20-24 years old) and represents less than 0.7%. For the year 2010 the lack of formal education is reported to be almost nonexistent in this age group.

On the left side of Figure 4 the femininity index is shown. This index expresses how many women there are for each men in a given age and educational group. In 1990, women prevailed in the groups with no and basic education. They had a small value, but still greater than 1 (which would mean gender equity) among the youngsters with higher education. However, in the rest of the groups, women were subrepresented. For 2010, the shape of the pyramid is less unequal between the base and the top groups, but the educational advance in both medium and high education as well as gender equity becomes evident. It is also interesting to note that among the younger groups, the disparity between men and women has been reverted.

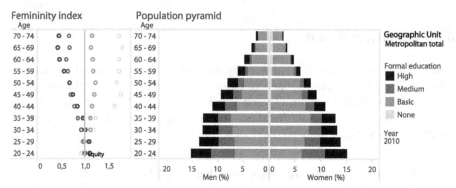

Fig. 4. Population pyramid showing educational achievement and femininity index for 2010 (source: authors' elaboration). Viewable at:
`http://public.tableausoftware.com/views/Piram_edad/Dashboard1?:e`
`mbed=y&:display_count=no`

In 2010, the population of Mexico was around 111 million inhabitants. The vast majority of them did not change their place of residence during the period between 2005 and 2010. However, almost 4 million people did move from one state to another, without taking into account another million people that entered the country from abroad. This implies that the flow of people could, in principle, give rise to a respectable sized city all by itself.

Digging deeper into detail, the municipal division level can be reached and for the year 2010 includes 2456 territorial units. This is translated into 6 million potential flows. Municipalities are the political-administrative units with the largest censual information available and for which it is possible to extract an origin-destination matrix. Table 1 summarizes the information about the percentage distribution of intermunicipal flows, with municipalities classified either as metropolitan or non-metropolitan.

This table allows catching a glimpse of a fundamental difference among recent migration patterns of highly qualified human capital with respect to the general population migration, consisting in the preeminence of flows between metropolitan areas. These flows represent a challenge for future planning: provision of public services, housing, and equipment. There is evidence pointing towards the main explanatory factors being changes in family composition, changes in home income and strategies to deal with day-to-day mobility, as pointed out by Suárez-Lastra and Delgado-Campos in their study analyzing mobility in Mexico City [27].

Table 1. Summary of the 2005-2010 internal migration[1] (source: authors' elaboration)

Municipality	Population ages 5+	University level	Graduate level	University + Graduate level	Knowledge-intensive occupations
Non- metropolitan	846,397	83,220	6,164	89,384	42,871
	(13.35%)	(8.45%)	(5.45%)	(8.14%)	(8.95%)
Moving into a metropolitan area	986,021	141,692	10,354	152,046	49,901
	(15.55%)	(14.39%)	(9.16%)	(13.85%)	(10.42%)
Moving out of a metropolitan area	868,576	112,845	11,729	124,574	55,646
	(13.70%)	(11.46%)	(10.38%)	(11.35%)	(11.62%)
Between metropolitan areas	**1,190,802**	**270,326**	**36,375**	**306,701**	**123,793**
	(18.78%)	(27.46%)	(32.18%)	(27.94%)	(25.86%)

6.2 Age Composition of Migrating Population

A specialization index by age was built to compare the age structures of interesting migrating groups, taking the structure of the general migrant population ages 20 to 74 as a reference. Figure 5 shows the specialization index in the horizontal axis, values higher than 1 mean an overrepresentation of a group and gender with respect to the same group and gender of the reference population.

The population pyramid of migrants usually has a larger base than rest of the population. That is, younger people usually changes residence more often than the rest. The specialization index reveals that among groups of highly qualified human capital there is also an overrepresentation of the first age groups, from 20 to 50 years old in men, and preeminence among young women with values of the index higher than men. It is also interesting to note that in the case of women with graduate degrees, the index reaches twice the previous value. Not only highly qualified population moves between metropolitan areas, but the general population does as well. Their age equality structure shows a specialization in relatively younger ages. This can be linked to two processes: the absence of an educational offer in smaller metropolitan areas – thus creating the need of younger people to change their residence to continue with education – and the pressure of ejection from the larger metropolitan areas where the labor market can be saturated – as is the case of Mexico City and Guadalajara – towards other regions in which they can insert themselves into the labor force.

The fact of subrepresentation of migrants in the older age groups of highly qualified population could imply that once people are 50 years or older, there is higher work stability.

---

[1] Non-metropolitan flow occurs between municipalities that are not classified as metropolitan. Moving into a metropolitan area implies the migration from a non-metropolitan to a metropolitan municipality. Moving out is the inverse flow.

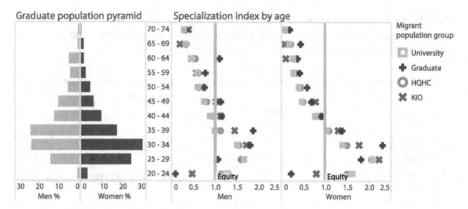

Fig. 5. Graduate population pyramid and Specialization index by age (source: authors' elaboration). Viewable at:
`http://public.tableausoftware.com/views/Estructuraetrea/`
`Dashboard1?:embed=y&:display count=no`

6.3   Gross Migration, Currents, Countercurrents and Centrality

The 8 largest metropolitan areas encompass 50% of the gross migration of population with university or graduate degrees. The net balance for Mexico City is negative for all the groups of interest although the highest disparity is found between immigrants and emigrants with graduate degrees. On the other hand, the volume of highly quali-fied population in the country's capital city has an important impact in the rates, even if these have the same sign. They reach very small values – about 8 migrants and 6 immigrants for every 1000 person years. The rate instability is also evident in small metropolitan areas, resulting in very high values contrasting with Mexico City. The ten cities with the highest net population migration balance in the university/graduate degree level are distributed across the country: Querétaro, Pachuca, Toluca and Cuer-navaca in the central region, Monterrey and Saltillo in the north, Mérida and Cancún in the southeast and León and Aguascalientes in the central-northwest part of the country.

Another interesting thing to note is the behavior of the immigration rate of highly qualified human capital and the general population aged 25 or more: the first is sys-tematically higher than the second. In other words, the mobility of population with university or graduate degree is much larger, relatively speaking.

In order to explore some results of this study, a web application was built to show the flows between different metropolitan areas. An example can be seen in Figure 6. Flow data is handled in JSON format and expressed as a one-to-many relationship, that is, outgoing flows go from one node to many. Each node is identified with an ID, name and geographical coordinates and thus can be put on a map to aid in the study of the spatial interaction between flows. The Leaflet library was used to overlay flows on maps and jQuery was used to handle user interaction, adapting Bostonography's Hubway Trip Explorer [28, 29, 30].

The functionality of this visualization is as follows: the user can click on any metropolitan area (dots on the map) and the migration flows will be shown. Arcs are displayed for incoming and outgoing flows, thus showing the diversity of connections in each network that gives rise to the migrant dominant currents and countercurrents. Clear dashed arcs represent emigrant flows from a hovered node to the selected one, while dark solid ones represent immigrant flows to the selected node. The line width is proportional to the gross migration of the selected metropolitan area.

Migration flows can be explored for people with university degree, graduate degree, university and graduate degree and people in knowledge-intensive occupations. It is also possible to take a look at the pattern of all the flows and get a glimpse of all the possible connections in the network.

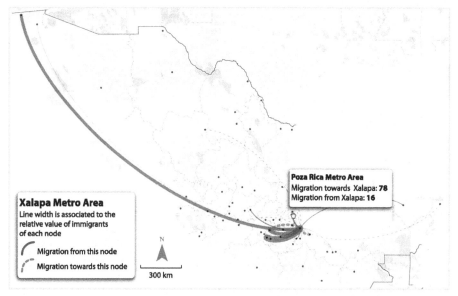

Fig. 6. Migration flows in a web interface (available at
http://rhac2.centrogeo.org.mx/reportes/migracion/migracion.html)

Another element that can be explored is network centrality. Networks were characterized according to their intermediation, their closeness and the degree of entry and then cross-compared with the typology resulting from the analysis of knowledge-intensive occupations and educational achievement [4]. Figure 7 shows scatter plots of these parameters. Circle diameters are proportional to the gross migration value. The betweenness centrality shows an exponential behavior in relation to the degree of entry and it also shows a good correlation with gross migration. It is important to remember that betweenness is a parameter of the global structure of the network, thus, it changes the position of metropolitan areas according to the population under observation. In the case of the network for the population of 5 years and older, Guadalajara is the metropolitan area with the highest values, while for the rest of the networks, Mexico City always comes first.

The value of *betweenness* centrality for Mexico City's metropolitan area is separated considerably from the rest of the metropolitan areas in the 'Graduates' network. That is, almost any shortest path in that network must go through the country's capital city.

The *closeness centrality* has an inverse relationship, thus a smaller value indicates that the distance from that node to the rest of the network is smaller, therefore making it more central. It can also be seen in Figure 7 that some metropolitan areas under the 'Mean' category are overlapped and these also happen to have a higher centrality value. This result can imply that these metropolitan areas constitute 'in-transit' regions for the formation of human capital, that is to say, they both receive population in their formation years and expel population looking to join the labor market.

Centrality - Graduate network

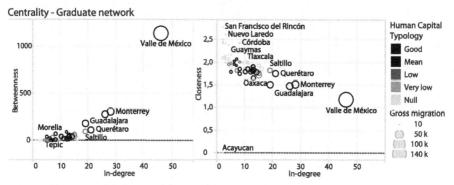

Fig. 7. Scatter plots of the centrality parameters with human capital typology[2] (source: authors' elaboration). Viewable at:
`http://public.tableausoftware.com/views/Rolesymontos/`
`Parmetrosdecentralidad?:embed=y&:display_count=no)`

6.4   Community and Brokerage Role Detection

The initial step to determine which role each metropolitan area plays is to classify them in terms of the flow network (absolute values) or immigration rates (relative values). In as much as possible, the number of communities was kept alike to be able to compare their composition in terms of the number of members and their geographic distribution.

The quality of the partition (the communities) turned out to be diverse, ranging from a minimum value of 0.44 for the communities of immigration rates of population having university degree, to a maximum value of 0.66 for the communities of absolute values of population having a graduate degree, implying that some elements could change their membership. This in turn is associated with high values of the betweenness centrality parameter. Some community detection algorithms are based on the elimination of the nodes with the highest betweenness values, resulting in communities with no members playing the liaison role.

---

[2]   The typology captures the concentration of knowledge-intensive occupations and educational achievement for each metropolitan area (Martínez-Viveros et al. [31]).

Another important element is that in many cases, the dominant role of a metropolitan area does not coincide with any of the five types previously mentioned. This means that the node, or nodes, only interact with members of their own community, and even more, they do not act as coordinators. This happens, in general, with the less connected metropolitan areas.

A common element in migration studies is that there is an important spatial component in the flows that is reflected in the communities. This means that, even though some metropolitan areas are distributed along different regions of the country, they have many interactions among them and can be joined in as a community. This happens in the central region of the country, where almost all of the metropolitan areas were classified as part of the same group and is also observed in other parts of the country, as can be seen in Figure 8.

Absolute values, HQHC

Fig. 8. Communities and dominant brokerage roles (source: authors' elaboration). Viewable at: http://public.tableausoftware.com/views/Rolesymontos/ Comunidadesyrolesdominantes?:embed=y&:display_count=no

As it was previously mentioned, roles depend on the position of the node in terms of their relationship with other nodes in the same or in a different community. The liaison role implies an intermediation relationship with two communities, that is, each node is important as it keeps brokerage relations among three communities. In the case of the 'University' network for both the absolute and relative values, this is the dominant role for the majority of its members (the exception being Cuernavaca, which for the immigration rates acts as a gatekeeper).

This result ties in directly with the difficulty to create a "hard" classification, because metropolitan areas with very high values of both degree and betweenness centrality have relationships with many other metropolitan areas. The counterpart of liaisons can be either gatekeepers or representatives, the latter being the dominant roles with higher frequency in the communities that are not in the central part of the country.

Only in the 'Graduate' and 'Knowledge-intensive occupations' networks there are metropolitan areas with dominant coordinator roles. This is relevant because it shows a higher degree of cohesion within the group, or endogamy. There are nodes having more interactions between members of their community than with nodes of foreign groups, which can be an indication that the labor market and the graduate studies offer promote flows to the interior of the communities. Despite that the geographical component is weaker in the 'Graduate' networks this result also ties in together with the size of community "1", which is the community where the highest frequency count of coordinating nodes is found (24 for the absolute values and 22 for the rates). In this scenario, the coordinating metropolitan areas are usually smaller in size and their interactions obey a small scale dominion. That is, they can have a limited regional importance.

This is an initial exercise of an exploratory analysis of social networks that has allowed demonstrating its potential to analyze migration flows. On the one hand, a distinctive feature of internal migration has been exposed, and it is the existence of an important spatial component. The unequal development produces a hierarchy of cities that maintain relatively intense local relations (the exception being the migration network of population with graduate degrees, in which community distribution is more dispersed). Community detection allowed revealing functional distance relationships given a specified migrating group, a key concept when trying to understand the redistribution processes of highly qualified population in Mexico.

Further exploration of some particular metropolitan areas makes it possible to get an idea of the importance of each and one of them by the roles they play in each migrating group. Mexico City's metropolitan area, for instance, has the highest values in all of the centrality parameters thus, its dominant role is as liaison for the 'University' and the 'University+Graduates' networks both in the absolute values and rates. In the case of immigrant rates in the 'Graduates' network, its role changes to gatekeeper as it receives migrants from many metropolitan areas outside its own community and, at the same time, has many relations with other cities within its own group.

7   Conclusions

In this study only one of the demographical components of a population has been explored: internal migration. Zelinsky's hypothesis about transition to mobility has aided explaining some vital transition patterns and their correlation with evolution in education and social growth in metropolitan areas: the higher the education, the higher the mobility. This also explains the stagnation of some metropolitan areas with higher population in the country, such as Mexico City or Guadalajara, which have a stabilized average educational level of the general population, while in medium-sized or small metropolitan areas an improvement in this indicator of human capital is observed. This is relevant because, after acknowledging the deficiencies on the data sources used, with the censual definition of 5-year difference for the regular place of residence, many intermediate movements are lost. These would provide much more detail on the characteristics of the migration of highly qualified population. Also, there is no information on the motivations of migration.

With respect to network analysis, an important conclusion is that the population with graduate degree has a weaker geographical component in its behavior than the rest of the groups. It will be interesting to try and find factors that explain these results. The betweenness centrality parameter looks promising as a migration predictor, except their relationship is not linear. This is not surprising as in many networks the degree distribution can be best described as a power law, yielding scale-free networks.

Another contribution from the network analysis standpoint, one that is not obtainable with the classical analysis tools, is that besides taking into account the aggregate indicators, it offers a way to obtain clusters in terms of their interactions. This exercise has shown its utility when combining results from network analysis —impervious to *geographical distance* but not to *functional distance*— with the visualization of results in a map. In this way flows are shown to have an important spatial component for some groups. In the case of the 'Graduates' network, it can be seen that since communities are dispersed throughout the country, it implies there is a larger breach between geographical and functional distance in the migration flows of this population group.

Future research includes the combination of network indicators with other social, economic and educational performance indicators to carry out multivariate analyses that would allow associating these factors with the migration process of highly qualified human capital.

Acknowledgements. The authors would like to thank Elvia Martínez-Viveros for fruitful discussions and comments on previous versions of this manuscript. This research paper is part of the project "Aporte de los recursos humanos altamente calificados a las capacidades locales de innovación. Un estudio con enfoque territorial" funded by the Consejo Nacional de Ciencia y Tecnología.

References

1. Askar, D., House, T.: Complex patterns of multiscale human mobility in United Kingdom. Working paper. University of Warwick (2010)
2. Maier, G., Vyborny, M.: Internal migration between US-states. A social network analysis. Working paper. University of Viena (2005)
3. Moreno, J.: The Sociometry Reader. The Free Press of Glencoe, New York (1960)
4. Martínez-Viveros, E., Caudillo-Cos, C., Coronel, A.B., García, R., Lesdesma, M., López, F., López, J.L., Chapela, J.I., Morales-Gamas, A., Ortega, A., Pérez-Esteva, M.C., Tapia-McClung, R.: Aporte de los recursos humanos altamente calificados a las capacidades locales de innovación. Un estudio con Enfoque Territorial. Primera etapa. Working paper (2012)
5. CONAPO, INEGI: Delimitación de las zonas metropolitanas de México 2010. SEDESOL-INEGI (2013)

6. Bender-deMoll, S.: Potential Human Rights Uses of Network Analysis and Mapping. A report to the Science and Human Rights Program of the American Association for the Advancement of Science (2008)
7. Annuska, D., Henke, R., Vanna, L.: Review of a Decade of Research on Trafficking in Persons. Center for Advanced Study / The Asia Foundation, Cambodia (2006)
8. Massey, D.S.: Economic Development and International Migration in Comparative Perspective. Population and Development Review 14(3), 383–413 (1988)
9. Marconi, G.: Not just passing through: international migrants in cities of 'transit countries'. SSIIM Paper Series, vol. (6) (2010)
10. Freeman, L.C.: Centrality in social networks: I. Conceptual clarification. Social Networks 1, 215–239 (1979)
11. Borgatti, S.P.: Centrality and network flow. Social Networks 27, 55–71 (2005)
12. Borgatti, S.P., Everett, M.G.L.: A graph theoretic perspective on centrality. Social Networks 28, 466–484 (2006)
13. Borgatti, S.P., López-Kidwell, V.: Network Theory. In: Carrington, P., Scott, J. (eds.) Handbook of Social Network Analysis, Sage, London (2011)
14. Blodel, V.D., Jean-Loup, G., Lambiotte, R., Lefebvre, E.: Fast unfolding of communities in large networks. Journal of Statistical Mechanics: Theory and Experiment (2008)
15. Fortunato, S.L.: Community detection in graphs. Complex Networks and Systems Lagrange Laboratory. Report (2010)
16. Gould, R.V., Fernandez, R.M.: Structures of mediation: A formal approach to brokerage in transaction networks. Sociological Methodology 19, 89–126 (1989)
17. deNooy, W., Mrvar, A., Batagelj, V.: Exploratory Social Network Analysis with Pajek. Cambridge University Press (2005)
18. Gephi, an open source graph visualization and manipulation software: http://gephi.org/
19. Elridge, H.T.: Primary, Secondary and return migration in the United States, 1955–60. Demography 2, 444–455 (1965)
20. Ravenstein, E.: The Laws of Migration. Journal of the Statistical Society of London 48(2), 167–235 (1885)
21. Zelinsky, W.: The hypothesis of the mobility transition. Geographical Review 61(2), 219–249 (1971)
22. Abel, J.R., Deitz, R.: Do colleges and universities increase their region's human capital? Journal of Economic Geography 12, 667–691 (2012)
23. Groen, J.A.: The effect of college location on migration of college-educated labor. Journal of Econometrics 121, 125–142 (2004)
24. Tableau software. http://www.tableausoftware.com/public
25. CONAPO: La población de México en el nuevo siglo. CONAPO, México (2001)
26. CONAPO: Dinámica demográfica de México 2000–2010. In: CONAPO: La situación demográfica en México, México (2011)
27. Suárez-Lastra, M., Delgado-Campos, J.: Patrones de movilidad residencial en la Ciudad de México como evidencia de co-localización de población y empleos. EURE 107(36), 67–91 (2010)
28. Leaflet, a JavaScript library for mobile-friendly maps: http://leafletjs.com/
29. jQuery: http://jquery.com/
30. Bostonography's Hubway Trip Explorer: http://bostonography.com/hubwaymap/
31. Martínez-Viveros, E., Tapia-McClung, R., Caudillo-Cos, C.: Geographic distribution of talent in urban Mexico and its expression in patterns of specialization and / or diversification. Paper presented at the XXVII IUSSP International Population Conference, Busan (2013)

Automatic Point of Interests Detection
Using Spatio-Temporal Data Mining Techniques
over Anonymous Trajectories

Anahid Basiri[1] and Pouria Amirian[2]

[1] Nottingham Geospatial Institute, the University of Nottingham, UK
Anahid.basiri@nottingham.ac.uk
[2] Department of Computer Science, National University of Ireland, Maynooth, Ireland
pouria.amirian@nuim.ie

Abstract. Location Based Services (LBS) are evolving very fast but despite this are still at an early phase of their evolution. The future LBS services will be more anticipatory of users needs and will exploit a broader range of information on users. In order to anticipate users' needs, LBS providers should be able to understand users' behaviours, preferences and interests automatically and without asking the user to specify them. Then using the user's current situation and previously extracted behaviours, interests and dislikes, the user's needs can be predicted at the relevant moment and provide the most appropriate sets of services. This paper shows the application of data mining techniques over anonymous sets of tracking data to recognise mobility behaviours and preferences of users such as Point of Interest (PoI). Tracking data are first anonymised then stored in a spatio-temporal database. Then, using data mining techniques, rules, models and patterns are recognised. Such knowledge, patterns and models are subsequently used for intelligent navigation services as input making suggestions.

Keywords: Spatio-temporal data mining, Trajectory, Knowledge extraction.

1 Introduction

Location Based Services (LBS) has become a part of our daily lives. According to different LBS market analysis and reports, such as the market report produced by The European GNSS Agency (GSA) (The European GNSS Agency (GSA), 2013) Location Based Services will be more and more intelligent in future. Users look for a mobile service/application which can monitor or understand their current situations, predict their needs and demands, provide the most appropriate set of services in most comfortable way. Users do not want to get distracted to specify their current situations and send queries to get relevant response, they would like a service which can detect the current situations (such as current location, travel mode, day time, interests, mood, etc.) automatically and anticipate their needs and demands and provide relevant information and service (such as navigating them to their preferred destination through

B. Murgante et al. (Eds.): ICCSA 2014, Part IV, LNCS 8582, pp. 185–198, 2014.

less-traffic streets, finding and giving the address of the nearest ATM nearby, finding a vacant parking space near to their workplace/subway station and navigating them to get there).

In order to have a more intelligent system, the system needs to learn the patterns and rules (such as detecting travel mode based on speed of movement, finding interests and dislikes of a user based on history of travels and places visited, etc.) then it can automatically find matching information/service user may need according to the current situation observed. Data mining techniques help to find out patterns, rules, similarities, abnormities and clusters over quite large set of input data. This paper focuses on recognising some of patterns existing in travel and urban mobility and also Point of Interest (PoI) over large set of anonymous trajectories. These patterns may be used to make navigational suggestions.

In this regard, tracking data, without any reference to moving user's identification, should be stored. This can be done by using different anonymizers such as K-anonymity (Kalnis, 2007). Trajectories as a kind of spatio-temporal data can be captured from different resources, such as CCTVs, GPS embedded in mobile phones or vehicles, accelerometers and so on. Considering this huge amount of input data from all these resources, then it is possible to apply an inference engine in order to detect anomalies, cluster and classify data (Baiget, 2008). Such inference system helps us to find some patterns and recognise rules (Kuntzsch, 2013), (Zhang, 2013) which can be used in path finding and routing decision making or making navigational suggestions.

For example, if we could get tracking data of different users within a quite long period of time and see that in special time interval, e.g. on 8 a.m. to 10 a.m., many of users' trajectories fit into street networks and also considering their average speed of movement, the first thing to infer would be the mode of their movement; driving. This can be considered as rule to be learnt by the system as if the average speed of movement is more than 50 km/h and trajectory is matched by street network, then the travel mode is car or if it matched with bus lines and there are stop (speed for a short period of time, like 30 seconds, becomes zero) at the bus stops then travel mode can be bus. Such rules can be used in the phase of "recognising user's current situation".

The inference can be more complicated than simple if-then rules. It is possible to find the pattern over input dataset using data mining techniques. By analysing input trajectories captured from different types of users, it is possible to find out some similarities and patterns. For example, if analysis shows a cluster of trajectories arrive to a specific area, and then the behaviour of movement changes, e.g. trajectories are then going through a park and also the average speed of movement is reduced. Then it is possible to infer that users have parked their cars and then they walk to get their destination. This pattern may help us to find possible parking spaces without having any further knowledge of traffic and urban features. Then for a new user, needing navigational services, it is possible to use such information in navigational suggestion making process; e.g. suggest the user to park his/her car in that area an continue the route walking (if it takes less than 10 minutes walking to get the destination) since it might be hard to find a parking space very near the destination (if it was a parking lot, other could park their car but based on learnt pattern there isn't!).

Take an art gallery as another example. It is possible to track users using indoor positioning technologies, such as wireless networks, RFID network, CCTVs, etc. By analysing trajectories as input data which have been stored over quite long period of time, it is possible to find some points where many of users visited and also they have stay in those points for some minutes (possibly to see an artistic object). After finding/extracting such points, which can possibly show an artistic feature such as sculpture, painting, etc., then it would be possible to suggest new users to go to those points since there might be an interesting feature too or provide navigational instructions or find best paths which pass such points as well. Next section discusses nature of input data. Then in section 3 theoretical aspects are explained and section 4 describes implementation of proposed framework.

2   Tracking Data

In order to extract patterns and rules of movements, there is a need to analyse quite large sets of input trajectory data which is called training data. Training data is used for pattern recognition and rule learning purposes. However, there should be another set of input data which is called control data and is used to control how learnt rules and recognised pattern can fit into this set of data. Usually at the beginning we have a very large dataset and then it is randomly divided into two sets of training data and control data. After analyzing and finding patterns on the training data, inference system will employ the extracted patterns on control data to see up to what degree input control data and estimated results are similar. If very similar it is possible to say we have found the pattern and any new data can be analyze using that pattern. However, in cases of small or incomplete data sets, the selection of the test data may disturb its patterns. Thus, it is better to use only a small amount of data for the purpose of testing.

Larger data sets as input data may lead the inference system to a more realistic results since abnormalities will have less weight in pattern recognition process. There should be a framework in which a large volume anonymous tracking data sets can be stored. Since there is not a universal positioning technique which can get locations of users seamlessly indoors and outdoors with a quite acceptable degree of accuracy, the anonymous tracking data can be collected from different source such as GPS receivers in vehicles, GPS receivers embedded in mobile phones, RFIDs tags and readers, video cameras, Bluetooth networks . Then trajectories are stored in a spatio-temporal database. In this section there is a short review of some of widely used positioning and tracking methods and technologies and their issues such as coverage, accuracy, privacy, etc.

Many mobile phones and vehicles are now employing Global Positioning System (GPS) receivers to get their position. The GPS technology provides location and time information in all weather, anywhere on or near the Earth, where there is an unobstructed line of sight to four or more GPS satellites. GPS often can achieve sub-meter accuracy. In addition to acceptable level of spatial accuracy provided by GPS, finding position using GPS is for free. In addition to GPS, other satellite systems are being used or under development based on the same concepts. The Russian GLObal

NAvigation Satellite System (GLONASS), the European Union Galileo positioning system, the Chinese Compass navigation system and the Indian Regional Navigational Satellite System are some of the most important space-based positioning systems. These technologies and techniques cannot provide reliable position for indoors, however based on The European GNSS Agency (GSA) report in 2013 they will remain as the most widely used approach to get position of users in future.

One of the most important and widespread technique of indoor positioning uses a Wireless Local Area Network (WLAN). It is also one of the most accurate positioning techniques for positioning but its accuracy depends on the signal strength received by the mobile phones. WLAN is not specifically designed and deployed to provide positioning. However, measurements of strength of the signal transmitted by either an access point (AP) can be used to calculate the location of a mobile user. There are some factors which may have an effect on accuracy of Wi-Fi positioning techniques that are more related to the mobile device such as its speed, metal casing causing radio interference and the device scan interval (Kushki and Venetsanopoulos 2007). All in all it is one of the most widely used techniques to determine the position of users both indoors and outdoors but there is still a need for it to be improved to be able to use that as a tracking technology.

E-OTD is another positioning/tracking technique which uses the data received from surrounding cell base stations in a mobile network to measure the time difference it takes for the data to reach the mobile phone. This requires that the base station positions are known and that the data sent from different sites is synchronized. The positioning calculation can then either be done in the terminal or over the network. It can be used as a seamless positioning technique but E-OTD accuracy is not usually sufficient for indoor positioning.

Cell-ID, another network based technique, uses the identity of each cell (coverage area of a base station) to locate the user. It is often complemented with the Timing Advance (TA) information. TA is the measured time between the start of a radio frame and a data burst. This information is already built into the network and the accuracy is good when the cells are small. The accuracy is dependent on the cell size and varies from 10 meters (a micro cell in a building) to 500 m (in a large outdoors macro cell).

There are also video/camera-based techniques and Quick Response (QR) Codes (Basiri, 2014) to get the position of a user. The main concept is to provide images which are processed and exteriorly oriented to track or calculate the position of objects in those images. In order to get the position of a mobile user within an image, firstly, based on image processing techniques, users or their minimum bounding boxes are detected. In this step the cells, in which the user is located, are determined. There are some real-time tracking and navigation applications based on camera positioning systems (Jaenen, 2012), (Brown and Silva 2011). One of the most important reasons based on which many tracking and navigation have been implemented based on video/camera positioning systems, is having the infrastructure and devices available since safety and security issues in large public open spaces are often ensured using surveillance cameras (Basiri et al., 2012). It is possible to use the same infrastructure in order to get tracking data (Gall, 2010).

After getting tracking data, it is very important to make it anonymized using different anonymizers such as K-anonymity (Gruteser, 2003). Most existing anonymizers on tracking data adopt a K-anonymity. In order to do so, the location of the user got by a query and K to the anonymizer, which is a trusted third party (Kalnis , 2007) (Mokbel, 2006) in centralized systems or a peer in decentralized systems (Ghinita, 2007) . The anonymizer removes the ID of the user and cloaks the exact user location. Then anonymizer sends the location and query to the spatial database or location based services sever.

Now the anonymous tracking data can be stored in a spatio-temporal database. A spatiotemporal database captures spatial and temporal aspects of data and deals with geometry changing over time and/or location of objects moving over invariant geometry (Güting, 2005). Since tracking data has got both spatial and temporal aspects to be considered it is very important to be store in a spatio-temporal database.

3   Pattern Recognition and Knowledge Extraction

In order to find patterns over users' data and extract knowledge from it, we applied some of data mining techniques. Data mining approaches and methods enables us to find patterns over available data to extract some knowledge, such as interesting features in the city to be visited, and also find users which are not following the pattern and have unusual behaviour.

Data mining is a field at the intersection of computer science and statistics, and is the process that attempts to discover patterns in large data sets. It utilizes methods at the intersection of artificial intelligence, machine learning, statistics, and database systems. The overall goal of the data mining process is to extract information from a data set and transform it into an understandable structure for further use. Data mining involves six common classes of tasks:

- Anomaly detection (Outlier/change/deviation detection) – The identification of unusual data records, that might be potential errors or anomalies
- Association rule learning (Dependency modelling) – Searches for relationships between variables. Using association rule learning, one can determine which variable are more related to another. This is sometimes referred to as market basket analysis.
- Clustering – is the task of discovering groups and structures in the data that are in some way or another "similar", without using known structures in the data.
- Classification – is the task of generalizing known structure to apply to new data.
- Regression – Attempts to find a function that models the data with the least error.
- Summarization – providing a more compact representation of the data set, including visualization and report generation

Spatio-temporal data mining is some computational techniques for the analysis of large spatio-temporal databases. Both the temporal and spatial dimensions add substantial complexity to data mining tasks.

First of all, the spatial relations, both metric (such as distance) and non-metric (such as topology, direction, shape, etc.) and the temporal relations (such as before and after) are information bearing and therefore need to be considered in the data mining methods.

Secondly, some spatial and temporal relations are implicitly defined and they are not explicitly encoded in a database. These relations should be extracted from the data. However there is always a trade-off between pre-computing them before the actual mining process starts (eager approach) and computing them on-the-fly when they are actually needed (lazy approach). Moreover, despite much formalization of space and time relations available in spatio-temporal reasoning, the extraction of spatial/temporal relations implicitly defined in the data introduces some degree of certitude that may have a large impact on the results of the data mining process.

Thirdly, working at the level of stored data, that is, geometric representations (points, lines and regions) for spatial data or time stamps for temporal data, is often undesirable. Therefore, complex transformations are required to describe the units of analysis at higher conceptual levels, where human-interpretable properties and relations are expressed.

Fourthly, spatial resolution or temporal granularity can have direct impact on the strength of patterns that can be discovered in the datasets. Interesting patterns are more likely to be discovered at the lowest resolution/granularity level. On the other hand, large support is more likely to exist at higher levels.

In this research, firstly, unusual data records has been detected and stored in a dataset. Such data, which may be either unusual behaviour or just simple errors, are found according to some statistical rules and also predefined constrains. For examples non-moving users, whose calculated speeds of movements are less than 0.1 m/second for a quite long period of time, can be stored in the database as unusual behaviour or errors. This example shows how predefined rules can find anomalies or errors. In addition to predefined rules and constrains, it is possible to use statistical methods to find anomalies and errors. For example if in a large enough input data set, there is an area which ladled as public property but there is no (or only a few) trajectory passing through it, that may means that area has changed its landuse to a private property, or the area might not be safe enough for pedestrians to passed through, etc.

After finding anomalies and errors, classification and clustering methods can be applied to categorise data into different data sets. According spatial attributes of trajectories, such as location, shape, topological relationship between trajectories and streets networks and also trajectories' bounding rectangles, also according to calculated speed for each segment, and finally temporal characteristics, it is possible to categorise them into different categories. As it is explained in implementation section in more detail, such classes and categories may identify the intention of users. For example, it is possible to deduce from topological relationship between a trajectory and streets network, time of the day and week, and also average speed of movement of user, that user is going to/ getting back from his/her work place or he/she is just visiting the city as a tourist.

Although it is possible to deduce many useful pieces of information from trajectories using above mentioned methods, there is always a possibility to have miss-matching and inaccuracy. It is possible to categorise a trajectory into a business travel category while it is a touristic sightseeing. On the other hand it is possible not to categorise a trajectory at all since it is not matched with none of above mentioned rules. In order to have a better understanding of how accurate the recognised patterns and rules are, it is better to have dataset with large number of trajectories. Then divide input data randomly into two sets of training data and control data. After analyzing and finding patterns and rules (such as anomaly detection, error detection, clustering and classification, etc.) on the training data, the pattern should be employed on control data to see up to what degree input data and estimated results are similar. Control data is the data set for which the class and category and in general rules and patterns is already available. If very similar it is possible to say we have found the pattern and any new data can be analyze using that pattern. However, in cases of small or incomplete data sets, the selection of the test data may disturb its patterns. Thus, it is better to use only a small amount of data for the purpose of testing. This mode is called leave-one-out method.

The final step of knowledge discovery from data is to verify that the patterns produced by the data mining algorithms occur in the wider data set. Not all patterns found by the data mining algorithms are necessarily valid. It is common for the data mining algorithms to find patterns in the training set which are not present in the general data set. This is called overfitting. To overcome this, the evaluation uses a test set of data on which the data mining algorithm was not trained. The learned patterns are applied to this test set and the resulting output is compared to the desired output. The accuracy of the patterns can then be measured from how many trajectories are correctly classified. A number of statistical methods may be used to evaluate the algorithm, such as ROC curves. If the learned patterns do not meet the desired standards, then it is necessary to re-evaluate and change the pre-processing and data mining steps. If the learned patterns do meet the desired standards, then the final step is to interpret the learned patterns and apply them for new input data or turn them into knowledge.

In order to test and implement above mentioned methods, an ArcGIS add-in has been developed to represent, store, analyze and extract patterns of pedestrians' trajectories based on spatio-temporal data mining methods. Next section shows the implementation of such system.

4 Implementation

One of the widely used applications of ambient services is smart rooms and in general indoor ambient services. In this regard, a conference room has been considered to install cameras. In the room there are 5 ceiling mounted cameras whose three dimensional coordinate are measured accurately. These cameras are allocated in the way that all points and corners of the room are covered. Figure 1 shows one of the cameras' coverage areas.

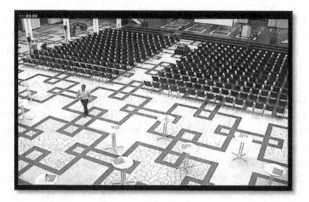

Fig. 1. One of mounted cameras' coverage area

In camera installation phase, the final goal was maximizing the room coverage and also having more overlapping area, which is covered by more than one camera. Having overlapping areas is very important since all analysis is doing over anonymous data and users are identified using a random number (Object ID) assigned. So in order to follow the user roaming from one camera's coverage area to another's, it is very important to have an area which is covered by two cameras to find the corresponding absolute position of a user in overlapping area.

After installation phase, we need to store location and time for every person detected by image processing module. Then detected user's location and corresponding time are stored in a document. Based on such document it is possible to visualise trajectories. In order to do so, an ArcGIS add-in was developed to visualise, analyse and mine the data. As it is illustrated in figure 2, a trajectory analyser douckable window is added to ArcMap. Its first tab Creates a feature class by reading the input XML file. It also can add two columns to the created feature class which calculate distance between every point and the point next to it (length of each segment) and speed of movement of user passing that segment.

The very last button (Create Distance Graph) in Creation tab generates graphs and diagram to visualise some statistical information, such as average speed of movement. Figure 3 shows a graph for speed of movement. As it can be seen in figure 3, minimum, maximum, average and standard deviation are calculated and shown. This graph may help to rule out some of errors and also exclude redundant data. For example if the speed of movement of the user between two points is much more than normal speed of a pedestrian (1.5 m/s), then it is possible to take that segment as error and store it in another dataset since a pedestrian cannot move with such high speed (e.g. 20 m/s). Also it is possible to find points where user has had no movement (speed is near zero) and replace them with one single point with a description of time interval during which user's speed is zero.

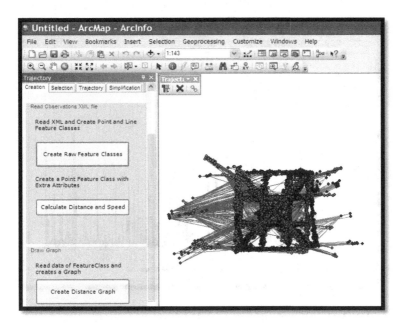

Fig. 2. Trajectory analyser duckable window (developed ArcGIS add-in)

Using such statistical analysis, it is possible to exclude some of errors, which may occur because of image processing and feature matching challenges (such as detecting user's reflection instead of user, etc.), and also manage the data in a better way.

Another challenge in this project is redundant data. It is possible to have more than one trajectory for each user at the same time interval, with different shapes and sizes, since users can be viewed by different cameras synchronously. If two or more cameras detect a person at the same time in their overlapping area, then we will have two or more trajectories stored for that person, as it is shown in figure 4. The simplest policy is to consider all trajectories ignoring this fact that some of them may not show different users' movements since they are redundant data for the same user moving in a same period of time. This policy may lead in having higher weight for trajectories located in overlapping areas. In order to handle this, we need to find only one trajectory which shows the user's movement as detail as possible. This become more complex where there is no link between user and trajectory. Because of privacy issues, there is no link between users and their trajectory. That means, it is not single user ID assigned to user and based on which the corresponding trajectories (got from different cameras) can be identified. In this regard trajectories belong to a single movement made by single user should be identified, and then they should be transformed into one polyline to show user's movement over a period of time.

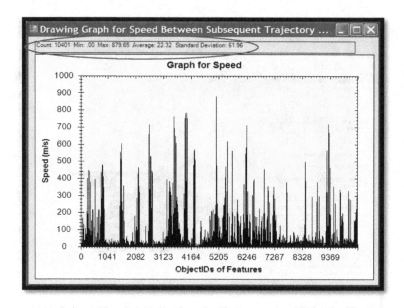

Fig. 3. Graph for speed of movement between every two sequent points

In order to find the trajectories to be matched and aggregated into one trajectory, we need to find correlated vertices, which represent one object captured by different cameras, and replace them with one vertex. In order to do this, the simplest policy can be finding vertices which are spatially and temporally near to each other. Because the vertices which are recorder spatially and temporally near to each other, are more likely to represent one user and the small differences between them can be because of camera synchronisation drift, calibration and instrumental errors. However this approach may be the simplest approach to find trajectories representing a single user's movement, it has got some issues. First of all, and the most important one, is mismatching issue. It is possible to find many trajectories which satisfy the conditions (captured within quite the same temporal interval and also spatially near) while some of these trajectories belong to another user and they represent another user's movements. In addition, finding the best time interval and also spatial buffering radios in which trajectories are considered to be matched is quite tricky. Figure 4 shows selection tab in the developed ArcGIS-Add in which allows selecting vertices captured spatially and temporally close to each other. Then a column will be added to the attribute table to show which segments should be matched. As it is shown in figure 4, spatial and temporal thresholds should be given or previously set, for example in figure 4 vertices whose distance from surrounding vertices were less than 2 centimetres and also have been captured in a 100 milliseconds time interval are selected. Finding the best temporal and spatial threshold depends on the applications, movements' characteristics, experts' comments, equipments' configurations and settings, etc.

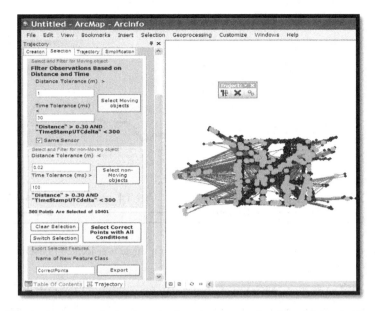

Fig. 4. Finding matched trajectories based on temporal and spatial thresholds

This might be the simplest approach, however because of mismatching and also being so experience and application dependant, makes it unreliable. In this regard, this paper proposes using data mining techniques to find matching segment. This means, introducing fixed grid windows in which segments whose patterns are quite same considered to be matched. In contrast with previous approach which only considers spatial and temporal proximity to find matching segments, this approach considers pattern of movements. Figure 5 shows how two trajectories are found as matching trajectories using this approach. The green trajectory is captured by camera1 and the red trajectory is captured by camera2. As cameras locations, confutations and settings are different they may capture different trajectories with different numbers of vertices, as it shown in figure 5. However the proposed approach considers pattern of movement within a temporal and spatial window. For example the green and the red trajectories in figure 5 can be found as matching trajectories since the general trends and pattern of movement in the predefined window, including four time intervals and a spatial bounding box, is quite same. Those very small noises (which are differences of the trajectories from the general trend) are ignorable according to the predefined settings.

Both approaches, finding near vertices based on a spatial and temporal threshold and also the pattern-based approach, have been implemented in the developed ArcGIS add in to make choices and also make easier to understand what would be results of both approaches on the same dataset. Figure 6 shows the matching tab, called simplification since it matches segments to reduce redundant data.

Now the data has been pre-analysed and it is possible to define some rules based on which Points of Interests (PoIs) can be extracted. This may be very helpful in tourist guidance or any indoor navigation services. Since cameras (CCTVs) are usually available indoors, especially in galleries and museums. In order to find PoIs, it is

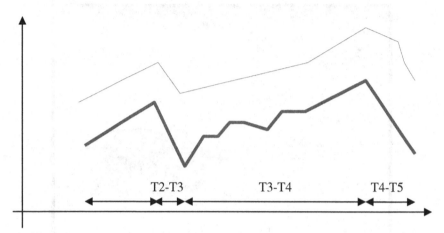

Fig. 5. Two trajectories generated by two cameras for a single user located in overlapping area of cameras' coverage

possible to select vertices where many users stay for a period of time (probably to see an interesting feature, such as sculpture, painting, etc.). In order to find such points, easily one can use selection tab and select non-moving users. If the number of users who stays in this point (or very close to this point) was more than a threshold, then we can export that point (or area) as a new feature class, called Point of Interests. Again spatial and temporal threshold and also number of users are selected based on the application and experts comments. These PoIs can be used to make navigational suggestions for new comers.

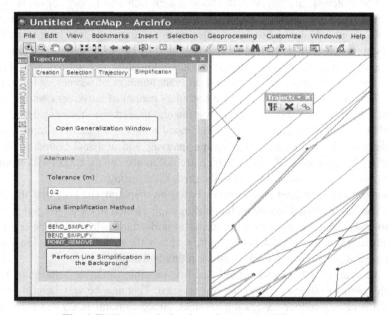

Fig. 6. Finding matched trajectories tab, simplification tab

5   Conclusion

The future LBS services will be more anticipatory of users needs. Users look for a mobile service/application which can monitor or understand their current situations, predict their needs and demands, provide the most appropriate set of services in most comfortable way. This paper aimed at detecting and extracting existing patterns over pedestrian trajectories The trajectories of anonymised tracked users in surveillance of cameras are considered as sequences of time-discrete observations of users' positions. These trajectories are processed to find patterns over pedestrians' movements. For example, it is possible to find attractive/important points or features in a gallery, based on users' trajectories; i.e. where there is no movement over a period of time for many of tourists. These points can be considered as point of interests to be applied in giving navigational instructions or suggestions to others; i.e. it is possible to recommend a new user to go to that point since many people have visited that. In order to extract patterns over spatio-temporal data of pedestrians including time and position of users, an ArcGIS add-in has been developed to represent, store, analyze and extract patterns of pedestrians' trajectories based on spatio-temporal data mining methods.

Acknowledgement. Research presented in this paper was funded by a Strategic Research Cluster grant (07/SRC/I1168) by Science Foundation Ireland under the National Development Plan. The authors gratefully acknowledge this support.

This work was financially supported by EU FP7 Marie Curie Initial Training Network MULTI-POS (Multi-technology Positioning Professionals) under grant nr. 316528.

References

1. Asahara, A., Maruyama, K., Sato, A., Seto, K.: Pedestrian-movement prediction based on mixed Markov-chain model. In: Proceedings of the 19th ACM SIGSPATIAL International Conference on Advances in Geographic Information Systems (GIS 2011), pp. 25–33. ACM, New York (2011)
2. Ashbrook, D., Starner, T.: Using GPS to learn significant locations and predict movement across multiple users. Personal & Ubiquitous Computing 7, 275–286 (2003)
3. Baiget, P., Sommerlade, E., Reid, I., Gonzàlez, J.: Finding prototypes to estimate trajectory development in outdoor scenarios. In: Proceedings of the First International Workshop on Tracking Humans for the Evaluation of their Motion in Image Sequences, THEMIS 2008 (2008)
4. Basiri, A., Amirian, P.: A Winstanley, The Use of Quick Response (QR) Codes in Landmark-Based Pedestrian Navigation. International Journal of Navigation and Observation (2014)
5. Basiri, A., Amirian, P., Winstanley, A.C., Kuntzsch, C., Sester, M.: Uncertainty handling in navigation services using rough and fuzzy set theory. In: Proceedings of the Third ACM SIGSPATIAL International Workshop on Querying and Mining Uncertain Spatio-Temporal Data, pp. 38–41 (2012)

6. Browarek, S.: High resolution, Low cost, Privacy preserving Human motion tracking System via passive thermal sensing, Master Thesis. Dept. Electrical Engineering and Computer Science, MIT (2010)
7. Elhayek, C., Stoll, N., Hasler, K.I., Kim, H.-P., Seidel, C.: Theobalt, Spatio-temporal Motion Tracking with Unsynchronized Cameras. MPI Informatik (2012)
8. Feuerhake, U.: Prediction of Individual's Movement based on Interesting Places. ISPRS Annals of Photogrammetry, Remote Sensing and Spatial Information Sciences I-2, 31–36 (2012)
9. Fidaleo, D., Nguyen, H., Trivedi, M.: The networked sensor tapestry: A privacy enhanced software architecture for interactive analysis and sensor networks. In: ACM 2nd International Workshop on Video Surveillance and Sensor Networks (2004)
10. Gall, J., Rosenhahn, B., Brox, T., Seidel, H.-P.: Optimization and filtering for human motion capture – a multi-layer framework. IJCV 87, 75–92 (2010)
11. Ghinita, G., Kalnis, P., Skiadopoulos, S.: Privé: Anonymous location-based queries in distributed mobile systems. In: WWW 2007 (2007)
12. Gruteser, M., Grunwald, D.: Anonymous Usage of Location-Based Services through Spatial and Temporal Cloaking. In: MobiSys (2007)
13. Güting, R.H., Schneider, M.: Moving Objects Databases. Academic Press (2005) ISBN 978-0-12-088799-6
14. Jaenen, U., Feuerhake, U., Klinger, T., Muhle, D., Haehner, J., Sester, M., Heipke, C.: QTrajectories: Improving the Quality of Object Tracking using Self-Organizing Camera Networks. ISPRS Annals of Photogrammetry, Remote Sensing and Spatial Information Sciences I-4, 269–274 (2012)
15. Kalnis, P., Ghinita, G., Mouratidis, K., Papadias, D.: Preventing Location-Based Identity Inference in Anonymous Spatial Queries. In: IEEE TKDE 2007 (2007)
16. Kuntzsch, A.: A Framework for On-line Detection of Custom Group Movement Patterns. In: Krisp, J.M. (ed.) Progress in Location-Based Services. Lecture Notes in Geoinformation and Cartography, pp. 91–107. Springer, Heidelberg (2013)
17. Kushki, A., Plataniotis, K.N.: Kernel-Based Positioning in Wireless Local Area Networks. IEEE Transactions on Mobile Computing 6(6) (2007)
18. Makris, Ellis, T.: Spatial and probabilistic modeling of pedestrian behavior. In: Proc. Brit. Machine Vision Conf., Cardiff, U.K, vol. 2, pp. 557–566 (2002)
19. Mokbel, M.F., Chow, C.-Y., Aref, W.G.: The new casper: Query Processing for Location Services without Compromising Privacy. In: VLDB 2006 (2006)
20. Monreale, A., Pinelli, F., Trasarti, R., Giannotti, F.: WhereNext: A Location Predictor on Trajectory Pattern Mining. In: Proceedings of the 15th ACM SIGKDD International Conference on Knowledge Discovery and Data Mining, 2009, pp. 637–646 (2009)
21. Yavas, G., Katsaros, D., Ulusoy, O., Manolopoulos, Y.: A data mining approach for location prediction in mobile environments. Data and Knowledge Engineering 54(2), 121–146 (2005)
22. Zhang, L., Dalyot, S., Sester, M.: Travel-Mode Classification for Optimizing Vehicular Travel Route Planning. In: Progress in Location-Based Services, pp. 277–295. Springer, Berlin (2013)

Prospecting the Evolution of a Winegrowing Region through MAS Modelling

Giovanni Fusco[1,2] and Matteo Caglioni[1,2]

[1] Université de Nice Sophia Antipolis, UFR Espaces et Cultures
[2] Centre National de la Recherche Scientifique, UMR 7300 ESPACE
{giovanni.fusco,matteo.caglioni}@unice.fr

Abstract. The Bandol winegrowing region (South-Eastern France) is presently confronted with growing urban pressure. This paper prospects the evolution of winegrowing in the region through scenario simulation in a Multi-Agent System model. The model is confronted with the difficulties of combining real data with expert knowledge. Entry data are the cadastral map, land-use and land-use plans, economic data on winegrowers and several external parameters. Spatial strategic foresight is exploratory and does not determine a certain future for the study area. The starting point is the simulation of a trend scenario. Alternative scenarios are later elaborated by modifying coherently the parameter set of the geosimulations. Their goal is to understand the possible consequences of departures from trend dynamics on the study area and its winegrowing economy. The paper shows the interest of using a MAS model, developed from real-world data, in order to simulate scenarios for spatial strategic foresight. Under the constraint of a set of hypotheses, the MAS model produces spatial forms for possible futures of the Bandol region and shows the overall resilience of its winegrowing economy.

Keywords: Multi-Agent Systems, Geosimulation, Spatial Strategic Foresight, Winegrowing, Urban Pressure, Land-Use.

1 Introduction

French rural landscapes are traditionally marked by the presence of winegrowing. Urban sprawl around French cities is nowadays exerting growing pressure on the vineyards in several regions. This is the case of the winegrowing region of Bandol, on the Mediterranean coast. As many French wines, Bandol is a controlled origin label, implying wine production within a precise geographical perimeter, from certain grape varieties and using determined production procedures. Stretching over almost 10 000 ha, the Bandol perimeter was established in 1941 and never modified thereafter. Within this perimeter, competition between wine-growing and land development for tourism and the residential sector was already observed in the sixties [6]. The main threat identified back then was the impossibility of expanding winegrowing as natural land was steadily being developed.

Local authorities and winegrowers are increasingly aware of the risks associated with uncontrolled continuation of present trends. But interactions among winegrowing and

B. Murgante et al. (Eds.): ICCSA 2014, Part IV, LNCS 8582, pp. 199–214, 2014.
© Springer International Publishing Switzerland 2014

new urbanization are at present even more complicated and encompass several factors such as land ownership, social networks among winegrowers, entrepreneurial demography, production constraints, capital transfers from land development to wine-growing and road accessibility. Prospecting the evolution of winegrowing in the Bandol region in face of urban pressure is thus a challenging task, which can greatly benefit from geo-simulation techniques. Expert knowledge on interaction rules within the winegrowing economy and availability of reliable databases made the development of a multi-agent system for the Bandol region an interesting option.

Multi-Agent Systems (MAS) are already known to be a powerful modelling technique of the interactions among agents in geographic space, allowing the simulation of emerging structures [1], [4, 5], [7]. These structures interact with the agents, modifying their behaviours and decisions. At the same time, parameters shaping agents' interaction rules and external constraints can be modified to simulate different scenarios of the evolution of the study area. Confronted with the political goals and strategic options of local decision-makers, these scenarios could be used to inform a strategic spatial foresight of the winegrowing region [2], [10].

The MAS model of the Bandol region was developed on the NetLogo platform [11]. Its entry data are the cadastral map, land-use and land-use plans, economic data on winegrowers and several external data (the price of wine, housing demand, developing land price, etc.). The Bandol perimeter is another main constraint for winegrowing in the study area: only non-Bandol vineyards are allowed outside the perimeter. Thanks to these data, and a few necessary approximations, the initial state of the Bandol model is known for the year 2010. A few key parameters are entered by the modeller as hypotheses. It is thus possible to simulate the interaction among the agents of the Bandol region within a time horizon of 40 years.

Spatial strategic foresight is exploratory and does not determine a certain future for the Bandol winegrowing system. The starting point is the simulation of a trend scenario, where urban pressure, winegrowing economic parameters and planning constraints take the most plausible values according to expert knowledge (the risk taken here is to extrapolate these values over 40 year simulations). Alternative scenarios, partially departing from the trend, are later elaborated by modifying coherently the parameter set of the geosimulations. Their goal is to understand the possible consequences of departures from trend dynamics on the study area, in general, and on its winegrowing economy, in particular. During the simulations, the MAS model produces charts for the evolution of key parameters (land-use, economic structure and spatial distribution of winegrowing agents, landscape ecology indicators). It is thus possible do obtain a fairly clear description of possible futures for the winegrowing landscape of Bandol. Knowledge of these possible (and more or less plausible) futures will be precious to local decision-makers (public authorities, winegrowers, land-developers) in order to identify strategies to attain a desired evolution of the territorial system.

The MAS model and the scenarios are clearly specific to the territorial functioning identified in the Bandol region. The modelling and prospective approach can nevertheless be applied to other geographic contexts, whenever the goal is to simulate, with a parcel grain, the evolution of the interaction between the winegrowing economy and land development. Modellers will have to specify the interaction rules and the parameter values which are typical of their study area. By introducing more important structural changes, the approach could be used to model other agricultural regions under urban pressure. The next section will present the main characteristics of the MAS model for the winegrowing region of Bandol. Section 3 is a description of the study area and of its initial state in 2010. Section 4 will show the main geosimulation results, both for the trend scenario and for alternative scenarios. A last section will give an overall view of geosimulation results and perspectives of further research.

2   The MAS Model

The model developed for the geosimulation of the Bandol region has peculiar characteristics which are worth mentioning. Overall, it is a multi-agent system confronted with the difficulties of describing the functioning of a given study area with a certain realism. Real world data are thus used to initialize the model and to calibrate the behavioural functions of its agents. Realistic description is focused on the dynamics of winegrowing, allowing coarser (or even exogenous) modelling of other phenomena (non-winegrowing agriculture, land development, land-use planning). A one-to-one description of the real world had clearly to be avoided both for the modelling effort required and for computational reasons.

The semantics of space reflect this choice (Fig. 1). Land-use is described at 25 m resolution through twelve classes. Seven land-use classes can evolve during the simulation: Bandol vineyards, other vineyards, other agricultural use, agricultural wasteland, natural space, discontinuous urban fabric (suburban developments) and scattered individual housing. Five additional land-uses are considered fixed: continuous urban fabric, main roads, highways, quarries, recreational facilities and water surfaces. Some land-uses exert land-development pressure on the neighbouring space (within a certain distance and respecting property boundaries, but with specificities that depend on land ownership): urban fabric (of all kinds), main roads (but not highways) and scattered individual housing. These are morphogenetic land-uses for future urban growth. The latter can take two different forms: suburban subdivisions by land-developers or scattered individual housing by households and farmers (including winegrowers). Land-use plans also shape future urban growth, dividing land parcels in three groups: urbanisable, non-urbanisable and protected natural areas (where even new agricultural activities are forbidden). The last two categories protect parcels from urban pressure. Plans can evolve, during simulations, under the pressure of land-developers and winegrowers, reflecting their relative strength in every municipality.

As far as agents are concerned, the model has a four-tiered structure:

1. The first tier is the one of farmer agents: these can be non-winegrowing farmers or one of three different kinds of Bandol winegrowers (owners of an independent winemaking domain, big or small cooperative winegrowers, cultivating more or less than 2 ha of vineyards, respectively). Age, family structure, land ownership, land tenure and eventual sharecropping agreements are known for all farmer agents. Bandol winegrowers are also characterised by their amount of Bandol

cultivation rights delivered by the wine Syndicate of Bandol (they can destroy the vineyard in a given parcel and plant it somewhere else, but within the limits of their cultivation rights).

2. Land parcels (plots) are also MAS agents, characterised by several attributes: surface, owner, exploiting agent (which could not be the owner), Bandol vineyards surface, list of neighbouring parcels, level of urban pressure, land-plan status, etc. A single parcel can have different land-uses. Urban parcels are not modelled.

3. 25 m squared pixels are the agents (NetLogo "patches") through which land-use is modelled. They make up both parcels and landscape ecology patches and have several attributes: parcel membership, exploitation membership, land-use, urban pressure, etc. Pixels belonging to urban parcels are assigned to a fictitious parcel.

4. Landscape ecology patches (not to be confused with NetLogo patches) are groups of contiguous pixels with the same land-use and are the last kind of agents in the model. They are mainly used to calculate landscape ecology metrics, which are not dealt with within this paper.

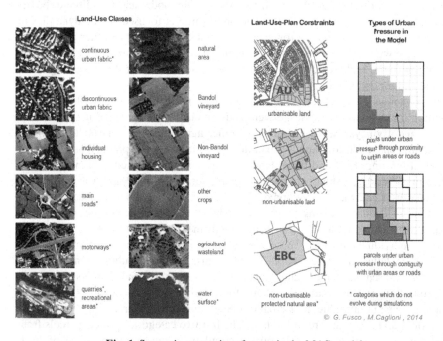

Fig. 1. Semantic categories of space in the MAS model

Several global variables also influence the agents and are exogenous to the model: the price of the Bandol wine (which is increasingly determined on international markets), the price of land for urban development, the quantity and the typology of housing demand (individual housing / subdivisions), a few key parameters of the winegrowing economy (overhead costs, exploitation costs, etc.). Different utility functions characterise the four kinds of farmer agents and several social relations structure their interaction (parentage, contract farming, sharecropping agreements).

Though differently parameterised, the four utility functions share the following overall structure:

$$\Delta K = \Sigma_i \, [\alpha_i S_i - f\,(S_i, d_{ci})] - f\,(K_b, t) - f(c)$$

Where ΔK is the annual capital variation for the agent, $\alpha_i S_i$ is the net value of agricultural production of parcel i (α is the net productivity of 1 ha of agricultural land, S is the exploited surface), $f\,(S_i, d_{ci})$ is the logistic cost of the exploitation of parcel i (d_{ci} being the distance of i from the logistic centre), $f\,(K_b, t)$ is the cost of borrowed capital (t being the interest rate) and $f(c)$ is the fix cost of living for every category of farmer. The net value of agricultural production depends from the cultivated produce (Bandol or non-Bandol grapes, other crops) and from the winegrowing category (winemakers produce wine and not just grapes and keep more profitable parts of the value chain).

The interaction among agents is governed by rules which are implemented as procedures in NetLogo. The model core is thus composed of more than twenty procedures which call each other during the simulations and can be broadly grouped in three categories: general procedures, procedures of the urban growth and procedures of the winegrowing economy. Auxiliary procedures (initialisation, file reading/writing, debugging) add to these three categories.

A few procedures are worth particular attention. Land sale, for example, takes place in different contexts. Land developers can bid for all land parcels under urban pressure, individual households only on land that is not in the hands of independent domains. At the same time, suburban developments need a minimum required surface sold from the same land-owner and is hindered from extreme ownership fragmentation. Extra-profits generated from land sale to urban development can be reinvested in winegrowing whenever economic and regulatory constraints allow it. Land sale among farmer agents happen whenever an agent is in need of capital or retires (without children taking over), unless he withholds land under urban pressure in the expectation of future sale to land developers or individual households. When agricultural land is sold, a Vickrey auction system [9] is modelled within the MAS, with the highest bidder paying the second highest bid. Particular attention is also given to the modelling of generational passage in the winegrowing business, with children being more or less inclined to taking over their parents' activity, according to the overall economic profitability of Bandol winegrowing.

Overall, the MAS model of the Bandol region contains 363 000 pixels, 14 700 parcel agents and 4 000 farmer agents. Simulations are thus relatively computing intensive. NetLogo being unable to manage multithread calculus, they also become time consuming. A 40 year simulation of our study area takes around 20 hours on a PC with Pentium i5 CPU at 3.30 GHz.

3   The Winegrowing Region of Bandol in 2010

The Bandol wine-growing region is situated on the Mediterranean coast of France, near the city of Toulon and stretches over eight municipalities (Fig. 2): Saint-Cyr-sur-Mer, La Cadière-d'Azur, Le Castellet, Le Beausset, Bandol, Sanary-sur-Mer, Ollioules and Evenos. The MAS model was implemented only on the first four municipalities, making up for around 95% of Bandol vineyards in 2010. Land-use data entered into the model are extremely precise as long as vineyards are concerned. These are derived from the local vineyard cadastre of 2010 (distinguishing Bandol and non-Bandol vineyards), from airborne imagery and from terrain surveys, and have a metric precision. Data are coarser for other land-uses, based on CORINE Land Cover 2006. Modelled through 25 m pixels (Fig. 3), the study area contains 1650 ha of Bandol vineyards and 990 ha of other vineyards in 2010. Vineyards often adjoin other agricultural land (1410 ha) and urban areas. The main natural areas (forest or garigue, making up 6800 ha) are on the outskirt of the study area (the big recreational area in the north is a car race track). Two big Bandol vineyards are in the north of the Cadière-d'Azur municipality. The highway is the main transportation infrastructure, linking the urban centres of the study area to nearby Toulon (to the south-east) and Marseilles (to the north-west).

The local vineyards cadastre and the numeric cadastre (BD MAJIC III) are also necessary to reconstruct land ownership and land tenure within the study area. Fig. 4 shows agricultural and forest land ownership among the model agents. In 2010, 53 independent winemaking domains and 120 big cooperative winegrowers already control a hefty swath of agricultural land within the Bandol perimeter (3300 parcels out of 7400). 3700 parcels are the ownership of other agricultural or forestry agents (which are more than 3500 in the study area). 129 small cooperative winegrowers own almost 400 small to medium-sized parcels within the Bandol perimeter.

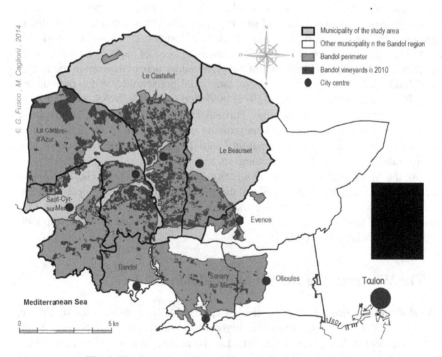

Fig. 2. The Bandol region and the study area of the MAS model

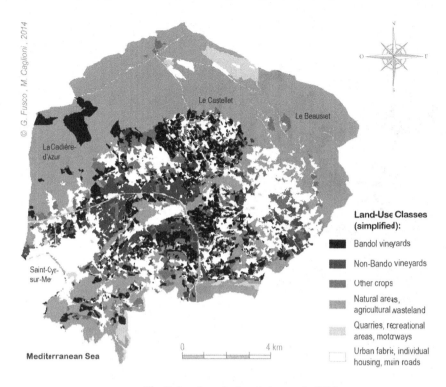

Fig. 3. Land-use in the study area in 2010

Winegrowers and winemakers also own land out of the Bandol perimeter (1250 parcels out of 7300), namely in the central part of the study area (often non-Bandol vineyards) or in the wooded areas in the north. Planning constraints are finally known from the municipal land-use plans into force. According to these plans, areas of potential urban growth are still considerable in 2010. If 2933 ha are already urbanised (including quarries and recreational areas), urbanisable agricultural or natural land sums to 1042 ha. Urbanisable land is scarcer along the coast: only 116 ha can host future urban growth in Saint-Cyr-sur-Mer, where coastal natural land is protected by the French Coastline Law. Most vineyards and other agricultural land are presently protected from urbanisation, but by-laws often allow farmers to build their own houses on their land. Forests in the north (but not around the car race track of Le Castellet) and along the coastline are protected even from farmers' individual housing developments. Strong legal constraints prevent municipalities from transforming these areas in urbanisable land. For the rest, land-use plans can evolve during the simulations. The MAS model makes the opening of new agricultural and natural areas to urbanisation depend on two factors in every municipality: the residual urbanisation potential (the smaller the potential, the more pressure will be exerted by land developers to open new areas to urbanisation) and the relative force of the winegrowing economy.

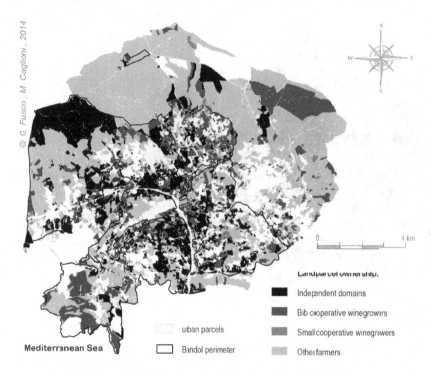

Land parcel ownership:
- Independent domains
- Big cooperative winegrowers
- urban parcels
- Small cooperative winegrowers
- Bandol perimeter
- Other farmers

Mediterranean Sea

Fig. 4. Land-ownership in the study area in 2010

4   Simulated Scenarios of Development over 40 Years

The trend scenario is characterised by a strong growth of urbanisation, as it was already observed in the four decades preceding the 2009-2013 crisis. The demand of individual housing developments is higher than the one for housing in subdivisions or in small collective buildings (8.75 and 3.75 ha/year, respectively). The final retail price of a litre of Bandol wine stays at around 10 Euros (as observed in the last decades) and agricultural activity, except for winegrowing, is particularly weak. Land-use plans keep on evolving under the pressure of land developers, but still protect remarkable natural areas. This context does not penalize winegrowing and the generational taking over process is almost always successful among the Bandol winegrowers.

Over the 40 year horizon (land-uses are shown in Fig. 5), the trend scenario sees progressive leapfrogging of urban growth in agricultural and natural land, often in the form of individual housing. Large residential subdivisions appear in the north of the municipality of Le Castellet, along main roads (concerned parcels differ among simulations). Vineyards don't reduce their surface: Bandol-vineyards are quantitatively stable; non-Bandol vineyards grow 50 ha. But vineyards are not spatially stable: they tend to move away from urbanised areas and colonise agricultural and natural land further away. Other crops thus pass from 1410 to 1140 ha; natural land loses 275 ha.

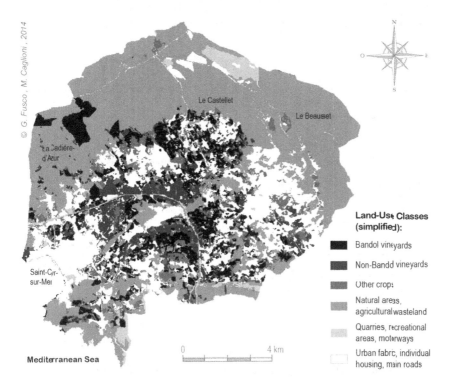

Fig. 5. Land-use in 2050 according to the trend scenario

As far as Bandol winegrowing agents are concerned (Fig. 6), small cooperative winegrowers increase significantly their number, the surface of their Bandol vineyards and the surface of the parcels owned. At the same time, their share in the Bandol vineyards total surface remains modest, passing from 5% to 9%. Small and big cooperative winegrowers benefit greatly from financial transfers from individual housing developments, whereas independent domains sell their parcels only for larger subdivision developments. The plantation of new vineyards is nevertheless blocked by the presence of large, protected wooded parcels in the north of the municipalities of Beausset and Castellet. Continual capital accumulation from big cooperative winegrowers allows some of them to build their own winemaking equipment and to evolve into independent domains. This explains why their number slowly decreases, whereas domains pass from 53 in 2010 to 70 in 2050. Within this context, the eventual splitting of domains due to property transfer between generations is quickly resorbed: availability of funds (thanks also to property value gains when land is sold to developers) lets heirs without winemaking facilities to buy them quickly. Bandol domains can thus continue their strategy of expanding vineyards, although the total Bandol vineyards surface stays almost constant (due to constraints in the availability of cultivation rights). The evolution of prices in land transactions shows the bigger value of parcels within the Bandol perimeter, even if a temporary price decline is registered halfway through the 40-year simulation.

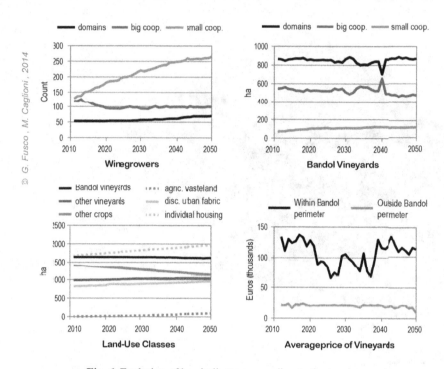

Fig. 6. Evolution of key indicators according to the trend scenario

Thanks to the MAS model, several alternative scenarios could be simulated and analysed for the Bandol wine-growing region. Five of them were finally validated with experts of the Bandol economy. They include a slowing down of urbanisation, increased urbanisation in the form of individual developments, increased urbanisation in the form of subdivisions, a wine price crisis and a stable increase of wine prices with weakening of the protection of wooded areas. Maps of simulated land-use and land-ownership for the six scenarios (including the trend scenario) have been published in a thematic atlas of wine- and olive-growing in Mediterranean France [3]. In this paper, we will focus our attention on land-use changes for two contrasting scenarios: the one of increased land development through new subdivisions (Fig. 7 and 8) and the one of declining wine prices leading to a wine-growing crisis (Fig. 9 and 10).

In the scenario of increased land development through new subdivisions, urban pressure increases on big parcels of agricultural land and on groups of contiguous small parcels from the same owner, whenever they are urbanisable and connected to the main road network or close to existing urban areas. The increased power of land developers is a key ingredient for this scenario. Despite the greater compactness of the developments, the total demand of urbanisable land increases compared to the trend scenario and sums to 14,5 ha/year, three quarters of which in the form of new subdivisions. Attractiveness and profitability of other agricultural activities is lower than in the trend scenario, farmers avoiding new plantations in areas under urban pressure in the expectation of future land sale to developers.

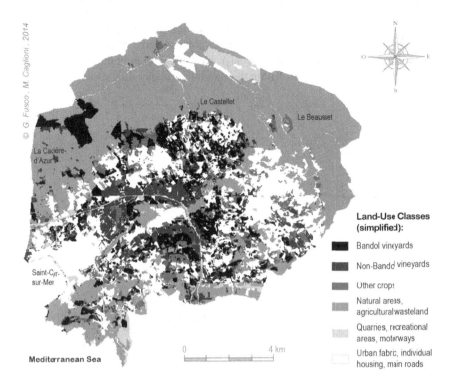

Fig. 7. Land-use in 2050 according to the scenario of increased land development

Of course, the MAS simulation translates these modelling hypotheses in a strong growth of discontinuous urban fabric, whose surface overtakes at the end of the simulation period both agricultural land and non-Bandol vineyards. Large residential subdivisions appear in the north of the municipality of Castellet (on parcels which were not always owned by wine-growers) as well as small- to medium-sized subdivisions in the centre of the study area. Here land was previously owned by independent domains or big cooperative wine-growers and winegrowing is thus financially boosted by capital transfers from land developers. New urbanization takes over agricultural land, with a correlative creation of agricultural wasteland of a speculative kind (120 ha) and a strong contraction of other crops (which lose 315 ha). On the contrary, winegrowing can adapt to growing urban pressure through a progressive migration towards more peripheral areas, whenever they are not protected by environmental constraints (a spatial dynamics already detectable in the trend scenario). Bandol vineyards are thus quantitatively (but not spatially) stable, other vineyards grow 70 ha.

Agricultural land prices within the Bandol perimeter stay high all along the simulation. The scenario is particularly favourable to independent domains, which grow from 53 to 71. The surface of their Bandol vineyards grows 42 ha (to a total of 906 ha) and most of the increase of non-Bandol vineyards takes place within their parcels. Domains extend their control (through ownership or contract farming) over large swaths of natural areas in the north of the municipality of Castellet, as well as on the coastal areas south of Saint-Cyr-sur-Mer. Big cooperative winegrowers diminish in number and in controlled surface of Bandol vineyards, as some of them evolve into independent domains. Small cooperative winegrowers increase in number and surface of vineyards, but their share of land within the Bandol perimeter diminishes.

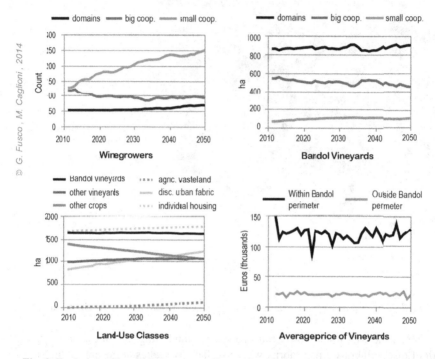

Fig. 8. Evolution of key indicators according to the scenario of increased land development

By changing a few external parameters, fairly different land-use structures emerge from the MAS model simulations. We thus kept the same annual demand of housing as the trend scenario, but we lastingly reduced by a factor two the retail price of Bandol wine and the attractiveness of the winegrowing business for the new generations, and increased, at the same time, the attractiveness and the profitability of other agricultural activities. This scenario results in a marked crisis of the winegrowing activity in the Bandol region. Simulations show a completely reversal of roles between vineyards and other crops, when compared to the trend scenario. Non winegrowing agricultural activities stabilize their surfaces (around 1400 ha), although they experience a spatial migration towards more peripheral areas. Adaptive capacity and resistance to new urbanisation from agriculture results in a contraction of around 310 ha from natural areas (which is higher than in the trend scenario).

Agricultural resistance and new urbanisation take place at the expense of vineyards, as well. Bandol vineyards lose 130 ha, non-Bandol vineyards lose around 20 ha. Financial transfers from land development benefit in the first place to other agricultural activities. Consequently, the average price of land sales within the Bandol perimeters tends to diminish as winegrowers are less able to bid for land. Indeed, the number of winegrowers does not diminish as much as Bandol vineyards surface.

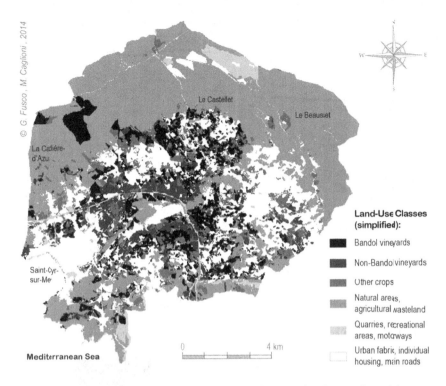

Fig. 9. Land-use in 2050 according to the scenario winegrowing crisis

Small cooperative winegrowers, whose living does not depend solely on grape production, increase in number. This is the case for independent domains as well (even if much less than in other scenarios) and only the number of big cooperative winegrowers diminishes. At the same time, independent domains and big cooperative winegrowers, who were in control of much agricultural land at the beginning of the simulations, tend to diversify their activities by growing different crops on their land. Strong urban pressure, profitability of other agricultural activities and lasting drop of wine prices make up the only scenario of deep crisis of winegrowing in the Bandol region, despite the adaptive capacity of winegrowing agents. Other scenarios with different characteristics of urban growth or with more favourable wine prices always result in the reinforcement and in the spatial development of the winegrowing economy in the Bandol region.

5   Results Overview and Conclusions

Overall, the six scenarios selected during the research were obtained through coherent combination of three main factors in the MAS model simulations: the intensity and the characteristics of urban pressure, the retail price of Bandol wine and the dynamism of other agricultural activities. Fig. 11 organises the scenarios in a global scheme shaped by these three factors and synthesises the main spatial dynamics of vineyards in the study area.

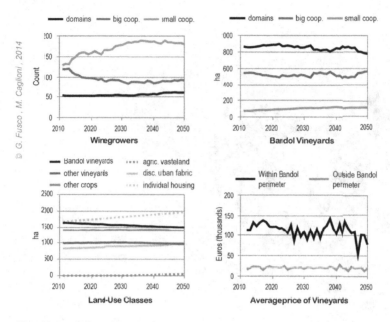

Fig. 10. Evolution of key indicators according to the scenario of winegrowing crisis

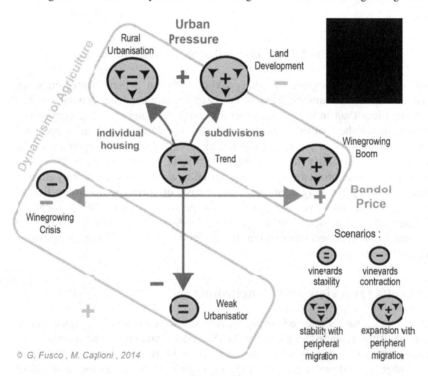

Fig. 11. Scenarios overview

The external factors constrain the interactions among the agents of the model, and from these interactions land-use and land-ownership patterns, as well as economic organisation of winegrowing, emerge. In most scenarios, winegrowing (including both Bandol and non-Bandol vineyards) shows unsuspected robustness, and is able to adapt and/or to resist to external pressure. This is possible thanks to financial transfers (which are never null) from land development to winegrowers, and to the complementary strategies of cooperative winegrowers and independent winemakers. Vineyards migrate during the simulations to more peripheral natural land, or substitute other agricultural activities in the central areas. The only crisis scenario for winegrowing sums the drop of wine prices to strong urban growth and attractiveness of other agricultural activities. But even in this scenario, winegrowing just shrinks marginally and the number of winegrowers even increases, thanks to agricultural diversification. The analysis of different scenarios of landscape evolution in the Bandol region should also take into account the dynamics of other agricultural activities and of natural areas, even if they are not the focus of the MAS model. Other crops show evolutions which are at the opposite of vineyards. Weak link in the local land-use system, they tend to shrink in all scenarios (but in the one of winegrowing crisis) and in some scenarios they lose up to a third of their 2010 surface. At the same time, speculative agricultural wasteland is produced in previously cultivated areas under urban pressure. Natural areas diminish in all scenarios, as well. Their maximum loss attains 610 ha in the scenario of winegrowing boom and weakening environmental protection. But these losses never affect their overall size considerably, as this is of another order of magnitude (6800 at the beginning of the simulations).

In conclusion, this paper showed the interest of using a MAS model, developed from real-world data, in order to simulate scenarios within a spatial strategic foresight approach. Under the constraint of a set of hypotheses, the model produces spatial forms for possible futures of the winegrowing economy of the Bandol region. In order to better assess advantages and shortcomings of the model, it is important to stress what it could and what it could not do. The model did show the dialectics between agent micro-behaviours and emergence of meso- and macro- spatial structures. It could also highlight the role of basic social and economic interaction among winegrowers, other farmers, land developers and local authorities. It also pointed to the role of land ownership structure and demographic variables within winegrowing and urban development. It finally identified the spatial impacts of urban pressure, land-use plans and exogenous economic variables on winegrowing landscapes.

At the same time, the model did not integrate with the same realism other important aspects of the study area. Urban dynamics are thus exogenous, and the modelling of the evolution of land-use plans is relatively simplistic. Other agricultural activities and forestry were very poorly modelled, as well. Topological and accessibility properties of road networks were not modelled (only contiguity to main roads was a catalyst of urban growth, regardless of other location parameters). Finally, landscape modelling is limited to land-use quantification (excluding qualitative aspects like perceptions, heritage conservation, or 3D rendering). Foreseen developments of the research are at present limited to analysing landscape ecology metrics of scenario outcomes.

Most of these shortcomings could be eliminated by further developing the model. Refinements could include the integration of a DTM for the development of both urban growth and new vineyards. The road network could become dynamic (or at least its development could contribute to scenarios) and accessibility within the network could be integrated in agents' rules. A more precise description of other agricultural activities could also be developed (as long as data are available for calibrating such refinements). With a more important structural development of the model, some external parameters could be internalised, namely housing demand.

But we think that all these improvements should not be carried out losing sight of the role of a MAS model in spatial strategic foresight. MASs are not prediction models, but tools to explore in a spatially explicit way the possible futures of the landscape under coherent sets of hypotheses. They allow the integration of socioeconomic and spatial processes, going beyond the black box of cellular automata modelling. But they also present the intrinsic danger of over-complexifying the modelling of spatial processes, by suggesting the false goal of a one-to-one model of reality. MAS models are precious tools to understand the role of key variables and policies in a spatially informed strategic foresight, but they should limit the modelling effort to the minimum necessary in order to cover the field of interest. This way, MAS models become useful companions for scenario building, as well. Within our research, scenarios were developed through a mix of expertise on external parameter and simulation of spatial and economic consequences of agents' interactions. An inductive approach to scenario building through search in parameter behaviour space could also be considered [8]. Problems of sampling in parameter space should then be addressed, as well as computing power issues, given the computing intensity of MAS simulations.

References

1. Benenson, I., Torrens, P.M.: Geosimulation: Automata-Based Modeling of Urban Phenomena. John Wiley & Sons, London (2004)
2. Fusco, G.: Démarche géo-prospective et modélisation causale probabiliste. Cybergéo 613 (2012), http://cybergeo.revues.org/25423
3. Fusco, G., Caglioni, M.: Quel avenir pour les paysages de la vigne? Éléments d'une géoprospective. In: Anglès, S. (ed.) Atlas des paysages de la vigne et de l'olivier en France méditerranéenne. Editions QUAE, Versailles (2014)
4. Helbing, D. (ed.): Social Self-Organization: Agent-Based Simulations and Experiments to Study Emergent Social Behavior. Springer, Heidelberg (2012)
5. Heppenstall, A.J., Crooks, A.T., See, L.M., Batty, M. (eds.): Agent-Based Models of Geographical Systems. Springer, Heidelberg (2012)
6. Masurel, Y.: La vigne dans la Basse-Provence Orientale. Louis-Jean, Gap (1967)
7. Rabino, G.: Processi Decisionali e Territorio nella Simulazione Multi-Agente. Esculapio, Bologna (2005)
8. Swarts, P.G., Zegras, C.: Strategically robust urban planning? A demonstration of concept. Environment and Planning B: Planning and Design 40, 829–845 (2013)
9. Vickrey, W.: Counterspeculation, Auctions, and Competitive Sealed Tenders. The Journal of Finance 16(1), 8–37 (1961)
10. Voiron-Canicio, C.: L'anticipation du changement en prospective et des changements spatiaux en géoprospective. l'Espace Géographique 2012-2, 99–110 (2012)
11. Wilensky, U.: NetLogo: Center for Connected Learning and Computer-Based Modeling. Northwestern   University,   Evanston,   Evanston   (1999), http://ccl.northwestern.edu/netlogo/

Land Use Change and Externalities in Poland's Metropolitan Areas

Application of Neighborhood Coefficients

Piotr Werner[1], Piotr Korcelli[2], and Elżbieta Kozubek[3]

[1] University of Warsaw, Faculty of Geography and Regional Studies, Warsaw, Poland
peter@uw.edu.pl
[2] Stanisław Leszczycki Institute of Geography and Spatial Organization,
Polish Academy of Sciences, Warsaw, Poland
korcelli@twarda.pan.pl
[3]Mazovian Office for Regional Planning, Warsaw, Poland
ekozubek@mbpr.pl

Abstract. Interdependence between spatial interactions and land use change generates various external effects. These, economic, social, cultural and environmental externalities may in turn have an impact on land use dynamics. The perception of such phenomena is related to the level of their acceptance which can be measured in psychological categories of comfort and discomfort. In this paper an attempt is made to compare the dynamics of land use with changes in the intensity of spatial interactions, using GIS (Geographic Information Systems) – especially map algebra (MA), cellular automata (CA), and the population potential model. According to the main hypothesis, interrelations between the observed land use change and the intensity of spatial interactions assume alternative forms, these giving rise to specific kinds of externalities.

Results of the research project: "Spatial interactions and external effects of land use changes in metropolitan areas in Poland", granted by Polish National Science Centre (UMO-2011/01/B/HS4/05194), carried on at the Institute of Geography and Spatial Organization, Polish Academy of Sciences.

Keywords: land use, metropolitan area, population potential, entropy, external effects.

1 Introduction

The aims of the study are identification, measurement and evaluation of interdependencies between socio-economic interactions and land use transition in selected metropolitan areas of Poland, i.e., the areas in which the phenomena in question, as well as externalities resulting from land use change are the most intense. The present day, dynamic land use changes may have sudden implications – economic, social, cultural and environmental and, in turn, they may influence land use changes. The inhabitants, visitors, decision-makers first perceive these phenomena in psychological categories of comfort or discomfort and, subsequently, accept or not the new situation.

B. Murgante et al. (Eds.): ICCSA 2014, Part IV, LNCS 8582, pp. 215–226, 2014.
© Springer International Publishing Switzerland 2014

In this study the physical models, which are used extensively in geographical studies, have been applied with the aim to specify the above mentioned interdependencies, as well as their external effects. The research is based on population potential model, entropy model and methods of cellular automata, comparison of dynamics of land use changes with the intensity of changes of spatial interactions using GIS (Geographic Information Systems), especially map algebra (MA), CA (cellular automata) and tools for measurement of spatial interactions (i.e. population potential). Use of GIS aims at extended detection and spatial exploration of probable unexpected phenomena (i.e. externalities) resulting from unbalanced relations between the observed land use changes and trends in spatial interactions. The specific original methods integrated into the model involve concepts of geography, cartography and geostatistics.

The spatial framework of research is regular geometric grid (raster), which consists of basic spatial units i.e. cells. Population data for statistical units (NUTS) are transformed into grid pattern.

The detailed research objectives focus on three problems directly related to the development of metropolitan areas. The first target is identification of the degree of land use change; the second target involve estimation of the significance of spatial interactions and, consequently, intensity of spatial interactions. The third research question, and the main question tackled in the study is the automatic detection and estimation of spatial extent and location of possible external effects which result from lack of balance between the intensity of spatial interactions and the land use dynamics.

This is especially important since over the last two decades Poland's urban agglomeration have been subject to far-going functional, as well as spatial transformation. Major aspects of the latter process are represented by suburbanization an peri-urbanization trends [9, 13, 14].

The study involved 12 metropolitan areas identified according to Poland's Spatial Development Concept till 2030 [12]. With regard to their spatial extent, delineation used for the purpose of Urban Audit project (Eurostat, 2012[1]) was used. Outer boundaries of metropolitan areas were kept constant for the whole period of analysis, i.e. 1990-2006. Within this time span two sub-periods were distinguished: 1990-2000 and 2000-2006. This was due to the accessibility of Corine LU/LUC data for the years mentioned.

2   Externalities

Agglomeration economies and externalities are essential elements of spatial economy, especially when a dynamic perspective is chosen [3]. Agglomeration spillovers (spatial non-market interactions or externalities) are also recognized in the analysis of residential choice [5, 6, 7]. The land market, therefore, is assumed to reflect the effect of these neighborhood externalities (ibid.). Non-market externalities are defined as material and immaterial products of human activity. They emerge as the spontaneous

---

[1]   http://epp.eurostat.ec.europa.eu/portal/page/portal/
region_cities/introduction

result of spatial interactions, or may be the result of directed activity of government or local authorities.

Externalities are material and non-material products which are received from the neighborhood, without bearing the cost of their production, assuming that the recipient is not able to control and influence them [23]. Local externalities, i.e. non-market goods associated with location are the result of preferences and choices of residents, or other actors [3].

There are two sets of externalities:

— Environmental (e.g. open spaces, green urban areas, forests, water reservoirs);
— Social (e.g. networks of services, social interactions, accessibility to education, public transport facilities).

Assessment of the impact of externalities depends on evaluation of size, shapes, density and distribution of clusters. These can be treated as results of individual decisions.

3   Population Potential

Spatial interactions represent flows of activity between locations in geographical space [1]. These may be flows of people, represented by population potential values. Population potential is a measure of nearness, or accessibility of a certain aggregation of people to a given point. The term is derived from social physics, and the concept is closely related to that of the gravity model, in that it relates mass (population) to distance [17, 18]. Whereas the gravity model deals with separate relationships between pairs of points, population potential encompasses the influence of all other points on a particular one [8].

Population potential is sometimes defined as the possibility of interactions, i.e. total ability of places (regions) to generate spatial interactions. In this sense, the value of the potential at a given point is a sensitive indicator of changes, including land-use changes that occur in the geographical space, particularly in urbanized areas. It should be noted that similar interpretation may pertain to changing population density values [2].

The population potential has been used in urban and regional studies in Poland [4], *inter alia* in modeling of commuting, transport networks [15] and simulations of ICT networks [19]. In the context of economic development level, and land use changes, population potential values were interpreted in: [10].

In this study population potential is treated as an appropriate measure of accessibility, and as a tool for assessment of the volume of interaction involving alternative locations. The general model of potential assumes that interactions pertaining to a given location decline proportionally with distance from the origin. The larger the population potential in target location, the greater its ability to generate spatial interactions. Population potential model encompasses both population size of places (nominator) and distances (denominator) involved; distance expresses the friction of space. The nominator represents the total number of people capable to overcome the resistance of geographical space, or the number of potential trips.

The model here applied includes a term corresponding to the Zipf's rule of rank and population size of spatial units [19], namely:

$$V_i = \sum_{\substack{j=1 \\ i \neq j}}^{n} \left(\frac{P_j}{d_{ij}}\right) h_{ij}^{-1} \tag{1}$$

where:
V_i – the population potential of ith municipality
P_j – the population of jth municipality
d_{ij} – the geographical distance in km between the municipalities i and j
h_{ij} – the difference in hierarchy between the municipalities i and j.

The values of population potential, as of 1990, 2000 and 2006 at the level of individual municipalities have been calculated for the territory of Poland as a whole, but only the subset of municipalities that fall within the boundaries of metropolitan areas have been considered in the analysis.

In the next step quotients of population potential for the defined subset of metropolitan municipalities: [2000/1990] and [2006/2000] have been examined. These procedures made it possible to assess changes in the intensity of spatial interactions. The quotients of population potential (indices of dynamics of spatial interactions) are in the range $(0; \infty)$. For value equal to one there is no change (stability); value less than one indicates a decline of spatial interactions; values greater than 1 are interpreted as growing intensity of spatial interactions.

Values of population potential for individual municipalities have been transformed into grid to enable comparison with more detailed information on land use changes.

4   Neighborhood Coefficients of Land Use (NBC)

Cellular automata are mathematical models for depicting complex, natural systems containing large numbers of simple, identical components with local interactions. The concept of neighborhood coefficient (NBC) is based on the combination of map algebra with two-dimensional cellular automata. NBCs [11, 20] are calculated on the basis of a mathematical formula which contains the numbers describing land use classes and the consecutive numbers of the cells in Moore neighborhood (3x3, eq.1). NBC is reversible. It is possible to reconstruct (recalculate) the original input land use classes in Moore neighborhood (ordinal numbers) on the basis of the value of the central cell's NBC.

$$NBC_c = \sum_{i=0}^{8} k_i n^i \tag{2}$$

where
i – number of a cell,
k – ordinal - class of land use,
n – total number of land use classes

NBC is the one (huge) integer which encodes nominal values of land use classes of all the fields in Moore neighborhood. The method is universal, and the only limit are computer resources. The main idea lying in the construction of NBC index is analogous to the situation of observer of view shed (360°), and counts the number of identified, distinguished classes of land use. The distance of sightseeing is limited in case of Moore neighborhood. The observer moves along the grid lines from left to right (W→E) and from up to down (N→S), touches each separate field and calculates its NBC. Neighborhood coefficients are the starting point to evaluation of entropy of land use in the neighborhood.

The only assumption made consisted of a condition on the identical sequence of cells in NBC, and the reverse neighborhood cell value calculations. The reverse algorithm of NBC entailed the set of divisions as shown in eq. 3, [20]:

$$k_i = \left(\frac{NBC_c}{n^i}\right) \bmod n \tag{3}$$

where: $k \in \{0, 1, \ldots, n\}$
i – the consecutive number of a cell in Moore neighborhood
k – the consecutive number describing the class of land use in an ordered nominal scale
n – the total number of land use classes.

This, in fact, is coding the sequence of land use nominal classes within the defined neighborhood as the one integer based on numeral system of total number of land use classes. The advantage of using this coding system is the possibility to create feasible cartographical visualizations and derive the appropriate transition rules and spatial patterns.

5   Entropy of Land Use Differentiation

Entropy is a universal measure of dispersion. The entropy index is not vulnerable to its operationalization [21, 22]. It is used in analysis of different systems (including socio-economic systems) to obtain measures of concentration and divergence of random variables (ibid.). The Shannon formula of entropy was applied for the purpose of this study, and the input data were the shares of individual land use types (classes). The structure of land use of the metropolitan areas in Poland was generalized into 16 classes[2]. Then, the set of indices of entropy have been calculated – for each neighborhood separately for 1990, 2000 and 2006, using the following formula:

$$E = -\sum_{i=1}^{n} p_i \log_2(p_i) \tag{4}$$

---

[2]   These are the following: continuous urban fabric, discontinuous urban fabric, industrial or commercial units, transport areas, airports, mineral extraction and dump sites, construction sites, green urban areas, sport and leisure facilities, agricultural areas, pastures, forests, natural and semi-natural vegetation, inland marshes and peat bogs, water courses and bodies, sea and ocean.

where:

p_i – probability of given land use class in neighborhood (share of land use class in % of whole area of neighborhood)

i – given land use class (nominal order)

n – total number of land use classes in study (n=16)

E – value of entropy index of land use.

Entropy is a continuous function with non-negative values, independent of the position of component elements. Entropy index takes values in range $[0, \log_2 n]$, with minimum, zero for the whole area covered with only one type (class) of land use, and maximum (equal 4 in the study) for the equal area patches of all possible land use classes.

Growth of entropy over time should be interpreted as a measure of increase of spatial differentiation of land use (a greater number of land use classes) in a neighborhood. Conversely – decline of entropy over time implies increasing homogeneity of the studied area.

Following the calculation of entropy indices, their quotients were estimated for analogous points of time, as in the case of population potential. This procedure made it possible to evaluate the dynamics of land use. Quotients of entropy equal to one indicate no changes in shares of individual land categories. Values greater than one show entropy increase and differentiation of land use (growth of disorder). Values of entropy change less than one describe tendency toward unification of land use, i.e. greater order of structure and growing dominance of a particular land use class.

Table 1. Example of land use and its Coefficient of Moore Neighborhood: decimal, hexadecimal and calculated entropy

Nominal values of land use classes (decimal) in Moore neighborhood			Nominal values of land use classes (hexadecimal) in Moore neighborhood			NBC (decimal and hexadecimal) in Moore neighborhood	Entropy (based on area shares)
3	8	15	3	8	F	48941694851	
15	6	2	F	6	2	NBC (HEX)	2,7255
5	6	11	5	6	B	B6526FF83	

6   Analytical Framework

According to the working hypothesis, as formulated in the introductory section, along with changing intensity of human interaction in space (which here is expressed by the population potential index values), the observed patterns of land use evolve accordingly, so as to become more diversified, or more uniform over space. At the same time, there is positive correlation between rates of change of the indicators used. Verification of the hypothesis relied on confrontation of the quotients of population potential (δV) and entropy (δE) for corresponding time periods. There are four (theoretical) possibilities of the pattern of change (see table 2):

1. **(A)** $\delta V > 1$ and $\delta E > 1$: growing potential for spatial interaction is accompanied by increasing differentiation of land use patterns. This type of change is normally associated with progressing peri-urbanization, and can be defined as **metropolitan expansion**.

2. **(B)** $\delta V \leq 1$ and $\delta E > 1$: decreasing volume of potential interaction in space corresponds with the transition of land use that leads to its differentiation. This can occur as an effect of dispersion of settlement in certain zones, with accompanying development of green and blue infrastructure, and can be linked with the **urban conservation** perspective [16].

3. **(C)** $\delta V > 1$ and $\delta E \leq 1$: increasing population potential values are positively correlated with decreasing differentiation of land use patterns, as recorded at the level of individual municipalities. Such type of relationship may be related to the **urbanization containment** perspective.

4. **(D)** $\delta V < 1$ and $\delta E < 1$: decreasing potential for spatial interaction is associated with growing homogeneity of land use. This suggests the phenomenon of urbanization shrinkage, although it can also imply a **land use consolidation** process, following an earlier phase of urban expansion.

7   Results of Spatial-Temporal Analysis

Some results of the study for the 12 metropolitan areas are presented in table 3, according to the categories identified in table 2. As these maps indicate, there are considerable differences among the 12 metropolitan areas concerning the observed dynamics of land use and spatial interactions. Three MAs are characterized by distinctly higher-than average level of anthropopressure in the metropolitan core. In this respect, Warsaw MA (1) represents a special case. Warsaw is in fact the only European – level metropolitan area in Poland, which attracts considerable numbers of in-migrants from other Polish regions, as well as foreign migrants. Lublin MA (7) was in a somewhat similar position during the 1990s, since the city was the first target of migration movement from Eastern Europe, first of all from Ukraine. Poznań, in turn, was the first Polish metropolitan area to be linked, before the year 2000, to the European motorway network.

Almost all the metropolitan areas are characterized by a ring of suburbanization , and peri-urbanization phenomena around the centers, which at the same time experience the decreasing population pressure, i.e. urbanization containment (except for

Table 2. Comparison of theoretical values of spatial-temporal quotients of entropy of land use and quotients of spatial interactions (population potentials, see explanation in text)

A	Metropolitan Expansion	$\delta E > 1$ $\delta V > 1$	
B	Urban Conservation	$\delta E > 1$ $\delta V < 1$	
C	Urbanization Containment	$\delta E < 1$ $\delta V > 1$	
D	Land Use Consolidation	$\delta E < 1$ $\delta V < 1$	

Table 3. Cartographical (iconized) presentation of spatial-temporal comparison of dynamics of spatial interactions and entropies in 12 Metropolitan Areas (MA) in Poland (main roads from 2005 year)

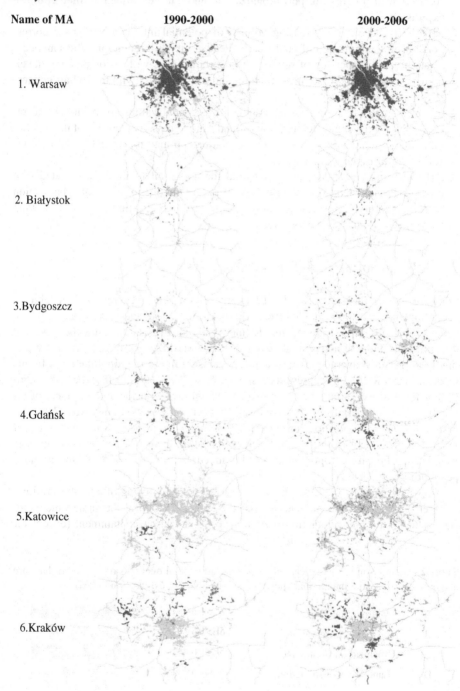

Name of MA	1990-2000	2000-2006
1. Warsaw		
2. Białystok		
3. Bydgoszcz		
4. Gdańsk		
5. Katowice		
6. Kraków		

Table 3. (*Continued*)

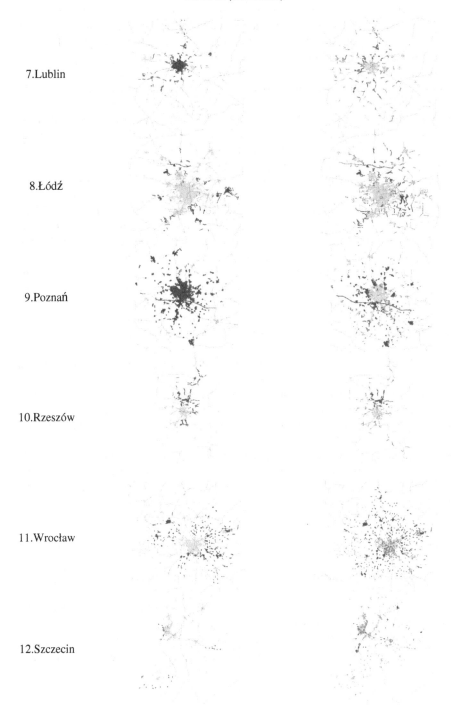

7.Lublin

8.Łódź

9.Poznań

10.Rzeszów

11.Wrocław

12.Szczecin

the three MAs listed above). On the other hand, there are smaller, compact zones of urban conservation, and of urban expansion, inside the metropolitan core areas. These phenomena seem to be the most important indicators of the location of possible externalities stemming from unbalanced relation between the rate of change of land use and the dynamics of spatial interaction.

In general, the observed patterns of land use change are compatible with local spatial development plans. One has to notice, however, that only some parts of the metropolitan areas are covered by the valid land use plans. When such plans are not available, the land development can still take place on the basis of the so-called development conditions, i.e. building permits that follow the good neighborhood rule, as well as the nature conservation requirements.

The existing transportation networks constitute one of the main factors determining specific land development trends. These areas that represent the patterns of change herewith defined as metropolitan expansion, and urban conservation, are rather scattered, frequently of peripheral situation. The dominant processes are those of land use consolidation, and urbanization containment. This shows that land use tends to become less differentiated at the municipality level, irrespective of the trends concerning population change. On the other hand, the areas characterized by high population potential values feature a relative stability of land use. These are, as a rule, the metropolitan core areas, as well as inner suburban ring zones. The larger the distance from the centre, and the more discontinuous the urban fabric, the greater is the observed intensity of land use transition. Small patches of land situated within high population potential zones tend to coalesce so as to form urbanized clusters. This phenomenon typically occurs along the major roads. The presence of numerous, rather small areas undergoing land use transition, situated relatively far from the urban core (in the metropolitan areas of Poznań, Gdańsk, Wrocław, or Białystok) signal peri-urbanization phenomena that affect former rural settlements.

When identifying the patterns of land use and population change in reference to the general analytical framework used, one has to take into account the specificity (including some external factors of local range) of the individual metropolitan areas. Thus, for instance, in the case of Szczecin MA, the growth of peri-urban areas during the 1990 – 2000 period was certainly influenced by the development of cross-border trade along the Polish – German boundary. In the case of Wrocław MA, the atypical pattern of land use change, falling under the urban conservation type, was a consequence of extensive land use reclamation measures taken in the aftermath of the "centennial flood" of 1997.

8   Concluding Remarks

In the next steps of the analysis the focus will be put on identification of more specific location of negative, as well as positive externalities, their description and interpretation. According to the analytical framework applied we expect that external effects occur first of all in those areas, in which the land use transition types of characterized as urban conservation and urbanization phenomena are mainly observed.

References

1. Batty, M.: Spatial Interaction. In: Encyclopedia of Geographic Information Science, pp. 417–419. SAGE, Thousand Oaks (2007- 2012), SAGE Reference Online, Web (January 10, 2012)
2. Bourne, L.S.: On the Complexity of Urban Land Use Change: or, What Theoretical Models Leave in the Dust. Papers in Regional Science 41(1), 75–100 (2005)
3. Caruso, G.: Integrating Urban Economics and Cellular Automata to Model Periurbanisation. These Presentee en vue de l'obtention du Grade de Docteur en Sciences. Universite catholique de Louvain, Louvain-la-Neuve (2005) (Access: December 16, 2012)
4. Chojnicki, Z.: Podstawy metodologiczne i teoretyczne geografii. Bogucki Wydaw. Naukowe, Poznań (1999), Methodological and theoretical fundamentals of geography
5. Durlauf, S.N.: Neighborhood effects. In: Henderson, J.V., Thisse, J.-F. (eds.) Handbook of Urban and Regional Economics. Elsevier-NorthHolland (2003)
6. Fujita, M., Thisse, J.F.: Agglomeration and growth with migration and knowledge externalities. In: The Economics of Political Integration and Desintegration, May 24-25. Center for Operations Research and Econometrics, Louvain-la-Neuve (2002)
7. Ioannides, Y.M.: Residential neighbourhood effects. Regional Science and Urban Economics 32, 145–165 (2002)
8. Johnston, R.J.: The dictionary of human geography. Wiley-Blackwell, Oxford (2000)
9. Korcelli, P.: The urban system of Poland. Built Environment 1978-, 133–142 (2005)
10. Korcelli, P., Grochowski, M., Kozubek, E., Korcelli-Olejniczak, E., Werner, P.: Development of Urban-Rural Regions: From European to Local Perspective. IGSOPAS, Warsaw (2012)
11. Kozubek, E., Werner, P., Ney, B.: The method of research and the prognosisof Land Use changes in Poland using Cellular Automata and Map Algebra, Report of the research project /No 1403/13/T02/2008/35/. Inst. of Geodesy and Cartography, Warsaw (2011)
12. KPZK: Koncepcja Przestrzennego Zagospodarowania Kraju 2030. Ministerstwo Rozwoju Regionalnego, Warszawa (2011)
13. OECD iLibrary, Organisation for Economic Co-operation and Development: OECD Urban Policy Reviews. ParisOECD (2011)
14. Parysek, J.J.: Miasta polskie na przełomie XX i XXI wieku: Rozwój i przekształcenia strukturalne. Bogucki Wydawnictwo Naukowe, Poznań (2005)
15. Ratajczak, W.: Modelowanie sieci transportowych (Modeling of transportation networks) Wydawn. Naukowe UAM, Poznań (1999)
16. Ravetz, J., Fertner, C., Nielsen, T.: The Dynamics of Peri-Urbanizationin. Peri-urban futures: Scenarios and models for land use change. In: Europe, N.K., Pauleit, S., Bell, S., Aalbers, C., Sick Nielsen, T.A. (eds.) Peri-urban futures: Scenarios and Models for Land use Change, pp. 13–44. Springer, Berlin (2013)
17. Stewart, J.Q.: Empirical Mathematical Rules Concerning the Distribution and Equilibrium of Population. Geographical Review, Vol 37, 461–486 (1948)
18. Warntz, W.: Geography of Prices and Spatial Interaction. Papers in Regional Science 3(1), 118–129 (2005)
19. Werner, P.: Geograficzne uwarunkowania rozwoju infrastruktury społeczeństwa informacyjnego w Polsce. UniwersytetWarszawski, Wydz. Geografiii Studiów Regionalnych, Warszawa (2003), (Geographical conditions of the information society development in Poland)

20. Werner, P.A.: Neighbourhood Coefficients of Cellular Automata for Research on Land Use Changes with Map Algebra. Miscellanea Geographica-Regional Studies on Development 16(1), 57–63 (2012)
21. Wilson, A.G.: Entropy in urban and regional modelling. Pion, London (1970)
22. Wilson, A.G.: Complex spatial systems: The modelling foundations of urban and regional analysis. Pearson Education, New York (2000)
23. Zarządzanie Rozwojem Przestrzennym Miast. Miasto, Metropolia, Region. Gdańsk: Urbanista (2010), Management of Spatial Development of Cities

Modeling Morbid Geographical Risk Exposure

Methodological Proposition and Application of EstimGRE Algorithm to Epidemiological Data

Stéphane Bourrelly

PhD student, University of Nice, UMR 7300 (ESPACE) - CNRS, Nice, France
s.bourrelly@hotmail.fr

Abstract. To address the priorities of the Public Healths, in particular those set by French's Cancer Plans, it is necessary to develop spatial tools able to identify environmental risk factors. The emergence of numerous databases provides access to many environmental parameters that describe geographical living environments. However, those ones are only interesting if there are crossed with health indicators. Epidemiological databases contain spatiotemporal references but are not suited to the geographic modeling.

The EstimGRE method provides two morbid spatiotemporal indicators (m.st.i) adapted to the analysis of interactions between health and environment. It also leads to construct a third m.st.i which characterizes spaces with morbid Geographical Risk Exposures (GRE). Therefore, it enables to develop medical solutions and public policies to improve the environmental health of populations, in line with the sustainable development objectives. Propositions are applied to the LEA cohort. The EstimGRE algorithm is the name of the method proposed.

Keywords: Geography, Health, Spatial modeling algorithm, Epidemiological data, Sustainable management.

1   Introduction

The increasing life expectancy due to medical progress and the access to effective treatments, *has positioned post-cancer and children's health at the core of socio-political concerns in Europe and in France*. The analysis of health states confers, *for several decades now, a special importance to the environmental quality* [1].

The introduction of human sciences, such as geography, to enable searching of environmental factors for *disease emergence* has extended the epidemiological approach, where environment *was considered partially*, to a more comprehensive approach [2]. Now, *all the items of the living environment are taking into account*: biological, physical, chemical, social, economic or cultural and *all scale are studied*, from individuals to the climatic zones, passing by the workplaces and the administrative geographical units. The health geographer looks *at all relevant environmental factors*, i.e. that have morbid effects documented *proven or only suspected* [3]. Also,

B. Murgante et al. (Eds.): ICCSA 2014, Part IV, LNCS 8582, pp. 227–242, 2014.
© Springer International Publishing Switzerland 2014

he takes into account the geography of individual and medical characteristics of the targeted populations, *otherwise the analysis is biased* [4] - hence the advantage to use cohort data.

Combining health informations with spatial and socio-economic features - when epidemiological data are lacking - has the advantage to help decision makers to plan effective public health policies and to ensure his primary purpose: *protecting individual health and promoting public health* [3]. Healthy human societies are suitable to the necessary dynamics for the sustainable territorial development [5].

The *emergence of heterogeneous databases* opened access to many parameters characterizing spaces [6]. Physical geography and human geography provide a lot of methods to model the factors describing the environmental quality of living environments. However, these indicators are inadequate to assess the adverse effects of the environment whenever they are not crossed with reliable health indicators [7]. In addition, Geographical Information System (GIS) tools can spatialize all epidemiological data without knowing any concept of geography. This paper propose a new kind of morbid spatiotemporal indicators (m.st.i) whose purpose is: (i) modeling and analyzing the geography of Morbid Phenomena (MP); (ii) characterizing as morbid Geographical Risk Exposures (GRE).

The EstimGRE method focuses on the analysis level of *environmental health interactions* - i.e. the potential health effects of various environmental factors which characterize living environments [1]. It is adapted to epidemiological data, such as cohorts. The algorithm EstimGRE: Models geography of particular states of health through two m.st.i $z'^{j}_{U_k,c}$ and $z'^{j}_{U_k,q}$; Characterizes spaces by morbid GRE with a synthetic m.st.i $z^{GRE,j}_{U_k}$ - suitable to sustainable management of territories.

The m.st.i proposed incorporate two dimensions: *the horizontal* - i e. space, and *the vertical* - i e. the time, as well stochastic uncertainties linked to the complexity of this spatiotemporal approach. Herein epidemiology, statistics and geography are coupled within *a phenomenological approach of the analysis of health facts*. The interactions between humans and their environment are studied in globality, and from *experimental data* [8]. Consequent, this research is conducted at the scale of the French administrative level: the *commune*.

2   Material and Methods

2.1   Specification of the Data and Objectives

EstimGRE algorithm is applied to epidemiological data of the LEA cohort (Leukemia in Childhood and Adolescence). This is an epidemiological database that repertories the persons treated for childhood leukemia, in France since 1980.

Leukemia is a cancer that affect the bone marrow. Over time, the survivors develop sequelae, their number and severity have an impact on their quality of life. In this paper the focus is on cataracts (CATA) and thyroid tumors (THYR) [9]. Boolean variables $y^j_i = \{YES \cup NO\}$, which allow whether the patient "i" developed the sequela "j". The only geographical information available in the database is the Postal Code (PC) x_i^{PC} of the last medical consultation. Patients included in this analysis are those of the base LEA 2010 and whose x_i^{PC} corresponds to a geographical area situated in France. Sample is composed of 747 individuals: 417 boys and 330 girls. The incidence of sequelae is: 13.0% for CATA and 10.2% for THYR.

More specifically the EstimREG method objectives are to return m.st.i, which should be able: To model the geography of the MP; To characterize territories by a type of GRE; And also, to be adapted to identify adverse environmental factors, focusing on the spatial and temporal dimension issues.

The different steps of the algorithm: (i) spatializing patients in the *communes*, (ii) constructing a *geographical fuzzy metric*, which reduces the *spatiotemporal uncertainties* due to the inputs used, (iii) modeling the geography of the MP studied by m.st.i, (iv) characterizing the geographical spaces by morbid GRE through a third synthetic m.st.i adapted to sustainable management of territories.

2.2   Geographic Modeling Morbid

To begin, EstimGRE algorithm computes two m.st.i $z'^j_{U_k,c}$ and $z'^j_{U_k,q}$, then a third $z_{U_k}^{GRE,j}$, which is obtained by fusing the first two. Those m.st.i incorporate a spatiotemporal weighting system, in order to achieve a more accurate representation of the geographical morbid reality and to be better suited to the spatial analysis of health-environment interactions.

The weighting system is composed of: (i) a *geographical fuzzy metric* [10] - that contains spatial uncertainty weights π_i^j - whose goal is to *transversally* incorporate the space [8] by injecting expert epidemiological, geographical and statistical (Epi-GeoStat) knowledge to reduce spatiotemporal uncertainties brought by the quality of morbid and community data used; (ii) a patient-years conversion [11], which allows to integrate *vertically* the time, within the process [8], through the exposure times at the environment ete_i^j of the patients [11].

The first m.st.i proposed is quantitative $z'^j_{U_k,c}$. It represents spatial prevalence of sequelae, which are EpiGeoStat weighted and converted to patient-years.

The second m.st.i is qualitative $z'^j_{U_k,q}$. This one expresses the susceptibility to develop sequelae. It pays special attention to: The spatial EpiGeoStat quality data; The exposure times at the environment; The knowledge of the patient's medical history.

The third m.st.i $z_{U_k}^{GRE,j}$ is obtained by merging the information contained by the first two. The estimation strategy is based on a geographical elasticity threshold φ_j^*. This one integrates: Statistical characteristics of the first two (m.st.i); Expert knowledge; Consequences of the modeling process. This m.st.i characterizes the communes U_k, by four kinds of morbid Geographical Risk Exposures (GRE) : PROBABLE; POSSIBLE; UNPROVABLE; WEAK.

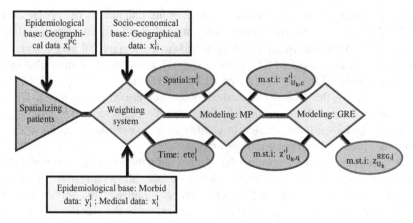

Fig. 1. Synoptic of EstimGRE algorithm: estimation processes of the three m.st.i

EstimGRE algorithm is programmed in Visual Basic [12]. It is autonomous if the inputs are correctly implemented. Then, the health outcomes are interfaced to a GIS - that used is ArcGIS.10 [13]. It can perform cartographies, statistics, as well as spatial analysis. The $z_{U_k}^{GRE.j}$ estimation involves a bootstrap process, which requires using the R software [14].

To illustrate the heuristic proposals EstimGRE algorithm is applied to the LEA data - sequelae CATA and THYR. Methods and principles specified below can be reproduced to all other sequelae and extended to any disease.

2.3   Principle of Spatialization

Patients are spatialized in the GIS units U_k. GIS base used is geofla2003. Each U_k represents a French administrative area: *a commune*. As shown in the Fig. 2, the PC involves spatial uncertainty. It may not be adapted to *communal* boundaries. They are often more coarse. They may cover several *communes* and straddle on several others.

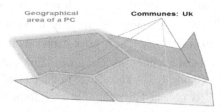

Fig. 2. Schematic of a PC overlay imprecise with several U_k

Spatialization hypothesis: patients living in the most populated *commune* which is covered by the geographical area of his Postal Code.

Advantages and limits: This strategy is inexpensive because it does not require to call the patients back. However to validate it, several patients whose x_i^{PC} spreads over wide geographic areas have nevertheless been randomly contacted. The hypothesis was correct in 90% of cases. But when this was not the case, the *commune* of residence was roughly in spatial contiguity. In addition, when EstimGRE makes a mistake, the verification has shown that patients roughly spatialized were practicing a social, professional or sport activity in the *commune* in which they were spatialized. Consequently patients undergoes, at least partially, the environmental factors of the most populated U_k- which combines many functionalities.

2.4   Principle of the Spatiotemporal Weighting System

Geographical Fuzzy Metric

Geographical fuzzy metric allows to associate spatiotemporal quality notion to the sequelae variables y_i^j. Estimating spatial EpiGeoStat uncertainty weights π_i^j (of this matrix), assumes that during aggregation of y_i^j in the U_k, some YES tend to NO and conversely. Values of π_i^j are proportional to the uncertainties associated with y_i^j. It is great for a patient who has not been seen since a long time, who is spatialized alone in a rural U_k and whose PC can match with several other *communes*.

<u>Heuristics propositions:</u> The *fuzzy set theory,* allows to inject and to *merge expert knowledge through mathematical functions*. Then, incorporate the π_i^j, in the estimation process of the m.st.i [10].

<u>Estimation principle:</u> (i) EpiGeoStat uncertainty weights cannot be calibrated because the morbid geographical reality is unknown. (ii) They have a sign, such as $\xi_i^j = -1 + 2 \cdot \mathbb{1}_{y_i^j = YES}$. (iii) π_i^j are divided into three fragmentary weights of uncertainty: epidemiological $\pi_i^{j,epi}$, geographical $\pi_i^{j,geo}$ and statistical $\pi_i^{j,stat}$. (iv) Under *the Law of Large Numbers* [15]; Assuming *the existence of an environmental effect* [9] ; and to ensure that m.st.i remain interpretable, then weight are defined as follows.

$$|\pi_i^j| = \frac{\xi_i^j}{3} \cdot \left|\left(\pi_i^{j,epi} + \pi_i^{j,geo} + \pi_i^{j,stat}\right)\right| \ll \frac{1}{2} \qquad (1)$$

With the upper limit was fixed *from expert knowledge* [16].

Epidemiological uncertainty weight

The $\pi_i^{j,epi}$ are defined by uncertainty Epidemiological Score (ES), which is introduced into an exponential function, linearly attenuated, and bounded to remain interpretable:

$$\pi_i^{j,epi} = \left\{1 - \zeta \cdot ES_i^j + e^{\left(-\hat{\lambda} \cdot ES_i^j\right)} \middle| \{\hat{\zeta}, \hat{\lambda}\} = \operatorname{argmin}_{\forall \{\zeta, \lambda\} \in \mathbb{R}_*^2} \sum_{l=1}^5 \left(\pi_l^{j,epid} - \pi_l^{j,exp}\right)^2\right\} \qquad (2)$$

The scores ES_i^j are used to estimate spatiotemporal uncertainties of the epidemiological data. Two cases can be distinguished, Fig.3 :
<u>The patient developed the sequela</u> $y_i^j = 1$, this information is safe, so $ES_i^j = 0$

The patient did not develop the sequela $y_i^j = 0$, this information contains several spatiotemporal uncertainties due to: Strategy filling gaps, systematically brought to a sequela absence $l_i^j = \|_{\{y_i^j=?\}}$; Quality monitoring, patients who received aggressive treatment are systematically screened $s_i^j = \|_{\{monitoring_i^j=NO\}}$; Capacity to develop sequelae, dead patients are no longer exposed to the environment, $m_i = \|_{\{death_i=YES\}}$; Temporal information significance, which is linearly proportional to ti_i, the time between the last visit and the database used $Q(ti_i)$, such as $ES_i^j = \left(l_i^j + s_i^j + m_i + Q(ti_i)\right)$

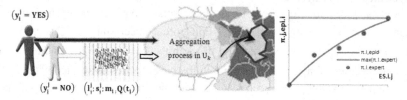

Fig. 3. Synoptic estimation of the Epidemiological uncertainty weight $\pi_i^{j,epi}$

The mathematical function, the upper limit, the expert weights $\pi_i^{j,exp}$: were assessed on medical opinions [16].

Geographical uncertainty weight
It is defined by the sum of three fragmentary, functions spatial or temporal

$$\pi_i^{j,geo} = g_1(\cdot) + g_2(\cdot) + g_3(\cdot) \tag{3}$$

The spatialization uncertainty: It is generated by the inaccuracy of the PC and estimated by a *piecewise constant function* [15], noted $g_1\left(q_{o,i}\middle|\hat{F}_n(q_{o,i})\right)$. It is calibrated from the *empirical distribution function* $\hat{F}_n(q_{o,i})$ - with $q_{o,i}$ is the number of U_k that can be matched with each patient, knowing its x_i^{PC}. This uncertainty is shown in green. The color is darker in proportion to the likelihood that patients are wrongly spatialized in red U_k, Fig. 4.

The uncertainty of geographical scale used: this uncertainty is based on the variability of environmental exposures in micro-scale. It depends on daily travel patients. The distances are assumed to be proportional to the log. of the geographical area commune $S_{(i|U_k)}$ [17], Fig.4. Therefore, $g_2\left(S_{(i|U_k)}; \hat{\alpha}, \hat{\beta}\right) = \hat{\alpha} \cdot \ln\left(1 + \hat{\beta} \cdot S_{(i|U_k)}\right)$. With α an upper bound to avoid outliers, and $\hat{\beta}$ a shape parameter defined by the surface's quantile, such as $\hat{\beta} = (e^1 - 1)/\hat{Q}_3\left(S_{(i|U_k)}\right) - \hat{Q}_1\left(S_{(i|U_k)}\right)$.

The uncertainty of life trajectories: It is assumed to be proportional to the probability that patients have moved from his *communal* residence U_k during T_i, the time between the date of first treatment received and the last consultation Fig. 4. The reconstitution of residential mobility is not possible from LEA data. It was made from, the only available source, *the INSEE survey housing 2006* [17]. These community data used to link the intentions to relocate idd_t , with the durations of house occupation dol_t - assimilated to T_i - and a maximum duration housing occupancy dol^i_{max}, from which the likelihood is stabilized. The idd_t are adjusted to a gamma distribution, whose parameters $\hat{\Theta}_m = (\hat{C}; \hat{\theta}; \hat{k})$ are specified by *the ordinary least squares criterion* and *two expert coefficients* are added $\tilde{C}$ and $idd^\sim_{max}$, to *avoid outliers* [15], as follows.

$$g_3\left(T_i; \hat{\theta}; \hat{k}; \tilde{C}, idd^\sim_{max} \,\middle|\, \hat{\Theta}_m\right) = \tilde{C} \cdot T_i^{\hat{k}-1} \cdot \frac{e^{-(T_i/\hat{\theta})}}{\Gamma(\hat{k}) \cdot \hat{\theta}^{\hat{k}}} \cdot \|_{\{T_i \le dol^i_{max}\}} + idd^\sim_{max} \cdot \|_{\{T_i > dol^i_{max}\}} \quad (4)$$

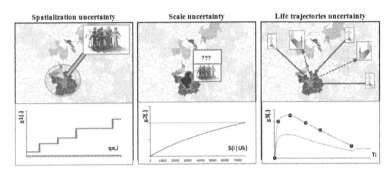

Fig. 4. Synoptic estimation of the Geographical uncertainty weight $\pi_i^{j,geo}$

Statistical uncertainty weight

Values $\pi_i^{j,stat}$ are affected by a *piecewise constant function*. They depend on the concept of *statistical inconsistence* [15], ie they are inversely proportional to the number of patients spatialized by *communes*. The parameters of $\pi_i^{j,stat}$ are estimated from the *empirical distribution function* $\hat{F}_n(n^I_{(i|U_k)})$, Fig. 5 - such as:

$$\pi_i^{j,stat} = s_1\left(n^I_{(i|U_k)} \,\middle|\, \hat{F}_n(n^I_{(i|U_k)})\right) \quad (5)$$

With $n^I_{(i|U_k)}$ the number of individuals spatialized with the patient "i", in its U_k.

Transformation Patient Years

Heuristic hypothesis: The *geographical fuzzy metric* enables to introduce the space. However, *it doesn't put the time into the shade* [8], Fig. 5.

Estimation principle: In spatial epidemiology, the conversion patient-years allows to take into account the patient's exposure times at the geographical environment, noted ete_i^j. This fundamental strategy increases population size spatialized, dissociates individuals at risk - those who have not developed the disease - from that in which it was detected, considers also the latency of the disease [11].

The participation time tp_i, is the duration between the date on which the LEA study starts - where all individuals are supposed healthy - and the date on which the study is stopped.

The exposure time at the environment ete_i^j, corresponds to tp_i for healthy individuals and to the duration between the start study and the sequaele diagnosis for the others.

The ratio of participation-exposure rpe_i^j is the division between the participation time and the exposure time at the environment, rounded to use higher number.

$$\text{rpe}_i^j = \lceil \text{tp}_i / \text{ete}_i^j \rceil$$

This one equals 1 when the sequela has not been diagnosed. It is greater than one when the patient has developed it. This value is even higher than the sequela early.

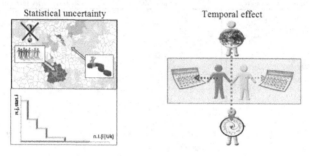

Fig. 5. Synoptic estimation of the: Statistical uncertainty weight $\pi_i^{j,\text{stat}}$, right; Exposure time at the environment ete_i^j, left.

2.5   Modeling of Morbid Phenomena

Estimation of the Spatiotemporally Weighted Morbid Prevalence
This m.st.i, is a classical quantitative variable. It models, for each U_k, the geography of MP by the sum of y_i^j weighted by spatial EpiGeoStat uncertainty and reduced by patient's exposure time at the environment.

$$z'^j_{U_k,c} = \frac{1}{n_{U_k}^{\text{ete}_i^j}} \cdot \Sigma_{i=1}^n (y_i^j + \pi_i^j) \cdot \mathbb{1}_{\{x_i^{CP} \supset U_k\}} \tag{6}$$

With $n_{U_k}^{\text{ete}_i^j}$ is the number of patients-years spatialized in each U_k.

Estimation of the Spatiotemporally Weighted Morbid Propensity
This m.st.i, is an original qualitative multi-class variable. It models the geographic susceptibility of population to develop a disease. It is obtained by maximizing the *information sources effect* [18]. In other words, it preserves the qualitative nature of inputs and includes: The spatial EpiGeoStat quality: the patient's exposure time at the environment, the knowledge of his medical history. It assigns the most likely modality among three: YES, the individuals have a high tendency to develop the disease; NO, this is not the case; UNCERTAIN, the y_i^j EpiGeoStat quality is too poor.

$$z'^{j}_{U_k,q} = \begin{cases} \text{YES} & \text{if} \quad \mathbb{P}_{F_n}\left(zq'^{j}_{\{.|U_k\}} = \text{YES}\right) \geq \left(k^j \cdot \bar{y}^j \wedge 0,5\right) \\ \text{UNCERTAIN} & \text{if} \quad \mathbb{P}_{F_n}\left(\{zq'^{j}_{\{.|U_k\}} = c^j\}\right) = \phi \\ \text{môd}\left(zq'^{j}_{\{.|U_k\}}\right) & \text{if} \quad \text{Otherwise} \end{cases} \tag{7}$$

With: $zq'^{j}_{\{.|U_k\}}$ the subset of patients spatialized in U_k, whose the y^j_i are repeated in proportion of the rpe^j_i values. The qualitative m.st.i: takes the value YES if the proportion of reliable sequaele is k^j times higher than the prevalence estimated on all the patients spatialized $\bar{y}^j$. With k^j a factor set to 2, *from expert advice* [16].

The value y^j_i is considered spatiotemporally significant if $|\pi^j_i|$ exceeds the Epi-GeoStat reliability threshold ψ^j_π, that estimated as follows.

$$\psi^j_\pi = \left| \text{môy}\left(|\pi^j_.|\right) - t_{(\alpha)} \cdot \frac{\hat{\sigma}\left(|\pi^j_.|\right)}{\sqrt{n}} \right| \tag{8}$$

With: $t_{(\alpha)} \xrightarrow[n \to +\infty]{} N \sim \mathcal{N}(0,1)$, and $\mathbb{P}\left(N \leq t_{(\alpha)}\right) = \{\alpha = 5\%\}$.

The $z'^{j}_{U_k,q}$ indicator assesses the certainty that the $z'^{j}_{U_k,c}$ is to take high or weak values. By integrating differently the EpiGeoStat quality sources and simultaneously the variables ete^j_i and tp_i, it identifies the significant values of $z'^{j}_{U_k,c}$ – i.e., those ones are adapted to health-environment spatial analysis.

2.6   Intermediary Results

The m.st.i $z'^{j}_{U_k,c}$ and $z'^{j}_{U_k,q}$ are primary variables to estimate the third m.st.i adapted to the sustainable management of territories. However, their joint analysis leads to interesting findings. The following maps show the values taken by the m.st.i in the *communes* of the PACA region (southeast of France), where more than 50% of patients are spatialized. The name is displayed for: Geographical extrema, i.e. outliers when $z'^{j}_{U_k,c} \geq \left(\hat{Q}_3(z'^{j}_{.,c}) + 1,5 \times I\hat{Q}R(z'^{j}_{.,c})\right)$; Or $z'^{CATA}_{U_k q} = \text{YES}$. Statistical tables provide: *average and standard deviation* of $z'^{j}_{U_k,c}$, and the *frequencies* of $z'^{j}_{U_k,q}$ [15].

Cartographic and Statistic Presentations

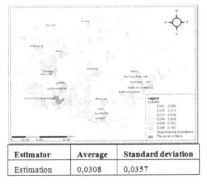
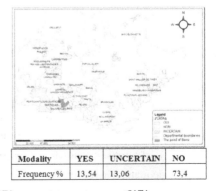

Estimator	Average	Standard deviation
Estimation	0,0308	0,0357

Modality	YES	UNCERTAIN	NO
Frequency %	13,54	13,06	73,4

Fig. 6. Sequela CATA: values of $z'^{CATA}_{U_k,c}$, right; And values of $z'^{CATA}_{U_k,q}$, left

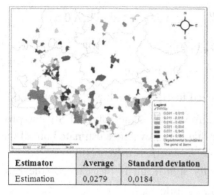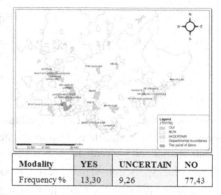

Estimator	Average	Standard deviation
Estimation	0,0279	0,0184

Modality	YES	UNCERTAIN	NO
Frequency %	13,30	9,26	77,43

Fig. 7. Sequela THYR: values of $z'^{THYR}_{U_k,c}$, right, and values of $z'^{THYR}_{U_k,q}$, left

Analyzis and Remarks

Sequela CATA: The spatiotemporally weighted prevalence average is $\bar{z}'^{CATA}_{.,c} =$ 0,0308. In the Southeast of France, the communes which are both $z'^{CATA}_{U_k,c}$ extrema and $z'^{CATA}_{U_k,q} = $ YES are numerous, there are: Port-Saint-Louis-Du-Rhone, Lambesc, Lancon Provence, Cavaillon, Sheds Bedouin Modragon, Andon, Biot, Saint-Vallier-de-Thiey. In these U_k, patients seem to have a curious tendency to develop cataracts – Fig. 6.

Sequela THYR: The spatiotemporally weighted prevalence average is $\bar{z}'^{THYR}_{U_k,c} =$ 0,0279. It is close to the one of CATA. Contrastingly, no commune takes an $z'^{THYR}_{U_k,c}$ extreme value on the maps although many are characterized as $z'^{THYR}_{U_k,q} = $ YES – Fig.7.

Simultaneous analysis of the sequelae CATA and THYR: typically the $z'^{j}_{U_k,q} = $ INCERTAIN involve rural communes - where the spatiotemporal quality of sources is poor - in these ones the extreme values of $z'^{j}_{U_k,c}$ - weak or strong - are unusable. Some are characterized by a high outlier of $z'^{j}_{U_k,c}$ and $z'^{j}_{U_k,q} = $ NO - or vice versa - those geographical antagonisms complicate the interpretation results.

Remarks:
The cross-spatial analysis of the i.st.m, independently conducted on each sequela, shows strong spatial variability. This finding suggests an environmental morbid tendency. When the analysis is carried out simultaneously by combining the $z'^{j}_{U_k,q} = $ YES, the extreme values of $z'^{C\,ATA}_{U_k,q}$ and high values (not extremes) of $z'^{THYR}_{U_k,q}$ – Spatial simultaneities are observed for U_k close to each other, or located in small geographical areas, when they are not directly contiguous. This finding assumes a contributive environmental adverse effect, which adds to the treatments received.

These remarks prove the importance of using both pieces of information contained in the m.st.i. and also show the analyze limits when it is led this way.

The $z'^{j}_{U_k,q}$ are used to characterize the significance of the $z'^{j}_{U_k,c}$. But the use of extrema is inadequate. Therefore, we must set a threshold beyond which the $z'^{j}_{U_k,c}$ may be considered *abnormally high*. Thus, they can be easily crossed with $z'^{j}_{U_k,q}$.

In addition, the use of two m.st.i, is not synthetically sufficient to develop public policies to improve environmental health.

2.7   Modeling of Morbid Geographical Risk Exposures

The aim is to propose a geographical elasticity threshold φ_j^* to identify *abnormally high values* of $z'^{j}_{U_{k,c}}$, and to merge wisely the informations of $z'^{j}_{U_{k,c}}$ and $z'^{j}_{U_k,q}$, in order to characterize the U_k by morbid Geographical Risk Exposures (GRE).

Principe of estimation:

$$
z^{GRE.j}_{U_k} = \begin{cases} \text{PROBABLE} & \text{if } \left\{z'^{j}_{U_k,c} \geq \varphi_j^*\right\} \cap \left\{z'^{j}_{U_k,q} = \text{YES}\right\} \\ \text{POSSIBLE} & \text{if } \left\{z'^{j}_{U_k,c} \geq \varphi_j^*\right\} \cap \left\{z'^{j}_{U_k,q} = \text{NO}\right\} \\ \text{UNPROVALE} & \text{if } \left\{z'^{j}_{U_k,c} \in \mathbb{R}_+\right\} \cap \left\{z'^{j}_{U_k,q} = \text{UNCERTAIN}\right\} \\ \text{WEAK} & \text{if } \left\{z'^{j}_{U_k,c} < \varphi_j^*\right\} \cap \left\{z'^{j}_{U_k,q} = \text{NO}\right\} \end{cases} \quad (9)
$$

Preliminary remarks:
  (i) In geography is difficult to develop a strategy *starting from expert knowledge to assess a mathematical parameter that has a real geographical significance* [8].
  (ii) When $z^{REG.j}_{U_k}$ = PROBABLE the sanitary quality of the communes is impli-cated. So φ_j^* should be sufficiently large so that this number is not equal to that of $z'^{j}_{U_k,q}$ = YES. The modality POSSIBLE tempering the PROBABLE kind of disease exposure ; UNPROVABLE is used to not conjecture wrongly a GRE.

Hypothesis: If the morbid phenomenon is completely random, ie independent of envi-ronmental factors, then under the *Law of Large Numbers*, the $z'^{j}_{U_k,c}$ should tend to the average of overall weighted sequalae spatialized $\overline{y}'^{j}$ [15]. But this is not case.

Estimation strategy of the geographical elasticity threshold:

$$
\varphi_j^* = \left\lceil \xi^j_{geo} \cdot \left(\overline{z}'^{j}_{(U_k),c} + b^{*,j}_{1-\alpha} \cdot \frac{\hat{\sigma}\left(z'^{j}_{U_k,c}\right)}{\sqrt{N(U_k)}} \right) \right\rceil \quad (10)
$$

With: $\lceil \cdot \rceil$ rounded the higher number function ; $b^{*,j}_{1-\alpha}$ et ξ^j_{geo} two parameters which cannot be calibrated.

Principe to estimate the bootstrap risk level:
The parameter $b^{*,j}_{1-\alpha}$, amplifies the estimation of the $z'^{j}_{U_k,c}$ average. Usually, the quan-tile $b_{1-\alpha}$ is pre-determined in assuming that if $z'^{j}_{U_k,c}$ is standardized then it follows standard normal distribution such that values will exceed $b_{1-\alpha}$ have a probability α. Here, to make a better use of information contained in the epidemiological data and to

dispense with assumption on the probability law, $b_{1-\alpha}^{*,j}$ *is estimated by bootstrap.* This process is *semi-parametric* because it depends on *the level of risk* α^j [19]. This value is set in conversely proportional to the *standard deviation estimator* $\hat{\sigma}(\cdot)$ and *under statistic paradigms established* [15], as $\alpha^j \in [\![5\% \; ; \; 10\%]\!]$.

Principe to estimate the geographical elasticity threshold:
The efficiency of α^j was very modest. The parameter ξ_{geo}^j, allows moderated the class PROBABLE by POSSIBLE. His value is specified from, expert knowledge acquired on the MP and proportionally to the geographical distortions carried by the combined action of the π_i^j and the ete_i^j. Moreover, if the upper geographic average is adequately rugged then their values are not random. Experience has shown that $\xi_{geo}^j \in [\![1 \; ; \; 2]\!]$.

3   Results

The entire principles proposed were programmed in EstimGRE algorithm. It has been used to characterize the geographic units U_k (communes) through morbid Geographical Risk Exposure (GRE). The sequelae studied are cataracts (CATA), Fig. 8 and thyroid tumors (THYR), Fig. 9. The parameters used to estimate φ_j^* are :

Table 1. Values used for the geographical elasticity thresholds

Sequela	CATA	THYR
α^j	10%	5%
ξ_{geo}^j	1.10	1.50

Cartographic representations: the communes are those situated in the PACA region and around. Labels are displayed when $z_{(U_k)}^{REG.j} = $ PROBABLE

Statistical Results: the bar graph of the feature class of $z_{U_k}^{GRE.j}$ gives the frequencies estimated on all U_k where patients are spatialized.

Cartographic and Statistic Presentations

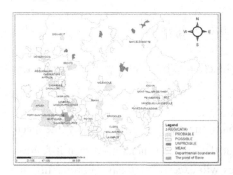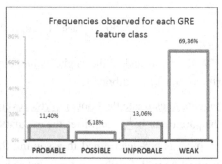

Fig. 8. Sequela CATA, i.st. $z_{U_k}^{REG(CATA)}$: Map values, right; And bar graph, left

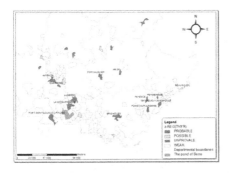 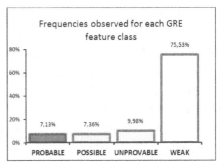

Fig. 9. Sequelae THYR, i.st. $z_{U_k}^{REG(THYR)}$: Map values, right; And bar graph, left

3.1   Analyzis and Remarks

The objective of spatial analysis of the Geographical Risk Exposures (GRE) is to bring general trends. The GRE: PROBABLE characterizes 11.4% of *communes* for the sequela CATA, and 7.1% of communes for the sequela: THYR - this proportion reached 14.5% of U_k when also considering the POSSIBLE kind. Then, 13.1% of municipalities have a GRE (CATA) described as UNPROVABLE. This impossibility to characterize the communes by a GRE concerns 10.0% of U_k, for THYR.

The $z_{U_k}^{GRE.j}$ = UNPROVALE, are related to U_k where few patients are spatialized and where the CP describes large geographic areas or whose the spatiotemporal quality sources is lacking. The U_k affected by a GRE: UNPROVABLE are located mainly in the North of France. In return, GRE: PROBABLE are essentially assigned to the U_k located in the PACA region, where more than 50% of patients are spatialized.

GRE (THYR) is PROBABLE for 13 communes and is POSSIBLE for 6 others. GRE (CATA) concerns 27 communes in PACA and 5 U_k are characterized by POSSIBLE

The joint analysis of sequelae CATA and THYR shows that the GRE is always PROBABLE for 7 communes in PACA: Mandelieu-la-Napoule, Puget-sur-Argens, Brignoles, Lancon-en-Provence, Avignon Cabannes and Lambesc. And when GRE is extended to the POSSIBLE modality, then 11 *communes* are concerned: la Valette-du-Var, Bandol, Isle-sur-la-Sorgue, Pelissanne, are added.

The results highlight differences and similarities to the attribution of GRE and morbid geographic patterns emerge. The GRE modality: PROBABLE gives an adversely affect to the environmental health policies, for the communes concerned. However, many uncertainties must be considered for the interpretation of GRE. Moreover, U_k where GRE is UNPROVABLE should be treated with great consideration. Indeed, these U_k can not be considered as communes where no patient would be spatialized because epidemiological data are well available.

4   Discussion

EstimGRE method allows: Modeling geography MP - by two m.st.i $z^j_{U_k,c}$, prevalence of diseases weighted EpiGeoStat converted into patient-years and $z^j_{U_k,q}$, EpiGeoStat propensities to develop the disease which include the quality of medical knowledge and exposure times at the environment ; Characterizing the territories by morbid Geographical Risk Exposures (GRE) - through $z^{REG.j}_{U_k}$. This last one assigns a modality among four GRE: PROBABLE, POSSIBLE, UNPROVABLE or WEAK.

Applying EstimGRE algorithm to the LEA data, the geography of particular sequelae can be modeled. The analysis of spatial contiguity-has shown that the phenomena under study are not randomly located - according to *the First Law of Geography* [20], and suggests an obvious environmental effect. Then characterization U_k by identical GRE highlighted *morbid clusters,* which strengthen the hypothesis that environmental exposure to risk factors could contribute to morbidity observed.

Sequelae developed are partly determined by the aggressiveness of treatment. The adverse effect of some contributory environmental factors, suggested by the geographical approach, strengthens epidemiological *assumptions of the studies carried out on LEA* [9], i.e with individual-centric patterns.

The i.st.m $z^j_{U_k,c}$ and $z^j_{U_k}$, are adapted to the spatial analysis of health-environment interactions. They can be crossed with indicators modeling dangerous environmental factors found in living environments, like: *The geography of individual characteristics, of targeted populations* [21]; The *geographic access to health care, which clearly influences sequelae* [22]; The *socioeconomic contexts that induce morbid predispositions* [23]; Exposure *with adverse physicochemical agents,* eg radioactivity and other toxic substances [24].

Uncertainties caused by the complexity of this spatiotemporal approach are numerous: imprecision of patient spatialization, ambiguities of their life trajectories, epidemiological variables with conflicting qualities... Therefore, it is necessary to consider the significance of the i.st.m proposed. Indeed, spatial indicators often modelize different phenomena than those for which are designed - it is impossible to identify this gap [25]. Therefore, in geography *the risk exposure can be expressed only as a possible relationship between the environment and human* [26].

Assuming that the i.st.m proposed, roughly model the reality of health states; then the $z^{REG.j}_{U_k}$ are well suited to sustainable management of territories. In fact, because they can be computed at different times, they allow us to estimate changes in health statements and they can be used to assess the efficiency of public policies related to limit the noxious environmental exposures.

5   Conclusion

The *geographical medico-metric aims to build spatial indicators.* When they are interfaced with a GIS, it can *integrate interdisciplinary dimensions*: epidemiological, economical, sanitary, social, psychological ... that serves as space media to develop effective public health policies [5].

The challenges of GRE modeling involve at different socio-spatial levels. At the individual scale, giving medics the power to propose solutions to reduce environmental risk exposures. And, assuming that has been established for a predisposed population, it can be generalized. So, at the community scale, they are suitable to develop policies measures devoted to improve the environmental health. *When public health policies take account of local population's needs*, i.e. morbid GRE and at the same time, *their desires to improve particular territorial structures* rather than others - *the political control remains under the social control* [3].

Healthy humain societies are more active, more involved in the socioeconomic and sanitary development of the spaces. And, when jointing: health considerations, and popular desires, that creates dynamic conducive to sustainable development with the social resonances necessary to an overall improvement of their quality of life [27].

Acknowledgments. I want to thank: especially the Prof. Auquier P. which funded my research and for his advices in epidemiology, the Prof. Voiron C. for the spatial weighting system of the geographical fuzzy metric, and Lauby P. for the illustrations.

References

1. Ministère de l'Enseignement Supérieur et de la Recherche; Ministère de la Santé et des Sports : Pan Cancer 2009-2013. Institut National du Cancer (INCa). Boulogne-Billancourt, France (2009)
2. Grmek, M.-D.: Le concept de maladies émergentes. In: History and Philosophy of the Life Sciences, Napoli, Italy, vol. 15, pp. 281–296 (1993)
3. Salem, G.: Géographie de la santé, santé de la géographie. In: Espace, Population, Société, Paris, vol. 1, pp. 25–30 (1995)
4. Abramson, J.-H., Abramson, Z.-H.: Making Sense of Data: A Self-Instruction Manual on the Interpretation of Epidemiological Data. Oxford University Press, Oxford (1988)
5. Bailly, A.: La géographie du bien-être. Presses Universitaires de France, Paris (1981)
6. Zeitouni, K.: Analyse et extraction de connaissances des bases de données spatio-temporelles. In: Habilitation à Diriger des Recherches. Université de Verseilles Saint-Quentin-en-Yvelines, Paris (2006)
7. IPCS. Glossary of key exposure assement terminology. In: International pro-gramme on Chemical Safety, Harmonization Project Document, num. 1 (2004)
8. Peguy, C.-P.: L'horizontal et le vertical, RECLUS, Montpellier, France (1996)
9. Michel, G., Bordigoni, P., Simeoni, M.-C., Curtillet, C., Hoxha, S., Robitail, S., Thuret, I., Pall-Kondolff, P., Chambost, H., Orbicini, D., Auquier, P.: Health status and quality of life in long-term survivors of childhood leukaemia, the impact of haemato-poietic stem cell transplantation. Bone Marrow Transplantation 40(9), 897–904 (2007)
10. Dubois, D., Prades, H.: On the use of aggregation operations in information fusion processes. In: Fuzzy Sets and Systems, vol. 42, pp. 144–161. Elsevier (2004)
11. Bernard, P.-M., Lapointe, C.: Mesures statistiques en épidémiologie. Presse de l'Université du Québec, Quebec (2003)
12. Micosoft. Microsoft office, http://www.office.microsoft.com

13. ESRI. Esri solution SIG - ArGis.10, `http://www.esrifrance.com`
14. Institute for Statistics and Mathematics - R, `http://www.cran.r-project.org`
15. Saporta, G.: Probabilité, analyse des données et Statistique. Technip, Paris (2006)
16. Michel, G., Auquier, P.: Elaboration of an epidemiological weighting system to model sequelae geography. Public Health Laboratory, Marseille (2013)
17. Couet, C.: La mobilité résidentielle des adustes: Existe-t-il des "parcours type"? Dans portrait social. In: INSEE, Paris, pp. 159–179 (2006)
18. Marcotte, D.: Cours: GML6402, `http://geo.polymtl.ca/~marcotte`
19. Chernick, M.R.: Boostrap methods: A practitioner's guide. Wiley, New-York (1999)
20. Tobler, W.: A Computer Movie Simulating Urban Growth in the Detroit Region. In: Economic Geography, pp. 234–240. Clark University, Worcester (1970)
21. Abramson, J.-H., Abramson, Z.-H.: Making Sense of Data: A Self-Instruction Manual on the Interpretation of Epidemiological Data. Oxford University Press, Oxford (1988)
22. Barnett, S., Roderick, P., Martin, D., Diamond, I., Wrigley, H.: Interrelations between three proxies of health care need at the small area level: an urban/rural comparison. Journal of Epidemiol Community Health 56(10), 754–761 (2002)
23. Chaix, B., Merlo, J., Chauvin, P.: Comparaison of spatial approach with the multi-level approach tor investigating place effects on health: the example of halthcare utilisation in France. Journal of Epidemiology and Community Health 59(6), 517–526 (2005)
24. Brucker-Davis, F.: Effects of environmental sythetic chemical on thyroid function. Thyroid 1, 827–856 (1998)
25. Brook, R.H., Lohr, K.N., Chassin, M., Kosecoff, J., Fink, A., Solomon, D.: Geographic variations in the use of services: do they have any clinical significance? Health Affairs 3, 64–73 (1984)
26. Bailly, A., Beguin, H.: Introduction à la géographie humaine. Armand Colin, Paris (2005)
27. Dumont, G.-F.: Les Populations du monde, 2nd edn. Armand Colin, Paris (2004)

Cokriging Areal Interpolation for Estimating Economic Activity Using Night-Time Light Satellite Data

Dimitrios Triantakonstantis and Demetris Stathakis

Laboratory of Spatial Analysis, GIS and Thematic Mapping,
Department of Urban and Regional Planning Engineers, University of Thessaly,
Pedion Areos, 38334 Volos, Greece
trdimitrios@gmail.com, dstath@uth.gr

Abstract. There is a strong correlation between economic activity, which can be measured by Gross Domestic Product (GDP) and the night-time light emissions data. Gross Domestic product is usually available in large aggregate units and therefore the disaggregation has become a necessity in urban and regional development. The night-time light data obtained by Defence Meterological Satellite Program – Operational Linescan System (DMSP-OLS) are supplementary information for measuring GDP in disaggregated units. Cokriging areal interplation was used in this present study in order to disaggregate the GDP contained in 51 Greek administrative divisions NUTS 3. The final disaggregated units are the 1035 municipal divisions. The supplementary night-time light emission data were used as additional variable (proxy variable) to cokriging method. The results showed high performance because cokriging incorporates the spatial aurocorrelation and cross-correlation between the examined variables.

Keywords: Cokriging, Gross Domestic Product, DMSP-OLS, areal interpolation, Greece.

1 Introduction

There is a significant need to link the socioeconomic with the environmental variables in order to better understand earth systems. It is a necessity to integrate socioeconomic datasets with other sources of environmental data. The socioeconomic data are usually available at a national level, which in many cases it is inconvenient for further spatial analysis. Therefore, the disaggregation of socioeconomic datasets has become a necessity.

GIS and Remote Sensing technologies can provide us not only with physical variables of the earth surface but also with many characteristics – variables associated with human activities. The following applications are some examples of these human related variables: measurement of anthropogenic CO_2 (Doll et al. 2000), measurement urban population density (Wu an Murray 2005) and estimation of Gross Domestic Product (GDP) (Elvidge et al. 1997; Sutton et al. 2007).

GDP is the market value of goods and services produced in a country within a specific time period. It is a direct indicator of the economy in a country, assisting policy

B. Murgante et al. (Eds.): ICCSA 2014, Part IV, LNCS 8582, pp. 243–252, 2014.

makers to formulate and evaluate development plans. It also helps individuals such as investors to make the appropriate decisions (Bhandari and Roychowdhury 2011). However, maps showing the distribution of GDP are usually available at national level. This is a limitation when we integrate GDP maps with other environmental data, which usually have higher spatial resolution. This paper is based on the disaggregation of socioeconomic data (e.g. Gross Domestic Product) using environmental datasets (e.g. Night-Time Light Emission) into smaller units.

The relationship between night-time light emissions with GDP was first investigated by Elvidge et al. (1997) and expanded by Doll et al. (2000), where a first map of disaggregated GDP based on satellite data estimated the global economy. Moreover, the correlation of GDP with night-time light emissions has been utilized by many studies (Ebener et al. 2005; Sutton et al. 2007; Ghosh et al. 2009; Henderson et al. 2009).

Cities like cats will reveal themselves at night (Rupert Brooke, 1887-1915). The evident that lights could be detected at night by remote sensors is one of the most important benefits of remote sensing. Defense Meteorological Satellite Program - Operational Linescan System (DMSP-OLS) provides night-time imagery since 1994. DMSP-OLS is a program generated by National Oceanic and Atmospheric Administration's National Geophysical Data Center (NOAA/NGDC).

Geostatistics is based on the theory of regionalized variables. Any variable which describes a phenomenon spreading in space and exhibiting a certain spatial structure is defined as regionalized variable (Matheron, 1971). Interpolation using geostatistical methods such as kriging or cokriging has the great advantage of considering the spatial autocorrelation and cross-correlation between examined variables (Kyriakidis 2004; Yoo and Kyriakidis 2006; Liu et al. 2008; Nagle et al. 2011). This asset has a unique value in environmental and socioeconomic phenomena.

Interpolation techniques are usually applied in point datasets. When we need to predict a variable in an areal unit, such as administrative city boundaries, then the Modified Area Unit Problem (MAUP) appears (Openshaw and Taylor 1979). The problem is called "the modified area unit" because the boundaries of many geographical units can be changed through time. Political and administrative boundaries are subject to be organized. These changes produce many inconsistences (Wong 2004). The MAUP consists of two components, which cause variations in statistical results: scale and zoning effects. The scale effect occurs when different results are obtained from the same data at different spatial resolutions. The zoning effect occurs when the scale of analysis remains constant but areas are aggregated in different zones (Wrigley et al. 1996).

The GDP in Greece in 2010 for NUTS 3 is available by Eurostat and used for the needs of this current study. The NUTS classification (Nomenclature of territorial units for statisics) is a hierarchical system for dividing European Union for statistical purposes. The NUTS 3 divisions refer to small regions for specific diagnoses. Our method describes the utilization of cokriging areal interpolation method in order to estimate the GDP in smaller spatial units, i.e. municipal administrative divisions. The supplementary information used in cokriging method was the DMSP-OLS night-time light emissions. The results of this research may be useful in governmental and private sectors for covering the needs of estimating the local economy and designing appropriate applications for development.

The objectives of this paper are a) to estimate GDP in disaggregate units using night-time light emissions and b) to evaluate a more "intelligent" technique, i.e. cokriging areal interpolation in a socioeconomic variable. Because cokriging usually is applied to environmental variables, its application to socioeconomic variable such as GDP marks a significant novelty.

2   Materials and Methods

2.1   Study Area and GDP Data

The study area is Greece, which contains 51 NUTS-3 administrative divisions and 1035 municipal divisions. According to the 2011 census, the total resident population is 10.815.197 inhabitants. The Greek economy, having achieved high growth by 2008, showed signs of recession in 2009, as a result of global financial crisis. Since 2010 the recession intensified due to fiscal imbalances. The strict incomes policy and the drastic reduction of public expenditure during the last three years negatively influenced the GDP, as expected. GDP declined by 4,9% in 2010, 7,1% in 2011 and 6,4% in 2012 (constant prices 2005).

It is estimated that the GDP will continue declining by 4,4% in 2013, while the economy will return to growth of 0,6% in 2014 (Eurostat, 02/2013). The impact of GDP decrease in unemployment was very important, reaching 27% in 2013 (Figure 1). In particular, youth unemployment which exceeds 50% is one of the greatest problems of the economic crisis.

GDP data are at spatial level of NUTS 3 regions and these values are in purchasing power standards (PPS). PPS are a fictive currency unit which eliminates the differences occurred in purchasing power, i.e. different price levels between countries.

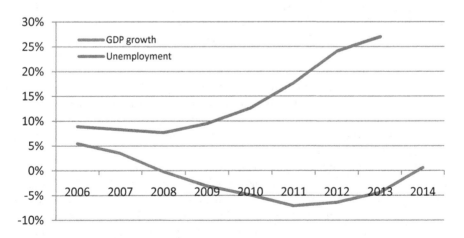

Fig. 1. Gross Domestic Product growth and unemployment during economic crisis period

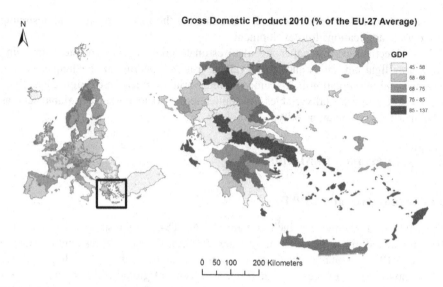

Fig. 2. Study area: Gross Domestic Product per inhabitant in purchasing power standards (PPS) by NUTS 3 regions (2010) (% of the EU-27 average, EU-27 = 100)

A GDP above the EU27 average is observed in Athens and in adjacent regions (Figure 2), i.e. divisions of Voiotia (national industrial pole) and Cyclades Islands (national touristic pole).

2.2   Night-Time Light Satellite Data

The National Geophysical Data Center (NGDC) of the National Oceanic and Atmospheric Administration (NOAA) provides Defense Meteorological Satellite Program - Operational Linescan System (DMSP-OLS) data since 1994. Each OLS has a swath width of 3000 Km and 14 orbits per day and therefore is able to cover the globe in a 24 h period. The night-time imagery (night time pass: 20.30 - 21.30 local time each night) is available in a number of different products (Elvidge 2002).

The nighttime lights product for 2010 named "Average Lights X Pct" was used for the purposes of this study. This product, which has been already processed, is derived from the average visible band digital number (DN) of cloud-free light detections multiplied by the percent frequency of light detection. The percent frequency of light detection normalizes the DN. The spatial resolution of the original night-time lights data is 2,7 Km. After geolocation to 30 arc-second grids, the resolution become approximately 1 km at the equator.

Night-time light imagery has been used in numerous applications. Among them, most common application is the mapping of urban areas. More sophisticated applications include mapping of socioeconomic variables. Population density, economic activity, greenhouse gas emissions, light pollution, disaster management due to fires are some examples where night-time light data have played a significant role (Doll 2008).

2.3   Areal Interpolation Using Cokriging

DMSP-OLS night-time light emission data can be used as supplementary information for GDP areal interpolation. GDP mapping usually based on relationships between GDP and night-time light data using regression analysis (Doll et al. 2006; Ghosh et al. 2010). A problem that arises is the assumption of spatial independency of GDP using regression. This problem can be addressed using areal interpolation such as cokriging, where spatial autocorrelation and cross-correlation between GDP and night-time light data are taken place. Cokriging is usually applied when the supplementary information is abundant and the estimated variable is not (Wu and Murray 2005).

The spatial dependency between neighborhood data is described by the semivariogram, which is a main tool in cokriging. The semivariogram indicates the relationship between the semivariance, given in equation (1) and the distance between sampling points (Journel and Huijbregts 1978).

$$\gamma_{uv}(h) = \frac{1}{2N(h)} \sum_{i=1}^{N(h)} \{(Z_u(x_i) - Z_u(x_i + h)\}\{(Z_v(x_i) - Z_v(x_i + h)\} \tag{1}$$

Where $\gamma(h)$ is the semivariance, $N(h)$ is the number of sampling pairs, h is the distance between sampling points, $Z(x_i)$ and $Z(x_i+h)$ are the value of the variable Z at geographic locations x_i and x_i+h. The u is the primary variable, e.g. GDP and the v is the covariate variable, e.g. the night-time light emissions. The semivariogram is the plot of semivariance with the distance h. Several types of theoretical models with best fit can be applied to the experimental semivariogram. Among them, spherical, exponential and Gaussian are some examples. The formula of the spherical model is given in Equation 2.

$$\gamma(h) = \begin{cases} C_0 + C \left(\frac{3h}{2a} - \frac{1}{2}\left(\frac{h}{a}\right)^3 \right), & 0 < h \le a \\ C_0 + C, & h > a \\ 0, & h = 0 \end{cases} \tag{2}$$

Where C_0 is the nugget variance representing spatially dependent variation occurring in small distances as well as measurement errors, $C+C_0$ is the sill representing the maximum value obtained, a is the range which is the maximum distance where pairs of observations remain correlated and h is the distance.

A cokriging estimation of variable u at a block B is a linear function of N_u measured values $Z_i(x_i)$ and N_v measured values $Z_j(x_j)$. Hence, the cokriging estimation will be (Equation 3):

$$Z_u(B) = \sum_{i=1}^{N_u} \lambda_{ui} z_i(x_i) + \sum_{j=1}^{N_v} \lambda_{vj} z_j(x_j) \tag{3}$$

Where λ_{ui}, and λ_{vj} are weight factors chosen to meet the Equation 4.

$$\sum_{i=1}^{N_u} \lambda_{z_i} = 1, \sum_{j=1}^{N_v} \lambda_{z_j} = 0 \qquad (4)$$

The cokriging variance is given in Equation 5.

$$\sigma^2 = \gamma_{z_u}(0) + \mu_{z_u} - \sum_{i=1}^{N_u} \lambda_{z_i} * \gamma_{z_u}(x_{z_i} - x_0) - \sum_{j=1}^{N_v} \lambda_{z_j} * \Gamma_{z_u z_v}(x_{z_j} - x_0) \quad (5)$$

Where μ_{z_u} is the Lagrangian multiplier used to satisfy the first part of Equation 4, γ_{z_u} is the semivariance of $z_i(x_i)$ and $\Gamma_{z_u z_v}$ is the cross-semivariance of $z_i(x_i)$ and $z_j(x_j)$.

3   Results

Cokriging areal interpolation was used to disaggregate the GDP contained in 51 Greek NUTS 3 administrative divisions using DMSP-OLS night-time light data. The final spatial disaggregate units produced are the 1035 Greek municipal divisions. Cokriging generates a geostatistical layer where each pixel represents the GDP and is further processed in order to aggregate the information to municipal units. The results are graphically presented in Figure 4.

Table 1. Parameters of spherical semivariograms

Sill (GDP)	Sill (lights)	Sill (cross)	Range	No lags	Lag distance
892,53	134,01	99,22	83,25 Km	12	87,69 Km

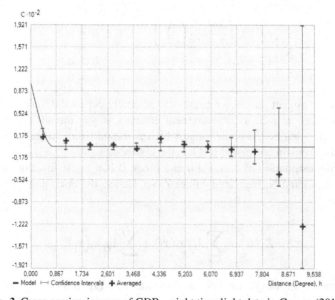

Fig. 3. Cross-semivariogram of GDP – night time light data in Greece (2010)

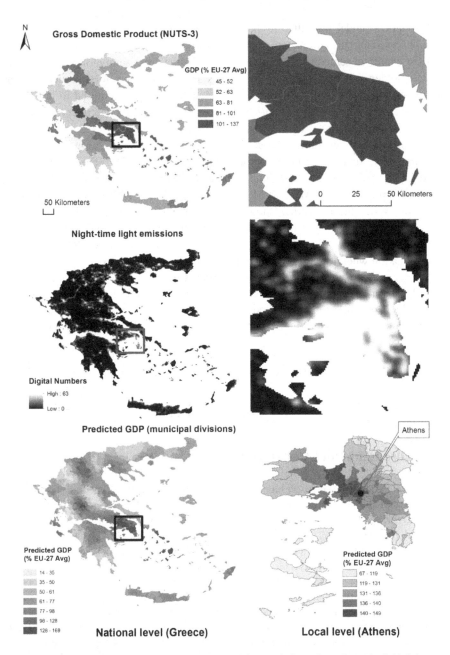

Fig. 4. Disaggregation method using cokriging areal interpolation where from the initial dataset (GDP at NUTS 3 in 2010), the predicted GDP at municipal divisions is produced. The DMSP-OLS night-time light emissions data (2010) are used as supplementary information into the model.

A spherical model was fitted to the experimental semivariograms. No anisotropies were observed and the parameters of the semivariograms are given in Table 1.

Most empirical semivariances (blue crosses) for the cross-semivariogram fall within the confidence intervals (Figure 3). This is an evidence of a good accuracy of the model. Although the semivariogram curve (blue curve) is not required to fall within the confidence intervals, in our case it falls within the confidence interval which is another evidence of a good fit.

The accuracy of the cokriging areal interpolation was achieved using the following measurements: Mean Error (ME), Root Mean Square Error (RMSE), Mean Standardized Error (MSE), Root Mean Square Standardized Error (RMSSE) and Average Standard Error (ASE). If ME and MSE are close to 0 and RMSE is small, then the model fits well. ASE and RMSE should have close values. If ASE>RMSE, the model overestimates the primary variable, while underestimation occurs when ASE<RMSE. RMSSE should have value close to 1. Underestimation occurs when RMSSE>1 and overestimation when RMSSE<1 (Burrough and McDonnell, 1998).

The accuracy results showed a quite significant performance (Table 2). More specifically, the method applied indicates a low underestimation of GDP. This probably occurs due to night-time lights product, which contains detection from fires and a variable amount of background noise. These limitations should be considered in a future research.

Table 2. Accuracy results of the cokriging areal interpolation

ME	RMSE	MSE	RMSSE	ASE
-1,49	21,02	-0,09	1,47	16,71

4   Discussion and Conclusions

The global economic recession has led many people to poverty. Especially in Greece, the effects of the economic recession can be seen in GDP decline as well as in high unemployment rates. The need to understand the spatial distribution of wealth is essential for undertaking appropriate policies in order to take the path to economic recovery. The map of GDP could assist in better understanding the causes and consequences of economic activity.

In this paper the disaggregation of GDP, initially measured at NUTS 3 level, is applied in Greece using cokriging areal interpolation method. The supplementary information in cokriging is the DMSP-OLS night-time light emission data. The aim is to disaggregate the GDP into smaller spatial units, i.e. municipalities. The areal interpolation using geostatistical methods, such as kriging and cokriging in estimating socioeconomic variables is quite new in the literature. Therefore, this study is one of the few examples where socioeconomic variables are interpolated using remote sensing data by geostatistics.

The advantage of using geostatistical method in areal interpolation is that spatial autocorrelation and cross-correlation between variables are considered. This is a powerful characteristic in estimating a socioeconomic variable because conventional interpolation methods such as regression violated the assumption of spatial independency of the data. Therefore, inconsistencies in prediction results are limited.

GIS and Remote Sensing technologies appear to highly contribute in environmental and socioeconomic studies. The ability to support with data in high spatiotemporal resolution makes them a unique resource of data and methods not only for research but also for everyday needs of urban and regional planning and decision making.

Finally, the proposed methodology could assist all the involved bodies responsible for development plans in national and regional perspective. Governmental and private sectors may benefit from knowing the economic activity in disaggregate units because they could make the appropriate decisions for recovering. Development strategies can be effectively supported by analyzing the GDP and therefore, the contribution of this study is highly important.

Acknowledgements. The research presented here was supported by Greek State Scholarships Foundation (IKY Fellowships of Excellence for Postgraduate Studies in Greece – Siemens Program, contract number: SR22020/13). Also, authors would like to thank NOAA and Eurostat for making public night-time light data and GDP data.

References

1. Burrough, P.A., McDonnell, R.A.: Principles of Geographical Information Systems. Oxford Univ. Press, Oxford (1998)
2. Doll, C.N.H., Muller, J.P., Elvidge, C.D.: Nighttime imagery as a tool for global mapping of socio-economic parameters and greenhouse gas emisssions. Ambio 29(3), 157–162 (2000)
3. Doll, C.N.H., Muller, J.P., Morley, J.G.: Mapping regional economic activity from night-time light satellite imagery. Ecological Economics 57, 75–92 (2006)
4. Doll, C.N.H.: CIESIN Thematic Guide to Night-time Light Remote Sensing and its Applications. Center for International Earth Science Information Network of Columbia University, Palisades (2008); http://sedac.ciesin.columbia.edu/tg/ (accessed on January 21, 2014)
5. Ebener, S., Murray, C., Tandon, A., Elvidge, C.D.: From wealth to health: modeling the distribution of income per capita at the subnational level using nighttime light imagery. International Journal of Health Geographics 4, 1–17 (2005)
6. Elvidge, C.D.: Global observations of urban areas based on nocturnal lighting. LUCC Newsletter 8, 10–12 (2002)
7. Ghosh, T., Powell, R.L., Elvidge, C.D., Baugh, K.E., Sutton, P.C., Anderson, S.: Shedding light on the global distribution of economic activity. The Open Geography Journal 3, 148–161 (2010)
8. Ghosh, T., Anderson, S., Powell, R.L., Sutton, P.C., Elvidge, C.D.: Estimation of Mexico's informal economy and remittances using nighttime imagery. Remote Sensing 1(3), 418–444 (2009)
9. Journel, A.G., Huijbregts, C.J.: Mining geostatistics. Academic Press, New York (1978)
10. Henderson, J.V., Storeygard, A., Weil, D.N.: Measuring economic growth from outer space. NBER Working Paper 15199. National bureau of economic research. Cambridge (2009)

11. Kyriakidis, P.C.: A Geostatistical Framework for Area-to-Point Spatial Interpolation. Geographical Analysis 36, 259–289 (2004)
12. Liu, X.H., Kyriakidis, P.C., Goodchild, M.F.: Population-density estimation using regression and area-to-point residual kriging. International Journal of Geographical Information Science 22(4-5), 431–447 (2008)
13. Matheron, G.: The Theory of Regionalized Variables and Its Applications. In: Cahiers du Centre de Morphologic Mathematique de Fontainebleau, vol. 5, Ecole National Superieure des Mines (1971)
14. Nagle, N.N., Sweeney, S.H., Kyriakidis, P.C.: A Geostatistical Linear Regression Model for Small Area Data. 一种适用于小区域数据的地统计线性回归模型. Geographical Analysis 43, 38–60 (2011)
15. Openshaw, S., Taylor, P.: A million or so correlation coefficients: Three experiments on the modified area unit problem. In: Wrigley, N. (ed.) Statistical Applications in the Spatial Sciences, London, Pio, pp. 127–144 (1979)
16. Sutton, P.C., Elvidge, C.D., Ghosh, T.: Estimation of gross domestic product at sub-national scales using nighttime satellite imagery. International Journal of Ecological Economics & Statistics 8, 5–21 (2007)
17. Wong, D.W.S.: The Modifiable Areal Unit Problem (MAUP). In: WorldMinds: Geographical Perspectives on 100 Problems, pp. 571–575 (2004)
18. Wrigley, N., Holt, T., Steel, D., Tranmer, M.: Analysing, modelling and resolving the ecological fallacy. In: Longley, P., Batty, M. (eds.) Spatial Analysis: Modelling in a GIS Environment, pp. 25–41. Wiley, Chichester (1996)
19. Wu, C., Murray, A.T.: A cokriging method for estimating population density in urban areas. Computers, Environment and Urban Systems 29, 558–579 (2005)
20. Yoo, E.Y., Kyriakidis, P.C.: Area to point Kriging with inequality-type data. Journal of Geographical Systems 8(4), 357–390 (2006)

Comparative Analysis of Models of Location and Spatial Interaction

Mauro Mazzei and Armando Luigi Palma

National Research Council,
Istituto di Analisi dei Sistemi ed Informatica "Antonio Ruberti",
Viale Manzoni 30, I-00185, Rome, Italy
mauro.mazzei@iasi.cnr.it, palma@arpal.it

Abstract. The transformations of urban and regional systems are the result of complex interactions between the physical environment and socio-economic environment differentiated in space and time. These systems are very complex and make it difficult to understand the phenomena of urban and territorial transformation and their causes; therefore it is necessary to reduce their complexity to make it easier to effect analyses, forecasting, optimization and control of these local systems. In order to simplify reality it is necessary to present models that represent all the most significant factors that allow an understanding of the systems studied. This work is focused on the comparative analysis of the various gravity models applied. Localization models of activity (Hansen, Lakshmanan), and spatial interaction models (Lowry and Lowry-Garin). Finally, we propose a model for the optimal location of a large shopping center in the province of Taranto.

Keywords: GIS, urban models, spatial data analysis, spatial statistical model.

1   Introduction

In everyday language a "system" is designed as a set of connected parts that can interact so as to form a functional unit. We give here a more precise definition that will be useful in considering the "system" as a collection of parts that communicate with each other, the activity of which is aimed at achieving a result. The communication between the parts of the "system" is intended to implement the necessary coordination of the activities that allows one to achieve the results expected. The communication between parts of the system involves different types of interaction such as the transmission (transport or spatial transformation) of material goods, of people, of messages.

The activities of the parts of a system are always connected to a place, the same thing happens for communications. The first step is therefore to define a system to recognize the activities related to communications. A system can be graphically represented by a connected graph; the nodes are representative of the parts of the system, the connecting arcs are representative of the existing channels of

B. Murgante et al. (Eds.): ICCSA 2014, Part IV, LNCS 8582, pp. 253–267, 2014.

communication between parts of the system. These channels can be constructed connections, such as roads, trails, railways, canals, pipelines, telephone cables, or natural connections such as rivers, air corridors, the summit of the mountains or the bottom of the valleys. In order to achieve the expected result the activities of the parts of a system should be subject to a controlled adjustment of the error. This adjustment (feedback) is implemented through a control mechanism that provides information on the actual state of the system, compared with the desired state.

In our case, the city is considered to correspond to the system that we want to check; we can assess what the actual state of the system is through various forms of investigation: population, employment, services, housing [1]. The city can be influenced by adding, removing or altering any parts, components or connections between components, i.e. influencing land use and communications [6][7]. This result can be obtained in two different ways.

Firstly, through public intervention relating to hospitals, schools, residence, services, roads, transport on road and rail, parking, etc.

Secondly, indirectly, by adjusting the modifications brought about by private intervention through appropriate control processes and developmental regulation [8]. It is clear that we must have the means to predict the effects of our actions, public or private, for the time in which they occur.

Sometimes the system can go beyond the limits set by a corrective action plan and it may be too late. The means that allow us to predict the effects of our actions are the simulation models that increase and expand the experience of the planner, indicating the need for corrective action and allowing the testing of the different forms of intervention.

In the simulation models variable time can be introduced; you can also check out the various aspects of the actions in the short, medium and long term.

2   Case Study and Data

The case study is located in the province of Taranto in southern Italy. We examined the 29 municipalities included in the province of Taranto. The statistical data of these municipalities have been provided by the Italian Institute of Statistics (ISTAT). Usually the statistical data are organized by census areas and census year. In this case study we used the data referring in the census of 2001 and 2011. For greater clarity the data were collected for each municipality. The aggregate data are shown in Tables 1 and Table 2.

Below are shown the thematic maps relating to the ISTAT data processed. These thematic maps show the variables used during the comparison of different urban models. In order to facilitate the reading of thematic maps, only four classes have been used in the process of data visualization.

Table 1. DATA 2001 - ISTAT

ZONES	BASIC EMPLOYMENT	SERVICE EMPLOYMENT	TOTAL EMPLOYMENT	TOTAL POPULATION	RESIDENTIAL AREA
1	245	507	752	7303	3442
2	55	303	358	6070	1997
3	369	1650	2019	17393	5501
4	268	1397	1665	12973	3345
5	171	189	360	3513	1139
6	109	280	389	5639	1983
7	1081	2013	3094	22146	6607
8	635	2186	2821	31894	7093
9	242	1534	1776	14996	3143
10	44	357	401	5810	4200
11	76	568	644	10195	5287
12	502	2754	3256	31747	19916
13	3848	5331	9179	48756	15866
14	64	357	421	5386	6944
15	1425	2865	4290	30923	6080
16	49	212	261	5199	1755
17	63	242	305	4277	962
18	13	129	142	2411	798
19	389	1248	1637	16575	4073
20	112	495	607	7483	1559
21	313	1022	1335	15815	3601
22	132	962	1094	10240	5470
23	4	62	66	1756	622
24	561	1139	1700	15613	2820
25	43	502	545	8830	3578
26	344	1465	1809	16163	7992
27	234	515	749	14585	2773
28	19317	32491	51808	202033	17581
29	199	284	483	4082	4923

Table 2. DATA 2011 - ISTAT

ZONES	BASIC EMPLOYMENT	SERVICE EMPLOYMENT	TOTAL EMPLOYMENT	TOTAL POPULATION	RESIDENTIAL AREA
1	131	679	810	6964	3442
2	135	469	604	6963	1997
3	194	1806	2000	17075	5501
4	226	1348	1574	13646	3345
5	225	345	570	3558	1139
6	200	500	700	5345	1983
7	876	2556	3432	22555	6607
8	694	3362	4056	32544	7093
9	954	1926	2880	15316	3143
10	109	677	786	7873	4200
11	107	781	888	10192	5287
12	531	3825	4356	30795	19916
13	2996	7856	10852	48958	15866
14	35	506	541	5355	6944
15	1315	4230	5545	32548	6080
16	786	300	1086	5530	1755
17	69	309	378	4037	962
18	5	117	122	2410	798
19	639	1626	2265	16127	4073
20	48	537	585	7829	1559
21	231	1525	1756	16111	3601
22	145	1194	1339	11221	5470
23	11	93	104	1797	622
24	571	1753	2324	15480	2820
25	66	746	812	9237	3578
26	299	1834	2133	16343	7992
27	286	823	1109	14055	2773
28	16491	35806	52297	198728	17581
29	193	372	565	4222	4923

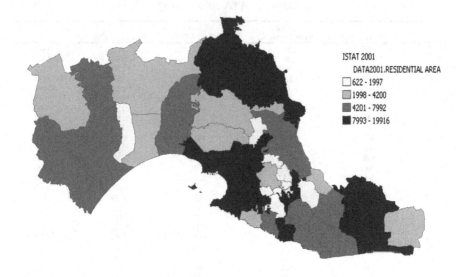

a)

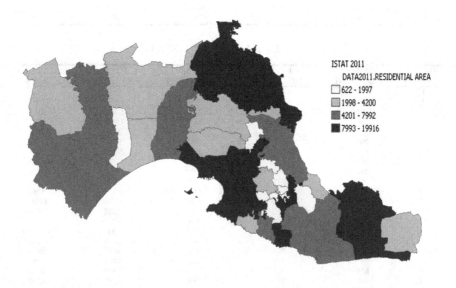

b)

Fig. 1. Residential area. (*a*) 2001 (*b*) 2011.

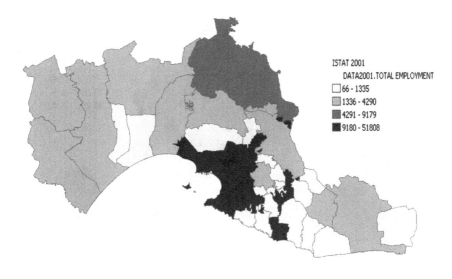

a)

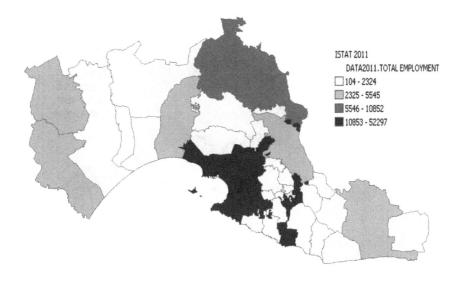

b)

Fig. 2. Total employment. (*a*) 2001 (*b*) 2011.

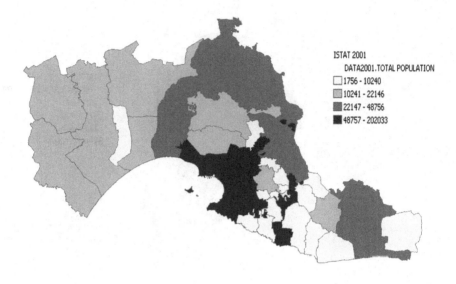

a)

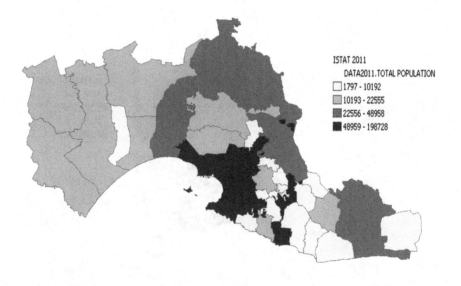

b)

Fig. 3. Total population. (*a*) 2001 (*b*) 2011.

3   Gravity Models

In order to study the spatial interactions of urban aggregates (flows of people, goods, money, etc.), the gravity models are the ones most commonly used as a form of mathematical model. According to these models (traffic flows, money, goods, etc.), the force of attraction that is exerted between two urban centers is directly proportional to the product of their "masses", they are measured as the number of residents, jobs, bank deposits and inversely proportional to a power of the distance between them [5]. This law can be expressed in general as:

$$T_{i,j} = k \; O_i \; D_j \; d_{ij}^{-\alpha}$$

where $T_{i,j}$ is the intensity of interaction from urban centers i and j, O_i and D_j express the intensity of attraction exerted by the two centers as places of origin and destination α is the exponent of the distance between the centers of the i and the center j (not necessarily equal to 2) and k is a proportionality constant that ensures the internal consistency of the model [13][14].

In this general formulation of the gravity model, you can add the formulation of the gravity model in the source and destination bound:

$$T_{i,j} = A_i \; O_i \; B_j \; D_j \; d_{ij}^{-\alpha}$$

where A_i and B_j are two separate sets of scale factors defined as follows:

$$A_i = 1 / (\Sigma_j \; B_j \; D_j \; d_{ij}^{-\alpha})$$

$$B_j = 1 / (\Sigma_i \; A_i \; O_i \; d_{ij}^{-\alpha})$$

below is the formulation of the gravity model for specific purposes:

$$T_{i,j} = W_i \; B_j \; D_j \; d_{ij}^{-\alpha}$$

with

$$B_j = 1 / (\Sigma_i \; W_i \; d_{ij}^{-\alpha})$$

$$\Sigma_i \; T_{i,j} = D_j$$

that W_i is an estimate of the attraction exerted by the urban center as i-th origin. Finally, we report the formulation of the gravity model in origin bound:

$$T_{i,j} = A_i \; O_i \; W_j^{*} \; d_{ij}^{-\alpha}$$

with

$$A_i = 1 / (\Sigma_j W_j^* d_{ij}^{-\alpha})$$
$$\Sigma_j T_{i,j} = O_i$$

that W_j^* is an estimate of the attraction exerted by the urban center as the seat of the j-th destination.

3.1   The Gravity Model or Potential Model of Hansen

The Hansen model is a localization model to predict the location of the population. It is based on the assumption of accessibility to workplaces. The accessibility is given by the time required on a given road network to move from the residence to the workplace. This is the biggest determinant of the location of residences [16][17].

This model is not a gravity model in the strictest sense since it is based on the interaction between the areas, and it is more correct to define it as a model "potential", in fact regards the interaction potential on the accessibility of the various zones [18].

The expression of the accessibility of the area in relation to the i-th in j-th area is given by:

$$A_{ij} = E_j d_{ij}^{-\alpha}$$

where $A_{i,j}$ is the accessibility index of the i-th area in relation to the area j-th, E_j are the total employment in the j-th, $d_{ij}^{-\alpha}$ is the distance between the area of the i-th and j-th area high exponent α. The overall accessibility to the j-th zone is the sum of the values of all indices of accessibility compared to other areas and is given by:

$$A_i = \Sigma_j E_j d_{ij}^{-\alpha}$$

In addition to accessibility, one of the major factors that determine the attraction of population to a certain area is the amount of land available in the area for residential use, Hansen called "residential growth potential" of the area. Accessibility A_i e la incremental residential growth H_i of a certain area can be combined in an index of development potential D_i given by the product of the index of accessibility A_i for the ability to incremental residential growth H_i. It has therefore:

$$D_i = A_i H_i$$

The potential for development D_i can be considered a measure of the attractiveness of each zone, based on accessibility to workplaces and on the availability of building land. The population is attributed to the i-th zone on the basis of its development potential of each of the zones, that is, the development potential of each area divided by the total potential of all areas:

$$(A_i \, H_i) \, / \, (\, \Sigma_i \, A_i \, H_i \,)$$

In other words, the portion of the increase in total G_t of the population (model input) which will be attributed to any zone, depends on the attraction of this area compared to other, to which the portion of the increase of population G_i going to the i-th zone, will be:

$$G_i = G_t \, (A_i \, H_i) \, / \, (\, \Sigma_i \, A_i \, H_i \,)$$

that is:

$$G_i = G_t \, (D_i \, / \, \Sigma_i \, D_i)$$

This model is tool that ids easy to use to locate the population, as a result of alternative distributions of employment, in accordance with changes in travel time resulting from different projects roads, and the different possibilities in the various zones of housing residential growth.

3.2   Lackshmanan/Hansen Models

This is a gravity model used to source frequently used consumer purchases to shopping centers. The model describes the flows of costs between the residential areas and shopping centers and estimate sales in each center by adding the cash for purchases from all areas to each of the centers [10][11].

The model states that the sale of the shopping center is directly proportional to the attraction of the shopping center, and inversely proportional to its distance from residential areas and the competitiveness of other centers. The only difference of this model from the basic model in origin bound is the addition of an exponent, *a*, the variable, from their surface, which measures the attractiveness of each center. This became necessary after it was experimentally verified that large shopping centers attracted proportionally more than their size would seem to indicate. The analytical formulation of the model is the following:

$$S_{ij} \, = C_i \, A_i \, F_j^a \, d_{ij}^{\,-\alpha}$$

where:

$S_{ij} =$ shopping from the residential area to the shopping center $_j$;

$C_i =$ total shopping in the residential area i;

$F_j^a =$ size attraction or of the shopping center j;

$A_i = (\, \Sigma_j \, F_j^a \, d_{ij}^{\,-\alpha})^{-1}$;

$d_{ij} =$ distance from the residential area to the shopping center j

a e $\alpha =$ exponents to be determined in the calibration of the model.

This model requires input data on the purchases made by each residential area, and the sales area (size) of the shopping centers, as well as the matrix distance / time of each residential area by each shopping center (Fig. 4).

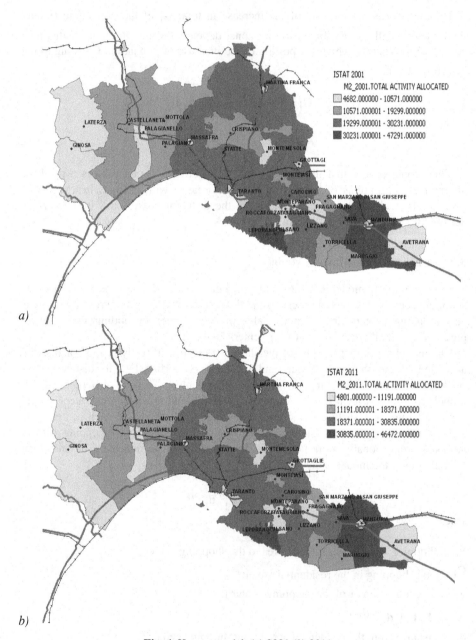

Fig. 4. Hansen model. (*a*) 2001 (*b*) 2011.

3.3   Lowry Model

This model integrates the theory of the economic basis with the theory of gravity models. The theory of the economic basis divides the area into two sectors of activity: the basic sector and the services sector. The basic sector, or export sector, produces goods that are consumed mainly outside the area and its growth is mainly linked to the national economy. The growth of service industries, that is not for export, depends on the increase in the population of the area, because the population that provides the workforce for the basic industries creates a services demand such as shops, transport, banking, etc. In the Lowry model the distinction between the population dependent employment base and population dependent on employment services is fundamental to the predictions of localization.

The assumptions of the method of the economic basis can easily be expressed in mathematical form. We represent with **E** total employment o fan area, with **B** employment base, with **S** employment service, we have that:

$$E = B + S$$

We can also consider that at some time a given number of total employment **E** maintains a certain population **P**. If we represent the number of persons to be retained by a single job with the symbol α, called dependency index, we have that:

$$P = \alpha\, E$$

where α is a factor that expresses the proportion of the total population employed. We have that:

$$\alpha = P/E$$

The central point of the Lowry model, however, concerns the distinction between the population dependent on the occupation of the base and that which depends on the occupation of service. Therefore, the population can be expressed in terms of employment base and employment service, assuming that the total employment is equal to the sum of basic employment and employment services (Fig. 5).

$$E = B + S$$

from which

$$P = \alpha\,(B + S)$$

$$P = \alpha\, B + \alpha\, S$$

The second assumption for the operation of the Lowry model is that the level of employment service is determined by the level of the population. Therefore, as we have said that the population P is a function of total employment E, we can also say that the employment service S is a function of the total population. That is, we place:

$$S = \beta P$$

β can be considered a service rate of the population, a factor that expresses the amount for employment as the service request or supported by a given population **P**. It also has:

$$\beta = S/P$$

From the basic assumptions, the Lowry model is able to estimate the distribution of the resident population and of jobs in the service sector from a given distribution of jobs in the sector of basic, through an iterative process. In fact, the model calculates the distribution of residents from the distribution of employment in the area of base by a gravity model for a specific purpose

$$T_{i,j} = W_i \, B_j \, D_j \, d_{ij}^{\ -\alpha}$$

that W_i measures the attraction of the area as the location for residence, a measure of the building areas obtained by subtracting from the whole area the equipment industry base and the area of land that cannot be used, while D_j represents the number of jobs in the area **j** . In the first iteration of the model, the jobs D_j are those of the base sector, while in later iterations the values D_j represents the successive increments of employment in the service sector. At this point, we calculate the number of moves that makes the resident population to go to the facilities of the service sector, using a gravity model to origin bound.

$$T_{i,j} = A_i \, O_i \, W_j^{\ *} \, d_{ij}^{\ -\alpha}$$

where $W_j^{\ *}$ is a measure of the attraction of the area as the location of the j-th equipment industry services; in principle, it is expressed by the area occupied by the equipment industry already located in the area. Once we have obtained the estimate of employment in the service sector relative to the first iteration, these values are used in the model for a specific purpose to determine the increase in the distribution of residents. The proceedings are repeated until we reach the equilibrium condition given by a very small increase in population and employment service and therefore not significant.

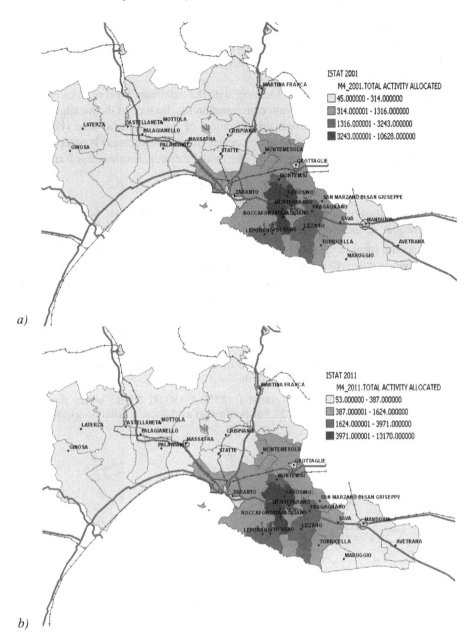

Fig. 5. Lowry model. (*a*) 2001 (*b*) 2011.

4   Conclusions

Hansen's model, compared with the Lowry model for the area of Taranto in 2001 and 2011 with various of calibration tests did not give satisfactory results. In fact, the correlation coefficient between observed and calculated values is 0.23 in 2001 and 0.288 in 2011. We have obtained the best correlation coefficients with the model Lakshmanan-Hansen 0.956 in 2001 and 0.942 in 2011 and Garin-Lowry 0.984 in 2001 and 0.971 in 2011.

We propose a model for the best location, with the maximum level of sales of a big shopping center. The model takes into consideration the distance of the location of the big shopping center by the various urban centers of the considered area [18]. The distance of the location of the big shopping center from all urban areas is perceived by users as an impediment to the movement. The model therefore assumes this form:

$$T_i = \Sigma_j \, d_{ij}^{-\alpha} \, E_j \, R \, \varepsilon$$

Where the exponent α can assume values between 1 and 2, the factor **R** is the average income of the employed **E**$_j$ and the factor ε represents the percentage of monthly income spent in the shopping center. Put $\alpha = 1$, **R**=1800, $\varepsilon = 0.1$ the City 24° corrisponding to "San Giorgio Ionico" (Fig. 6), is the best potential location in relation to sales, with a value of purchases for the year 2001, equal to 7408.0*180,00 EUR / month, equal to annual sales of 7408*180,00*12 whereas localization in 1° City, corrisponding to "Avetrana" (Fig. 6), with a potential value of minimum annual turnover equal to 2698.6*180,00*12 EUR, would result in a difficult survival of a large shopping center.

Finally, it should be noted that the use of the model with census data of 2011 confirmed that "San Giorgio Ionico" is the best localization, with a value of T_{24} = 8550.3 and with monthly sales potential 8550.3*180,00 EUR, while the City "Avetrana", has a value of T_1 = 3120.5 is confirmed as the worst possible location.

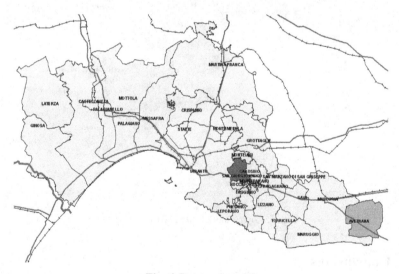

Fig. 6. Proposed model

References

1. Batty, M.: An activity allocation model for the Notts/Derby sub-region. Regional Studies 4(3) (1970)
2. Broadbent, T.A.: Zone size and spatial interaction. Centre for Environmental Studies, Working Note 106 (1969)
3. Carroll, J.D., Bevis, H.W.: Predicting local travel in urban regions. Papers and Proceedings of Regional Science Association 3 (1957)
4. Hayes, C.: Retail Location Models. Centre for Environmental Studies, Working Paper 16 (1968)
5. Cripps, E.L.: Limitations of the Gravity Concept, in Styles (1968)
6. Cripps, E.L., Carter, E.: The Empirical Development of a Disaggregated Residential Location Model: Some Preliminary Results. Urban Systems Research Unit, University of Reading, Working Paper 9 (1971)
7. Freeman, F.: Wilbur Smith and Assoc: London Traffic Survey, vol. II. Greater London Council (1966)
8. Hansen, W.G.: How accessibility shapes land use. Journal of the American Institute of Planners, maggio (1959)
9. Isard, W.: Methods of Regional Anaiysis. MIT Press (1960)
10. Lakshmanan, T.R., Hansen, W.G.: A retail market potential model. Journal of American Institute of Planners, maggio (1965)
11. Lewis, J.P.: The invasion of planning. Journal of the Town Planning Institute, maggio (1970)
12. McLaughlin, J.B., et al.: Regional Shopping Centres in North West Englattd, Part 11. University of Manchester (1966)
13. Shcneider, M.: Gravity models and trip distribution theory. Papers and Proceeding of the Regional Science Associution 5 (1959)
14. Styles, B.J. (ed.): Gravity Models in Town Planning. Lanchester Polytechnic (1968)
15. Tanner, J.C.: Sotne Factors afecting the Amoicnt of Travel. Road Research Laboratory Paper No. 58 (1961)
16. Wilson, A.G.: The use of entropy maximising methods in the theory of trip distribution. Journal of Transport Econoinics and Policy 3(1) (1969a)
17. Wilson, A.G.: Disaggregating Elementary Residential Models. Centre for Environmental Studies, Working Paper 37 (1969b)
18. Wilson, A.G.: Entropy in Urban and Regional Modelling. Centre for Environmental Studies, Working Paper 26 (1969c)

Modelling Impact of Morphological Urban Structure and Cognitive Behaviour on Pedestrian Flows

Marija Bezbradica and Heather J. Ruskin

Centre for Scientific Computing Research and Complex Systems Modelling (Sci-Sym),
School of Computing, Dublin City University, Dublin, Ireland
{mbezbradica,hruskin}@computing.dcu.ie

Abstract. A novel, discrete space-time model of pedestrian behaviour in real urban networks is presented. An agent-based approach is used to define characteristics of individual pedestrians, based on spatial awareness and cognition theories, combined with preferential choices of different social groups. Behaviour patterns are considered incorporating rules of movement along pedestrian routes and for intermediate decision and conflict points. The model utilises dynamic volunteered geographic information system data allowing analysis of arbitrary city networks and comparison of the effect of grid structure and amenity distribution. As an example, two distinctive social groups are considered, namely 'directed' and 'leisure', and their interaction, together with the way in which flow congestion and changes in network morphology affect route choice in central London areas. The resulting stress and flow characteristics of the urban network simulations as well as the impact on individual agent paths and travel times, are discussed.

Keywords: GIS, agent-based modelling, urban spatial-temporal modelling, pedestrian behaviour, geovisualisation.

1 Introduction

As urban environments expand, routine travel to work or other destination becomes more complex causing increase in commuter stress. From the pedestrian viewpoint, familiarity with and ease of navigation on the urban grid can facilitate rapid transit to daily destinations, such as schools, offices, parks and entertainment venues. This in turn can influence lifestyle choice by making walking or cycling both more attractive and efficient than driving.

Although pedestrian behaviour has been studied for more than several decades [1,2], the research has intensified since the early nineties due to improved computational models and increased availability of computing power. Pedestrian dynamic studies have mostly focused on self-organisation and interaction of pedestrian flows [3,27]. A typical application of such models is the prediction of evacuation patterns from enclosed spaces, such as buildings, underground stations and public venues [4,5,6]. Additionally, models were developed to address

B. Murgante et al. (Eds.): ICCSA 2014, Part IV, LNCS 8582, pp. 268–283, 2014.

large-scale problems, notably evacuation in the context of naturally-occurring and man-made disasters, such as hurricanes and terrorist attacks, [7,8].

Notably, following early use of discrete methods (such as cellular automata), agent-based modelling (ABM) has gained considerable popularity for representation of individual interactions. The approach has several key advantages, the most important being the expressive and intuitive nature of the modelling language, its suitability to high-performance execution environments, adaptability to inclusion of heterogeneous behaviour and incorporation of stochasticity [9,10]. The origins of application of ABM to pedestrian modelling lie in simulations of social behaviour and decision-making, [12], introduced in detail in [13]. From early models, where agents of two distinct types populated a simple grid, [11], use has expanded to representation of complex real-world situations and social behaviour involving millions of entities (TRANSIMS, [14]).

In simulating crowd and group dynamics, ABM enables exploration of force effects at different crowd densities by using discrete grid cells with assigned force vectors, [15], and demonstration of local patterns for random pedestrian walks, emphasising the importance of both micro- and macro-simulations, [17]. In addition, ABM can be linked to geographic information systems (GIS), combining spatial and temporal aspects in an effective geo-simulation tool to enable interpretation of urban environments, [18,16]. Nevertheless, models using both separate crowdsourced GIS and ABM are relatively unexplored [19] and, in order to link these, investigation of social behaviour patterns is required, [20]. Studies report that, rather than individual movement, interactions inside and between groups lead to formation of typical walking patterns [21]. Visual perception and route choice is shown to depend on the configuration of the urban street network with focus on cognitive understanding of spatial complexity, and the way in which directional change, rather than distance, impacts the route choice [22,23].

While motorised (and non-motorised) road-using vehicles are constrained by traffic rules, signalisation and street orientation, pedestrian flows are subject to fewer rules and exhibit more flexibility and randomness of choice at every time point [30,24]. Depending on real-time assessment of congestion, route choice can be readily adapted. Further, pedestrian behaviour is much more diverse with each individual permitted flexible options for movement through crowds or definition of 'optimal' route.

These properties motivate the need for bottom-up modelling of pedestrian movement, with the agent basis providing a flexible tool for analysis of complex social behaviour [25]. In this context, the ABM paradigm allows simulation of individual actions by representing pedestrians as agents with active awareness of their environment (traffic, neighbouring pedestrians and the street network).

In this paper, we introduce a discrete, behaviour-driven space-time framework, allowing pedestrian movement to be modelled on a real urban network. The main focus here is on exploring the potential of the approach through example scenarios and investigation of simple hypotheses of pattern evolution as a foundation for later extension. We consider pedestrian movement originating from three main 'cognitive features' [28,23]: (i) walking strategy, (ii) spatial awareness and (iii) knowledge of the urban grid.

The paper is organised as follows: in Section 2 we present the core behavioural model by defining the agent state machine and types of pedestrian behaviour. We also describe the GIS from which the data is sourced for the models. In Section 3 we describe several different behavioural scenarios with respect to different city locations and analyse the key movement patterns observed. Finally, in Section 4 we summarise findings and possible future questions of interest.

2   Model

2.1   Agent State Machine

In order to create direct mapping between pedestrian behaviour in a particular simulation scenario and the urban network through which an individual will move, we decompose pedestrian movement into three basic states (Figure 1). Each of these corresponds to a 'mode of thinking' of an idealised pedestrian:

- *Decision state.* This state describes the options open to an individual agent, positioned at an intersection (node) between two or more streets (edges). Each agent, depending on its behavioural type, can choose to move in one of several directions (towards the next decision point or intersection). This decision is made based on a number of factors (see Section 2.2), distinguished mainly by the individual's knowledge of the 'optimal' route. An agent, having made a choice, is then in *Transition*.
- *Transition state.* In this state, the agent attempts to move from the origin intersection in the direction chosen. The agent will move in discrete steps (one per iteration) along the connecting edge, with speed (number of iterations) determined by its behavioural type. Before each step, a search is made for any agents occupying the next 'point' (space) on the connecting edge in the direction chosen. Movement is then evaluated, based on the average throughput of street type and the number of street spaces occupied by other agents. Depending on the outcome, the agent can make one of two state transitions: if there is adequate space, the agent moves forward one step, and stays in *Transition* state, or, if the road ahead is blocked, the agent pauses and enters the *Waiting* state for one time step. If the next step requires a further decision, e.g. direction choice at an intersection, the agent enters a new *Decision* state.
- *Waiting state.* If the agent movement is blocked due to congestion, it is assigned a 'wait counter' representing the maximum number of iterations it is 'patient' enough to wait. The agent attempts to move until the counter expires, at which point it 'recalculates' the decision to proceed by the current route. Effectively, it removes the congested edge from its decision tree and enter a new *Decision* state.

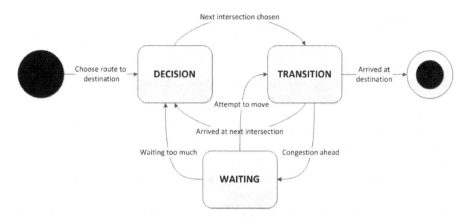

Fig. 1. Behavioural state machine representing transitions between three possible pedestrian states - decision, transition and waiting

2.2   Types of Pedestrian Behaviour

The *Destination* state of the core state machine presented above allows for arbitrary pedestrian cognitive behaviour to be incorporated into the model, on the basis that the next point along the pedestrian path to the ultimate destination is the outcome of such behaviour. We note here that, at present, we consider only path selection behaviour. Another important category, 'crossing' behaviours [24] has not been considered. Given current model scale and the fact that purposeful choice is constrained by road usage in urban environments, paths between intersections tend to be the rational choice and random crossovers are limited. Inclusion of crossings may lead to some impact on pedestrian congestion at intersections.

The model implements a number of core cognitive behaviours, divided into two broad groups:

Pedestrians with Partial or no Knowledge of the Urban Network. This first group represents 'leisure' pedestrians, mostly tourists or casual walkers, with a vague, or even no, knowledge of the configuration of streets between their current location and the intended destination. In [33] authors have emphasised the importance of this group to overall pedestrian traffic in urban environments. These pedestrians can be represented by agents, whose node choice behaviour is driven by minimisation of the following function:

$$f(n) = \delta \cdot d + \omega \cdot o \tag{1}$$

where n represents the node descriptor, d - the straight line distance between the node and the pedestrian's destination (the measure of perceived proximity) and o - current occupancy of the node (the measure of perceived crowding); δ and ω are variable weight factors which reflect the particular pedestrian preference when it comes to route selection. Certain behavioural types will ignore crowding and attempt strictly to approach the destination as directly as possible. However, more cautious individuals will select what they perceive to be a sub-optimal route, if it means that congestion can be avoided and time minimised, even if distance is slightly longer.

Based on the possible options for minimisation of Equation 1, we define three fundamental types of 'leisure' agents in our model:

- **Aggressive** - This type of agent has high δ factor, and low ω factor, and aims consistently to transition to the perceived nearest node;
- **Cautious** - This type of agents has high ω factor, and tends to avoid crowds. However the δ factor value is still high, with node proximity always an important driver of the agent behaviour;
- **Random** - A certain proportion of pedestrians will have few or no route preferences, but are content to wander randomly about the urban grid. This category is useful for simulating urban areas of high attraction to tourists where a single destination is not dominant (or even apparent), or for simulating the effect of spontaneous reaction of other pedestrian types to 'background noise' present in a given street layout.

Pedestrians with Full Knowledge of the Urban Network. Unlike the previous major group, which makes decisions based on 'local' knowledge i.e. the next node only, pedestrians who are completely familiar with the street layout (e.g. local residents or daily office workers) generally attempt to follow a well-known, predetermined route, which they take every day from origin to destination. The behaviour of these agents is based on the previous work [23], which relates route choice to spatial awareness of the street layout. Before embarking on a trip, each agent computes a desired path (a series of nodes connected via street edges) using Dijkstra's algorithm, where edge 'cost' can take one of the following three forms:

- **Least cumulative angle change between the streets.** In this mode, an agent computes Dijkstra's shortest path that considers edge cost as the angle formed by edges connecting at the next node, Figure 2-II. The agent attempts to find the minimum of the following sum:

$$d(p) = \sum_k \alpha_k, \alpha_k = \pi - \delta_k \tag{2}$$

where d is the total perceived distance of a subset of nodes $p = \{n_1, n_2, ...n_n\}$ along the determined path, with $alpha_k$ representing the cost of travelling between the two edges and δ_k being the lesser of two angles between the lines constructed by geometric coordinates of the nodes n_{k-1}, n_k and n_{k+1}, respectively, with $\alpha_0 = \alpha_n = 0$.
- **Least cumulative number of turns between the source and destination.** In this mode, an agent computes Dijkstra's shortest path such that edge costs depend on the number of turns (direction changes) between

origin and destination nodes, (Figure 2-III), making this a binary form of Equation 4:

$$d(p) = \sum_k A_k, A_k = \begin{cases} 1, & \text{if } \alpha_k \neq 0 \\ 0, & \text{if } \alpha_k = 0 \end{cases} \tag{3}$$

where A_k represents the binary metric of turn or no turn with boundary values $A_0 = A_n = 0$.

- **Shortest distance between source and destination.** Finally, a certain number of pedestrians will know the shortest metric distance (the geographical distance) to their destination, even though it might not be the most obvious route, Figure 2-I:

$$d(p) = \sum_k d_k, d_k = |\mathbf{x}_k - \mathbf{x}_{k-1}| \tag{4}$$

where $\mathbf{x}_k$ is the vector containing the geographical coordinates of node n_k, and $d_0 = 0$.

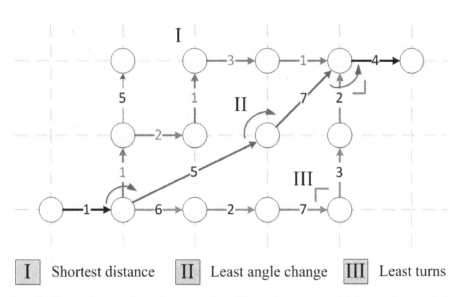

| **I** | Shortest distance | **II** | Least angle change | **III** | Least turns |

Fig. 2. Shortest path algorithm examples. The nodes represent intersections, and the numbers a distance metric representing cost. Three possible route choices based on the type of edge costs are shown.

The appropriate distance function is recalculated each time congestion occurs, by removing the congested path from an agent's network view, and permitting it to re-compute its route.

2.3 Modelling Workflow

In order to analyse and evaluate assumptions of the model proposed with respect to real urban environments, (including both street network layout and the location of main transportation, business and leisure zones of interest), we have built a framework that sources urban map data on demand from a suitable, external GIS. A rise in popularity of volunteered geographic information (VGI) has recently been observed [19]. The data, obtained from such systems, provide several advantages for macroscopic pedestrian modelling:

- Open data format, allowing for easier consumption and exchange of data between applications
- Street layout details, which compare favourably with commercial solutions, including street types and venue information
- Ability to make local edits of the layout data, allowing experimentation and flexibility with respect to infrastructure changes.

The framework sources data from OSM datasets, readily available from the OpenStreetMap APIs, for the coordinates provided. Once the dataset, along with the main pedestrian source and destination locations is loaded into our framework, the street network is automatically extracted and converted to graph representation. The resulting grid is populated with pedestrians of the various behaviour types. During simulation, the framework enables visualisation of individual pedestrian movement and keeps tracks of the paths used togheter with flow and density information. As well as the grid-level visualisation, the framework outputs detailed density maps, flow graphs and average time information that can be used for further analysis, (Figure 3). The sparsity of data, inherent in early versions of crowdsourced systems, has been alleviated in recent years [34] with a surge in the number of contributors to the GIS dataset. Super-urban areas, such as London, attract a very high level of detail, adequate for modelling applications and readily combined with available census data.

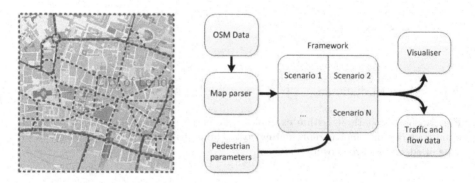

Fig. 3. Example transport graph parsing and model workflow schematics

3 Simulation and Results

3.1 Model Outputs

In order to evaluate the pedestrian traffic patterns resulting from our model, we focus on: (a) sections of the urban grid with good connectivity and presence of alternative routes between the nodes which, according to [23] are expected to show similar performance for non-congested and congested scenarios, and (b) sections where few alternative routes exist, which are expected to lead to significant bottlenecks as the traffic increases.

As density of pedestrian traffic along a given street varies with time of the simulation [13], peaking for high congestion periods, such as the morning rush, and mimicked by the simulation, we estimate the average density of pedestrians at a given geolocation by making use of the bivariate normal *kernel density function* and plot the mean of the density against the map:

$$f_h(\mathbf{x}) = \frac{1}{n} \sum_{i=1}^{n} K_H(\mathbf{x} - \mathbf{x}_i) \tag{5}$$

here, $\mathbf{x}$ and $\mathbf{x}_i$ are vectors of agent positions, $\mathbf{K}_H$ is the smoothing bandwidth function applied to a standard bi-variate normal distribution curve and n is the number of agents passing through the given measurement point.

To measure the velocity of pedestrians along the grid, we also calculate the flow through the map by measuring the number of pedestrians passing a given intersection over time [26]:

$$J = \frac{1}{\langle \Delta t \rangle}, \langle \Delta t \rangle = \frac{1}{N} \sum_{i=1}^{N} (t_{i+1} - t_i) = \frac{t_{N+1} - t_1}{N} \tag{6}$$

where J denotes flow, and $\langle \Delta t \rangle$ the average of time taken for two consecutive pedestrians to pass through the intersection. N denotes the total number of measurements (total number of pedestrians passing through).

Finally, we observe and chart the paths taken and the average travel distances and times between source and destination nodes, and compute the average speed of pedestrians of given type within the context of the total urban network load.

3.2 Scenario Matrix

In order to perform robust *sensitivity analysis* of the relationships between model inputs and outputs, several behaviour scenarios were simulated for two different urban environments within London. For one location, we chose the area of the inner financial district within the city of London, with the pedestrian traffic of interest being prevailingly directional (office workers and daily commuters). For the other, we chose London's West End, where the pedestrian groups mainly consist of tourists and people attending leisure venues, such as theatres, restaurants or pubs.

Table 1. Direct/leisure pedestrian distribution for different simulation runs

Scenario	Agent numbers	Leisure	Direct
1	1000	0%	100%
2	2000	50%	50%
3	3000	66%	33%
4	7000	70%	30%
5	10000	50%	50%
6	15000	33%	66%

For directional traffic, public transport, specifically underground stations, were taken as points of origin, and a central location as the destination, while leisure traffic destinations were taken to be equally likely all entertainment venues present in the given map segment. Points of origin of leisure traffic varied between public transport points to start points at random positions on the map.

For each location, we simulated scenarios for the relationship between the size of the agent population (from 1000 to 15000 pedestrians) and the behavioural profile of different groups. In terms of output, we monitored the flow and density effects as well as average transit times for each group. Table 3 presents the matrix of explored scenarios.

Table 2. Scenario list

#	Area	Direct behaviours	Leisure behaviours
1	Financial district	Distance, turn or angle	Random
2	Financial district	Distance, turn or angle	Cautious
3	Financial district	Distance, turn or angle	Aggressive
4	West End	Distance, turn or angle	Random
5	West End	Distance, turn or angle	Cautious
6	West End	Distance, turn or angle	Aggressive

Physical dimensions of each agent were taken to be 50cm width and 30cm breadth [31], with the maximum throughput of street sidewalks averaged using estimates on street size from OSM data. Base speed of an individual agent was taken to be 1.5 m/s [28]. Street sizes used are outlined in Table 3, with data on standard sidewalk sizes in urban London sourced from [32]. The number indicates the maximum number of people standing abreast on the sidewalk, at either side of the street. Non-pedestrian transport routes (such as railways, metro lines and river transport routes) were removed from the map for the current exploratory work, although these form obvious designated points of origin of pedestrian traffic for the future extensions of the framework. Also, local pedestrian to road traffic interactions (including cross-walks and traffic lights) were omitted from consideration in order to focus on the primary effects of the global street layout. This can be additionally incorporated at later stages. Each simulation was run for 3000 iterations, representing 50 minutes of real time.

Table 3. Pedestrian throughput of different street profiles

Type	Throughput
Footways and walkways	3
Steps and stairways	2
Residential area streets	4
Pedestrian paths in parks and green areas	6
Major thoroughfares (highways, embankments)	12
Main streets (primary roads)	8
Side streets (secondary and tertiary roads)	6

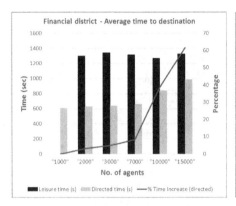 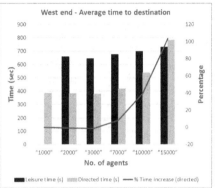

Fig. 4. Average walking times for various pedestrian ratios (directed vs. leisure). The red line indicates relative change in travel time of directed agents, with the inflection point marking the point where the network starts experiencing significant congestion.

3.3 Results

Figures 5 and 6 represent the density and flow of the two major pedestrian categories for scenarios 1 - 6 from Table 2. The results help gauge the ability of the inner city network to accommodate pedestrian traffic densities. For public transports origin points in both areas, we have used the nearby London Underground stations.

From the graphs and model visualisations several marked phenomena can be observed:

– **Lane formation.** As an individual agent's waiting period typically allows the agent in front to move one step, pedestrians naturally form lanes of traffic going from origin to destination. Only at intersections where multiple flows merge, does an agent wait long enough to consider taking an alternate route. For directed flows, lanes are mostly formed closer to origin point, while for leisure traffic these tend to occur at intersections where multiple flows meet. For single origin traffic (e.g. pedestrians originating from a single underground station), the fan out degree (number of distinct lanes) of lane formation is fairly small, with primary roads having highest flow. Side streets are much less used as alternative means of navigating to the destination.

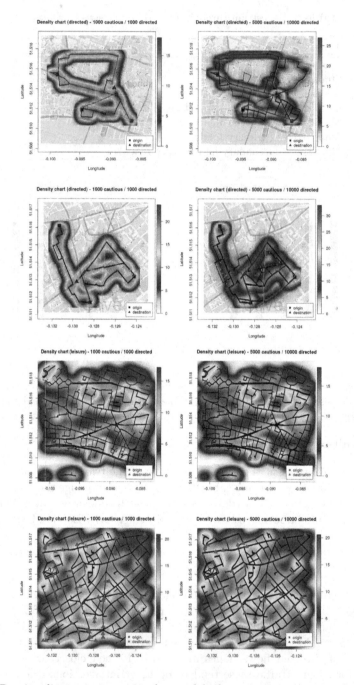

Fig. 5. Density diagrams comparing low and high congestion scenarios for directed (top two rows) and leisure traffic (bottom two rows) in London city vs. West End. The colours indicate approximate density per $9m^2$ (a single step, unit time interval of 1 second, in any direction from the agent's current location).

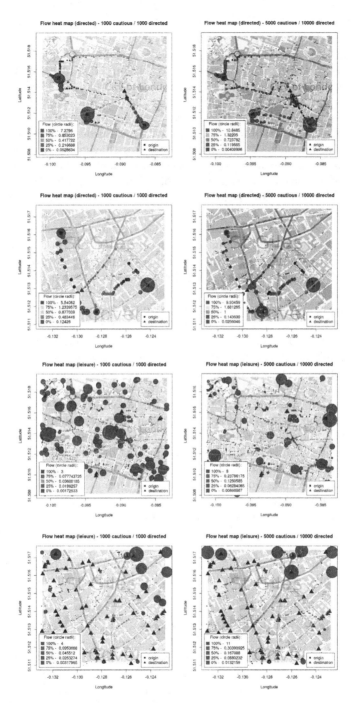

Fig. 6. Flow diagrams comparing low and high congestion scenarios for directed (top two rows) and leisure (bottom two rows) traffic in London city vs. West End. The circle size indicates flow rate of pedestrians per unit time.

- **Jamming.** In situations where the destination node has a limited number of approaches (Figure 5 top row), traffic merging on a given approach will cause local congestion, with agents attempting to re-route to a second approach but being blocked by traffic streaming in that direction. A given number of agents will take a sub-optimal (longer) route to avoid this block (Figure 5 top row, right.), but the majority stays at the intersection until the flow of traffic subsides as preceding agents reach the destination.
- **Local congestion avoidance.** As shown by the pedestrian path traces in Figure 5, pedestrians of both behaviour types (having both local and extensive knowledge) exhibit only localised congestion avoidance by preferring shortest routes around the congestion point instead of seeking alternative approaches *ab initio* to destination (i.e. these are reactive rather than planned). In this scenario the utilisation of side streets becomes significant. In extreme cases, this local congestion avoidance will be repeated as congestion occurs at every proximate side street intersection causing the network of side streets to become crowded in turn. Layouts, such as the one of Figure 5 second row, where central intersections have large interconnection degree and better side street connectivity, allow agents to switch from one major road to another by traversing side streets and generally give rise to lower travel times.
- **Venue dispersion effect.** Venue grouping has a *significant effect*, both on travel time and total network congestion. In the financial district layout, available leisure venues were grouped in several distinct places on the map, requiring lengthy paths from any given point on the grid. This extends the time interval for which pedestrians are present on the network, increasing congestion pressure. Layouts where the distribution of venues is more uniform across the map exhibit lower congestion, and consequently, shorter travel times, (note here, of course, that pedestrians rarely choose the 'nearest' map venue as their destination, instead picking distant ones causing paths to intersect across the network).
- **Network inflection points.** As shown in Figure 4, and in flow graphs in Figure 6, both urban networks are large enough to contain the the total number of pedestrians in the simulations. However, each network grid exhibits an 'inflection point' at which traffic levels start to exponentially impact on directed traffic, which is unable to find routes allowing free flow, thus leading to jamming behaviour. Again, networks with smaller interconnection rates exhibited inflection points for lower pedestrian traffic numbers.

4   Conclusions and Future Work

In this paper we have shown the potential for modelling macroscopic pedestrian flows using the agent based paradigm, combined with cognitive behaviour characteristics of different pedestrian groups. The main objective was to build an initial, general framework enabling adaptable combinations of elementary behaviours giving rise to more complex scenarios. We have demonstrated the ability of the framework to mimic important flow phenomena and to facilitate comparison of the effect of different urban grid structures. By using dynamically sourced map data, we were able to compare arbitrarily chosen city layouts. Of future interest is the expansion both in number of points of origin as well as behavioural features, to include addition of connected groups (such as families with children) rather than individual pedestrians. Additionally, network characteristics, such as "closeness" and "betweenness" metrics of nodes [23], and their correlation with traffic flows could be investigated, in order to provide a fuller picture of morphological effects. Validation against experimental observations of pedestrian traffic behaviour is obviously crucial, but dependent on detailed publicly available sources with social media recording (typically) only endpoints. Utilisation of census data is however a realistic objective for refinement of pedestrian group profiles.

Acknowledgement. Financial support from the ERA-Net Complexity Project, P07217, is gratefully acknowledged.

References

1. Carstens, R.L., Ring, S.L.: Pedestrian capacities of shelter entrances. Traffic Engineering 41, 38–43 (1970)
2. O'Flaherty, C.A., Parkinson, M.H.: Movement on a city centre footway. Traffic Engineering and Control 13, 434–438 (1972)
3. Couzin, I.D., Krause, J.: Self-organization and collective behavior in vertebrates. Advances in the Study of Behavior 32, 1–75 (2003)
4. Kirchner, A., Schadschneider, A.: Simulation of evacuation processes using a bionics-inspired cellular automaton model for pedestrian dynamics. Physica A: Statistical Mechanics and its Applications 312(1-2), 260–276 (2002)
5. Pan, X., Han, C.S., Dauber, K., Law, K.H.: A multi-agent based framework for the simulation of human and social behaviors during emergency evacuations. AI & Society 22(2), 113–132 (2007)
6. Augustijn-Beckers, E.W., Flacke, J., Retsios, B.: Investigating the effect of different pre-evacuation behavior and exit choice strategies using agent-based modeling. Procedia Engineering 3, 23–35 (2010)
7. Lu, Q., George, B., Shekhar, S.: Capacity Constrained Routing Algorithms for Evacuation Planning: A Summary of Results. Department of Computer Science and Engineering, University of Minnesota (2005)
8. Moussaïd, M., Helbing, D., Theraulaz, G.: How simple rules determine pedestrian behavior and crowd disasters. Proceedings of the National Academy of Sciences 108(17), 6884–6888 (2011)
9. Gilbert, N., Troitzsch, K.G.: Simulation of the social scientist. Open University Press (1999)
10. Fiedrich, F., Burghardt, P.: Agent-based Systems for Disaster Management. Communications of the ACM 50(3), 41–42 (2007)
11. Schelling, T.C.: Dynamic Models of Segregation. Journal of Mathematical Sociology 1, 143–186 (1971)
12. Bonabeau, E.: Agent-Based Modelling: Methods and Techniques for Simulating Human Systems. Proceedings of the National Academy of Sciences of the United States of America (PNAS) 99(3), 7280–7287 (2002)

13. Castle, C.J.E., Crooks, A.T.: Principles and Concepts of Agent-Based Modelling for Developing Geospatial Simulations. CASA Working Papers, Centre for Advanced Spatial Analysis, UCL (2006)
14. Cetin, N., Nagel, K., Raney, B., Voellmy, A.: Large-scale multi-agent transportation simulations. Computer Physics Communications 147, 559–564 (2002)
15. Henein, C.M., White, T.: Agent-Based Modelling of Forces in Crowds. In: Davidsson, P., Logan, B., Takadama, K. (eds.) MABS 2004. LNCS (LNAI), vol. 3415, pp. 173–184. Springer, Heidelberg (2005)
16. Crooks, A.T., Castle, C.J.E.: The Integration of Agent-Based Modelling and Geographical Information for Geospatial Simulation. In: Heppenstall, A.J., Crooks, A.T., See, L.M., Batty, M. (eds.) Agent-Based Models of Geographical Systems, pp. 219–251. Springer, Netherlands (2012)
17. Longley, P., Batty, M.: Advanced Spatial Analysis: The CASA book of GIS. ESRI Press, Redlands (2003)
18. Batty, M.: Cities and Complexity: Understanding Cities with Cellular Automata, Agent-Based Models, and Fractals. MITPress, Mass (2005)
19. Andrew, T.C., Sarah, W.: GIS and agent-based models for humanitarian assistance. Computers, Environment and Urban Systems 41, 100–111 (2013)
20. Batty, M.: Predicting where we walk. Nature 388, 19–20 (1997)
21. Moussaïd, M., Perozo, N., Garnier, S., Helbing, D., Theraulaz, G.: The Walking Behaviour of Pedestrian Social Groups and Its Impact on Crowd Dynamics. PLoS ONE 5(4), e10047 (2010)
22. Duckham, M., Kulik, L.: "Simplest" paths: Automated route selection for navigation. In: Kuhn, W., Worboys, M.F., Timpf, S. (eds.) COSIT 2003. LNCS, vol. 2825, pp. 169–185. Springer, Heidelberg (2003)
23. Hillier, B., Iida, S.: Network and Psychological Effects in Urban Movement. In: Cohn, A.G., Mark, D.M. (eds.) COSIT 2005. LNCS, vol. 3693, pp. 475–490. Springer, Heidelberg (2005)
24. Papadimitriou, E., Yannis, G., Golias, J.: A critical assessment of pedestrian behaviour models. Transportation Research Part F: Traffic Psychology and Behaviour 12(3), 242–255 (2009)
25. Crooks, A., Castle, C., Batty, M.: Key challenges in agent-based modelling for geospatial simulation. Computers, Environment and Urban Systems 32(6), 417–430 (2008)
26. Schadschneider, A., Klingsch, W., Klüpfel, H., Kretz, T., Rogsch, C., Seyfried, A.: Evacuation Dynamics: Empirical Results, Modeling and Applications. In: Meyers, R. (ed.) Encyclopedia of Complexity and Systems Science, pp. 3142–3176. Springer, Berlin (2009)
27. Schwandt, H., Huth, F., Bärwolff, G., Berres, S.: A Multiphase Convection-Diffusion Model for the Simulation of Interacting Pedestrian Flows. In: Murgante, B., Misra, S., Carlini, M., Torre, C.M., Nguyen, H.-Q., Taniar, D., Apduhan, B.O., Gervasi, O. (eds.) ICCSA 2013, Part V. LNCS, vol. 7975, pp. 17–32. Springer, Heidelberg (2013)
28. Antonini, G., Bierlaire, M., Weber, M.: Discrete choice models of pedestrian walking behavior. Transportation Research Part B: Methodological 40(8), 667–687 (2006)
29. Blecic, I., Cecchini, A., Congiu, T., Pazzola, M., Trunfio, G.: A Design and Planning Support System for Walkability and Pedestrian Accessibility. In: Murgante, B., Misra, S., Carlini, M., Torre, C.M., Nguyen, H.-Q., Taniar, D., Apduhan, B.O., Gervasi, O. (eds.) ICCSA 2013, Part IV. LNCS, vol. 7974, pp. 284–293. Springer, Heidelberg (2013)

30. Helbing, D., Molnár, P., Farkas, I.J., Bolay, K.: Self-organizing pedestrian movement. Environment and Planning B: Planning and Design 28(3), 361–384 (2001)
31. Oberhagemann, D.: Static and Dynamic Crowd Densities at Major Public Events. Technical Report vfdb TB 13-01 (2012)
32. Design Manual for Roads and Bridges. Pavement Design and Maintenance. Published by the Highways Agency 7, Section 2(5) (2001)
33. Kowald, M., Frei, A., Hackney, J.K., Illenberger, J., Axhausen, K.W.: Collecting data on leisure travel: The link between leisure contacts and social interactions. Procedia - Social and Behavioral Sciences 4, 38–48 (2010)
34. Haklay, M.: How good is volunteered geographical information? A comparative study of OpenStreetMap and Ordnance Survey datasets. Environment and Planning B: Planning and Design 37, 682–703 (2010)

Generation of Routes with Points-of-Interest to Help Users in Their Trip

Ana María Magdalena Saldaña, Jose Giovanni Guzmán, Marco Antonio Moreno,
Miguel Torres, Rolando Quintero, and Oleksiy Pogrebnyak

Centro de Investigación en Computación, Instituto Politécnico Nacional
asaldana_a12@sagitario.cic.ipn.mx,
{jguzman1,marcomoreno,mtorres,rquintero,olek}@cic.ipn.mx

Abstract. Many research areas are developing applications that use the Global Positioning System (GPS) to improve human's life. In routing systems most of the studies are centered in time improvements. Despite they notify users the instructions that they should follow in their travels, they do not include support elements that could be useful as land marks or points of interest. We propose the Generation of Routes with Points of interest (GRP), a methodology based on an application ontology, that will be used to describe points of interest as shops, hospitals and schools, among others, located on the roadways of study area. The GRP gives users the instructions to go from one place to another combining geospatial analysis tools, web and mobile technologies. The routes generated begin in the mobile position and after a geospatial process, the user gets a map with the route, points of interest and instructions to change the direction to help him on his trip.

Keywords: Ontology, point-of-interest, routing, geospatial processing, mobile platform.

1 Introduction

Routing with GPS data provides programmers opportunities to build routes to go from one geographic point to another. The use of Geographic Information Systems (GIS) has increased the GPS data precision. Despite of their low cost and developing time, GPS routing devices have some limitations; one of the most important is the use of directional and longitudinal measurements without objects to help users in their trip [5].

Several works try to include visual elements in their routes; most of them are focused on the study of route distance and travel time. The methodology GRP proposed through GPS data generates the route to go to a specific place, showing landmarks and points of interest located on the route. It is used an ontological model of roads classification to get points of interest.

Through this paper, we will mention the GRP previous published works, followed by a description of GRP components and stages, then our results are shown and finally we present the investigation conclusions.

B. Murgante et al. (Eds.): ICCSA 2014, Part IV, LNCS 8582, pp. 284–298, 2014.

2   Related Work

Going to a specific place in a city can be a difficult situation, especially if either the area or precise destination is unknown, or if the instructions to get there are not clear [14]. Currently, there are at least three options to solve the problem: 1) it is possible to ask for indications from someone who knows the area [24], 2) use a local map, or 3) use of a GPS navigator.

2.1   Routing Algorithms for GPS

If a person has chosen a GPS navigator, that person will be working with a GPS signal processor. This kind of device works efficiently just when it is in an unobstructed area, it means without elements that wear out the signal of GPS system. In an ideal area the GPS processor should have free transmission in all directions and 15 elevation grades from the antenna [6]. There are two ways for configure a GPS receiver in a car. The first is to coordinate the processor to work with the personal data assistant that some vehicles possess, where the user has to input some information about travel, passengers and city where is driving. In the second, the GPS processor works passively, it does not need interaction with users, because it gets routes information automatically [4].

The algorithms that work in the GPS devices have changed at the same time that GPS has evolved. For example, they have raised their efficiency in routes calculation, exactitude of instructions and have decreased the energy consume. These algorithms are classified according to their cyclic functions, their metric distribution, the accuracy of the information they deliver to users, scalability and robustness [22]. Algorithms with elevated scalability and robustness are capable of working dynamically with the communication network and they also have a higher information backrest. Among the most important algorithms used in GPS devices are the Map Matching Algorithms (MM) which obtains deterministic information from roads and arcs represented in GPS. In spite the fact MM consider arcs continuity and connectivity, they do not guarantee the full usability of routes [2]. The Multiple Hypotheses Technique (MHT) connects the route segments designed by the MM algorithm to a geographic database at real time increasing the accuracy of the data. Other algorithms save the starting point of the route in a search engine of the GPS with a group of candidate routes. Each of them has a calculated score that takes into account attributes such as distance, speed and the number of points and arcs presented in the route. The route with the highest score is then chosen and displayed [21]. The use of modeling algorithms to generate sets of potential routes was one of the first uses in mobile GPS routing. This algorithms use topological measures of the area, and temporal information like time and speed to calculate the probability where each route point matches with the device data, while the user follows the route.

Most of published reports related to GPS data routing focus on time-longitude relation optimization and energy saving, but few of the use landmarks to help users. To define an element as a landmark or point of interest, it is necessary to consider some aspects such as the place it is representing, if the point of interest is easy to identify by people, its structure and its relevance. The relevance depends directly from the person perception [25]. Some GIS try to include support elements in their routes as obstacles, but these are not user references [3].

Building personal maps from GPS data is a system proposed by Lin et al. [9]. It extracts places, recognizes activities related to them and determines the place where user wants to go by using the GPS data. The system learns specific route mistakes, mobility and transportation patrons by a hierarchical model. To get map personalization, the project defines significant places; it means those where users frequently spend long time. This make possible the time measuring and GPS coordinates saving at a cluster.

Developing an algorithm for generating weighted routes semantically, project studied by Luna [10], generates a set of routes based on users requirements as traffic lights, crime rate and traffic. Finally, with a semantic function applied to road ontology, choses the routes in more accordance to the requirements of the user giving him some options. The algorithm is programmed in a web environment.

2.2   GPS Routing Applications

There exist some commercial applications that provide routing service for mobile applications, despite their characteristics as precision and environment they do not use points of interest as references to help users, they give their instructions by using measures and orientation terms that in some cases are difficult to understand.

Waze is a mobile application designed to show users the traffic in their city according to its community user's data. The system has been programmed for users that are in a vehicle and its routing process gives the route with less traffic, not only the shortest one. Each Waze member is able to edit maps and help some other users to find new ways to avoid car accidents and traffic jams.

Google Maps for Android is the implementation of Google Maps for web in mobile devices. This program shows users the route to go from one point to another in a determined area, lets user know its position on the map, can show traffic at real time and by its social network allows members to interchange comments about places and routes. One important characteristic of Google Maps is that makes possible to user search information about places that are interesting for him while routing.

3   Methodology

The GRP methodology shows landmarks, called points of interest or reference points, located in a route calculated using GPS data. This methodology is organized in three stages, Acquisition, Routing and Ontological process. The Acquisition determines the beginning and the end of the route by a mobile application. These data are used in the Routing to generate the trip applied to the study area roadways by a set of geospatial operations. The Ontological process searches points of interest as schools, shops, hospitals and some other business from application ontology designed for GRP, and shows users the obtained results. Fig. 1 shows methodology general schema.

3.1 Acquisition

This stage function is get user position coordinates and the name place where it wants to go. For this purpose, an Android [1] application of two modules has been programmed. It has been used Android because nowadays, is one of the most important mobile platforms, it is easy to manipulate for users and is the operative system in many mobile devices with GPS integrated.

The first module is connected to GPS device to get the user actual position. It has a textbox where user writes the destiny name in; the textbox has an autocomplete function to avoid user's mistakes. Additionally, it is possible to deactivate the GPS device and send the data obtained to the GRP server.

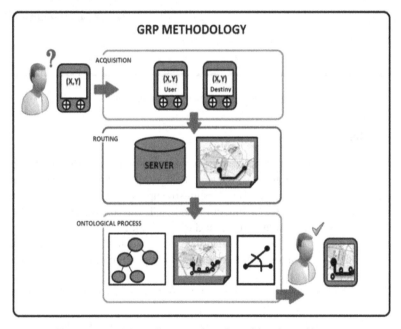

Fig. 1. Methodology diagram of routing with points of interest

The second android module is related to the third methodology stage. It shows a map with instructions to arrive to the earmark, the device coordinates and the points of interest on the route. Mapsforge an open source library to visualize OpenStreetMap cartography in mobile devices without a permanent internet connection, was used to show the map at the application [11], Mapsforge is in continuous developing; its use requires the change of an OpenStreetMap file [13], from *.osm* to a *.map* format. Merkaartor OpenStreetMaps editor was used to download the study area map and edit its components [12]. The study area map in format *.osm* is shown in the Fig. 2.

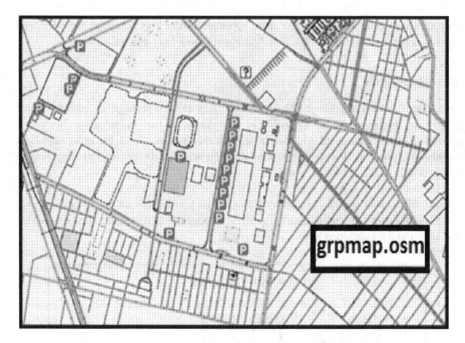

Fig. 2. Study area map in .osm format

3.2   Routing

To make routing possible, the study area roadway shapefile was imported to project database allocated in the database manager Postgresql [18], by its spatial database extendert tool Postgis [17] that adds to Postgresql support for geographic objects. It is possible to apply a set of spatial functions and SQL queries on data to make them available to work with Pgrouting [16] after the importation.

Pgrouting is the Postgis tool that calculates routes with Postgresql data. By the Pgrouting A star routing algorithm, the methodology gets the route between the user's position and the final target.

To have a link between the mobile application and Postgresql database, a web server has been programmed by a PHP [15] code. This code obtains the mobile application information and save it into the database where it is manipulated. The data interchange is possible using the JSON data structure [7].

The spatial functions are buffers and intersections applied if necessary, according to the GRP methodology which is described below.

1. Coordinates from the mobile application are obtained.
2. Point A=user. User coordinates are represented as a point.
3. Coordinates of name destiny are obtained from server.
4. Point B=destiny. Destiny coordinates are represented as a point.
5. Intersection 1(Point A, roadways). Intersection between user position and roadways is calculated.

     a. If (Intersection 1==null), then:
         i. While (Intersection 1== null)
         ii. Point A1 = increase the Point A are by using a buffer.
         iii. Intersection 1(Point A1, roadways).
         iv. End While.
     b. End If.
6. Intersection 2 (Point B, roadways). Intersection between target and roadways is calculated.
     a. If (Intersection 2==null), then:
         i. While (Intersection 2== null)
         ii. Point B1 = increase the Point B area by using a buffer.
         iii. Intersection 2(Point B1, roadways).
         iv. End While.
     b. End If.
7. Obtain route between Intersection 1 and Intersection 2, by A star algorithm usage.
8. Calculate angle between route segments.
9. Identify direction change on route by angles value to define instructions.
10. Consult ontology from mobile application.
11. Get street name and points of interest name and coordinates from ontology.
12. Draw route and map ontology data in map.
13. Write direction change instructions with points of interest located on each change.
14. Show map in the mobile application.

First, destiny latitude and longitude coordinates are searched and represented as spatial points by a Postgresql's function in the server. This gave a table called Points, as result. After that, the origin route and end points are intersected with the roads data

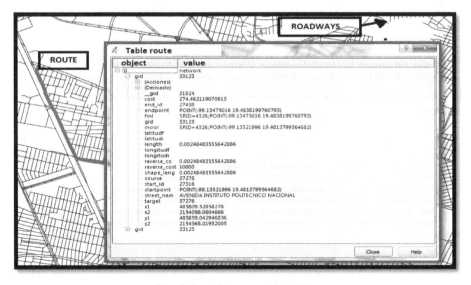

Fig. 3. Route's segment data table

to identify its beginning and ending geometric points. In case of a null result from the intersections, the point area will be increased until the intersection returns a positive value. The buffer to increase the areas is created by a Postgresql function, which is executed in server and is represented on a table.

Once the geometric beginning and ending are known, it is possible to apply the Pgrouting A star algorithm to get the shortest route between points. The result gives a set of segments with attributes as id, latitude, longitude, reverse cost, shape and street name of the road represented, Fig. 3.

3.3   Ontological Process

Direction Change Instructions
To determine direction changes that user has to follow to arrive to his destiny, the angle between route consecutive segments is calculated, and according to the [-14,14] interval, GRP determines if it is necessary to notify a directional change or not. This interval is needed because segments are not completely lineal and each street is divided in multiple lines. Relation between angles is shown in Fig. 4.

To calculate angles we use equation (7), which components are calculated by applying equations (1) to (6).

It is necessary to deduct A latitude coordinates from B latitude coordinates, and take A longitude coordinates from B longitude coordinates. Subtractions are shown in equations (1) and (2).

$$Ry = coordBx - coordAx \tag{1}$$

$$Rx = coordBy - coordAy \tag{2}$$

The Angle between A and B points results by applying equation (3).

$$\theta 1 = atanRy/Rx = \tan{-1}Ry/Rx \tag{3}$$

To get the angle between C and A points, the arctangent of the deduction of A coordinates from C coordinates is done as equation (6) shows. Equation (4) shows the latitude coordinates subtraction; equation (5) shows the longitudinal coordinates subtraction.

$$Sy = coordCx - coordAx \tag{4}$$

$$Sx = coordCy - coordAy \tag{5}$$

$$\theta 2 = atanSy/Sx = \tan{-1}Sy/Sx \tag{6}$$

Finally to determine the direction's change, the deduction of θ_1 from θ_2 is done, as shown in equation (7).

$$\theta = \theta 2 - \theta 1 \tag{7}$$

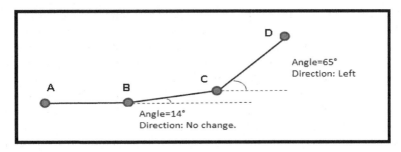

Fig. 4. Diagram that show a direction change to the left

The coordinates where the direction change is needed are saved; in case of a positive angle higher than 14, the direction change will be consider to the left; on the other hand, if a negative result smaller than -14 is obtained, the direction change will be consider to the right.

Ontology Model

By the ontology model usage, points of interest located near to the direction change point are added to instructions.

Methontology, a methodology for ontology's development based on IEEE 1074 standard [23], was used to generate the GRP ontology model. Lars Kulik work [8], mentions the relation between polygonal curves, their semantic weight at a particular context and their polygonal and geometric attributes, this characteristics have been added to the ontology structure.

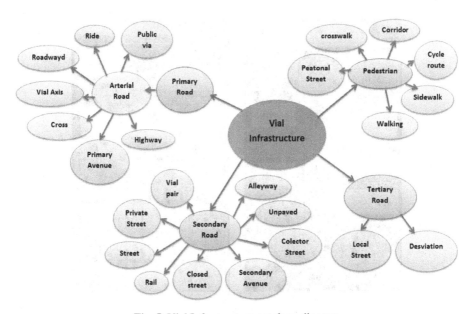

Fig. 5. Vial Infrastructure ontology diagram

The GRP ontology is formed by the main class Vial Infrastructure, divided in four subclasses, composed by entities. Each entity has its own instances, which information is based on *Secretaría de Transportes y Vialidades* (SETRAVI) data, from Mexico; the ontology implementation was done in *rdf* format in Protégé, an ontology editor [19]. The ontology general schema is shown in Fig. 5.

The instances have a hierarchical level, and three attributes: business' name, latitude and longitude coordinates. By using the latitude and longitude coordinates, the ontological element is represented in the GRP map as a point.

By a semantic query made from the mobile application to the ontological model, the ontology´s searches the street name, and sends as a result to the mobile application the attributes of the entity. The business obtained by the query and which location is in an fifteen meters interval from the direction changes coordinate, are mentioned in the instructions to help user during his trip.

One of the GRP ontology advantages is that it can be expanded to be used in some other related systems or to increase the information of some other ontologies.

4   Results

In this section, an example of GRP methodology usage is described.

Suppose that a visiting researcher is at the Investigation on Computation Center in Mexico (CIC), after a congress he decides to go to a restaurant (Vips). When he asks to the mobile application GRP for a route to get his objective, the GRP recovers his actual position with the device´s GPS, and the endpoint name as shown in Fig. 6. These data are sent to the web server to initialize the GRP methodology.

 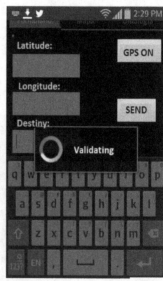

Fig. 6. GRP mobile application interphase to get users coordinates

shape_leng double precision	street_name character varying	reverse_co double precision	the_geom geometry	id serial	campo geometry	latitud double precision	longitud double precision
0.000156115678888:	EJE 4 NTE	10000	0105000020E61	4	0101000020E61	19.4833900762214	-99.13104
0.0007710486776283	EJE 4 NTE	10000	0105000020E61	5	0101000020E61	19.4834900806753	-99.13115988
0.000921956031126;	EJE 4 NTE	10000	0105000020E61	7	0101000020E61	19.4846200078117	-99.13365
0.000191132923830!	EJE 4 NTE	10000	0105000020E61	8	0101000020E61	19.4848200151199	-99.13455
0.00430958466204B	EJE 4 NTE	10000	0105000020E61	9	0101000020E61	19.484840049794	-99.13474008
0.000417411863306!	EJE 4 NTE	10000	0105000020E61	10	0101000020E61	19.4856100245649	-99.13898016
0.000414617240677!	EJE 4 NTE	10000	0105000020E61	11	0101000020E61	19.4856899930664	-99.13938984
0.0020367732837371	EJE 4 NTE	10000	0105000020E61	12	0101000020E61	19.4858300651859	-99.13978008
0.001510954740318;	EJE 4 NTE	10000	0105000020E61	13	0101000020E61	19.4861599557662	-99.14178996
0.000136054641802!	AVENIDA LAZARO CA	10000	0105000020E61	14	0101000020E61	19.4866299170857	-99.14321988
0.000497828885987;	AVENIDA LAZARO CA	10000	0105000020E61	15	0101000020E61	19.4867599711666	-99.14325984
0.000869188119108(	AVENIDA LAZARO CA	10000	0105000020E61	16	0101000020E61	19.4872200833196	-99.14344992
0.000551520066785(	AVENIDA LAZARO CA	10000	0105000020E61	17	0101000020E61	19.4880199282711	-99.14379012
0.000791300658122;	AVENIDA LAZARO CA	10000	0105000020E61	18	0101000020E61	19.4885299525545	-99.144
0.000946991307938!	AVENIDA LAZARO CA	10000	0105000020E61	19	0101000020E61	19.4892700281757	-99.14428008
0.000206108952222!	AVENIDA LAZARO CA	10000	0105000020E61	20	0101000020E61	19.4901499989314	-99.14463
0.000752810294673;	AVENIDA LAZARO CA	10000	0105000020E61	21	0101000020E61	19.4903399823354	-99.14470992

Fig. 7. Postgresql's table with GRP route data

When the route has been calculated, it is represented as a table in Postgresql database as shown in Fig. 7.

The processed data can be consulted in server by using QGIS [20] a GIS that can be connected to Postgresql server, Fig. 8. The GRP calculates the angle between the route´s segments to determine the direction changes, and searches in the Vial Infrastructure ontology the business located on the route.

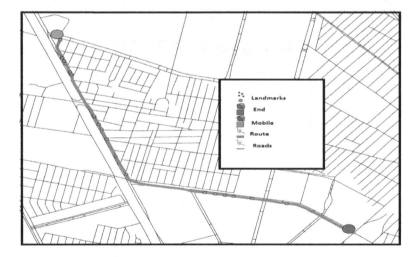

Fig. 8. Visualization in QGIS of GRP data

The instance "Axis North 4", one of the route streets, is searched by its classification as Road Axis in the model. As it is shown in Fig. 9, Road Axis is an entity of Arterial Road, a Primary Road subclass.

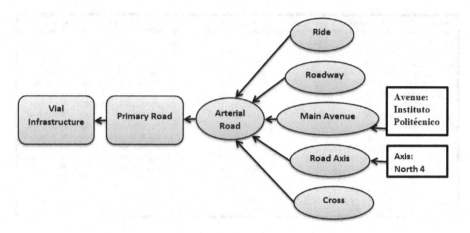

Fig. 9. Vial Axis entity ontological classification

In the map, points of interest are represented as blue and red circles, the route is drawn as a dashed line, and the instructions to the user are written at top of the screen, in the mobile application like in Fig. 10.

When user touches the button located in the right corner of the screen, instructions will be given to him one by one each time he is walking near to the direction change. In Fig. 11 and Fig. 12, shot screens with the individually instructions given are shown. In case of internet connection lost, user will be able to see the map with the route, interest points, and instructions of the last request done.

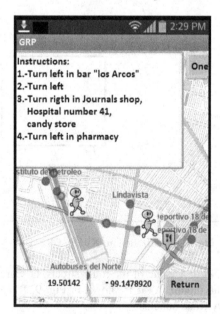

Fig. 10. Mobile application results

Instructions are short but try to be useful for user, giving the business located near to the direction changes; in case of no business near, the instruction just indicate the change of direction, represented by the green icon in the map.

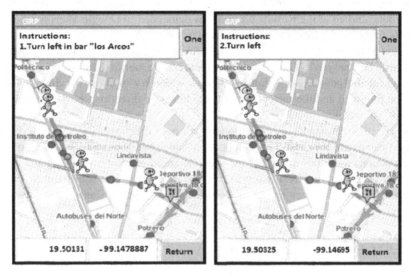

Fig. 11. Shot screens of instructions 1 and 2 given one by one

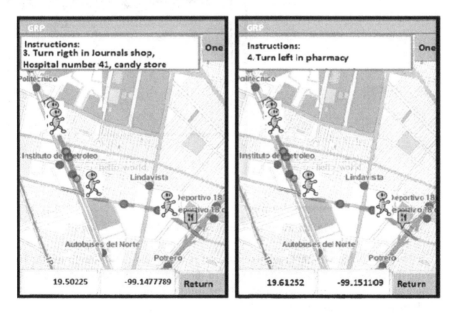

Fig. 12. Shot screens of instructions 3 and 4 given one by one

5   Comparison

To prove the routing with points of interest GRP methodology, some comparison between the obtained results and Google Maps for Android service where compared.

The route studied at the Results section was demanded to Google Maps for Android, the routing done by this application is shown at Fig. 12.

Google Maps give as result two routes, in this case the one that is like the GRP resultant will be analyzed. The instructions to users are notified by using the streets names where the person is passing by, arrows which try to show the direction that should be followed and the meters before that change of direction; the duration of the trip is also showed.

The use of street names at instructions is not ever functional because in some cities street names are not correct or are not at a visible place, so if a person is following that instructions and does not know the area, he could get lost or walking following the instructions in an incorrect roadway. For a walking person it is difficult to have a real perception of distances so the use of distances on instructions is not always a good idea.

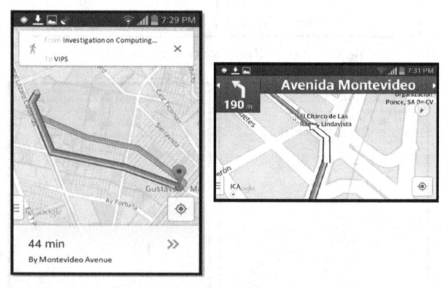

Fig. 12. Results of Google Maps for Android

GRP considers these points and gives to user two ways to get instructions, the first showing the route with icons at each change of direction, the points of interest near to the route and all the instructions that should be followed, notifying the ones located at each direction change as in Fig. 13; the second way, showing the entirely route and giving the instructions one by one as the person is approaching to each change preserving the points of interest at the instructions.

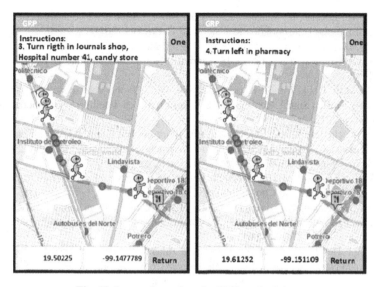

Fig. 13. Instructions given by GRP methodology

6   Conclusions

The methodology for generate routes with points of interest GRP, is an option to provide routes with descriptive instructions and points of interest by using an application ontology and mobile technologies.

GRP combines many tools of geospatial processing, web and mobile development software in its performance.

The ontology model designed for GRP to get the interest points is the application ontology of Vial infrastructure that classifies roads into primary, secondary, tertiary and walking classes, each class is also divided in entities as streets, avenues, road axis, etc., with attributes like name, business, and geographic coordinates.

GRP has a mobile application programmed in Android that has two modules; the first gets the mobile coordinates and the name of the place where the user wants to go, the second shows a final map with the route and instructions the person should follow to go from one place to another, and the points of interest located on the route.

The data used to generate routes, belongs to delegation *Gustavo A. Madero*, a territorial division of *Distrito Federal* in Mexico City, and the information for the business and interest points is part of the statistics directory of economic units (DENUE) of Statistics and Geography National Institute (INEGI) in Mexico.

GRP is an option of the ontology's usage in mobile technologies to process geospatial information.

References

1. Android Mobile Platform, http://www.android.com
2. Bierlaire, M., Newman, J., Chen, J.: A probabilistic map matching method for smartphone GPS data. Transport and Mobility Laboratory, Ecole Polytechnique Fédérale de Lausanne, Switzerland. Transportation Research Part C. Elsevier (2012)

3. Castro, M., Iglesias, L., Sánchez, J.A., Ambrosio, L.: Sight distance analysis of highways using GPS tools. Transportation Research Part C. Elsevier (2011)
4. Findley, D.J., Zegeer, C.V., Sundstrom, C.A.: Finding and Measuring Horizontal Curves in a Large Highway Network A GIS Approach. North Carolina State University, Raleigh (2011)
5. Hunter, M.P., Kook Wu, S., Kyoung, K., Suh, W.: A Probe-Vehicle-Based Evaluation of Adaptive Traffic Signal Control. IEEE Transactions on Intelligent Transportation Systems 13(2) (2012)
6. Imran, H., Hassan, Y.: GPS-GIS-Based Procedure for Tracking Vehicle Path on Horizontal Aligments. Computer Aided Civil and Infrastructure Engineering, 383–394 (2006)
7. JavaScript Object Notation, http://www.json.org
8. Kulik, L., Duckham, M., Egenhofer, M.J.: Ontology-Driven Map Generalization. Journal of Visual Languages and Computing 16(3), 245–267 (2005)
9. Liao, L., Patterson, D.J., Fox, D., Kautz, H.: Building personal maps from GPS Data. Department of Computer Science and Engineering, University of Washington Seattle, Washington 98195, USA (2006)
10. Luna, V.: Desarrollo de un algoritmo para rutas semánticamente ponderado. Instituto Politécnico Nacional, Centro de Investigación en Computación. México (2012)
11. Mapsforge, https://code.google.com/p/mapsforge/
12. Merkaator, http://merkaartor.be/projects/merkaartor
13. OpenStreetMap, http://www.openstreetmap.org
14. Orellana, D., Bregt, A.K., Ligtenberg, A.: Exploring visitor movements patterns in natural recreational areas. Wageningen University, Centre for Geo-Information Science and Remote Sensing, Países Bajos. Tourism Management. Elsevier (2011)
15. Php, http://www.php.net
16. PgRouting, http://www.pgrouting.org
17. Postgis, http://www.postgis.net
18. Postgresql, http://www.postgresql.org
19. Protégé, http://protege.stanford.edu/overview/portege-owl.html
20. Quantum GIS Project, http://www.qgis.org
21. Quddus, M.A., Noland, R.B., Ochieng, W.Y.: Validation of map matching algorithms using high precision positioning with GPS. The Journal of Navigation 58(02), 257–271 (2005)
22. Stojmenovic, I., Giordano, S., Blazevic, L.: Position Based Routing Algorithms for Ad Hoc Networks a Taxonomy. University of Ottawa, Ontario Canadá (2008)
23. Suárez, M.C., García, R., Villazón, B., Gómez Pérez, A.: Essential. In: Ontology Engineering Methodologies, Languages, and Tools. Ontological Engineering State of the Art. Ontology Engineering Group, Universidad Politécnica de Madrid (2011)
24. Winter, S., Truelove, M.: Talking About Place Where it Matters. Department of Infrastructure Engineering, Universidad de Melbourne, Parkville, Australia. Lecture Notes in Geoinformation and Cartography. Springer, Heidelberg (2011)
25. Wu, Y., Winter, S.: Interpreting Destination Descriptions in a Cognitive Way. Department of Infrastructure Engineering, University of Melbourne. Parkville, Australia (2011)

Study of Effect of Urban Green Land on Thermal Environment of Surrounding Buildings: A Case Study in Beijing, China

Qingzu Luan, Caihua Ye, Yonghong Liu, and Shuyan Li

Beijing Municipal Climate Center, Beijing, China
{qzluan,caihuaye,yonghongliu,shuyanli}@gmail.com

Abstract. We analyzed dominant landscape feature parameter of Urban Green Land (UGL) that impact thermal environment of buildings, and researched the spatial distance that UGL could affect surrounding thermal environment and correlation between landscape coefficients of UGL and cooling range caused by UGL. In the scale of 100m spatial resolution, most UGL patches have a role in cooling effect on their surrounding buildings within 100m range. All UGL patches over 0.5 km^2 could exert significant cooling effect on surrounding buildings within 100m, with temperature difference ranging from 0.46 ℃ to 0.83 ℃, averaged at 0.72 ℃, while UGL patches below 0.5 km^2 with high vegetation cover could play a cooling role, but nonsignificant cooling effect with low vegetation coverage. In addition, UGL patches' perimeter, area, shape index and vegetation coverage have no significant correlation with cooling range of their surrounding buildings.

Keywords: Urban green land, Buildings, Thermal environment, Remote sensing, Landscape coefficients.

1 Introduction

With the acceleration of urbanization in China, a series of urban climate change takes place as a result of the expansion of urban impervious surface area and the release of anthropogenic heat caused by effluent and exhaust gases in city run [1,2], producing urban heat environmental problem which becomes an important issue to be considered during urban development and planning. Numerous studies indicate that UGL, wind, and water are the key factors alleviating urban thermal environmental effect [3], for which rational planning and UGL increase are effective means [4,5].

Currently practical observation is widely used in the studies of the impact of UGL on the heat environment, most of which focus on the mitigation of UGL to the heat island effect. For example, Chang [6], Lee [7], Li yanming [8], and Cheng chengqi [9], by means of observation, researched temperature difference between the UGL and its surrounding environment, cooling effect of green land, and heat island mitigation by green land. In recent years, the development of remote sensing technology

B. Murgante et al. (Eds.): ICCSA 2014, Part IV, LNCS 8582, pp. 299–314, 2014.

promotes the studies about the effect of UGL on heat environmental impact. References [10,11,12,13,14,15] used information derived from remote sensing images, together with meteorological data and green land statistic, to study how UGL mitigate and reduce the heat island effect, as well as their interrelation, indicating that the application of remote sensing technology to the study of urban heat environment is feasible and efficient.

UGL is obviously effective in improving the urban heat environment, but there were few studies referring to the effect of UGL on the surrounding building from the perspective of the spatial distribution of green land. We used remote sensing technology and GIS, set the surrounding buildings of UGL as the main objects, and studied the relationship between the UGL spatial distribution and the heat environment of surrounding buildings, in the expectation of applying spatial analysis method to identify the effect and spatial scale of the effect, which will ultimately guide the planning of UGL construction, and build theoretical foundation for quantitative evaluation of UGL construction.

2   Data and Method

2.1   Research Area and Data

Research was carried out in six central districts of Beijing, including Dongcheng district, Xicheng district, Chaoyang district, Haidian district, Shijingshan district, and Fengtai district, with distribution shown in Figure 1.

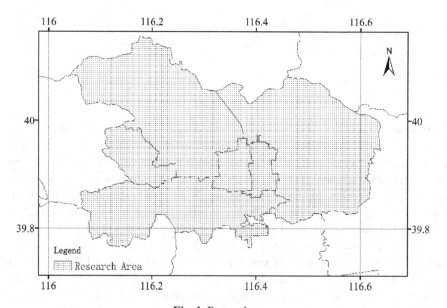

Fig. 1. Research area

We used remote sensing image of U.S. Landsat TM (track number 123/32) as the basic data for the extraction of UGL and the retrieval of land surface temperature. About the image, spatial resolution of visible channels was 30 meters, and spatial resolution for thermal infrared channel (Band 6) was 120 meters. Transit time of the image was at 2:21 p.m. on August 2nd, 2008, which is the day with highest temperature reaching 35.5 ℃ monitored by Beijing Meteorological Bureau, as high temperature condition is conducive to comparative analysis. Basic survey data came from the topographic vector data of scale 1:2000, provided by Beijing Institute of Surveying and Mapping in 2009.

2.2   Retrieval of Land Surface Temperature

Calibration of Landsat TM data was conducted according to the band gain parameters of TM sensor provided by NASA, and radiance of different spectral bandwidths was calculated by formula (1) [16,17] . Based on the vector data of topographic maps of scale 1:2000 and digital elevation model (DEM), orthorectification was done using the improved polynomial model [18], with error controlling in one pixel or less.

$$L_\lambda = \alpha \cdot DN + \beta \tag{1}$$

where, L_λ is surface radiance, α is gain of each band of TM5 sensor provided by NASA, β is band offset, and DN is the original pixel value of image.

Land surface temperature retrieval was referred to the retrieving formula (2) of TM image temperature given by Schneider etc. [19].

$$T = \frac{K_2}{ln(K_1 / L_\lambda + 1)} \tag{2}$$

Where, T is thermodynamic temperature, K_1 and K_2 are calibration constant, and L_λ is radiance of thermal infrared channel (band 6) calculated by formula (1).

To facilitate subsequent calculations and statistical analysis, data obtained from land surface temperature retrieval was resampled to the spatial resolution of 100 meters.

2.3   Extraction of UGL

In order to precisely extract the spatial information of UGL, we manually mapped multi-scale urban green patches, with the smallest scale 3 * 3 pixel. In the process, only patches of green land that don't contain or only contain small amount of water body were chosen, so as to avoid the cooling effect produced by water. Due to different landscape features of green land patch, there is difference in thermal environment effect among various types of green land. According to the definition of landscape parameter in landscape ecology [21,22], we selected area, perimeter, shape index (ratio of perimeter to area), and vegetation coverage as the four landscape feature parameters of UGL. Thereamong, vegetation coverage was calculated using vegetation index, referred in formula (3), which was retrieved from NDVI vegetation index using equal density model in mixed pixels.

$$f = \frac{NDVI - NDVI_{\min}}{NDVI_{\max} - NDVI_{\min}} \tag{3}$$

In formula (3), f is vegetation coverage, $NDVI$ is the NDVI value of each pixel, $NDVI_{\max}$ is the maximum of NDVI value in the research area, and $NDVI_{\min}$ is the minimum.

2.4   Flowchart of the Model

Technological route was shown in figure 2, as the research was conducted by four steps.

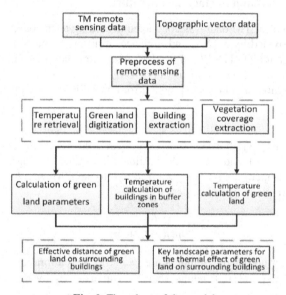

Fig. 2. Flowchart of the model

First, preprocess remote sensing data, including calibration and orthorectification of original TM remote sensing data. Secondly, extract patches of green land and surrounding buildings from preprocessed remote sensing images through artificially digitizing, and retrieve land surface temperature and vegetation coverage using method mentioned in section 2.2 and 2.3. Thirdly, analyze the statistical data. This step consists of three parts. (1)Spatially overlay the distribution of UGL patches with the temperature of city's landmarks, recording maximum, minimum and average temperature corresponding to every green land patches. (2) Obtain area, perimeter, and shape index (ratio of perimeter to area) of every artificially digitized UGL patch, and analyze the spatial feature of each patch. (3) Set multi-scale buffer zones for each green land patch, extract building pixels at buffer areas of different scales, and calculate average top roof temperature of buildings. Fourthly, analyze the effect of UGL on thermal environment of surrounding building. We studied the effect from two aspects.

On one hand, use spatial statistical analysis method and isotherm perimeter-temperature curve's slope breakpoint method [23] to study the distance of effect of UGL on thermal environment of surrounding buildings. On the other hand, make separate correlation analysis between cooling range of buildings in the buffer area, and the main landscape parameters, such as the area, perimeter, shape index and vegetation coverage of UGL patch, resulting in quantitative measure of the effect.

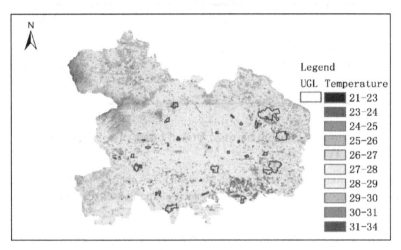

Fig. 3. Land surface temperature and green land digitizing distribution in Beijing by the retrieval of remote sensing data on August 2nd, 2008 (Unit: ℃)

3   Analysis

3.1   Retrieval of Land Surface Temperature

Calculated by formula (1) and (2), landmark temperature in the central district of Beijing retrieved was shown in figure 3. Land surface temperature of main districts of Beijing ranges from 21.86 ℃ to 34.00 ℃. Higher temperature areas are majorly distributed at industrial and residential areas with less vegetation coverage in the southeast, while lower at shallow mountain areas in the northwest.

3.2   Extraction of UGL

Through the calculation of landscape parameters, it is found that area, perimeter, and shape index vary significantly amongst green lands. There are 16 pieces of green land patches with area ranging from 0 to 0.5 km^2, 3 pieces with area ranging from 0.5 to 1.0 km^2, 3 pieces with area ranging from 1.0 to 2.0 km^2, 2 pieces with area ranging from 2.0 to 5.0 km^2, 2 pieces with area ranging from 5.0 to 8.0 km^2. The average area of green land is 1.12 km^2. Spatial analysis shows that, 5 patches of green land with the lowest temperature are located in the north of the city, and 5 patches with the highest temperature are located in the south. The regularity, higher in the south and lower in the north, shown in the spatial pattern of thermal environment of UGL in Beijing, is coherent with that of the whole city.

Table 1. Statistics of UGL

Index	Average	Maximum	Minimum
Area (km^2)	1.12	7.65	0.08
Perimeter (km)	5.29	20.22	1.55
Shape index	0.0093	0.0203	0.0026
Vegetation coverage	0.77	0.92	0.58
Temperature (℃)	26.80	28.97	25.58

3.3   Analysis of Relationship between UGLs' Landscape Parameters and Average Temperature

We separately studied the correlation between UGLs' average temperature and four landscape parameters, including area, perimeter, shape index, vegetation coverage of each green land patch.

1. In general, there is a rather weak relation between the average temperature and area of UGL. But among the patches of green land whose area ranges from 0.26 km^2 to 1.03 km^2, temperature decreases non-uniformly with increasing area. Among those patches with area ranging from 0.08 km^2 to 0.22 km^2, there is no regularity in the area-temperature curve.
2. Analysis in the figure 3 shows that, there is no significant correlation between the perimeter of green land patch and the temperature of thermal environment.
3. According to Wiens' research [24] in 1993, the shape index is an important parameter to reflect the shape of patch. For a patch of green land with higher value of shape index, implying a more complex surface shape, it is easier to exchange the internal energy, material and information with the surrounding environment. So we developed correlation analysis between shape index and temperature of green land patch. Result shows that, the lowest value of shape index is 0.0026, located at green land surrounding China Aviation Museum, corresponding to the temperature of 26.9 ℃, while the highest value of shape index of 0.0203 at Sihui park corresponds to 28.1 ℃. However, as shown in figure 3, there is weak correlation between the two in the absence of obvious trend shown in the curve.
4. The average temperature of UGL decreases as the vegetation coverage increases, with the correlation coefficient of -0.50, which has passed the F test. Sihui park, with the lowest vegetation coverage of 0.58, corresponds to the temperature as high as 28.1 ℃, and Wangjing street green land, with the highest vegetation coverage of 0.92, corresponds to the lowest temperature of 25.9 ℃. Therefore, there is a certain negative correlation between vegetation coverage and temperature of urban thermal environment, that is, the higher the degree of vegetation coverage, the

lower the temperature. So the vegetation coverage significantly influence the cool-ing effect.

It is necessary to specify that, researches by Wangxue [25] and Ganlin [26] showed that temperature has negative correlation with area and perimeter of UGL, yet has positive correlation with shape index, which is not consistent with the result in this research. One of the important reasons is the specialty of our samples. For exam-ple, green land of Zizhuyuan park contains some water body, and in some other parks, there are impervious surface and a small number of buildings. Therefore, the different land surface cover of parks will result in thermal environmental difference even among parks of the same area. The other possible key reason is the significant differ-ence in inner land surface cover of green land patch. For instance, temperature of Linglong park with area of 0.08 km^2 is 26.8 ℃, whereas temperature of Ditan park with area of 0.8 km^2 is 26.6 ℃, close to the former one. Without considering other sources of heat, the main reason is that the vegetation coverage of Linglong park reaches as high as 0.83, while the vegetation coverage of Ditan park is 0.69.

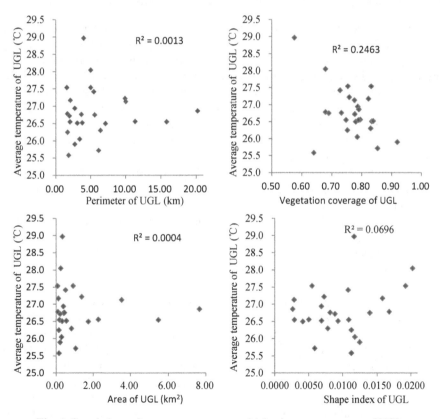

Fig. 4. Correlations of average temperature with landscape parameters of UGL

4   Thermal Environment Effect of UGL on Surrounding Buildings

In order to make quantitative research in spatial mode on the thermal environment effect of UGL on its surrounding building, we constructed buffer zone of different distance for each green land patch. Based on spatial distance, the area outside each green land patch was divided into four buffer zones, including 0-100m, 100-200m, 200-300m, and 300-400m buffer zones. Buildings in the buffer zones were extracted and overlaid with land surface temperature distribution to obtain the statistics of thermal environmental temperature of the four levels of buffer zones. We compared the average temperature of every buffer zone, quantitatively measured the local effect of UGL on the surrounding buildings, and made correlation analysis between average temperature of buildings impacted by green land and parameters of UGL, including area, perimeter, shape index, and vegetation coverage.

4.1   Thermal Effective Distance of Green Land on Surrounding Buildings

Spatial Statistics Analysis. From the theoretical analysis, temperature of inner green land patch is lower than that of non-vegetation underlying surface such as surrounding buildings, resulting in certain cooling effect on thermal environment. If roof temperature difference between the buildings in the buffer area of green land and common buildings is obviously above a certain threshold, it implies that buildings within the distance are impacted by green land. In most cases, the further away from UGL, the higher temperature of building. Herewith, referring to research result by Huang [27], we defined cooling range as follows. If temperature variation of buildings from one buffer zone to its adjacent zone was no more than 0.3 ℃, the buffer zone was selected as the normal building buffer zone that was not impacted by the green land. Then cooling range of green land was the average temperature difference between each buffer zone and that of the normal buffer zone.

According to the definition, in order to identify the thermal effective distance of green land on surrounding buildings, we calculated average temperature difference separately between buffer zones of 100m and 100-200m, 100-200m and200-300m, 200-300m and 300-400m, as shown in table 2.

Comparing the building's temperature difference between 100m and 200m buffer zones among parks, we can find as follows. (1) There are only 5 parks with temperature difference less than or equal to 0, which is greater than 0 at the remaining 21 areas. (2) Among all the parks, Huaxiang park has the greatest cooling range, reaching 2.37 ℃. There is a large area of road and construction at the 100-200m buffer zone, with little green area at these impervious areas. At the same time, the green area of Huaxiang park reaches as high as 2.27 km^2. Therefore, the cooling effect of the park is the most obvious. As a consequence, we consider that the above six pieces of green land are greatly influenced by the background environment, and remove them at the consequent analysis.

Table 2. Statistic of temperature of green land and buildings in buffer area (Unit: ℃)

UGL Name	Area	Avg. temp.	A	B	B-A	C	C-B	D	D-C
Linglong park	0.08	26.78	28.06	28.82	0.76	28.77	-0.05	28.49	-0.28
Wanfangting park	0.10	27.54	27.4	28.21	0.81	28.08	-0.13	27.79	-0.29
Chaoyang golf	0.13	27.17	28	28.43	0.43	28.37	-0.06	28.22	-0.15
Jingshan park	0.15	26.25	27.28	26.79	-0.49	27.51	0.72	27.34	-0.17
Tiancun vegetable field	0.16	25.58	NO Builds	28.67	—	28.47	-0.2	28.42	-0.05
Ritan park	0.18	26.55	28.06	28.06	0	28.03	-0.03	28.05	0.02
Wangjing street	0.21	25.9	28.18	28.69	0.51	28.31	-0.38	28.07	-0.24
Side park	0.22	26.72	28.07	27.57	-0.5	27.56	-0.01	28.05	0.49
Sihui park	0.25	28.05	28.01	28.79	0.78	29.09	0.3	28.41	-0.68
Zizhuyuan park	0.29	26.05	27.97	28.25	0.28	28.15	-0.1	28.22	0.07
Tiancun street park	0.33	26.51	28.18	27.17	-1.01	28.47	1.3	28.24	-0.23
Taoyuan park	0.34	28.97	28.13	29.04	0.91	28.27	-0.77	28.94	0.67
Sontlindao park	0.39	26.94	27.74	28.18	0.44	28.14	-0.04	28.4	0.26
Ditan park	0.40	26.75	29.11	28.47	-0.64	28.78	0.31	28.72	-0.06
Xinglong park	0.45	26.76	28.02	28.33	0.31	28.72	0.39	28.75	0.03
Kandanjiaoye park	0.50	27.42	28.68	28.84	0.16	28.86	0.02	28.38	-0.48
Wanliu golf club	0.55	26.52	26.99	27.71	0.72	27.68	-0.03	27.75	0.07
Jiangfu park	0.82	26.3	28.37	29.2	0.83	29	-0.2	28.71	-0.29
Hongbo park	0.91	27.54	28.42	29.18	0.76	28.82	-0.36	28.9	0.08
Yuanmingyuan park	1.04	25.72	27.2	28.02	0.82	28.21	0.19	28.27	0.06
Yamenkou	1.37	27.22	28.1	28.86	0.76	28.95	0.09	28.99	0.04
Tiantan park	1.73	26.5	27.86	28.58	0.72	28.76	0.18	28.49	-0.27
Huaxiang park	2.27	26.56	26.64	29.01	2.37	28.76	-0.25	28.98	0.22
Sanjianfang	3.52	27.13	28.46	29.2	0.74	29.62	0.42	29.65	0.03
Nanhuayuan	5.48	26.55	27.98	28.44	0.46	28.66	0.22	28.3	-0.36
Aviation museum	7.65	26.86	28.66	29.36	0.7	29.25	-0.11	29.44	0.19

Remark: A is the average temperature of buildings within the 100m buffer zone, B is the average temperature of buildings within the 100-200m buffer zone, C is the average temperature of buildings within the 200-300m buffer zone, and D is the average temperature of buildings within the 300-400m buffer zone. B-A is the temperature difference between the 100-200m buffer zone and the 100m buffer zone, and so on by such analogy.

For the remaining green land, we drew a line diagram between the temperature variation of buffer zones and the area of green land patch (Figure 5). From the given line plot, together with the statistics of table 2, we can find information as follows. First, temperature increased significantly at zones far away from the green land patch, and the temperature variation of the three level buffer zones decreases as the distance increases. In table 2, data mostly follows the regularity showing (D-C) < (C-B) < (B-A) , which indicates that the cooling effect of green land on buildings reduces as the distance increases. Secondly, more than 90 percent of the temperature difference between the 200m and the 100m buffer zone of green land exceeds 0.3 ℃, and there are only 2 green land patches with temperature difference of the two buffer zones lower than 0.3. However, more than 90 percent of the temperature difference between the 300m and the 200m buffer zone of green land and more than 95 percent of the temperature difference between the 400m and the 300m buffer zone of green land is less than 0.3 ℃. The two green land patches with temperature difference of the 200m and 100m buffer zones lower than 0.3 ℃ are located at Kandanjiaoye park and Zizhuyuan park. The difference of Zizhuyuan park is 0.28 ℃ approximating 0.3 ℃. And the variation of Kandanjiaoye park is 0.16 ℃, where the green area is 0.5 km², but its shape index as an irregular rectangle, reaches as high as 0.0108, and the vegetation coverage is as low as 0.29, so its cooling effect is not as significant as other green lands. Thirdly, temperature variation of the green land patches over 0.5 km² wholly exceeds 0.3 ℃, ranging from 0.46 ℃ to 0.83 ℃, averaged at 0.72 ℃, which indicates significant cooling effect on surrounding buildings within 100m. For the green land smaller than 0.5 km², those with high vegetation coverage play a cooling role, which is not significant among those with low vegetation coverage.

Therefore, we infer that, in the scale of 100m spatial resolution, thermal effective distance of green land on surrounding buildings is 100m, that is, green land only affected the temperature of buildings within 100 meters. And green land doesn't influence the temperature of buildings at the 100 to 200m buffer zone, of which temperature can be chosen as the background temperature. Furthermore, the temperature difference between the 200m and 100m buffer zone is the cooling range impacted by green land on buildings within 100 meters.

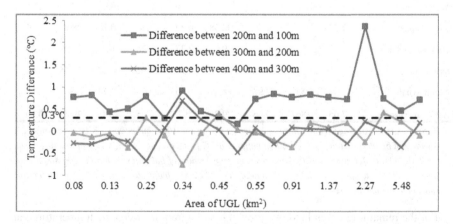

Fig. 5. Relationship between the temperature variation of buildings in buffer area and the area of green land

Isotherm Perimeter-Temperature Curve Slope Breakpoint Method. From the theoretical analysis, UGL can result in certain cooling effect on its surrounding environment, which can be characterized by the distribution and density change of isotherm around the urban green land, so it is feasible to use the isotherm around each green land patch to quantitatively measure their thermal effect. Jia Liuqiang [12] used the perimeter-temperature diagram of isotherm around green land to find out the slope breakpoint, so as to obtain the cooling scope and range of green land. Figure 6 shows the isotherm of Yuanmingyuan ruins park, with interval of 0.1 ℃, which was derived from the land surface temperature.

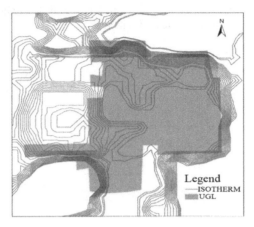

Fig. 6. Map of Isotherm with Interval of 0.1 ℃ around Yuanmingyuan Ruins Park

As shown in figure 6, 21 closed isotherms are distributed from inside to outside, reflecting the thermal effect of green land patch on surrounding landmarks. As the perimeter of isotherm expands, with land surface temperature increases, the cooling effect of green land reduces, converging toward the surrounding surface temperature. Scope of cooling effect can be considered as the area from the boundary of patch to the closed isotherm, from which temperature is consistent with that of outside area. And cooling range is the temperature difference between the given isotherm and the average temperature of the green land patch.

We further analyzed the curve concerning the perimeter of 21 isotherms and their temperature, fit by polynomial, in which the determination coefficient, R^2, is 0.97 at the 0.05 level of confidence. There are two slope breakpoints of the curve, 25.9 ℃ and 26.7 ℃. The former one corresponds to the edge of the green land, some located inside the green land, which is the start of thermal effect. The latter one corresponds to the critical point of the thermal effect, from which the temperature does not significantly change.

Using the isotherm distribution chart to check the two slope break points of the curve above, it is found that the cooling scope reflected by the curve is consistent with regularity interpreted from the distribution of isotherm around the green land patch. Temperature changes significantly near the edge of the green land of Yuanmingyuan ruins park, forming the first slope change point, corresponding to the 25.9 ℃ iso-therm. As the perimeter of isotherm increases, the land surface temperature gradually increases until reaching the second slope change point, corresponding to the 26.7 ℃ isotherm, outside of which the trend of temperature increase disappears, converging to the surrounding landmarks. So it is the outer boundary of the cooling scope of the green land. Therefore, the cooling range of Yuanmingyuan ruins park is 0.8 ℃, which is subtracted from 26.7 ℃ to 25.9 ℃, result of which is basically the same as 0.82 ℃ calculated by the method of buffer area mentioned above. Although the en-closed area between the two isotherms is quite different with the 100m buffer zone, the minimum interval distance of the two isotherms is still less than 100 meters, indi-cating the effect of green land less than 100 meters.

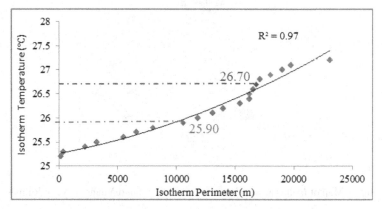

Fig. 7. Slope breakpoints of the Isotherm perimeter-temperature curve

It should be noted that, there is also defect concerning the isotherm perimeter-temperature curve slope breakpoint method. If the land use of underlying surface at the green land area is complex, or the green land is with a complex shape, the thermal environment around the green land will be affected by multi factors, and the isotherm will also become irregular, which will not be the regular closed isotherms centered by the green land from inside to outside, but will be diverging and extending complex curves.

4.2   Correlation between Parameters of Green Land Patch and Cooling Range

The spatial distance of thermal effect of green land on surrounding environment is 100 meters, so we mainly analyzed the parameters of green land patch and cooling range from 100m to 200m buffer zone, so as to find out the regularity of relieving effect of green land on thermal environment. From the scatter diagram (Figure 8), there is no obvious correlation between the cooling range and parameters of green land, such as area, perimeter, shape index, and vegetation coverage. Research by Jia Liuqiang indicates that, cooling range of green land is inveterately influenced by the vegetation, area, perimeter, and shape index of green land patch, in which the area and the perimeter are the most important, with the correlation coefficient of 0.76 and 0.72. We selected surrounding buildings of green land as the research objectives. On one hand, the buffer area of one green land patch is an outward expanding area with the similar shape like the patch. But in actual, impacted by landform, environment, and wind direction, it is difficult to create a green land cooling affected area with the similar shape. On the other hand, besides influenced by the environment, temperature of buildings is also significantly impacted by its own structure and inner heat consumption with strong uncertainty. Furthermore, the small number of samples of urban green space may also affect the outcome. Therefore, there is certain uncertainty in using quantitative method based on suffer area analysis to study the thermal effect of green land on its surrounding environment and define the cooling range of green land.

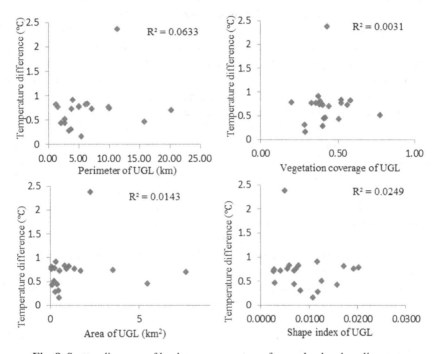

Fig. 8. Scatter diagrams of landscape parameters of green land and cooling range

5   Conclusion and Discussion

We took the green land inside the fifth circle road in the central city of Beijing in 2008 and the surrounding buildings as the research object, discussed the regularity of relieving the thermal environment of surrounding buildings by green land, studied UGL's average temperature's relation with vegetation coverage and spatial characteristics parameters of green land, including area, perimeter, and shape index of green land, and quantitatively analyzed the relationship between these parameters and the effective scope and effective degree, and validated the scope of effect based on isotherm perimeter-temperature curve's slope breakpoint method. Conclusions are obtained as follows.

1. Average temperature of the green land patch is weakly related with the area, perimeter, and shape index of patch, but is negatively correlated with vegetation coverage, with the correlation coefficient of -0.50, indicating that vegetation cover provides cooling benefits.

2. In the scale of 100m spatial resolution, most green lands in the research only affected the temperature of buildings within 100 meters. Some patches with higher vegetation coverage may have the effective distance more than 100m. This result also implies that, no matter how large the green land is, its cooling effect on surrounding buildings is limited in certain spatial scope. Therefore, it is better to construct decentralized green land than centralized large land

3. Green land patches over 0.5 km^2 has obvious cooling effect on buildings with 100m, with temperature decrease ranging from 0.46 ℃ to 0.83 ℃ , averaged at 0.72 ℃. But for the green land patches smaller than 0.5 km^2, those with high vegetation coverage play a cooling role, which is not significant among those with low vegetation coverage. Based on the result, during the planning and designing of green land, it is suggested that area of green land patch reach more than 0.5 km^2, to gain the maximum and optimization of the thermal effect of green land, and increase the vegetation coverage as much as possible, which is positive for cooling effect.

4. Cooling scope and range of green land on surrounding buildings, affirmed by the method of isotherm perimeter - temperature curve's slope breakpoint, is partially consistent with the result above. It indicates that this method aids in quantifying the thermal effect of green land. Under the condition of simple shape of green land, and not complex land use of underlying surface, this method can more accurately determine the effective scope and cooling range of each green land patch on its surrounding environment. At the same time, to some extent, this method avoid error caused by artificially unified buffer distance, which ensures the independence of the data acquisition concerning effective scope and cooling range, and improves the reliability of conclusion.

5. Statistics of samples selected in the paper shows that, cooling range of surrounding buildings is not significantly related with the area, perimeter, and shape index of green land, and its vegetation coverage. Due to the complex underlying condition, the buildings' structure and the strong uncertain inner heat consumption, there is some detect in the analysis method based on buffer area.

Additionally, scale is an important issue in research. Given a benchmark data source, conclusion will be various if the scale of study is different. Limited by the spatial resolution of remote sensing image, conclusion of this study is still relatively rough, which needs more precise, accurate quantitative analysis of green land's thermal effect. In the future, it is needed to study UGL's thermal effect in combination of green land's temperature and thermal infrared remote sensing images with higher spatial resolution.

References

1. Oke, T.R.: Urban climatology and its applications with special regard to tropical areas. In: WMO, Mexico, vol. 652 (1984)
2. Peng, S.L., Zhou, K., Ye, Y.H., Su, J.: Research progress in urban heat island. Ecology and Environment 14(4), 574–579 (2005)
3. Taha, Akbari, Rosenfeld: Heat island and oasis effects of vegetative canopies: Micrometeorological field-measurements. Theoretical and Applied Climatology 44(2), 123–138 (1991)
4. Akbari, Rosenfeld, Taha: Summer heat islands, urban trees, and white surfaces. Winter meeting of the American Society of Heating. Refrigerating and Air-Conditioning Engineers 45(19), 359–367 (1990)
5. Jauregui: Influence of a large urban park on temperature and convective precipitation in a tropical city. Energy and Buildings 15(3-4), 457–463 (1990)
6. Chang, L.: A preliminary study on the local cool-island intensity of Taipei city parks. Landscape and Urban Planning 80(4), 386–395 (2007)
7. Lee, Lee, Jin, et al: Effect of an urban park on air temperature differences in a central business district area. Land Scape and Ecological Engineering 5(2), 183–191 (2009)
8. Yanming, L.: Research of Ecology Effect of Municipal Gardens Landscape Architecture in Beijing (in Chinese). Urban Management Science & Technology 56(1), 24–27 (1999)
9. Chengqi, C., Ning, W., Shide, G., et al.: A Study on the Interaction Between Urban Heat Island and Vegetation Theory, Methodology, and Case Study (in Chinese). Research of Soil and Water Conservation 78(3), 172–174 (2004)
10. Mei, Z.H., Hu, Z.C., Qiang, G.W., et al.: The Surveying on Thermal Distribution in Urban Based on GIS and Remote Sensing (in Chinese). Acta Geographica Sinica 56(2), 189–197 (2001)
11. Hongmei, Z., Yang, G., Weiqiang, G., et al.: The Research on the Relationship Between the Urban Expansion and the Change of the Urban Heat Island Distribution in Shanghai Area (in Chinese). Ecology and Environment 17(1), 163–168 (2008)
12. Liuqiang, J., Jian, Q.: Study of Urban Green Patch's Thermal Environment Effect with Remote Sensing: A Case Study of Chengdu City (in Chinese). Chinese Landscape Architecture (12), 97–101 (2009)
13. Jian, Q., Liuqiang, J., Yong, W.: Spatial Correlation between Heat Island and Green Space in Qingdao City Based on Remote Sensing (in Chinese). Journal of Southwest Jiaotong University 43(4), 427–433 (2008)
14. Weng, O., Schubring, J.: Estimation of land surface temperature-vegetation abundance relationship for urban heat island studies. Remote Sensing of Environment 89(4), 467–483 (2004)
15. Streutker, D.R.: Satellite-measured growth of the urban heat island of Houston, Texas. Remote Sensing of Environment 85(3), 282–289 (2003)
16. Yunhao, C., Jing, L., Xiaobing, L.: Urban Spatial Thermal Environment Analysis Based on Remote Sensing-Process Simulation and Impacts (in Chinese). Wangping, Science Publication, Beijing (2004)
17. Chander, G., Helder, D.L., Markham, B.L., et al.: Landsat-5 TM reflective band radiometric calibration. IEEE Transactions on Geo-science and Remote Sensing 42(12), 2747–2760 (2004)
18. Qingzu, L., Huiping, L., Zhiqiang, X.: Comparison between Algorithms of Ortho-rectification for Remote Sensing Images (in Chinese). Remote Sensing Technology and Application 22(6), 743–747 (2007)

19. Schneider, K., Mauser, W.: Processing and accuracy of Landsat Thematic Mapper Data for lake surface temperature measurement. International Journal of Remote Sensing 17(11), 2027–2041 (1996)
20. Liu, A.J., Cameron, G.N.: Analysis of landscape patterns in coastal wetlands of Galveston Bay, Texas. Landscape Ecology 16, 581–595 (2001)
21. Hulshoff, R.M.: Landscape indices describing a Dutch landscape. Landscape Ecology 10, 101–111 (1995)
22. Strahler, A.H.: The use of prior probabilities in maximum likelihood classification of remotely sensed data. Remote Sensing of Environment 10(2), 135–163 (1980)
23. Liuqiang, J.: Research on the Spatial Characteristics of the Urban Green Space's Function of Mitigating Urban Heat Island (in Chinese). Jia Liuqiang, Southwest Jiaotong University, Chengdu (2009)
24. Wiens, J.A., Stenseth, N. C., Van, H.B., et al.: Ecological mechanisms and landscape ecology. Oikos (1993)
25. Xue, W.: A Remote Sensing Study of Urban Green Space Distribution and Its Thermal Environment Effect (in Chinese). Wangxue, Beijing Forestry University, Beijing (2006)
26. Lin, G.: Planning Strategies of Beijing's Greenbelt: Oriented to the Mitigation of Surface Thermal Effect (in Chinese). Ganlin, Tsinghua University, Beijing (2011)
27. Huang, J., Akari, H., Taha, H., Rosenfeld, A.: The potential of vegetation in reducing summer cooling loads in residential buildings. Journal of Climate and Applied Meteorology 26, 1103–1106 (1987)

Predicting Land Cover Change in a Mediterranean Catchment at Different Time Scales

Hari Gobinda Roy, Dennis M. Fox[*], and Karine Emsellem

UMR 7300 CNRS ESPACE, Université de Nice Sophia Antipolis, BP 3209,
06204 Nice cedex 3, France
roy.hari.gobinda@etu.unice.fr,
{fox,Karine.Emsellem}@unice.fr

Abstract. Land cover has been changing rapidly throughout the world, and this issue is important to researchers, urban planners, and ecologists for sustainable land cover planning for the future. Many modeling tools have been developed to explore and evaluate possible land cover scenarios in future and time scales vary greatly from one study to another. The main objective of this study is to test land cover change prediction at different time scales in a Mediterranean catchment in SE France. Land cover maps were created from aerial photographs (1950, 1982, 2003, 2008, and 2011) of the Giscle catchment (235 Km^2) and surfaces were classified into four land cover categories: forest, vineyard, grassland, and built area. Explanatory variables were selected through Cramer's coefficient. Different time scales were tested in the study: short (2003-2008), intermediate (1982-2003), and long (1950-1982). To test the model's accuracy, Land Change Modeler (LCM) of IDRISI was used to predict land cover in 2011 and predicted images were compared to a real 2011 map. Kappa index and confusion matrix were used to evaluate the model's accuracy. Altitude, slope, and distance from roads had the greatest impact on land cover changes among all variables tested. Good to perfect level of spatial and perfect level of quantitative agreement were observed in long to short time scale simulations. Kappa indices ($K_{quantity} = 0.99$ and $K_{location} = 0.90$) and confusion matrices were good for intermediate and best for short time scale. The results indicate that shorter time scales produce better predictions. Time scale effects have strong interactions with specific land cover dynamics, in which stable land covers are easier to predict than cases of rapid change and quantity is easier to predict than location for longer time periods.

Keywords: Time scale, Land cover change modeling, Mediterranean Europe, Land change Modeler (LCM).

1   Introduction

1.1   Land Cover Change Modeling

Land cover is changing rapidly throughout the world, and it has become an important issue for urban planners, ecologists, economists, and resource managers to evaluate

---

[*] Corresponding author.

B. Murgante et al. (Eds.): ICCSA 2014, Part IV, LNCS 8582, pp. 315–330, 2014.
© Springer International Publishing Switzerland 2014

environmental change and establish sustainable development planning [7, 10, 17]. Land cover change models are able to identify location and quantity of change, predict land cover change considering past changes, test explanatory variables, and simulate management policies. For this reason, many interdisciplinary research projects have been initiated for land cover change modeling, measuring regional and global land cover change, forecasting future conditions, and planning for sustainable development [28]. As a result, researchers have created a large set of operational modeling tools to implement prediction and exploration of possible land cover change trajectories and land cover planning and policy in recent years [29]. Moreover, land cover change, urban growth, and spatial modeling have drawn considerable interest in the last two decades due to better computing power, availability of spatial data, and the need for innovative planning tools for decision support [7]. Advanced urban and land cover change modeling techniques have been included in many GIS software package.

1.2   The Role of Time Scale in Land Change Prediction

The selection of prediction and validation time intervals has a great impact on prediction accuracy [6]. Prediction accuracy can depend on the rate and process of transitions in both time intervals. Modeling of land cover change using a coarser temporal scale may fail to understand landscape change patterns properly and can hamper model performance [2], so most studies on future land cover change use short to intermediate historical time scales (5–15 years). Many studies on urban land cover change modeling use short time scales that achieve better prediction [1, 11, 18, 24]. Some studies use intermediate time scales [13, 14, 15, 20, 25, 26, 27] and very few studies use long time scales to simulate urban land cover [4] and multiple land cover change [10, 21]. Average historical and prediction time periods are about 10 and 12 years, respectively, analyzing 25 recent studies on land cover change using CA-Markov and Multi-Layer Perceptron (MLP).

Very few studies were found on the comparison of the impact of historical time periods on land cover prediction using different time scales. To investigate the impact of time interval on prediction accuracy in Gorizia-Nova Gorica (Italy), urban area was predicted for different years (2005 to 2010) from initial conditions in 1985 and 2004 [5]. The authors found that prediction accuracy increased with decreasing prediction time period.

1.3   Objectives

The objective of this paper is to explore the impact of temporal scales on land cover change modeling for predicting land cover change in a Mediterranean catchment in SE France. Land cover maps of 2011 were predicted from different time scales (1950-1982, 1982-2003, and 2003-2008) and compared with the digitized land cover map of 2011 to measure model accuracy. The study is part of a larger program to evaluate the impacts of land cover change on runoff and soil erosion at the catchment scale.

2   Methods

Study area, land change modeling steps, and data are discussed in this section.

2.1   Site Description

The study area (about 235 km²) is situated in the Var department of SE France near the Gulf of St. Tropez. The western part of the watershed (about 70% of the catchment) is forest (mostly pine and oaks), and the topography is uneven with the highest elevation at about 650 m. The lower part of the catchment is a gently sloping alluvial plain. The catchment area is characterized by a Mediterranean climate with hot dry summers, and cooler rainier winters. Average temperatures range between 22°C to 26°C in summer and 5°C to 10°C in winter. The mean annual rainfall is about 900 mm, and the main rainy season is from October to January [9]. Several tributaries flow into the Giscle main channel, including the Môle, the Grenouille, the Tourre, and the Verne. Three main municipalities are located within the catchment: Cogolin, Grimaud, and La Môle.

2.2   Land Change Modeling Procedure

Land Change Modeler (LCM) in IDRISI [8] was originally designed to manage impacts on biodiversity, and analyze and predict land use and land cover changes. Only thematic raster images with the same land cover categories listed in the same sequential order can be inputted in LCM for analysis, and background areas must be identified on maps coded with 0. LCM evaluates land cover changes between Time 1 (initial time) and Time 2 (second time). It calculates the changes, and displays the results with various graphs and maps. Finally, it predicts future (Time 3) land cover on the basis of relative transition potential maps. LCM was used in this study to identify explanatory variables, create transition potentials, and predict future land cover maps.

Digital Data and Land Cover Categories
Land cover maps were digitized from grey scale ortho-rectified aerial photographs of 1950 and 1982, and color ortho-photos of 2003, 2008, and 2011. Spatial resolution for all aerial photographs was reduced to 1 m from 0.5 m to facilitate data manipulation during digitization. Surfaces were initially characterized into five categories: forest (F), vineyard (V), grassland (G), urban (U) and suburban (S), but the last 2 categories were collapsed into a single built area (B) class to improve category attribution as described below. Methods of land cover digitization, classification, and characteristics of land cover classes were discussed in [23]. Land cover classification was facilitated by numerous field visits, and validation was carried out through a group of 15 third year Geography students of the University of Nice Sophia Antipolis. Each student was provided with a sample of 20 selected cells to identify land cover class; each sample had a roughly equal number of cells in each category, and there were 5

students for each year (1950, 1982, and 2003). This was the students' first contact with digital air photos, so the validation is considered a worst case scenario.

Slope was created from a 25 m Digital Elevation Model (DEM). Road and stream networks were screen digitized from the aerial photographs of 2008. Only major roads were taken into account, so road network was considered constant for all time periods. In order to make the land cover maps compatible with the explanatory variables, celle size was converted to 25 m.

Explanatory Variables and Constraints

Topographic and distance variables have been used to simulate land cover change studies throughout the world [16, 18, 19, 27]. In an earlier study [23], major topographic and distance variables were identified. These include the following: slope, altitude, distance from roads, distance from built area (initial year), and distance from streams. In addition, three constraints and incentives (forest to built area, vineyard to built area, and grassland to built area) were included in the prediction process. These were created from the "Plan Local d'Urbanisme" (PLU) and "Schéma de Coherence Terrtoriale" (SCOT). The PLU is the local urban plan in France; it determines land use guidelines. The SCOT integrates different policies regarding urban planning: social and private housing, communication infrastructure and public transport, commercial infrastructure, and environment protection. Constraints and incentives are multiplied by the corresponding transition potential during modeling. In this study, values of 0 on the map were used to define absolute constraint, and 1.1 was used for incentives to emphasize the expansion of built areas in suitable selected zones for development according to the regional plan. In addition, distance from streams was also added with above mentioned constraints. Disincentive areas situated within a distance from streams of 0-25 m, and 25-50 m were defined by values of 0.6 and 0.8, respectively to maintain the historical trend of less urbanization near stream networks in the study area according to [23].

Selection of Explanatory Variables

The simulation of multiple categories of land cover change depends on several explanatory variables [18]. Explanatory variables that were drivers of past land cover change are expected to be an influential force in future changes and are selected based on available data and their explanatory abilities. DEM, slope, and distance from road represent the accessibility of a neighborhood, and distance from built area highlights the proximate location of urbanization. The significance of explanatory variables was tested using Cramer's V which measures the strength of association between two categorical variables based on Chi-square statistics [21]. In this study, land cover change in a historical time period and explanatory variables are taken into account to test Cramer's V for a particular variable. LCM calculates Cramer's V automatically and displays the association level of explanatory variables with land cover categories. Variables with greater values are considered more important than other variables. Cramer's V values of ≥ 0.4 and ≥ 0.15 are considered good and useful, respectively; and values <0.15 should be removed from the model [8].

Transition Potentials
Transition potential maps were created for each transition possibility (F to V, F to G, F to B, V to F, V to G, V to B, G to F, G to V, and G to B) based on historical changes and selected explanatory variables. The Multi-Layer Perceptron Neural Network (MLPNN) algorithm of IDRISI [8] was employed to create transition potentials. Each transition potential was modeled individually using the same explanatory variables, but only transition potentials with an accuracy rate greater than 70% were utilized for land cover prediction.

Land Cover Prediction and Time Scales Test
Land cover change prediction has two aspects: the quantity of change is provided by the Markov change model matrix and the spatial distribution of change is given by MLPNN. LCM provides the quantity of change by evaluating the Markov matrix comparing the initial (T1) and second land cover (T2), and then predicts the future land cover (T3) using a transition probability matrix for the future. The transition probability matrix displays the probability of each land cover category changing into another category. A value close to 0 indicates a low conversion probability, and 1 indicates a high conversion probability for the target land cover. Land cover maps were predicted for 2011 using transition potential maps from several historical time periods (1950-1982, 1982-2003, 2003-2008) (Table 1). The same variables and constraints were incorporated in all simulations.

Table 1. Historical time periods, prediction and validation dates for different scales

Historical time period	Prediction date	Historical time interval	Validation time interval
1950-1982	2011	32	29
1982-2003	2011	21	8
2003-2008	2011	5	3

Land Cover Prediction Validation
Validation of a model is needed in order to assess its accuracy. To do this, simulated land cover maps of 2011 created using different time scales were compared with a digitized map of the same year. Kappa indices and error matrix analysis were used in the study for model validation. The standard 'Kappa index' is a comparative analytical process that measures spatial and non-spatial aspects between predicted and reference maps [8]. Kappa values were characterized as excellent over 0.75, 0.40 to 0.75 as fair to good, and below 0.40 as poor [8].

Several components of Kappa indices are described in [22]: Kappa standard ($K_{standard}$), Kappa for location ($K_{location}$), and Kappa for quantity ($K_{quantity}$). They [22] define *"$K_{standard}$ as an index of agreement that attempts to account for the expected agreement due to random spatial reallocation of the categories in the comparison map, given the proportions of the categories in the comparison and reference maps, regardless of the size of the quantity disagreement"*. $K_{quantity}$ is a ratio of quantitative difference between the categories in the comparison map and reference map, and $K_{location}$ is the spatial allocation agreement between them.

The confusion matrix was analyzed using the ERRMAT module of IDRISI [8] to assess the fitness of spatial cell allocation between predicted and true values. ERRMAT outputs an error matrix containing a tabulation of the number of cells found in each possible combination of true and mapped categories and a summary of statistics [8]. Error of omission estimates the proportion of the area of a particular land cover that is omitted by the model. Error of commission represents the proportion of wrongly attributed land cover of a particular category that is overestimated by the model for each category.

3   Results

3.1   Land Cover Change Analysis during Different Time Periods

The classification validation procedure revealed that classifying land cover into five categories was difficult from grey scale photographs and simpler for the 2003 color air photos. For 1950, classification error was 27%, and sources of error were either a confusion between vineyard and grassland or urban and suburban. The classification error decreased to 20% when urban and suburban were collapsed into a single built category. For 1982, category error was 10% and 20% for 4 and 5 categories, respectively. Finally, for 2003, the error was only 4% for 4 categories, down from an initial 15% due to confusion between urban and suburban classes (by one student). It should be noted that the exercise was for unexperienced undergraduates just introduced to digital air photos. The actual classification was carried out by an experienced user over several months and verified thoroughly by a second experienced user, so the actual classification accuracy can be considered much greater than the values cited above.

Fig. 1a-d show land cover maps (1950, 1982, 2003, and 2008) digitized from the air photos. Most of the land cover changes occurred in the alluvial plain (East), where most of the vineyard, grassland and built areas are concentrated.

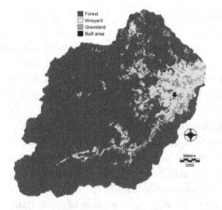

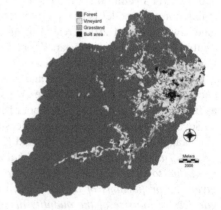

Fig. 1a. Land cover map of 1950   **Fig. 1b.** Land cover map of 1982

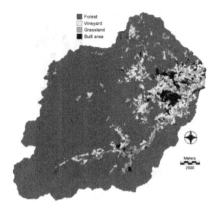

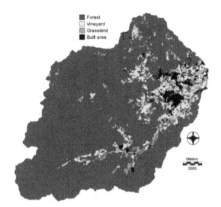

Fig. 1c. Land cover map of 2003   **Fig. 1d.** Land cover map of 2008

Fig. 2 a-d present land cover changes (ha) in all categories of the study area, and Table 2 shows the percentage of total surface area of each land cover category in different years. Two general trends can be identified in land cover change since 1950: forest and vineyard decreased while grassland and built area increased. Some changes in forest occurred in 1982-2003 as it lost about 120 ha (Fig. 2 a). A marked decrease was observed in vineyard (28% of the initial year) that lost 854 ha between 1950 and 2003 (Fig. 2 b). Then, it increased 67 ha in 2003-2008 and resumed its decreasing trend in the last time period 2008-2011. Vineyard was 10.4% of the catchment in 1950 and decreased to 6.6% in 2003 and then remained more or less stable till 2011. Grassland increased from 3.4% to 5.4% of the catchment in 1950-2003 and decreased slightly to 4.9% in 2011. It increased greatly (383 ha) in 1982-2003, decreased 122 ha in the next time period (2003-2008) but resumed the increasing trend again in 2008-2011 (Fig. 2 c). Built area remained a minor component of the catchment, and increased rapidly from only 0.1% to 3.2% of the catchment during the study period (Table 2).

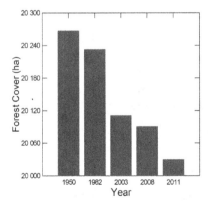

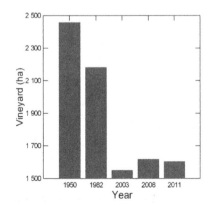

Fig. 2a. Forest change in 1950-2011   **Fig. 2b.** Vineyard change in 1950-2011

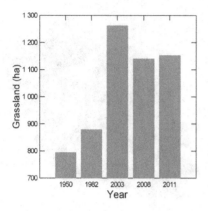 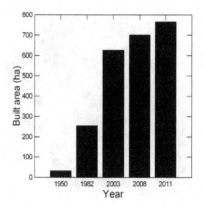

Fig. 2c. Grassland change in 1950-2011   **Fig. 2d.** Built area change in 1950-2011

Table 2. Percentage of the catchment area for each category

	Total surface area (% of the catchment)				
	1950	1982	2003	2008	2011
Forest	86.1	85.9	85.4	85.3	85.1
Vineyard	10.4	9.3	6.6	6.9	6.8
Grassland	3.4	3.7	5.4	4.8	4.9
Built area	0.1	1.1	2.7	3.0	3.2

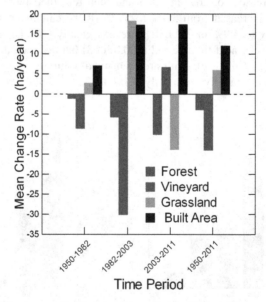

Fig. 3. Mean rates of land cover change (ha) in different time periods

Fig. 3 summarizes the mean rate of change of each land cover category in the different time periods. Forest loss was -1.1 ha yr^{-1} and -5.8 ha yr^{-1} in 1950-1982 and 1982-2003, respectively, it lost -10.1 ha yr^{-1} in the recent time period 2003-2011. The average forest depletion rate was -3.9 ha yr^{-1} in 1950-2011. The greatest rate of vineyard loss was -30.1 ha yr^{-1} in 1982-2003, and the average overall rate of vineyard depletion was -14 ha yr^{-1}. The rate of grassland expansion was 2.7 ha yr^{-1} in 1950-1982; it increased to 18.2 ha yr^{-1} in 1982-2003, and then to 13.8 ha yr^{-1} in 2003-2011. Grassland gained an average of 5.9 ha yr^{-1} in the study period. The rate of built area expansion was 7 ha yr^{-1} in 1950-1982 and increased to 17.6 ha yr^{-1} in the recent time period 2003-2011. So, the average rate of built area expansion was 12 ha yr^{-1} in 1950-2011.

3.2   Selection of Explanatory Variables

The association level between explanatory variables and land cover types in different time periods is shown in Table 3. It is measured through Cramer's V. All variables have a Cramer's V value ≥ 0.15 with all land cover types except forest in the long time period (1950-1982).

Table 3. Cramer's V coefficient (relationship between land cover change and explanatory variables). Values ≥ 0.40 are highlighted in bold

Time period		Altitude	Slope	Dist. Road	Dist. Built area	Dist. stream
	Forest	0.20	0.15	0.31	**0.40**	0.12
1950-1982	Vineyard	**0.69**	**0.65**	**0.59**	**0.46**	**0.41**
	Grassland	**0.52**	**0.50**	**0.44**	0.33	0.32
	Built area	0.39	0.36	0.28	0.22	0.20
	Forest	0.30	0.22	**0.49**	**0.60**	0.16
1982-2003	Vineyard	**0.67**	**0.63**	**0.59**	**0.59**	**0.41**
	Grassland	**0.40**	**0.40**	0.36	0.33	0.27
	Built area	**0.44**	**0.42**	0.30	0.30	0.25
	Forest	0.30	0.22	**0.49**	**0.64**	0.16
2003-2008	Vineyard	**0.67**	**0.62**	**0.59**	**0.60**	**0.41**
	Grassland	**0.41**	**0.41**	0.36	0.34	0.27
	Built area	0.39	0.38	0.27	0.29	0.25

The strongest explanatory variable is altitude, which has a good association level (Cramer V ≥ 0.40) with all land covers except forest for all time periods. A good association level is also observed in slope with all land covers in all time periods, especially with vineyard and grassland. Distance from roads shows a high association level with vineyard in all time periods, and has good association level with forest and grassland in the intermediate (1982-2003) and long (1950-1982) time periods, respectively. Distance from built area also has a good association level with forest and vineyard in all time periods. Distance from streams is the weakest variable; it shows comparatively limited association with existing land covers and has only a good level of association with vineyard in all time periods. The lowest association is observed for forest with all variables except distances from road and built area, indicating that the dominant forest category (about 85%) is less influenced by topographic variables.

3.3 Transition Potentials

Transition potentials for different time periods present similar patterns and the same explanatory variables were used in all simulations for the different time scales. Table 4 presents the accuracy rate of all transition potentials for different time periods. Accuracy rate represents the agreement between a particular transition and selected explanatory variables. A high accuracy rate is observed for several transitions in all time periods: forest to all other categories, and vineyard and grassland to built area. Transition from vineyard to forest in 2003-2008 also shows high accuracy. Therefore, transition potentials from forest to all and vineyard and grassland to built area are good. All transitions from vineyard and grassland to other land covers except built area have low to intermediate accuracy rate.

Table 4. Accuracy rate (%) of transition potentials in different time periods (F-Forest, V-Vineyard, G-Grassland, B-Built area)

Time period	Accuracy rate (%)								
	F-V	F-G	F-B	V-F	V-G	V-B	G-F	G-V	G-B
1950-1982	85	86	99	64	58	97	63	58	97
1982-2003	83	81	97	64	60	85	62	57	83
2003-2008	91	97	98	100	63	85	63	64	82

3.4 Validation of Predicted Land Cover

Simulations for 2011 were executed using transition potentials from 1950-1982, 1982-2003, and 2003-2008, respectively. Simulated and actual land cover maps of 2011 are presented in Fig. 4a-d. Dissimilarities are observed mainly in the plain land

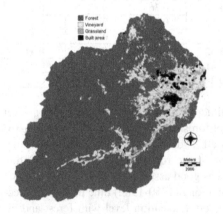 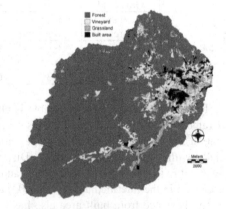

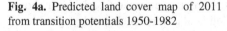

Fig. 4a. Predicted land cover map of 2011 from transition potentials 1950-1982   **Fig. 4b.** Predicted land cover map of 2011 from transition potentials 1982-2003

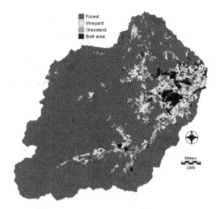 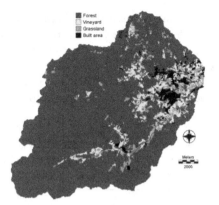

Fig. 4c. Predicted land cover map of 2011 from transition potentials 2003-2008

Fig. 4d. Land cover map 2011 (actual)

of the eastern part of the catchment where most of the conversion took place as described in [23]. Visual interpretation (Fig. 4 a-c) suggests the simulated maps from intermediate (Fig. 4 b) and short (Fig. 4 c) time scales are reasonably similar to the actual map of that year (Fig. 4 d).

Kappa Indices for Predicted Land Cover from Different Time Periods

The summary of the Kappa indices at different time scale simulations is presented in Table 5. These indices are acquired from the VALIDATION module of IDRISI [8] and can also be obtained using the Pontius matrix following [22]. Results show that all Kappa components increase with decreasing time scale up to the near perfect level of agreement for the short time scale. However, simulation from long time scale also achieved a perfect level for $K_{quantity}$ and a reasonable level of agreement for $K_{location}$, and $K_{standard}$.

Values of $K_{quantity}$ were observed in the perfect level of agreement in all three simulations, and these values increased a little from 0.95 to 1.00 for long to short time scale simulations. $K_{location}$ gives the overall spatial accuracy of a simulation. Spatial accuracy was difficult to achieve from the long time simulation. Values of $K_{location}$ varied greatly from long to short time scale though the simulation for the long time scale also had a good level of agreement (0.75); this increased to 0.87 and 0.94 for intermediate and short time simulations, respectively. The greatest changes were also observed in $K_{standard}$ for different time scales which increased from 0.66 to 0.94 with decreasing time scale.

Table 5. Summary of Kappa indices

	Initial time period		
	1950-1982	1982-2003	2003-2008
$K_{quantity}$	0.95	0.99	1.00
$K_{location}$	0.75	0.90	0.94
$K_{standard}$	0.66	0.87	0.94

Error Matrix Analysis for Predicted Land Cover from Different Time Periods
Table 6 presents the error matrix analysis of the actual land cover map 2011 (column) against predicted land cover (row) for different time scales. The table contains three 6 x 6 matrices for the 1950-1982, 1982-2003, and 2003-2008 time periods. In addition to overall errors, this table also shows where errors occur. For example, 158 ha of vineyard was wrongly attributed to forest, and 438 ha of vineyard was omitted that should be forest.

Table 6. Error matrix analysis of actual land cover map 2011 (column) against predicted (row) land cover from transition potentials for different time periods. Values are expressed in hectares (ha) and errors of commission and omission are expressed in % and in bold.

Initial time period		Forest	Vineyard	Grassland	Built area	Total	Error of commission (%)
	Forest	19,277	158	236	113	19,784	**2.6**
	Vineyard	438	1,305	488	156	2,387	**45.3**
1950-1982	Grassland	295	113	403	118	930	**56.6**
(long)	Built area	20	27	23	378	450	**16.0**
	Total	20,030	1,603	1,152	765	23,550	
	Error of Omission (%)	**3.8**	**18.6**	**65.0**	**50.6**		**9.3**
	Forest	19,716	45	52	51	19,864	**0.7**
	Vineyard	68	1,413	80	30	1,590	**11.2**
1982-2003	Grassland	204	119	965	37	1,326	**27.2**
(interme- diate)	Built area	42	26	54	647	770	**15.9**
	Total	20,030	1,603	1,152	765	23,550	
	Error of Omission (%)	**1.6**	**11.9**	**16.2**	**15.4**		**3.4**
	Forest	19,953	30	45	27	20,055	**0.5**
	Vineyard	16	1,496	94	15	1,621	**7.7**
2003-2008	Grassland	44	68	997	17	1,127	**11.5**
(short)	Built area	16	9	16	706	747	**5.4**
	Total	20,030	1,603	1,152	765	23,550	
	Error of Omission (%)	**0.4**	**6.7**	**13.4**	**7.7**		**1.69**

Errors for all land covers decreased with decreasing time scales. The lowest commission and omission errors were observed in forest for all time scales and these decreased slightly with decreasing time scales. Errors of commission and omission were 2.6% and 3.8%, respectively, for forest in the long time scale prediction, and these decreased to 0.7% and 1.6% in the intermediate and 0.5% and 0.4% in the short time scale predictions, respectively. High error of commission (45.3%) was observed in vineyard in the long time scale where the greatest amount of vineyard (1,082 ha) was wrongly attributed, and commission error decreased markedly in intermediate and short time scales. However, error of omission was relatively low in the long time scale simulation for vineyard. The highest errors of commission and omission were observed in grassland in all time scale simulations, particularly the long time scale where errors of commission and omission were 56.6% and 65%, respectively. Errors for this land cover also decreased greatly with decreasing time scale (Table 6). Considerable amounts of vineyard and grassland were wrongly attributed to forest, and considerable areas of vineyard and grassland were omitted by the model in the long time scale simulation; this occurred mainly due to high swapping of these land covers with forest. For this reason, high errors of commission and omission were generated for vineyard and grassland in the long time scale; errors decreased considerably in the intermediate and short time scale simulations. In long time simulation, errors of commission of built area were lower than for vineyard and grassland due to its small coverage in the catchment, and it was wrongly attributed 72 ha of other land covers. However, high error of omission was observed in the same simulation because much built area (388 ha) was omitted.

4   Discussion

Land cover dynamics and changes in individual land covers have an important impact on land cover simulation. As it is described in the results, forest is easy to predict, and it obtains the best level of agreement and the lowest error in all simulations using different time scales due to its dominant coverage in the study area. It is the least probable to change in all transition potentials of forest to other land covers, so $K_{quantity}$ is better for all time scales.

Simulations of vineyard and grassland are extremely difficult to predict: accuracy is lower and errors greater due to the dynamic changes in different time periods and high swapping between these covers. Hence, high commission and omission errors are observed in vineyard and grassland simulations, particularly in the long time scale. These errors may occur due to different rates of change in initial and prediction time periods and the selection of transition potentials where transition potentials from vineyard to forest and grassland, and grassland to forest and vineyard were avoided due to their limited accuracy rate (<70%). Simulations of vineyard and grassland may improve using constraints for vineyard and grassland. Vineyard fields belonging to the wine making "domaines" tend to remain stable and convert to other covers less [23], so a "domaine" layer could be used as a constraint for vineyard. This information, however, was not available in this study. In addition, fire breaks, horseback riding, and other tourism related activity zones that are classified as grassland could perhaps be taken as a constraint for grassland.

Accurate prediction of urban expansion is difficult due to the complexity in urbanization which depends on several spatial variables, urban planning, and land use demand [12]. The rapid relative rate of urban growth impacted the urban prediction. For example, the model predicts (for 2011) about 40% less built area than the actual map of 2011 using the long time scale because the rate of built area expansion increased by more than double in the latter time period (1982-2011) compared to the initial period (1950-1982) (Fig. 3). However, intermediate and short time periods perform better since increasing trends in the initial time periods are about the same as in the prediction time periods (2003-2011 and 2008-2011). In addition, several scattered urban areas are developed exceptionally far away from existing built area in the recent year, and these remain difficult to predict because the model is based on historical trends. Earlier trials showed the use of constraints for the transitions to built area from other land covers reduced error in built area in all simulations.

Time scales have a significant impact on land cover simulation. Quantity was predicted better than location, probably due to the dominant forest cover in the study area. Therefore, $K_{quantity}$ is nearly perfect in all time scales. However, complex land cover changes and swapping between land covers generate less perfect levels of agreement for $K_{location}$ than $K_{quantity}$, and values increase with decreasing time scales.

Although different indexes are used, there is a general trend for Shorter time scales to Produce better prediction results [1, 15, 16, 20, 21, 24, and 27], as found in this study was. The values of $K_{quantity}$ and $K_{location}$ are in acceptable ranges for different time scales in this study. Maximum commission and omission errors observed in crops and grassland [27] were also noted in this study since complex changes in grassland and vineyard are difficult to simulate.

5   Conclusion

Studies of the temporal and spatial distribution of land cover change have become an important issue due to the rapid conversion of land cover and its impact on environment change. Time scale has a significant impact on prediction. Near perfect quantitative accuracy was achieved in all time scales but spatial accuracy varied with different time scales. High quantitative and location accuracy were found in forest prediction due to its large surface area, in which changes are relatively small and swapping does not impact prediction. Prediction of vineyard and grassland were difficult due to high swapping with one another and forest, and prediction of built area was complicated by the dramatic relative growth that increased in the recent time periods and the emergence of urban lots far from historic centers. Cell size and catchment area may also impact land cover change simulation and this is under study now.

References

1. Ahmed, B., Ahmed, R.: Modeling urban land cover growth dynamics using multi-temporal satellite images: A case study of Dhaka, Bangladesh. ISPRS International Journal of Geo-Information 1, 3–31 (2012), doi:10.3390/ijgi1010003
2. Álvarez-Martínez, J.M., Suárez-Seoane, S., Luis Calabuig, E.D.: Modelling the risk of land cover change from environmental and socio-economic drivers in heterogeneous and changing landscapes: The role of uncertainty. Landscape and Urban Planning 101, 108–119 (2011), doi:10.1016/j.landurbplan.2011.01.009
3. Araya, Y.H., Cabral, P.: Analysis and Modeling of Urban Land Cover Change in Setúbal and Sesimbra, Portugal. Remote Sensing 2, 1549–1563 (2010), doi:10.3390/rs2061549
4. Bohnet, I., Pert, P.L.: Pattern, drivers and impacts of urban growth- A study from Cairns, Queensland, Australia from 1952 to 2031. Landscape and Planning 97, 239–248 (2010), doi:10.1016/j.landurbplan.2010.06.007
5. Chaudhuri, G., Clarke, K.C.: Temporal accuracy in urban growth forecasting: A study using the SLEUTH model. Transactions in GIS 18, 302–320 (2014), doi:10.1111/tgis.12047
6. Chen, H., Pontius Jr., R.G.: Diagnostic tools to evaluate a spatial land change projection along a gradient of an explanatory variable. Landscape Ecology 25, 1319–1331 (2010), doi:10.1007/s10980-010-9519-5

7. Dietzel, C., Clarke, K.: The effect of disaggregating land use categories in cellular automata model calibration and forecasting. Computers, Environment and Urban Systems 30, 78–101 (2006), doi:10.1016/j.compenvurbsys.2005.04.001
8. Eastman, J.R.: IDRISI Selva Help System. Clark Labs, Clark University, Worcester (2012)
9. Fox, D.M., Witz, E., Blanc, V., Soulié, C., Penalver-Navarro, M., Dervieux, A.: A case study of land cover change (1950-2003) and runoff in a Mediterranean catchment. Applied Geography 32, 810–821 (2012), doi:10.1016/j.apgeog.2011.07.007
10. Guan, D., Li, H., Inohae, T., Su, W., Nagaie, T., Hokao, K.: Modeling urban land use change by the integration of cellular automata and Markov model. Ecological Modelling 222, 3761–3772 (2011), doi:10.1016/j.ecolmodel.2011.09.009
11. He, C., Okada, N., Zhang, Q., Shi, P., Zhang, J.: Modeling urban expansion scenarios by coupling cellular automata model and system dynamic model in Beijing, China. Applied Geography 26, 323–345 (2006), doi:10.1016/j.apgeog.2006.09.006
12. He, C., Okada, N., Zhang, Q., Shi, P., Li, J.: Modelling dynamic urban expansion processes incorporating a potential model with cellular automata. Landscape and Urban Planning 86, 79–91 (2008), doi:10.1016/j.landurbplan.2007.12.010
13. Huang, Q., Cai, Y.: Simulation of land use change using GIS-based stochastic model: The case study of Shiqian County, Southwestern China. Stochastic Environment Research Risk Assessment 21, 419–426 (2007), doi:10.1007/s00477-006-0074-1
14. Jenerette, G.D., Wu, J.: Analysis and simulation of land -use change in the central Arizona Phonix region, USA. Landscape Ecology 16, 611–626 (2001)
15. Kamusoko, C., Aniya, M., Adi, B., Manjoro, M.: Rural sustainability under threat in Zimbabwe-Simulation of future land use/cover change in the Bindura district based on the Markov-cellular automata model. Applied Geography 29, 435–447 (2009), doi:10.1016/j.apgeog.2008.10.002
16. Khoi, D.D.: Spatial modeling of deforestation and land suitability assessment in the Tam Dao National Park region. University of Tsukuba, Vietnam. Ph. D. thesis (2011)
17. Lambin, E.F., et al.: The cause of land-use and land-cover change: moving beyond the myths. Global Environment Change 11, 261–269 (2001)
18. Li, X., Yeh, A.G.O.: Neural network based cellular automata for simulating multiple land use change using GIS. International Journal of Geographical Information Science 16, 323–343 (2002), doi: 10.108 0/13658810210137004
19. Mas, J.F., Pérez-Vega, A., Clarke, K.C.: Assessing simulated land use/cover maps using similarity and fragmentation indices. Ecological Complexity 11, 38–45 (2012), doi:10.1016/j.ecocom.2012.01.004
20. Mhangara, P.: Land use/ cover change modeling and land degradation assessment in the Keiskamma catchment using remote sensing and GIS. University of Nelson Mandela Met-ropolitan. Ph. D. thesis (2011)
21. Pérez-Vega, A., Mas, J.F., Ligmann-Zielinska, A.: Comparing two approaches to land use /cover change modeling and their implications for the assessment of biodiversity loss in a deciduous tropical forest. Environmenta Modelling & Software 29, 11–23 (2012), doi:10.1016/j.envsoft.2011.09.011
22. Pontius Jr., R.G., Millones, M.: Death to Kappa: birth of quantity disagreement and allocation disagreement for accuracy assessment. International Journal of Remote Sensing 32, 4407–4429 (2011), doi:org/10.1080/01431161.2011.552923
23. Roy, H.G., Fox, D.M., Emsellem, K.: Spatial dynamics of land cover change in a Mediterranean catchment (1950-2008). Journal of Land use Science (2014)

24. Sang, L., Zhang, C., Yang, J., Zhu, D., Yun, W.: Simulation of land use spatial pattern of towns and villages based on CA-Markov model. Mathematical and Modelling 54, 938–943 (2011), doi:10.1016/j.mcm.2010.11.019

25. Silva, T.S., Tanliani, P.R.A.: Environment planning in the medium littoral of the Rio Grand do Sul coastal plain - Southern Brazil: Elements for coastal management. Ocean and Coastal Management 59, 20–30 (2012), doi:10.1016/j.ocecoaman.2011.12.014

26. Tewolde, M.G., Cabral, P.: Urban sprawl analysis and modeling in Asmara, Eritrea. Remote Sensing 3, 2148–2165 (2011), doi:10.3390/rs3102148

27. Valdivieso, F.O., Sendra, J.B.: Application of GIS and remote sensing techniques in generation of land use scenarios for hydrological modeling. Journal of Hydrology 365, 256–263 (2010), doi:10.1016/j.jhydrol.2010.10.033

28. Verburg, P.H., de Koning, G.H.J., Kok, K., Veldkamp, A., Bouma, J.: A spatial explicit allocation procedure for modellling the pattern of land use change based upon actual land use. Ecological Modelling 116, 45–61 (1999)

29. Verburg, P.H., Schulp, C.J.E., Witte, N., Veldkamp, A.: Downscaling of land use change scenarios to assess the dynamics of European landscapes. Agriculture, Ecosystem and Environment 114, 39–56 (2006), doi:10.1016/j.agee.2005.11.024

Measuring Territorial Vulnerability?
An Attempt of Qualification and Quantification

Florent Renard and Didier Soto

Université Jean Moulin Lyon 3, UMR 5600 CNRS Environnement Ville Société,
Lyon, France
florent.renard@univ-lyon3.fr

Abstract. Risk management traditionally addresses the control of hazards. However, this field of research is now more focused on reducing territorial vulnerability. This requires prior accurate knowledge of this territorial vulnerability. The present work, applicable to any location, proposes a GIS-based methodology for vulnerability assessment using the greater Lyon, France, as a model. The study is based on a spatial decision-making method that awards priority to vulnerability using expert judgments. Vulnerability is defined as the combination of the sensitivity of assets facing hazards and their strategic importance in the functioning of a city. This approach can be applied to multiple layers of hazards for a specific envisioned risk.

Keywords: Vulnerability, risk assessment, decision making, analytic hierarchy process, hazard, GIS.

1 Introduction

The greater Lyon has a population of 1.3 million people in an area of 515 km² (Fig. 1). With such a concentration of assets, policy makers and local elected officials attach great importance to the management of rainfall risks, particularly in the light of the inherent characteristics of the area. Indeed, flooding of river systems and run-off are the most frequent and costly natural hazard in the greater Lyon [24] as in the majority of the countries in the world on a regular basis [10], [33]. Thus, flood risks are numerous and can occur nearly anywhere in the greater Lyon because of very frequent intense rainfall in the summer, the presence of two major rivers, many torrential streams terrain with steep slopes conducive to the accumulation of water at lower elevations and certain impervious areas that cause heavy runoffs and flash floods [4], [14, 15], [17], [30] (Fig. 1).

Hazards associated with flood risks in the greater Lyon have been the subject of numerous analyses [24]. However, links between atmospheric patterns and precipitation have not been established, unlike the case Mediterranean region further south [16], [18], [34]. Studies on the vulnerability of the greater Lyon are rare and have common shortcomings that call their validity into question [5], [25]. This is regrettable because the assessment of the vulnerability of urban systems is as important as that of heavy rainfall hazards. Moreover, acquiring a precise knowledge of vulnerability is the first

B. Murgante et al. (Eds.): ICCSA 2014, Part IV, LNCS 8582, pp. 331–343, 2014.

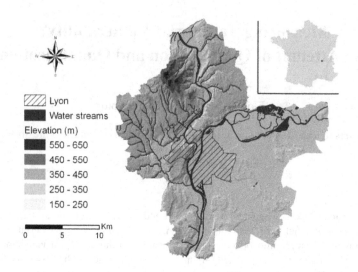

Fig. 1. Topography and hydrology of the greater Lyon and location in France

step in building more flexible and resilient cities (ability to absorb disturbance and recover functions after a disturbance [13], [29], [32].

The risk can be seen as a function of the hazard and the vulnerability of assets in the area studied [12], [19, 20, 21]. This definition has also been used in recent major reports on risk and climate change impacts [9], [33]. Urban vulnerability can be the degree to which a city is susceptible to a hazard or unable to cope with it [7,8]. Flash floods are an example of such a hazard.

Vulnerability assessment is in two parts in this article. The first is the sensitivity of the assets facing hazards such as urban run-off or flash floods, i.e. their propensity for damage and the degree to which they are affected during a flood. The second factor is their strategic importance (or priority) in the functioning of the city.

Indeed, not all assets make an equal contribution to the function of the greater Lyon. For example, damage to street furniture, such as bus shelters, seems less harmful to the activities of the urban area than that of the overall transport infrastructure [23]. We therefore assess the vulnerability of the greater Lyon urban area to hydrological hazards and especially a flood depth of about 25 cm. The greater Lyon is a very complex assembly of human, environmental and equipment assets, as shown in figure 2.

The first step is to identify all the assets, and then classify them in a vulnerability structure, i.e. a hierarchical breakdown [1,2], [6], [11], [31] from global to site-specific. Multicriterion decision-making is also needed to evaluate the two distinct components of vulnerability: the sensitivity of an asset (human, environmental and equipment [26] and the contribution it makes to the functioning of the community, i.e. its strategic importance or strategic priority.

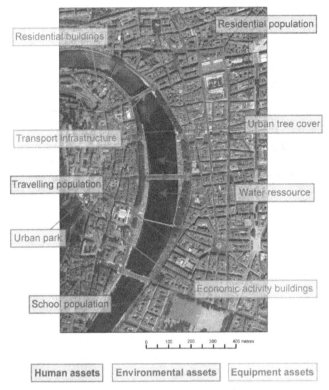

Fig. 2. Some different components of the human, environmental, and equipment assets of the greater Lyon. The map also shows the need for early and exhaustive examination prior to evaluation.

2   A Vulnerability Structure to Evaluate the Sensitivity and Strategic Importance of Urban Assets

2.1   Building a Vulnerability Structure of the Various Assets of the Greater Lyon

A hierarchical vulnerability structure was devised in collaboration with engineers and elected officials of the greater Lyon. This structure combines the three categories of assets (human, environmental, and equipment), which are then separated, as shown in figure 3.

Targets that share the same rank in this structure do not necessarily have the same sensitivity to a flood hazard, nor do they have the same strategic importance for the region as a whole. The second stage therefore defines the sensitivity and strategic importance of the different assets based on the judgments of experts.

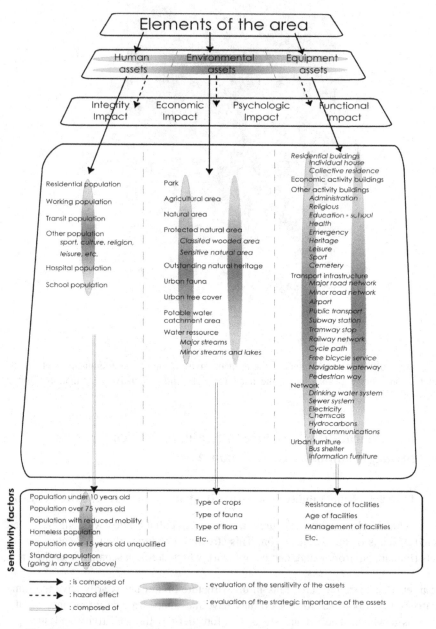

Fig. 3. Hierarchical vulnerability structure of the assets of the greater Lyon urban area

2.2   A Multicriterion Method of Decision Support to Evaluate the Sensitivity and Strategic Importance of the Issues

Multicriterion decision methods are used to rank features and have been adapted to the modelling of the sensitivity and the strategic importance of the issues of the greater Lyon [3], [27]. The methods are used to assess the priorities of different components in a given situation. Thomas Saaty's analytic hierarchy process (AHP) is the approach most frequently used and was chosen for reasons of accessibility, flexibility and adaptation and because it fully matches the modelling required for this situation [28].

This was the basis for the evaluation of the sensitivity and the strategic importance of assets. These evaluations are based on the opinions of 38 experts from different sectors, such as local authorities (32%), universities (53 %), engineering firms (10%) or associations (5%). They are all involved in the professional field of flood risk management. The experts gave verdicts on assets that seem to be of relatively greater importance for the urban community than others, using the Saaty semantic scale (table 1).

Table 1. Scale of binary comparison used to assess the importance and the sensitivity of the assets (adapted from Saaty, 1980)

Degree of importance or sensitivity	Definition
1	Equal importance (sensitivity) of two elements
3	Weak importance (sensitivity) of an element in comparison to the other one
5	Strong importance (sensitivity) of an element in comparison to the other one
7	Certified importance (sensitivity) of an element in comparison to the other one
9	Absolute importance (sensitivity) of an element in comparison to the other one
2, 4, 6, 8	Intermediate values between two appreciation
1/2, 1/3, 1/4, 1/5, 1/6, 1/7, 1/8, 1/9	Reciprocal values of the previous appreciation

Semi-structured interviews of were conducted and binary comparison made of the assets with the same vulnerability hierarchical structure (Fig. 3). They used the same method to determine sensitivity stakes for a flood risk of about 25 cm. Finally, the consistency of ratings given by experts was validated by calculating a coherence ratio with the results aggregated to give strategic importance functions and sensitivity functions. The two latter categories of function were then reciprocally combined to obtain general functions of vulnerability.

3   Using a GIS to Map the Risk Associated with Hydrological Hazards

3.1   Vulnerability Functions to Quantify Urban Vulnerability

The AHP enabled the researchers to define the strategic importance of a wide range of assets in the greater Lyon and to determine their sensitivity to hydrological hazards. The global vulnerability function (Equation 1 and Fig. 4) was obtained from the combination of the sensitivity (Equation 2 and Fig. 4) and the strategic importance functions (Equation 3 and Fig. 4), based on a reciprocal weight between these two functions. It shows the major importance of human issues (75%). The proportions of vulnerability of environmental assets and equipment assets are about 12% each. However, it can be seen from equations 2 and 3 that the environmental and equipment stakes do not have the same weight in the strategic importance and sensitivity functions. Indeed, the environmental issues seem more important than the equipment ones (17% for the environmental stakes whereas only 5% for the equipment stakes in the strategic importance function) but are less sensitive to hydrological hazard (6% for the environmental issues and 20% for the equipment issues in the sensitivity function). Similarly, the priority functions and the sensitivity functions were established for all assets at all levels of the vulnerability structure.

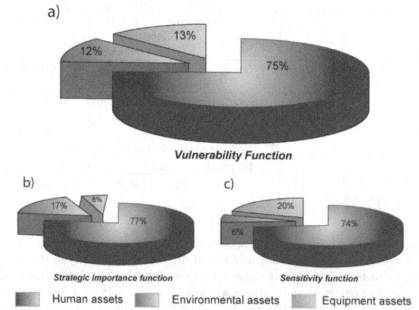

Fig. 4. Results of the 38 semi-structured interviews with reference to the global vulnerability function (a), the strategic importance function (b) and the sensitivity function (c)

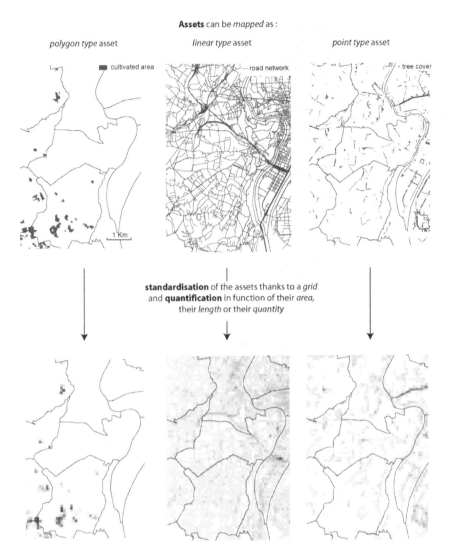

Fig. 5. Examples of standardisation frameworks for polygon (area) type assets (cultivated area), line type assets (road network) and point type assets (urban tree cover) in mesh form, according to their quantity in a mesh

Equation 1- global vulnerability function:

Global Vulnerability = 0.753 x Human Vulnerability + 0.118 x Environment Vulnerability + 0.128 x Equipment Vulnerability   (1)

Equation 2 - sensitivity function:

Global Sensitivity = 0.735 x Human Sensitivity + 0.064 x Environment Sensitivity + 0.201 x Equipment Sensitivity   (2)

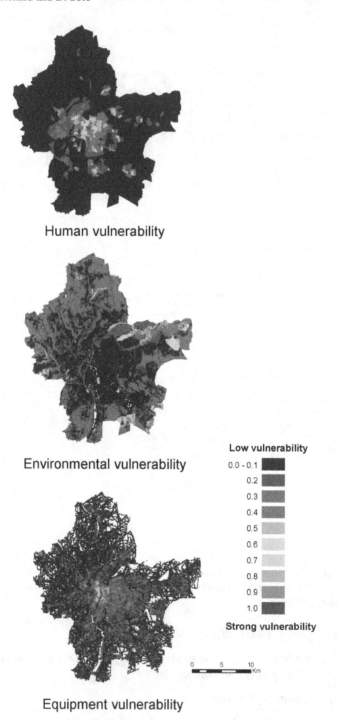

Fig. 6. Human, environmental and equipment vulnerabilities of the greater Lyon

Equation 3 - strategic importance function:

Global Priority = 0.772 x Human Priority + 0.173 x Environment Priority + 0.055 x Equipment Priority   (3)

The spatial framework and the heterogeneous nature of the data must be standardized to make comparison possible. The stakes were converted into a mesh form 100 metres wide (fig. 5). This method gave a detailed, consistent vulnerability appraisal of the territory of the greater Lyon (fig. 6). Human vulnerability is centred in the core of the greater Lyon and follows the decreasing population gradient from the centre to the outskirts. Equipment vulnerability displays similar location features. Logically, the opposite case applies for environmental factors, with vulnerability located at the periphery of the greater Lyon. Overall vulnerability (fig. 7) is also centred on the core of the region, given the strong weight of human vulnerability (75% - Equation 1 and Fig. 6) in the overall vulnerability function.

The degree of risk is determined by combining hazard and vulnerability. For example, figure 8 shows the risk of flooding in the greater Lyon area caused by the two major rivers, and figure 9 shows the risk of flooding caused by run-off. In the first kind of risk (Fig. 8), the flooding hazard affects the most vulnerable areas and thus forms a major risk for the city, whereas in the second case (Fig. 9), the run-off hazard does not harm the most vulnerable parts of the city.

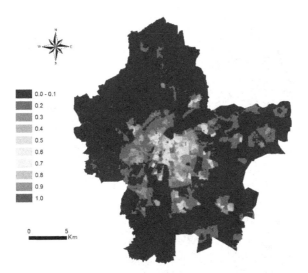

Fig. 7. Overall vulnerability of the greater Lyon

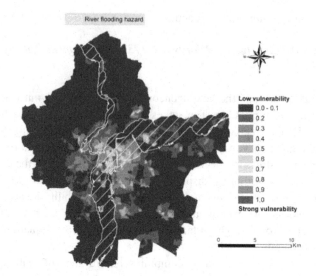

Fig. 8. Flooding risk from the main streams (Rhône and Saône) (a). It can be seen that the risk (relation between vulnerability and hazard) caused by the same hazard may be completely different according to urban vulnerability.

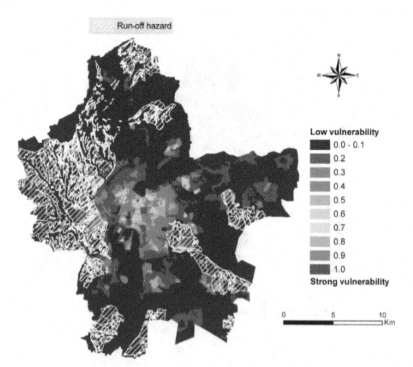

Fig. 9. Run-off risk (relation between vulnerability and hazard) in the greater Lyon (to be compared to fig. 9)

4 Conclusion

Risk management has long been focused entirely on the control of hazards. The management tools used have shown their limits as a result of the increase in population and asset concentrations in cities coupled with climate change expected to lead to more severe storms and rainfall events that will increase river discharges and peak flows. Consequently, the concept of vulnerability has emerged in the last decade and, more recently, the theory of resilience. This situation is typical in the territory of the greater Lyon, which is threatened by numerous natural hazards. The first studies on the vulnerabilities of the greater Lyon were not sufficient and certain conceptual and technical limits prevented them from being used. This study is therefore aimed at overcoming these problems by using a diverse set of assets (human, environmental and equipment) in a coherent framework enabling their comparison using a multicriterion method for decision support, based on expert appraisals. The study demonstrated the strong preponderance of human vulnerability in the overall vulnerability arena. This leads to a decreasing gradient centred on the urban core but with certain highest vulnerability points in the periphery particularly influenced by environmental vulnerability. From this perspective, the analysis shows the need for managers to take very accurate account of local vulnerability as it displays very strong spatial variability. In conclusion, this work, whose methodology is universal, can provide a detailed, precise knowledge of urban vulnerability, and should be one of the first steps in studying the resilience of cities.

References

1. Barroca, B., Bernardara, P., Mouchel, J.M., Hubert, G.: Indicators for identification of urban flooding vulnerability. Natural Hazards and Earth System Sciences 6, 553–561 (2006)
2. Fedeski, M., Gwillian, J.: Urban sustainability in the presence of flood and geological hazards: The development of a GIS-based vulnerability and risk assessment methodology. Landscape Urban Plan 83, 50–61 (2007)
3. Fernandez, D.S., Lutz, M.A.: Urban flood hazard zoning in Tucuman Province, Argentina, using GIS and multicriteria decision analysis. Eng. Geol. 111, 90–98 (2010)
4. Gaume, E., Valerie, B., Pietro, B., Newinger, O., Barbuc, M., Bateman, A., Blaskovicova, L., Bloschl, G., Borga, M., Dumitrescu, A., Daliakopoulos, J., Garcia, J., Irimescu, A., Kohnova, S., Koutroulis, A., Marchi, L., Matreata, S., Medina, V., Preciso, E., Sempere-Torres, D., Stancalie, G., Szolgay, J., Tsanis, I., Velasco, D., Viglione, A.: A compilation of data on European flash floods. J. Hydrol. 367, 70–78 (2009)
5. Grand Lyon: Cahier Risques majeurs du Référentiel environnemental du Grand Lyon (2004)
6. Griot, C.: Des territoires vulnérables face au transport de matières dangereuses. Proposition d'un outil d'aide à la gestion de crise. Géocarrefour 82, 51–63 (2007)
7. IPCC (Intergovernmental Panel on Climate Change): Climate Change 2001: Synthesis report, World Meteorological Organization, UN Environment Programme (2001a)
8. IPCC (Intergovernmental Panel on Climate Change): Climate Change 2001: Impacts, Adaptation and Vulnerability. Contribution to the Working Group I to the third assessment report of the Intergovernmental Panel on Climate Change (IPCC). Cambridge University Press, Cambridge (2001b)

9. IPCC: Managing the Risks of Extreme Events and Disasters to Advance Climate Change Adaptation, A Special Report of Working Groups I and II of the Intergovernmental Panel on Climate Change. Cambridge University Press, Cambridge (2012)

10. Jongman, B., Kreibich, H., Apel, H., Barredo, J.I., Bates, P.D., Feyen, L., Gericke, A., Neal, J., Aerts, J.C.J.H., Ward, P.J.: Comparative flood damage model assessment: towards a European Approach. Nat. Hazards Earth Syst. Sci. 12, 3733–3752 (2012)

11. Kienberger, S., Lang, S., Zeil, P.: Spatial vulnerability units – expert-based spatial modelling of socio-economic vulnerability in the Salzach catchment, Austria. Nat. Hazards Earth Syst. Sci. 9, 767–778 (2009)

12. Kron, W.: Flood risk = hazard × exposure × vulnerability. In: Wu, B., Wang, Z., Wang, G., Huang, G., Fang, H., Huang, J. (eds.) Flooddefence, pp. 82–97. Science Press (2002)

13. Lhomme, S., Serre, D., Diab, Y., Lagagnier, R.: Les réseauxtechniques face aux inondations ou comment définir des indicateursde performances de ces réseaux pour évaluer la résilienceurbaine. Bulletin de l'association des Géographes Français 4, 487–502 (2010)

14. Llasat, M.C., Llasat-Botija, M., Prat, M.A., Porc'u, F., Price, C., Mugnai, A., Lagouvardos, K., Kotroni, V., Katsanos, D., Michaelides, S., Yair, Y., Savvidou, K., Nicolaides, K.: High-impact floods and flash floods in Mediterranean countries: the FLASH preliminary database. Adv. Geosci. 23, 47–55 (2010)

15. Marchi, L., Borga, M., Preciso, E., Gaume, E.: Characterisation of selected extreme flash floods in Europe and implications for flood risk management. J. Hydrol. 394, 118–133 (2011)

16. Martinez, M.D., Lana, X., Burgueno, A., Serra, C.: Spatial and temporal daily rainfall regime in Catalonia (NE Spain) derived from four precipitation indices, years 1950–2000. Int. J. Climatol. 27, 123–138 (2007)

17. Montz, B.E., Gruntfest, E.: Flash flood mitigation: recommendations for research and applications. Environ. Hazards 4, 15–22 (2002)

18. Nuissier, O., Joly, B., Joly, A., Ducrocq, V., Arbogast, P.: A statistical downscaling to identify the large-scale circulation patterns associated with heavy precipitation events over southern France. Q. J. Roy. Meteorol. Soc. 137, 1812–1827 (2011)

19. Poussin, J.K., Ward, P.J., Bubeck, P., Gaslikova, L., Schwerzmann, A., Raible, C.C.: Flood Risk Modelling. In: Climate Adaptation and Flood Risk in Coastal Cities, pp. 93–121. Earthscan (2012)

20. Propeck-Zimmermann, E., Saint-Gérand, T., Bonnet, E.: Nouvelles approches ergonomiques de la cartographie des risques industriels. Mappemonde 4, 19 (2009)

21. Reghezza, M.: Géographes et gestionnaires face à la vulnérabilité métropolitaine. Quelques Réflexions Autour du cas Francilien. Annales de Géographie 669, 459–477 (2009)

22. Renard, F.: Le risque pluvial en milieu urbain. De la caractérisation de l'aléa à l'évaluation de la vulnérabilité: le cas du Grand Lyon. PhDthesis, Université Lyon 3, 528 (2010)

23. Renard, F., Chapon, P.M.: Une méthode d'évaluation de la vulnérabilité urbaine appliquée à l'agglomération lyonnaise. L'Espace géographique 1, 35–50 (2010)

24. Renard, F., Chapon, P.M., Comby, J.: Assessing the accuracy of weather radar to track intense rain cells in the Greater Lyon area, France. Atmospheric Research 103, 4–19 (2012)

25. Rufat, S.: L'estimation de la vulnérabilité urbaine, un outil pour la gestion du risque. Approche à Partir du cas de l'agglomérationlyonnaise. Géocarrefour 82, 7–16 (2007)

26. Ruin, I., Creutin, J.D., Anquetin, S., Lutoff, C.: Human exposure to flash floods – relation between flood parameters and human vulnerability during a storm of in Southern France. J. Hydrol. 361, 199–213 (2008)

27. Saaty, T.L.: The Analytic Hierarchy Process. McGraw-Hill, New York (1980)

28. Saaty, T.L., Hu, G.: Ranking by eigenvector versus other methods in the Analytic Hierarchy Process. Applied Mathematics Letters 11, 121–125 (1998)
29. Schelfaut, K., Pannemans, B., van der Craats, I., Krywkow, J., Mysiak, J., Cools, J.: Bringing flood resilience into practice: The FREEMAN project. Environ. Sci. Policy 14, 825–833 (2011)
30. Tarolli, P., Borga, M., Morin, E., Delrieu, G.: Analysis of flash flood regimes in the North-Western and South-Eastern Mediterranean regions. Natural Hazards and Earth System Sciences 12, 1255–1265 (2012)
31. Tixier, J., Dandrieux, A., Dusserre, G., Bubbico, R., Mazzarotta, B., Silvetti, B., Hubert, E., Rodrigues, N., Salvi, O.: Environmental vulnerability assessment in the vicinity of an industrial site in the frame of ARAMIS European project. Journal of Hazardous Materials 130, 251–264 (2006)
32. Tromeur, E., Menard, R., Bailly, J.-B., Soulie, C.: Urban vulnerability and resilience within the context of climate change. Nat. Hazards Earth Syst. Sci. 12, 1811–1821 (2012)
33. UNISDR, 2011. Global Assessment Report on Disaster Risk Reduction. Revealing Risk, Redefining Development. United Nations International Strategy for Disaster Reduction Secretariat (2011)
34. Vicente-Serrano, S.M., Begueria, S., Lopez-Moreno, J.I., El Kenawy, A.M., Angulo Martinez, M.: Daily atmospheric circulation events and extreme precipitation risk in Northeast Spain: the role of the North Atlantic Oscillation, Western Mediterranean Oscillation, and Mediterranean Oscillation. J. Geophys. Res. Atmos. 114 (2009)

Planning Support Tool for Rural Architectural Intensification

Roberto De Lotto, Tiziano Cattaneo, Elisabetta Maria Venco, Susanna Sturla, and Sara Morettini

DICAr – University of Pavia, via Ferrata 3, 27100 Pavia, Italy
{uplab,roberto.delotto}@unipv.it

Abstract. The paper proposes results derived from a national research project (PRIN, Progetti di Ricerca di Interesse Nazionale - Research Projects of National Interest) named "Rural Architecture Intensification" developed in the local unit of the University of Pavia. Authors developed the research considering the different themes and disciplines that are involved in planning and design actions aimed to improve, restore and re-functionalize existing minor settlements in rural-urban context. The entire process is a typical planning and design algorithm. In the paper authors focus on the treatment of geographical data and on the integration of the data sets that have dissimilar origin, diverse formats (they may be not only digital) and different meaning value. They usually belong to various disciplines and together they compose the information set from which it is possible to deduce specific knowledge. Authors first expose the logical framework of the whole process, then describe the specific theme and finally show some related spatial analysis applications.

Keywords: Spatial Decision Support System; Spatial Analysis; Inter-scalar design process; Rural and rurban settlements.

1 Introduction

When approaching a composite territorial problem that involves different scales and disciplines (considering also systems and subsystems belonging the disciplines), it is necessary to establish a precise logical framework. Every planning or design activity is an iterative process applied to a complex system; not-linear relations among the entities that compose the system are numerous and it is problematic to spell out them.

Authors develop an approach that can help planners to analyze problems and needs of a specific region taking into account landscape and socio-economic aspects at local and regional level. This approach can bring back to the trend of Ecological Planning [1,2].

Authors developed a framework that has a hybrid structure in which Spatial Decision Support Systems (SDSS) [3,4,5], Knowledge Discovery and Data Mining (KDD) [6] and Expert Systems (ES) converge.

The method is not completely automatic and there is a continuous interaction between user and system.

B. Murgante et al. (Eds.): ICCSA 2014, Part IV, LNCS 8582, pp. 344–355, 2014.

Spatial analysis is the main theme, and the proposed system uses models and techniques such as: Multi- criteria Analysis (MCA) [7,8], Cluster [9,10,11] and Spatial analysis [12,13,14], Map Overlay, best practices database collection, semantic indexing, etc. Used technological instruments are: ArcGIS, QGIS, Excel and Visual Basic, Hyperlinks.

2   Procedure

The main aim of the entire research group who participated to PRIN[1] research, was to define an informed methodology for decision makers, stakeholders and public bureaus who have (or want to) face the problem of improving and intensifying insediative activities in minor centers located in rural-urban context.

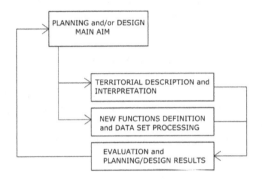

Fig. 1.   General framework

The related conceptual scheme of the main phases is quite similar to a classic planning or design problem: the process described here is an adaptation of conventional planning processes[2] [1]. Starting from a main objective (that could be a well-defined decision or a scenario proposal defined by stakeholders or public bureaus) the territory is specifically interpreted extracting knowledge from the multiple available data. The entire process involves elements with different scales (territorial scale and building scale) [1,15,16] and expertise coming from different disciplines [17] (regional planning,

---

[1] PRIN 2009 "Architecture as Heritage: innovative instruments for the tutelage and the improvement of the local border systems". University of Pavia participated the research theme on "Regeneration and renewal of rural landscape. Building strategies in the surroundings of new urban centers". This issue aims to the regeneration of the rural landscape, with particular focus on landscape and architecture design which can become the instrument for promoting the value, the memory and the environmental quality of the rural landscape as cultural heritage. Enhancing rural architecture, small towns, farmsteads and ancient relics is one of the main components for the regeneration of the countryside. It is a strategy with a positive outcome, even only if it has been supported simultaneously by the possibility of creating more business, but which nevertheless is planned taking into account the improvement of the perceived aesthetic structure of the countryside. (http://www.raintensification. com/#).

[2]   I.e. Steiner quotes Hall 1975; McDowell 1986; Moore 1988; Stokes et all. 1997.

architecture, landscape planning, social sciences, economics) involving different scale as well.

The basic conceptual framework considers the relation among the following different aspects: scale, theme, data set, techniques (and models) [18]. This scheme could be valid for every action that involves both territorial and building scale, and both regional planning and architectural design.

Everything is organized throughout technical and information system instruments.

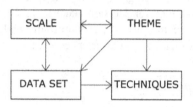

Fig. 2. Basic conceptual structure

Scale issues have influence on, and itself influences: data set and theme.

Theme influence all the other issues; techniques are influenced (chosen) in respect of thematic choices and data set availability.

Data set collecting and systematization depend on chosen theme and planning and/or design scale.

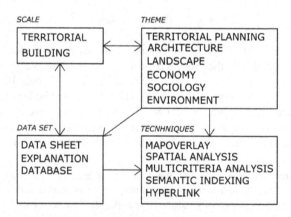

Fig. 3. Specific conceptual structure

For the specific application here presented, Rural Architectural Intensification, each cluster is formed by a list of precise issues (Fig. 4).

The list is changeable considering particular applications or main goals, such as the interrelations among the different precise issues.

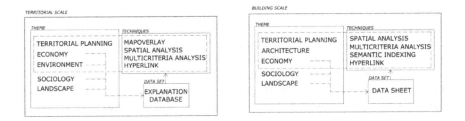

Fig. 4. Territorial and Building scale: semantic correlation

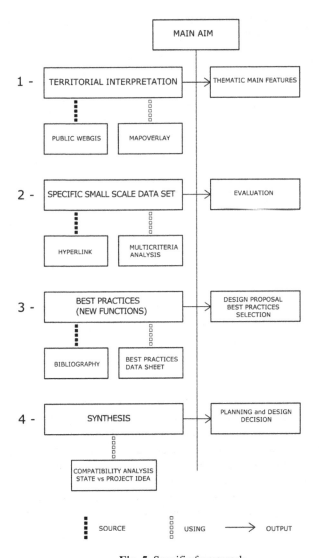

Fig. 5. Specific framework

In Fig. 4 and 5 for each specific scale themes, techniques and data set are selected and connected to underline the logical structure of the whole process.

In particular, in Fig. 5 there is the sequence of operations that a planner has to follow; this sequence is inserted in a circular complex process[3] [1]

Connections among specific issues may vary considering peculiar cases.

Finally, the considered process of landscape intensification in rural context is as sequence of defined operations that may be described as Fig. 6 shows.

The main aim determines the specific outputs while each phase (from 1 to 4) gets information from a primary origin and it uses specific techniques or models.

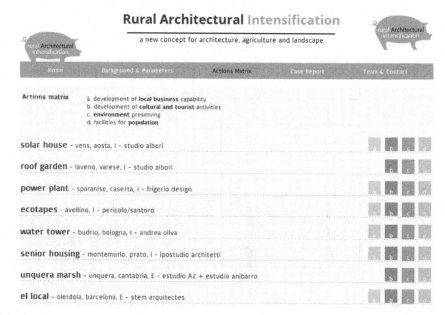

Fig. 6. Source: http://www.raintensification.com/
#!actions-matrix/c1d3f

3   Theme

The selected theme of the research is "Rural Architectural Intensification" (RAI). At the moment a website, http://www.raintensification.com/#!130/cfmf (built by authors), illustrates partial results of the bibliographic research at the building scale.

The output is a collection of semantically indexed sheets that represent applications and realized projects concerning four specific pillars: development of local business capability, development of cultural and tourist activities, environment preserving, facilities for population. These four pillars came from a synthesis of different Regional and European Rural Development Plans and from documents such as: Urban Act 1, Urban Act 2, Lipsia Charter.

---

[3]   See [1] Fig. 1.1 p. 9 (Italian edition, 2004).

Fig. 7. Source: `http://www.raintensification.com/#!136/cb4s`

Selected pillars are considered as assessment criteria or parameters for indexing a best practices list. In particular, as it is described in the website, the parameters are expressed as follows:

1. development of local business capability. Through local business skills, seen as their ability to increase the region's endogenous productive resources and to encourage the diversification of economic activities (also working from the existing architectural heritage). Into this parameter compare specific qualitative indicators: business and tourism, diversification into non-agricultural activities, diversification and innovation in agriculture, cultivate the landscape, cooperation and short chain, growth of the bio-economy, business and infrastructure.

2. cultural and tourist activities. Through the region's attractiveness, seen as its ability to boost innovative tourism and cultural activities, which will enhance the local resources in an integrated and ecological manner (the architectural and environmental heritage together). Into this parameter compare specific qualitative indicators: tourism and architectural heritage, tourism and environment, small-scale tourism services, countryside vs sea and mountain, tourism and water, tourism and infrastructure, tourism in less-favorable areas.

3. environment preserving. Through the quality and quantity of the natural environment present, of its resources, biodiversity or agro-forestry systems with a high nature value, regional green systems, ecological corridors. Into this parameter compare specific qualitative indicators: environment and biodiversity protection , environment as heritage, environment and water, soil and environment, environment and animals, environmentally sustainable operations, limit consumption of the environment, bioenergy, environmental reservoirs, environment and urban space, environment and infrastructure, environment and waste, environmental risk, environment in disadvantaged areas, diversified environmental redevelopment.

4. facilities for population. Through the local population's quality of life, as well as the essential public services and infrastructures. Into this parameter compare specific qualitative indicators: population and employment: tourism, population and

employment: diversification of agricultural activities, essential services to the population, country-city, population: energy saving, young population, population: infrastructure, cooperation in development, safe population, population and environment.

4   Spatial Analysis Applications

Following the previous conceptual schemes, the entire process is a logical sequence of passages to solve a complex problem, so it is basically an algorithm.

The system, GIS based, has been tested in the territory between the Italian cities Milano and Pavia [19].

Many web sources provide an exhaustive quantity of descriptive geo-referenced information. All used shapefiles (such as: DUSAF - Destinazione d'Uso dei Suoli Agricoli e forestali[4], ERSAF - Ente Regionale per i Servizi all'Agricoltura e alle Foreste[5], and territorial plans) come from the Lombardy Region website of territorial data set: WebGIS, Open Data Italia, PTR - Piano Territoriale Regionale, PTCPWEB - Piano Territoriale di Coordinamento Provinciale online[6] and PGTWEB - Piano di Governo del Territorio online[7], trans. Planning of Territory Government We. [1][8]

The availability of a comprehensive geo-information set guarantees the possibility to extract specific contents and gain knowledge.

The analyzed rural land is characterized by a strong presence of typical rural settlements (named *cascine*) and clear signs of sprawl phenomena. Many minor communities are widespread in a quite isotropic way. Major attention has been given to *cascine* settlements.

In the following examples, authors focus on geographical data treatment; with reference to the previous frameworks it is the Territorial Interpretation phase. [20,21,22,23,24,25]

4.1   Example 1

Main Aim: creation of widespread communities with assorted facilities.

Territorial interpretation: each *cascina* has been treated as a unique point with proper attributes. The first objective was to divide the whole set into groups basing on spatial and physical characteristics. Throughout a cluster analysis the results were variable (there was no convergence), throughout a grouping analysis 5 groups have been defined.

To find appropriate locations for diverse collective facilities it was necessary to find a sufficient built surface available for a change of destination (from rural to facility). So sub-clusters made of 3 or more *cascine* with the minimum distance

---

4   Forest and agricultural land use
5   Regional Bureau for services to agriculture and forests
6   WebGIS of Provincial Territorial Plan
7   WebGIS of Planning of Territory Government
8   See [1] Fig. 1.2 p. 12 and Table 1.1 p. 13 (Italian edition, 2004).

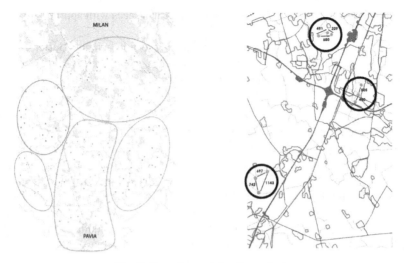

Fig. 8. Grouping and density analysis

among them (in order have the possibility to reach by feet every destination from the barycenter or from a parking area) were found throughout ArcGIS density operations (Fig. 9).

Specific small scale data set: each *cascina*, considered as a unique element, is provided of a hyperlink to Google Maps and Google Street View. Hyperlink has a purely visual configuration, and it is useful for a detailed analysis of visual resources and then for a visual evaluation [26].

In this way it is possible to verify landscape details and architectural features of each building.

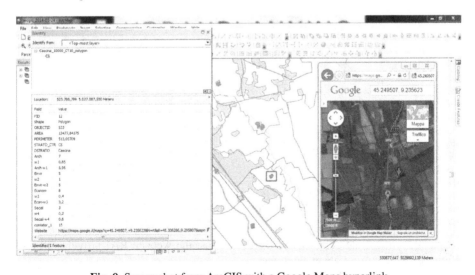

Fig. 9. Screen shot from ArcGIS with a Google Maps hyperlink

Best practices: the website RAI provides all the references with appropriate evaluations.

Synthesis: once specific facilities have been defined for each sub-cluster, using an hyperlink to "case report" section of RAI website a compatibility among existing conditions and design solutions has been found.

4.2   Example 2

Main Aim: creation of eco-commercial clusters.

Territorial interpretation: considering an isotropic consumer base and the Main Aim, a sequence of ArcGIS operation was defined: for each *cascina* (that had a minimum dimension of 30.000 m^2) it was firstly identified the proximity and the accessibility to the main infrastructure net (500 m). Then it was identified the possible presence of some environmental or protective restrictions excluding the settlement located in such fields. Finally, to avoid environmental impacts on residential settlements (due i.e. to traffic) the *cascine* inside small urban agglomeration were excluded (Fig. 10),

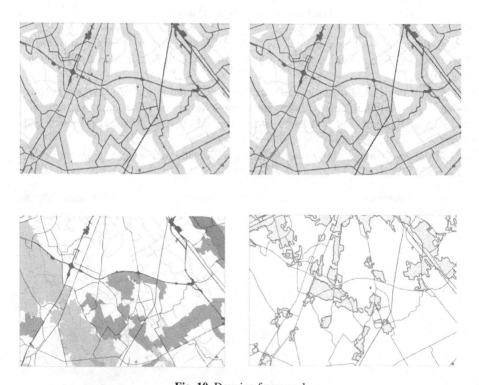

Fig. 10. Drawing framework

Specific small scale data set and Best practices: the website RAI provides all the references with appropriate evaluations.

Synthesis: using the hyperlink to "case report" section of RAI website a compatibility among the selected *cascine* and design solutions has been found.

4.3   Example 3

Main Aim: possibility to locate a decentralized eco-hotel.

Territorial interpretation: a decentralized eco-hotel is composed by different locations not farer than 2 km from each other. It should be preferably located close to the most relevant environmental components and reasonably far from the main urban settlements. In the specific context, the distance has been fixed as 1 km and the dimension of the settlement has been fixed as more than 200.000 m^2. Moreover, in order to have good environmental conditions (low pollution and low noise) the eco-hotel must be at least 500 m far from the main road net.

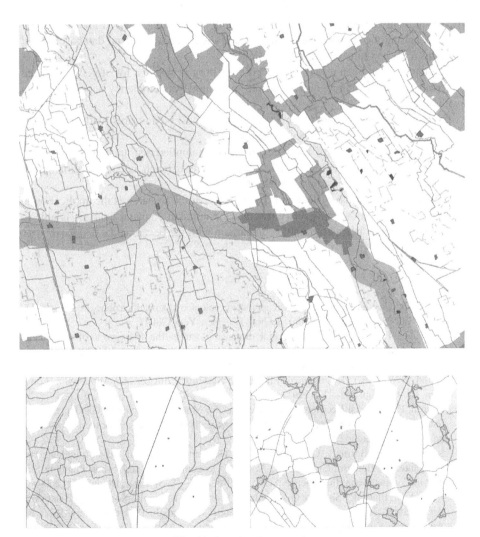

Fig. 11. Drawing framework

In ArcGIS an analysis of the environmentally meaningful elements such as woods, rows and hedge, environmental value charts, RER (Rete Ecologica Regionale, trans. Regional Ecological Network REN), hydric network, carried to define specific fields that, crossed with a buffer of small settlements and a buffer of road net, delimited the appropriate location area (Fig. 11).

In such area there are not *cascine* so close to create a real unique structure, so the hypothesis is not spatially affordable.

5   Conclusions

At the moment the proposed system is not completely computerized and more connections among the spatial analysis results and the semantic indexed database of best practices must be improved.

The method has been tested in a specific territory in which a lot of geo-referenced information were available, and the whole spatial analysis efficiency depends on data availability.

Every considered element (such as the *cascine*) must be described with appropriate attributes in order to provide effective analyses.

The logical framework of the procedure can be applied also to bigger minor settlements (such as small villages) and to wider contexts (such as entire regions). The only necessary condition is data availability.

This is the limit of the system and the next step of the research is the application to different contexts in order to verify whether the automatic spatial analysis is able to produce results that permit a reliable comparison with the best practice database.

References

1. Steiner, F.: Costruire il paesaggio. Un approccio ecologico alla pianificazione. In: Palazzo, D., Treu, M.C. (eds.), Italian, 2nd edn. McGraw-Hill, New York (2004)
2. Blaschke, T.: The role of the spatial dimension within the framework of sustainable landscape and natural capital. Landscape and Urban Planning 75(3-4), 198–226 (2006)
3. Densham, P.J.: Spatial Decision Support Systems. In: McGuire, D.J., Goodchild, M.S., Rhind, D.W. (eds.) Geographical Information Systems: Principle and Application, Longman, England, pp. 403–412 (1991)
4. Keenan, P.B.: Spatial Decision Support Systems. In: Mora, M., Forgionne, G., Gupta, J.N.D. (eds.) Decision Making Support Systems: Achievements and Challenges for the New Decade, pp. 28–39. Idea Group (2003)
5. Murgante, B., Borruso, G., Lapucci, A.: Geocomputation and urban planning. In: Murgante, B., Borruso, G., Lapucci, A. (eds.) Geocomputation and Urban Planning Studies in Computational Intelligence. SCI, vol. 176, pp. 1–18. Springer, Heidelberg (2009)
6. Fayyad, U., Piatetsky-Shapiro, G., Smyth, P.: From Data Mining to Knowledge Discovery: An Overview. In: Fayyad, U., Piatetsky-Shapiro, G., Smyth, P., Uthurusamy, R. (eds.) Advances in Knowledge Discovery and Data Mining. AAAI Press/The MIT Press (1996)
7. Voogd, H.: Multicriteria evaluation for urban and regional planning. Pion, London (1983)

8. Malczewski, J.: GIS and Multicriteria Decision Analysis. John Wiley & Sons Inc., New York (1999)
9. Jambu, M., Lebeaux, M.O.: Cluster Analysis and data analysis. North Holland Publishing Company, Amsterdam (1983)
10. Spath, H.: Cluster Analysis Algorithms for data reduction and classification of objects. Ellis Horwood Limited, Chichester (1980)
11. Anderberg, M.R.: Cluster Analysis for applications. Academic Press, Inc., New York (1973)
12. Getis, A., Ord, J.K.: The analysis of spatial association by use of distance statistics. Geographical analysis 24(3), 189–206 (1992)
13. Murgante, B., Danese, M., Las Casas, G.: Analyzing Neighbourhoods Suitable for Urban Renewal Programs with Autocorrelation Techniques. In: Burian, J. (ed.) Advances in Spatial Planning. InTech (2012)
14. Murgante, B., Danese, M.: Urban versus Rural: The decrease of agricultural areas and the development of urban zones analyzed with spatial statistics. Special Issue on Environmental and Agricultural Data Processing for Water and Territory Management. International Journal of Agricultural and Environmental Information Systems (IJAEIS) 2(2), 16–28 (2011)
15. Johnson, A.H.: Guest editorial: Human ecological planning-methods and studies. Landscape Planning 8(2), 107–108 (1981)
16. Meentemeyer, V.: Geographical perspective of space, time and scale. Landscape Ecology 3, 163–173 (1989)
17. Carta, M.: Pianificazione territorial ed urbanistica. Dalla conoscenza alla partecipazione, Laboratorio di Pianificazione Territoriale, Medina, Palermo (1996)
18. Kaiser, E., Godschalk, D., Chapin Jr, F.S.: Urban Land Use Planning, 4th edn. University of Illinois Press, Urbana (1995)
19. Multiplicity.lab, Centro Studi PIM, Comune di Milano: Le cascine di Milano verso ed oltre Expo 2015; un sistema di luoghi dedicati all'agricolture, all'alimentazione, all'abitare e alla cura (2009)
20. Frampton, K.: In search of the Modern Landscape. In: AA.VV., Denatured Visions. Landscape and Culture in the Twentieth Century. Museum of Modern Art, New York (1991)
21. Seddon, G.: Landscape planning: A conceptual perspective. Landscape and Urban Planning 13, 335–347 (1986)
22. Spaziante, A., Murano, C.: Rural development programs and strategic environmental assessment: Towards a sustainable rural territory. International Journal of Agricultural Resources, Governance and Ecology 8(2), 205–223 (2009)
23. Van der Vaart, J.H.P.: Towards a new rural landscape: Consequences of non-agricultural re-use of redundant farm buildings in Friesland. Landscape and Urban Planning 70(1), 143–152 (2005)
24. Fuentes, J.M.: Methodological bases for documenting and reusing vernacular farm architecture. Journal of Cultural Heritage 11(1), 119–129 (2013)
25. Cano, M., Garzon, E., Sanchez-Soto, P.J.: Historic preservation, GIS, & rural development: The case study of Almeria province, Spain. Applied Geography 42, 34–47 (2013)
26. Zube, E.H., Sell, J.L., Taylor, J.G.: Landscape perception: research, application and theory. Landscape Planning 9, 1–33 (1982)

Application of an Integrated Transport Land Use Model: An Empirical Approach

Georgia Pozoukidou

Department of Planning and Development,
Aristotle University of Thessaloniki, Thessaloniki, Greece
gpozoukid@plandevel.auth.gr

Abstract. Spatial allocation of households and regional employment are the most significant data inputs in land use forecasting development and transportation modeling. Therefore obtaining and validating socioeconomic data and forecasts are becoming crucial tasks for any planning organization that wants to utilize land use forecasting models as planning analysis tool. This paper presents the results of an ongoing research investigating the application of an integrated transport land use model in small and medium sized metropolitan planning organizations. To this purpose it uses FHWA's freely available Transportation Economic Land Use Model for Ada and Canyon counties of Boise, ID. The paper concludes with the strengths and limitations of the model in the context of data requirements and calibration process.

Keywords: Calibration, integrated transport land use models, metropolitan planning organizations, land use model application, urban models.

1 Introduction

The significance of the relationship between land use change and transportation infrastructure has been recognized in the relative literature for some decades now. Therefore integrated transport land use models should be an integral part of the decision making process in planning. On the other hand literature review on computer use in planning suggests that there is a continuous failure to use complex planning methods, like integrated land use transport models, in planning practice [1, 2].

Focusing on the use of such models in U.S.A., we should not forget to mention that the enactment of three major federal actions in the early 1990s added momentum to the integration of transportation and land use planning. Clean Air Act amendments, Intermodal Surface Transportation Efficiency Act and Transportation Equity Act for the 21st Century, brought new requirements and responsibilities to elected officials regarding transportation decision-making. Therefore Metropolitan Planning Organizations (MPOs) became responsible for transportation and land use planning decisions, and were required to assess the impact of their transportation policies on land use development using appropriate modeling tools [3].

B. Murgante et al. (Eds.): ICCSA 2014, Part IV, LNCS 8582, pp. 356–371, 2014.
© Springer International Publishing Switzerland 2014

Federal's Highway (FHWA) freely available transportation land use model-TELUM, predicts employment and household spatial allocation in a region using a spatial interaction approach. TELUM is part of a larger decision support system that was initiated and funded by U.S. Department of Transportation, with Rutgers University and North Jersey Planning Authority being responsible for designing and developing the system. Today the system is copyrighted and every MPO is eligible to use TELUM at no cost.

In this context, the paper presents the results of an ongoing research that examines how an integrated transport land use model like TELUM, can be used by small and middle-sized MPOs. Therefore land use and transportation data for Ada and Canyon counties of Boise Metropolitan Statistical Area were used with the model. The paper starts by providing a short description of the employment and household allocation models emended in TELUM, the required data inputs and a description of the calibration process. Then a detail report of TELUM application for Ada and Canyon counties is provided. The paper concludes with a reference to the problems occurred during data acquisition and calibration process and highlights the importance of ease implementation of these tools in order to achieve higher utilization percentages in planning practice.

2   Forecasting with TELUM

Basic parts of TELUM land use model are DRAM and EMPAL, a residential and an employment location model emended with other auxiliary modules in one system. Following is a brief description of these two basic parts of TELUM, focusing mainly on the structure of the equations that models utilize. These equations are the final product of an extensive 40 year research of employment and residential location models that has been developed by Dr. S.H. Putman, Professor at the Department of City and Regional Planning, University of Pennsylvania [4].

2.1   The Employment Location Model – EMPAL

TELUM-EMPAL is a modified version of the standard singly-constrained spatial interaction model. There are three modifications: 1) a multivariate, multiparametric attractiveness function is used, 2) a separate, weighted, lagged variable is included outside the spatial interaction formulation, and 3) a constraint procedure is included in the model, allowing zone and/or sector specific constraints [5].

TELUM-EMPAL model normally uses for 4-8 employment sectors. The equation structure used for TELUM- EMPAL and for this project is as follows:

$$E_{j,t}^k = \lambda^k \sum_i P_{i,t-1} A_{i,t-1}^k W_{j,t-1}^k c_{i,j,t}^{\alpha^k} \exp(\beta^k c_{i,j,t}) + (1-\lambda^k) E_{j,t-1}^k \tag{1}$$

where
$$W_{j,t-1}^k = (E_{j,t-1}^k)^{a^k} L_j^{b^k} \tag{2}$$

and
$$A_{i,t-1}^k = \left[\sum_{\ell} (E_{\ell,t-1}^k)^{a^k} L_\ell^{b^k} c_{i,\ell,t}^{\alpha^k} \exp(\beta^k c_{i,\ell,t}) \right]^{-1} \tag{3}$$

where

$E_{j,t-1}^k$ employment (place of work) of type k in zone j at time t-1

$E_{j,t}^k$ employment (place of work) of type k in zone j at time t

L_j total area of zone j

$c_{i,j,t}$ impedance (travel time or cost) between z ones i and j and time t

$P_{i,t-1}$ total number of households in zone I at time t-1

$\lambda^k, \alpha^k, \beta^k,$ empirically derived parameters
$a^k, b^k,$

2.2 The Residential Location Model - DRAM

TELUM-DRAM is also a modified version of a singly – constrained spatial interaction model. There are two major modifications: 1) a multivariate, multiparametric attractiveness function is used, 2) a consistent balanced constraint procedure is included in the model, allowing zone and/or sector specific constraints. The model is normally used for 3-5 (the current maximum is 8) household categories, whose parameters are individually estimated. A more detail description of model's structure is available in texts written by Pr. Putman [5]. The equation structure that is as follows:

$$N_i^n = \sum_j Q_j^n B_j^n W_i^n c_{i,j}^{\alpha^n} \tag{4}$$

where
$$Q_j^n = \sum_k a_{k,n} E_j^k \tag{5}$$

and
$$B_j^n = \left[\sum_j W_i^n c_{i,j}^{\alpha^n} \right]^{-1} \tag{6}$$

and
$$W_i^n = (L_i)^{q^n} (x_i)^{r^n} (L_i^r)^{s^n} \prod_{n'} \left(1 + \frac{N_i^{n'}}{\sum_n N_i^n} \right)^{b_{n'}^n} \tag{7}$$

where

E_j^k employment of type k (place of work) in zone j

N_i^n households of type n residing in zone i

L_i^v vacant developable land in zone i

x_i 1.0 plus the percentage of developable land already developed in zone i

L_i^r residential land in zone i

$a_{k,n}$ regional coefficient of type n households per type k employee

$c_{i,j}$ impedance (travel time or cost) between zones i and j

$\alpha^n, q^n,$ empirically derived parameters

$r^n, s^n, b_{n'}^n$

2.3   Required Data Inputs for the Forecasting Procedure

The study area is divided in appropriate analysis zones which usually are the prede-fined transportation analysis zones. The forecasting procedure starts with TELUM-EMPAL. The model uses four to eight employment types or sectors and needs the following input data: employment by type in all zones, population by income per zone, total area per zone, zone to zone travel cost or travel time between all zones.

Next step is the residential location forecast which is performed using the DRAM module of TELUM. The model uses four to six household types, which represent different income groups i.e. high income, low income etc., the parameters of which will be individually estimated. To forecast residential location TELUM-DRAM needs the following input data: residents of all types in all zones at time t, land use for resi-dential purposes in each zone at time t, the percentage of developable land that has already been developed in each zone at time t, the vacant developable land in each zone at time t, zone to zone travel cost, employment of all types in all zones at time t+1. Residential location forecasts are then used as inputs to generate and distribute trips, split trips by mode and then assign vehicle trips to the transportation network.

2.4   Calibration

Calibration is the process of fitting TELUM-DRAM and TELUM-EMPAL models into the real world by estimating the parameters for each locator type (i.e. high in-come households, manufacturing etc.), which will be used in models' equation. These parameters will be the ones that best fit in the structure of the dataset and will minim-ize the discrepancies between the model results and the real data. The calibration process used by CALIB module of TELUM is based on the maximization of the likelihood function and employs a gradient search method [5].

Using CALIB we calculate partial derivatives (or estimate parameters) for each one of the locator types. Each locator type (government employment, low-income households etc.) in TELUM EMPAL and DRAM models will have different "locating behavior" in a particular region. At the same time a particular locator type may also exhibit different "locating behavior" in different regions. Because of this, it is neces-sary to estimate the equation coefficients of the model equations separately for each locator type in each region. The process of estimating the equation coefficients is called model calibration. For each locator type CALIB runs are performed. It may take one or several CALIB runs for each locator type's full calibration.

One should examine the results of the calibration process in order to evaluate if the estimated partial derivative values are reasonable and acceptable as inputs in TELUM

DRAM and EMPAL models. In order to do this one or more indicators of goodness of fit are used [5]. These are:

R-Square. R-square value is an indicator of how well the model fits the data. The smaller the variability of the residual values around the regression line relatively to the overall variability the better is our prediction.

Best-Worst Likelihood Ratio. The "best fit" of a model is when the difference between the model's estimate of the depended variable and the observed values in the calibration dataset is as small as possible. The value of this ratio range from 1 to 0, were 1 is the best/perfect fit and 0 is the worst fit. The B/W likelihood ratio takes the following equation form:

$$\varphi = \frac{L - L_w}{L_b - L_w} \tag{8}$$

Mean Absolute Percent of Error (MAPE). Of the several statistics, which are often used to test the results of forecasting models the Mean Absolute Percent of Error is one of the most appropriate measures of goodness of fit. MAPE examines the distribution of the residuals (or errors) between the observed data and model's current best-fit estimates. More specifically it is the average of the absolute values of percent of error between the observed and the estimated by the model values.

When using MAPE as a measure of goodness of fit we should be aware that it does not take into account the size of the zones (population and employment wise). This can create distortions especially when we have large percentages of errors in small zones. In order to avoid such misinterpretations it is wise to examine MAPE indicator for the biggest 25% and the smallest 25% of the zones and explore if we are likely to get mistakes because of zone sizes. If this is the case then another indicator could used for the same purposes, such as MARMO.

MARMO. MARMO is very similar to MAPE. It also expresses the average of the absolute values of percent of errors between the observed set of data and the data estimated by the model, but it is weighted by the size of the observation (actual count of population or employment). A 20% to 30% percent usually represents a good MARMO.

Regional Location Elasticities. Location elasticities measure the sensitivity of household and employment location to changes in the attractiveness variables of DRAM and EMPAL models. Location elasticities are defined for each one of the employment and residential zones. For instance for a 1% increase in an attractiveness variable in a specific zone, the location elasticity measures the resulting percentage of change in the number of households and employees in that zone.

3   TELUM Application for COMPASS MPO

Community Planning Association of Southwest Idaho (COMPASS) is the metropolitan planning organization responsible for Ada and Canyon counties which compose our study area. Ada and Canyon counties are two out of the five counties of Boise

Metropolitan Statistical Area, which is currently the third largest metropolitan area in the Pacific Northwest region of US, after Seattle and Portland [6]. Boise MSA includes Idaho's three largest cities and nearly half (40%) of Idaho's total population lives in this area.

In order to better understand the physiognomy and future trends of the study area, following is a brief presentation of basic demographic and economic facts for Ada and Canyon County. Numbers for Idaho State and Boise MPO are also provided for comparison purposes.

3.1   Geographic and Demographic Facts

Boise MSA lies in an area of 11,738 sq. miles and as of 2010 a total of 616,561 people were living in this area. With a national population growth rate of 8.8 percent and a state of 21 percent for 2000-2010, Boise MSA growth rate is way above the average (table 1).

Table 1. Population Number and Percent of Change

Area	2000	2010	% Change
Idaho	1,293,953	1,567,582	21.1%
Boise MSA	464,840	616,561	42.6%
Ada C.	300,904	392,365	30.3%
Canyon C.	131,441	188,923	43.7%

Source: US Census Data, 2010.

According to 2010 census data Ada's County population increased by 30%, when Canyon's County population experienced an even larger increase by almost 44%. It is very important to note that both in Ada and Canyon County spatial distribution of population growth is inversely proportional to the distance from its respective city centers (Boise, Garden, Nampa and Caldwell), indicating that there has been a conti-nuous process of urban sprawl undergoing in the metropolitan area [7].

As far as age composition, relative data (table 2) indicates a "healthy" population distribution among the different age groups. This information is especially significant since it indicates the size of future employment based on current population at the age group 0-17 years old. Median age for Ada and Canyon counties is 34.8 and 31.6 years old respectively, which indicates a very young population.

Table 2. Population Age Structure: Percent of total Population/age group

	Idaho	Boise MSA	Ada C.	Canyon C.
Total Population	1,567,582	616,561	392,365	188,923
0- 17 years	27.37	26.6	26.4	31.4
18-64 years	60.21	59.4	63.1	57.71
65+ years	12.42	13.8	10.4	10.08
Median Age	34.6	38.74	34.8	31.6

Source: US Census Data, 2010.

In terms of racial composition the study area is pretty homogeneous with the white representing a high percentage in both counties (almost 90%). It should be noted that in the category "white" Hispanic race is included. More detail information in regard to racial composition of the study area is provided by the US Census General Demographic Profile Characteristics tables [7].

As far as housing occupancy rates it seems that both counties have high occupancy rates that are also higher than Idaho's and Boise's MSA (table 3). This information is extremely valuable since it indicates the available housing stock, which will partly determine if the city will further expand or if it is more likely to use its existing housing stock.

Table 3. Housing Occupancy

Area	Total Housing Units	% Occupied Housing Units	%Vacant
Idaho	667,796	86,8	13,2
Boise MSA	245,962	83,6	16,4
Ada County	159,471	93,1	6,9
Canyon County	69,409	91,6	8,4

Source: US Census Data, 2010.

Table 4 shows the median household income for 2008-2012. The numbers reveal that the state of Idaho, Boise MSA and Canyon County are below the U.S. average. In contrast Ada County has a relative high median household income overcoming national average and Idaho State.

Table 4. Household Income

Area	Median Household Income
U.S. National	53,046
Idaho	47,015
Boise MSA	42,570
Ada County	55,499
Canyon County	42,691

Source: US Department of Commerce, 2012.

3.2   Employment

Boise's metropolitan area labor force grew by over 10,000 in both 2005 and 2006 before slowing in 2007 and ultimately declining in 2009. The labor force rebounded in 2010 and is now 12,000 above its recession low. According to Idaho Department of Labor, total employment for Boise MSA is 283,012 (2012). Total civilian labor force is 307,591 and unemployment rate is 7.3% which is at the same level as that of Idaho's (6.6%) and U.S. national (7.9%).

Boise's MSA employment composition is similar to what it had been over the past decade with "Trade, utilities and transportation" still dominant, occupying almost 20% (49,364 employees) of the employment. At the same time there has been a shift away from manufacturing, which went from 15% to 9% of jobs in 10 years, towards education and health care, which now makes up 15% of total employment rather than 11% in 2002.

As far as Ada and Canyon counties they follow the trends diagnosed for the metropolitan area. Ada County labor force grew significant between 2002 and 2012, to a total of 191,379 employees in 2012 (table 5). Education and health sector increased the most during this period when construction and manufacturing lost significant percent of jobs. Canyon County exhibited same trends in terms of construction and manufacturing job losses. A detail analysis in regard to sectorial and spatial employment distribution for these two counties is provided in the section of calibration.

Table 5. Employment per Sector for Ada and Canyon Counties

	Canyon County			Ada County		
	2002	**2012**	**%Change**	**2002**	**2012**	**%Change**
Total	47,298	50,861	7.5%	182,311	194,714	6.8%
Agriculture	2,824	2,800	-0.8%	604	764	26.5%
Mining	46	22	-52.2%	113	84	-25.7%
Construction	3,604	2,811	-22.0%	12,898	9,104	-29.4%
Manufacturing	10,821	7,499	-30.7%	23,615	14,909	-36.9%
Trade, Utilities & Transportation	9,340	11,043	18.2%	34,845	37,260	6.9%
Information	652	620	-4.9%	3,270	3,590	9.8%
Financial Activities	1,469	1,541	4.9%	9,459	10,800	14.2%
Professional and Business Services	2,811	3,631	29.2%	28,878	33,338	15.4%
Educational and Health Services	5,200	7,460	43.5%	19,800	30,616	54.6%
Leisure and Hospitality	3,085	3,452	11.9%	16,722	19,101	14.2%
Other Services	940	1,451	54.4%	5,127	5,214	1.7%
Government	6,508	8,531	31.1%	26,980	29,935	11.0%

Source: Idaho Department of Labor, 2012.

4   Model Calibration

4.1   Data Availability

Following is a description of the data used in order to perform the calibration procedure for TELUM DRAM and EMPAL models.

Zones. The study area is composed of 471 Transportation Analysis Zones (TAZ). TAZ is the most commonly used spatial level to generate land used forecasts. As mentioned above the study area consists of two counties, Ada and Canyon with 285 and 187 transportation analysis zones respectively.

Population. Total population for all zones according to population estimations census data for COMPASS MPO for 2008 was 589,720 people. Total population for 2012 was estimated to be 590,070 which indicates a very small increase within 5 years.

Spatial distribution of population in two counties for 2008, shows a population concentration mainly around the four major cities of Boise, Garden, Nampa and Caldwell. Distribution of population density (Fig.1) verifies the concentration trend. It should be noted that population concentration patterns remain the same for the 2012 estimated population.

Spatial distribution of percent of change between 2008 and 2012 indicates that the immediate to Boise city zones are experiencing a decrease in their population numbers. The trend is opposite for the TAZs close to Garden city and northeast boundaries of Ada county.

Households. Household spatial distribution for 2008 verifies that the 212,896 households have, as expected, similar spatial pattern as population since household distribution is directly related to population distribution.

It should be noted that the available household data is limited to the total number of households per zone. That means that necessary information to execute calibration runs is missing. In order to overcome this problem we made the assumption that the total number of households is equally divided between four household types of income groups. These are: High Income Households, High-Middle Income Households, Low-Middle Income Households and Low Income Households.

Employment. Total employment and employment by type is available, for each transportation analysis zone for both 2008 and 2012. Total employment number for 2008 is 267,732 and for 2012 is 274,303. For the calibration process six employment types were used; Agriculture, Education, Government, Industrial, Office and Retail. These six employment types is a consolidation of the 12 standard employment types provided by US Census Bureau when categorized by "Industry".

Employment distribution for Ada and Canyon Counties for the six employment sectors (in 2008) shows that employment is heavily concentrated at the "office" sector with almost 45 percent. Another 35% of employment is distributed amongst retail (27%) and government (8%), which both constitutes third employment sector. A comparison between total employment for 2008 and the projected employment for 2012 shows that there is an increase of 3 percent in five years. Employment distribution between the six different employment types for 2012 remains almost the same.

In terms of employment distribution for 2008, it seems that there is a concentration near the four major cities of the study area (fig.2). A series of maps showing employment distribution for each sector were also created but due limited space they were not included in the paper.

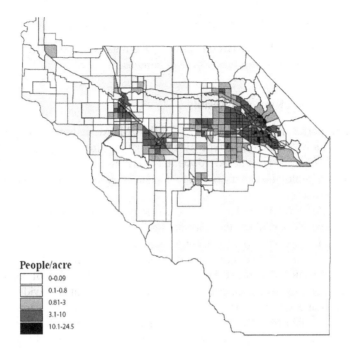

Fig. 1. Spatial Distribution of Population Density, 2008

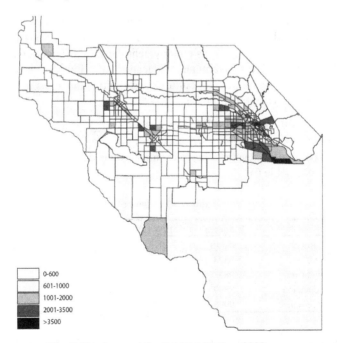

Fig. 2. Employment Spatial Distribution, 2008

Land Use Data. Land Use data was not available for the study area.

Zone to Zone Travel Cost (time). Travel time data was available. The travel time matrix contains zone to zone travel time in minutes.

4.2   Calibration Runs and Results

In order to perform calibration with TELUM-EMPAL it is necessary to have employment data by employment type by zone for two time points, five years apart. In this case 2012 was considered to be the current time point and 2008 the lag time point. Employment data is the only data required for two time points in the calibration process.

For TELUM-DRAM calibration it is necessary to have employment data by employment type by zone, for the current time point. TELUM-EMPAL current year should match TELUM-DRAM current year, which in our case is 2012. It is also necessary to have household data for one time point that will match current employment year. TELUM-DRAM can use lag household data in the form of total number of households per zone five years earlier. Finally, household data is divided by type (4 to 8 categories) and zones in current year. As mentioned earlier this case study is peculiar, due to the fact that we do not have a breakdown of households in different income categories but the total number of households in each zone. In order to be able to run calibration we divided the total number of households into four different categories by equally distributing households in each one of them. As a result four different household types were included in the calibration runs. These were: High Income Households, High-Middle Income Households, Low-Middle Income Households and Low Income Households. As far as employment the five different employment types used were: Retail, Office, Government, Education, and Industry. It should be noted that the employment sector of "Agriculture" was omitted due to its insignificance as a percentage in employment distribution.

Employment Calibration Results. The measures of goodness of fit achieved from the calibration process were very satisfactory. Following is the table showing a summary of the calibration results. The four indicators as described above are presented with their values (table 6).

Table 6. Goodness of Fit

	R-Square	B/W LR	MAPE	MARMO
Retail	.98346	.9735	17.259%	6.740%
Office	.9823	.9725	21.575%	6.976%
Industry	.9848	.9889	12.167%	7.549%
Education	.9849	.9879	14.457%	5.637%
Government	.9998	.9994	4.561%	1.219%

R-Square and B/W LR are really high indicating that employment data almost perfectly fit TELUM-EMPAL model. At the same time MAPE values also indicate that the estimated parameters for TELUM-DRAM and TELUM-EMPAL models' equation are very close to "best fit" since the percentage of error between observed and model's current best fit estimates for each one of the locator types are within the acceptance range.

For reasons explained earlier in this paper (see section 2.4) a more detail examination of Mean Absolute Percent of Error (MAPE) for the dataset was made. Table 7 shows the values of MAPE for the smallest 25% and the largest 25% for each one of the employment types.

Table 7. Mean Absolute Percent Error (MAPE)

	Min Observed Value	Max Observed Value	MAPE for Smallest 25% of Zones	% of smallest zones of region total	MAPE for Largest 25% of Zones	% of largest zones of region total
Retail	1	3568	.000%	.18%	7.756%	82.25%
Office	1	4382	462%	.15%	6.452%	81.64%
Industry	1	6534	28.3%	.35%	6.302%	81.94%
Education	1	1541	.000%	.12%	7.985%	84.48%
Gov/ment	1	3509	.000%	.34%	6.852%	82.33%

For most employment types MAPE has values within an acceptable range for the 25% smallest zones. Exception is the case of office employment sector were MAPE for the 25% smallest zones is 462%, which actually inflates the value of MAPE for the dataset. For that MARMO is a more secure indicator for our purposes. MARMO values as shown in table 6 are within an acceptable range.

Examination of lambda values is another way to evaluate calibration results. Lambda is a weighting factor or a parameter related to the extent to which a specific employment type is oriented to past locational determinants or to current intra-urban location of the population [5]. Locational determinants are related to parameters like the value of the land, the accessibility patterns (i.e. proximity to a major rail line), or even to the agglomeration effects of some economic activities.

In this case the value of lambda for retail, as shown in table 8, indicates the high importance of the prior retail location and high spatial interaction. The low λ values for the rest of the employment types suggest that these employment groups are rather static and is most likely to be found in zones were they were previously concentrated.

Table 8. EMPAL Estimated Parameter Values

	Retail	Office	Industry	Education	Government
α	.0163	3.6854	3.6855	3.4265	3.4148
β	-.1278	-1.632	-1.628	-1.648	-1.693
EMP	.9351	0.999	.9901	.9906	.9999
LAND	.0054	.0000	.0796	.0853	.0022
λ	.0261	.0001	.0001	.0100	.0004
1-λ	0.973	0.999	0.999	0.99	0.999
R^2	.98346	.9823	.9848	.9849	.9998

Regional location elasticities is another sensitivity indicator for employment location of a specific employment type to the changes in the attractiveness variables. For instance regional location elasticities for "Retail" shows that one percent increase in land will cause a subsequent increase in employment by 0.97 percent. Results are similar for the rest of employment types.

Household Calibration Results. Due to the peculiarity of the household data we needed to run calibration for TELUM-DRAM equation just for one of the four income categories. As the results show (table 9, 10, 11) household data almost perfectly fits TELUM-DRAM equations. MAPE though has a relative high value, indicating that further analysis is needed. A closer look of MAPE indicator shows that it has low values for the 25% of the largest zones of the study area and high values for the 25% of the smallest zones, which actually inflates the overall MAPE value. In conclusion we could say that household data almost perfectly fits TELUM-DRAM equations. It should be noted that the parameter values shown at the tables below are referring to all household types since we did not have individual data for each household type.

Table 9. Goodness of Fit

	Households
R-Square	.9432
B/W LR	.9267
MAPE	25.338%
MARMO	10.696%

Table 10. DRAM Estimated Parameter Values Hholds

α	.8389
β	-1.269
VACDEV	.1958
PERDEV	.4560
LIHH	1.0173
LAGHH	.0633

Table 11. Mean Absolute Percent Error MAPE for Households

Min Observed Value	0
Max Observed Value	702
MAPE for Smallest 25% of Zones	60.529%
% of smallest zones of region total	3.63%
MAPE for Largest 25% of Zones	9.348%
% of largest zones of region total	58.63%

Calibration Residuals. A more detail examination of residual's distribution will give us a better idea about data quality and validity of calibration results. Also an examination of their locational distribution might show interesting spatial patterns.

Calibration residuals represent the unobserved factors that influence the relative employment attractiveness. Calibration process creates a set of residuals that indicate how accurate is the calibration and consequently how accurate is the forecasted population or employment values for each zone. Residuals are calculated by dividing the difference of the observed and predicted values by the observed. In other words residuals are the difference between the actual/observed number of employees or households and predicted number of employees or households in each zone divided by the observed values of employees or households in the same zone.

Actually their spatial distribution shows interesting patterns. For instance in the case of Retail (fig. 3) it seems that a number of zones have been underestimated by 28 percent to 97 percent. At the same time the darker colored zones shows that these zones have been overestimated by the relative percentage. Zones that they do not have any color are within the range of plus/minus five percent, which is an acceptable error.

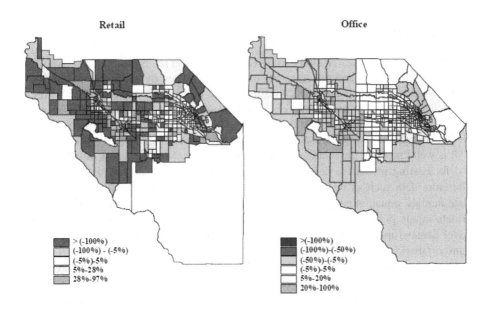

Retail   Office

> (-100%)   >(-100%)
(-100%) - (-5%)   (-100%)-(-50%)
(-5%)-5%   (-50%)-(-5%)
5%-28%   (-5%)-5%
28%-97%   5%-20%
                                                          20%-100%

Fig. 3. Calibration residual for Retail and Office employment types

Validity of Household and Employment Projections. Despite the fact that the measures of goodness of fit seem to be very satisfactory and indicate that the estimated parameters for TELUM-EMPAL and TELUM-DRAM model equations are very close to the best fit model there is a suspicion that the data for 2012 is just a linear extrapolation of the 2008 data. If this is the case then the calibration results presented above are unreliable. A simple regression between 2008 and 2012 population confirms that the projected 2012 data is a linear projection of the 2008 data. The chart (fig. 4) confirms the relationship between the two datasets. The same analysis was done for employment and showed a linear relationship between the actual (2008) and projected (2012) data. It is obvious that such data is inadequate for our research purposes.

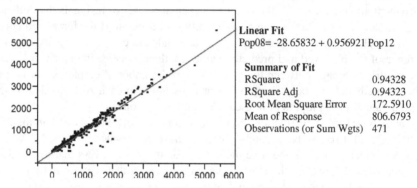

Fig. 4. Population- Bivariate Fit of Households 2008 by Households 2012

5   Conclusions

The purpose of this research study was to examine how a relative simple, in terms of its data requirements, land use transport model can be used by small and middle-size MPOs to address their planning responsibilities. In order to do that COMPASS MPO was selected as a case study and TELUM land use model was used. TELUM was specifically selected since it is U.S. Federal's Highway Administration freely available land use model, designed especially for use by MPOs.

It is well known in the relative scientific community, that the success of calibration and forecasting procedure depends, to a large extend, on the quality of data inputs. Therefore data such as future population and employment distribution used to estimate models' equation parameters becomes extremely crucial. In addition, the quality of data inputs in the land use model affects the validity of results in the subsequent travel demand model, setting the quality of data as a key factor in the land use and transportation forecasting process.

Current study demonstrates that the quality of data used for COMPASS MPO was insufficient to produce statistically reliable and replicable forecasts for agency use. At this point it is very important to note that in order to perform a full iteration of calibration and forecasting, TELUM requires very few and basic data, in comparison to other types of land use models like micro-simulation model UrbanSim [8].

Previous efforts to apply TELUM [9, 10] using as test bed different MPOs had several issues in regard to data availability and subsequently the validity of calibration results. In addition a comprehensive usability evaluation for TELUM using several criteria [3], such as system requirements, transparency, ease of use, user expertise etc. confirms that one of the factors causing major bottlenecks in TELUM's application were data requirements. It is worth noting that similar studies dealing with the usability and applicability of planning support systems, highlighted the issue of data availability as a serious bottleneck in the application of such systems [11].

It is becoming quite clear that inadequate data and data availability could be a serious bottleneck towards the application of land use models in planning practice. This paper demonstrates that even simple, in terms of its data requirements, land use models could be inapplicable in practice. Therefore, if we want for land use models to have practical use as policy analysis tools, then scientific community should focus not only on the improvement of model formulation and predictability issues but also on data requirements and data availability issues.

References

1. Brail, R., Klosterman, R.: Planning Support Systems, Integrating geographical information systems models and visualization tools. ESRI, California (2001)
2. Pozoukidou, G.: Utilisation of Urban Modeling Tools in Decision Making Processes. The TELUM Case Study. In: Psycaris, Y., Skayannis, P. (eds.) The Context, Dynamics and Planning of Urban Development: A Collection of Papers. University of Thessaly Press, Volos (2008)
3. Pozoukidou, G.: Increased usability of urban and land use models. The role of knowledge based systems. In: Facilitating Land use Forecasting to Planning Agencies. Doctoral Dissertation. University of Pennsylvania, Philadelphia (2005)
4. Putman, S.: DRAM residential location and land use model: 40 years of development and application. In: Pagliara, F., Preston, J., Simmonds, D. (eds.) Residential Location Choice, pp. 61–76. Springer, Heidelberg (2010)
5. Putman, S.: Integrated Urban Models 2. Pion Limited, London (1992)
6. Community Planning Association of SW Idaho, http://www.compassidaho.org
7. U.S. Census Bureau, http://data.spokesman.com/census/2010/idaho/
8. Patterson, G., Kryvobokov, M., Marchal, G., Bierlaire, M.: Disaggregate models with aggregate data: Two UrbanSim applications. Journal of Transport and Land Use 3(2), 5–37 (2010)
9. Pozoukidou, G.: TELUM Land Use Model: An Investigation of Data Requirements and Calibration Results for Chittenden County MPO, U.S.A. International Journal of Environmental, Earth Science and Engineering 8(3), 1–10 (2014)
10. Pozoukidou, G.: Predicting the outcome of Plans. Retrospect and prospects of the use of models in planning. In: Brebbia, C.A., Hernandez, S., Tiezzi, E. (eds.) Urban Regeneration and Sustainability 2010, pp. 83–93. WIT Press, Boston (2010)
11. Vonk, G., Geertman, S., Schot, P.: Bottlenecks blocking widespread usage of planning support systems. Environment and Planning A 37, 909–924 (2005)

Modelling Spatially–Distributed Soil Erosion through Remotely–Sensed Data and GIS

Antonello Aiello[1,*], Maria Adamo[2], and Filomena Canora[3]

[1] Department of Sciences, University of Basilicata, Potenza, Italy
antonello.aiello@unibas.it
[2] Institute of Intelligent Systems for Automation, National Research Council of Italy, Bari, Italy
adamo@ba.issia.cnr.it
[3] School of Engineering, University of Basilicata, Potenza, Italy
filomena.canora@unibas.it

Abstract. Estimation of soil erosion using common empirical models has long been an active research topic. Nevertheless, application of those models at basin scale is still a challenge due to data availability and quality. In this study, the Revised Universal Soil Loss Equation (RUSLE) and the Unit Stream Power–based Soil Erosion/Deposition (USPED) were applied and compared to determine the spatial distribution of soil erosion of a coastal watershed in Basilicata, southern Italy. A comprehensive approach that integrates ancillary data, digital terrain model, products derived from satellite remote sensing (multi–temporal Landsat imagery) and GIS techniques was adopted to identify major factors influencing soil erosion. Soil loss and soil erosion/deposition maps were produced. The study provided a reliable prediction of soil erosion rates and definition of erosion–prone areas within the watershed.

Keywords: Soil erosion, RUSLE, USPED, Remote sensing, GIS.

1   Introduction

Erosion includes a broad range of processes that involves soil detachment and transport due to drivers that act upon Earth surface [1]. Soil erosion is a complex and dynamic phenomenon. It represents one of the most serious land degradation problems all over the world, and denotes a main source of environmental worsening. It also leads to decline in soil fertility, agricultural productivity, water quality, reservoir capacity, etc. [2,3]. Mediterranean areas are particularly prone to soil erosion because they are subjected to extended dry period followed by heavy erosive rainfalls, which fall on steep slopes characterized by fragile soils [4,5,6]. According to [7], soil loss of more than 1 ton ha^{-1} yr^{-1} can be considered as irreversible within a time span of 50–100 years. As stated by [8], losses of 20–40 ton ha^{-1} yr^{-1} in individual storms, which may happen once every two or three years, are measured regularly in Europe with losses of more than 100 ton ha^{-1} measured in extreme events.

---

[*] Corresponding author.

B. Murgante et al. (Eds.): ICCSA 2014, Part IV, LNCS 8582, pp. 372–385, 2014.
© Springer International Publishing Switzerland 2014

Different models and relations for predicting soil erosion/deposition and for estimating sediment yield have been developed and applied by several researchers [9,10,11,12]. Modeling can provide a quantitative and consistent approach to estimate soil erosion under a wide range of conditions [13]. Over the last decades, estimation of soil erosion using empirical models has long been an active research topic [14,15,16,17,18], [7], [19]. Nevertheless, their application over large areas is still a challenge due to data availability and quality [20]. Spatial and temporal patterns of soil erosion and deposition are caused by several key processes such as rainfall; surface, subsurface and ground water flow; vegetation growth; soil detachment, transport and deposition. The assessment of soil erosion requires the specific knowledge of soil parameters (i.e., soil susceptibility to erosion, soil protection, etc.), as well as physical parameters (precipitation, temperature, slope, surface area, etc.), all highly variable in space and time. Although the various methods are based on the key factors of soil erosion process, they differ on the processing of data and on the accuracy of results. Moreover, the specific characteristics of catchment basins and the accurate estimation of each parameter can deeply affect the final results. Successful monitoring can be realized with the integration of ancillary data and remotely–sensed data within a GIS environment. Remote sensing offers a privileged point of view in detecting environment abrupt and phenological changes, and in retrieving parameters. GIS offers challenging opportunities in evaluating the potential effects of single factors on the overall result, in processing huge amount of difference data sources, and in simulating future trend. Its synoptic spatial information can provide important additional information layers for designing scientifically–based strategies that must be oriented to protection and to sustainable planning/use of the environment.

The purpose of this study was to compare and to evaluate the suitability of the Revised Universal Soil Loss Equation (RUSLE) and Unit Stream Power–based Soil Erosion/Deposition (USPED) models in assessing soil erosion distribution at watershed scale. The first one is a detachment capacity limited model. It predicts the spatial distribution of soil erosion (Section 3) [17], [1]. The second one is a transport capacity limited model. It predicts the spatial distribution of net erosion and deposition rates for a steady state overland flow with uniform rainfall excess conditions (Section 3) [18], [1].

In order to assess the rate of soil erosion, the two models were applied to the Basento river basin and to the sub–basin subtended by the Camastra Dam (Fig. 1). To this end, digital terrain model, products derived from satellite remote sensing (multi–temporal Landsat imagery), soil texture maps and rainfall data were integrated and processed in a GIS. To test the models, the computed soil erosion rates over the sub–basin subtended by the Camastra Dam were compared with the dam silting value provided by an interregional authority responsible for its management (Section 4). The two models proved to be effective in quantifying soil loss at watershed scale. The study provided a reliable evaluation of soil erosion rate and a good definition of erosion–prone areas within the Basento basin.

2   Study Area

RUSLE and USPED were applied on the Basento basin of Basilicata, southern Italy (Fig. 1). The study area covers a surface of 1,530 km². The climate is typically Mediterranean, characterized by cool–to–mild wet winters and warm–to–hot dry summers. Mean annual rainfall ranges between 450 to 1,350 mm per year. Rainfalls are concentrated between October and March, and sharply decrease during the following period of the year. The Basento River has an approximate stream length of 150 km, starts from the Apennine Chain on the northwest side of the river basin and reaches the *Jonian* Sea after crossing the Bradanic Trough. The river initially runs from west to east and then changes course to southeast, until it is perpendicular to the coastline [21] (Fig. 1). The morphology of the river basin is characterized by a mountain sector, followed by hills areas and then flat valleys. It exhibits a wide and steep basin in the upper part, which becomes quickly narrow as a consequence of lithological features; subsequently, the riverbed expands, showing characteristics of meandering rivers especially in the lower part (Fig. 1).

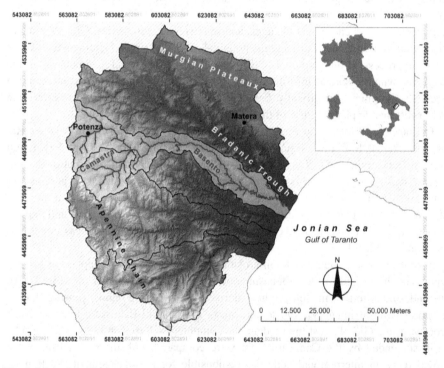

Fig. 1. Map of the Basento watershed and the sub–basin subtended by the Camastra Dam (Basilicata, southern Italy), showing also the other four main watersheds reaching the *Jonian* Sea, the digital terrain model (DTM) of the relative hydrographic basins, and major cities.

The mean altitude is equal to 600 m, about 21% of the basin has an altitude up to 200 m, about the 42% of the basin has an altitude ranging between 200 and 800 m, and about 37% has an altitude ranging between 1,000 and 1,800 m. The average slope is about 21%, and the 20% of the area has a slope more than 30%. In the upper part of the basin, flyschoid formations prevail, with calcareous–siliceous lithology, conglo-meratic–arenaceous succession, arenaceous and argillaceous lithotypes. In the middle sector, the basin flows almost totally over clastic Plio–Pleistocene deposits from the Bradanic Trough. The middle and lower parts of the basin are dominated by Plio–Pleistocene conglomerate, sand and clay, while towards the *Jonian* coast the river flows through recent alluvial sediments [22]. Geologic features and geomorphic li-neaments of the middle and lower parts of the river valley show several orders of fluvial and marine terrace deposits. During the Quaternary, the Bradanic Foredeep area was subjected to progressive uplift due to isostatic adjustment after thrust em-placement [23]. The series of marine terraces were assumed to be shaped as a result of the interference between tectonic uplift and glacio–eustatic fluctuations [24,25,26]. The study area is characterized by a mountainous landscape dominated by forests, and hilly and flat areas dominated by widespread cultivation of subsidized cereals. The improvement in earth–moving machinery has led to a general increase in field area, removal of terraces and change slope morphology. As a result of this landscape re-modeling, slope lengths are also greatly increased determining a spatial coupling of soil erosion processes [27].

3   Materials and Methods

The data sets collected and used in this study are summarized in Tab. 1. All datasets were geo–referenced in the WGS84 ellipsoid/UTM zone 33N projection system, and stored in a GIS database. A digital terrain model (DTM), provided by the National Institute for Environmental Protection and Research of Italy (ISPRA), was used to generate the topographic parameters such as slope–length, slope–steepness, aspect,

Table 1. Dataset collected and used in this study

Datasets	Description
DTM	Raster format with 20 m spatial resolution
Landsat 7 ETM+ images	Two scenes of Landsat 7 ETM+ images, acquired at Jan 3rd 2001 and Jul 13th 2001, with 30 m spatial resolution
Rainfall data	Daily rainfall data between 1991 and 2011 from local rain gauge stations
Soil texture map	Vector format with five soil texture types and 1:250,000 scale
CORINE land cover 2006	Vector format at 1:100,000 scale

drainage network, upslope contributing area. The soil texture map, provided by the Basilicata Region, was used to generate the soil thematic layers. Two Landsat 7 ETM+ images, acquired at 03/01/2001 and 13/07/2001, were used in order to monitor the vegetation cover of the study area (Fig. 2). The CORINE land cover from 2006, provided by ISPRA, was used to take into account the impact of different support practices on the average annual erosion. Since different data sources had various data formats, they were resampled to the same pixel size of 20 m by 20 m. This study was performed using ArcGIS 10.0 (ESRI), ENVI 5.1 (Exelis VIS), IDL 8.3 (Exelis VIS) and MATLAB R2013a (MathWorks) software.

Choosing between different models, RUSLE and USPED were applied in order to estimate the spatial distribution of soil loss and net soil erosion/deposition. RUSLE is a revision and update of the widely used Universal Soil Loss Equation (USLE) and it is suitable for GIS–based soil erosion modelling [7]. It has a number of improvements for its application at watershed scale, in contrast to the USLE designed for agricultural fields [28]. RUSLE uses the same factorial approach employed by the USLE, but different equations to obtain the same parameters: each factor has been either updated with recent information, or new factor relationships have been derived based on modern erosion theory and data. RUSLE can be expressed as

$$A = R \cdot K \cdot L \cdot S \cdot C \cdot P \tag{1}$$

where A is the computed annual soil loss per unit area (ton ha^{-1} yr^{-1}); R is the rainfall–runoff erosivity factor (MJ mm ha^{-1} h^{-1} yr^{-1}); K is the soil erodibility factor (ton h ha^{-1} MJ^{-1} mm^{-1}), L is the slope–length factor, S slope–steepness factor, C is the cover and management factor, and P is the support practice factor. L–, S–, C– and P– factors are dimensionless.

RUSLE is a detachment capacity limited model. It assumes that water flow can transport an infinite amount of sediment, and the amount of eroded soil is limited only by the capacity of water to detach soil particles. Thus, it is not able to predict soil deposition [1]. On the other hand, USPED does not only predict soil erosion rates, but also the spatial distribution of sediment deposition at watershed scale for a constant state overland flow with uniform rainfall–excess conditions. As a transport capacity limited model, the USPED assumes that the water flow can transport a limited amount of sediment given by the transporting capacity of water flow. Moreover, it supposes that the amount of sediment carried by water is always at its full transporting capacity [1]. The USPED is based on the theory originally proposed by [29], and the net erosion/deposition rate is estimated as the change in sediment flow rate expressed by the divergence in sediment flow [16], [18], [1]. It assumes that the sediment flow rate $q_s(r)$ corresponds to the sediment transport capacity T(r):

$$q_s(r) = T(r) = K_t(r) \cdot [q(r)]^m \cdot [sin \ \beta(r)]^n \tag{2}$$

where q(r) is the water flow rate (m^3 m^{-1} s^{-1}); $K_t(r)$ is the transportability coefficient, which is dependent on soil and cover; $\beta(r)$ is the slope (deg); m and n are parameter that vary according to type of flow and soil properties. According to [1], RUSLE R–, K–, C– and P–factors can be used into the USPED equation to incorporate the approximate impact of soil and cover and obtain a relative estimate of net erosion/deposition. Thus, USPED assumes that sediment flow at sediment transport capacity is [28], [1]:

$$T = R \cdot K \cdot C \cdot P \cdot L \cdot S \tag{3}$$

$$L \cdot S = A^m \cdot (sin\beta)^n \tag{4}$$

where A is the upslope contributing area per unit contour width (m), m and n are parameters depending on prevailing rill or sheet erosion. Thus, to incorporate the impact of flow convergence, λ was replaced by the upslope contributing area per unit width A. Then the net erosion/deposition rate (ED) is estimated as:

$$ED = \nabla (T \cdot s) = \frac{d}{dx}(T \cdot cos\alpha) + \frac{d}{dy}(T \cdot sin\alpha) \tag{5}$$

where s is the unit vector in the steepest slope direction, and α is the aspect of the terrain surface (deg).

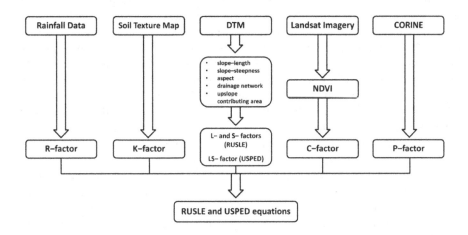

Fig. 2. RUSLE and USPED factors generation flowchart

According to [30], the rainfall–runoff erosivity factor (R) represents the input that drives the sheet and rill erosion process, and it is considered one of the most influencing factors in soil loss. In the present study, daily rainfall dataset of 20 years (1991–2011) for 20 rain–gauge stations were collected from the Regional Agency for Environmental Protection of Basilicata (ARPAB) and from the Agriculture Development and Innovation Agency (ALSIA) and were used for the R–factor calculation. A number of formulas have been developed to estimate the R–factor [14], [31,32,33,34,35,36,37,38], [7]. It was adopted the Modified Fournier Index (MFI) with the approach suggested by [38] for the area of Sicily (Italy):

$$MFI = \frac{1}{N}\sum_{j=1}^{N}\sum_{i=1}^{12}\frac{P_{ij}^2}{P_j} \qquad (6)$$

$$R = 0.6120 MFI^{1.56} \qquad (7)$$

where P_{ij} is rainfall depth for i–th month and j–th year (mm) and P_j is the rainfall total for j–th year (mm). According to [38], [7], the MFI is highly correlated with the rainfall erosivity and, thus, with the soil erosion. It was chosen for taking into account the monthly rainfall distribution during each year for a period of N years. Dixon's Q test was performed in order to identify and reject of possible outliers, which were not found. Subsequently, a continue surface was produced from the point data of the 20 rain–gauge stations with a tension spline function. Compared with other interpolation functions, it proved to be the most effective for producing the R–factor map. The rainfall erosivity factor for the 20–year period, calculated on a monthly bases, ranges between 450 MJ mm ha^{-1} h^{-1} yr^{-1} and 1360 MJ mm ha^{-1} h^{-1} yr^{-1}.

The K–factor is an empirical measured of the inherent erodibility of a soil as affected by its properties [30], [7]. According to [14], a soil type becomes less erodible with the decrease in silt fraction, regardless of whether the corresponding increase is in the sand fraction or in the clay fraction. In the present study, the K–factor was estimated by means of the soil texture map provided by the Basilicata Region at a scale of 1:250,000. The soil texture map categorizes the soil formations into 5 different major classes: coarse sandy loam, sandy loam, silty clay loam, silty clay, and silty loam. According to [14], [2], [39], the estimated K values vary from 0.08 ton h ha^{-1} MJ^{-1} mm^{-1} for coarse sandy loam, 0.12 ton h ha^{-1} MJ^{-1} mm^{-1} for sandy loam, 0.26 ton h ha^{-1} MJ^{-1} mm^{-1} for silty clay loam, 0.28 ton h ha^{-1} MJ^{-1} mm^{-1} for silty clay, and 0.30 ton h ha^{-1} MJ^{-1} mm^{-1} for silty loam.

The L– and S–factors are related to the slope length and the slope gradient, and represent the role of topography on soil erosion. As slope gets steeper, the higher is the velocity of overland flow, thus increasing the shear stress on the soil particles. Furthermore, as slope length increases, the overland flow and flow velocity also progressively increase, leading to greater erosion forces applied to the soil surfaces [19], [39]. In equation (1), L– and S– calculations are based on the following expressions of [40]:

$$L = \left(\frac{\lambda}{22.13}\right)^m \qquad (8)$$

$$m = \frac{\beta}{1+\beta} \qquad (9)$$

$$\beta = (\sin\theta)/[3\cdot(\sin\theta)^{0.8}+0.56) \qquad (10)$$

$$S = 10.8\cdot\sin\theta + 0.03\,\theta < 9\% \qquad (11)$$

$$S = 16.8\cdot\sin\theta - 0.5\,\theta \geq 9\% \qquad (12)$$

where λ the slope length (m), m is a variable length–slope exponent, β is a factor that varies with slope gradient, and θ is the slope angle (deg).

The C–factor represents the effects of the presence of ground, tree and grass covers on reducing soil loss in natural and agricultural contexts. As vegetation cover increases, soil loss decreases. The procedure used for estimating the C–factor involves the use of the Normalized Difference Vegetation Index (NDVI). The NDVI represents an effective remote sensing derived indicator of the green vegetation distribution by determining spectral reflectance difference between near infrared (NIR) and red (Red) bands of the electromagnetic spectrum [41], with values ranging between –1.0 and +1.0. Since the C–factor used in RUSLE and USPED reflects the role of vegetation cover on annual soil erosion loss, a single C–factor layer calculated from a single remotely–sensed image cannot fully represent the overall condition due to the vegetation phenology or the agricultural practices. In addition, several disturbing factors can influence the Top of Atmosphere (ToA) reflectance values. For those reasons, the overall C–factor distribution able to better represent the overall yearly conditions was calculated from two available images, which were selected on the base of the phenological and water seasonality of species. Two Landsat 7 ETM+ images were considered, acquired at 03/01/2001 (winter, w) and 13/07/2001 (summer, s), and with a spatial resolution of 30 m. Those images consisted of band 1: 0.45–0.52 mm, band 2: 0.52–0.60 mm, band 3: 0.63–0.69 mm, band 4: 0.76–0.90 mm, band 5: 1.55–1.75 mm and band 7: 2.08–2.35 mm [42]. Each band was radiometrically calibrated and pixel values were converted in ToA reflectance. For a Landsat 7 ETM+ image the NDVI is computed utilizing band 3 (Red) and band 4 (NIR) as follows:

$$NDVI = \frac{NIR-RED}{NIR+RED} \tag{13}$$

After the production of the NDVI for each image, the following formula was used to generate the C–factor surface from NDVI values [43]:

$$C = 0.431 - 0.805 \cdot NDVI \tag{14}$$

The relation was obtained comparing several NDVI profiles derived from Landsat 7 ETM+ imagery with *in situ* data relative to (semi–) natural vegetation [43]. C–factor values range between 0 and 1.

The support practice factor (P) takes into account the impact of different support practices on the average annual erosion [17]. The value of P–factor varies between 0 and 1. Commonly, P–factor is assigned based on land cover distribution [20], [2]. In this study, the P–factor was estimated by means of the CORINE land cover map from 2006 provided by the National Institute for Environmental Protection and Research of Italy (ISPRA) at a scale of 1:100,000, which is available online on http://www. sinanet.isprambiente.it/. Areas with no conservation practices were assigned with a P–factor equal to 1. On the other hand, areas considered to be less prone to erosion were assigned lower values according to expert opinion (Tab. 2). Thus, higher P values indicate poorer performance in soil conservation.

Each factor involved into the equations (1) and (3), (5) represents a specific thematic layer over the study area. Those layers were generated with a spatial resolution of 20 m, in accordance with the DTM, and then multiplied. The results obtained by RUSLE and USPED represent the annual soil loss (Fig. 3) and the net erosion/deposition (Fig. 4), respectively.

Table 2. Land use/land cover in the Basento watershed with P values

Land use/land cover classes	P value
Built–up land; water bodies	P = 0
Coniferous forest	P = 0.1
Wetlands; sclerophyllous vegetation	P = 0.4
Broad–leaved forest; mixed forest; transitional woodland–shrub	P = 0.5
Permanent crops	P = 0.8
Heterogeneous agricultural areas	P = 0.9
Arable lands; pastures; open spaces with little or no vegetation	P = 1

4   Results and Discussions

RUSLE and USPED have the ability to predict long–term average annual rate of soil loss and net erosion/deposition at catchment scale, using topography, rainfall, soil type, cover and management practice. Potential annual soil loss and net erosion/deposition rates were estimated from the input factors (L, S, R, K, C, P) by means of an overlay analysis of the thematic layers, which represent the geo–environmental scenario of the study area, in a GIS.

RUSLE outputs were classified using the following 4 classes (Tab. 3): 1. Extreme erosion; 2. High erosion; 3. Moderate erosion; 4. Low erosion. USPED results were classified according to the following 8 classes (Tab. 4): 1. Extreme erosion; 2. High erosion; 3. Moderate erosion; 4. Low erosion; 5. Low deposition; 6. Moderate deposition; 7. High deposition; 8. Extreme deposition.

Fig. 3 shows the spatial distribution of soil loss after the application of RUSLE at the Basento river basin. According to the RUSLE estimation (Tab. 3), about the 44.70% of the basin is at low risk of erosion, about the 29.25% is at moderate risk of erosion, about the 24.00% of the basin is at high risk of erosion and the 2.05% is at extreme risk of erosion.

As stated in Section 3, RUSLE is a detachment capacity limited model and it is not able to predict soil deposition. Conversely, the USPED is a transport capacity limited model, and it does not only predict soil erosion rates, but also the spatial distribution of sediment deposition at watershed scale for a constant state overland flow with uniform rainfall–excess conditions. Fig. 4 shows the spatial distribution of net soil erosion/deposition after the application of the USPED at the study area. According to the USPED estimation (Tab. 4), about 25.80% of the basin is stable (low erosion and low deposition), about 12,5% is at moderate risk of erosion, about 18.60% is at high risk of erosion, 10.75% is at extreme risk of erosion, about the 9.40% experiences moderate deposition, about the 12.65% experiences high deposition and 10.3% experiences extreme deposition. Moderate to low soil erosion occurred into the plain parts of the basin, while high to extreme soil erosion occurred into the hillslope sectors.

Table 3. RUSLE estimation

RUSLE classes (ton ha^{-1} yr^{-1})	% surface
Low erosion (0 ÷ -10)	44.70
Moderate erosion (-10 ÷ -50)	29.25
High erosion (-50 ÷ -200)	24.00
Extreme erosion (> -200)	2.05

Table 4. USPED estimation

USPED classes (ton ha^{-1} yr^{-1})	% surface
Extreme deposition (> 200)	10.30
High deposition (50 ÷ 200)	12.65
Moderate deposition (10 ÷ 50)	9.40
Stable (-10 ÷ 10)	25.80
Moderate erosion (-10 ÷ -50)	12.50
High erosion (-50 ÷ -200)	18.60
Extreme erosion (> -200)	10.75

Due to the lack of measured data of soil erosion in the study area, the application of both models was useful to compare the erosion rates. In particular, USPED can be considered an important tool in estimating the deposition process, especially at basin scale, to further understanding the complex interaction of involved variables, and accurately predicting the net erosion phenomenon. Furthermore, to validate the results with measured data, RUSLE and USPED models were tested in the sub–basin sub-tended by the Camastra Dam (Fig. 1), and compared with the dam silting value supplied by an interregional authority responsible for its management [44]. As shown in Tab. 5, the positive match of those values proved that RUSLE and USPED are relia-ble in quantifying the soil erosion at watershed scale. In particular, estimated rates obtained by USPED are greater than those achieved by RUSLE as USPED replaces the local slope length λ, considered in RUSLE, by the upslope contributing area per unit contour width A, which better reflects the impact of convergence/divergence of flow on increased erosion.

The studies revealed the important impact of each factor into the process modeling, showing that topography and rainfall regime were the main drivers of the phenome-non in the upper part of the basin. Despite, in topographically gentle areas, vegetation covers and soil types were more relevant factors than the others.

Table 5. Comparison between RUSLE and USPED computed soil erosion rates integrated over the Camastra sub–basin surface and Camastra dam silting value provided by an inter–regional authority responsible for its management.

Basin	Measured silting data [44]	RUSLE	USPED (Erosion)
Camastra	ton yr^{-1} 665,000	ton yr^{-1} ~520,500	ton yr^{-1} ~666,650

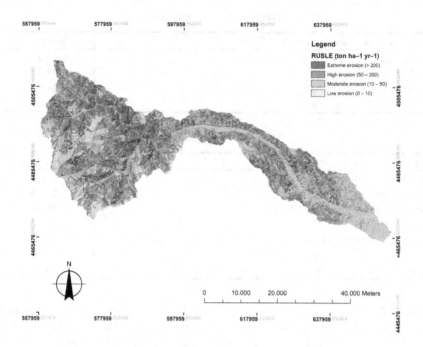

Fig. 3. Spatial distribution of soil loss rate estimated with RUSLE

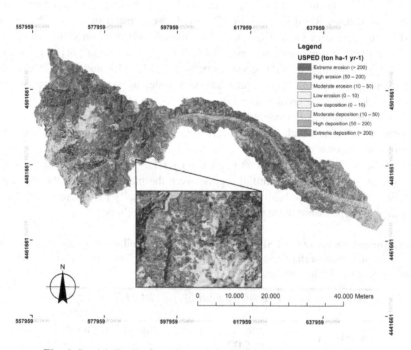

Fig. 4. Spatial distribution of net erosion/deposition rates with USPED

5   Conclusions

Estimation of soil erosion for a large river basin can often being a difficult task. This study demonstrated the feasibility to apply the RUSLE and USPED models to estimate spatial distribution of soil loss and net soil erosion/deposition at a watershed scale. Spatially soil erosion distribution highlighted that rainfall and slope play an important role in soil loss in the upper sector of the Basento basin, whilst vegetation covers and soil properties are dominant in the plain areas.

The validation of both RUSLE and USPED outputs was not possible over the entire Basento basin due to the lack of measured data. Thus, the rates of soil erosion were verified at the sub–basin scale by comparison with the Camastra reservoir silting value. Both models showed good accuracy and reliability of the estimations.

The application of those methods provided useful data to further understanding the soil erosion phenomenon, in which different variables such as climate conditions, topography, land use and land cover, etc., contribute in determining the various interacting processes of land degradation.

Moreover, the results obtained in this study can be a useful reference to further identifying prone areas, soil conservation practices, environmental planning and management of the Basento river basin.

References

1. Mitasova, H., Barton, M., Ullah, I., Hofierka, J., Harmon, R.S.: 3.9 GIS-Based Soil Erosion Modeling. In: Shroder, J., Bishop, M.P. (eds.) Treatise on Geomorphology, vol. 3, pp. 228–258. Academic Press, San Diego (2013)
2. Prasannakumar, V., Vijith, H., Abinod, S., Geetha, N.: Estimation of soil erosion risk within a small mountainous sub-watershed in Kerala, India, using Revised Universal Soil Loss Equation (RUSLE) and geo-information technology. Geoscience Frontiers 3, 209–215 (2012)
3. Guo, J., Niu, T., Rahimy, P., Wang, F., Zhao, H., Zhang, J.: Assessment of soil erosion susceptibility using empirical modeling. Acta Meteorol. Sin. 27, 98–109 (2013)
4. van der Knijff, J.M., Jones, R.J.A., Montanarella, L.: Soil erosion risk assessment in Italy. European Soil Bureau, vol. EUR 19022 EN. Office for Official Publications of the European Communities, Luxembourg (1999)
5. Capolongo, D., Pennetta, L., Piccarreta, M., Fallacara, G., Boenzi, F.: Spatial and temporal variations in soil erosion and deposition due to land-levelling in a semi-arid area of Basilicata (Southern Italy). Earth Surf. Process. Landforms 33, 364–379 (2008)
6. Terranova, O., Antronico, L., Coscarelli, R., Iaquinta, P.: Soil erosion risk scenarios in the Mediterranean environment using RUSLE and GIS: An application model for Calabria (southern Italy). Geomorphology 112, 228–245 (2009)
7. Kouli, M., Soupios, P., Vallianatos, F.: Soil erosion prediction using the Revised Universal Soil Loss Equation (RUSLE) in a GIS framework, Chania, Northwestern Crete, Greece. Environ. Geol. 57, 483–497 (2008)
8. Morgan, R.P.C.: Soil erosion in the northern countries of the European Community. In: EIW Workshop Elaboration of a Framework of a Code of Good Agricultural Practices, Brussels, pp. 21–22 (1992)

9. Lal, R.: Soil degradation by erosion. L. Degrad. Dev. 12, 519–539 (2001)
10. Merritt, W.S., Letcher, R.A., Jakeman, A.J.: A review of erosion and sediment transport models. Environ. Model. Softw. 18, 761–799 (2003)
11. Aksoy, H., Kavvas, M.L.: A review of hillslope and watershed scale erosion and sediment transport models. Catena 64, 247–271 (2005)
12. Vrieling, A.: Satellite remote sensing for water erosion assessment: A review. Catena 65, 2–18 (2006)
13. Bhattarai, R., Dutta, D.: Estimation of Soil Erosion and Sediment Yield Using GIS at Catchment Scale. Water Resour. Manag. 21, 1635–1647 (2006)
14. Wischmeier, W.H., Smith, D.D.: Predicting rainfall erosion losses: A guide to conservation planning. Agriculture Handbook, vol. 537. U.S. Dept. of Agriculture, Washington, DC (1978)
15. Williams, J.: Sediment routing for agricultural watersheds. JAWRA Journal of the American Water Resources Association 11, 965–974 (1975)
16. Mitasova, H., Hofierka, J., Zlocha, M., Iverson, L.R.: Modelling topographic potential for erosion and deposition using GIS. Int. J. Geogr. Inf. Syst. 10, 629–641 (1996)
17. Renard, K.G., Foster, G.R., Weesies, G.A., McCool, D.K., Yoder, D.C.: Predicting soil erosion by water: A guide to conservation planning with the revised universal soil loss equation (RUSLE). Agriculture Handbook, vol. 703. U.S. Dept. of Agriculture, Washington, DC (1997)
18. Mitas, L., Mitasova, H.: Distributed soil erosion simulation for effective erosion prevention. Water Resour. Res. 34, 505–516 (1998)
19. Ranzi, R., Le, T.H., Rulli, M.C.: A RUSLE approach to model suspended sediment load in the Lo river (Vietnam): Effects of reservoirs and land use changes. J. Hydrol. 29, 422–423 (2012)
20. Huang, J.Q., Lu, D.S., Li, J., Wu, J.S., Chen, S.Q., Zhao, W.M., Ge, H.L., Huang, X.Z., Yan, X.J.: Integration of Remote Sensing and GIS for Evaluating Soil Erosion Risk in Northwestern Zhejiang, China. Photogramm. Eng. Remote Sens. 78, 935–946 (2012)
21. Aiello, A., Canora, F., Pasquariello, G., Spilotro, G.: Shoreline variations and coastal dynamics: A space–time data analysis of the Jonian littoral, Italy. Estuar. Coast. Shelf S. 129, 124–135 (2013)
22. Tropeano, M., Sabato, L., Pieri, P.: Filling and cannibalization of a foredeep: The Bradanic Trough (Southern Italy). In: Jones, S.J., Frostick, L.E. (eds.) Sediment Flux to Basins: Causes, Controls and Consequences, Spec. Publ., vol. 191, pp. 55–79. Geological Society, London (2002)
23. Ciaranfi, N., Ghisetti, F., Guida, M., Iaccarino, G., Lambiase, S., Pieri, P., Rapisardi, L., Ricchetti, G., Torre, M., Tortorici, L.: Carta neotettonica dell'Italia meridionale. Progetto Finalizzato Geodinamica, vol. 515. CNR, Rome (1983)
24. Boenzi, F., Digennaro, M., Pennetta, L.: I terrazzi della valle del Basento (Basilicata). Rivista Geografica Italiana 85, 396–418 (1978)
25. Cotecchia, V., Ricchetti, E., Polemio, M.: Studio delle caratteristiche morfoevolutive del fondovalle del F. Basento fra Pisticci e la foce, finalizzato all'ottimizzazione dell'intervento antropico. Mem. Soc. Geol. It. 47, 587–608 (1991)
26. Boenzi, F., Capolongo, D., Gallicchio, S., Di Pinto, G.: Morphostructure of the Lucania Apennines front between the Basento and Salandrella rivers (Southern Italy). J. Maps 10, 478–486 (2014)
27. Clarke, M.L., Rendell, H.M.: The impact of the farming practice of remodellinghillslope topography on badland morphology and soil erosion processes. Catena 40, 229–250 (2000)

28. Garcia Rodriguez, J.L., Gimenez Suarez, M.C.: Methodology for estimating the topographic factor LS of RUSLE3D and USPED using GIS. Geomorphology 175-176, 98–106 (2012)
29. Moore, I., Burch, G.: Modelling erosion and deposition: Topographic effects. Transactions of the ASAE 29, 1624–1640 (1986)
30. Renard, K.G.: Computerized Calculations for Conservation Planning. Agr. Eng. 73, 16–17 (1992)
31. Arnoldus, H.M.J.: An approximation of the rainfall factor in the Universal Soil Loss Equation. In: De Boodt, M., Gabriels, D. (eds.) Assessment of Erosion, pp. 127–132. Wiley, Chichester (1980)
32. Renard, K.G., Freimund, J.R.: Using Monthly Precipitation Data to Estimate the R-Factor in the Revised USLE. J. Hydrol. 157, 287–306 (1994)
33. Yu, B., Rosewell, C.J.: An assessment of a daily rainfall erosivity model for New South Wales. Aust. J. Soil Res. 34, 139–152 (1996)
34. Yu, B., Rosewell, C.J.: Rainfall erosivity estimation using daily rainfall amounts for South Australia. Aust. J. Soil Res. 34, 721–733 (1996)
35. Yu, B., Rosewell, C.J.: A robust estimator of the R-factor for the universal soil loss equation. Transactions of the ASAE 39, 559–561 (1996)
36. Ferro, V., Porto, P.: A comparative study of rainfall erosivity estimation for southern Italy and southeastern Australia. Hydrolog. Sci. J. 44, 3–24 (1999)
37. Angulo-Martínez, M., Beguería, S.: Estimating rainfall erosivity from daily precipitation records: A comparison among methods using data from the Ebro Basin (NE Spain). J. Hydrol. 379, 111–121 (2009)
38. Ferro, V., Giordano, G., Iovino, M.: Isoerosivity and Erosion Risk Map for Sicily. Hydrolog. Sci. J. 36, 549–564 (1991)
39. Alexakis, D.D., Hadjimitsis, D.G., Agapiou, A.: Integrated use of remote sensing, GIS and precipitation data for the assessment of soil erosion rate in the catchment area of "Yialias" in Cyprus. Atmos. Res. 131, 108–124 (2013)
40. McCool, D.K., Foster, G.R., Mutchler, C.K., Meyer, L.D.: Revised slope length factor for the Universal Soil Loss Equation. Transactions of the ASAE 32, 1571–1576 (1989)
41. Rouse, J.W., Haas, R.H., Schell, J.A., Deering, D.W.: Monitoring vegetation systems in the Great Plains with ERTS. In: Stanley, C., Mercanti, E.P., Becker, M.A. (eds.) Proceedings of the Third Earth Resources Technology Satellite Symposium, NASA Special Publication, vol. 351, pp. 309–317. NASA, Washington, DC (1973)
42. Chander, G., Markham, B.L., Helder, D.L.: Summary of current radiometric calibration coefficients for Landsat MSS, TM, ETM+, and EO-1 ALI sensors. Remote Sens. Environ. 113, 893–903 (2009)
43. de Jong, S.M., Brouwer, L.C., Riezebos, H.T.: Erosion hazard assessment in the Peyne catchment, France. In: Working Paper DeMon-2 Project, Dept. of Physical Geography. Utrecht University (1998)
44. Regione Basilicata: Stato conoscitivo preliminare e primi strumenti operativi del Piano Regionale di Tutela delle Acque in Basilicata (PRTA/Bas). In: Bollettino Ufficiale della Regione Basilicata, Potenza, vol. 25 (2004)

Assessing Groundwater Vulnerability to Pollution through the DRASTIC Method

A GIS Open Source Application

Lia Duarte[1,2], Ana Cláudia Teodoro[1,2], José Alberto Gonçalves[1,3],
António J. Guerner Dias[1,4], and Jorge Espinha Marques[1,4]

[1] Department of Geosciences, Environment and Land Planning,
Faculty of Sciences, University of Porto, Porto, Portugal
[2] Geo-Space Science Research Center, Faculty of Sciences,
University of Porto, Porto, Portugal
[3] Interdisciplinary Centre of Marine and Environmental Research, University of Porto
[4] Geology Centre of the University of Porto, Porto, Portugal
liaduarte@fc.up.pt

Abstract. Groundwater pollution is a constant concern. Geographical Information Systems (GIS) provide useful tools to manipulate the variables that can be used prevent/minimize these issues. This article presents the development of a tool to produce maps under a GIS open source environment. The application was developed through Quantum GIS (QGIS) software. The tool is developed based on DRASTIC method and incorporates some procedures under a plugin. The Drastic method comprises several steps and several maps: Depth to groundwater (D), Net Recharge (R), Aquifer media (A), Soil media (S), Topography (T), Impact of Vadose Zone (I) and Hydraulic Conductivity (C). These maps are produced according to indexes defined by Aller et al (1987), [2]. One of the main advantage of this application is the easiness to use. The user can generate the maps according to his perception regarding field conditions. The application is free for the institution or user and presents a great contribution to predict the intrinsic vulnerability pollution through GIS open source.

Keywords: groundwater contamination, drastic map, open source software, Quantum GIS.

1 Introduction

1.1 Groundwater Pollution Vulnerability

Groundwater is a natural resource of great economic importance since it is an essential input for many human activities. In fact, groundwater is used for drinking and domestic water supply, agriculture, industry, balneotherapy (mineral water), geothermal energy, among other [7], [10].

Most of the world populations rely on fresh groundwater as the main source of domestic and drinking water. Due to the pollution of surface water resources the demand for groundwater is growing every year [4], [10].

B. Murgante et al. (Eds.): ICCSA 2014, Part IV, LNCS 8582, pp. 386–400, 2014.
© Springer International Publishing Switzerland 2014

Foster (1987), employs the expression "aquifer pollution vulnerability" to represent the intrinsic characteristics which determine the sensitivity of the groundwater system to being adversely affected by an imposed pollutant load [8]. This author also differentiates pollution vulnerability from pollution risk, because the last one results both from the natural vulnerability of the aquifer and from the pollution load that is, or will be, applied on the subsurface environment as a result of human activity. Consequently, an aquifer may have high vulnerability but no pollution risk, due to the absence of significant pollution loading, and vice versa. NAP (1993), defines groundwater vulnerability to pollution as the tendency or likelihood for pollutants to reach a specified position in the groundwater system after introduction at some location above the uppermost aquifer [13].

Given the importance of sustainable aquifer management, groundwater vulnerability assessment and mapping has become a central subject. Groundwater pollution involves many factors, for example, geological, geomorphological, climatic and biological conditions. Some of these factors are particularly important for vulnerability mapping, such as geological materials, landforms, unsaturated zone, aquifer hydraulic features and land use. Hence, it will be necessary create several models with different vulnerability indexes, different statistical methods, process-based methods and overlay and index methods [23].

Geographic Information Systems (GIS) applications are often used to estimate groundwater vulnerability due to the easiness and efficiency to manipulate, analyze and incorporate geographic data, such geologic and hydrogeological data [23]. DRASTIC index is one of the methods used to estimate groundwater vulnerability [1]. Some GIS software was already used to analyze DRASTIC index. For instance, ArcGIS® software was used to calculate DRASTIC index, in a case of study in North Dakota, USA [11]. Also in Isparta, Turkey [22] and in Portugal, ArcGIS® software was used to mapping the vulnerability in thermal waters of Nisa [12]. A Quantum GIS plugin was already developed, in the scope of a research, in order to evaluate the groundwater vulnerability with DRASTIC method [5]. This research allowed producing a map with a single window where the user defines the input rasters and respective weights. However, is not possible to change the indexes of the model. This possibility is relevant so the user may evaluate and analyze the results obtained.

The main objective of this work was to create an Open Source GIS application for assessing groundwater vulnerability to pollution through the DRASTIC method. The developed application presents some advantages in opposition to the maps creation under various steps and tools, through different GIS software. With this application, the user can obtain the results in a few minutes and can assess in real time the consequences of applying different indexes and weights. The user can also change the values of the indexes and their corresponding weights by trial and error until the adequate/expected result is reached. This tool is provided as free open source software, with permission for other users to change the source code and improve it, according to user needs. This is an important advantage of the open source concept.

1.2   DRASTIC Index Parameters

DRASTIC index is composed by seven parameters and it is a result of weighted average of these seven values [2]. These parameters can be defined as hydrogeological indicators and are described as Depth to Goundwater (D), Net Recharge (R), Aquifer Media (A), Soil Media (S), Topography (T), Impact of Vadose Zone (I) and Hydraulic Conductivity (C). The DRASTIC index is calculated according to equation (1).

$$DRASTIC = Dp \times Di + Rp \times Ri + Ap \times Ai + Sp \times Si + Tp \times Ti + Ip \times Ii + Cp \times Ci \quad (1)$$

Where i and p correspond to the index and weight for each parameter, respectively. Each parameter is reclassified by the indexes and multiplied by the weight. The final value of the index is obtained from the sum of these results. Each parameter is divided in different classes, and an index is assigned to each class.

Depth to Groundwater
Depth to groundwater parameter represents the depth from the ground surface to the water table [1]. The material thickness that a pollutant has to go through to reach the aquifer is conditional to aquifer depth [15]. Table 1 presents the index assigned to the parameters according to Aller et al. (1987), [2]. A deeper water table implies a less vulnerable aquifer [1]. Usually the depth to the water table is measured *in situ* in wells or piezometers.

Table 1. Depth parameter indexes (according to Aller et al (1987) [2])

Depth (meters)	< 1,5	1,5 – 4,6	4,6 – 9,1	9,1 – 15,2	15,2 – 22,9	22,9 – 30,5	>30,5
Index	10	9	7	5	3	2	1

Net Recharge
The net recharge presents the quantity of water that is infiltrated in the ground and reaches the water table. A high recharge value, corresponds a high groundwater level of pollution, because the recharge transports the pollutants [1]. The indexes and intervals were used according to Aller et al. (1987) [2]. Recharge values are used in millimeters per year. Usually the precipitation data is obtained from meteorological stations. Other type of information such as surface runoff and evapotranspiration combined with the precipitation could also be used.

Aquifer Media
The aquifer media is characterized by the aquifer capacity to attenuate the pollutants effects. For instance, a rock more fissured has less attenuation capacity [16]. This information is based on geological description. The descriptions are defined according to each study area. The indexes, defined according to Aller et al. (1987), are presented in the Table 2 [2]. The index column contain intervals of values that can be assign to the aquifer type respectively and the typical index presents the usual index value used in DRASTIC method.

Table 2. Aquifer parameter indexes (according to Aller et al (1987) [2])

Aquifer	Index	Typical index
Shale, argillite	1-3	2
Igneous/ metamorphic rock	2-5	3
Altered igneous/ metamorphic rock	3-5	4
Till	4-6	5
Sandstone, limestone and shale	5-9	6
Sandstone	4-9	6
Limestone	4-9	6
Sand and ballast	4-9	8
Basalt	2-10	9
Karstified Limestone	9-10	10

Soil Media
The soil type has an important role, because it controls the amount of recharge that can infiltrate and the movement of pollutants in the vadose zone [1]. The main reason to his importance is its potential to attenuate pollution [15]. Aller et al (1987) defined the indexes that were used by default in this application [2]. Such as aquifer media, the soil information is related with geological information.

Topography
Topography is related to slope. In this case, the slope determines the probability of a pollutant to remain at the surface without infiltrating [15]. Areas where the slope is low tend to retain the water for a long time, causing infiltration and recharge and a potential to pass contaminants [1]. Slope is calculated in percentage from the Digital Terrain Model (DTM). The indexes assign to slope intervals were used according to Aller et al. (1987) [2].

Table 3. Impact of the Vadose Zone parameter indexes (according to Aller et al (1987) [2])

Vadose Zone	Index	Typical index
Confined layer	1	1
Clay/ Silt	2-6	3
Shale clay, mudstone	2-5	3
Limestone	2-7	6
Sandstone	4-8	6
Sandstone, limestone and shale	4-8	6
Sand and ballast with significant amount of silt and clay	4-8	6
Igneous /metamorphic rock	2-8	4
Sand and ballast	6-9	8
Basalt	2-10	9
Limestone	8-10	10

Impact of the Vadose Zone

Vadose zone has a great impact in groundwater vulnerability. This parameter controls the communication between the contaminant with the saturated zone [1]. The vadose zone material determines the contact time with the pollutant and several phenomena happens in this period, such as biodegradation, neutralization, chemical reactions and dispersion among others [15]. Such as aquifer and soil, vadose zone information is obtained through geological maps. Table 3 presents the parameter indexes.

Hydraulic Conductivity

The hydraulic conductivity values are obtained *in situ* with several tests and correspond to the ability of the aquifer to transmit water [16]. This capacity controls the groundwater flow. An aquifer with high hydraulic conductivity is vulnerable to substantial contamination [1]. The hydraulic conductivity is expressed in centimeters per second (cm/s). The parameter indexes were adopted by Aller et al. (1987), [2].

Parameters Weights

Weight values are assigned to all parameters reflecting their relative importance related to the groundwater vulnerability. After the indexes assignment, the final map is multiplied by the respective weight. These values were used, according to Aller et al (1987) and range from one and five (Table 4) [2].

Table 4. Parameter weights (according to Aller et al (1987) [2])

Parameter	D	R	A	S	T	I	C
Weight	5	4	3	2	1	5	3

2   Methodology

2.1   Open Source GIS Software

The Quantum GIS (QGIS) desktop software was used to develop the DRASTIC application due to its characteristics and many advantages. The project objectives and the authors experience and skills [25] were taking into account to choose the software. QGIS was started in 2002 with Gary Sherman and it is licensed under a GNU GPL license. QGIS is an open source software, so it respects the Stallman four freedoms, freedom to run the program for any purpose, freedom to study how the program works and modified if you want so, freedom to redistribute copies and freedom to distribute copies of modified versions [24]. The main advantage of QGIS relies on the easiness and quickness on developing new plugins, using python language [18],[25]. QGIS presents their own APIs (Application Programming Interface), such as QGIS API, Gdal/ OGR API, PyQt4 API that were used to develop the application. These APIs have functions, classes and modules which interact with geographic information. QGIS is developed in C++ language and has extensions, named plugins, developed in Python language. The QGIS plugin developed in this work presents a great advantage due to its effectiveness and quickness to use. The drastic application was developed in Python language.

2.2   Programming Language

Python is an easy, simple, quick and efficient programming language. It is used in a variety of applications domains. Python is compared to other powerful languages, such as JAVA, Tcl, Perl, Scheme or Ruby. It runs in Windows, Linux/ Unix, OS2, Mac and others and it comes with documentation, tutorials and helpful documents that a new developer can use. Python language is licensed under an open source license. In order to increase the user productivity in the scope of this application, was used Python.

Plugin Development
QGIS plugins developing rules are different in the last QGIS version, (QGIS 2.0 *Dufour*) when compared to the last ones. According to structure model to develop plugins presented in QGIS 2.0 official page [19], some steps have to be followed:

1. *Idea*: Have an idea of the main purpose of the new plugin.
2. *Create files*: QGIS have a specific files structure that the user has to follow.
3. *Write code*: Write the code in the main program.
4. *Test*: Some tests have to be done during the developing.
5. *Publish*: In the end, the plugin can be distributes through QGIS official repository or through a personal repository.

The directory structure of a plugin must respect the following files.

```
PYTHON_PLUGINS_PATH/Plugin/
    __init__.py
    plugin.py
    metadata.txt
    resources.qrc
    recources.py
    form.ui
    form.py
```

The developed plugin comprises twenty one classes. The first one *(__init__.py)* is the starting point of the plugin and corresponds to the DRASTIC toolbar. The *plugin.py* presents the main code of the plugin. This code contains all the information about the plugin actions configurations, it creates the toolbar and adds the toolbar button and the menu item to the DRASTIC window, which is defined in *Drastic_window.py* program. In this application the code is named *DrasticToolbar.py*. The *form.ui* and the *form.py* are not necessarily just one program. It can be several programs with different names. In Drastic application eighteen classes were defined, nine with *form.py* and nine with *form.ui*. The *form.ui* points to the feature interface, windows, edit lines, combo boxes, buttons and many other interface options and the *form.py* presents the code associated to that feature. In these classes are presented the *Drastic_window.py* and *Ui_Drastic_window.py* which defines the application window. The first one sets the window title, DRASTIC, and sets the map canvas and the canvas color (in this case white). This class adds the menus to the window and connects the menus to the corresponding classes. The actions to file menu are also added in this class. The last one matches window interface and defines the plugin window, gives the resize, define a menu bar, a menu file, the menu DRASTIC and the menu Help. Programming in python to develop a QGIS plugin implies to use some API which contains a set of programming codes that allow to developed applications [25]. Drastic application import QGIS API [20], which contains QGIS Core and QGIS Gui modules, PyQt4 API [17], which includes the Core and Gui modules and Numpy API [14], to use in mathematical operations between raster files. Some libraries were used, such as GDAL (Geospatial Data Abstraction Library) and OGR Simple Feature Library, in order to manipulate vector information and raster files.

3   Results

3.1   DRASTIC Window

The application is composed by a window where the spatial objects can be presented. This window allows the user to analyze the result and modify the input parameters. The DRASTIC window is composed by a map canvas, a menu bar containing a *File* menu, the *DRASTIC* menu and the *Help* menu. The first one is composed by two buttons that allows a user to add a vector or a raster file (*Add Vector File* and *Add Raster File* respectively). The DRASTIC menu is composed by eight buttons, *Depth to Groundwater (D), Net Recharge (R), Aquifer Media (A), Soil Media (S), Topography (T), Impact of the Vadose Zone (I), Hydraulic Conductivity (C)* and *DRASTIC* as shown in figure 1.

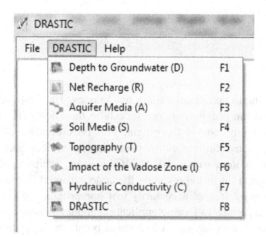

Fig. 1. DRASTIC menu

Each one of these buttons creates a raster file according to its name. The last button, *DRASTIC*, creates the DRASTIC index map through the seven raster files previously generated. The *Help* menu provides feature information and basic information about the application capabilities. All feature interfaces are composed by an input field, an attribute field, a cell size definition, a weight definition, the indexes table and an output directory field. It was necessary to import some *SEXTANTE* modules [19], such as GRASS through *v.surf.idw*, *v.to.rast.attrib* [9] and SAGA through *reclassify-gridvalues* [21]. To read the text fields of the attribute table some functions of PyQGIS API, such as *QgsVectorLayer (ogr* type), were used.

Depth (D) Creation
Depth feature could comprise different interpolation methods to estimate the depth groundwater map. However, in this application the interpolation method used is the *Inverse Distance Weighting* [3]. This feature allows interpolating data point with the depth values into a raster file. The input data is a point file. Afterwards, interpolation raster is reclassified according to the indexes definition. The weight value was defined as presented in Table 4. The multiplication result is saved in an output path defined by the user.

Net Recharge (R) Creation
The three input files are converted to raster files by using the first module. The feature is composed by two methods to determine the recharge map. The user can choose depending on user available files. If the user has precipitation, surface runoff and evapotranspiration data (mm/year), these files can be defined as input files. The procedure includes the conversion these files to raster files. The recharge is obtained by subtracting the precipitation raster to the surface runoff and by subtracting again the result obtained by the evapotranspiration raster. On the other hand, if the user only has the precipitation data, he can use this data as input file. Through this method, the recharge is defined as 10% of precipitation values. In both cases, the resulting recharge raster was reclassified with the indexes values defined before by the user. The user can modify the indexes in the feature table, by adding or removing classes. The weight values can be also modified. By default, the weight is defined according to Aller et al (1987) and presented in Table 4. In the end, this value is multiplied by the final raster, producing the recharge raster [2].

Aquifer media (A) Creation
The user has the possibility to modify the attribute table description. For example, if the vector file attribute table did not have the descriptions used by the model, the user has to adjust the descriptions to that model. Therefore, the feature assigns the respective indexes to each description. These indexes are added to the vector file attribute table in a new column. After, these values were used to convert the vector file to a raster file. The result is a raster file with the same vector file extent and cell size, where each cell has the respective index number correspondent to the description. The result is automatically added to the canvas.

Soil map (S) Creation

This feature considers the soil map with the description according to the application requirements. This feature works identically to the aquifer feature, it creates a new column in vector file attribute table with the assigned indexes. Then the vector file is converted to raster format. The result is saved in the output file and added automatically to the canvas.

Topography (T) Creation

Topography is related to terrain surface slope. As cited in Ferreira et al (1995) slope values between 0 and 2 percent provide the greatest opportunity for a pollutant to infiltrate [6]. The topography feature is composed by five main fields. In this feature, the weight by default is 1. The reclassified map is added to the map canvas.

Impact of the Vadose Zone (I) Creation

This feature creation is also identical to the aquifer feature. In this feature, the weight field is 5 by default, according to Aller et al (1987) [2]. A new column is created in the geological file attribute table, filled with the indexes values according to the geologic description. Then, the vector file is converted to raster format through the new column, resulting in an impact vadose zone map.

Hydraulic Conductivity (C) Creation

The hydraulic conductivity values are obtained through the user who can introduce the values in attribute table of the geological vector file. Usually, the hydraulic conductivity values are derived from the geologic information. The vector file is reclassified and converted to raster file according to the indexes values assigned. In the end, the result is added to the map canvas.

DRASTIC Map

The last feature, DRASTIC index, corresponds to the final map, which is composed by the seven maps created and explained above. As mentioned in section 1.2, the DRASTIC map is the result of a sum of the seven maps multiplied by the corresponding weights.

3.2   DRASTIC Map: A Case Study in the Region of Castelo Branco - Portugal

In order to assess the DRASTIC GIS tool a test was computed with real data. The study case described refers to a region in the Tagus basin, near to Castelo Branco city, Portugal (figure 2). The groundwater system of the study area covers approximately 1500 hectares with a maximum depth of 35 meters. The spatial resolution of the raster files for the analysis was chosen as 20 meters. This spatial resolution is sufficient for analyzing the spatial variation of the variables involved. All data are in ETRS89 PTTM06 coordinate system.

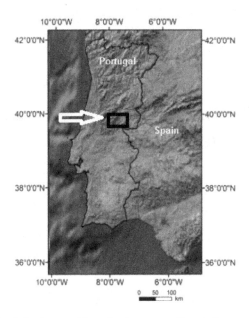

Fig. 2. Study area: Tagus basin, near to Castelo Branco city

The Depth to groundwater map was obtained from a point vector file with the depth values of study area. Through IDW interpolation, the Depth Groundwater map was created and reclassified which is present in figure 3. The recharge map was obtained through precipitation data (mm), using 10% of that. The Aquifer map was generated through a geological vector file. The classifications used in the vector file were described in the application tool with an index assign. Therefore, the user is free to modify the base model according to geological vector file nomenclature. The aquifer map was multiplied by the weight. The application creates a soil map reclassified by the indexes and then multiplies by the weight which, by default, is 2. The study area topography was obtained through slope map in percentage and multiplied by the weight, which, by default is 1. This map was created from geological map and the classes and respective indexes were modified according geological nomenclature, such as aquifer map. The result was multiplied by 5 (weight value by default). The Impact of Vadose Zone was created from geological data. The Hydraulic Conductivity is expressed in centimeters by seconds (cm/s) and was created based on conductivity values associated to the respective area. Finally, the DRASTIC map was created (figure 4).

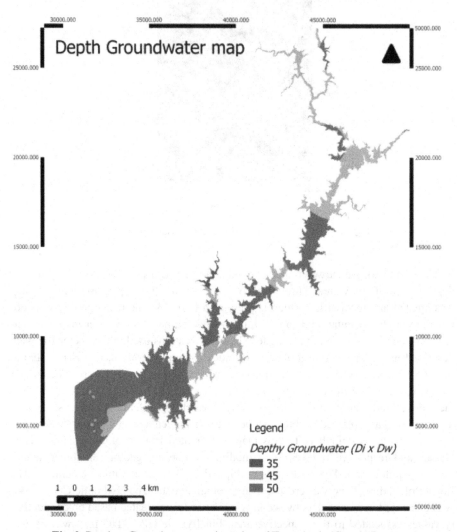

Fig. 3. Depth to Groundwater map in region of Tagus basin, Castelo Branco

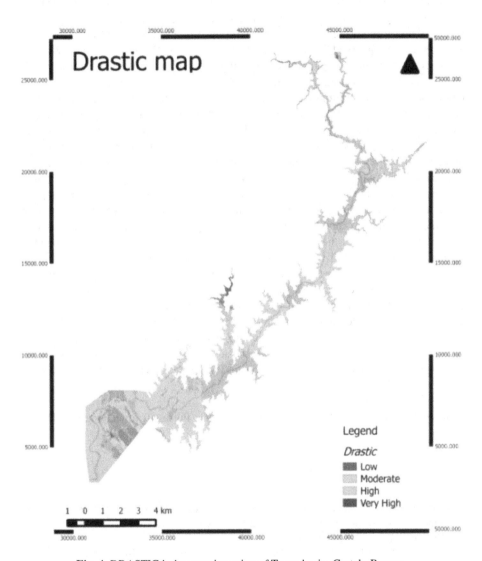

Fig. 4. DRASTIC index map in region of Tagus basin, Castelo Branco

4   Discussion and Conclusions

This objective of this research was to develop an open source application in QGIS environment which allows creating DRASTIC maps. To test the application was used a study case in Portugal. The DRASTIC index values vary between 100 and 181. These results suggest that, in this study area, the groundwater vulnerability can be moderate to high in some zones. The available data were not enough to evaluate more precisely the groundwater vulnerability. Despite this, the DRASTIC index was obtained based in data available based in Aller et al. (1987) DRASTIC model [2]. This application offers the advantage of generating several procedures, giving the possibility of analyzing and change the final DRASTIC map. Some features presents different methods to obtain the respective maps, such as net recharge parameter where the user can choose between two methods based on data available. This is a crucial advantage when few data are available. This application can be very helpful to the user due to the easiness and quickness to producing maps. The user has the advantage of change the indexes until get the better interpretation to his study area. The weights can be modified in order to obtain a better result. The fact of being developed through open source software makes it easier to obtain and use. In the future we attempt to create dynamic maps in QGIS environment to automatize DRASTIC maps creation with the possibility of modifying indexes and weights of each parameter and in real time assess the result of these changes. Also, the application will allow to change the aquifer, soil and impact of vadose zone descriptions according to other nomenclatures adopted. The DRASTIC index was created in the 80's of the last century. However, this index is worldwide used for the hydrologist's community. Nevertheless, we are currently developing an improved version of this application. The future application aspires to associate the aquifer description, defined by the user, with soil and impact of vadose zone parameter. In case of depth to groundwater parameter, the application needs to use an interpolation method. The basic model presented in this research considers the *Inverse Distance Weighting*, but in the future, other interpolation methods will be available.

Aknowledgements. We thank to EDP (Energias de Portugal) for having provide the test data.

References

1. Al-Rawabdeh, A.M., Al-Ansari, N.A., Al-Taani, A.A., Knutsson, S.: A GIS-Based Drastic Model for Assessing Aquifer Vulnerability in Amman-Zerqa Groundwater Basin. Jordan Engineering 5, 490–504 (2013)
2. Aller, L., Lehr, J.H., Petty, R., Bennet, T.: Drastic: A standardized system to evaluate graoundwater pollution potential using hydrogeologic settings. National Water Well Association, Worthington (1987)
3. Bonham-Carter, G.F.: Geographic Information Systems for Geoscientists, Modelling with GIS. Computer Method in Geosciences 13, 152–153 (1994)

4. Borevsky, B., Yazvin, L., Margat, L.: Importance of groundwater for water supply. In: Zektser, I.S., Everett, L.G. (eds.) Groundwater Resources of the World and Their Use. IHP-VI, Series on Groundwater, vol. (6), pp. 20–24. UNESCO, Paris (2004)
5. Coulibaly, N.: An Opensource GIS tool for Integrated Water Resources Manage-ment (IWRM) in a basin. Indian Institute of Technology Bombay, Computer Sciences Engineering Department (CSE), Laboratory Geospatial Information Science and Engineering Laboratory, GISE-Lab (2011)
6. Ferreira, J.P.C.L., Oliveira, M.M., Ciabatti, P.: Desenvolvimento de um inventário das águas subterrâneas de, vol. I. Laboratório Nacional de Engenharia Civil, Departamento de Hidráulica, Grupo de Investigação de Águas Subterrâneas (1995)
7. Fetter, C.W.: Applied Hydrogeology, 4th edn., p. 598. Prentice Hall, New Jersey (2001)
8. Foster, S.S.D.: Fundamental concepts in aquifer vulnerability, pollution risk and protection strategy. In: Van Duijvenbooden, W., van Waegeningh, H.G. (eds.) Proceedings and Information on Vulnerability of soil and Groundwater to Pollutants, vol. 38, pp. 69–86. TNO Committee on Hydrological Research (1987)
9. GRASS GIS, 2013, The world's leading Free GIS software, http://grass.osgeo.org/ (accessed January, 2014)
10. Job, A.C.: Groundwater Economics, p. 650. CRC Press, Boca Raton (2010)
11. Li, R., Merchant, J.W.: Modeling vulnerability of groundwater to pollution under future scenarios of climate change and biofuels-related land use change: A case study in North Dakota, vol. 447, pp. 32–45. Science of the Total Environment, USA (2013)
12. MotaPais, M.A.A., Antunes, I.M.R.H., Albuquerque, M.T.D.: Vulnerability mapping in a thermal zone, Portugal – a study based on DRASTIC index and GIS, Mul-tidisciplinary Research on Geographical Information in Europe and Beyond. In: Proceedings of the AGILE 2012 International Conference on Geographic Information Science, Avignon, April 24-27, pp. 978–990 (2012) ISBN: 978-90-816960-0-5
13. Ground water vulnerability assessment - contamination potential under conditions of un-certainty, p. 189. NAP (National Academy Press), Washington D.C (1993)
14. Numpy API, 2013, Numpy Reference, http://docs.scipy.org/doc/numpy/reference/ (accessed November 2013)
15. Oliveira, M.M.: Cartografia da vulnerabilidade à poluição das águas subterrâneas do con-celho de Montemor-o-novo utilizando o método Drastic, Laboratório Nacional de Engen-haria Civil, Departamento de Hidráulica, Grupo de Investigação de Águas Subterrâneas (2002)
16. Paralta, E.A., Oliveira, M.M., Batista, S.B., Francés, A.P., Ribeiro, L.F., Cerejeira, M.J.: Aplicação de SIG na avaliação da vulnerabilidade aquífera e cartografia da contaminação agrícola por pesticidas e nitratos na região do Ribatejo. A hidroinformática em Portugal (2001)
17. PyQt4 API, 2012, PyQt Class Reference, http://pyqt.sourceforge.net/Docs/PyQt4/classes.html (accessed November 2013)
18. Python, 2011, Python Programming Language, http://python.org/ (accessed February 2014)
19. QGIS, 2013, Quantum GIS Project, http://www.qgis.org/ (accessed November 2013)
20. QGIS API, 2013, Quantum GIS API Documentation, http://www.qgis.org/api/ (accessed November 2013)
21. SAGA, 2013, System for Automated Geoscientific Analyses, http://www.saga-gis.org/ (accessed January 2014)

22. Sener, E., Davraz, A.: Assessment of groundwater vulnerability based on a modified DRASTIC model, GIS and an analytic hierarchy process (AHP) method: The case of Egirdir Lake basin (Isparta, Turkey). Hydrogeology Journal 21, 701–714 (2013)

23. Shirazi, S.M., Imran, H.M., Akib, S., et al.: GIS-Based DRASTIC method for groundwater vulnerability assessment: A review. Journal of Risk Research 5(8), 991–1011 (2012), doi:10.1080/13669877.2012.686053

24. Stallman, P.: Why 'Open Source' misses the point of free software, GNU Operating System (2007), http://www.gnu.org/philosophy/open-sourcemisses-the-point.html (accessed, December 2013)

25. Teodoro, A.C., Duarte, L.: Forest Fire risk maps: A GIS open source application – a case study in Norwest of Portugal. International Journal of Geographic Information Science 27(4), 699–720 (2013)

A New Data-Driven Approach to Forecast Freight Transport Demand

Massimiliano Petri[1,2], Giovanni Fusco[3,4], and Antonio Pratelli[5]

[1] Tages s.c., Pisa, Italy
[2] University of Pisa, Pisa, Italy
petri@tages.it, m.petri@ing.unipi.it
[3] Université de Nice Sophia Antipolis, UFR Espaces et Cultures
[4] Centre National de la Recherche Scientifique, UMR 7300 ESPACE
giovanni.fusco@unice.fr
[5] University of Pisa-Dept. Civil and Industrial Engineering, Pisa, Italy
a.pratelli@ing.unipi.it

Abstract. Transport modeling, in general, and freight transport modeling, in particular, are becoming important tools for investigating the effects of investments and policies. Freight demand forecasting models are still in an experimentation and evolution stage. Nevertheless, some recent European projects, like Transtools or ETIS/ETIS Plus, have developed a unique modeling and data framework for freight forecast at large scale so to avoid data availability and modeling problems.

Despite this, projects that had multi-million Euros funding, using these modeling frameworks, have provided very different results for the same forecasting areas and years, giving rise to serious doubts about the results quality, especially in relation to their cost and development time. Moreover, many of these models are purely deterministic.

The project here developed tries to overcome the above-mentioned problems with a new easy-to-implement freight demand forecasting method based on Bayesian Networks using European official and available data. The method is applied to the Sixth European Freight Corridor.

Keywords: Freight Demand Model, Bayesian Networks, European Freight Corridor.

1 Introduction and Goals

Transport modeling, in general, and freight transport modeling, in particular, are becoming important tools for investigating the effects of investments and policies, involving large amounts of resources. However, freight demand forecasting models are still in an experimental and evolution stage (Inaudi et al. 2013) for the following reasons:

B. Murgante et al. (Eds.): ICCSA 2014, Part IV, LNCS 8582, pp. 401–416, 2014.
© Springer International Publishing Switzerland 2014

- lower seniority (about 10 years) than the respective passenger models;
- high number of decision-makers to consider (companies, shippers, carriers, logistics operators, port operators, deposits, etc.);
- variety of products transported (in terms of categories, dimensions, weight, value, etc.);
- high variability in decision-making processes;
- limited availability of information (data often aggregated, dated, partial, heterogeneous, etc.).

Recently, European projects like Transtools (Burgess et al., 2008) or ETIS/ETIS Plus (NEA Transport research and training BV, 2005; Chen, 2011) have developed a unique modeling and data framework to forecast freight flows at large scale so to avoid data availability and modeling problems (Albert and Schafer, 2013)   (Fig. 1).

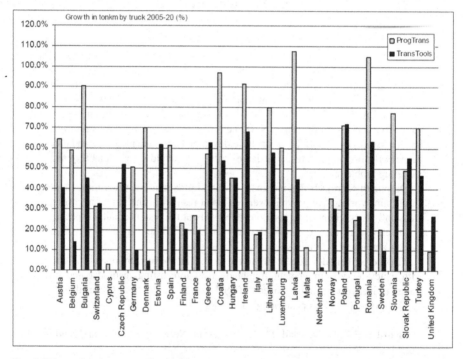

Fig. 1. Divergence between the results of the two projects Prog-Trans and TransTools for truck flows (Source: FreightVision Project-Deliverable 4.3, p.17)

Despite this, projects that had multi-million Euros funding, using these modeling frameworks, have provided very different results for the same forecasting areas and years, giving rise to serious doubts about the results quality, especially in relation to their cost and development time.

Moreover, many of these models are purely deterministic in results, giving no information about their estimation errors or the probability of the occurrence of forecast values. Other problems include forecasting different scenarios with very long-term simulations. We think that projects of national/European importance would benefit from the contribution of probabilistic data-driven models that take into account the uncertainties and variability of attributes and scenarios, especially for long-term estimates, in order to have more truthful decision-support.

There are a lot of freight demand models (Chase et al., 2013), with some methods similar to the one adopted here like the use of Trend Analysis/Time series or Neural Networks (NCFRP, 2010), but Bayesian Networks have the advantage to allow the introduction in the model of expert knowledge and the possibility to verify the results that are in form of an easy-to-understand oriented causal graph among variables and not complex or black-box relations, like with Neural Networks (Floreano et al., 1996).

Our goals are to understand quantitative and qualitative aspects of future traffic demand and evaluate possible future scenarios according to most relevant and influencing variables (Meersman and Van de Voorde, 2013) of the freight market. We also want to overcome the above-mentioned problems with a new freight demand forecasting framework based on Bayesian Networks and using European official and available data. The model has to be easy to implement, not onerous and give probabilistic results in less time, with an estimation error similar to the more complex methods. It should be capable of giving order of magnitude of forecasted freight flows for strategic decision making at a very early phase of policy development, and be complementary to more traditional, more precise, but much more expensive freight models for later stages of analysis.

2   A New Methodology for Freight Demand Forecast

Within our study, we applied the general demand forecast methodology to freight flows within the Sixth European Rail Freight Corridor. The flow chart in Fig. 2 summarizes the key elements of the different activities to be performed during demand forecast. In the present paper we will describe only the first three steps because they are where the major innovation elements lie. These steps are namely:

- the input data analysis;
- the Decision Tree Induction model analysis (Witten and Frank, 2000);
- the final freight demand forecasting by Bayesian Network models (Pearl 2000, Jensen 2001, Korb and Nicholson 2004, and more particularly Fusco 2010 and 2012 for their use as spatial strategic forecasting tools).

These steps are logically connected. The input data analysis allows to know how each input variable influences the actual freight flow dynamics in terms of relative growth (i.e. percentage variation between reference years 2005 and 2010) so to understand which variables are directly (with the uni-variate analisys) or indirectly (bi-variate and tri-variate analysis) related to it.

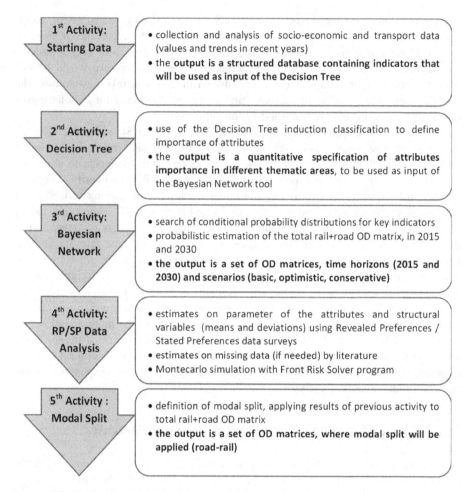

Fig. 2. The Overall Methodology (activities 1 to 3 are described in this paper)

The Decision Tree classification refines this preliminary analysis with a complex multi-variate elaboration having as target variable always the freight flow dynamics (the evolution of the freight flows between 2005 and 2010). Finally, the Bayesian Network models use as input data only the most influencing variables in order to avoid irrelevant data in the model, resulting in errors and reduction in the forecasting capacity.

The Bayesian Networks models were finally used to forecast freight flows in different scenarios. More precisely, the final traffic forecasts were carried out according to three different estimates of GDP growth for the study area: basic, optimistic, conservative. The demand forecasting models were developed with reference to two different geographic areas: at first, the analysis were conducted with reference to the mobility data of whole European O/D Matrix, later it was decided to focus only on the area interested by Corridor 6 and to calibrate the model accordingly in order to obtain more reliable estimates.

2.1 The Preliminary Data Analysis

The initial data analysis was carried out on the whole road and rail ETIS Origin-Destination Freight Flows Matrix in Europe for 2005 and 2010 years. Origins and destinations in this database are known at the NUTS 2 level. The original road 2005 O/D matrix has thus about 134.000 O/D pairs while the corresponding 2010 matrix has only 102.000 O/D pairs. 88.000 O/D couples are common to the two matrices. Taking into account only these common data (88.000 O/D pairs), we lose around 4% of total flows (containing also flows not interesting directly the Corridor 6). For each O/D couple an evolution rate between 2005 and 2010 could thus be calculated.

Table 1. List of the Initial Variables

ID	Indicator	Starting year	Forecast year 1	Forecast year 2
1	GDP of NUTS2i	Δ 2005-10	Δ 2010-15	Δ 2010-30
2	GDP of NUTS2j	Δ 2005-10	Δ 2010-15	Δ 2010-30
3	Population of NUTS2i	2010	2010	2010
4	Employment of NUTS2j	2010	2010	2010
5	Gross Capital Formation of NUTS2j	Δ 2005-10	Δ 2010-15	Δ 2010-30
6	Production of Manifactured Goods sold of NUTS2j	Δ 2005-10	Δ 2010-15	Δ 2010-30
7	Production Value by industry of NUTS2j	Δ 2005-10	Δ 2010-15	Δ 2010-30
8	Import of Good of NUTS0j	Δ 2005-10	Δ 2010-15	Δ 2010-30
9	Export of Goods of NUTS0i	Δ 2005-10	Δ 2010-15	Δ 2010-30
10	Total freight flows between NUTS2i and NUTS2j	Δ 2005-10	Δ 2010-15	Δ 2010-30
11	Minimum distance between NUTS2i and NUTS2j	2010	2010	2010
12	Macroregion Name of NUTS1i	2010	2015	2030
13	Macroregion Name of NUTS1j	2010	2015	2030
14	NMF – Net Migration flows of NUTS0j	Δ 2005-10	Δ 2010-15	Δ 2010-30
15	NMF – Net Migration flows of NUTS0i	Δ 2005-10	Δ 2010-15	Δ 2010-30
16	Unemployment rate on NUTS0i	2010	2010	2010
17	Transport taxation revenues of NUTS0i (million of €)	Δ 2005-10	Δ 2010-15	Δ 2010-30
18	Transport taxation revenues of NUTS0j (million of €)	Δ 2005-10	Δ 2010-15	Δ 2010-30
19	Diesel price of NUTS2i (€/litre)	Δ 2005-10	Δ 2010-15	Δ 2010-30
20	Diesel price of NUTS2j (€/litre)	Δ 2005-10	Δ 2010-15	Δ 2010-30

Together with freight flows, the starting data include twenty variables belonging to different fields like economy, geography and transportation and are summarized in Table 1. This general data analysis phase explores the freight flow dynamics and its correlation with the main variables, some of which are normally used in Transport Distribution Models (like distance, population and GDP) while others are not included in these models but can be used in data-driven Bayesian Network learning (for example unemployment rate, the variation of origin export and destination import or binary variables like the belonging to the European Union) (Caplice et al., 2010).

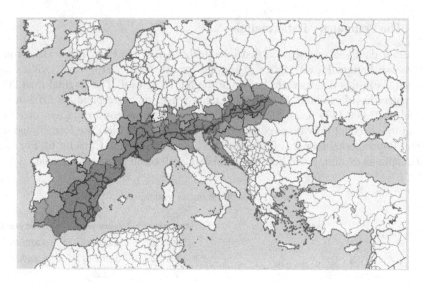

Fig. 3. The Corridor 6 Catchment Area

The starting data analysis is divided in three parts of increasing complexity: orthogram, bi-variate and tri-variate analysis. The following analyses concern only road and rail freight flows because they are the most interesting for Corridor 6 study area (see Fig. 3).

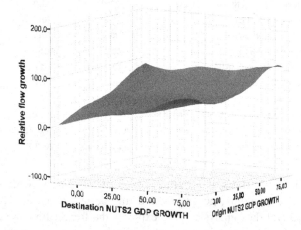

Fig. 4. Smoothing interpolation of 3D Scatterplot between Origin and Destination GDP growth and freight flow variation

The orthogram analysis allows studying the correlation between different kinds of variables (categorical and numeric for example). The analysis shows that 2005-2010 freight flow variation is correlated with the belonging or not of each area to the EU: there is a clear distinction between areas belonging to the EU and other areas.

EU Countries have more stable freight flows while non-EU Countries have opposite behaviors with some showing a big increase of freight flows and others a considerable decrease. These bi-modal behaviors are difficult to model with classical Transport Distribution Models. This first analysis already shows the interest of using different, more exploratory methods, like Decision Tree Induction and Bayesian Network modeling.

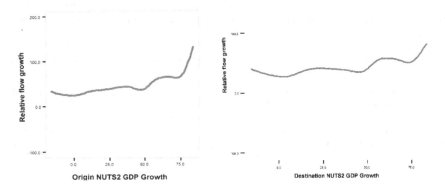

Fig. 5. Smoothing interpolation of 2D Scatterplots between Origin and Destination GDP growth and freight flow variation

The bi-variate analysis shows that correlation of freight flow dynamics is practically absent both with the distance between Origin and Destination (measured in kilometers on the transportation networks), with Origin Population, with unemployment rate at the Origin or with the Origin Export Variation and with the Destination Import Variation (Import and Export variations are known at the country level).

The tri-variate analysis correlates simultaneously the freight flow variations with Origin and Destination GDP variations. The 3D scatterplot, with a smoothing interpolation effect (see Fig. 4) indicates an overall positive correlation between these three variables with more specific local trends. To better understand this last point, two 2D scatterplots are extracted from the 3D diagram (see Fig. 5). The correlation is similar for the destination and origin GDP variations. Curiously a positive flow growth characterizes even negative GDP Variations, showing for some countries an inverse correlation, which could indicate a profound restructuring of the economy following the integration in the European market. More than linear flow growths are to be observed beyond 75% of GDP increase rate (more evident for Origin than Destination).

2.2   The Decision Tree Induction Classification

Decision Tree models are useful multivariate classification instruments allowing analysis of data correlation on the base of a target variable, O/D freight flow relative growth. Moreover, instead of regression models where we need to hypothesize a shape of the correlation (linear, cubic, exponential, etc.), Decision Tree models don't require any assumption and give more than one type of correlation. Finally the IF THEN framework is very useful and understandable for users and Decision Tree models can be used as a preliminary phase for the Bayesian Network modeling in order to understand the most influential variables to simulate the target one. Decision Trees Induction is an inductive classificatory technique belonging to the Data-Mining and to the Knowledge Discovery in Databases fields. It will be applied to the complete list of variables (Table 1), keeping the O/D freight flow relative growth as target variable.

The extracted classifier has a percentage of Correctly Classified Instances of about 38%, which appears relatively inaccurate. However, the analysis shows two main points:

- the classification ability is higher for the first and last flows variation classes and for the class nearest to zero;
- once again, distance (DIST_2010) between the individual Origins and Destinations does not have a relevant influence.

The analysis suggests introducing new variables so to add detail in the information (GDP at NUTS 2 Level, Internal, Belonging to EU and others) and to add interaction between territorial dimensions at NUTS2 and NUTS0. The new variables are:

- Internal (indicates if an O/D couples belong to the same country);
- No_EU (indicates if an O/D couples belong to EU countries or not);
- Delta GDP 2010-2005 at NUTS2 level;
- Flow 2005 (to indicate flow level before the 2008 economic crisis);
- EU15_CH_NO (indicate whether a flow belongs to the 15 EU member states before 2004 plus Switzerland and Norway);
- Weight of the exit flow for a given origin = $F_{ij}/F_{i.}$;
- Weight of the entry flow for a given destination = $F_{ij}/F_{.j}$;
- Weight of exports to Country J from i = $F_{iJ}/F_{i.}$;
- Weight of imports from Country I to j = $F_{Ij}/F_{.j}$

Where F_{iJ} means total flows from NUTS2 i to all Country J while $F_{.j}$ means total exit flows to NUTS2 j.

Introducing these new variables, the extracted Decision Tree identifies the variable "weight of the exit flow" as the most important one and shows the relatively chaotic evolution of flows for non-EU countries. Decision Trees results for the whole ETIS O/D Matrix describe a non-unique freight traffic evolution, with different variables explaining flow growth for each country and mainly different from countries belonging or not to the early EU member states. The only shared important variable is the weight of the exit flow (Fij/Fi) showing the relative importance of the economic relation between the origin and destination areas with respect to all the exit flows.

The Decision Tree extracted from the same variables but including only O/D flows belonging to the area of interest for Corridor 6 (Fig. 3) shows clearly two different dominant behaviors:

- the first is related to the countries with more stable economy and freight market where the only element that explains the freight dynamic is the actual weight of outgoing flows (this concerns more than 50% of total flows);
- the second is the already noted bi-modal behavior.

2.3   The Bayesian Network Forecasting Model

The Decision Tree technique produces knowledge only for the pre-processing phase. This limit of this technique is mainly due to the difficulty of the application of the rules extracted from the sample to the whole population:

- first, it is possible that a combination of conditional attributes never occurred in the extracted rules (IF part), whereas it can be present in the prevision dataset; the problem would then be to compute the relative conditional probability distribution;
- secondly, it could also be possible not to find a rule exactly identical (in the IF part) with the record to be classified: this problem can be solved only with the search of an attribute set close enough to the one to be classified.

Due to these possible situations, the extracted influent variables were used as input variables to implement a Bayesian Network. Bayesian Networks are more suitable to predict phenomena due to their robustness (they can couple statistical robustness from data-mining to expert knowledge directly implemented in the model, whereas Decision Trees are only based on data frequencies) and the possibility to make probabilistic inference so to have a probability values attached to predictions. Even in the absence of expert knowledge (as in our application), prior probabilities in the network initialization produce non-null probabilities for combination of attributes that are not present in the learning data-base. Through Bayesian learning algorithms from data (Jensen 2001, Korb and Nicholson 2004), the model links the variables in acyclic and directional graphs, showing their reciprocal influence in a cause-effect relationship between "parent" and "child" nodes. Finally, a conditional probability table is calculated for each dependent variable (with incoming link in the node), detailing the probabilistic relationship between the values of the "parent" and "child" variables. Unconditional probability tables are calculated for independent variables (without incoming links in the node). Learning algorithms search for the best possible combination of structure (links among nodes) and parameters (probability values in the tables) within a subspace of possible solutions. The best solution is found through likelihood maximization, knowing the empirical data.

Different Bayesian Network models were calculated from data covering the whole ETIS O/D Matrix, or just the area of interest for Corridor 6. Continuous variables were discretized in eight classes of equal frequencies (other discretizations were also attempted). Each model allows probabilistic inference of O/D freight flow relative growth between 2005 and 2010 from 2005 and 2010 data. Under the assumption of model stationarity, the probabilistic relationships embedded in the model can be used to infer O/D freight flow relative growth between 2010 and 2015 (end hence 2015 freight flows) from 2010 data and scenarios on 2015 data. A more problematic stationarity assumption was also used in order to forecast 2030 freight flows.

The Forecast for the Whole ETIS O/D Matrix

The final model for the whole ETIS O/D Matrix (Fig. 6) shows that the most important variables in order to forecast freight flows relative growth are the GDP national growth in the country of origin (NUTS2 GDP growth had too many missing data to produce statistically significant links in the model) and the relative importance of the outflow for the origin (weight of the exit flow Fij/Fi.). The mutual information analysis (resumed by the position of each node within the model) shows a clear clustering of economic (with internal circle in dark grey) and geographic (without internal circle) variables.

A first validation of the extracted Bayesian Network concerns its predictive power in inferring the value of the target variable of flow relative growth knowing the other variables. The resulting confusion matrix shows that the model can predict values of the target variable with a total precision of 25%, when considering the prediction of the exact variation class, but of more than 50% when considering prediction of the right class or of the two (eventually one) nearest ones (flow growth rates are discretized in cight classes). The second validation tests the model generalizability (or presence of over-fitting problem) through a ten-fold cross validation (that's to say the iterative use of 9/10 of the total O/D data to build the network and 1/10 of the total O/D data to validate it). Results of the cross validation are very similar to the initial model, which leads us to the conclusion that the model doesn't have particular over-fitting problems. During the cross validation, another validation of the model regards the stability of its network structure (called confidence analysis) and relative variable dependencies (represented from the arc connections) in the ten simulated networks. The arcs directly connected with the target variable (flow_growth) remain always the same (see Fig. 7) and are present in all the networks produced within the cross validation (100%).

A first problem of this methodology arises when we need to use the probabilistic results of the Bayesian Network inside the Discrete Choice model (Ben-Akiva and Lerman, 1985) that is based on deterministic values of total demand and, based on Revealed Preferences / Stated Preferences interviews (RP/SP, Danielis and Rotaris, 1999), elaborates probabilistic results on the modal split. In our application, modal split predictions are carried out using the weighted average of the median value of each flow variation class. An example is shown in Fig. 8 with a probability distribution for the target variable freight flow growth. For each of the eight classes, the central value is reported in the right column and is used to calculate the expected mean value (-26,17% in the example) as the weighted average (on the predicted probabilities) of the mean class values.

Once the Bayesian Network model is calibrated for 2010 (base year), scenario values can be defined for 2015 and 2030 for the main economic variables. Subsequently, the most probable values of freight flow growth can be inferred through the Bayesian Networks for very O/D couple in 2015 and 2030. The scenarios for the economic variables are as follows:

- Base scenario: 2015 and 2030 forecast baseline (natural development of the market from the current situation);
- Optimistic scenario: GDP growth forecast increased by 30%;
- Conservative scenario: GDP growth forecast decreased by 30%.

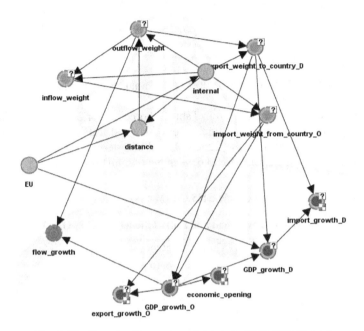

Fig. 6. The Bayesian Network model (whole ETIS O/D Matrix)

Arc	Arc Frequency	Inverted Arc Frequency	Edge Frequency	Total Frequency
internal -> distance	0%	0%	100%	100%
EU -> distance	0%	0%	100%	100%
EU -> internal	0%	0%	100%	100%
outflow_weight -> flow_growth	100%	0%	0%	100%
GDP_growth_O -> flow_growth	100%	0%	0%	100%
internal -> outflow_weight	100%	0%	0%	100%
outflow_weight -> inflow_weight	90%	10%	0%	100%
internal -> inflow_weight	100%	0%	0%	100%
import_weight_from_country_O -> GDP_growth_O	100%	0%	0%	100%
export_weight_to_country_D -> GDP_growth_D	100%	0%	0%	100%
GDP_growth_O -> export_growth_O	100%	0%	0%	100%
import_weight_from_country_O -> export_growth_O	100%	0%	0%	100%
export_weight_to_country_D -> GDP_growth_D	100%	0%	0%	100%
GDP_growth_O -> GDP_growth_D	100%	0%	0%	100%
EU -> GDP_growth_D	100%	0%	0%	100%
GDP_growth_D -> import_growth_D	100%	0%	0%	100%
export_weight_to_country_D -> import_growth_D	100%	0%	0%	100%
outflow_weight -> export_weight_to_country_D	100%	0%	0%	100%
internal -> export_weight_to_country_D	100%	0%	0%	100%
inflow_weight -> import_weight_from_country_O	100%	0%	0%	100%
internal -> import_weight_from_country_O	100%	0%	0%	100%
GDP_growth_O -> economic_opening	100%	0%	0%	100%
distance -> outflow_weight	90%	0%	0%	90%
distance -> inflow_weight	-10%	0%	0%	-10%

Fig. 7. Bayesian Network (whole ETIS O/D Matrix): confidence analysis of the extracted model

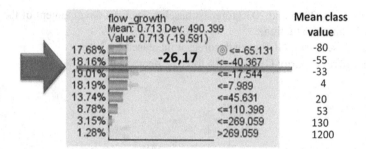

Fig. 8. Bayesian Network (whole ETIS O/D Matrix): evaluation of the mean flow prediction

The model results are summarized in Table 2. Forecasts for 2015 show freight flow dynamics practically stable with a final level of general traffic in 2015 below 2005 levels. Forecasts for 2030 show, instead, a general long-term increase of traffic flows, with growth rates ranging from 20% to 61% in the optimistic scenario.

Table 2. Demand Forecast (whole ETIS O/D Matrix) for 2015 and 2030 Scenarios

Year	Freight flows (road and rail) of the whole ETIS O/D Matrix	
2005	17.752 millions of tons	
2010	16.229 millions of tons	
2015	16.367 ÷ 17.037 millions of tons	$\Delta_{2010 \div 2015}$: 0%÷+5%
2030	19.530 ÷ 26.167 millions of tons	$\Delta_{2010 \div 2030}$: +20%÷+61%

The Forecast for the Corridor 6 Study Area

A second Bayesian Network model was developed more specifically for the area concerned by Corridor 6. Flows are grouped as follows:

- Internal, with Origin AND Destination in Corridor zones;
- Exchanges, with Origin OR destination in Corridor zones;
- Transits, with Origin AND Destination outside of Corridor zones.

Once again, under 5-year and 20-year stationarity assumptions, freight flows were inferred for 2015 and 2030, using the most probable values of flow relative growth. The forecasts for Corridor 6 flows (see Table 3 and Fig. 9) show that the flows variation in 2015, relative to 2010 base year and considering the three scenarios, lie between -1% (conservative scenario) and +10% (optimistic scenario), with very low probability of having total flow decrease and high probability of having total flow increase, although small in quantity. The results of the demand forecast for 2030 show a general long-term increase of traffic flows with high percentage variation from the conservative scenario, with a 27% of increase to a 96% of increase for the optimistic one. It is very difficult to verify these results. We thus tried to compare our results with those produced by a recent work by the French Ministry of the Environment (2005). This is one of the few comparable works to ours, in terms of geographical extension of the study area.

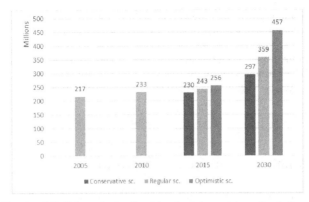

Fig. 9. Road and Rail flows in the Corridor 6 catchment Area (including transit)

Table 3. Evolution of freight flows concerning Corridor 6 catchment Area (including transit)

Year	Freight flows (road and rail) of Interesting O/D couples	
2005	217 millions of tons	
2010	233 millions of tons	.. $\Delta_{2005 \div 2010}$: +7.3%
2015	230 ÷ 256 millions of tons	$\Delta_{2010 \div 2015}$: -1% ÷ +10%
2030	297 ÷ 457 millions of tons	$\Delta_{2010 \div 2030}$: +27% ÷ +96%

Table 4. Comparison of evolution of freight flows through Pyrenees between the french study and the ETIS real values (values in mill. of tons)

ETIS 2005	Study on freight flows through Pyrenees - 2010 estimates		ETIS 2010
	Low scenario	High scenario	
10,86	12,5	13,5	11,68

The study on freight flows through the Pyrenees predicts the following annual average freight flow growth rates between the Iberian Peninsula and the rest of Europe between two scenarios: 2,9% (low scenario) and 4,5% (high scenario). By applying these growth rates to the observed 2005 freight flows within the area of interest for Corridor 6 (data derived from the 2005 ETIS O/D matrix), the estimated 2010 road and rail freight flows from the Iberian Peninsula to the rest of the catchment area of Corridor 6 would be much higher than the ones actually recorded within the ETIS 2010 O/D Matrix (Table 4).

Table 5 provides a comparison between 2015 and 2030 forecasted freight flows in the two studies (the study of freight flows through the Pyrenees actually provides estimates for 2025, but due to the hypothesized linearity of the evolution, it was possible to determine the "most probable forecast" for 2030).

Table 5. Comparison of prevision of freight flows through Pyrenees between the French study and our results (values in mill. of tons)

Year	Freight transport of Pyrenees' study - Estimates for 2010		Our study on Corridor 6		
	Low scenario	High scenario	Conservative	Regular	Optimistic
2015	14,8	16,9	11,5	11,9	13,6
2030	22,7	32,6	16,3	17,3	24,4

3   Conclusions and Future Developments

The data-driven methodology applied within this work seems to be very promising from many points of view. First of all, the data it needs are easy to find from official European level sources (even if more complete economic data-bases at the NUTS 2 level could have improved the performance of our models). Secondly, the methodology, because of its simplicity, is applicable in the short term, through model updating by incremental learning or new model development; it will thus be possible to update forecasts as new data are available and to follow multi-temporal economic dynamics. Moreover, the Bayesian Network framework adopted allows the recognition of different flow evolutions (which is similar to having multiple transport distribution system equations with different calibrated parameters) and their application in the forecasted scenarios. In addition, a comparison of the results with some official studies shows that our results are acceptable estimates.

The starting database for this first application covers two base years, namely 2005 and 2010, which are a very particular period for the European economy (arrival of new member states in 2004 and deep economic crisis after 2008), with some peculiar correlations and dynamics among economic, transportation and social variables. Availability of the 2015 version of the ETIS database will allow data-driven model development over the 2005-2015 period, which should produce more reliable results. Of course, the development of new infrastructures or geo-economic dynamics (entrance of new member states in the EU) will always be exogenous to the model, and the use of time- or cost-distances could be used instead of km-distance to better model the impact of transportation networks on the study area. Finally, the stationarity hypotheses on the links between economic, geographic and transportation variables are much more appropriate for short-term forecast (5 years) than for long-term ones (20-30 years).

A further point to be developed is the link between the total demand forecast and the following modal split scenarios. The use of average prediction values necessary for this further methodological step involves the loss of the richness of the Bayesian Network results, that is the probability distribution of the estimated flows demand. We are presently trying to use Montecarlo simulation approaches (Train, 2009) in order to extract a large number of possible deterministic demand values from the demand probability distribution. Subsequently, a modal choice probabilistic distribution will be derived from each of these values. It will then be possible to estimate an overall probability distribution for flows by mode and the results will be expressed in terms of values accompanied by statistical parameters such as mean, variance, and quartiles. The methodology is similar to that used in Mixed Discrete Choice Models.

Another option would be to develop the entire demand forecast, that is the generation and the modal distribution of freight flows, within the Bayesian Network framework. It will then be possible to preserve a consistent probabilistic approach for flows estimation by transport mode.

References

1. Albert, A., Schafer, A.: Demand for Freight Transportation in the U.S: A High-Level View. Journal of Transportation Statistics, (2013) (in print)
2. Ben-Akiva, M., Lerman, S.R.: Discrete choice analysis. MIT Press (1985)
3. Burgess et al.: Final Report TRANS-TOOLS (TOOLS for TRansport Forecasting ANd Scenario testing) Deliverable 6. Funded by 6th Framework RTD Programme. TNO Inro, Deft, Netherlands (2008)
4. Caplce, C., Phadnis, S.: Driving Forces Influencing Future Freight Flows-NCHRP, web-only document 195. Transport Research Board, Washington D.C. (2010)
5. Chase, K., Anater, P., Pelan, T.: Freight Demand Modeling and Data Improvement - Second Strategic Highway Research Program. Transport Research Board, Washington (2013)
6. Chen, M.: ETIS and TRANS-TOOLS v1 Freight demand. In: CTS-seminar –European and National Freight Demand Models, Stockolm (March 1, 2011)
7. Danielis, R., Rotaris, L.: An analysis of freight transport demand using stated preference data: A survey and a research project for the Friuli-Venezia Giulia region. Transporti Europei (13) (1999)
8. Floreano, D., Mattiussi, C.: Manuale sulle Reti Neurali. Mulino Editrice, Bologna (1996)
9. Fusco, G.: Handling Uncertainty in Interaction Modelling in GIS: How will an Urban Network Evolve? In: Prade, H., Jeansoulin, R., Papini, O., Schockaert, S. (eds.) Methods for Handling Imperfect Spatial Information, pp. 357–378. Springer, Berlin (2010)
10. Fusco, G.: Démarche géo-prospective et modélisation causale probabiliste. Cybergeo: European Journal of Geography, Systems, Modelling, Geostatistics, document 613 (2012), http://cybergeo.revues.org/25423, doi:10.4000/cybergeo.25423
11. Inaudi, D., De Jong, G., Arnone, M.: Un modello matematico per la valutazione degli scenari di sviluppo del sistema del trasporto merci nel Nord-Ovest. In: XXXII Italian Conference of Regional Science (2013)
12. Jensen, F.V.: Bayesian Networks and Decision Graphs. Springer, New York (2001)
13. Korb, K.B., Nicholson, A.E.: Bayesian Artificial Intelligence. Chapman & Hall / CRC, Boca Raton (2004)
14. Meersman, H., Van de Voorde, E.: The Relationship between Economic Activity and Freight Transpot. In: Ben-Akiva, et al. (eds.) Freight Transport Modelling. Emerald Group Publishing (2013)
15. Ministère de l'Écologie, du Développement durable et de l'Énergie - Direction des Transports Terrestres-Bureau d'Informations et de Prévisions Economiques: Analyse et évolution des flux de transport de marchandises à travers les Pyrénées, Issy, Edition BIPE (2005)
16. National Cooperative Freight Research Program: Freight-Demand Modeling to Support Public-Sector Decision Making. Transport Research Board, Washington D.C (2010)
17. NEA Transport research and training BV: Core Database Development for the European Transport policy Information System (ETIS), Final Technical Report v1(2005)

18. Pearl, J.: Causality – Models, Reasoning and Inference. Cambridge University Press, Cambridge (2000)
19. Train, K.E.: Discrete Choice Methods with Simulation, 2nd edn. Cambridge University Press, USA (2009) ISBN: 9780521766555
20. Witten, I.H., Frank, E.: WEKA – Machine Learning Algorithms in Java. University of Waikato, Morgan Kaufmann Publishers (2000)

An Ontology Framework for Flooding Forecasting

Annalisa Agresta[1], Grazia Fattoruso[1], Maurizio Pollino[2], Francesco Pasanisi[3],
Carlo Tebano[3],Saverio De Vito[1], and Girolamo Di Francia[1]

[1] UTTP/Base Material and Devices Dept., ENEA Portici Research Centre, P.le E. Fermi,
1 - 80055 Portici (NA) - Italy
grazia.fattoruso@enea.it
[2] UTMEA/Energy and Environmental Modeling Dept., ENEA Casaccia Research Centre, Via
Anguillarese, 301 - 00123 Roma - Italy
[3] UTTP/CHIA Dept., ENEA Portici Research Centre, P.le E. Fermi,
1 - 80055 Portici (NA) – Italy

Abstract. Floods can cause significant damage and disruption as they often
affect highly urbanized areas. The capability of knowledge using and sharing
is the main reason why the ontologies are suited for supporting the phases of
forecasting in (near-) real time disastrous flooding events and managing
the flooding alert and emergency. This research work develops an ontology,
FloodOntology for floods forecasting based on continuous measurements of
water parameters gathered in the watersheds and in the sewers and simulation
models. Concepts are captured across the main involved domains i.e. hydrolog-
ical/hydraulic domains and SN-based monitoring domain. Classes hierarchies,
properties and semantic constraints are defined related to all involved entities,
obtaining a structured and unified knowledge-base on the flooding risk forecasting, to be integrated in expert systems.

Keywords: Flooding Forecasting, Ontology, Semantic Web Technology, Sensor Network, Modeling.

1 Introduction

Establishing a viable flood-forecasting system for communities at risk requires heterogeneous knowledge that encompasses fundamentally three main domains i.e. hydrological, hydraulic and sensor networks ones.

Following an integrated approach, flooding forecasting is realized by (near) real
time and continuous monitoring based on sensor networks and by simulations of the
involved hydrological/hydraulic processes such as surface runoff and discharge propagation through the drainage system network of pipes and channels.

As regards floods monitoring, up-to-date information is crucial. Geo-sensors ranging from water gauges and weather stations to stress monitors attached to dams or
bridges are generally used in network to gather such information [1], [2]. But they
return huge amounts of data and in different formats and semantics [3]. The semantic
heterogeneity of data observations has been identified by scientific community as a
key issue for data sharing between different domains [4], [5], [6] and overcome by

B. Murgante et al. (Eds.): ICCSA 2014, Part IV, LNCS 8582, pp. 417–428, 2014.
© Springer International Publishing Switzerland 2014

using Semantic Web methods and technologies as a means to enable the interoperability for sensors and sensing systems. In fact, an ontology of the sensor networks, namely SSN (Semantic Sensor Network) [7] has been developed, realizing a unique knowledge management base for sensor networks.

As regards floods simulations, an advanced component–based modeling approach is required where a unified model framework is used, primarily composed of hydrological and hydraulic components interlinked through data exchange. But for using this approach, more issues have to be overcome, especially the semantic heterogeneity across the disciplines [8] due to the variety of terminology used to describe the involved physical processes and related mathematical equations as well as variables, parameters, and units within models [9]. An important step in overcoming such issue is to build an ontology that specifies and organizes the concepts and terminologies used in the different involved domains [10]. An ontology is an explicit conceptualization of human knowledge that focuses on shared understanding by defining vocabularies that represent and communicate knowledge about a specific domain [11]. Establishing a shared understanding of concepts aids in eliminating conceptual and terminological confusion [12, 13].

Gruber [11] specified as primary characteristics of an ontology the flexibility to merge with other ontologies and the extensibility to accommodate any required future modifications.

In this research work, the effort has hence been to develop an ontology, FloodOntology, that provided a unified and structured knowledge-base for flooding risk forecasting by building and merging ontologies related to the involved SN domain (on the basis of the existing SSN ontology) and the hydrological and hydraulic domains.

This paper is structured as following: Section 2 discusses the methodology used to create FloodOntology. Section 3 describes in detail the developed ontology. The paper concludes with a summary and ongoing works.

2   The Methodology

For building FloodOntology, we have used the widely accepted skeletal methodology described by Uschold and Gruninger (1996) [14], which have been successfully applied for building much ontology. (e.g. [7], [15], [16], [17]).

According this methodology, an ontology is built by three phases: (i) concept capture, (ii) coding, and (iii) integration with complementary ontologies. These steps are necessary to define the concepts used within the community, determine the method of presenting these concepts, and benefit from prior efforts in building related ontologies, respectively.

The objective of our research work has been to develop a floods ontology realizing the interoperability across hydraulic and hydrologic domains and data monitoring framework. This aim has been achieved by modelling and integrating all components of these three domains.

By the concept capture phase, the basic ideas, relationships, and terms corresponding to each domain have been identified and defined. The classes and their definitions

Altitude
As a general definition, altitude is a distance measurement, usually in the vertical or "up" direction, between a reference datum and a point or object.

Aquifer
An underground water-bearing formation that is capable of yielding water. Aquifers have specific rates of discharge and recharge. As a result, if groundwater is withdrawn faster than it can be recharged, the underground aquifer cannot sustain itself. (WFL)

Aquifer (Unconfined)
An aquifer whose upper water surface (water table) is at atmospheric pressure, and thus is able to rise and fall.

Artesian water
Ground water that is under pressure when tapped by a well and is able to rise above the level at which it is first encountered. It may or may not flow out at ground level. The pressure in such an aquifer commonly is called artesian pressure, and the formation containing artesian water is an artesian aquifer or confined aquifer. See flowing well.

Artesian well
An artesian well is a pump less water source that uses pipes to allow underground water that is under pressure to rise to the surface.

Base Flow
The fair-weather or sustained flow of streams; that part of stream discharge not attributable to direct runoff from precipitation, snowmelt, or a spring. Discharge entering streams channels as effluent from the groundwater reservoir. Also referred to as *Groundwater Flow*. (NALMS)

Capillary action
The means by which liquid moves through the porous spaces in a solid, such as soil, plant roots, and the capillary blood vessels in our bodies due to the forces of adhesion, cohesion, and surface tension. Capillary action is essential in carrying substances and nutrients from one place to another in plants and animals.

Catchment
The area contributing surface water to a point on a drainage or river system, which may be divided into sub-catchments.

Fig. 1. A segment of FloodOntology Glossary

have been collected in a glossary (Figure 1); a class hierarchy has been derived and properties and semantic constraints have been defined and attached to the classes.

In literature, three approaches have generally used for accomplishing the knowledge organization and the restructuring task: top-down, middle-out, or bottom-up [14]. Top-down starts with the highest concept and moves down for details while, logically, bottom-up begins with details and tries to generalize them. The middle-out approach conserves a balance in terms of the level of details and has the advantage over the other approaches that it allows higher-level classes to arise naturally. In this study, we have used the middle-out approach on the recommendation from prior studies[14].

The process of coding the ontology represents the domain metaontology (i.e., information about ontology components) gathered in the concept capture phase in a formal specification using an ontology coding language. The advantage of using a formal language is to establish the information about the concepts and their relationships in the form of axioms. Different languages are used in the coding of ontologies, but we have used OWL (Ontology Web Language) [18] because it is appropriate when the information needs to be processed by an application, rather than just for the purpose of presentation of information. OWL is very useful in representing the meaning (i.e., semantics) of terms and relationships between terms, and hence developing ontologies.

The most commonly used tools for creating OWL documents are OntoEdit, OilEd, and Protégé. Protégé is an ontology editor and a knowledge-based editor as well as an open source Java tool that provides an extensible architecture for creation of customized knowledge-based applications [19]. Using the Protégé OWL Plugin, we can load and save OWL and Resource Description Framework (RDF) ontologies, edit and visualize classes and properties, define logical class characteristics as OWL expressions, execute reasons such as description logic classifiers or edit OWL individuals for Semantic Web markup. Protégé also enables an ontology structure to be defined by automatically generating forms that facilitate knowledge acquisition. For this reasons Protégé is been used in this work. A segment of OWL code related to our ontology automatically generated by Protégé editor is shown in the Figure 2.

```
    <rdfs:range rdf:resource="http://www.co-ode.org/ontologies/ont.owl#Object"/>

</owl:ObjectProperty>



<!-- http://www.semanticweb.org/paola/ontologies/2013/11/untitled-ontology-29#has_postcondition -->

<owl:ObjectProperty   rdf:about="http://www.semanticweb.org/paola/ontologies/2013/11/untitled-ontology-
29#has_postcondition">

    <rdfs:label xml:lang="en">has postcondition</rdfs:label>

    <rdfs:label xml:lang="it">ha postcondizione</rdfs:label>

    <rdfs:comment rdf:datatype="&xsd;string">Direct succession applied to situations.
E.g., &apos;A postcondition of our Plan is to have things settled&apos;.</rdfs:comment>

    <rdfs:domain rdf:resource="http://www.co-ode.org/ontologies/ont.owl#Situation"/>

    <rdfs:range rdf:resource="http://www.co-ode.org/ontologies/ont.owl#Situation"/>

    <rdfs:subPropertyOf       rdf:resource="http://www.semanticweb.org/paola/ontologies/2013/11/untitled-
ontology-29#directly_precedes"/>

</owl:ObjectProperty>
```

Fig. 2. Segment of OWL code related to FloodOntology

Ontologies are built to be reused in different applications [20]. One common way ontologies are reused is through integration with other ontologies. Not only does this helps in speeding up ontology construction, but it also allows reuse of definitions and properties already built and tested, providing consistency and robustness to new ontologies. In this work, segments of different ontologies have been reused such as, just as examples, measurement units by SWEET (Semantic Web for Earth and Environmental Terminology) ontology [21] and the abstract class by SSN (Semantic Sensor Network ontology) [7].

3   Building FloodOntology

The objective of FloodOntology is to model the knowledge-base characterizing the floods scenarios, by building and merging the knowledge-bases of the involved domains, to be integrated into urban hydraulic risk safety and management systems.

FloodOntology has been built on the basis of the concepts captured across the main hydrological, hydraulic and SN based monitoring domains that characterize floods modeling and monitoring. Hierarchies, properties and semantic constraints have been defined and modeled among the entities, obtaining a structured knowledge on the flooding risk, to be integrated in an expert system [22].

3.1   Hydrological and Hydraulic Ontologies

A flood event is a natural event of great complexity. Numerical modeling is amply used for studying the flood phenomenon, assessing the flood risk at the current situation and after suggested changes in the flood prone areas, as well as forecasting also in (near-) real time disastrous flooding events and managing the flooding alert and emergency. All these scenarios require an advanced multi-component-based modeling framework, where the hydrological and hydraulic components, interlinked through data exchange, are identified as primary. In fact, for simulating flood events, the physical phenomena to be modeled are basically: (1) how flow is routed over the land surface after falling as rain; (2) the behavior of water once it is channeled into rivers and sewers. In particular, hydrological models are used to simulate the surface runoff at rainfall events of different magnitudes, while hydraulic models use this information to transform flows into water levels that can be used to estimate inundation extent (Figure 3).

The mathematical equations related to hydrological phenomena require information about the catchment elevations and gradients, volumes of water entering the catchment as rain and parameters that account for flow resistance and runoff. To characterize the catchment land surface, topographic data are required in the form of a Digital Elevation Model (DEM), a raster grid where cells of a given size are assigned an elevation value. The water input to the equations is given by point-based rain gauge records. A number of parameters must also be specified in order to perform the flow routing calculations and to account for processes that are not explicitly represented (e.g. losses to infiltration).

As regards the hydraulic phenomena, they are modelled by energy and flow equations. The involved parameters are related to the geometry   and topology of the water network, characteristics of the channels such as their slope,   roughness, hydraulic sections and forth on, in order to provide an estimate on water levels, flows and velocities along a river as well as a sewer system.

The hydraulic and hydrological characteristics of the watershed and urban drainage system have been captured and modeled in our ontology as classes and subclasses. In the following figures, some classes hierarchies are shown.

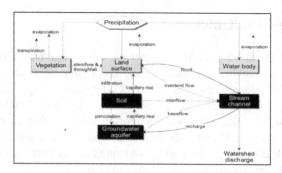

Fig. 3. Conceptual model of the rainfall/ranoff phenomemon by [23]

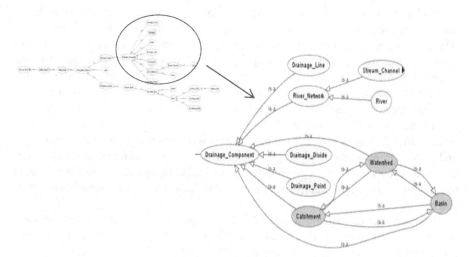

Fig. 4. Segment of FloodOntology: sub-classes of *Drainage-Component* class

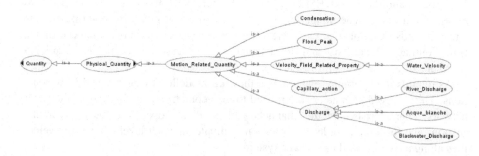

Fig. 5. Segment of FloodOntology: hydraulic characteristics classes

3.2   Sensor Network Ontology

Observations, and the sensors that gathered them, are at the core of any flooding risk management application.

Due to their various purposes, sensor networks employed within flooding risk management take many forms and cover various geographical and temporal scales. They range from remote sensing satellites, providing invaluable images of large regions, through instruments installed on the Earth's surface to sensing devices situated along drainage networks (rivers and sewers) and within drainage basins, providing highly-detailed point-based information from single sites. Data from sensor networks useful for flooding risk forecasting and management purposes are summarized in the Table 1 by [24].

Table 1. Summary of data from sensor networks useful for risk management purposes

REMOTE SENSING	IN SITU
Large scale meteorological information (e.g. cloud cover) from satellite images and aerial photographs	Meteorological information (e.g. temperature, rainfall, humidity, pressure and wind speeds and directions) from weather stations
Geological maps	Hydrometric data (on the level of rivers and streams and the discharge rates)
Digital elevation and terrain models (through, e.g., satellite images)	Geological maps
Information on the environment and population density (from, e.g., aerial photographs and satellite images)	Water well logs
Information on the soil roughness (from, e.g., aerial photographs and satellite images)	Rain radars for short-medium range rain forecasting and measurements of rain density.
SAR PRI images (weather invariant) Multi-spectral images for land cover evaluation and flood mapping	
Land cover map from satellite images.	

The use of sensing devices and networked sensing devices means huge volume of data, as well as heterogeneity of data and devices, data formats, and measurement procedures. In the last years, the semantic heterogeneity issues have been addressed by using the Semantic technologies [7], such as the ontologies, that assist in managing, querying, and combining sensors and observation data.

In this research work, sensors and networked sensors for monitoring meteorological data (e.g. temperature, rainfall, humidity, pressure and wind speeds and directions) by weather station networks as well as hydrometric data (on the level of rivers and streams and the discharge rates) have been described in terms of capabilities, measurement processes, observations and deployments, by building a SN ontology (Figure 6), re-using some segments of the SSN ontology [7].

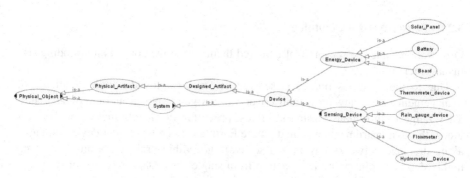

Fig. 6. Segment of the sensor network ontology

3.3   Integrating *FloodOntology*

Generally, in an integration process, source ontologies are aggregated, combined, assembled together, to form the resulting ontology, possibly after the reused ontologies have suffered some changes, such as, extension, specialization or adaptation.

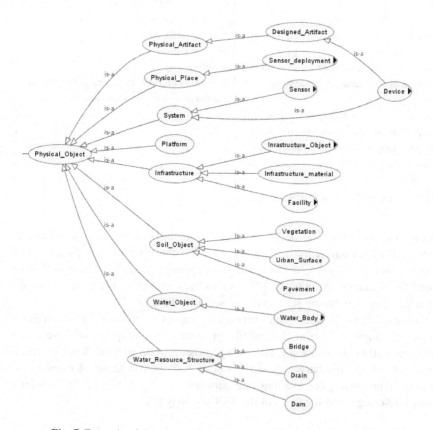

Fig. 7. Example of domain and sensor concepts integration in FloodOntology

This research work applies the ontology integration concept to the hydrological/hydraulic and SN-based monitoring ontologies, separately developed, building an unified ontological framework for flooding forecasting.

The adopted integration approach argues for an integrated view, instead of purely sensor-centric methods [25]. Sensor concepts (i.e., how observations are performed) are related to domain concepts (i.e., observed properties and their associated hydrologic/hydraulic entities), thus, describing the features (e.g., river, watershed, and sewer) as well as the properties (e.g., rainfall) measured by sensors (e.g. pluviometers) (Figure 7).

Hierarchies, properties and semantic constraints are defined and modeled among the entities (Figure 8).

Fig. 8. Example of properties defined in FloodOntology

An example of relation is shown in Figure 9 where the property measure links the Sensing Device class to the Meteorological Quantity class. Equivalent classes, i.e. different classes that are the same entities, have also been identified and defined. Just as example, in Figure 9, equivalent classes are shown as orange classes.

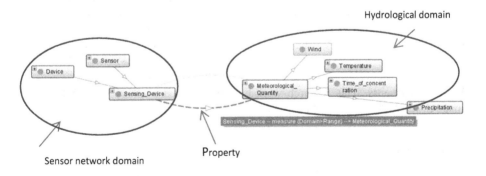

Fig. 9. Integrating the ontologies: the *measure* property links the *Sensor Device* class to *Meteorological Quantity* class

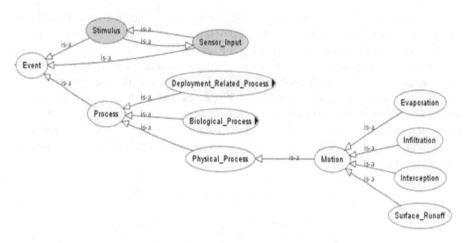

Fig. 10. Example of equivalent classes

4   Conclusion

Building FloodOntolgy, the effort has hence been to develop a unified and structured knowledge-base to be integrated in flooding forecasting and alert and emergency managing systems. Ontologies related to the main involved domains, i.e. the SN-based monitoring domain and the hydrological and hydraulic domains, have been developing and merging. FloodOntology constitutes a headway in the capability of inferring new knowledge, re-using and sharing it for risk forecasting and management. Through FloodOntology, local and national authorities involved in flooding risk forecasting and management as well as criticizes may use a standardized knowledge framework, enhancing the analysis potential into the flooding management processes.

Building centralized or distributed flood-forecasting and management applications on this ontology, recognized issues related to knowledge heterogeneity within and across the involved domains may be overcome. Though other domains (e.g. geographical domain, authority regulation domain and so on) have to be conceptualized and integrated with the developed ones in order to make available to the communities at risk and decision makers a shared and structured understanding of overall flooding risk domain.

Acknowledgments. This research work has been funded by PON R&C 2007-2013 Smart Cities/AQUASYSTEM Project.

References

1. Lai, S., Zoppi, C.: An Ontology of the Strategic Environmental Assessment of City Masterplans (2011)
2. Montenegro, N., Gomes, J.C., Urbano, P., Duarte, J.P.: A Land Use Planning Ontology: LBCS. Future Internet 4(1), 65–82 (2012), doi:10.3390/fi4010065.

3. Bröring, A., Beltrami, P., Lemmens, R., Jirka, S.: Automated Integration of Geosensors with the Sensor Web to Facilitate Flood Management.....
4. Elag, M., Goodall, J.L.: An ontology for component-based models of water resource systems. Water Resources Research 49(8), 5077–5091 (2013)
5. Jirka, S., Bröring, A., Stasch, C.: Applying OGC Sensor Web Enablement to risk monitoring and disaster management. In: GSDI 11 World Conference, Rotterdam, Netherlands (June 2009)
6. Laurini, R.: Pre-consensus Ontologies and Urban Databases. In: Teller, J., Lee, J.R., Roussey, C. (eds.) Ontologies for Urban Databases. SCI, vol. 61, pp. 27–36. Springer, Heidelberg (2007)
7. Compton, M., Barnaghi, P., Bermudez, L., García-Castro, R., Corcho, O., Cox, S., Taylor, K.: The SSN ontology of the W3C semantic sensor network incubator group. Web Semantics: Science, Services and Agents on the World Wide Web 17, 25–32 (2012)
8. Arara, A., Laurini, R.: Formal contextual ontologies for intelligent information systems. In: Proceedings (WEC 2005) 3rd World Enformatika Conference, vol. 5, pp. 303–306 (2005)
9. Barnaghi, P., Meissner, S., Presser, M., Moessner, K.: Sense and sens' ability: Semantic data modelling for sensor networks. In: Conference Proceedings of ICT Mobile Summit 2009 (June 2009)
10. Murgante, B., Garramone, V.: Web 3.0 and knowledge management: Opportunities for spatial planning and decision making. In: Murgante, B., Misra, S., Carlini, M., Torre, C.M., Nguyen, H.-Q., Taniar, D., Apduhan, B.O., Gervasi, O. (eds.) ICCSA 2013, Part III. LNCS, vol. 7973, pp. 606–621. Springer, Heidelberg (2013)
11. Gruber, T.R.: A translation approach to portable ontology specifications. Knowledge Acquisition 5(2), 199–220 (1993)
12. Beran, B., Piasecki, M.: Engineering new paths to water data. Computers & Geosciences 35(4), 753–760 (2009)
13. Uschold, M., Gruninger, M.: Ontologies: Principles, methods and applictions. The knowledge Engineering Review 11(02), 93–136 (1996)
14. Zerger, A., Wealands, S.: Beyond modelling: linking models with GIS for flood risk management. Natural Hazards 33(2), 191–208 (2004)
15. Wang, J., Huang, Q., Liu, Z.P., Wang, Y., Wu, L.Y., Chen, L., Zhang, X.S.: NOA: A novel Network Ontology Analysis method. Nucleic Acids Research 39(13), e87 (2011)
16. Taylor, K., Leidinger, L.: Ontology-driven complex event processing in heterogeneous sensor networks. In: Antoniou, G., Grobelnik, M., Simperl, E., Parsia, B., Plexousakis, D., De Leenheer, P., Pan, J. (eds.) ESWC 2011, Part II. LNCS, vol. 6644, pp. 285–299. Springer, Heidelberg (2011)
17. Ahmedi, L., Jajaga, E., Ahmedi, F.: An Ontology Framework for Water Quality Management.
18. McGuinness, D.L., Van Harmelen, F.: OWL web ontology language overview. W3C Recommendation 10, 2004-03 (2004)
19. The Protégé Ontology Editor and Knowledge Acquisition System, http://protege.stanford.edu
20. Fernández-López, M., Gómez-Pérez, A., Juristo, N.: Methontology: From ontological art towards ontological engineering (1997)
21. Raskin, R.: Semantic Web for Earth and Environmental Terminologies (SWEET) Ontologies (2005), http://sweet.jpl.nasa.gov/ontology
22. Gruber, T.R.: The Role of Common Ontology in Achieving Sharable, Reusable Knowledge Bases (1991)

23. Robinson, M., Ward, R.: Principles of Hydrology. McGraw-Hill Publishing Company, Berkshire (2000)
24. Douglas, J., Usländer, T., Schimak, G., Esteban, J.F., Denzer, R.: An open distributed architecture for sensor networks for risk management. Sensors 8(3), 1755–1773 (2008)
25. Pinto, H.S., Martins, J.P.: A methodology for ontology integration. In: Proceedings of the 1st International Conference on Knowledge Capture. ACM (2001)

A Sequence-Encoded Relocation Scheme for Electric Vehicle Transport Systems

Junghoon Lee and Gyung-Leen Park[**]

Dept. of Computer Science and Statistics,
Jeju National University, Republic of Korea
{jhlee,glpark}@jejunu.ac.kr

Abstract. This paper designs a resource distribution scheme for city-wide electric vehicle (EV) transport systems, evaluating its performance via a prototype implementation. With the help of computational intelligence and future demand forecasts, the resource distributor tries to enhance the service ratio of EV sharing systems. A genetic algorithm is designed for reasonable response time, focusing on how to encode a relocation schedule so as to represent not just relocation pairs but also operation sequences. The genetic operators are customized for the encoding scheme, while the fitness function estimates relocation distance considering the encoded vector and the number of service men. The experiment result shows that the proposed scheme reduces the resource distribution overhead for the given parameter set and fully benefits from potential operation concurrency, improving the relocation distance by up to 56.9 %, compared with vehicle-by-vehicle moves.

Keywords: Electric vehicle, vehicle sharing system, sequence encoding, service ratio enhancement.

1 Introduction

Electric vehicles, or EVs in short, require a variety of well-developed information technologies for their wide penetration, including taxis, rent-a-cars, shared vehicles, and so on. Vehicle sharing services are now being deployed to overcome high price and maintenance difficulty of EVs in many cities [1]. In this service, users rent out and return EVs from and to sharing stations. Particularly, in one-way rental systems, rent-out and return stations for each transaction can be different [2]. Here, a sharing request can be accepted only when the sharing station from which a user wants to take has enough EVs. However, due to uneven demand patterns, which can be easily found especially in commute hours and evening time, some stations are highly likely to lack EVs. Accordingly, it is necessary to explicitly relocate EVs according to the current distribution and the future demand prediction [3].

[**] Prof. Gyung-Leen Park is the corresponding author.
This work was supported by the research grant from the Chuongbong Academic Research Fund of Jeju National University in 2013.

B. Murgante et al. (Eds.): ICCSA 2014, Part IV, LNCS 8582, pp. 429–437, 2014.

For EV relocation, the number of vehicles that can be moved simultaneously is limited even with highly expensive towing cars, making it reasonable and economic to relocate EVs by human service staff [4]. Here, for a two-man team, two members go to an overflow station in their service vehicle. One drives an EV to an underflow station while the other follows in the service vehicle. Then, both go to another overflow station. If there are n service men in a relocation team, $(n-1)$ EVs can be moved simultaneously if the overflow station has more than $(n-1)$ EVs to move to the same underflow station. The distance taken by the service vehicle is the most important criteria for the performance of the relocation procedure and affects the length of out-of-service time. We will call it relocation distance from now on. It is decided by how to match overflow and underflow stations and how to create a relocation order between each matched pair. An efficient relocation mechanism can cut down the hiring cost in the sharing business.

Relocation scheduling is a very complex problem demanding tremendous execution time, especially when there are many EVs to move and multiple staff members [5]. Hence, it is necessary to design a suboptimal scheme, such as a genetic algorithm, for practical application [6]. The relocation distance is consist of not only the distance between overflow and underflow stations in each relocation pair (intra-pair distance) but also the distance between each pair, that is, how far the service vehicle goes to the next overflow station after the completion of a pair (inter-pair distance). Our previous work first finds the relocation matching to reduce the intra-pair distance first and then inter-pair distance [7]. However, a schedule having the smallest intra-pair distance does not always lead to the smallest relocation distance. After all, our observation finds out the essentiality of orchestrating them in a single schedule in traversing the search space [8].

Accordingly, this paper designs a sequence-encoded relocation planner based on genetic algorithms. To this end, the encoding scheme is designed to represent a relocation schedule by an integer-valued vector. Each schedule includes not only the set of relocation pairs, each of which is consist of an overflow station and an underflow station, but also the sequence the service vehicle processes each pair. In addition, our fitness function evaluates the relocation distance for a schedule considering both intra-pair and inter-pair distances to improve the quality of population generation by generation. In a chromosome, relocation pair and sequence parts are separated and handled differently, while the more than 2 service men can constitute a relocation team. It prefers a schedule having many identical pairs to better benefit from simultaneous relocation.

As an extended version of our previous work [9], which have presented just a separated encoding scheme, this paper includes much more detailed description of the main idea, that is, sequence-encoded relocation scheduling, and extensive performance measurement results. Our scheduler can provide its service on an EV telematics service framework embracing ubiquitous communication channels. The rest of this paper is organized as follows: Section 2 and Section 3 describe our scheduler design in detail, focusing on how to encode a relocation schedule as well as how to evaluate its fitness. Section 4 conducts extensive performance measurement via prototype implementation, with main performance metric being relocation distance. Finally, Section 5 summarizes and concludes this paper with a brief introduction of future work.

2   Encoding Scheme

The first step of an evolutionary algorithm is to encode a feasible solution, namely, a relocation schedule in our topic, by an integer-valued vector, or a chromosome. This paper proposes that each one consist of matching and sequence parts as shown in Figure 1. Inherited from our previous work [10], the matching part accounts for an unordered set of relocation pairs. The index in the vector points to the location in the overflow station list while the vector element points to the location in the underflow station list. In the overflow (underflow) station list, a station appears as many times as the number of surplus (lacking) EVs, which is decided by current and the target EV distributions. As an example, S_0 is an overflow station having 3 surplus EVs. It appears 3 times in the overflow index. In the encoded vector, number 1 is placed at location 2. It denotes the relocation from S_0 to S_1, as the 1-st entry of the overflow index is S_0 and 2-nd entry of the underflow index is S_1.

Next, the sequence part specifies the execution order of each pair. In the example of Figure 1, the relocation number 2, which corresponds to a move from S_0 to S_1, is carried out first. By this encoding, we can estimate the relocation distance straightforwardly. The intra-pair distance is retrieved after identifying the overflow and the underflow station in each vector element. In addition, the inter-pair distance can be calculated by accumulatively adding the distance between underflow and overflow stations in the next pair. After all, for the given encoded vector of Figure 1, the final relocation distance is shown in the bottom line. Here, solid arrows denote the intra-pair distance and dotted arrows the inter-pair distance, respectively. The sequence part is the main difference from our previous work.

3   Genetic Algorithm Adaptation

It must be mentioned that the number of elements in the matching part is equal to the number of EVs to move. The number of elements in the sequence part is, too. Hence, the vector length is different relocation plan by plan. The genetic operators, such as selection, reproduction, and mutation, are applied to each of two parts independently. Only selection operations handle two parts as a single entity, evaluating the fitness of a solution. In reproduction operations, two crossover points are selected in each part and substrings are swapped. The reproduction operator creates duplicated genes in a vector. The duplicated genes will be replaced by disappearing ones. In addition, mutation operators are triggered when duplicated chromosomes are found to make them different. Here, two chromosomes are different when at least one of two parts is different. As a result, each chromosome has sequence-encoded relocation pairs.

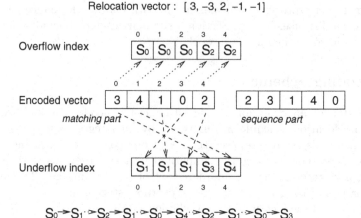

Fig. 1. Encoding scheme

In addition, in the example of Figure 1, 0 at 3 and 2 at 4 are identical pairs. S_2 is their overflow station while S_1 is their underflow station. For the case of 3 or more-man team, the second one will be removed, as 2 or more EVs can be moved simultaneously. Hence, the fitness function first scans the sequence to find identical pairs and merges into one up to the number of simultaneously relocatable EVs. Actually, hiring more service men generally reduces the relocation distance. In case the number of EVs to move gets larger for the small number of stations in a sharing system, we can expect more identical pairs will be generated. However, if there is not enough number of identical pairs, it is not so cost-effective. We can extend our scheme so as to evaluate the cost factor in the fitness function.

4   Experiment Result

This section implements the prototype version of the proposed scheme to measure its performance. As for genetic algorithm-related parameters, the population size is set to 64 and the number of iterations is to 1,000, beyond which further improvement can hardly be expected in most cases. In addition, the ratio of mutation is not explicitly specified. According to our observation, not a few identical chromosomes are generated. Hence, each time they are found, mutation operations convert them to new ones. In genetic evolution, our implementation exploits Roulette wheel selection and random initial population. Besides, the number of EVs in the sharing system is 50 while the number of stations is 10. 5 out of 10 stations are set to be underflow stations. Inter-station distance exponentially distributes with the average of 3.0 km. Finally, the experiment assumes that a single team moves all EVs, and this restriction can be eliminated by combining clusters.

The first experiment measures the relocation distance according to the number of moves, and the result is plotted in Figure 2. The figure includes two curves, one for the 2-man team and the other for the 3-man team, respectively. With 3-man team, two EVs can be moved simultaneously, significantly reducing the relocation distance especially for relocation pairs having long intra-pair distance. To begin with, in both cases, the relocation distance linearly increases according to the increase in the number of moves. However, over 25 moves, the planner finds relocation pairs having smaller intra-distance, as there are more options in relocation matching. In addition, the performance gap between 2-man team and 3-man team gets larger according to the increase in the number of moves, reaching 56.9 %, when the number of moves is 25. This result indicates that our scheme better exploits concurrent relocation operations, giving precedence to a schedule having many identical pairs.

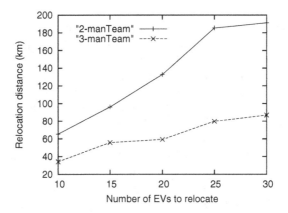

Fig. 2. Relocation distance measurement

Next, Figure 3 shows the relocation distance according to population size, the most important genetic algorithm-related parameter. Here, the number of moves is set to 10. A larger population size does not always bring smaller relocation distance, as it follows different evolution streams. Figure 3 discovers that the 3-man team is less affected by the change in the population size than the 2-man team. The maximum differences within each curve are 10.4 % and 10.0 %, respectively, while the 3-man team outperforms the 2-man team by up to 53.9 % when the population size is 40. According to this result, we can find out that the relocation distance is not significantly influenced by the population size and we can achieve reasonable quality schedule within a short response time, as a sorting procedure mainly dominates the genetic loop for the entire population. Here, the evaluation step, consisting of a sequence of arithmetic operations, brings just a small overhead.

Figure 4 measures the effect of the number of service men in a relocation team. The number of service men, ranging from 2 to 5 in this experiment, is physically

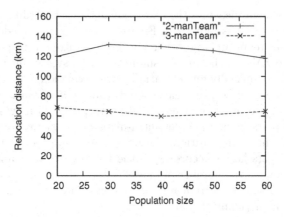

Fig. 3. Effect of population size

limited by vehicle capacity and practically by the management budget. In a 5-man team, 4 EVs can be moved simultaneously, significantly reducing the relocation distance, only if such identical relocation pairs exist. Practically, in cluster-based relocation systems [10], the number of stations is not so large and the number of EVs can get larger, generating many identical pairs. Figure 4 shows 2 curves, one for the case of 10 moves and the other for the case of 20 moves. For the case of a 2-man team, where no simultaneous relocation is expected, the relocation distance of 20 moves is twice as much as that of 10 moves. However, for the 5-man team case, the relocation distance of 20 moves is just 34.0 % of that of 10 moves. Apparently, the 20-move case better benefits from more service men, significantly reducing the relocation distance from 124.1 km to 34.6 km. This result can give system designers a guideline on how many service men are hired to meet a system goal.

In addition, Figure 5 plots the relocation distance according to the number of stations from 7 to 11 for the cases of 2-man team and 3-man team. Here, the number of EVs to move is fixed to 20. With more stations, there can be more options to select underflow stations, which can possibly reduce the intra-pair distance. However, the number of identical pairs is likely to decrease. As shown in Figure 5, the second factor has a stronger effect, as the relocation distance gets longer according to the increase in the number of stations. The case of a 3-man team is more affected, its relocation distance increasing from 33.8 to 71.3 km, which corresponds to the overhead increase by 110 %. On the contrary, the 2-man team case shows only 23.4 % increase. Hence, the number of stations is one of the most critical factors in assigning human staff members to a cluster, in addition to the terrain effect.

Finally, the relocation distance is measured according to the change in the average inter-station distance. Due to the inherent characteristics of exponential distributions, some of inter-station distances will be extraordinarily long. To cope with this situation, 20 sets are generated and their results are averaged. The experiment changes the average inter-station distance from 0.5 km to 5 km

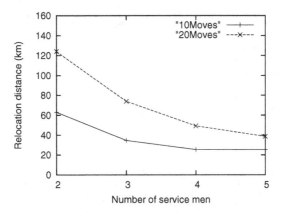

Fig. 4. Effect of the number of service men

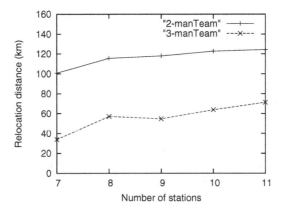

Fig. 5. Effect of the number of stations

with the step of 0.5 km. For the 2-man team case, the relocation distance is roughly linear to the average inter-station distance, as each EV must be moved one by one. The relocation distance increases by 9.63 times according to the 10 times increase of the inter-station distance, as shown in Figure 6. For the 3-man team case, the increase in the inter-station distance by 10 times leads to the increase in the relocation distance by 9.81 times. In addition, the curve of 3-man team case looks much smoother, as the number of relocation pairs is reduced.

The prototype is run on the average-performance personal computer equipped with Intel Core2 Duo CPU and 3.0 GB memory as well as installing Windows 7 operating system. It takes less than 1 second to obtain a final schedule with the proposed scheme. Actually, it is not possible to compare with an optimal schedule, which can be obtained after vast time amount. This responsiveness allows our relocation planner to be implemented as an EV information service in combination with EV tracking, charger reservation, and the like. We can create a new sophisticated service by incorporating them and sharing core information.

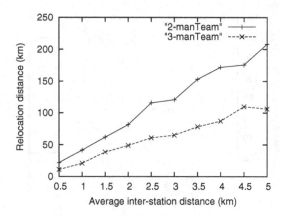

Fig. 6. Effect of inter-station distance

In addition, there is a trade-off between the number of moves and the relocation period. The relocation procedure can be triggered either during non-operation hours or on-demand basis, for example, when the operator senses significant drop-down of the service ratio of the whole system [11].

5   Concluding Remarks

Information services are extending their support to the smart grid system to make it more intelligent and achieve energy efficiency. For EV systems, it is important to prepare EV resources for the future demand forecast. In this paper, we have designed an EV resource distribution scheme for better serviceability based on genetic algorithms, considering relocation matching and the sequence at the same time. Relocation pairing is based on two extra lists, namely, one for overflow stations the other for underflow stations, creating an unordered set of relocation pairs. The sequence part determines the processing order of a relocation team, making it possible for a fitness function to estimate the relocation distance. Here, the operation concurrency is accounted for by the number of service men. Genetic operators are tailored for this sequence-encoded relocation vector, independently handling two parts.

We have implemented a prototype version for further revision and extensive performance measurement according to various system parameters such as the number of stations, underflow station occupation, the number of service men, and the like. The experiment result reveals that our scheme improves the relocation distance by up to 56.9 % for the given range of the number of moves, compared with vehicle-by-vehicle moves based on the scheduling efficiency. Particularly, the resource distributor works better when more identical relocation pairs are likely to take place, for example, there are many EVs to move, and there are fewer stations. In addition, it shows a relatively stable behavior even in the change in the average inter-station distance, eliminating relocation pairs having too long intra-pair distance.

The proposed scheme is currently targeting at short-term rent-a-car systems, or EV sharing systems. However, it can also work for long-term daily rent-a-car services. Such services are instances of optimization problems, which can be handled by genetic algorithms, simulated annealing schemes, and many other heuristics. With reasonably short response time, it can provide services combined with other EV telematics components. As future work, we are planning to apply our resource sharing solution to diverse EV systems, especially developed in Jeju city, which tries to build a city-wide EV information service framework to prompt the penetration of EVs in various transport. It includes taxis, rent-a-cars, and the like, mainly focusing on the environmental viewpoint. It will involve a large volume of data collected from EVs and charging facilities [12].

References

1. Cepolina, E., Farina, A.: A New Shared Vehicle System for Urban Areas. Transportation Research Part C, 230–243 (2012)
2. Lue, A., Colorni, A., Nocerino, R., Paruscio, V.: Green Move: An Innovative Electric Vehicle-Sharing System. Procedia-Social and Behavioral Sciences 48, 2978–2987 (2012)
3. Wang, H., Cheu, R., Lee, D.: Logistical Inventory Approach in Forecasting and Relocating Share-use Vehicles. In: International Conference on Advanced Computer Control, pp. 314–318 (2010)
4. Correia, G., Antunes, A.: Optimization Approach to Depot Location and Trip Selection in One-Way Carsharing Systems. Transportation Research Part E, 233–247 (2012)
5. Weikl, S., Bogenberger, K.: Relocation Strategies and Algorithms for Free-Floating Car Sharing Systems. In: IEEE Conference on Intelligent Transportation Systems, pp. 355–360 (2012)
6. Sivanandam, S., Deepa, S.: Introduction to Genetic Algorithms. Springer (2008)
7. Lee, J., Park, G.: Planning of Relocation Staff Operations in Electric Vehicle Sharing Systems. In: Selamat, A., Nguyen, N.T., Haron, H. (eds.) ACIIDS 2013, Part II. LNCS (LNAI), vol. 7803, pp. 256–265. Springer, Heidelberg (2013)
8. Wang, H., Cheu, R., Lee, D.: Dynamic Relocating Vehicle Resources Using a Microscopic Traffic Simulation Model for Carsharing Services. In: International Joint Conference on Computational Science and Optimizations, pp. 108–111 (2010)
9. Lee, J., Park, G.: Sequence-Encoded Resource Relocation Scheme for Electric Vehicle Information Systems. In: International Conference on Advanced Computing and Services (2013)
10. Lee, J., Park, G.-L., Lee, I.-W., Park, W.K.: Relocation matching for multiple teams in electric vehicle sharing systems. In: Pathan, M., Wei, G., Fortino, G. (eds.) IDCS 2013. LNCS, vol. 8223, pp. 260–269. Springer, Heidelberg (2013)
11. Kek, A., Cheu, R., Meng, Q., Fung, C.: A Decision Support System for Vehicle Relocation Operations in Carsharing Systems. Transportation Research Part E, 149–158 (2009)
12. Fang, X., Yang, D., Xue, G.: Evolving Smart Grid Information Management Cloudward: A Cloud Optimization Perspective. IEEE Transactions on Smart Grid 4, 111–119 (2013)

The Relationships between Depression Spatial Clusters and Mental Health Planning in Catalonia (Spain)

Maria Luisa Rodero-Cosano[1], Jose Alberto Salinas-Perez[1],
Juan Luis Gonzalez-Caballero[2], Carlos R. Garcia-Alonso[1], Carolina Lagares-Franco[2],
and Luis Salvador-Carulla[3] for the GEOSCAT group.

[1] Universidad Loyola Andalucía, C/Escritor Castilla Aguayo, Córdoba, Spain
{mlrodero,jsalinas,cgarcia}@uloyola.es
[2] Universidad de Cádiz, Cádiz, Spain
{juanluis.gonzalez,carolina.lagares}@uca.es
[3] University of Sydney, Sydney, Australia
luis.salvador-carulla@sydney.edu.au

Abstract. This study aims to analyse potential risk factors which could influence the occurrence of hot spots of depression. They cannot only be explained through municipal socio-demographic characteristics and which is why causes at catchment area level should also be studied. Indicators at both spatial levels were analysed by a multi-level regression model. The analysis included various socio-demographic, geographical and service allocation indicators. According to scientific literature, unemployment and rurality were identified as risk factors for depression and, therefore, for hot spots. On the other hand, low educational levels and poor accessibility showed little relationship here while other studies indicated otherwise. Preliminary results described diverse risk factors at two levels which were related to a high likelihood of hot spots, although more in-depth analysis will be needed.

Keywords: Spatial cluster, risk factor, multilevel analysis, catchment areas, mental health.

1   Background

Spatial data can be analysed through a large and growing number of spatial statistical methods for detecting spatial patterns in their geographical distribution [1]. Spatial clustering analysis usually has two stages [2]: first, to identify spatial clustering by an exploratory spatial data analysis; second, to relate clusters to other factors in order to find their original causes by some multivariate method.

Epidemiological research uses these analyses intensively to study spatial patterns in diseases [3,4]. In this area, Multi-objective Evolutive Algorithms were used (MOEA), especially designed for solving multi-objective spatial problems [5] in order to identify hot spots and cold spots of treated prevalence of depression at municipality level in Catalonia (Spain) [6]. This research aimed to locate these significant spatial clusters, describing them together with some characteristics of mental health catchment areas;

B. Murgante et al. (Eds.): ICCSA 2014, Part IV, LNCS 8582, pp. 438–447, 2014.

it did not seek, however, to link them to socioeconomic risk factors of depression or other factors potentially related to variations in medical practice.

Scientific literature has related depression to different socio-economic indicators. For instance, unemployment [7] and rurality [8] are considered to be among its major risk factors. In addition to these, other socio-economic indicators such as poverty, belonging to minorities, gender and low incomes have been related to such prevalence and hospital admissions for depression [9].

Thus, socio-economic variables are key but there may be other factors involved. In a previous article, mental health catchment areas were described through several indicators related to specialised service allocation such as their accessibility, availability and adequacy. This last concept is the smallest geographical healthcare area where different studies have found the existence of variations in health care delivery that could be useful for health planning [10,11]. In mental health, the smallest specialized care area is usually the catchment area served by Mental Health Community Centres.

Thus, socio-economic and demographic characteristics of the population and indicators in health service allocation, at two geographical scales or levels (municipality and mental health catchment area), may be related to spatial clusters of very high or low depression cases. Hierarchical linear models are specially designed to analyse data at different levels, in this case geographic scales. These models often use health fields [12,13,14,15,16,17,18], covering many of the multilevel models described in the theory.

The main objective of this article is to analyse potential risk factors which may influence the occurrence of hot spots or spatial units where treated prevalence scores for depression in Catalonia are significantly high. Depression clusters have been identified and described in a previous paper by a MOEA. The relationships are studied by two multilevel models which use several socio-economic and service provision indicators at municipal scale and in catchment areas.

2   Methods: Multilevel Logistic Regression Models

In general, univariate multilevel models [19,20] assume that there is a hierarchical data set, with one single outcome or response variable (Y) that is measured at its lowest level, and explanatory variables at the rest of existing levels. If the outcome is a dichotomous or binary variable representing the presence or absence of a characteristic, our interest is in predicting the probability (percentage) of subjects that present the characteristic ; then the common approach is to use generalized linear models [21,22], in particular the logistic regression model. In this case, Raudenbush & Bryk [23] or Goldstein [19] describe the multilevel extension of generalized linear models that give explanatory variables (x_{ij}) for the different levels. The probability of the desired outcome $P(Y = 1|x_{ij}) = \pi_{ij}$ is predicted using the logit function $\text{Logit}(\pi_{ij}|x_{ij}) = \log[\pi_{ij}/(1 - \pi_{ij})]$ in two steps: firstly, modelling the linear regression equation at the lowest level with the explanatory variables of the logit function of π_{ij} and, later, using the inverse of the logit function; thus we can predict the probability π_{ij} by explanatory variables at different levels.

The OR obtained by the model represents the odds that an outcome will occur given a particular value x_{ij}^*, compared to the odds of the outcome occurring with another value x_{ij}. Special interpretation is obtained when an explanatory variable is binary, representing whether there has or has not been exposure to a factor to evaluate the association between the exposure to the factor and its outcome. OR represent the ratio between the odds for the presence and absence of the exposure factor, obtained by e^β where β is the specific coefficient of the explanatory binary variable.

HS was modelled to allow the differentiation of relationships. The dependent variable is whether there is (1) or is not (0) a HS (binary). The independent variables for catchment areas and municipality levels are explained above.

With this amount of variables at both levels, and since our hypothesis is not based on a strong theory, we have used an exploratory procedure to select the most parsimonious model. The strategy to construct the linear model for the logit component has been from the bottom up, starting with a single model and proceeding to add parameters which are tested using the Wald test for significance after they have been added. Due to that, sometimes there was recoding of some initially numerical variables into categorical variables using the percentiles, or reducing categories in initially categorical variables.

We have started by building [20] the intercept-only model, then the best model was fitted with fixed lower-level explanatory variables, and finally the higher-level explanatory variables were added. All models were built nested, using full maximum likelihood via adaptive Gaussian quadrature [24].

3   The Study Case

3.1   Scope of the Study

Catalonia is a Spanish region located in the Northeast of the Iberian Peninsula with 7.5 million inhabitants. The Catalonian Public Health System is universal, as in the rest of the country, and the provision of care includes both public and private organisations under contract agreements with the health administration. Regarding mental health care, Mental Health Community Centres for adults are assigned to large city districts or whole municipalities forming catchment areas or small mental health areas [25]. Catalonia is divided into 74 catchment areas, which include a total of 946 municipalities. Catchment area levels were reduced to 60 because several areas in populated cities corresponded to infra-municipal level and so they were grouped into 7 units which coincide with municipal levels.

3.2   The Dependent Variable: Depression Hot Spot in Catalonia

Depression clusters were identified by a MOEA, located and described in a previous article [6]. The multivariate spatial problem was to find groups of close municipalities (spatial unit of analysis) with significantly high prevalence through the maximization of the treated prevalence of depression, minimizing the standard deviation of prevalence and minimizing the distance between spatial units. The MOEA's solutions were evaluated by means of four fitness functions selecting the most frequent municipalities. Thus, this type of problem does not have a unique solution. Finally, the results were mapped in a Geographic Information System (GIS).

The MOEA found six hot spots formed by 39 municipalities which were included within 13 catchment areas. The first is situated in the Seu d'Urgel region located in Lleida, which is mainly a rural area whose activity focus is on agriculture. The second is Baix Berguedà near the Seu d'Urgel region which lies in transition to the plains of the Catalan Central Depression. This region has always been sparsely populated and here agriculture, cattle ranching and forestry have proven generally complementary to one another and compatible with tourism. The third is in the Catalonia central area in Barcelona including the North Anoia region with its important paper-making industry, along with the neighbouring municipalities of La Segarra, a grain-growing region. The fourth is located in Barcelona province within the Vallès Oriental region, which is an urban area whose main economic activity focuses on the industrial and services sectors. The fifth, as in the case mentioned above, is a rural zone with a major wine sector. Finally the last hot spot is located in Borges Blanques, mainly a rural area.

Municipality hot-spots have an average of 8.16 cases per 1,000 inhabitants: the median is 7.85 and its standard deviation is 6.49. The highest value is 35.80 while the lowest is 1.74 per 1,000. The main statistical values are shown in table 1.

Table 1. Basic statistics of the dependent variable

Su	Nº mun.	Catchment areas	Mean	Median	Standard deviation
HS1	3	Bages	9.9	8.3	3.4
HS2	8	Bages, Berga and Osona	12.0	9.0	10.1
HS3	7	Anoia and La Segarra	11.6	9.6	6.6
HS4	11	Vallès Oriental and Osona	4.6	3.3	3.2
HS5	9	Alt Penedès, Garraf, Gavà, Martorell and Sant Feliu	5.2	5.4	2.0
HS6	1	Borges Blanques	13.9	13.9	

Thus, the hot spot municipality is a binary variable that divides municipalities into two classes depending on whether they have not been detected as hot spots (category 0) or they have been (category 1), located and described above. Therefore, the dependent variable corresponded to level 1 in our multi-level analysis.

3.3   Independent Variables

Level 1: Municipalities.
Scientific literature has related depression to different socio-economic indicators, such as unemployment, poverty, rurality etc. Consequently it is also important to analyse the relationship between the spatial cluster with high values of depression-treated prevalence and socio-economic indicators. Based on studies on the spatial distribution of mental-illness prevalence, the indicators selected are: population density, unemployment, income and university studies.

Depression-treated prevalence (PRE) is a variable which was calculated using the gender, age, municipality of residence and main diagnosis through the direct method [26] that took into consideration the population of Catalonia and is measured in cases per 1,000 inhabitants.

The population density (DENS) is a binary variable related to rurality [27]. The municipalities were separated into two groups: those whose population density is lower than 45 inhabitants per square kilometer (0) indicating municipalities with very low density and, therefore, highly rural; and those whose population is greater or equal to 45 inhabitants per square kilometer (1) which indicates moderately rural municipalities.

The unemployment rate (UNE) measures the prevalence of unemployment and its percentage is calculated by dividing the number of unemployed individuals by the active population currently between 16-64 years of age [28].

Income (INC) is a continuous variable that allows the economic poverty level to be evaluated by measuring the income per inhabitant [29].

University studies (UNI) is a proxy indicator of the educational level of municipality populations and was calculated using the number of inhabitants who had finished their degrees [28]. Municipalities were classified in three categories: values below 7.39 (0), values between 7.39 and 9.86 (1) and values above 9.86 (2).

The main statistics of these variables can be studied in Table 2.

Table 2. Municipal level variables initially considered independent in the analysis

							Percentiles				
	Type	N	Mean	SD.	Mín	Máx	25h	33.3h	50h	66.7h	75h
PRE	Num.	946	2.57	2.9	0	35.8	0	1.1	1.99	2.97	3.61
DEN	Num.	946	437.47	1562.9	0.93	18871.88	13.01	20.42	44.99	110.28	182.74
UNE	Num.	946	7.15	3.18	0	21.14	4.81	5.6	6.93	8.46	9.46
INC	Num.	943	17840.53	4192.9	7403	40109	15076	16157	17717	19030.33	19845
UNI	Num.	946	9.21	3.76	0.85	36	6.67	7.39	8.48	9.86	10.8

Level 2: Mental Health Catchment Areas.
Catchment areas in Catalonia have been previously described in four domains: urbanicity, service availability, accessibility to care and adequacy or appropriateness. These domains have been used in previous studies on the spatial distribution of mental-disorder prevalence. If hot/cold spots are spatially associated with specific catchment areas, it could be relevant to analyse if they are mainly rural or urban, if their accessibility is high or not, and so on.

The percentage of hot spots (PHS) is a numerical variable that indicates the percentage of hot spot municipalities in the catchment area.

Urbanicity (URB) was determined following a classification of the OECD [30], so that the variable was classified as being predominantly urban (0), significantly rural (1) and predominantly rural (2). The level 'predominantly urban' means 85% of the inhabitants reside in municipalities whose density is greater than 150 inhabitants/km^2, 'significantly rural' when this percentage is between 50% and 84%, and 'predominantly rural' when it is lower than 50%.

The accessibility (ACE) to the MHCC of each catchment area was assessed using a standard Geographical Information System (GIS) and was obtained from an article which studied accessibility to health services in Catalonia [31]. Accessibility was measured in minutes taken by car to the corresponding MHCC from the least accessible zone of the catchment area in intervals of fifty minutes, consequently selected intervals are: 0-15 min (0), 15-30 min (1), 30-45 min (3), 45-60 min (4), and >60 min (5).

MHCC availability (AVA) was measured by the rate of outpatient MHCC per 100,000 inhabitants. It indicates the relationship between the number of MHCC and the inhabitants in the catchment area. This indicator does not consider the differences in staff allocated to each MHCC. Therefore, the variable was classified as adequate when the values were within the range 1-2.5 (0) and inadequate when the values were outside this range, that is, <1 or >2.5 (1).

Lastly, the adequacy (ADE) of the provision of mental health services in the catchment areas was assessed by a group of PSICOST experts using information from the Mental Health Atlas of Catalonia [25]. This assessment includes all types of services such as hospitalization units, day hospitals, day centres, etc. Experts rated 7 levels of provision in every catchment area in (very high, high, medium high, medium, medium low, low and very low). This rating was represented in semaphore scale and agreed with official results from the Department of Health of Catalonia. Level 1 indicates that all types of services have been allocated to the catchment area and that most of them are located within it, while level 7 indicates that some types of services have not been allocated to it. For this research the variables are classified into two values: 0 when adequacy is very high or high and 1 when adequacy is medium, low or very low.

The main statistics of these variables can be studied in Table 3.

Table 3. Catchment area level variables initially considered in the analysis

Numerical Variables	N Municipalities	Mean	Sd.	Min	Max	Median 50th percentile
NUM (Number of municipalities in the catchment area)	946	15.77	17.33	1.00	73.00	10.00
PHS	39	3.34	7.83	0.00	29.41	0.00

4   Results

Taking into account the above explanation, we can obtain the model coefficients whose reduction allows us to assess the most fitting model. Based on the results of Model 1, we can see that there are differences in prevalence between areas of mental health, as well as between municipalities, although population differences at this level are more pronounced.

According to the results of the model (table 4), rurality, accessibility and adequacy are the three most important risk factors at catchment area level, and the unemployment rate shows a protective factor.

Table 4. Final estimation of fixed effects (full maximum likelihood via adaptive Gaussian quadrature)

Fixed Effect	Coefficient	Standard error	t-ratio	Approx. d.f.	p-value	Odds ratio	Confidence interval
For INTERCEPT, β_0							
INT, γ_{00}	-2.991	1.035	-2.890	58	0.005	0.050	(0.006,0.399)
URB, γ_{01}	-1.690	0.611	-2.766	58	0.008	0.184	(0.054,0.627)
For PRE slope, β_1							
INT, γ_{10}	0.288	0.111	2.584	878	0.010	1.333	(1.072,1.659)
ACC, γ_{11}	0.111	0.047	2.333	878	0.020	1.117	(1.018,1.226)
For DENS slope, β_2							
INT, γ_{20}	3.013	0.905	3.328	878	<0.001	20.352	(3.442,120.335)
ADE, γ_{21}	-2.651	1.208	-2.194	878	0.028	0.071	(0.007,0.756)
For UNE slope, β_3							
INT, γ_{30}	-0.577	0.144	-4.015	878	<0.001	0.562	(0.424,0.745)
PHS, γ_{31}	0.014	0.003	4.986	878	<0.001	1.014	(1.009,1.020)
ADE, γ_{32}	0.063	0.034	1.857	878	0.064	1.065	(0.996,1.138)
For UNI3 slope, β_4							
INT, γ_{40}	-0.646	0.291	-2.222	878	0.027	0.524	(0.296,0.927)

In short, the intercept in the model is the expected log-odds of HS for a municipality with zero value for all the predictor variables. In this case, this expected log-odds corresponds to a probability of $1/[1+\exp(2.991)]=0.047$, that is approximately 39/946.

In level 1, university studies directly reduce the probability of being a hot spot. This factor decreases the probability of being a hot spot when taking values 0, 1, 2. In this case, using as reference the municipalities whose percentage is under 7.39 of university, the probability of being a hot spot must be divided by 2 (OR=$e^{-0.646}$ = 0.524) for municipalities with a percentage between 7.39 and 9.86, and divided by 4 (OR=$e^{-0.646*2}$ = 0.275) for municipalities with a percentage over 9.86. The employment-rate factor may seem to act as a protective factor for a municipality hot spot, since the probability would have to divide it by 2 (OR=$e^{-0.577}$ = 0.561). However, this factor affects the variable percentage of hot spots and adequacy in level 2. This causes the probability of being a hot spot to increase for each unit increase in the percentage of the hot spot factor (OR=$e^{0.014}$ = 1.014) and increase in catchment areas with poorer adequacy (OR=$e^{0.063}$ = 1.065).

In principle, density values above 45 produce an increase in probability, multiplying it by 20 (OR=$e^{3.013}$ = 20.35). However, probability is roughly divided by 14 (OR=$e^{-2.651}$ = 0.071) in municipalities with density values higher than 45 and medium or low adequacy values. Finally, the prevalence factor increases the probability per each unit increase of prevalence (OR=$e^{0.288}$ = 1.333), and influences the accessibility variable increasing the probability of its being a hot spot.

In level 2, urbanicity affects in such a way that the more rural the municipality is (0 to 2), the less probability it has of being a hot spot. Regarding a predominantly urban municipality, the probability is reduced to almost one fifth (OR=$e^{-1.69}$ = 0.184) in the case of being significantly rural, and reduced almost 30 times (OR =$e^{-1.69*2}$ = 0.034) in the case of being predominantly rural. Moreover, for the same value of prevalence, the municipality may experience an increase in the probability of being a hot spot for each 15 minute increase of time regarding accessibility (OR=$e^{0.111}$ = 1.117).

5   Discussion

As mentioned in the introduction, scientific literature has related depression to different socio-economic indicators, among these highlighting unemployment [7] and rurality [8] as the most important risk factors. In fact, urban counties in the USA had higher hospitalization rates for depression than rural counties [9]. The results obtained are coherent with respect to this relationship.

Unemployment was the second most important socio-demographic risk factor for depression in Spain [32] in accordance with the literature [33]. High depression admissions in hospitals are related to a high unemployment rate [9]. In our research the employment rate factor may seem to act as factor which reduces the risk that a municipality is hot spot, although this factor affects the percentage of hot spot variables and adequacy at level 2, which coincides with the above study.

Focusing on the educational level, according to the scientific literature it was found to be related to the prevalence of depression and anxiety [34]. Patients who are better educated make significantly more medical visits due to depression than other patients [35]. However, hospital admission of patients with depression is not related to the educational level [9]. Nevertheless, the results show that a high level education reduces the risk of depression prevalence.

Finally, poor accessibility has appeared as a risk factor increasing depression prevalence in our research. However, other authors have supported that high accessibility has previously been associated with patients with depression making more mental health visits [35]. Therefore, this variable needs further study.

6   Conclusion

The spatial data analysis of depression, as well as schizophrenia, was included in the mental health atlas of Catalonia which is the first mental health report in Spain about integral mental health care. Spatial analysis has allowed the identification of geographical areas where the distribution of depression is not random and may be due to well-known but also unknown risk factors. This research endeavours to be an initial approach to the identification of these factors at two geographical levels, although deeper analysis must be carried out. These results may help planners and decision-makers in their search for efficiency, quality and equality in mental health care.

References

1. Shekhar, S., Evans, M.R., Kang, J.M., Mohan, P.: Identifying patterns in spatial information: A survey of methods. Wiley Interdiscip. Rev. Data Min. Knowl. Discov. 1(3), 193–214 (2011)
2. Wakefield, J., Kelsall, J.E., Morris, S.E.: Clustering, cluster detection, and spatial variations in risk. In: Elliot, P., Wakefield, J., Best, N., Briggs, D. (eds.) Spatial Epidemiology. Oxfor University Press, New York (2000)

3. Auchincloss, A.H., Gebreab, S.Y., Mair, C., Diez, A.V.: A Review of Spatial Methods in Epidemiology, 2000–2010. Annu. Rev. Public Health 33(1), 107–122 (2012)
4. Elliott, P., Wartenberg, D.: Spatial epidemiology: Current approaches and future challenges. Env. Health Perspect. 112(9), 998–1006 (2004)
5. García-Alonso, C.R., Salvador-Carulla, L., Negrín-Hernández, M.A., Moreno-Küstner, B.: Development of a new spatial analysis tool in mental health: Identification of highly autocorrelated areas (hot-spots) of schizophrenia using a Multiobjective Evolutionary Algorithm model (MOEA/HS). Epidemiol Psichiatr Soc. 19(4), 302–313 (2010)
6. Salinas-Pérez, J.A., García-Alonso, C.R., Molina-Parrilla, C., Jordà-Sampietro, E., Salva-dor-Carulla, L.: Identification and location of hot and cold spots of treated prevalence of depression in Catalonia (Spain). Int. J. Health Geogr. 11, 36 (2012)
7. Jefferis, B.J., Nazareth, I., Marston, L., Moreno-Kustner, B., Bellón, J.Á., Svab, I., et al.: Associations between unemployment and major depressive disorder: Evidence from an international, prospective study (the predict cohort). Soc. Sci. Med. 1982 73(11), 1627–1634 (2011)
8. Wang, J.L.: Rural–urban differences in the prevalence of major depression and associated impairment. Soc. Psychiatry Psychiatr. Epidemiol. 39(1), 19–25 (2004)
9. Fortney, J.C., Rushton, G., Wood, S., Zhang, L., Xu, S., Dong, F., et al.: Community-Level Risk Factors for Depression Hospitalizations. Adm. Policy Ment. Health Ment. Health Serv. Res. 34(4), 343–352 (2007)
10. Kelly, A., Jones, W.: Small area variation in the utilization of mental health services: Implications for health planning and allocation of resources. Can J. Psychiatry Rev. Can Psychiatr. 40(9), 527–532 (1995)
11. Mercuri, M., Birch, S., Gafni, A.: Using small-area variations to inform health care service planning: What do we «need» to know? J. Eval. Clin. Pract. 19(6), 1054–1059 (2013)
12. Barnett, S., Roderick, P., Martin, D., Diamond, I.: A multilevel analysis of the effects of rurality and social deprivation on premature limiting long term illness. J. Epidemiol. Community Health 55(1), 44–51 (2001)
13. Carey, K.: A multilevel modelling approach to analysis of patient costs under managed care. Health Econ. 9(5), 435–446 (2000)
14. Chandola, T., Clarke, P., Wiggins, R.D., Bartley, M.: Who you live with and where you live: setting the context for health using multiple membership multilevel models. J. Epidemiol Community Health 59(2), 170–175 (2005)
15. Eikemo, T.A., Bambra, C., Judge, K., Ringdal, K.: Welfare state regimes and differences in self-perceived health in Europe: A multilevel analysis. Soc. Sci. Med. 1982 66(11), 2281–2295 (2008)
16. King, M.T., Kenny, P., Shiell, A., Hal, J., Boyages, J.: Quality of life three months and one year after first treatment for early stage breast cancer: Influence of treatment and patient characteristics. Qual. Life Res. Int. J. Qual. Life Asp. Treat Care Rehabil. 9(7), 789–800 (2000)
17. Mitchell, R., Gleave, S., Bartley, M., Wiggins, D., Joshi, H.: Do attitude and area influence health? A multilevel approach to health inequalities. Health Place 6(2), 67–79 (2000)
18. Wong, I.O.L., Cowling, B.J., Lo, S.V., Leung, G.M.: A multilevel analysis of the effects of neighbourhood income inequality on individual self-rated health in Hong Kong. Soc. Sci. Med. 1982 68(1), 124–132 (2009)
19. Goldstein, H.: Multilevel statistical models. Arnold, London (2003)
20. Hox, J.J.: Multilevel analysis: Techniques and applications, 2nd edn. Routledge, New York (2010)

21. Gill, J.: Generalized linear models: A unified approach. Sage Publications Inc., Thousand Oaks (2001)
22. McCullagh, P., Nelder, J.A.: Generalized linear models, 2nd edn. Chapman & Hall/CRC, Boca Raton (1998)
23. Raudenbush, S.W., Bryk, A.S.: Hierarchical linear models: Applications and data analysis methods. Sage Publications, Inc., Thousand Oaks (2002)
24. Raudenbush, S.W., Yang, M.L., Yosef, M.: Maximum Likelihood for Generalized Linear Models with Nested Random Effects via High-Order, Multivariate Laplace Approximation. J. Comput. Graph. Stat. 9(1), 141–157 (2000)
25. Salvador-Carulla, L., Serrano-Blanco, A., García-Alonso, C.R., Fernández, A., Salinas-Pérez, J.A., Ruiz, M., et al.: Integral Map of Mental Health Resources of Catalonia 2010. Direcció General de Planificació i Recursos Sanitaris Generalitat de Catalunya (2013) http://www20.gencat.cat/docs/canalsalut/Home%20Canal%20Salut /Professionals/Temes_de_salut/Salut_mental/documents/pdf/mem oria_integral_atles_sp.pdf.
26. Rezaeian, M., Dunn, G., Leger, S.S., Appleby, L.: Geographical epidemiology, spatial analysis and geographical information systems: A multidisciplinary glossary. J. Epidemiol. Community Health 61(2), 98–102 (2007)
27. National Statistics Institute: Revision of the 2009 municipal register (2010), http://www.ine.es/jaxi/menu.do?type=pcaxis&file=pcaxis&path= %2Ft20%2Fe245%2Fp05%2F%2Fa2009
28. National Statistics Institute: Economically Active Population Survey (2010), http://www.ine.es/en/inebaseDYN/epa30308/epa_microdatos_en.htm
29. National Statistics Institute: Household budget survey, Base 2006 (2010), http://www.ine.es/en/prodyser/micro_epf2006_en.htm
30. OECD: Creating rural indicators for shaping territorial policy. Organisation for Economic Co-operation and Development, Paris (1994)
31. Olivet, M., Aloy, J., Prat, E., Pons, X.: Health services provision and geographic accessibility. Med. Clínica 131(suppl. 4), 16–22 (2008)
32. Gabilondo, A., Rojas-Farreras, S., Vilagut, G., Haro, J.M., Fernández, A., Pinto-Meza, A., et al.: Epidemiology of major depressive episode in a southern European country: Results from the ESEMeD-Spain project. J. Affect. Disord. 120(1-3), 76–85 (2010)
33. Fryers, T., Melzer, D., Jenkins, R.: Social inequalities and the common mental disorders: A systematic review of the evidence. Soc. Psychiatry Psychiatr. Epidemiol. 38(5), 229–237 (2003)
34. Sabes-Figuera, R., McCrone, P., Bogic, M., Ajdukovic, D., Franciskovic, T., Colombini, N., et al.: Long-Term Impact of War on Healthcare Costs: An Eight-Country Study. PLoS ONE 7(1), e29603 (2012)
35. Fortney, J.C., Rost, K., Zhang, M., Warren, J.: The impact of geographic accessibility on the intensity and quality of depression treatment. Med. Care 37(9), 884–893 (1999)

European Air Traffic:
A Social and Geographical Network Analysis[*]

Gabriella Schoier and Giuseppe Borruso

DEAMS – Department of Economic, Business, Mathematic and Statistical Sciences,
University of Trieste, Via A. Valerio, 4/1 – 34127 Trieste, Italy
{gabriella.schoier,giuseppe.borruso}@econ.units.it

Abstract. In this paper we consider methods to highlight clusters in a network structure. Network structures are to-date widely present in transport, telecommunication and, increasingly, in social networks and media. Big amounts of data are being created and methods to find patterns and scheme in the network structure and distribution are required. In this paper we use a sample dataset of passengers' air traffic flows between European countries to test social network analysis algorithm, comparing them to more classical geographical techniques as the nodal regional analysis.

Keywords: Clustering algorithms, social network analysis, Spatial data mining, GIS.

1 Introduction

Network structures are to-date widely used and recognizable in different fields, as transport, telecommunication networks, social media and networks. Basic structures as nodes and connections between them allow huge amount of flows to be located and draw patterns not immediately recognizable on them. The huge increase in the amount of spatially referenced data available on Internet has outstripped our capacity to meaningfully analyses such networks and run into significant computational barriers in large networks. This has induced the need for better analysis techniques to understand the various phenomena. Geographers have studied networks and their structure relying on the different methods and tools. To-date different classical social network theories can be used; among these the small-world literature has shown that there is a high degree of local clustering in the networks, this suggests that an approach for studying the structure of large networks would involve first the identification of local clusters and then the analysis of the relations within and between clusters. On the other hand papers on peer influence have shown that, based on an endogenous influence process, close units tend to converge on similar attitudes and thus clusters in a small-word network should be similar along multiple dimensions. Our aim here is to

---

[*] The paper derives from the joint reflections of the three authors. Giuseppe Borruso wrote paragraphs 2.2, 2.3, 3.1 and 3.2 while Gabriella Schoier realized the other paragraphs.

B. Murgante et al. (Eds.): ICCSA 2014, Part IV, LNCS 8582, pp. 448–462, 2014.
© Springer International Publishing Switzerland 2014

explore such phenomena, in terms of the presence of clusters in network data analysis and in terms of the connections between nodal elements, trying to understand the origins and motivations of such elements in space. In this paper the attention is focused on methods to highlight such clusters and see their impact on the network structure. Spatial clustering algorithms as well as network-based techniques will be evaluated and tested over a sample dataset.

2   Social Network Theory and Geographical Networks

2.1   Social Network Analysis

As it is well known social network analysis has become very popular in recent times. Much of this interest can be due to the appealing focus of social network on relationships among social entities not only on real life but on the Web. In general the network perspective can give formal definition to different aspects of the political, economical or social environment and, with the increasing of importance of the relations established on the web, to our life in its different aspects.

The basic idea of social network analysis is an assumption of the importance of relationships among interacting units. Theories, models, and applications are expressed in terms of relational concepts or processes." Along with growing interest and increased use of network analysis has come a consensus about the central principles underlying the network perspective" (Wassermann, 1994).

Social network analysis is a set of mathematical methods for the analysis of social structure. These methods bring towards an investigation of the relational aspects of the members of a group (efficiency, moral satisfaction, leadership). "The use of these methods, therefore, depends on the availability of relational rather than attribute data" (Scott, 1992).

The aim of positional analysis is to simplify the information and its relations as found in a database of subjects. Fundamental concepts in network analysis are actors, relational tie, relation, dyad, triad, subgroup, group and network.

As the main purpose of a network analysis is understanding the link between entities and the implications of these linkages the starting points are these entities are viewed as interdependent rather than independent, autonomous units. The entities are called *actors, node or point*. They can be individual, web pages, I.P. addresses, spatial entities etc.

Actors are linked to one another by *tie, edges, links* or *arcs*, the collections of ties of a specific kind among members of a group is called a *relation*. The relation can be single or multiple. Relational ties (linkages) between units are channels for transfer of resources. This transfer can be either material or nonmaterial depending on the relations and on the actors.

In the case of Web data the actors may be the visited pages while the relation between two pages   *sharing common users* (single relation) or the actors may be the I.P. addresses and the relation *having viewed a certain amount of pages in common and having stayed a certain amount of time over them* (multiple relation). In the spatial case the actors can be the airports and the relation *the flows in terms of passengers* among them (single relation) or *the flows in terms of passengers and goods* (multiple relation).

At basic level the relationship shows a tie between actors, many positional analyses are concerned with understanding ties between actors. The unit of analysis in network analysis is not the individual, but an entity consisting of a collection of individuals and the linkages among them. For this reason the methods used focus on dyads, triads, or larger systems subgroups or groups of individuals, or entire networks. A *dyad* consists of a pair of actors and the possible tie(s) between them. The analyses based on dyads are focused on pairwise relationships. The concept of dyad can be extended to a subset of three actors and the (possible) tie(s) among them, in this case we refer to a *triad*.

Dyads are pairs of actors and their ties, more in general a *subgroup* is subset of actors and all ties among them. A *group* is collection of all actors on which ties are to be measured this notion depends on the context in which it is used. For instance in the Web analysis it is connected with Internet characteristics while in Spatial Analysis it deals with spatial attributes which have to take into account in all the analysis.

Let $E = \{x_1, x_2, ..., x_n\}$ be a finite set of units or *actors*, these are related by binary relations: $R_t \subseteq E \times E$ (for $t = 1,...,r$) which determine a *network* (set of units and relation(s) defined over it) N: $N = (E, R_1, R_2, ..., R_r)$.

In the following we will consider a single relation R described by a corresponding binary matrix:

$$\mathbf{R} = \{r_{ij}\} \ i, j = 1,.., n \quad \text{where } r_{ij} = \begin{cases} 1 & \text{if } x_i \ R \ x_j \\ 0 & \text{otherwise} \end{cases} \tag{1}$$

There are two modes of representing a network. Through graphs or matrices.

A graph is a model used to represent relations between pairs of units in a network, thre are different types of graphs: non directional graphs, directional graphs, signed graphs, valued graphs, multivariate graphs, hypergraphs.

Usually data are stored in the so called adjacency matrix. Commonly, the [i,j] element of the adjacency matrix corresponds to the communication behavior of actor x_i to actor x_j The adjacency matrix is a *one-mode* matrix, in which the rows and the columns represent the nodes of the graph.

$$X = \{x_{ij}\} \ i, j = 1,...,n \quad \text{where } x_{ij} = \begin{cases} 1 & \text{if } n_i \text{ is adjacent to } n_j \\ 0 & \text{otherwise} \end{cases}$$

Social networks is characterized by a methodology which gathers different techniques such as graphs, matrices, structural and location properties with the notions of centrality, structural balance transitivity and cohesive subgroups, role and position with the notions of structural equivalence, blockmodeling (Schoier, 2002), relational algebra, network position and roles, different statistical methods, etc...

2.2   Geographical Networks

A geographical network recalls a well-defined geometric structure, where relations and liaisons between basic elements building the network itself are set. Nodes and segments (or points and lines) are the basic elements that build and characterise networks. Nodes represent a set of spatially distributed locations which contain a set of potential relations to be settled between each other. Segments (or lines) are the means by which nodes can be related and linked. Networks are therefore consistent means to analyse spatial interaction phenomena (Hagget and Chorley, 1968).

According with different authors a network is aimed to represent the relations a group set with a portion of space and is composed of points, nodal features and attributes. A network setting these relations over space and because of the creation of privileged means to set such relations lead us to talk about network space and therefore network geography. It must however be kept in mind that network phenomena are usually linked to some attributes as network's existence on space is not justified *per se* but as it allows different kinds of connections and contacts between elements over space.

Immaterial networks remind the set of relations that can be set between different points or nodes composing the network. These can be connections as well as commercial, industrial, cultural, occupational agreements or links among the locations considered. There can be particular proximity and affinity relations among two locations, maybe distant from each other on the geographic space.

We can refer to different network typologies when we define such relations and links between two or more nodes and represent such relations with linear features. These can present or not a precise reference to existing geographic space between the linked nodal features, according to the different kinds of networks considered.

2.3   Linkage System. The Basics

Connections as linkages between nodes can be used to draw a contribute in drawing hierarchical network systems. Authors as Nystuen and Dacey (1961) did that to visualize nodal regions in space by the exam of the dominance of some cities in terms of telephone calls traffic. Their theory is aimed at highlighting the 'backbone' structure of the dominant relations between cities (nodes) in space that are connected by some kind of flows, examining the "degree of association between [nodes] in terms of interactions that occur directly between two [nodes] or indirectly through one or more intermediary [nodes]" (p. 29). Largest outflows from subordinate nodes are defined as nodal flows and they draw the nodal structure of the region, portraying the functional association of nodes in a determined region (p.34). In the analysis the strongest flows between city (node) pairs are considered, although the overall flows to a city – point contribute to build a ranking of cities, and a hierarchy can be built according to the flows between city-pairs. There is a need to respect three properties: a node is dominant in the hierarchy when its largest outflow is directed to a smaller node; a subdominant node is one having its largest outflow to a larger node; nodes are part of a unique hierarchy. Nodes must be ranked by their size or some measure of magnitude or importance (Nystuen and Dacey, 1968, p. 34;).

Such method can be used to examine the direct association as well as the indirect contacts between nodes, or considering both direct connections between node a and b, but also those from a and b passing through c. This analysis of the indirect connections implies obtaining am adjacency matrix in order to consider also the indirect connections between the nodes. This can be done modifying the matrix of flows to obtain values, for each origin-destination nodes, proportional to the total association value of the main center in the region. A matrix Y is obtained, with each cell representing a percentage of the total association of the dominant node. That means that each cell will host a value between 0 and 1, while the total column value will have 1 as a maximum value.

In a second stage the matrix Y is raised to a power corresponding to the networks' diameter that means (the minimum number of passages needed to connect the farther node) the largest number of nodes which must be traversed in order to travel from one node to another (Rabino, 1998, Weisstein, 2011). In an air transport application, that means counting the maximum number of linkages needed to connect any two or regions. The different powers of the Y matrix (Y_2, ..., Y_n) consider the second and nth indirect linkages between nodes. The matrix Y and its powers can be summed in the matrix B:

$$B = Y_1 + Y_2 + \ldots + Y_n$$

The matrix B considers all the direct and indirect connections between node pairs. The column totals in matrix B can be used to rank the different nodes and identify the dominant and subdominant nodes in terms of major (direct and indirect) flows and therefore build and the 'skeleton' of the nodal regional association as it happens with the analysis of the direct links. A connectivity matrix can be obtained from matrix B to highlight the dominant linkages between nodes.

2.4   Small World Literature and the Peer Influence Model: The INWM Algorithm

Early social network theorists argued that the power of social networks lies in large-scale connectivity.

A tradition of work on the small-world problem similarly rests on large-scale connectivity, which has been shown to have potentially important consequences for information and large scale coordination.

Two classical social network theories provide insights that can help analyze large networks.

First, the small-world literature has shown that while most of our acquaintances tend to be acquainted with each other, short acquaintanceship chains (relative to the size of the network) link most pairs in the network. This high degree of local clustering suggests that a practical approach to studying the structure of large networks would involve first identifying local clusters and then analyzing the relations within or between clusters. Second, we know from work on peer influence that people tend to be similar to each other.

Based on an endogenous influence process, close friends tend to converge on similar attitudes (Friedkin, 1998) and thus clusters in a small-world network should be similar along multiple dimensions.

The analysis of high dimensional networks are performed in different phases: identification of local clusters, analysis of the internal structure of the clusters, analysis of the relations between clusters, individuation of the global structure of the network. In this paper the problem of identification of local clusters is considered.

There are different methods for the identification of local clusters

- Subgroups based on complete mutability.
- Subgroups based on reachability and diameter.
- Subgroups based on nodal degree.
- Subgroups based on matrices permutations.
- Subgroups based on classical statistical methods.
- Subgroups based on the peer influence model.

Moody (2001), on the base of the peer influence model, according to which close units tends to converge on similar attitudes (Friedkin, 1998), proposes an algorithm: the Recursive Neighbourhood Mean (RNM) algorithm. This algorithm tries to identify clusters of closely related units in large networks (see Moody (2001)) on the base of a matrix of influence. The starting point is a matrix of adjacencies that quantify the presence or absence of one or more relations among units.

A modification of Moody's algorithm has been presented by Schoier and Melfi. (2004): The Modified Recursive Neighbourhood Mean (MRNM) algorithm. Another version which use a more parsimonious structure is proposed by Schoier (2005). In the latter case an algorithm to take into account the weight of the relation(s) among different units using a more parsimonious structure: the Iterative Neighbourhood Weighted Mean (INWM) Algorithm is presented.

THE INWM ALGORITHM

Given $u_1,...,u_N$ units on the network on which $R_1,...,R_k$ relations are defined:

Step1. evaluate the matrix of weights on the basis of the relations among the units;

Step2. assign an uniform random number (between 0 and 1) to each unit for every of the m influence variables so to obtain the influence matrix $Y^{(0)}$:

Step3. re-assign to each element of the matrix $Y^{(t)}$, $t=1,...,n$, the weighted mean of the contacts on the basis of the matrix of weights. The elements of the $Y^{(t)}$ are:

$$Y_{ik}^{(t)} = \frac{\sum_{j \in L_i} Y_{ik}^{(t-1)} N_{ij}}{\sum_{j \in L_i} N_{ij}}$$

where:

L_i : set of units in relation with unit i, $i=1..N$;

N_{ij} :weight relative to the relation between units i and j;

k=1..; numbers of components of the influence matrix;
t=1,..,n; numbers of iterations.

Step4. Repeat Step3 n times.

The advantages in using this algorithm are:
- The previous versions need the adjacency list in which all the units with their contacts (relations) are included. If the network has a lot of units this may be heavy in terms of time of elaboration. At this point they execute some preliminary operations in order to obtain a sort of matrix with fixed number of rows (units of the network) and variable number of columns (according to how many units are in relation with a certain unit).
- Our version needs only the matrix of the weights so time of data transformation remain more or less the same with the increasing of the units of the network.

3   The Application

3.1   The European Space of Passengers National and International Air Traffic

The nodal regional analysis has been performed on passengers' flows between member states pairs, aimed at highlighting both national and international traffic and also the direct and indirect association (Knowles and Hall, 1998, Hoyle and Knowles, 1998). While direct association could be observed from the original flows matrix, indirect association implied considering also the steps necessary to link every country to another one. In this sense we considered a total of two steps needed, as the two farther countries could be reached with at maximum an intermediate country to be visited. The analysis is a continuation of what elaborated by Borruso (2013 a, b) on air traffic data at European level. Table 1 shows the original matrix used for the analysis.

From the original matrix the Y matrix was derived, where each cell hosts the ratio between the original corresponding cell values and the maximum column value (in this case the sum of inbound plus outbound flows of Spain, being this country the one having the highest value in direct plus indirect connections). Each cell value, and therefore each relation between country pairs, represents a percentage of the total association of the dominant country. Matrix Y is then powered to the diameter of the network, in this case 2, and summed to the original Y matrix. The resulting B matrix considers all the direct and indirect links between country pairs. From B matrix a connectivity matrix is derived, with the maximum row values showing the regional association between two countries. Ranking the countries by their column maximum allows discriminating between flows and therefore building the 'skeleton' network of major connections between countries. As suggested above in paragraph 2, it is not considered the case in which a maximum column value corresponds to a flow from a higher rank node to a lower one. In this sense the higher rank center corresponds to a nodal center (Table 2).

Table 1. Inbound plus outbound passengers flows between EU countries, 2011 (millions of passengers). Source: EUROPEAN COMMISSION, European Union Energy & Transport in Figures 2013, Office for Official Publications of the European Communities, Bruxelles, 2013)

Reporter	Belgium	Bulgaria	Czech Republic	Denmark	Germany	Estonia	Ireland	Greece	Spain	France	Italy	Cyprus	Latvia	Lithuania	Luxem bourg	Hungary	Malta	Nether lands	Austria	Poland	Portugal	Romania	Slovenia	Slovakia	Finland	Sweden	United Kingdom
Belgium	47.4	165.2	282.6	459.8	1.510,4	29.7	343.0	844.8	4.221,6	1.543,1	2.905,0	73.7	152.7	82.8	0.3	298.7	84.5	191.6	405.9	417.5	745.4	241.8	57.1	71.2	259.4	457.6	1.221,2
Bulgaria	156.7	204.9	256.7	102.7	1.223,5	9.5	40.8	98.6	276.1	228.9	346.1	73.9	9.1	12.5	18.8	136.3	13.8	185.3	401.3	224.2	2.9	33.2	7.6	83.6	60.1	47.0	831.8
Czech Republic	283.5	257.1	172.0	236.7	1.161,3	59.5	86.5	600.3	826.5	1.016,8	927.0	133.4	56.1	66.9	12.1	151.5	8.2	341.9	155.8	219.8	61.1	112.9	37.5	117.0	139.3	175.4	1.298,6
Denmark	460.1	98.5	237.3	2.365,9	2.344,4	133.2	199.5	666.0	2.195,6	1.311,1	947.5	72.8	159.3	177.9	56.9	176.1	58.7	1.138,1	420.8	562.4	225.4	21.3	17.7	0.2	774.8	1.658,5	2.562,7
Germany	1.301,2	1.223,3	1.139,2	2.551,8	24.418,4	238.6	1.361,5	4.543,1	21.973,1	7.110,1	11.170,2	384.0	586.0	285.5	286.4	1.495,8	520.6	3.301,9	6.418,6	2.843,6	2.841,5	1.292,1	198.5	1.4	1.680,1	2.720,0	11.588,9
Estonia	29.7	9.7	59.2	138.2	241.0	29.0	47.9	18.4	60.6	13.5	51.3	2.1	173.4	54.8	0.1	0.1	0.6	56.1	2.9	21.7	4.8	0.1	0.4	0.1	191.9	159.4	245.1
Ireland	342.5	44.5	86.5	200.4	1.571,4	48.1	113.9	56.6	2.863,2	1.513,8	905.4	12.9	122.5	183.5	11.9	173.0	53.1	746.5	124.0	949.1	598.2	74.9	1.9	89.1	20.3	129.8	9.655,8
Greece	859.4	97.8	594.8	688.0	4.305,5	17.9	56.2	3.658,1	613.2	1.801,3	2.507,6	1.129,9	46.9	64.2	50.4	269.9	15.5	1.257,4	656.2	700.0	30.7	259.0	66.2	101.8	422.8	776.2	4.749,7
Spain	4.249,5	278.3	831.1	2.198,5	22.292,4	60.9	2.867,6	625.4	37.966,5	9.166,3	11.783,5	41.1	134.7	89.5	247.0	199.4	175.6	5.817,2	1.415,2	1.147,5	3.228,1	1.218,6	15.0	198.9	1.140,2	2.442,7	31.656,8
France	1.514,5	217.0	1.040,8	1.301,9	8.210,5	13.5	1.505,1	1.826,5	9.965,7	27.718,7	9.241,0	92.9	112.5	51.8	100.7	527.5	205.9	2.596,2	1.098,9	885.2	3.320,0	690.0	142.6	72.2	561.3	1.067,9	10.793,5
Italy	2.891,5	359.9	923.7	947.4	11.069,3	50.9	900.6	2.334,2	11.773,0	9.218,2	32.009,6	82.5	516.7	150.2	107.3	602.7	619.3	3.245,0	1.209,3	1.112,6	1.285,4	1.309,5	2.6	184.5	516.0	871.4	10.251,8
Cyprus	74.5	74.1	130.0	72.7	376.8	1.1	12.9	1.112,8	59.9	90.0	65.9	0.1	0.0	0.0	6.1	78.6	31.4	94.8	148.7	66.6	1.7	165.3	2.7	9.8	69.5	212.5	2.667,5
Latvia	152.4	9.1	56.0	160.0	588.9	184.3	121.5	47.2	134.8	125.4	316.9	0.0	0.4	236.9	0.0	11.7	0.0	94.4	60.1	83.5	0.8	1.6	0.4	0.2	467.3	226.5	585.1
Lithuania	81.5	12.5	66.5	177.7	285.3	34.4	177.3	64.3	88.0	51.9	129.7	1.7	255.7	0.0	0.0	0.0	1.2	54.0	43.8	60.8	0.9	0.0	1.2	0.0	112.0	75.3	557.3
Luxembourg	0.4	22.1	12.7	57.9	285.3	0.0	11.8	56.9	251.8	103.9	111.8	0.0	0.0	0.5	0.0	3.2	107.6	57.7	0.5	193.3	0.3	0.0	0.0	0.0	0.0	0.4	253.9
Hungary	298.2	132.9	147.5	179.2	1.498,2	0.1	173.4	173.5	294.9	505.5	604.9	79.2	11.7	0.0	0.3	16.4	546.9	78.5	135.4	70.2	224.9	0.2	1.1	203.8	398.8	1.043,4
Malta	84.4	12.4	8.1	60.8	520.9	0.6	54.5	16.5	174.9	198.5	616.7	52.1	0.0	1.2	2.8	16.4	0.0	93.1	67.9	25.8	0.1	6.7	3.6	2.5	3.2	69.7	1.086,1
Netherlands	188.3	187.6	544.1	1.158,7	3.522,5	56.1	742.9	1.269,2	5.824,4	2.619,4	3.245,3	103.0	96.5	54.1	105.8	547.2	96.0	0.3	687.8	499.0	1.191,9	335.4	25.3	9.2	488.5	998.2	7.671,1
Austria	402.9	400.4	131.9	418.6	6.429,5	2.8	124.0	658.1	1.395,5	1.051,9	1.202,0	147.6	59.9	43.9	34.6	78.0	68.4	684.3	665.7	287.4	81.4	519.0	59.8	41.8	225.6	419.1	1.695,0
Poland	418.2	225.8	217.3	564.3	2.861,0	21.6	940.0	678.1	1.129,1	871.2	1.114,7	68.2	85.8	60.9	0.4	134.0	26.0	499.5	288.3	1.118,7	123.8	66.1	3.6	4.0	226.6	578.0	4.288,6
Portugal	753.9	4.3	68.7	225.8	2.664,2	4.8	598.4	31.7	9.276,8	3.209,1	1.290,4	1.9	0.8	189.9	57.0	0.2	1.188,4	85.4	125.9	2.898,0	14.5	0.4	3.7	161.4	143.2	5.368,3
Romania	233.7	38.1	111.5	21.3	1.236,7	0.1	72.0	251.7	1.144,8	659.4	2.154,0	138.9	1.7	0.0	0.3	176.6	6.2	330.7	501.7	70.7	13.3	736.1	0.2	0.0	32.1	0.0	661.3
Slovenia	56.4	7.7	37.4	18.1	199.2	0.5	1.9	67.0	15.0	144.1	1.9	5.1	0.4	1.2	0.0	0.0	3.6	15.3	60.0	5.5	0.0	0.2	0.1	0.0	12.1	5.1	126.0
Slovakia	70.1	66.2	104.5	0.1	2.2	0.4	87.3	101.0	201.9	72.5	185.9	100.0	0.1	0.0	0.0	0.1	2.5	9.1	45.1	3.2	2.8	0.1	0.0	34.1	0.6	5.7	463.0
Finland	257.5	59.9	158.9	772.4	1.682,9	195.6	10.3	417.4	1.119,0	962.5	512.8	67.7	444.4	112.1	0.0	202.2	8.3	489.0	125.3	227.3	158.5	29.8	21.8	0.6	1.754,0	1.365,0	1.003,4
Sweden	454.5	47.1	175.9	1.388,8	2.710,0	160.5	123.6	757.9	2.364,8	1.079,1	894.0	105.0	226.2	76.1	0.7	356.7	68.5	908.2	420.3	577.9	140.0	0.3	5.2	5.3	1.367,9	6.929,8	1.332,1
United Kingdom	1.235,5	878.1	1.194,8	2.565,5	11.626,5	244.1	9.670,6	4.778,4	31.487,8	10.840,6	10.254,7	2.658,9	584.2	564.0	251.7	1.037,3	1.091,1	7.515,5	1.704,2	4.260,1	5.329,1	717.1	126.3	468.8	1.009,5	2.539,2	20.920,6

Table 2: Matrix of dominant connections between EU countries according to nodal regional analysis – national and international flows between member states (our elaboration from EUROPEAN COMMISSION, European Union Energy & Transport in Figures 2013, Office for Official Publications of the European Communities, Bruxelles, 2013)

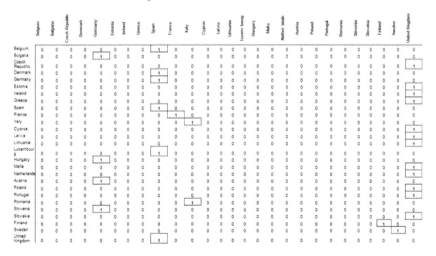

From the matrix of dominant connection deriving from the analysis the presence of a higher hierarchical center can be noticed and identified in Spain with reference to total movements and dominant for internal flows and for the international ones coming from Belgium, Denmark, Luxemburg and the UK. The UK, although being dominant over a wider amount of countries (Bulgaria, Estonia, Ireland, Greece, Cyprus, Latvia, Lithuania, Malta, The Netherlands, Poland, Portugal and Slovakia), is sub-dominant to Spain, having its larger flows towards Spain, higher in terms of major flows. Germany is also subdominant to Spain, and attracts traffic from Bulgaria, Hungary, Austria and Slovenia. Some countries like Spain, France, Italy, Finland and Sweden are also dominant on their own internal traffic.

Table 3. Matrix of dominant connections between EU countries according to nodal regional analysis –international flows between member states (our elaboration from EUROPEAN COMMISSION, European Union Energy & Transport in Figures 2013, Office for Official Publications of the European Communities, Bruxelles, 2013)

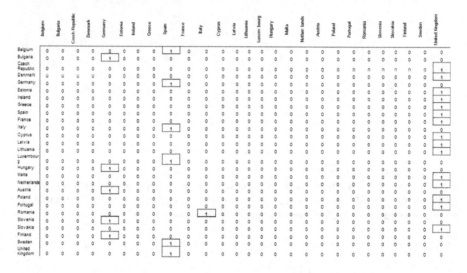

Analyzing the international traffic, that is eliminating the diagonal value in the original matrix, the level of international dominance can be observed. Here things change as the UK become dominant over a wide set of countries, including Spain. That is because the UK has a lower weight of the internal traffic if compared to Spain. On top of the hierarchical structure we have the UK dominant over a most of the countries including Spain. Spain is also dominant over several countries (Belgium, Germany, Italy, Luxemburg and Sweden), while a third level of ranking can be highlighted in Germany being dominant on countries like Bulgaria, Hungary, Austria, Slovenia and Finland, Italy dominant over Romania.

As a general conclusion of this part, the air traffic system at European level seems to be very clustered and organized in a hierarchical structure, with three tiers of dominance. UK and Spain are mutually interconnected and according to the analysis (total flows versus international ones) exchange their positions between the first and second tiers.

3.2 The Application of the Peer Influence Model to the analysis on National and International Air Traffic

We illustrate our approach on the basis of the data of Table 1 which contains the matrix of dominant connections between EU countries according to nodal regional analysis – national and international flows between member states (our elaboration from EUROPEAN COMMISSION, European Union Energy & Transport in Figures 2013, Office for Official Publications of the European Communities, Bruxelles, 2013).

In the following we will consider a single relation R " airport flows x_i is dominant with respect to airport flows x_j if the flow of passengers from x_i to x_j is greater"

$$\mathbf{R} = \{r_{ij}\} \ i, j = 1,..,n \quad \text{where } r_{ij} = \begin{cases} 1 & \text{if } x_i \ R \ x_j \\ 0 & otherwise \end{cases}$$

The one mode matrix given in Table 1 and the relation defined above allows to obtain the adjacency matrix reported in Table 2.

At this point we introduce a matrix Y , called influence matrix, of dimension (N x m) where N is the number of EU countries (in our case N = 27), and m represents the number of components describing the reciprocal influences.

A reasonable assumption is to set m = 5. This corresponds to assume that each user associated to an airport flows may be influenced by the behavior of five other airport flows .

In order to build the matrix Y we use a modifed version of the Recursive Neighbourhood Mean (RNM) algorithm , proposed by Moody (2001). Our algorithm has been implemented in SAS. The Iterative Neighbourhood Weighted Mean (INWM) consists in the computation of a suitable weighted mean by iteration, and generalises the RNM and the MRNM algorithm. This algorithm has been described in a previous paragraph.

The result of the INWM procedure are five positional variables reported in Table 4.

Table 4. Matrix of the positional variables airport flows

	pos1	pos2	pos3	pos4	pos5
Belgium	0,320492	0,262141	0,25486	0,11156	0,69179
Bulgaria	0,320492	0,262141	0,25486	0,11156	0,69179
Czech Rep.	0,320492	0,262141	0,25486	0,11156	0,69179
Denmark	0,320492	0,262141	0,25486	0,11156	0,69179
Germany	0,320492	0,262141	0,25486	0,11156	0,69179
Estonia	0,320492	0,262141	0,25486	0,11156	0,69179
Ireland	0,320492	0,262141	0,25486	0,11156	0,69179
Greece	0,320492	0,262141	0,25486	0,11156	0,69179
Spain	0,320492	0,262141	0,25486	0,11156	0,69179
France	0,304851	0,96163	0,198794	0,587503	0,306615
Italy	0,651282	0,381438	0,883215	0,578288	0,035082

Table 4. (*Continued*)

Cyprus	0,320492	0,262141	0,25486	0,11156	0,69179
Latvia	0,320492	0,262141	0,25486	0,11156	0,69179
Lithuania	0,320492	0,262141	0,25486	0,11156	0,69179
Luxembourg	0,320492	0,262141	0,25486	0,11156	0,69179
Hungary	0,320492	0,262141	0,25486	0,11156	0,69179
Malta	0,320492	0,262141	0,25486	0,11156	0,69179
Netherlands	0,320492	0,262141	0,25486	0,11156	0,69179
Austria	0,320492	0,262141	0,25486	0,11156	0,69179
Poland	0,320492	0,262141	0,25486	0,11156	0,69179
Portugal	0,320492	0,262141	0,25486	0,11156	0,69179
Romania	0,651282	0,381438	0,883215	0,578288	0,035082
Slovenia	0,320492	0,262141	0,25486	0,11156	0,69179
Slovakia	0,320492	0,262141	0,25486	0,11156	0,69179
Finland	0,225539	0,274004	0,716651	0,618815	0,398231
Sweden	0,423714	0,944501	0,738672	0,829675	0,893444
United Kin.	0,320492	0,262141	0,25486	0,11156	0,69179

On the basis of the five positional variables a Ward's minimum variances cluster analysis, carried out. In such a way we obtain a clear clustering that reveals a structure of three groups as one can see from Figure 3.

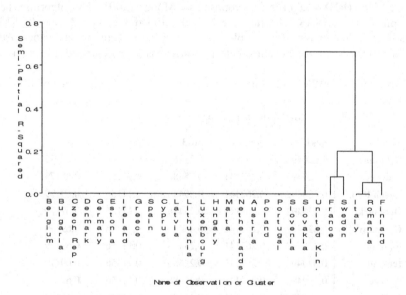

Fig. 3. Dendrogram of the cluster analysis

If no national air traffic is considered the result of the INWM procedure are five positional variables reported in Table 5.

Table 5. Matrix of the positional variables airport flows

	pos1	pos2	pos3	pos4	pos5
Belgium	0,336638	0,874356	0,352195	0,059614	0,280257
Bulgaria	0,093988	0,112068	0,637499	0,73621	0,195241
Czech Rep.	0,093988	0,112068	0,637499	0,73621	0,195241
Denmark	0,093988	0,112068	0,637499	0,73621	0,195241
Germany	0,336638	0,874356	0,352195	0,059614	0,280257
Estonia	0,093988	0,112068	0,637499	0,73621	0,195241
Ireland	0,093988	0,112068	0,637499	0,73621	0,195241
Greece	0,093988	0,112068	0,637499	0,73621	0,195241
Spain	0,093988	0,112068	0,637499	0,73621	0,195241
France	0,093988	0,112068	0,637499	0,73621	0,195241
Italy	0,336638	0,874356	0,352195	0,059614	0,280257
Cyprus	0,093988	0,112068	0,637499	0,73621	0,195241
Latvia	0,093988	0,112068	0,637499	0,73621	0,195241
Lithuania	0,093988	0,112068	0,637499	0,73621	0,195241
Luxembourg	0,336638	0,874356	0,352195	0,059614	0,280257
Hungary	0,093988	0,112068	0,637499	0,73621	0,195241
Malta	0,093988	0,112068	0,637499	0,73621	0,195241
Netherlands	0,093988	0,112068	0,637499	0,73621	0,195241
Austria	0,093988	0,112068	0,637499	0,73621	0,195241
Poland	0,093988	0,112068	0,637499	0,73621	0,195241
Portugal	0,093988	0,112068	0,637499	0,73621	0,195241
Romania	0,093988	0,112068	0,637499	0,73621	0,195241
Slovenia	0,093988	0,112068	0,637499	0,73621	0,195241
Slovakia	0,093988	0,112068	0,637499	0,73621	0,195241
Finland	0,093988	0,112068	0,637499	0,73621	0,195241
Sweden	0,336638	0,874356	0,352195	0,059614	0,280257
United Kin.	0,336638	0,874356	0,352195	0,059614	0,280257

On the basis of the five positional variables a Ward's minimum variances cluster analysis, carried out. In such a way we obtain a clear clustering that reveals a structure of ttwo groups as one can see from Figure 4.

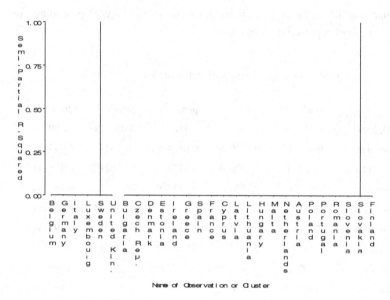

Fig. 4. Dendrogram of the cluster analysis

4   Conclusions

In this paper we have explored such phenomena, in terms of the presence of clusters in network data analysis and in terms of the connections between airports, trying to understand the origins and motivations of such elements in space.

The combination of social network analysis algorithms and more traditional analysis as the Nystuen and Dacey's nodal regional analysis allowed to find similar results and interesting information concerning the structure of flows in a network. The clustering technique in particular helped in overcoming some of the limitation of nodal regional analysis. This latter theory is in fact limited by the fact it considers only the dominant flows, highlighting in particular the backbone of flows in a network but with little consideration of the other not dominant flows. On the other end the network-based cluster analysis allowed going more in depth into the exam of the other relationship. It allowed confirming the results from the nodal regional analysis but providing more detailed information on the subdominant relationship between the other not dominant nodes.

References

1. Banfield, J.D., Raftery, A.E.: Model-based Gaussian and non-Gaussian clustering. Biometrics 49, 803–821 (1993)
2. Dempster, A.P., Laird, N.M., Rubin, D.B.: Maximum likelihood from incomplete data via the EM algorithm. Journal of the Royal Statistical Society, Series B (Methodological) 39(1) (1977)

3. Dasgupta, A., Raftery, A.E.: Detecting features in spatial point processes with cluster via model-based clustering. Journal of the American Statistical Association 93, 294–302 (1988)
4. Fraley, C., Raftery, A.E.: How many clusters? Which clustering method? Answers via model-based cluster analysis. The Computer Journal 41(8) (1998)
5. Fraley, C., Raftery, A.E.: Model-based clustering, discriminant analysis and density estimation. Journal of the American Statistical Association 97(458) (2002)
6. Fraley, C., Raftery, A.E.: MCLUST Version 4 for R: normal mixture modeling and model-based clustering, classification and density estimation, Technical Report no. 597. Department of Statistics, University of Washington (2012)
7. Fung, G.: A comprehensive overview of basic clustering algorithms (2001), http://pages.cs.wisc.edu/~gfung/ (cited October 2012)
8. Han, J., Kamber, M., Tung, A.K.H.: Spatial clustering methods in data mining. A survey (2001), http://www.cs.uiuc.edu/homes/hanj/ (cited December 2012)
9. Hoyle, B.S., Knowles, R.D.: Modern transport geography, 2nd edn. Wiley, Chichester (1998)
10. Knowles, R.D., Hall, D.: Transport deregulation and privatization. In: Hoyle, B.S., Knowles, R.D. (eds.) Modern Transport Geography, 2nd edn., Wiley, Chichester (1998)
11. Kochen, M.: The small World. Ablex Publishing Corporation, Norwood (1989)
12. Koperski, K., Han, J., Adhikary, J. : Mining Knowledge in Geographical Data (1998), http://www.cs.uiuc.edu/homes/hanj/pdf/geo_survey98.pdf (retrieved October 2011)
13. Miller, H.: A measurement theory for time geography. Geographical Analysis 37, 17–45 (2005)
14. Moody, J.: Peer influence groups: identifying dense clusters in large networks. Social Networks 23, 261–283 (2001)
15. Nystuen, J.D., Dacey, M.F.: A graph theory interpretation of nodal regions. Papers and Proceedings of the Regional Science Association 7, 29–42 (1968)
16. O'Sullivan, D., Unwin, P.J.: Geographic Information Analysis. Wiley, Chichester (2003)
17. Rabino., G., and Occelli, S.: Understanding spatial structure from network data: Theoretical considerations and applications. Cybergeo, 29 (1997) http://www.cybergeo.eu/index2199.html
18. Schoier, G.: Blockmodeling Techniques for Web Mining. In: Haerdle, W., Roenz, B. (eds.) Proceedings of Compstat 2002. Springer, Berlin (2002)
19. Schoier, G., Melfi, G.: Clusters d'ensemble de données larges dans le Web Log Mining. In: Attidel 10mes Rencontres de la Société Francophone de Classification, Neuchâtel Settembre., pp. 160–164 (2003)
20. Schoier, G., Borruso, G.: A Clustering Method for Large Spatial Databases. In: Laganá, A., Gavrilova, M.L., Kumar, V., Mun, Y., Tan, C.J.K., Gervasi, O. (eds.) ICCSA 2004. LNCS, vol. 3044, pp. 1089–1095. Springer, Heidelberg (2004)
21. Schoier, G., Melfi, G.: A Different Approach for the Analysis of Web Access Logs. In: Vichi, M., Monari, P., Mignani, S., Montanari, A. (eds.) New Developments in Classification and Data Analysis. Springer (2004)
22. Schoier, G.: An Algorithm for the individuation of clusters of units in large databases. In: Atti del Quarto Convegno su Modelli Complessi e Metodi Computazionali Intensivi per la Stima e la Previsione (S.CO. 2005), Bressanone (2005)
23. Scott, J.: Social Network Analysis. Sage, NewburyPark (1992)

24. Trasarti, R., Giannotti, F., Nanni, M., Pedreschi, D., Renso, C.: A Query Language for Mobility Data Mining. International Journal of Data Warehousing and Mining 7(1), 24–45 (2011)
25. Wasserman, S., Faust, K.: Social Network Analysis: Methods and Applications. Cambridge University Press, New York (1994)
26. Weisstein, E.W.: Graph Diameter, MathWorld–A Wolfram Web Resource, http://mathworld.wolfram.com/GraphDiameter.html (accessed January, 2011)
27. Borruso, G.: A nodal regional analysis of air passenger transport in Europe. International Journal of Business Intelligence and Data Mining 8(4), 377–396 (2013)
28. Borruso, G.: Polarization and Hierarchies in Air Passenger Transport in Europe. In: Murgante, B., Misra, S., Carlini, M., Torre, C.M., Nguyen, H.-Q., Taniar, D., Apduhan, B.O., Gervasi, O. (eds.) ICCSA 2013, Part IV. LNCS, vol. 7974, pp. 389–402. Springer, Heidelberg (2013)

Geomatics to Support the Environmental Impact Assessment in Renewable Energy Plants Installation

Emanuela Caiaffa[1], Alessandro Marucci[2], Flavio Borfecchia[1], and Maurizio Pollino[1]

[1] ENEA National Agency for New Technologies, Energy and Sustainable Economic Development, UTMEA Energy and Environmental Modeling Technical Unit, C.R. Casaccia, Via Anguillarese, 301 – 00123 Rome, Italy
{emanuela.caiaffa,flavio.borfecchia,maurizio.pollino}@enea.it
[2] Abruzzo Ambiente Srl, L'Aquila, Italy
marucci79@hotmail.it

Abstract. Projections to 2020 indicate that renewable energy sources (RES) could cover, from 20 to 30 percent of the world's energy needs. To implement an effective e-governance in this direction, it is necessary to implement new methodologies to support decision-making in the local energy planning. The environmental impact is one of the main concern existing at different levels, in addition to the growing soil consumption in Europe. A significant problem for some types of plants, mainly solar and wind power, is the interaction of the devices with the surrounding environment, with possible negative effects in terms of visual impact and soil consumption. It is, therefore, very important to define which weight can have different impacts and to consider all possible scenarios.

This work proposes a methodology to investigate, through some specific indicators, these impacts and to support territorial planning processes for the best choice of supply from RES. The processing of geospatial data and environmental information, collected at different scales, has been conducted through the use of GIS tools, allowing the implementation of a criticality index, as an indicator related to weight of the impacts that the construction of RES plants can have on the surrounding environmental features.

Keywords: GIS, renewable energy sources, land cover, landscape ecology, soil consumption.

1 Introduction and Objectives of the Work

As indicated in the 2009/28/EC Directive of the European Parliament and Council, ambitious energy and climate change objectives for 2020 have been stated: greenhouse gas emissions reduction for 20%, renewable energy sources (RES) increase for 20%, improvement in energy efficiency for 20% [1]. Projections to 2020 indicate that renewables could cover, for that time, from 20 to 30 percent of the world's energy needs. Existing tools, to implement an effective e-governance in this direction, are not many, so it is necessary to implement new methodologies to support decision-making in the local energy planning [2, 3].

B. Murgante et al. (Eds.): ICCSA 2014, Part IV, LNCS 8582, pp. 463–478, 2014.
© Springer International Publishing Switzerland 2014

Different forms of intervention on the territory, to produce energy from renewables, are certainly comparable to the construction of anthropogenic "linear and areal" infrastructures. As a consequence, the environmental impact is one of the main concern existing at different levels, in addition to the growing *soil consumption* in our Country. Soil consumption refers to land cover change processes, due to urbanization and increase of artificial surfaces over time, usually at the expense of rural or natural areas [4, 5].

This work aims to set up a methodology to support the territorial planning processes for the issues related to production and supply from Renewable Energy Sources (RES), focusing on the assessment of the possible environmental pressure due to RES plants installation. The methodology, by means of the realization of specific thematic maps, is conceived as a tool for the decision makers, able to understand the potentiality of a given area as opposed to its capability to bear RES installations. Therefore, trough the study of environmental conditions, we propose the definition of a criticality index, to be used both at large scale and locally, specifically for habitats that make up the natural mosaic under study. The reason of the latter aspect is that a wide-range assessment could not be exhaustive of all the peculiarities and the local situations.

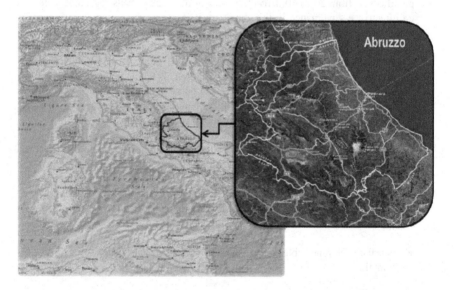

Fig. 1. Area of interest for the methodology developed: Abruzzo Region (Central Italy)

In this framework, during the last years, the design of new techniques for spatial data processing and the integration of geospatial data with other data sources (field surveys, thematic maps, cartography, etc.) have strongly improved environmental monitoring activities, in terms of quality, efficiency, timeliness and cost-effectiveness.

The approach proposed makes possible: a) to assess the distribution of criticalities existing on the territory, depending on different factors/parameters; b) to detect the presence of high-value natural elements; c) to individuate the ecosystems potentially more affected by possible modifications due to, for example, soil consumption. The area selected for this study is the Abruzzo Region, in the central Italy (Fig. 1).

2   Materials and Methods

The study is founded on the basic idea that there is a close relationship between the spatial complexity of the Land Cover (LC) and the heterogeneity of real environment. To cope with this goals, the evaluation of ecological values attributable to the LC features has been carried out exploiting the information provided by the Corine Land Cover thematic map (CLC 2006) [6], whose LC classes definition has been adopted and used.

A large number of different LC classes involves more contact surfaces between such classes and, then, a more complex environmental system, characterized by the presence of *ecotonal* areas [7, 8] (an ecotonal environment is a transition area between two ecosystems, and more in general between two different ecological systems). The ecotonal areas contain species peculiar of the surroundings areas and species exclusive of ecotonal area itself. Therefore, such areas are characterised by high biodiversity and richness and play a fundamental role, because through these structures occurs the connection between very different environments (from forests to meadows, from lakes to forests, from fresh water to salt water).

However, it has been noticed that, where there is a high variety of LC classes (in other words, scattering of LC), not necessarily it is possible to detect the presence of eco-systemic relations of high ecological value. The environmental "quality", expressed in terms of ecological value, is closely related to vegetation types and habitats in the area. A clear example is provided by the areas defined as "populated", where there are substantial diversity of LC types and of ecotonal areas (border line), but that do not express high values of environmental quality.

Experiences conducted during the development of several interdisciplinary researches (some ongoing) [9] have shown that, even if the concept of environmental continuity could be applied in terms of large scale, such pertinence is reduced as the territorial detail increases, which is the typical scale used in the framework of planning applications.

The present study has shown that there is a correlation among spatial scattering, land use classes and ecological value. This is related to specific conditions, such as:

- topological complexity: the units identified by the CLC polygons have a complex spatial distribution, with the presence of intrusion between the same units;
- possibility to assign an "ecological" value on the basis of the LC categories detected: this issue is very sensitive, because it is strongly subjective;
- presence of both of the above conditions: this increases the probability of finding environmental units with high ecological values.

So, the confluence of several factors having different nature creates the necessary conditions to assume a one to one correspondence between structure and ecological function. The reliability of such statements becomes more and more consistent when the spatial and structural analysis of the basic components is supported by the inclusion of empirical data (e.g., field surveys data), through which it is possible to delineate the factors really useful for demonstration of the thesis.

The amount of collected data, used to demonstrate the above mentioned correlation, cannot be easily reproduced on a large scale, because it requires a relevant time and resources consumption. Then, it is possible to set-up an approach aiming at the modelling of the landscape under investigation, by means the exploitation of specific indexes and environmental indicators. These values can be able to express the degree of area suitability, in order to assess the criticality level due to the location of RES plants and/or related devices.

Hence, to get to the definition of such criticality index it was followed a procedure based on the analysis of multiple variables involved in interaction assessment, by creating a data set that can be used to build impact scenarios, depending on the natural features of the landscape and on the structure/infrastructure here located.

The recourse to this approach is based on two main reasons:

1. to evaluate, on a local scale and in a planning phase, the variability of impacts produced by the installations, on the basis of the environmental features with which technological elements and devices could interfere;
2. to evaluate, on a larger scale, the entity of the interventions in terms of "environmental costs", considering the application of such index as valid on the whole area under investigation.

Moreover, to deal with the need of detailed and punctual data about vegetation (types and species), habitats and other environmental relevant features, covering the whole Abruzzo Region, it was introduced the use of an indicator based on CLC information summarising these salient features and characterising the landscape in terms of *environmental value*.

2.1   Basic Data

To study the landscape of the area of interest and to understand the structure and the patterns, RS data from different sources have been used as reference: Landsat-TM satellite scene (July 2010, available for download at GLCF) [10], digital aerial orthophotos (available for consultation and visualization at the Italian National Geoportal, NG) [11]. Then, by means a GIS approach, CLC 2006 data have been overlaid and analysed in combination with other spatial data, such as:

— Regional Technical Maps (RTM, 1:10,000 scale);
— Digital Terrain Model (DTM), spatial resolution 20 m (source: Italian Ministry for the Environment Land and Sea);
— Basic territorial layers: Administrative boundaries, Urban Areas, Delimitation of protected areas, Road network, Railways, etc., (DBPrior10K, 1:10,000 scale, source: Italian Ministry for the Environment Land and Sea);
— Hydrographic network (DBPrior10K, 1:10,000 scale, source: Italian Ministry for the Environment Land and Sea)

— Socio-economical and Census data (source: ISTAT, Italian National Institute of Statistics)

2.2   The Criticality Index: Describing the Sensitivity to Changes

The structural complexity of the ecosystems in the investigated area is the result of the evolution of the environment due to climatic conditions, as well others natural factors or anthropogenic causes.

As previously stated, the purpose of this study is to set-up a methodology able to understand if, through specific parameters, it is possible to estimate and map the spatial distribution of the criticality in the area of interest. To this end, has been chosen a level of detail to suitably test the level of plausibility of the initial assumptions. In this sense, high environmental values are the result of the contribution of several factors and, at the same time, indicate an ecosystem which could suffer in consequence of landscape alterations.

Thus, the Criticality Index (CI) is a value related to provide an estimate of the impacts, both real and potential, that the construction of RES production plants could have on the environmental features.

To achieve the CI implementation, it was tried to identify the factors able to describe how an area can be sensible to changes. Then, by means GIS procedure, these factors have been spatialized, to make them representable through suitable thematic maps.

In order to take into account all the factors involved in CI definition, a formula (1), synthetizing several territorial characteristics and features, has been implemented:

$$CI = D + EQ + LCI + AF + PI \qquad (1)$$

All terms of the CI formula (1) were normalized in order to have the identical variation range $(1 \div 10)$. The terms, representative of the factors involved in the definition of CI index, are the followings:

— D: *Density* (structure of *ecomosaic*). The presence of numerous habitats in a definite space determines the formation of further ecotonal areas, increasing the richness of species and the ecological relationships;
— EQ: *Ecological Quality*. These values derive from the LC texture, considering the specific features and the characteristics.
— LCI: *Landscape Conservation Index*. This index has been used to express the naturalness (meaning the conservation level of the natural status [12]) of the territory considered;
— AF: *Altitude Factor*. This index has been derived from the DTM, in order to highlight the relations with altitude.
— PI: *Positive Interference*. The degree of CLC quality classes, in relation to the surface, as an expression factor describing environmental continuity.

In the following paragraphs is reported the description of each single factor involved in the definition of the CI.

2.3   Density (D)

The presence of different habitats in a specific space determines the formation of further areas, in which the richness of species increases and consequently multiplies also the density of ecological relationships. To this end, we have introduced the evaluation of a Density factor D [13], calculated starting from the polygon vertices that represent the LC classes within the study area. Thus, the D factor takes into account the "geometric" articulation and the richness of the features vertices, correlating them with the LC typologies (based on CLC classes): the result is a layer containing information about the spatial density of such features within the study area. The larger are D values, the wider are the borders between LC classes characterised by high environmental values. In environmental terms, at the present analysis scale, high values of D factor outline the condition of greater *ecotonality* (Fig. 2), due to the presence of a large number of neighbouring natural and semi-natural areas.

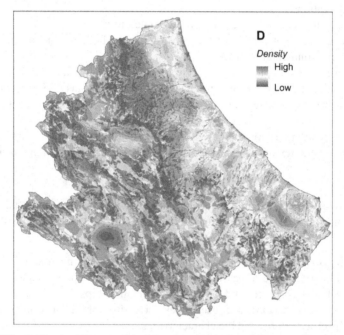

Fig. 2. Density (D) map, describing the complexity of the area of interest

2.4   Ecological Quality Attributable to Corine Land Cover (EQ)

Our territory, as we presently perceive it, is the result of deep changes due to human intervention on environment. During the years, natural ecosystems, such as forests, grasslands, wetlands, etc., have been converted to soils used as farmland, pastures or destined to urban settlements. The dynamics of urbanization and the conversion processes of natural areas are the main driver to a territory transformation [4, 5]. The European guidelines on territorial development [14] request to the planners to govern these transformation processes, by the definition of local development strategies, differentiated in relation to the different vocations of each territory and aimed at mitigating the inequalities, with a view to sustainable and balanced territorial development [15]. Hence, it is important to be able in analysing the transformation of the territory and in understanding the implications of ecological and environmental issues. In particular, the CORINE program, by means the CLC project [6], has produced and provided a valuable set of data, mapping the territory characteristics and focusing on land use and land cover. Since the early 90's, CLC data have been effectively exploited in the course of numerous environmental studies and, obviously, play a crucial role also in the framework research activities here described.

Table 1. CLC classes reclassification in accordance with the ecological quality (EQ)

CLC Nomenclature			EQ reclassification
2. AGRICULTURAL LAND	2.1.Arable crops	2.1.1. Non irrigated arable crops	*1*
		2.1.2. Arable crops in irrigated areas	*1*
		2.1.3. Rice fields	*1*
	2.2.Permanent crops	2.2.1 Vineyards	*3*
		2.2.2 Fruit trees and minor fruits	*3*
		2.2.3 Olive groves	*3*
	2.3. Permanent grass-land	2.3.1. Permanent grassland	*4*
	2.4. Heterogeneous agricultural areas	2.4.1. Annual crops associated with permanent crops	*3*
		2.4.2. Cropping systems and complex particle	*3*
		2.4.3. Areas mainly occupied by agricultural fields with significant areas of natural	*5*
		2.4.4. Agroforestry areas	*5*
3. FOREST AND SEMI NATURAL TERRITORY	3.1. Wooded areas	3.1.1. Deciduous forests	*9*
		3.1.2. Coniferous forests	*6*
		3.1.3. Mixed forests	*7*
	3.2. Areas with shrub vegetation and/or herbaceous	3.2.1. Areas of natural pasture and high-altitude meadows	*7*
		3.2.2. Moorlands and shrub	*6*
		3.2.3. Areas with sclerophyllous vegetation	*8*
		3.2.4. Areas of forest and shrubs vegetation in evolution	*6*
	3.3. Open areas with sparse or no vegetation	3.3.1. Beaches, dunes, sand (larger than 100 m)	*10*
		3.3.2. Bare rocks, cliffs, outcrops	*8*
		3.3.3 Areas with sparse vegetation	*6*
		3.3.4. Areas hit by fire	*10*
		3.3.5. Glaciers and permanent snow	*10*

The methodology to obtain the EQ factor starts from the classes defined in CLC 2006, for each of which it was assigned a value of "quality" considering the salient features of their biological function, naturalness, bio-permeability, level of alteration suffered. Then, LC categories, combined into vegetation types, were reclassified in order to assign them an "ecological connotation", according to a gradient of natural value, increasing from classes with high degree of transformation (e.g., artificial surfaces), to classes with high natural value. This classification has led to produce the map of the "ecological quality", that not only has allowed to describe the area in terms of naturalness, but also to detect areas with the highest values [16].

The objective was to express, in agreement with the assumptions relative to the scale of investigation and derived implications, the spatial distribution of natural and non-natural areas, according to their sensitivity or fragility and significance. As demonstrated by Clevenger et alii [17], such assumption represent the basic information layer, so that, for a large scale, we can define the structural composition of natural areas compared to populated areas and, as far as possible, give them a realistic ecological value. In Table 1 is reported a classification of CLC typologies according to the principles of environmental and ecological quality above described. The range spans from 0 to 10 (0 = minimum; 10 = maximum), on a scale to 11 classes of naturalness, expressing the ecological and functional value of LC categories with particular attention to water courses, because they represent a natural network of vital importance. Figure 3 shows the EQ values map, calculated on the basis of the specific reclassification of the CLC for the Abruzzo Region.

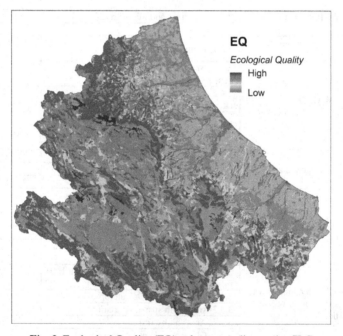

Fig. 3. Ecological Quality (EQ) values according to the CLC

The EQ factor acquires an important meaning when such analysis is carried out at a larger scale, where the outstanding features of the natural environment, through which we can deduce the environmental conditions, are represented by the connectivity and the continuity of natural systems. In fact, as emerges from Fig. 3, it is possible to identify, with good approximation, the areas with the highest ecological characteristics.

Among these, the areas along rivers become preferential axes traversed by the fauna to move from a natural zone to another.

Furthermore, the EQ map provides information on the sensitivity of the LC classes considered, at the external modifications due to human activities. Specifically, higher it is the value of EQ, higher it is the sensibility to potential damage of the ecosystem coming from the alteration of natural features. For example, through GIS analysis, it has been possible to observe that the coastal areas have a low value of EQ, because these zones were more interested by strong transformations during 80's and 90's.

2.5   Landscape Conservation Index (LCI)

The Landscape Conservation Index (LCI) [18, 19] can be seen as another useful indicator for the naturalness of an area. The LCI was calculated attributing the EQ value to each CLC class. The LCI spans from 0 (minimum conservation) to 10 (maximum conservation) and is obtained as the average of LCI classes values weighted on the surface extent of the same classes, according to the following formula (2):

$$LCI = \sum_{i=1}^{n} \left[\left(\frac{CLC_{area}}{Total_{area}} \right) \cdot \left(\frac{EQ}{n} \right) \right] \tag{2}$$

where n (with n = 11) is the number of EQ classes, CLC_{Area} is the ith class surface extent, and EQ is the relative value of ecological quality.

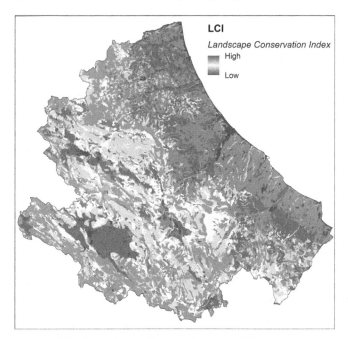

Fig. 4. Landscape Conservation Index (LCI) map

The final value, in relation to the investigation scale, provides important information for assessing the degree of conservation of the landscape. The most useful data, however, are those related to the observed values for the individual classes, that provide a clear representation of the components, interpreted according to the CLC, having a higher degree of conservation (Fig. 4).

2.6   Altitude Factor (AF)

With the increase of altitude, environments show higher naturalness and sensitivity to the transformations, due to both the morphological structure, both the climatic conditions. Therefore, the AF is important in the calculation of the final Criticality Index (CI), for instance to take into account that the Italian legislation consider a special protection for the areas with an altitude greater than 1,000 metres (Law n. 431, 1985, also known as "Galasso Law", and Decree n. 42, 2004).

The AF map (Fig. 5) was created from the DTM and included in the stratification of the critical factors in order to highlight the features related to altitude.

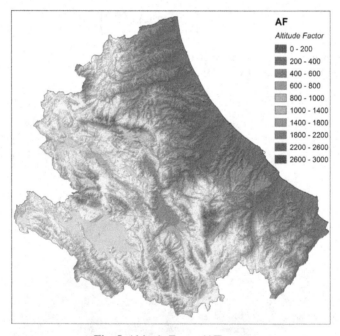

Fig. 5. Altitude Factor (AF) map

2.7   Positive Interference (PI)

Positive Interference has been introduced as factor expressing the environmental continuity, taking into account the quality level of CLC classes in relation to the surface extent. To parameterize the entity of disturbance different approaches have been proposed [19]. One of the most interesting is the calculation of the negative interference, due to the presence of urbanized areas and artificial infrastructure. It is based on the ratio between the surface extent of urbanized areas and their perimeter. It was decided to apply - in a complementary way - this calculation to natural areas, assessing, in terms of positive interference (PI), expressing the level of quality of these natural areas in relation to environmental continuity. The PI was not calculated for classes indicated in CLC map as "Artificial Surfaces" (typologies from 1.1.1 to 1.4.2, Level 3) because they are representative of urbanized/anthropized areas.

The values of PI were calculated as the ratio between the surface area and the perimeter of CLC polygons (LC classes), multiplied by their own EQ value. Higher values of PI indicate areas with large and compact surfaces, unusually or marginally affected by the fragmentation phenomena and characterized by a high environmental suitability [20, 21]. On the contrary, low values of PI indicate small areas, with possible high fragmentation, which are potentially more susceptible to degradation phenomena such as fringe/border effects (Fig.6).

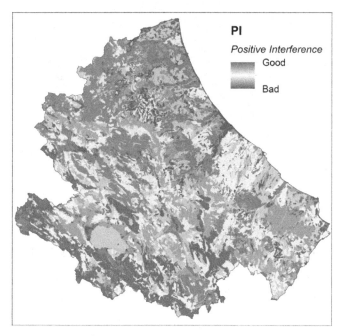

Fig. 6. Positive Interference (PI) map

3 Results

3.1 Criticality Index (CI) map

At this point, considered all the involved factors, the Criticality Index (Figure 10) has been calculated through the formula (1).

Interpolating the various indices was obtained a geographical information layer that synthetize, by spatializating, the main characteristics of the factors considered in this study (Fig. 7). The final CI is composed of ten classes, which express increasing values of land criticality in relation to the insertion of areal or linear structural elements (0 is the minimum, 10 is the maximum).

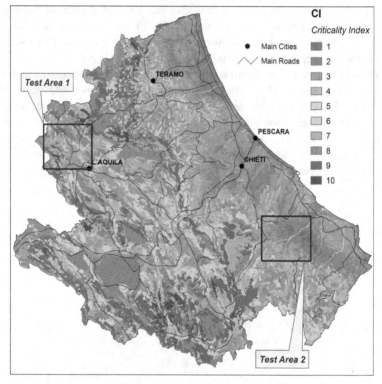

Fig. 7. Final Criticality Index (CI) map with Test Areas TA1 and TA2 (blue boxes)

3.2   Qualitative Environmental Impact Assessment: Examples of Applications

This paragraph is devoted to describe the application of the methodology developed to assess the impacts of a Wind Power Farm (WPF) installation and of a planned road infrastructure. The aim is to verify how the CI is able to support the analysis of environmental impacts generated on the surrounding environment by the development of such infrastructures. To this end, two distinct Test Areas have been considered (blue outlined boxes in Fig. 7).

The first one, Test Area 1 (TA1), concerns a real case: the project of a wind farm in the territory of Scoppito, in the province of L'Aquila (West of the town of Scoppito and North of Vigliano village). The area considered for the WPF does not fall inside protected areas, but is located close to the Site of Community Importance (SCI) named "Monte Calvo e Colle Macchialunga" (SCI identifier: IT7110208), where is planned to locate one of the wind turbines at about 600 m from the SCI.

The area is characterised by different territorial features (Fig. 8), such as urbanized areas (eastern side), hydrological bodies, forest areas, etc. Also in Figure 8 is reported the location of WPF elements: black points individuate the wind turbines, blue lines represent the relative road connections.

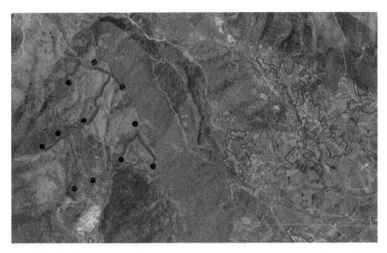

Fig. 8. Features characterizing TA1: territorial morphology, landscape texture, WPF turbines (black points) and road connections (blue lines), urban areas boundaries (black lines)

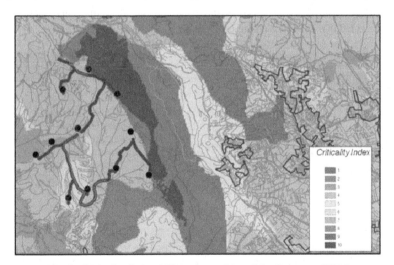

Fig. 9. Criticality Index (CI) map overlapped the features of TA1

By overlapping the CI map to the territorial and environmental features present within the area of interest (Fig. 9), it is possible to exploit our methodology to clearly highlight the impacts in terms of soil consumption. In fact, some Wind turbines fall very close to or inside areas characterised by high natural value, and for these reasons mapped with higher values of CI.

The developed methodology has been applied also to assess the environmental impacts due to the planning of a new road connection between two urban settlements, indicated in Test Area 2 (TA2). In Figure 10 are shown the urbanized areas (black contour) integrated in the context of a landscape characterised by natural features with high ecological sensitivity (rivers, fluvial habitats, forests, etc.).

Fig. 10. TA2 is characterized by several anthropized features, like settlements, roads, etc. (elements represented in overlap to a colour orthophoto)

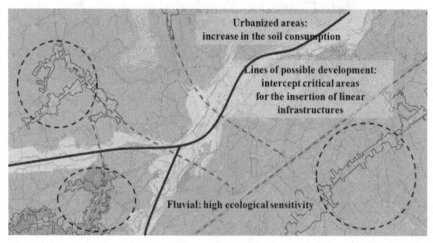

Fig. 11. Lines of possible development (dashed red lines), intercepting fluvial areas, characterised by an elevate degree of ecological sensitivity as represented in the CI map

In this case, it is planned to realize a new road connection (red dashed lines in Fig. 11) among the urbanized areas (delimited by the dashed black circle, also in Fig.11). The planned development will intercept areas defined as critical (CI = 6), especially in correspondence of the intersection between road and river (e.g., bridges). In fact, the presence of the river in the area constitutes a sort of natural path for a lot of bird species, which use this element to drive their migration from mountain to sea and vice versa; moreover, fluvial habitats are also important zone in which they can rest.

4   Conclusions

According to the proposed methodology and by means the production of a CI map, it has been possible to set-up a territorial classification in critical zones with different values. For example, coastal areas are, as expected, characterised by low values of criticality, because these areas were subjected to man-made changes. On the contrary, internal and mountain areas are characterised by higher criticality values.

On the whole, considering the checks carried out to test the effectiveness of the method, the results obtained from the analysis of the investigated areas can be considered reliable and summarized as follows:

- inland areas, due to the presence of several different factors, have CI values higher than coastal areas;
- in coastal areas it is possible to distinguish the main ecosystem of river courses, which represent some of the most important ecological connections;
- the level of detail of the indicators allows the analysis both at large and local scale;
- the critical intrinsic value of biological systems in respect to a transformation process is unrelated to the RES production and, therefore, the indicator provides a value for all the actions capable to affect an area and modify it.
- the *pulverisation* of the interventions on the territory, even if it can be considered as virtuous, because related to the production of clean energy, is a phenomenon to be avoided or at least to be carefully planned and controlled;
- the cumulative effect, combined with the pulverization, could represent a risk factor for natural habitats, because such phenomenon, uncontrolled, could affect the areas characterised by high naturalness as well as the marginal/fringe areas, still intact or partially intact.

The above mentioned typologies of disturbance, caused to the environment, can be compared to those which could derive from the urbanization processes.

In conclusion, criticality index can be used as a valuable and effective tool, at different scales, to perform an environmental impact assessment and to analyse the compatibility or not, of the RES infrastructure in the context of natural and semi-natural ecosystems.

References

1. European Union: Directive 2009/28/EC of the European Parliament and of the Council of 23 April 2009 on the promotion of the use of energy from renewable sources and amending and subsequently repealing Directives 2001/77/EC and 2003/30/EC (2009)

2. Caiaffa, E., Marucci, A., Pollino, M.: Study of sustainability of renewable energy sources through GIS analysis techniques. In: Murgante, B., Gervasi, O., Misra, S., Nedjah, N., Rocha, A.M.A.C., Taniar, D., Apduhan, B.O. (eds.) ICCSA 2012, Part II. LNCS, vol. 7334, pp. 532–547. Springer, Heidelberg (2012)

3. Borfecchia, F., Caiaffa, E., Pollino, M., De Cecco, L., Martini, S., La Porta, L., Marucci, A.: Remote Sensing and GIS in planning photovoltaic potential of urban areas. European Journal of Remote Sensing 47, 195–216 (2014)

4. Fichera, C.R., Modica, G., Pollino, M.: Land Cover classification and change-detection analysis using multi-temporal remote sensed imagery and landscape metrics. European Journal of Remote Sensing 45, 1–18 (2012)

5. Modica, G., Vizzari, M., Pollino, M., Fichera, C.R., Zoccali, P., Di Fazio, S.: Spatio-temporal analysis of the urban-rural gradient structure: an application in a Mediterranean mountainous landscape (Serra San Bruno, Italy). Earth System Dynamics 3, 263–279 (2012)

6. Corine Land Cover (2006),
   http://www.eea.europa.eu/publications/COR0-landcover

7. Meffe, G.K., Carroll, C.R.: Principles of conservation biology. Sinauer Associates Inc., Sunderland (1994)

8. Rastelli, F., Staffolani, L., Hruska, K.: Ecological study of the vegetal component in the terrestrial ecotones of central Italy. Journal of Mediterranean Ecology 4(2) (2003)

9. Odum, E.P.: Ecology: A bridge between science and society. Sinauer Associates Inc., Sunderland (1997)

10. GLCF - Global Land Cover Facility, Landsat Images, http://glcf.umd.edu/

11. NG, Italian National Geoportal,
    http://www.pcn.minambiente.it/GN/index.php?lan=en

12. Farina, A.: Principles and Methods in Landscape Ecology: Towards a Science of the Landscape. Springer, New York (2006)

13. Marucci, A.: Studio della sostenibilità delle fonti energeticamente rinnovabili attraverso tecniche di analisi GIS. Valutazione delle potenzialità territoriali tramite indicatori di sviluppo e sostenibilità. PhD Thesis, University of L'Aquila (Italy) (2012) (in Italian)

14. ESDP - European Spatial Development. Perspective Towards Balanced and Sustainable Development of the Territory of the European Union Agreed at the Informal Council of Ministers responsible for Spatial Planning in Potsdam (1999)

15. Council of Europe Committee of Ministers. Recommendation Rec, 1 of the Committee of Ministers to member states on the Guiding principles for sustainable spatial development of the European Continent (2002),
    https://wcd.coe.int/ViewDoc.jsp?id=257311

16. Bennett, A.F., Saunders, D.A.: Habitat fragmentation and landscape change. Oxford University Press (2010)

17. Clevenger, A.P., Wierzchowski, J., Chruszcz, B., Gunson, K.: GIS-generated expert based models for identifying wildlife habitat linkages and mitigation passage planning. Conservation Biology 16, 503–514 (2002)

18. Hanski, I.: Landscape fragmentation, biodiversity loss and the societal response. EMBO Reports 6(5) (2005)

19. Christensen, N.L.: Succession and natural disturbance: Paradigms, problems, and preservation of natural ecosystems. In: Agee, J.K., Johnson, D.R. (eds.) Ecosystem Management for Parks and Wilderness. University of Washington Press, Seattle (1988)

20. Pizzolotto, R., Brandmayr, P.: An index to evaluate landscape conservation state based on land-use pattern analysis and Geographic Information System techniques. Coenoses 1, 37–44 (1996)

21. Rutledge, D.: Landscape indices as measures of the effects of fragmentation: can pattern reflect process? DOC Science Internal Series 98. Department of Conservation, Wellington. 27 p. (2003)

Geomatics in Analysing the Evolution
of Agricultural Terraced Landscapes

Giuseppe Modica[1], Salvatore Praticò[1], Maurizio Pollino[2], and Salvatore Di Fazio[1]

[1]Dipartimento di Agraria, Università degli Studi Mediterranea di Reggio Calabria, Loc. Feo di
Vito, 89122 Reggio Calabria, Italy
{giuseppe.modica,salvatore.pratico,salvatore.difazio}@unirc.it
[2]ENEA National Agency for New Technologies, Energy and Sustainable Economic
Development, UTMEA-TER Lab, Via Anguillarese, 301 – 00123 Rome, Italy
maurizio.pollino@enea.it

Abstract. The present paper shows the first step of an ongoing research implemented for a dynamic characterisation and valorisation of a historical terraced landscape in Italy, the 'Costa Viola' landscape in the Calabria region. The Costa Viola dry-stone terraces characteristics and the Land Use/Land Cover (LU/LC) evolution between 1976 and 2012 were analysed. In order to better understand the evolution trends and dynamics, an intermediate step in 1989 was investigated. Taking into consideration the very steep slopes of the Costa Viola landscape and the need of precision in analysing the historical evolution of agricultural terraces, the use of precision tools coupled with in situ detailed surveys were implemented. As a result, a comprehensive picture of the evolutionary trends in nearly forty years has been provided also highlighting the dynamic of abandonment still ongoing despite the policies of valorisation implemented up to now.

Keywords: Agricultural Terraced Landscape; Geomatics; Landscape Evolutionary trends; Change Detection.

1   Introduction

The European Landscape Convention (ELC) [1] defines the landscape as 'an area, as perceived by people, whose character is the result of the action and interaction of natural and/or human factors'. This also means that landscape results from the composition of different cultural visions, corresponding to different perceiving subjects (farmers, tourists, ecologists, geographers, economists, planners, etc.) and experiences, which all refer to the same spatial reality [2]. As recognised by many scholars, the ELC considers the landscape as a common good [3]. In Italy the historic and traditional agricultural landscapes cover a large part of the national territory [4]. Although motivated by their productive function, they are today seen as an important cultural heritage. The terraced landscapes, in particular, have been attributed a number of cultural values, among which the scenic value, associated with the demand for landscape amenities, is the most appreciated by the wider public [5, 6]. As a consequence,

B. Murgante et al. (Eds.): ICCSA 2014, Part IV, LNCS 8582, pp. 479–494, 2014.

in many regions, the terraced landscapes have acquired a special status from the point of view of their management and planning, as it is in the case of the area of Cinque Terre (Liguria), which is a National Park and is included in the UNESCO World Heritage list. The terraced landscapes are a "living" matter, undergoing continuous change over time depending on the evolution of the relationship between the local population – with its economic activities, lifestyles, culture – and a given territory, with all its resources (natural/cultural and tangible/intangible) [7]. Today's increasing demand for conserving the quality of historic rural landscapes can only be answered by creating socioeconomic systems sustaining or increasing their resilience [4, 8]. For these reasons a strong public action is needed to tackle the present increasing urban pressure on the rural landscape, which is progressively leading to its depletion and to the loss of ecological stability. A nationwide initiative aimed at the inventory, characterization and continuous monitoring of the Italian historic and traditional agricultural landscapes, particularly the terraced ones, has been recently launched [4].

The terraced landscapes result from a social construction over time. On the one hand this implies that the whole community should be engaged in the protection and conservation of the terraced system; but, on the other, this last requires a more complex approach to its sustainable territorial management, since there must be compatibility between the constraints given by the conservation policies and the transformation needed by agriculture to acquire satisfactory economic competitiveness. While in the past agriculture with its terracing works could be considered as a very impactive factor of landscape transformation, today the permanence of or re-introduction of a suitable agricultural use in the historic terraced landscape is considered as crucial for assuring its physical elements (dry-stone walls and stairs, pathways, water drainage/channelling systems, etc.) a regular and effective maintenance; this, also thanks to the presence of farmers scattered on a wide area.

The above described change in the cultural point of view has been recently coupled by a change in the national strategies on the governance of the environmental/cultural resources and also on the valorisation and conservation of the landscape [9]. In Italy, according to the European policies on Rural Development, an acknowledgment of these indications can be found in the recent Italian National Strategic Plan for Rural Development, 2007-2013, which sets the agricultural and forestry reuse of redundant and derelict landscapes as a strategic objective of the agricultural policies. This approach also constitutes the reference basis of the forthcoming 2014-2020 European Rural Development Policy.

In the case of terraced historic landscape sustainable management/planning strategies need to be based not only on a detailed, updated, accessible and specific knowledge of their characteristic features and present state, but also on a very precise knowledge of landscape changes over time. Change detection should concern both the physical elements determining landscape configuration and functionality (retaining walls, channelling and drainage systems, pathways, stairs, etc.), and Land Use / Land Cover (LU/LC). This study is focused on an important historic terraced area of South Italy (Costa Viola, Calabria) in order to better understand its evolution trends and landscape dynamics. The multi-temporal analyses here presented mainly concern the dry-stone terraces characteristics and the Land Use/Land Cover (LU/LC) change in the period 1976-2012.

2   Materials and Methods

2.1   Study Area

Costa Viola is a coastal strip of land about 20 km long and 1 km wide, located in Southern Calabria (Italy). Altitude varies between 0 and 500 m a.s.l. From the administrative point of view, Costa Viola falls within the territory of five municipalities in the province of Reggio Calabria: Villa S. Giovanni, Scilla, Bagnara Calabra, Seminara and Palmi (Fig. 1). In the present research large part of the investigations were conducted in the municipality of Bagnara Calabra, where most of the terraced areas are located. The Strait of Messina separates Costa Viola from Sicily. The coast is particularly exposed to the west and south winds, faces the Tyrrhenian sea and is characterised by steep slopes by which Aspromonte (Aspermont, literally the "Harsh Mountain") reaches the sea, thus forming impressive cliffs and terraces. Altitude varies between 0 and 500 m. The land has a dramatic form presenting deep and narrow valleys excavated in time by the water of the many torrents (the so-called *"fiumare"*) characterising the region. Land morphology is very troubled also in the sample area characterized by very steep slopes formed by igneous and metamorphic rocks. Soils are subtle with acid or sub-acid reaction. Climate is Mediterranean with temperate winter and hot and drought summer with average precipitations of about 1000 mm·year^{-1}. The natural value of the territory is highlighted by the presence of a number of protected areas. Many flora and fauna species live in the various habitats still present in this area, which still keeps highly natural places. Agricultural terraces in the area had a great development between the 18th and the first half of the 19th century, a period of time in which they spread also in very steep sites mainly to allow the cultivation of vine, but also for creating small citrus and olive orchards. Until 1930s they were reported terraced vineyards presenting slopes over 250%; it means terraces 1 m large with earth retaining dry-stone walls being over 2.50 m high. In the average, slope was around 100% and there were agricultural terraced areas presenting over 200 consecutive terraces [10–12].

In the early 1930s in the area between Scilla and Bagnara the terraced vineyards extended over 600 hectares. They contributed to the economy of a local population then counting 20.000 inhabitants who lived on a mix of agriculture, trade and fishery. All these activities had complementarity and the uncertain incomes of the one was usually compensated by the others. Terraces not only had an agricultural function but also played an important role in assuring the safety of the towns and the villages below; in fact, by retaining the earth and controlling the rainwater flow through their complex drainage and channelling systems, they helped reduce the hydraulic and geologic risk, thus preventing landslides and floods.

From the 1930s up to now the terraced vineyards have undergone continuous abandonment followed by progressive deterioration and ruin. Today in Costa Viola the area covered by still cultivated terraced vineyards is roughly estimated around 200 hectares [13–15]. The decline of viticulture in the terraced areas is due to many reasons, among which the most relevant are: the general abandonment of agriculture and the decline of employment in agriculture; the ageing of the rural population; further

fragmentation of the land property, already high, as a consequence of heritage transmission in families with many children; increase of labour cost and of its relative
share of the total production cost; delays in the definition of policies and actions to
support agriculture in the terraced areas; lack of integrated actions for the recovery of
the dry-stone walls and the valorisation of: the regional landscape, local typical products, traditional crafts and culture, local distinctiveness [12, 13, 15, 16].

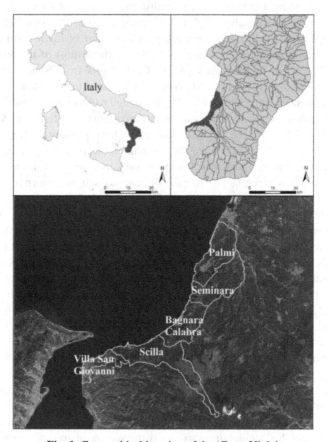

Fig. 1. Geographical location of the 'Costa Viola'

In Costa Viola traditionally the dry-stone walls were built by specialized crafts-
men. Today very few of them, mostly old aged, are still active, since the abandonment
of terraces has also caused a decline of this craftsmanship, so interrupting the chain of
transmission of the traditional skills and knowledge.

Agricultural mechanisation could greatly reduce the grape harvesting and transport
cost; although technically viable and supported in a recent past by the Regional Government with financial help, it has not been consistently applied, mainly because of
the high fragmentation of land property and the lack of co-operation between the wine
farmers. Co-operation, in fact, would have helped reducing the cost per plot of the
mechanic plants thus achieving economic viability. Abandonment is more relevant in

those vineyards which are less accessible, and therefore more difficult to cultivate, either because of the high altitude or the steep slopes. Progressive disuse of the terraced agricultural land, due to the subsequent lack of care and maintenance, has made the whole coastal area environmentally unbalanced and not safe, in particular from the hydraulic and geologic point of view [17]. The original asset of the dry-stone wall, water drainage and channelling systems has been greatly altered by undergoing progressive deterioration. The situation has also been worsened by the spontaneous renaturalisation of the agricultural terraces, where wild vegetation has regained little by little the space once occupied. This way the new configuration of the terraced landscape of Costa Viola differs from both the previous wild system and from the historic man-made landscape. Now it is rather a mix of both, evolving in such a way that is substantially unbalanced and uncontrolled. Recent studies concerning Costa Viola have shown the emblematic character of this case-study, and the close correlation between the abandoning of farming activities, the need to find new markets for the typical local products also in order to have strong economic reasons for maintaining the traditional terraced landscapes and, doing so, preventing soil erosion and landslides [18].

2.2   Land Use Mapping

As introduced, the present paper reports the first step of an ongoing research implemented for a dynamic characterisation and valorisation of the agricultural terraced landscape of Costa Viola. To detect the Land Use/Land Cover (LU/LC) dynamics of the agricultural terraces, still now subject to a continuous abandonment, a representative case-study was carried out. To this end, the municipality of Bagnara Calabra was chosen as sample area of the entire Costa Viola.

To implement the spatial analyses, the following base materials were acquired (Tab. 1):

— Black and White (B/W) digital georeferenced orthophotos of the "National Geo-portal" of the Italian Ministry of Environment (May 1989 and May 1994 flights);
— Colour digital georeferenced orthophotos of the "National Geoportal" of the Italian Ministry of Environment (May 1998 and May 2006 flights);
— Colour digital georeferenced orthophotos loaded from Bing Maps (May 2010 flight)
— Colour digital georeferenced orthophotos (May 2012 flight);
— Black and White (B/W) aerial frames, rasterized at 800 dpi, acquired from Italian Military Geographic Institute (IGMI) (May 1976 survey).

For each of the reference years (1976, 1989, 2012) under investigation a LU/LC vectorial layer was produced into a dedicated geodatabase. Prior to proceed the subsequent analyses, these files were subjected to topological error identification in order to eliminate holes, overlaps self-intersections, etc. In order to ensure a high level of detail, photointerpretation was carried out at a fixed scale of 1:900 with a Minimum Mapping Unit (MMU) of 0.20 hectares. These details allowed the identification of the different categories and subcategories of the agricultural terraces.

The referenced legend is derived by an implementation of CORINE Land Cover[1] (CLC) by European program CORINE (Coordination of information on the Environment). When needed, the CLC legend was implemented at IV hierarchical level. In more details, this was useful in providing the "terraced" attribute of the LU/LCs. Furthermore, for immediate identification of the terraced areas, a column providing information on the presence/absence of the terraces was added to the attribute table.

Table 1. Technical characteristics of the aerial photographs and orthophotos used

Frame data	Date	Spatial resolution [m]	Source
B/W digital aerial ortho-photos GIS server catalogue (WMS)	May 1989	1.0	National GeoPortal of the Italian Ministry of the Environment, Land and Sea National (www.pcn.minambiente.it)
	May 1994		
RGB digital aerial ortho-photos GIS server catalogue (WMS)	May 1998	1.0	
	May 2006	0.50	
RGB digital aerial orthophotos	May 2010	~ 0.50	Microsoft® Bing Aerial Imagery
RGB and IR digital aerial orthophotos	May 2012	0.50	Regional Agency for Agricultural Payments (ARCEA)
B/W aerial photos (average elevation 2700 m a.s.l.) scanned at 800 dpi	May 1976	1.0	Italian Military Geographical Institute (IGMI) (www.igmi.org)

2.3   Digital aero-Photogrammetry Techniques for 1976 Orthophotos Production

The tormented nature of the area and the high level of detail required made necessary the use of precision tools and detailed surveys. To this end, digital aero-photogrammetry techniques have been implemented and exploited.

Digital aero-photogrammetry techniques, using stereo-correlation algorithms and specific software, can provide a description of complex environments. This approach uses statistical algorithms for corresponding point-correlation and matching, which results in a Digital Terrain Model (DTM) generated mainly from interpolated points, and only a few matched points. Matched points are those for which the correlation was successful and a height value was determined from parallax displacements. For those points where correlation failed, the algorithm interpolates their heights from those of the nearest points matched. By using stereo-correlation algorithms and specific software (ERDAS Imagine® Suite), it has been possible to achieve the

---

[1]   www.eea.europa.eu/publications/COR0-landcover

description of a complex environment such as the study area. The correlation algorithm needs to be manually set up through a series of parameters, in order to obtain a coherent, suitable and efficient "strategy" to extract the DTM.

The following are the procedures implemented and applied for the test area:

1. Preliminary operations:

- High resolution scan and pre-processing of the aerial photographs;
- Ground Control Points (GCPs) survey.

2. Digital photogrammetry steps:

- Internal orientation and external orientation (triangulation);
- DTM processing strategy elaboration
- DTM extraction;
- Orthophotos generation;
- Mosaic of orthophotos creation.

In order to georeference the aerial photos, it is necessary to have an adequate number of GCPs with geographical coordinates (X, Y) and elevation (Z) known. In the present research, GCPs were surveyed and detected with a geodetic GPS/GNSS which, thanks to its RTK (Real Time Kinematic) connection, allows to have a precision < 1 cm. More than 20 GCPs were used for this work, chosen in correspondence of targeted point visible at the three time intervals considered and easily identifiable in field and cartography. They also were used a lot of points with unknown coordinate and elevation but that were identifiable in two or more photos to better allows the overlapping of the original frames (the so-called "tie points").

The processing steps outlined at no. 2 of the previous item (Digital Photogrammetry) were performed using LPS module, a photogrammetric add-on of the ERDAS IMAGINE® suite, that offers several interesting tools to work with aerial photos and create DTMs, orthophotos and mosaics starting from one or more pairs of aerial photos (stereo-pairs), belonging to the same strip or to contiguous strips. The georeferencing process starts with the study of the calibration certificate and the data strip of photograms, which show a number of useful data like model and characteristics of camera, focal length, lens distortions, flight average elevation and fiducial marks separation (used to set up the orientation of photograms). These data and parameters were exploited to carry out the processing steps performed by LPS software. Once provided these reference data, next step was to create an internal orientation properly inserting the four fiducial marks placed on the four sides of each photo. After this, to pass from internal to external orientation, it was necessary to provide the coordinates of GCPs (Fig. 2a) in order to assign to the frame X, Y and Z (elevation) coordinates in an unique Geographic Coordinate System (GCS): in this work the UTM WGS84 33N was adopted. Thus, the intermediate results were the creation of a georeferenced model (internal and external orientation) (Fig. 2b) and the extraction of the DTM.

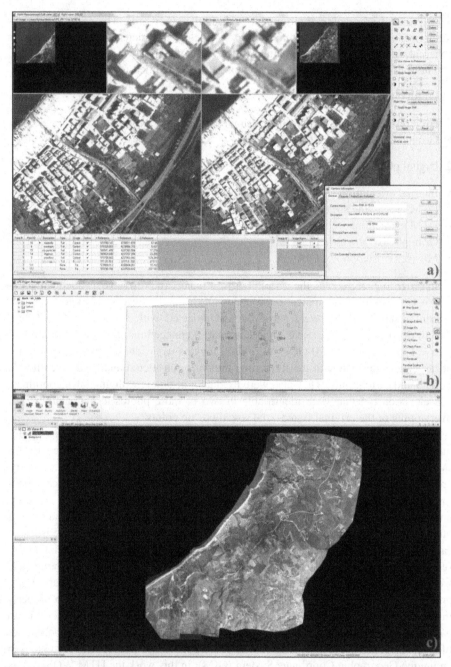

Fig. 2. Main steps of the digital aero-photogrammetry techniques for obtaining the orthophoto for year 1976: a) GCPs recognition and registration in the scanned aerial photographs frames and adding X, Y and Z coordinates for each of them; b) internal orientation and flight stripe reconstruction c) the obtained orthophoto after orthorectification, radiometric calibration and mosaicking.

Considering the study area, characterised by steep and undulating terrain, topography can distort the original aerial photos. DTM has therefore to be superimposed on the photos, in order to obtain, as an output, the orthophotos, in which every location of the aerial frame has been geometrically corrected (orthorectification) and located with its real geographic coordinates.

Finally, the orthophotos produced have been mosaicked together (Fig. 2c), to obtain a single georeferenced scene for the entire area, to be used to map LU/LC.

2.4   Change Detection

The evaluation of landscape patterns and dynamics is expression of the social context at a given time period [19]. Change detection is a process which allows to identify differences in the state of an object or phenomenon observed in different time ranges [20] and involves the application of multi-temporal datasets (maps, cartography, aerial photos, satellite images, in-situ surveys, etc.), to carry out a quantitative analysis of the effects, over time, of the phenomenon [21].

The second step of this work was to compare, by a spatial analysis, LU/LC changes within the time period examined. In order to avoid misregistration errors inducing false change alerts, a coregistration procedure between the three images was preliminarily carried out using GCPs as reference points.

This is a typical diachronic analysis: the assessment of the area situation in two different times. Then, through a thematic overlay it is possible to identify the changes occurred over the years in the study area. Such approach was conducted in GIS environment, by exploiting the *Geoprocessing* functions on vector data (maps of LU/LC produced for the different years considered) (Fig. 3).

This is a post-classification comparison technique that, starting from vector data, can provide a complete matrix of change dynamics [22], allowing to quantify the changes for each LU/LC class considered and to measure the areas subjected to change. Such approach enables to determine the difference between independently classified images from each time-interval analysed [23]. Where the change occurs, it is important to measure the areal extension, assess the spatial pattern and understand the reasons of change [24].

In this work 10 classes of LU/LC change were analysed and three cross-tabulation matrices (10x10) were produced. For each time-interval considered (1976÷1989; 1989÷2012) and for the overall one (1976÷2012) they were reported the LU/LC changes for the classes at 2nd hierarchic level of CLC, from time t_1 to time t_2. Moreover, only for the two extreme years of the examined time range were analysed 21 classes of change and was made a cross-tabulation matrix (21x21), reporting the LU/LC changes for the classes at 3rd hierarchic level of CLC. The rows contain the values (in ha) of changes occurred for t_1 categories and the columns show the amount of changes occurred for areas belonging to t_2 categories. The main diagonal expresses the *persistence areas*, which are those where no changes occurred.

Table 2. Transition matrix showing LU/LC changes for the main time interval (1976÷2012)

1976/2012	111	112	121	122	123	133	141	142	211	221	222	223	311	323	324	331	332	333	334	511	523	Sum
111	60.45	-	-	-	-	-	-	-	-	-	-	-	-	-	-	-	-	-	-	-	-	60.45
112	-	7.84	-	-	-	-	-	-	-	0.04	-	0.02	-	-	-	-	-	-	-	-	-	7.90
121	-	-	5.20	-	-	-	-	-	-	-	-	-	-	-	-	-	-	-	-	-	-	5.20
122	0.21	-	-	35.24	-	0.93	-	-	-	-	-	0.01	0.45	-	0.21	-	-	-	-	-	-	37.05
123	-	-	-	-	-	-	-	-	-	-	-	-	-	-	-	-	-	-	-	-	-	0
133	-	-	0.28	0.27	-	14.99	-	-	-	-	-	0.11	6.93	-	1.35	-	0.93	0.11	-	-	-	24.97
141	-	-	-	-	-	-	0.24	-	-	-	-	-	-	-	-	-	-	-	-	-	-	0.24
142	-	-	-	-	-	-	-	-	-	-	-	-	-	-	-	-	-	-	-	-	-	0
211	0.37	0.82	0.46	0.41	-	65.06	-	-	111.25	0.40	-	104.12	20.83	-	6.63	0.40	-	1.18	-	-	-	311.51
221	3.61	-	1.07	0.29	-	-	-	-	1.13	56.70	-	9.55	13.70	3.09	70.64	-	-	1.92	1.05	-	-	163.15
222	7.09	-	-	-	-	-	-	0.46	2.28	0.60	2.95	-	0.38	-	7.70	-	-	-	-	-	-	21.45
223	-	-	-	-	-	6.35	-	-	1.36	-	-	227.07	14.81	-	1.52	-	-	-	-	-	-	252.63
311	0.59	1.48	0.83	0.69	-	67.23	-	0.81	24.39	1.22	1.76	116.63	775.83	16.83	32.18	-	-	0.20	26.76	-	-	1050.4
323	-	-	0.49	1.62	-	0.92	-	-	-	-	-	-	20.35	106.98	8.22	-	-	1.91	1.42	-	-	138.37
324	4.83	-	0.48	0.58	-	1.04	-	-	1.11	5.38	0.52	15.22	66.23	-	204.24	0.18	0.40	15.61	-	-	-	334.87
331	-	-	-	-	-	-	-	-	-	-	-	-	-	-	-	17.19	-	-	-	-	-	17.27
332	-	-	-	-	-	-	-	-	-	-	-	-	-	1.18	1.35	-	11.18	-	-	-	-	13.88
333	-	-	-	-	-	-	-	-	-	-	-	-	-	-	0.13	-	-	26.31	-	-	-	26.41
334	-	-	-	-	-	-	-	-	-	-	-	-	-	-	-	-	-	-	-	-	-	0
511	-	-	-	-	-	-	-	-	-	-	-	-	-	-	-	-	-	-	-	7.78	-	7.78
523	-	-	-	-	1.76	-	-	-	-	-	-	-	-	-	-	-	-	-	-	-	-	1.76
Sum	77.16	10.13	8.82	39.11	1.76	156.50	0.24	1.34	141.52	64.34	5.23	472.73	919.51	128.07	334.32	17.76	12.50	47.24	29.22	7.78	0	2475.3

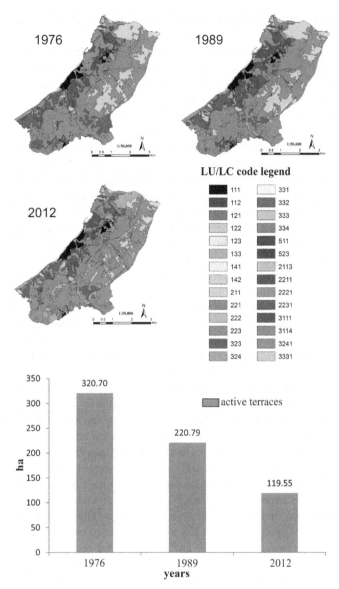

Fig. 3. The three LU/LC maps produced and the chart showing the amount of active agricultural terraces for years 1976, 1989, 2012

3   Results

Landscape changes have been studied by analysing the LU/LC changes in the territory of Bagnara Calabra. As reported in the methodological section, LU/LC changes were detected in two time intervals: from 1976 to 1989 and from 1989 to 2012, creating for each interval a cross-tabulation matrix. Moreover, they were created two matrices, one at the 2nd ad one at the 3rd hierarchic level of CLC, for the whole period considered, from 1976 to 2012.

Analysing the matrices it can be observed immediately that the LU/LC class 1.3 "mine, dump and construction sites" has greatly increased its extension at the expense of other classes, especially of classes 2.1 "arable land", 2.2 "permanent crops" and 3.1 "forests". This phenomenon can be ascribed to the construction sites for the modernization of the "A3 Salerno–Reggio Calabria" motorway, which have subtracted areas previously cultivated or wooded. In fact, the extension of class 1.3 passed from 25.17 ha in 1976 to 13.19 ha in 1989 and further expanded to 156.50 ha in 2012, increasing by about 1000%. As a further result, a loss of forests (about 130 ha) was identified, but it is important to clarify that this dynamic of regression is the consequence of two great wild fires recently occurred. Another important factor is represented by progress of LU/LC class 3.2 "shrub and/or herbaceous vegetation associations", that has a trend irregular and difficult to understand, because the above class includes: a) areas originating after the abandonment of cultivated terraces; b) transitional zones connected to natural dynamics of forest species and c) areas occupied by edaphic-climax formations, typical of the Mediterranean region. Thus, such increase, that can be a symptom of the abandonment of the terraces, could be underrated if it corresponds to an area decrease due to natural evolutionary dynamics of woodlands. An increasing of LU/LC class 3.3 "open spaces with little or no vegetation" of about 60 ha was caused, among other things, by landslides due, in the most cases, to the abandonment of cultivated terraces. An interesting element is represented by the transition of all areas of the class 5.2.3 to the class 1.2.3. This is due to the fact that in 1976 the port of Bagnara was not yet realised and surface was occupied by the sea. Thus it was used the class "sea and ocean" to respect the total of municipal area in the absence of "port areas" class. Another important factor is represented by the evolution of terraced areas. This parameter was analysed through the identification, in GIS environment, of visible terraced polygons at the end of both the time intervals analysed. Thus, it is possible to delineate an evolutionary pattern for these areas, comparing polygons that have terraced features at the time t_1 with polygons that have the same attribute at the time t_2. This method allows for a quantitative assessment concerning the extension of the terraced areas and their persistence, loss or recovery. The resulting data are, however, difficult to read because within the sample there are both cultivated and abandoned (inactive) terraces. Then, there are areas covered by crops and others abandoned or being abandoned, occupied by herbaceous or forestry formations. To better understand dynamics of actual recovery or loss, it is necessary to focus the research only on cultivated terraces (active) in order to obtain helpful information on the evolutionary dynamics occurring within the time interval considered. The total area occupied in 1976 by active terraces was about 320 ha. This amount, however, has suffered a decrease, coming about 220 ha in 1989 and finally about 120 ha in 2012, with a decrease of terraced vineyards from about 160 ha in 1976 to about 50 ha in 2012; terraced olive plantations decreased from about 135 ha in 1976 to about 55 ha in 2012 (Fig. 4). An interesting factor is the decrease of terraced orchards due, in addition to the natural dynamics of abandonment observed for all crops, to the urban expansion resulting from the construction of the "Marinella" neighbourhood.

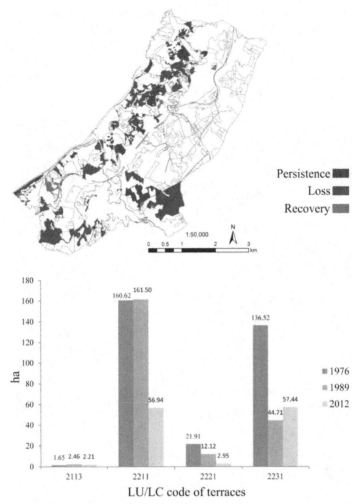

Fig. 4. Map showing the dynamic of the agricultural terraces in the study-area in the period 1976-2012.The chart at the bottom side depicts the amount of terraces for years 1976, 1989 and 2012, according to the different LU/LC classes.

4   Final Discussions

The results obtained in the framework of the present research has shown how GIS software and methodologies are able to effectively support not only the management of geographical data, but also the analysis carried out to describe and understand the landscape dynamics within the study area. These tools allow the planners to operate with the help of advanced spatial analysis models, thematic overlaps and maps of synthesis made "ad hoc" for the examined areas. Among these maps there are the LU/LC maps, considered as basic and fundamental information for territorial planning thanks to their capability to be easily read and understood by stakeholders and decision makers. What emerges from the evolutionary dynamics shown in this paper is that the territory of Bagnara Calabra and, more generally, the entire Costa Viola, is now at the mercy of a trend that is leading to a general abandonment of agricultural activities, primarily in those areas more difficult to reach because of the greater distance from the main road connections (Fig. 5).

―――― boundary of terraced vineyards in 1976

▨▨▨ loss of terraced vineyards in 2012

Fig. 5. An orthophoto detail (2012) showing the loss of agricultural terraces in the period 1976-2012 in a representative terraced area of Bagnara Calabra ('Caccipuju')

This is consequence of an increasing industrialisation pressure and of a socioeconomic change, that in rural areas lead to terraced crops abandonment in favour of more profitable crops [25]. All these factors involve not only an agricultural loss but also an increase of landslide risk, whose effects have already caused significant damages in the study area. But, fortunately, is also possible to notice the presence of a lot of people that, like "heroes" of landscape and agriculture preservation, are today trying in some terraced plots to tackle abandonment, by recovering the degraded areas, maintaining dry-stone walls and resuming crops. For all these reasons, it is important to implement and maintain a spatial database of terraces, and to keep under control the evolutionary dynamics of the terraced landscape; this, in order to constantly monitor the terraces state and to better calibrate the possible interventions for risk management, vulnerability mitigation and agriculture and rural development funding-allocation. Moreover, in an area such as the Mediterranean basin, a challenge still arising is to ensure a durable balance between development and the protection of valuable natural and cultural resources [26]. With the aim to support decision making, a future research direction will deal with the implementation of a free WebGIS platform suitable for favouring and improve the proactive e-participation of the local communities since the earlier stages of the planning process [27]; this, also by means of participative approaches such as collaborative mapping (Volunteered Geographic Information, VGI) [26–28].

Acknowledgements. The Authors are grateful to the Agency of Calabria Region for Agricultural Payments (A.R.C.E.A.) for providing RGB and IR digital aerial orthophotos of the study-area for year 2012. The research was partially funded by the LANDsARE project (Landscape Architectures in European Rural Areas: a new approach to the local development design Co-operation project).

References

1. Council of Europe: European Landscape Convention. Florence (2000)
2. Antrop, M.: The concept of traditional landscapes as a base for landscape evaluation and planning. The example of Flanders Region. Landsc. Urban Plan. 38, 105–117 (1997)
3. Pedroli, B., Antrop, M., Pinto Correia, T.: Editorial: Living Landscape: The European Landscape Convention in Research Perspective. Landsc. Res. 38, 691–694 (2013)
4. Agnoletti, M.: Italian Historical Rural Landscapes: Dynamics, Data Analysis and Research Findings. In: Agnoletti, M. (ed.) Italian Historical Rural Landscapes Cultural Values for the Environment and Rural Development, pp. 3–87. Springer, Dordrecht (2013)
5. Pinto-Correia, T., Barroso, F., Menezes, H.: The changing role of farming in a peripheric South European area – the challenge of the landscape amenities demand. In: Wiggering, H., Ende, H.-P., Knierim, A., Pintar, M. (eds.) Innovations in European Rural Landscapes, pp. 53–76. Springer, Heidelberg (2010)
6. Parris, K.: European agricultural landscapes supply and demand: implications of agricultural policy reform. In: Jongman, R.H.G. (ed.) The New Dimensions of the European Landscape, pp. 7–37. Springer, Dordrecht (2004)
7. Steiner, F.R.: The Living Landscape, Second Edition: An Ecological Approach to Landscape Planning. Island Press, Washington (2008)
8. Plieninger, T., Bieling, C.: Resilience and the cultural landscape: understanding and managing change in human-shaped environments. Cambridge University Press, Cambridge (2012)
9. De Montis, A., Caschili, S.: Nuraghes and landscape planning: Coupling viewshed with complex network analysis. Landsc. Urban Plan. 105, 315–324 (2012)
10. Bova, G.: Alcune tipiche sistemazioni dei terreni di Calabria. Tipografia Morello, Reggio Calabria (1934)

11. Gambi, L.: Calabria. UTET, Torino (1965)
12. Di Fazio, S.: I terrazzamenti viticoli della Costa Viola. Caratteri distintivi del paesaggio, trasformazioni in atto e gestione territoriale in un caso-studio in Calabria. In: I Georgofil. Quaderni 2008-II, pp. 69–92. Edizioni Publistampa, Firenze (2008)
13. Previtera, G., Zocali, A.: Descrizione delle zone viticole di Bagnara e Scilla secondo criteri di omogeneità nella giacitura, esposizione, vie di comunicazione, struttura fondiaria. Possibilità di recupero e riutilizzazione dei terreni abbandonati a vite o altre colture. Agric. Calabr. VI, 9–32 (1983)
14. Albanese, G.: Istituzione di Paesaggi Protetti nel territorio del "Basso Tirreno Reggino". Costa Viola e Piana degli Ulivi. Laruffa Editore, Reggio Calabria (2001)
15. Di Fazio, S., Malaspina, D., Modica, G.: La gestione territoriale dei paesaggi agrari terrazzati tra conservazione e sviluppo. Proceeedings of the National Congress of AIIA (Italian Society of Agricultural Engineers) Ingegneria agraria per lo sviluppo sostenibile in area mediterranea, Catania, Italy (2005)
16. Di Fazio, S., Modica, G.: The valorisation and characterisation of the agrarian terraced landscape. A case study in the Costa Viola area (Italy). In: International Conference of Agricultural Engineering CIGR-AgEng 2012, Valencia, Spain (2012)
17. Di Fazio, S., Modica, G.: Le pietre sono parole: letture del paesaggio dei terrazzamenti agrari della Costa Viola. Iiriti Editore, Reggio Calabria, Italy (2008)
18. Agnoletti, M., Cargnello, G., Gardin, L., Santoro, A., Bazzoffi, P., Sansone, L., Pezza, L., Belfiore, N.: Rural development: comparative study in three terraced areas in northern, central and southern Italy to evaluate the efficacy of GAEC standard 4.4 of cross compliance. Ital. J. Agron. 6, 121–139 (2011)
19. Antrop, M.: Why landscapes of the past are important for the future. Landsc. Urban Plan. 70, 21–34 (2005)
20. Singh, A.: Review Article Digital change detection techniques using remotely-sensed data. Int. J. Remote Sens. 10, 989–1003 (1989)
21. Lillesand, T.M., Kiefer, R.W., Chipman, J.W.: Remote sensing and image interpretation. John Wiley & Sons, Hoboken (2008)
22. Lu, D., Mausel, P., Brondízio, E., Moran, E.: Change detection techniques. Int. J. Remote Sens. 25, 2365–2401 (2004)
23. Fichera, C.R., Modica, G., Pollino, M.: Land Cover classification and change-detection analysis using multi-temporal remote sensed imagery and landscape metrics. Eur. J. Remote Sens. 45, 1–18 (2012)
24. Di Fazio, S., Modica, G., Zoccali, P.: Evolution Trends of Land Use/Land Cover in a Mediterranean Forest Landscape in Italy. In: Murgante, B., Gervasi, O., Iglesias, A., Taniar, D., Apduhan, B.O. (eds.) ICCSA 2011, Part I. LNCS, vol. 6782, pp. 284–299. Springer, Heidelberg (2011)
25. Dunjó, G., Pardini, G., Gispert, M.: Land use change effects on abandoned terraced soils in a Mediterranean catchment, NE Spain. Catena 52, 23–37 (2003)
26. Modica, G., Zoccali, P., Di Fazio, S.: The e-participation in tranquillity areas identification as a key factor for sustainable landscape planning. In: Murgante, B., Misra, S., Carlini, M., Torre, C.M., Nguyen, H.-Q., Taniar, D., Apduhan, B.O., Gervasi, O. (eds.) ICCSA 2013, Part III. LNCS, vol. 7973, pp. 550–565. Springer, Heidelberg (2013)
27. Pollino, M., Modica, G.: Free Web Mapping Tools to Characterise Landscape Dynamics and to Favour e-Participation. In: Murgante, B., Misra, S., Carlini, M., Torre, C.M., Nguyen, H.-Q., Taniar, D., Apduhan, B.O., Gervasi, O. (eds.) ICCSA 2013, Part III. LNCS, vol. 7973, pp. 566–581. Springer, Heidelberg (2013)
28. Goodchild, M.F.: Citizens as sensors: the world of volunteered geography. GeoJournal 69, 211–221 (2007)

Geomorphological Fragility and Mass Movements of the Archaeological Area of "Torre di Satriano" (Basilicata, Southern Italy)

Stefania Pascale[1,*], Jessica Bellanova[2], Lucia Losasso[1], Angela Perrone[2], Alessandro Giocoli[2,3], Sabatino Piscitelli[2], Beniamino Murgante[1], and Francesco Sdao[1]

[1] School of Engineering, University of Basilicata, 85100 Potenza (PZ), Italy
[2] CNR-IMAA, Tito, Potenza, Italy, 85100 Potenza (PZ), Italy
[3] ENEA, 00196 Rome, Italy
pascalestefania@gmail.com

Abstract. This paper describes the results of geomorphological and stability studies carried out in the archaeological site of Satriano di Lucania (Basilicata, Southern Italy), where an important sanctuary was built during the 4th Century B.C. This study is based on a mutidipliscinarity approach including accurate interpretation of aerial photos, geomorphological and geoelectrical surveys , and stability analyses. A description of the stability condition of the archaeological site with reference to the landslide that affects the sacred complex is provided in this work.

Keywords: Slope instability, Landslide hazard, Lucanian sanctuary, 4th Century B.C., Basilicata (South Italy).

1 Introduction

Basilicata Region (Southern Italy), due to geological, geomorphological, climatic and seismic characteristics, is one of the Mediterranean basin regions most subject to geomorphological instability, taking form by different kinds of landslides, evolution, mechanisms and processes of intense and selective erosion. In particular, in some areas of Basilicata Region, landslides are so intense to generate sometimes extensive and serious damages to people and properties. In the last years, recent geomorphological studies have shown that many archaeological sites of Basilicata Region, especially in the Apenninic areas, are characterized by widespread and intense landslides which damage valuable historical and archaeological remains.

Such situations can be observed, for example, in the archaeological area of Rossano di Vaglio ([1] and [2]), where the shrine of the goddess Mephitis falls within a large and ancient landslide; in the "Parco Storico Naturale delle Chiese Rupestri del Materano", in which there are numerous medieval testimonials characterized by a serious state of collapse due to landslides [3], [4] [5] and in the Satriano di Lucania area, which represents the study area of this article [6]. This work reports the main results of

---

* Corresponding author.

B. Murgante et al. (Eds.): ICCSA 2014, Part IV, LNCS 8582, pp. 495–510, 2014.

accurate geological, geomorphological and geophysical studies of "Torre di Satriano" archaeological area, supplemented by analysis and interpretation of aerial photos, which allowed the definition of the geomorphological instability and subsequent drafting of landslide inventory. Researches and surveys previously carried out both in the investigated area and the neighbour zones have been also considered [6]. In particular, it was analyzed the area on which a worship place arises and where there are the remains of the Lucanian Sanctuary structures [7]. Wall structures of the sanctuary, unearthed during various archaeological excavations, show progressive deformations which led the local population to repeatedly restructure the sacred building between the fourth and first centuries B.C.. Since its total abandonment a subsequent fossilization of the area has been caused by the deposition of detrital and colluvial materials. As a whole, the settlement of Lucanian age located downstream of the "Torre di Satriano", is configured as a space-time unity with specific lithological, morphological characteristics and well-define anthropogenic activities [8]. Finally, environmental and/or socio-cultural changes have influenced the local population to definitively abandon the place of worship during the first century A.D..

2   Historical Notes on the Satriano Archaeological Site

From the archaeological point of view, the Torre di Satriano represents one of the most interesting areas of the Basilicata Region, due to the anthropogenic presence, starting as early as the Bronze Age to the Middle Ages [6]. Over the years human settlement has been involved in a complex evolution characterized by abandonment and subsequent reorganization of the territory differently distributed in space. The first abandonment has been recognized between the Late Bronze Age and the Iron Age followed by a repopulation during the eighth century B.C., as evidenced by burials found in the area. The major phase of growth of the pre-Lucanian indigenous settlement took place from the sixth century B.C. that coincides with the implantation of a settlement with its necropolis, along the southern side of the Torre di Satriano. In the IV-III century B.C., the area was involved in great transformations corresponding with the arrival of people from Basilicata. It dates at this time the Lucanian Sanctuary which is located over a large area close to a spring and not far from a sheep track that favored the interaction between people, and represented a road network linking the Tyrrhenian and the Ionian sectors of the peninsula. The Sanctuary activity developed in a large enough time span, also characterized by quite long periods of scarcity of documentation, such as the period between the second century and the beginning of the first century B.C., until the complete abandonment occurred in the first century A.D. ([6] and [9]). During its period of activity, the Sanctuary was characterized by repeated renovations, especially between the third and the first century B.C., when, due to the damage of the seabed structures, the abandonment of the lower shelf, on which stood one of the buildings in favor of the upper one, occurred. The disappearance of the Lucanian settlement started at the end of the third century B.C., probably as a result of the Romanization phenomenon of the area that led to the emergence of new urban realities as *Potentia*. In the Middle Ages, finally, there was a restocking with the birth of *Satrianum* that became an important bishopric in the twelfth century A.D..

3   Geomorphological Fragility and Landslides of "Torre di Satriano" Archaeological Site

3.1   Geological Setting of the Study Area

The study area, located in the central sector of the Campano-Lucano Appennines (fig. 1), represents a high morphological and structural, standing as a small relief with a convex morphology and with steep slopes.

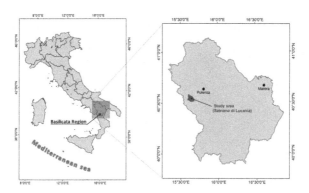

Fig. 1. Location of the study area

The ridge on which the ruins of Torre di Satriano stand, is a close alignment NW-SE oriented (fig. 2), which acts as both geographical and hydrographic watershed between two great morpho-tectonic depressions, respectively known as the Tito-Picerno intermontane basin at north and the Pergola-Melandro at south, that over the past two million years were filled by coarse clastic sediments [10]. The recent tectonic uplift rate occurred in this part of the chain during the last 730 ka, with speeds estimated in the order of 0.6-1 mm/yr, has increased erosional activity exerted by the water courses that has profoundly affected the hundreds of meters of clastic sediments, filling the intermontane depressions [10]. Lateral erosion, alternated to the vertical one, has produced a widening and, sometimes, a decrease in the slope of the river valleys, up to the current configuration of the landscape. From a geological point of view, in the examined area outcrop the terrains belonging to the Ceno-Mesozoic pelagic successions of the Lagonegro Units, the Tortonian pelitic-arenaceous-calcareous alternations of the "Monte Sierio" Formation (sensu [11]), and the middle-inferior Pliocene sands (Fig. 2).

The Lagonegro Units are, in particular, represented, from the bottom to the top, by "Monte Facito", "Calcari con Selce", "Galestri" and "Flysch Rosso" formations.

"Monte Facito" formation (Lower-Middle Triassic) is subdivided in a) a terrigenous (siliciclastic) unit, made up of a succession of quartz-rich calcarenites and mudstones, marls, shales and green micaceous sandstones, strongly deformed and frequently associated with gravels; and b) an organogenic unit, made up of grey massive limestones and heavily fractured black limestones and marls.

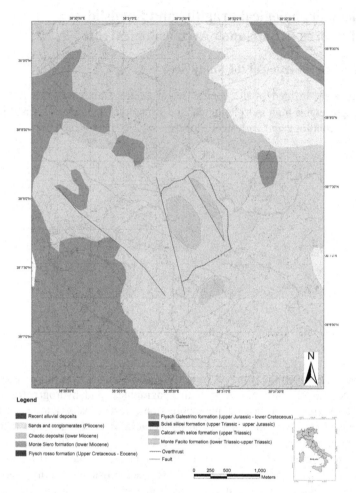

Fig. 2. Geological map of the study area (Geological Map of Italy at 1:100,000 scale, Servizio Geologico D'italia (1971) - Carta Geologica d'Italia scala 1:100.000. F. 199 "Potenza", Roma; modified by [6].

"Calcari con Selce" formation (Upper Triassic), conformably lying on the "Monte Facito" formation, consists of centimetres-to-decimetres-thick layers of grey limestones with beds and nodules of black and white chert, marls and calcareous marls, mudstones and marly clays. This succession is well layered and strongly fractured. This formation, showing a clear stratification and an intense cracking, constitutes the ridge oriented towards NW-SE on which the medieval site of "Torre di Satriano" is located.

"Scisti Silicei" formation (Jurassic), composed of thin – layered and fractured variegated cherty shales, partially cherty marls, green and red Radiolarites and graded pebbly limestones. This formation, well stratified and finely cracked, essentially emerges along the southern slopes of Mt. Caruso and near some reliefs in the area.

"Galestri" formation (Lower-Middle Cretaceous), consisting of clayey and marly thinly bedded rocks, extensively emerges near the north-western and southern sector of the examined area.

"Flysch Rosso" formation Auctt. (Upper Cretaceous – Eocene) is made up of marly clays and clayey marls finely scaly randomly, alternated with levels of mudstones, marly calcilutites, bioclastic calcarenites, cherty radiolarites, clays and marly clays. This geological unit, strongly fractured and severely deformed as a consequence of its severe tectonic history, is particularly prone to slope instability phenomena. This formation is present in the South-western sector of the examined area.

"Monte Siero" formation (lower Miocene) consists of an arenaceous-pelitic alternation in flysch facies and emerges in the NW of the study area and mostly in the orographic left of the Tito river.

Pliocenic conglomerates outcrop in discordant transgression on the Lagonegrese Unit. They are constituted by elements of different rocks and variable sizes, poorly stratified and poorly cemented. The rounded blocks, with a diameter ranging from some millimeters to some centimeters, are immersed in an abundant sandy matrix. These conglomerates outcrop in the sector of the study area, mainly on the left and right orographic of the Tito river.

These geological formations, finally, are capped by Pliocene wedge-top deposits and by Pleistocene-Holocene alluvial sediments.

Mass movements are widespread in the study area and play an important role in the present-day landscape evolution. In particular, very large ancient and recent landslide bodies can be observed and, mainly consist of calcareous-siliceous eterometric debris pieces in a yellow-brown sandy-clay matrix.

Furthermore, the thickness of the detrital accumulations, produced by landslide activity, that has always involved the considered area, varies from a few meters to a few dozen meters (landslide of Lucanian Sanctuary, [6]).

3.2   The State of Landslides of the Examined Area: Methodological Approach

The definition of the landslide state and the resulting geomorphological map are based on a methodological approach consisting in close synergy between the methods of applied and evolution geomorphology, the traditional techniques of landslides recognition (constituted of the analysis of aerial photos and satellite and of cartographic maps) and geomorphologic detection. Based on this methodology, the study has been divided into distinct and sometimes interconnected phases, summarized in fig. 3.

Studies have also taken into account the results obtained by geological, geomorphological and archaeological surveys previously carried out, concerning both the selected areas and the neighboring ones. For the definition of the geomorphologic and landslide characteristics of the area, accurate geological and geomorphological survey at 1:2.000 scale have been conducted in the years 2008 – 2013 and compared with aerial photos for the years 1956, 1977, 1982, 1991, 1994, 1997, 2004 and 2010. As far as the southern side of the relief of "Torre di Satriano" is concerned, and in particular for the area of the "Lucanian Sanctuary" and the nearby necropolis, also some

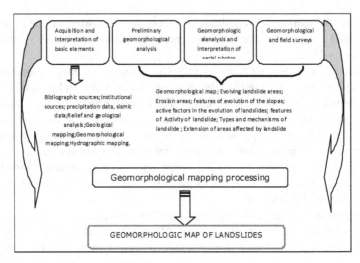

Fig. 3. Study methodology

archaeological and geophysical surveys have been conducted in recent years [6]. The current state of landslides has been schematically shown on a geomorphological map, in which each landslide has been reported and mapped.

For the drafting of such a map a thematic legend based on scheme proposed by [12] has been adopted and used, suitably adapted to local geomorphological conditions and already applied in other situations similar to the one in examination. In particular, for each landslide mapped the main geomorphological elements, the type and direction of movement and activity status have been shown.

4   Features and Distribution of Landslides of the Study Area

The intense and recurrent landslides characterizing the investigated area are well known and studied [6]. This geomorphological fragility, which is common in many areas of the Basilicata Region, is related to the peculiar lithological and geomechanical nature and to the widespread and intense cracking of the surface terrains (clayey-marly intensely fissured successions significantly predominate), the morphological configuration of the slopes, the applicants and intense event of rain, the frequent earthquakes. The analysis of landslide inventory map (fig. 4) shows that the area of Satriano has widespread and perfectly visible traces of a lively gravitational dynamics, which is still clearly evident and manifests through mass movements, of various types, size and degree of activity.

In the study area, 50 different landslides were recognized and defined, on a total area of 22 km^2 and about the 78% (17 km^2)was investigated. The most mapped landslides are related to complex types, where the retrogressive and roto-translational mass movements dominate, evolving, often, in earthflows. These landslides often involve the whole slope, from the top of the reliefs to the underlying watercourses of the Melandro and of the Tito-Picerno rivers which are characterized, for this reason,

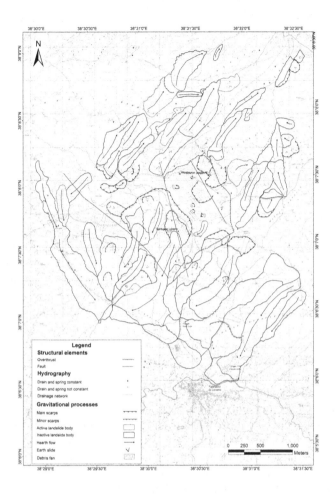

Fig. 4. Geomorphological map of the archaeological area of "Torre di Satriano"

by a markedly meandering configuration, with evident lateral deviation and partial occlusion of the riverbeds. The numerous mapped landslides show similar geomorphological features: often modeled or degraded scarps, connected with the surrounding relief; landslide channels with modeled edges and linked with the landslide bodies; bulge, morphological depressions and surface landslide; extended accumulation zones with partial landslide reactivations caused by the excavation made to the foot by Melandro River. The thickness of such landslides is variable and can reach a maximum value of about 20-25 m. Multi-temporal aerial photo interpretation and field surveys allowed to distinguish between active and inactive landslides. In particular, the landslides that showed signs of activity after 2013 were considered active (e.g. bright light colors on the aerial photos, landslide area with no vegetation, evident cracks in the source area and by a distinct bulge at the toe). Fig. 5 shows the distribution of landslides for each lithological class, according to the landslide density (ratio of landslide area of each type of landslide in every lithological complex "Af" and the total area of this last "Al"). The figure and the map show that landslides are present with varying intensity in each geological formations.

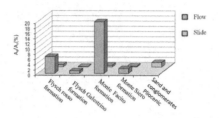

Fig. 5. Landslides distribution for lithological classes

The lithological formations most affected by landslides are clayey-marly ones belonging to the Lagonegro Unit and, in particular, to the formation of Monte Facito and Galestri. The same figure shows that the typology, geometry and evolutionary features of landslides in the examined areas are markedly affected by lithological and structural characteristics of the lithologies outcropping. Distribution and type of mass movements is closely related to the geological and morphological setting, including structural and stratigraphic factors (alternating weak and hard rocks), tectonic setting and fluvial undercutting ([13] and [14]). The interaction between landsliding and lithology showed that mass movements mainly affect slopes carved in the flyschoid rocks, sandy and conglomeratic deposits. In the areas, where flyschoid lithotypes crop out of various kinds and with different pelitic content, the landslide type is closely linked to the prevailing lithological nature and to their degree of fracturing. Moreover, a chaotic and a disorganized tectonic setting further promotes mass movements, because the structural discontinuities such as joints, faults, foliation and bedding planes form the pre-existing lines of weakness in the rocks and high local relief. These lines of weakness are likely to be areas with high permeability and deeply weathering of rocks, consequently, are areas with particularly high incidence of slope instability. Slide type mass-movements occur, where flyschoid rocks, sandy deposits and conglomeratic deposits crop out. Flow mass movements, that are closely related to concave surface topography and involved mainly a clayey-flysch bedrock covered by slope and/or old landslide deposits. The source area of many flow-type landslides mapped was branched in several coalescent landslide scarps and show elongated accumulation zones, ending with fan-shaped toes. Many shallow landslides are active within the source areas. Rock falls, are predominantly located in the zones where steep slopes and more resistant rocks crop out (e.g. carbonatic complex). Rock fall source areas are often connected to fault scarps, joints, or multiple cleavage systems. Rock falls may be represented by single boulder falls or events that move large volumes of rock. The falling material is deposited at the base of the cliffs and often produces typical debris cones. Among the various recognized landslides, the great landslide of "Lucanian Sanctuary" is particularly significant (figs. 4 and 5). This is a great, articulate and ancient mass movement, consisting of rototraslational slidings evolving to earth flows. This large mass movement has already been the subject of study in the recent past by [6]. Another important archaeological site is the Anaktoron that is located on a countersloping terrace site at an altitude of about 775 m above sea level delimited by a detachment developing between the units 740 m and 780 m above sea level, with a maximum length of 238 m and a width of about 230 m (fig. 4).

5   Geomorphological and Archaeological Analysis of the Lucanian Sanctuary

The area on which the Sanctuary was built is located along the southern foothills of the limy slope on which stands the Torre di Satriano, at around 1000 meters above sea level, and takes the form of a plateau with a lower slope than the rest of the slope located upstream which climbs to the summit from the SS 95. Downstream landscape, in a more or less constant way towards the watershed of a stream tributary orographic right of the Melandro River degrades with the exception of small flat areas - constituting the ancient stabilized landslide counterslope terraces - interrupting the physic continuity. In the area there are also water points that the emergence and development of the Lucanian settlement have favored and among them a particular significance has a spring that comes from the southern side and that is generated, in all likelihood, by the tectonic contact between the limestone and more permeable rocks forming the relief of the Tower which are in the bed of the fault and the clayey-marly top that serve as relative waterproof. This spring, which currently at the Land Registry of the water bodies in the Basilicata Region is collected and recorded, shows a flow of approximately 0.750 l/s measured at the end of the summer and that allows to hypothesize water supplies of greater magnitude in winter In morphological terms, the natural landscape has been in antiquity an ideal site for the construction of the Lucanian age Sanctuary; to this is added the strategic geographical location of the area in terms of connection between different cultures, placed along routes joining the Tyrrhenian coast to the Ionian coast [6]. The landforms related to exogenous recently dynamics, analyzed with a aerial photo analysis and with field surveys (fig. 4), have revealed that the territory of the Lucanian settlement is affected by widespread landslides both ancient and recent consisting mainly of rototraslational landslide that evolve to earth flows and evolving in clayey-marly lithologies belonging to different geological formations that outcrop in the area. In relation to the morphological conditions and the stability of the landslide were identified inactive landslide without obvious active signs of movement, dormant landslide characterized by an apparent stability but with well preserved forms of detachment and accumulation and characterized by active landslide movements and that often promptly the ancient landslide reactivate (fig.6) At a more detailed scale, the water outflow, along the southern slope of the place of Santuary, is divided into small streams with low degree of vertical depth and with an almost total absence of fluvial tributaries that determine above an elongated but not straightaway physiography with sinuous course and with small radius, due to the unstable conditions of the land. Throughout the southern slope, flat areas generated by the landslide terraces, locally present with signs of waterlogging forming small pools seasonally fed The entire slope is characterized by a large landslide (fig. 6) with complex kinematics in which roto-traslational movements, evolving to earth flows, are recognizable. The activation of these large landslides (fig. 6) precedes the colonization of the area by man and is likely to be placed at the post-glacial period, that is at the end of the last phase of the Wurm glacial expansion that starts about 10 ka due to the considerable supply of water resulting from melting snow. The sacred activity of the Santuary probably has not been continuously carried out but it has experienced more or less long periods of stasis - or of showy decrease ritual - witnessed by the absence of votive objects, as seems to be happening in the period between the second and first centuries B.C. [9]. After considering the environmental characteristics of the area (this term refers to all physical, biological and socio-cultural territory) the reasons that have produced this gap in the sanctuary activity, that was followed, in the first century A.D., by his final abandonment, is not entirely clear, or at least does not seem sufficient to justify these events only with the arrival of Roman culture that in the centuries following will grow with the birth of the ancient *Potentia*. It seems clear that the abandonment of the sacred place is not only to be related to socio-economic factors but also to biological and environmental factors that would have forced the local population, already prepared, to completely abandon the site.

6   The Large Landslide of "Lucanian Sanctuary"

The southern slope of Torre di Satriano is entirely covered by an ancient and complex rototranslational slide that evolved into a large earthflow (fig. 6) ([15] and [16]). It is approximately 2060 m long, from 150 to 725 m wide, and extends between 850 and 605 m a. s.l. with an average inclination of about 10°. As already highlighted, this large landslide has significantly influenced the historical evolution of this archaeological area [6] . In particular, the activation of this large landslide is certainly before the VIII century A.C., while the eastern sector remobilization of the rototraslational slide, occupied by the Lucan sacred Sanctuary settlement, took place after the renovation around the 1st century A.C.; this caused a rotation against Mount of approximately 13° of the same Sanctuary (fig. 7). The large landslide of "Lucanian Sanctuary" involves the formation of "Calcari con selce" (upper Triassic) and the Galestri formation (lower - middle Cretaceous); the landslide is essentially constituted by very degraded and plastic marly-clayey material; in the clay matrix calcareous and/or siliceous heterometric sharp-edged stone fragments can also be found. The alleged thickness of this landslide body, assessed both on geomorphological basis and indirect investigation (geoelectrical survey), varies from 20 to 25 m.

6.1   Geomorphological Survey

In situ geological and geomorphological surveys and aerial photo-analysis allowed the description of the main geomorphological features and the evaluation of the state of activity of the three different landslide areas (source, channel and accumulation). Along the main body of the landslide there are several secondary scarps, morphological depressions, and minor surface landsliding; in addition, a wide countersloping landslide terrace and creeping evidence can be observed. The latter are fed by surface water runoff coming from precipitations and by the discharge of the water source upstream which drain towards the south-east and probably in the past have been used for religious purposes of the sanctuary. The main source area is referable to a multiple and retrogressive rototranslational slide. It is mostly emptied and has a concave shape. The main scarp (fig. 6), at an elevation of about 850 m a.s.l., is affected by rockfalls and small rockslides. The source area is almost entirely covered by debris deposits of disjointed limestone and marl blocks immersed in a fine-grained matrix. Within this landslide there are morphological depressions and countorslope terraces, one of these holds the Lucan sanctuary (fig. 5): especially the latter terrace shows evident signs of recent rimobilition as confirmed by the presence of a clear cut plan highlighted by recent archaeological excavations (fig. 8).

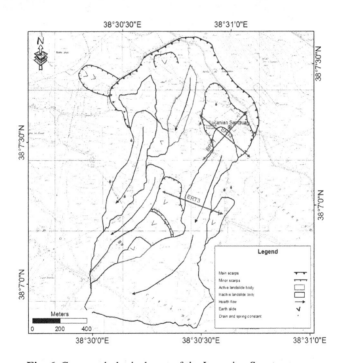

Fig. 6. Geomorphological map of the Lucanian Sanctuary

Since the area of the Lucanian Sanctuary is frequented by humans since the eighth century B.C., as documented by the discovery of some burials, it is evident that the landslide terrace on which it was realized, was already present at the time of realization of the settlement so the activation of the landslide is the oldest of the eighth century B.C. configures the phenomenon - along with a low degree of conservation of landslide morphologies that is the scarps, flanks and accumulation areas from the drainage network - as old and inactive.

The flow channel, which is probably placed on a preexisting drainage line, extends between 770 and 675 m a.s.l. and has an average slope of 6°. It is about 860 m long and the width varies from 150 m to 350 m. It is delimited by two evident flanks.

The accumulation zone shows a typical fan shape with an average slope of 5°. It is about 700 m long and 750 m wide. The landslide toe is located in the bed of the Cammarara stream, tributary of the river.

Fig. 7. Photo detail of the rotation against Mount of approximately 13° of the same Sanctuary [6]

Fig. 8. Site plan of the sanctuary area with planimetric (dotted line) of the detachment surface deforming the lower left corner of the bottom structure. In the photo the sliding surface of the landslide.

6.2   Geoelectrical Survey

The Electrical Resistivity Tomography (ERT) is an active, non-invasive and effective geophysical technique that provides 2D/3D images of the subsurface resistivity pattern. ERT is widely applied in landslide areas in order to collect indirect information about the geometrical features (lateral extention, sliding surface, thickness, etc.) of the mass movements ([17], [18] and [19]). The instrumental units necessary to carry out electrical resistivity measurements are a voltage generator, through which a train of square waves of current is injected in the subsoil, by means of the current electrodes A and B, and a receiving system constituted by a digital millivoltmeter connected to a computer, which stores the values of voltage across the electrodes M and N. Resistivity data may be acquired through different electrode configurations

(Wenner, Schlumberger, Dipole-Dipole, Pole-Pole, etc.) that are chosen considering: the array sensitivity to vertical and/or horizontal resistivity variations, signal/noise ratio, depth of investigation, horizontal data coverage, logistic problems. The final product of the data acquisition is the realization of a pseudo-section of resistivity, which represents a cross-section of the subsoil showing apparent resistivity values distribution as function of pseudo-depths. Using ad hoc modelling and inverting software, apparent resistivities and pseudo-depths are subsequently inverted in true resistivities and depths, producing the ERT. The high resolution obtained by this technique allows to discriminate effectively resistivity contrasts existing in the subsoil, providing, thus, more reliable information on the physical conditions of the rocks, the presence of subsurface discontinuities, the lithological-stratigraphic and/or structural setting (limits, faults, sliding surfaces, etc.), the presence of underground aquifers, etc. To characterize the large landslide of the "Torre di Satriano", three ERT were carried out (figs. 6 and 9); data were acquired through a multi-electrode system using a Syscal R2 (Iris Instruments) device and the apparent resistivity data were inverted by the RES2DINV software [20]. All the ERT were carried out using 48 electrodes with an electrode spacing of 10 m, having a total length of 470 m and reaching a maximum investigation depth of about 70 m. The ERT1 and ERT2 were performed across the main scarp of the landslide body, with longitudinal and transversal direction to the body, respectively. ERT 3 was carried out in WNW-ESE direction, near the SP12 road affected by the mass movement (figs. 6 and 9). Data inversion was performed obtaining 2D models with a RMS error variable between 2.5 (ERT 2) and 2.9 (ERT 3) after 3 iterations. The range of the electric resistivity values is between 8 and more than 81.2 Ω m. For the sake of brevity, only ERT 1, carried out with direction parallel to the main axis of the "Lucanian Sanctuary" landslide, is reported and analyzed (fig. 6). This tomography highlights the presence of both vertical and horizontal resistivity changes due to the different lithologies outcropping in the study area.

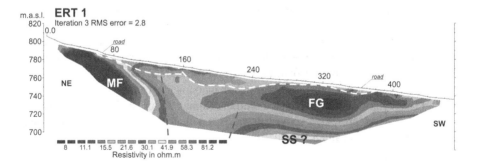

Fig. 9. Electrical resistivity image carried out longitudinally to the landslide body. Dashed white line represents the hypothetical sliding surface, whereas dashed red lines possible tectonic contacts (faults). MF: Monte Facito formation; FG: Galestri Formation; SS: Scisti Silicei formation.

In particular, the first 20 m of the ERT are characterized by relatively high resistivity values (15-20 Ω m) that could be associated with slide material (dashed white line). Between 0 and about 80 m from the origin of the profile it is possible to observe a sector relatively more resistive that can be associated with the Monte Facito formation (MF). In the central and SW sector, there is a low resistivity zone, about 40 m thick that can be interpreted as clayey of the Galestri Formation (FG) or a water-rich material zone. The vertical contrasts between the high-resistivity zone and the low-resistivity material could be related to the presence of tectonic contacts (dashed red lines). In the deeper part of the ERT, there is a relatively more resistive layer that could be associated with a lithological change, probably due to the presence of the Scisti Silicei formation (SS).

7   Conclusions

A multidisciplinary geological, geomorphological and geophysical study of the Lucanian Sanctuary area has been presented. The Sanctuary is located on the foot of the southern ridge of "Torre di Satriano", representing a sacred place from the fourth century B.C. up to the first century A.D. when it was completely abandoned The Sanctuary has been regularly damaged by natural disasters (i.e. landslides, earthquakes), The evolution of the landscape, in the study area, appears to be strongly conditioned by landslide activity. This generates small landslide counterslope terraces among which also the one where the sacred place was built. The presence in the north-eastern side of the site of a concave recess oriented in the direction of the Sanctuary, that drains water from the upstream, could have fed a small marshy area at the terrace of the landslide that generated a natural ideal pre- sanctuary for the realization of the sacred structure. Changes in the flow of the spring and/or small landslide reactivation could be also some of the reasons which have led the local population to achieve the surface drainage with the construction of canals recognized in the archaeological excavation The extreme vulnerability of the area and the lack of a continuous water supply in time are, in our opinion, the causes which have led the local population to abandon the place of worship in the first century A.D.. Since the last renovation of the Sanctuary took place around the middle of the first century B.C. seems realistic to assume that among the natural disasters, landslide activity has been particularly intense because it would have caused the counterslope tilt of the sanctuary structure in the top shelf, just when the abandonment of the area occurred. The advent of new religious cults associated with the Roman culture and the emergence of new rural realities represented socio-cultural issues that may have favored the final disappearance of the Lucanian Sanctuary.

References

1. Sdao, F., Chianese, D., Lapenna, V., Lorenzo, P., Perrone, A., Piscitelli, P.: Instabilità dei versanti in aree archeologiche della Basilicata: il caso del Santuario di Mephitis - Rossano di Vaglio (Basilicata). In: SIRIS. Studi e ricerche della Scuola di Specializzazione in Archeologia di Matera (2002-2003), vol. 4, pp. 119–131. EDIPUGLIA srl, Bari (2003)

2. Sdao, F., Simeone, V.: Mass movements affecting Goddess Mefitis sanctuary in Rossano di Vaglio (Basilicata, southern Italy). Journal of Cultural Heritage (2007)

3. Sdao, F., Lioi, D.S., Pascale, S., Caniani, D., Mancini, I.M.: Landslide susceptibility assessment by using a neuro-fuzzy model: a case study in the Rupestrian heritage rich area of Matera. Natural Hazards and Earth System Sciences 13, 395–407 (2013), doi:10.5194/nhess-13-1-2013, ISSN: 1561-8633

4. Sdao, F., Pascale, S., Rutigliano, P.: Instabilità dei versanti e controllo, mediante tecniche integrate di monitoraggio, delle frane presenti in due siti sacri del Parco Archeologico Storico Naturale delle Chiese Rupestri di Matera. In: SIRIS. Studi e ricerche della Scuola di Specializzazione in Archeologia di Matera, vol. 9, pp. 87–100 (2008)

5. Pascale, S., Pastore, V., Sdao, F., Sole, A.: Landslide Susceptibility in Archaeological and Natural Historic Park of Rupestrian Churches. In: Landslide Science and Practice, Second World Landslide Forum, Roma. Risk Assessment, Management and Mitigation, vol. 6, pp. 715–722. Springer-Verlag GmbH, Heidelberg (2013), doi:10.1007/978-3-642-31319-6_91, ISBN: 9783642313189

6. Giano, I.S., Sdao, F., Zotta, C.: Analisi archeoambientale del santuario di Torre di Satriano. In: Osanna, Sica (eds.) Torre di Satriano I. Il Santuario Lucano, pp. 466–473. Osanna Edizioni (2005)

7. Osanna, M., D'Alessio, A., Di Lieto, M., Ricci, A., Sica, M.M.: L'insediamento indigeno di Torre di Satriano (PZ): le nuove ricerche dell'Università degli Studi della Basilicata. In: Siris 3 - Studi e Ricerche della Scuola di Specializzazione in Archeologia di Matera (2000-2001), pp. 233–268. Adda Ed. (2002)

8. Schiattarella, M., Giano, S.I., Guarino, P.M.: Interazione Uomo-Ambiente e Sistemi Geoarcheologici. In: Albore Livadie, C., Ortolani, F. (eds.) Il Sistema Uomo-Ambiente tra passato e presente. CUEBC, Territorio storico e ambiente, vol. 1, pp. 181–184. Edipuglia (1998)

9. De Vincenzo, S., Osanna, M., Sica, M.M.: La lunga vita di un piccolo santuario lucano: Torre di Satriano in età Romana. In: Ostraka, Rivista di Antichità, Anno XIII, vol. 1, Loffredo Ed., Napoli (2004)

10. Schiattarella, M., Di Leo, P., Beneduce, P., Giano, S.I.: Quaternary uplift vs tectonic loading: a case-study from the Lucanian Apennine, southern Italy. Quaternary International 101-102, 239–251 (2003)

11. Castellano, M.C., Sgrosso, I.: Età e significato dei depositi miocenici della Formazione di M. Sierio e possibile evoluzione cinematica dell'Unità Monti della Maddalena nell'Appennino campano-lucano, pp. 239–249 (1996)

12. Canuti, P., Casagli, N.: Considerazioni sulla valutazione del rischio di frana. Estratto da "Fenomeni franosi e centri abitati". Atti del Convegno di Bologna del 27 Maggio 1994. CNR—GNDCI—Regione, Emilia Romagna (1994)

13. Conforti, M., Pascale, S., Pastore, V., Pepe, M., Sdao, F., Sole, A.: Geomorphology and GIS analysis for mapping landslide in the Camastra basin (Basilicata, South Italy). Rendiconti Online Della Società Geologica Italiana 21, 1152–1154 (2012)

14. Conforti, M., Pascale, S., Pepe, M., Sdao, F., Sole, A.: Denudation processes and landforms map of the Camastra River catchment (Basilicata - South Italy). Journal of Maps 9, 444–455 (2013), doi:10.1080/17445647.2013.804797, ISSN: 1744-5647

15. Cruden, D.M., Varnes, D.J.: Landslide types and processes. In: Turner, A.K., Schuster, R.L. (eds.) Landslides, Investigation and Mitigation: Transportation Research Board, pp. 36–75. US National Research Council, Special Report 247, Washington, DC (1996)

16. Keefer, D.K., Johnson, A.M.: Earth flows: morphology, mobilization, and movement. U.S. Geol. Survey Prof. Paper 1264, 56 p. (1983)

17. Perrone, A., Iannuzzi, A., Lapenna, V., Lorenzo, P., Piscitelli, S., Rizzo, E., Sdao, F.: High-resolution electrical imaging of the Varco d'Izzo earthflow (Southern Italy). J. of Applied Geophys. 56, 17–29 (2004)
18. Lapenna, V., Lorenzo, P., Perrone, A., Piscitelli, S., Sdao, F., Rizzo, E.: 2D Electrical Resistivity Imaging of some Landslides in Lucanian Apennine (Southern Italy). Geophysics 70(3), B11–B19 (2005), doi:10.1190/1.1926571
19. Naudet, V., Lazzari, M., Perrone, A., Loperte, A., Piscitelli, S., Lapenna, V.: Integrated geophysical techniques and geomorphological approach to investigate the snowmelt-triggered landslide of Bosco Piccolo village (Basilicata, southern Italy). Engineering Geology 98, 156–167 (2008), doi:10.1016/j.enggeo.2008.02.008
20. Loke, M.H.: Tutorial: 2-D and 3-D electrical imaging surveys. In: Course Notes for USGS Workshop "2-D and 3-D Inversion and Modeling of Surface and Borehole Resistivity Data", Storrs, CT, March 13-16 (2001)

Walkability Explorer:
An Evaluation and Design Support Tool for Walkability

Ivan Blečić, Arnaldo Cecchini, Tanja Congiu,
Giovanna Fancello, and Giuseppe A. Trunfio

DADU, Department of Architecture and Planning - University of Sassari, Alghero, Italy
{ivan,cecchini,tancon,gfancello,trunfio}@uniss.it

Abstract. Walkability Explorer is a software tool for the evaluation of urban walkability which, we argue, is an important aspect of the quality of life in cities. Many conventional approaches to the assessment of quality of life measure the distribution, density and distances of different opportunities in space. But distance is not all there is. To reason in terms of urban capabilities of people we should also take into account the quality of pedestrian accessibility and of urban opportunities offered by the city. The software tool we present in this paper is an user-friendly implementation of such an evaluation approach to walkability. It includes several GIS and analysis features, and is interoperable with other standard GIS and data-analysis tools.

Keywords: Walkability, evaluation, decision support, GIS, ELECTRE TRI.

1   Introduction

In this paper we present Walkability Explorer, a software tool for the evaluation of urban walkability.

Walkability of places is an important aspect of the quality of life in cities. Making cities more walkable does not merely improve the accessibility of places, it also is beneficial to the quality of the public use of space and the social climate in general. Ultimately, making places more walkable may expand capabilities of inhabitants, visitors and city-users, especially of those "week population" whose capabilities are curtailed by the predominant motorized practices of the use of space.

We use 'capability' here in specific sense of the so called capability approach [1]: a person's capabilities are valuable states of being that a person has effective access to. Thus, a capability is the effective freedom of an individual to choose between different things to do or to be that she has reason to value. In this conception, a capability constitutively requires two preconditions: (1) the ability, person's internal power, detained but not necessarily exercised, to do and to be, and (2) the opportunity, presence of external conditions which make the exercise of that power possible. A person is thus capable, has the capability to do or to be something, only if both conditions – internal and external, ability and opportunity – allow her to. The physical urban space – the city's hardware – influences capabilities primarily through the channel of the *opportunity* component of capabilities.

B. Murgante et al. (Eds.): ICCSA 2014, Part IV, LNCS 8582, pp. 511–521, 2014.
© Springer International Publishing Switzerland 2014

Many conventional approaches to the assessment of quality of life usually measure the distribution, density and distances of different opportunities in space. But distance is not all there is. If we want to reason in terms of capabilities, we should also take into account the quality of accessibility and the quality of urban opportunities. Besides the mere distance, it matters a great deal if a place can be reached also by foot or by bicycle, if the pedestrian route is pleasant and spatially integrated with the surrounding by good urban design, if it is brimful of urban activities, if it is well maintained and (perceived as) secure, if it is not submissive and surrendering to the car traffic whether by design or by predominant social practices of use of that space. At the same time we need to go beyond the simple presence of urban services, to understand their characteristics, if they are able to serve different categories of individuals, if their relevance is on the neighbourhood, urban or metropolitan/regional level, if there are possibilities of choice between two or more relevant places.

For Walkability Explorer, the software tool which is the focus of this paper, we have developed evaluation approaches which attempt to take into account the aforementioned facets of walkability. The assumption of an accessibility-enhancing perspective requires a very strict integration and collaboration between transportation planning, land-use planning and urban design. Walkability Explorer is therefore a milestone is our ongoing research to build evaluation models and a planning and design support tools that takes into consideration many of these concerns, and focuses on the quality of accessibility as an important factor for the extension of urban capabilities.

2   Evaluating Walkability

2.1   The Data

The evaluation of walkability is based on the exploration of how someone at different points in space can walk to destinations of interest in an urban area. A destination of interest is a place, service or facility which promotes an urban opportunity.

The concept of walkability pinpoints at features beyond the geometry of urban space. Besides mere presence of places of interest and their distances, factors related to the quality of pedestrian routes such as urban design and quality, track and road conditions, land-use patterns, building accessibility, degree of integration with the surrounding, safety and other features and practices of use of space, are all potentially relevant for walkability.

Therefore, for an operational evaluation of walkability, much richer spatial datasets are required. Our starting point are: (1) a detailed graph representation of the street network and (2) a detailed map of relevant places (destinations).

The street network graph is the cartographic base for the pedestrian route analysis. Besides their geometric properties, the edges hold relevant features for the walkability of a pedestrian route. In Table 1. we report an example list of edge attributes we used in our experimental runs of WE.

Table 1. Example of edge attributes

	Values Description
Urban design	
Building density	(qualitative) dense – rarefied – undeveloped	Describes the density of the urban fabric surrounding the edge.
Degree of integration	(qualitative) Integrated – filtered – separated	Describes how the pedestrian pathway is integrated with the surrounding buildings and areas. "Integrated" stands for complete integration and permeability; "filtered" means that the access is possible but "filtered" with specific points of access, pathways, etc.; "separated" stands for a complete separation (e.g. a wall or fence).
Street type	access – residential – crossing/bypass	The predominant type of the street: "access" to services, shops, offices, etc.; "residential"; or a "crossing/bypass"
Physical features	
Bicycle track	present – absent	
Number of car lanes	(number)	
Car speed limit (in km/h)	(number)	
One-way street	yes – no	
Car parking along the road	not allowed/practiced – allowed/practiced	Whether cars are parked/allowed to park along the motor lane
Footway width (in meters)	(number)	
Degree of maintenance	(qualitative) good – average – bad	A qualitative evaluation of the degree of maintenance (footpath, illumination, trash bins, flowerbeds, etc.)
Land-use pattern	
Commercial activities	(qualitative) predominant – present – absent	Whether commercial activities (shops, bars, restaurants, etc.) are predominant, present or absent
Services and offices	(qualitative) predominant – present – absent	Whether services, businesses and offices are predominant, present or absent

This, of course, is only an example and far from a complete list. Many other attributes could be useful to assess walkability, and we are surely failing to account for important aspects such as practices of use of space, social climate, perception of personal security, and many more. WE is a flexible tool and can import any set of attributes which scholars and users may consider of relevance for the evaluation of walkability in accordance to particular normative assumptions, empirical findings and available data.

The map of relevant places describes the spatial distribution of places, services and facilities and represent the information base for the analysis of particular attributes determinant for the promotion of urban opportunities. These attributes may in principle describe the quality of places, design of space, capacity to attract different categories of peoples at different times of the day, capacity to favour different uses in the space (play, meetings, study, …), and other features important for the accessibility of the space, intended as the possibility of appropriation of the urban space in respect to human needs. For the example runs of WE, we have classified destinations of interest in three categories: commercial (shops, bars, restaurants, etc.), services (schools, health services, libraries, etc.) and recreational and leisure areas (green areas, urban parks, sport facilities open to public). In Fig. 1 we show a screen capture of the maps with these three types of destinations.

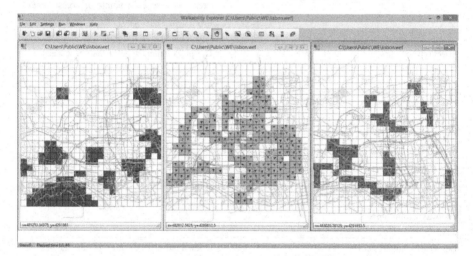

Fig. 1. A screen capture of WE representing the destination cells for different type of attractions and the destination node for each cell

2.2   Evaluation Models

In a previous paper [2] we described the an evaluation model based on a pseudo-utility functions, where for each category of destinations we define the pedestrian behaviour as an utility maximisation problem given the distance and the quality of pedestrian accessibility of destinations belonging to that category.

Here we concentrate to describe an alternative model based on the ELECTRE TRI rating procedure. In particular, for the purpose of rating urban opportunities we adapted the ELECTRE TRI approach in a particular nested procedure.

The aim of the evaluation model is to assign an "opportunity rating" to each point in space, that is to say, to put it in one among several classes of urban opportunity (one class for each among different types of urban opportunities). The core idea of the evaluation approach we propose is based on nesting several ELECTRE TRI evaluation procedures, one within another. So before laying down our "nested" model, let us briefly recall the basic general ELECTRE TRI model.

Among the methods for multiple criteria evaluation of ratings [3], the so called ELECTRE TRI model [4, 5] is a prominent classification approach. This rating approach possesses several desirable properties for our purposes: (1) it allows a complete classification, and the aggregation over multiple criteria is fairly flexible, permitting to account for (2) the importance (weights) of criteria, (3) coalitions (majority rule and threshold) and (4) possible veto powers. Besides, as it will be shown, out nesting ELECTRE TRI procedure allows a careful aggregation over criteria at each level of nesting in a controllable and meaningful way more in accordance with "natural" human reasoning.

The general ELECTRE TRI procedure works as follows. Given a set of objects, evaluated on a set of criteria $h_1 \ldots h_n$, to be assigned a rating class from a set of classes with ordinal property $C^1 \ldots C^m$, ELECTRE TRI first requires that the so called limiting profiles be defined for each class. That is to say, each class C^k is defined by a limiting profile π^k on m criteria: $\pi^k = \left(\pi_1^k, \ldots, \pi_n^k\right)$. To respect the ordinality of classes, the limiting profiles should be defined so that $\pi_i^k < \pi_i^{k+1}$ for every $i=1,\ldots,n$.

To assign an object a to a rating class we then apply the following two rules [3]:

— if the object a has the same or higher evaluation on the m criteria than π^k, it should at least belong to the class C^k;
— if π^{k+1} has the same or higher evaluation on the m criteria than the object a, then it should at most belong to class C^k.

Formally:

$$a \in C^k \Leftrightarrow a P \pi^k \wedge \pi^{k+1} P a \tag{1}$$

where P is the binary outranking relation meaning "belongs to the same or a higher class than".

The binary outranking relation P uses a crisp relation based on a concordance-discordance principle, that is to say, an object a outranks a limiting profile π^k if there is a "significant" coalition of criteria for which "a belongs to the same or higher class than π^k" (concordance principle) and there are no "significant opposition" against this proposition (discordance principle). In other words:

$$a P \pi^k \Leftrightarrow C\left(a, \pi^k\right) \wedge \neg D\left(a, \pi^k\right) \tag{2}$$

where:

— $C(a, \pi^k)$ means there is a majority of criteria supporting the proposition that a outranks ("is at least as good as") π^k;
— $D(a, \pi^k)$ means there is a strong opposition, that is to say a veto, to the proposition that a outranks ("is at least as good as") π^k.

Following Roy (1968), for two evaluation profiles x and y, we use the following definitions of $C(x, y)$ and $D(x, y)$:

$$C(x, y) \Leftrightarrow \frac{\sum\limits_{i \in H(x,y)} w_i}{\sum\limits_{j=1}^{n} w_j} \geq \gamma \qquad (3)$$

$$D(x, y) \Leftrightarrow \exists h_i : h_i(y) - h_i(x) > v_i \qquad (4)$$

where:

- h_i, $i = 1,\ldots,n$ are the criteria (the higher the value the higher the class);
- w_i are the importance coefficients (weights) associated to each criterion;
- $h_i(x)$ is the evaluation of x on the criterion h_i;
- $H(x,y)$ is the set of criteria for which x has the same or higher evaluation than y, that is, for which $h_i(x) \geq h_i(y)$;
- γ is the majority threshold;
- v_i is the veto threshold on criterion h_i.

After this recall of the basic ELECTRE TRI rating procedure, let us now lay down the specific nested procedure which we have developed for evaluating urban opportunities in space.

Let we define a set of ordinal opportunity classes from lowest to highest, $O^1 \ldots O^m$. Again, our objective is to assign to each point in space one and only one class by taking into consideration both (1) the quality and (2) the accessibility of destinations of interest from that point.

Each destinations may fall into one or more types of urban opportunity (e.g. green areas, retail, services, etc.). To represent this fact, each destination d is evaluated in terms of "quality" per each type of opportunity, which we will denote with $q_l(d)$, where l stands for the type of opportunity.

To evaluate the accessibility, we use a detailed graph representation of the street network. A path from an origin to a destination is a set of interconnected edges. Besides their length, edges are described with further attributes which shape the quality of pedestrian accessibility, with characteristics such as physical features, urban design, presence (or absence) of variety of urban activities, and so on (see Table 1. above for a example of edge attributes). Hence, in general terms, for every edge i in a path from one point in space to one destination, we have the edge's length l and a set of attributes $a_1,\ldots a_p$ which describe its characteristics.

Given such a configuration of definitions and available data, the "nested ELECTRE TRI" procedure we propose proceeds in four steps:

- Step 1: Assign a walkability class to each edge in the path;
- Step 2: Aggregate the walkability of edges in the path (from Step 1) to assign an overall walkability class to the entire path;
- Step 3: Combine the walkability class of the path (from Step 2) with its length to assign an accessibility class to the couple origin-destination

— Step 4: Combine the accessibility of all the destinations (of one type of urban opportunities) reachable from an origin, to assign an urban opportunity score/class to that origin (for that type of urban opportunities)

Step 1. Edge walkability rating. In this step we use ELETRE TRI to assign a walkability rating to each edge, using edge attributes as criteria. The step further requires that a corresponding set of criteria weights, possible veto thresholds, and the majority threshold be defined.

Step 2. Path walkability rating. Here, the ELECTRE TRI serves to assign a *walkability rating to the entire path*, by using the edges themselves as criteria. Their walkability classes (obtained in the Step 1) are used as criteria values, while their lengths are used as weights. So, this step only requires the definition of the majority threshold and possible vetos.

Step 3. Accessibility rating of each couple origin-destination. We now need to evaluate the overall *accessibility* of the destination from the origin. The accessibility should take into account both the quality of walk, i.e. walkability, and the distance. Therefore, for this purpose we again employ ELECTRE TRI, this time using two criteria: the walkability of the path (obtained in the Step 2) and its length. This step therefore requires to further settle the respective weights of the two criteria, as well as the majority and possible veto thresholds.

Step 4. Urban opportunity scores/rating. This is the final phase in which we assign the final urban opportunity ratings to the origin point in space. It combines the information about the quality of the destinations which are reachable from that origin with their accessibility rating (obtained in the Step 3). Therefore, this step may be performed only after all the accessibility ratings have been assigned to every couple origin-destination. Also, since different destinations are, as we said, relevant for different types of urban opportunity, we proceed separately and independently, calculating an opportunity score per each type of opportunity. The opportunity score of an origin $U(o)$ is obtained with:

$$U(o) = \sum_{i \in D} q(d_i) a(o, d_i) \tag{5}$$

— where D is the set of reachable destination relevant for the type of opportunity under assessment;
— $q(d_i)$ is the quality score of the destination i;
— $a(o, d_i)$ is the accessibility score of the destination i from the origin o; the accessibility scores are accessibility ratings (obtained in Step 3.) transformed into numeric factors $[0,1]$.

In the end, having calculated the urban opportunity scores for each type of opportunity, the final urban opportunity ratings are assigned by defining fuzzy thresholds on scores per each different type of urban opportunity.

3   A General Overview of Walkability Explorer

We are currently working on fully implementing the two evaluation models in Walkability Explorer (WE).

WE is an application running on Microsoft Windows whose user interface allows an easy assessment of the walkability. It furthermore allows a comparison in terms of walkability between the current situation and hypothetical projects concerning features relevant for the walkability, in terms of the evaluation model described above.

In Fig. 2 we show the standard workflow to perform a walkability evaluation in We.

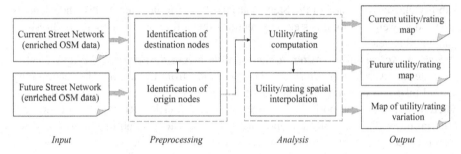

Input   Preprocessing   Analysis   Output

Fig. 2. The typical WE workflow. The required input data are the current and future street network in OSM with graph edges enriched with attributes relevant to the walkability (e.g. Table 1.). After the preprocessing and analysis phases, the main output is represented by the utility-score maps (if using the pseudo-utility evaluation model) or ratings (if using nested ELECTRE TRI procedure) at the desired resolution.

First, the user is asked to provide the road networks in the format defined by the Open Street Map (OSM) project (see screen capture in Fig. 3) OSM is a collaborative project for the creation of street maps that currently makes available a huge data base covering most part of the world. In addition to the availability of street network data, the advantage of using OSM for this application lies in the ease of introducing new attributes and topological changes that affect the graphs. For this purpose there are indeed several effective editing applications freely available. If the purpose is to compare the current situation with a future project, a further road network with the features modified by the project has to be provided.

Given the OSM data enriched with the set of edge attributes, the program identifies the areas of attractions using a regular grid of cells, according to a resolution set by the user, and constructs the sets of destination nodes (for an example see Fig. 1 above).

It is worth noting that the size of cells can be set independently for the different types of attractions. In particular, WE identifies the areas with prevalence of retail/commercial and service activities using the specific attributes attached to the edges in the OSM data. For the green/recreational attractions, the current implementation of WE exploits the polygons representing such urban areas, which are typically included in the OSM data. The program builds the set of destination nodes by finding for each attractive cell the node of the street network which is closer to its centroid (Fig. 4).

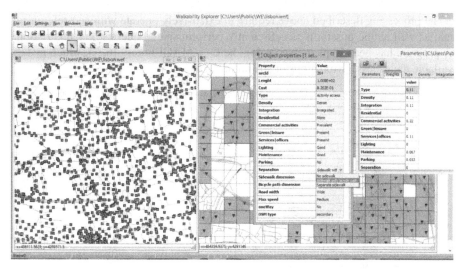

Fig. 3. A screen capture of WE representing the street network with the visualization of the edge parameters and an object property

Fig. 4. Origin nodes, attractive area and the corresponding destination node

WE determines the origin nodes for both the current and future street network. It is worth noting that, to increase the comparability of results, during the filtering process the program tries to make sure that the origin nodes of current and future road networks coincide. This is not possible in areas where there are geometrical and topological changes of the network.

The analysis run allows to calculate the utility-scores (if using the evaluation model based on pseudo-utility functions) or ratings (if using nested ELECTRE TRI procedure). The computation is carried out for the current and the future street network and for the each types of attraction. In order to shorten the run-time, WE exploits the available multi-core CPU computers implementing a parallel multi-thread approach.

The final output of the program are the georeferenced utility-score maps (e.g. Fig. 5) or ratings for both the current and future street networks and for all the types of attractions. Moreover, WE provides the map of utility/rating variation due to the project. All the above maps can be exported in a suitable GIS format for further elaborations.

The processing described above require to extensively operate with geo-referenced data, as well as the possibility to efficiently perform spatial queries. For this reason, the program has been implemented using the C++ MAGI library [6, 7], which makes available the necessary functions of spatial indexing.

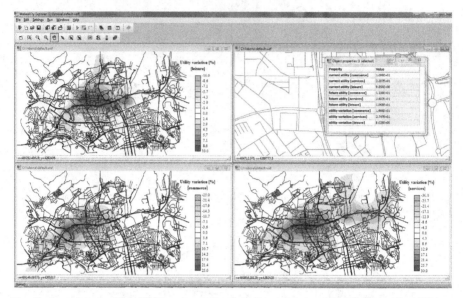

Fig. 5. A screen capture of WE representing georeferenced pseudo-utility maps for different types of attractions

Besides producing georeferenced maps, WE allows the results to be exported in the open csv format for further analysis in other GIS and statistical analysis tools. One such possible analysis is to calculate indicators for comparing the aggregate variation of walkability between alternative scenarios as well as its spatial distribution and dispersion in relation to the populations inhabiting the urban area under consideration.

4   Conclusions

Capability approach coupled with the analysis of accessibility provides a compelling theoretical framework for assessing relevant aspects of the quality of life in cities. The space and urban environment are an important constituent of certain human capabilities and is determinant for the individual life in cities. Among other dimensions of individual wellbeing (health, education, political participation, and so on), the way our cities and physical environment 'functions' – the way they are shaped, organized, and used by social practices – matters.

Architects, urban planners and policies makers could use urban capabilities to read and interpret the multiple relations between the individual and the city, to unveil the circumstances in which the city is an 'obstacle' to the needs and aspirations of its inhabitants, to better define and govern urban design processes which aim at removing these obstacles, to promote the right to the city [8,9,10] for all.

Such design attitude requires tools. Walkability Explorer is an attempt to implement evaluation models and to provide an user-friendly tool for assessing walkability which may prove useful for improving effectiveness, relevance, and inclusiveness of urban design and transport planning.

There is further work to be done and there are many areas in which we plan to extend WE's features. Foremost, to become a more complete decision support for assessing urban capabilities, besides walkability it should also be able to take into account the car and public transportation accessibility, and the way they interact with the pedestrian accessibility. Such incorporation of non-pedestrian mobility into WE would be an indispensable step to also take into account the quality of accessibility of not only neighborhood-level destinations, but also those on the urban and metropolitan/regional level, which of course also play a relevant role in shaping overall urban capabilities of people.

We intend to pursue these objectives in our future work.

References

1. Sen, A.: Capability and Well-Being. In: Nussbaum, M., Sen, A. (eds.) The Quality of Life, pp. 30–53. Oxford Clarendon Press, New York (1993)
2. Blecic, I., Cecchini, A., Congiu, T., Pazzola, M., Trunfio, G.A.: A Design and Planning Support System for Walkability and Pedestrian Accessibility. In: Murgante, B., Misra, S., Carlini, M., Torre, C.M., Nguyen, H.-Q., Taniar, D., Apduhan, B.O., Gervasi, O. (eds.) ICCSA 2013, Part IV. LNCS, vol. 7974, pp. 284–293. Springer, Heidelberg (2013)
3. Bouyssou, D., Marchant, T., Pirlot, M., Tsoukias, A., Vincke, P.: Evaluation and decision models with multiple criteria, Stepping stones for the analyst, vol. 86. Springer (2006)
4. Yu, W.: ELECTRE TRI: Aspects méthodologiques et manuel d'utilisation. Document du LAMSADE No. 74, Université Paris-Dauphine (1992)
5. Roy, B., Bouyssou, D.: Aide multicritère à la décision: méthodes et cas, Economica, Paris (1993)
6. Blecic, I., Borruso, A., Cecchini, A., D'Argenio, A., Montagnino, F., Trunfio, G.A.: A cellular automata-ready GIS infrastructure for geosimulation and territorial analysis. In: Gervasi, O., Taniar, D., Murgante, B., Laganà, A., Mun, Y., Gavrilova, M.L. (eds.) ICCSA 2009, Part I. LNCS, vol. 5592, pp. 313–327. Springer, Heidelberg (2009)
7. Blecic, I., Cecchini, A., Trunfio, G.A.: A general-purpose geosimulation infrastructure for spatial decision support. Transactions on Computational Science 6, 200–218 (2009)
8. Lefebvre, H.: Le droit à la ville. Anthropos, Paris (1968)
9. Harvey, D.: Social justice and the city. The University of Georgia Press, Athens (2009)
10. Soja, E.W.: Seeking spatial justice. University of Minnesota Press, Minneapolis (2010)

A Generic Approach for Analysis of White-Light Interferometry Data via User-Defined Algorithms

Max Schneider[1], Dietmar Fey[1], Kay Wenzel[2], and Torsten Machleidt[2]

[1] Friedrich-Alexander-Universität Erlangen-Nürnberg
[2] Gesellschaft für Bild- und Signalverarbeitung (GBS) mbH
{max.schneider,dietmar.fey}@cs.fau.de,
{kay.wenzel,torsten.machleidt}@gbs-ilmenau.de

Abstract. The non-destructive and automatic 3D surface analysis using white-light interferometry has very high demands on the required computing systems. This is due to both, the high volume of data that are collected and the tremendous computational power, required by algorithms which evaluate the data. Furthermore, material characteristics and environmental influences like temperature or ambient light have a not negligible effect on the scan procedure, such that the algorithmic approaches must be adjusted to assess the correct surface profile. Thus, to obtain the desired analysis results and to fulfill the computational requirements, the application developer has to consider the physical characteristics of the white-light interferometry process as well as the attributes of the computational platform used. This paper describes the generic computational module of a new framework for the white-light interferometry surface scanning procedure to address these problems. This framework allows a user-transparent automatic assessment and usage of available computational resources.

Keywords: White-light interferometry, High-performance computing, Function objects, GPGPU, Generic algorithms.

1 Introduction

This article describes results from an academic-industry project, which brings different disciplines together. The industrial partner, GBS mbH Ilmenau, develops software and hardware systems for digital image processing with focus on automation of quality assurance applications in manufacturing industry. One major branch in the product portfolio of GBS mbH is the development of 3D surface analysis systems based on white-light interferometry (WLI) measurement principle. The academic partner, chair of computer architecture of the Friedrich-Alexander-Universität Erlangen-Nürnberg, provides the knowledge required for developing highly parallel applications and effective utilization of available heterogeneous computational resources.

B. Murgante et al. (Eds.): ICCSA 2014, Part IV, LNCS 8582, pp. 522–537, 2014.
© Springer International Publishing Switzerland 2014

WLI is a principle derived from Michelson's method [1] and has been used for several decades for the measurement of three-dimensional object surfaces in the nanometer range. Today, WLI is a sophisticated and versatile method, that provides the capability of fast, contactless and high-precision 3D measurements of surface topology. The main application is the quality assurance in manufacturing processes, e.g. of components used in aerospace applications, integrated circuits and tools.

The basic white-light interferometry setup utilizes an optical interferometer, e.g. Michelson interferometer, as shown in Fig. 1a. The principle behind the measurement process is based upon the fact that spatially and temporally coherent lightwaves can be superimposed. To exploit this behaviour, the optical interferometer is equipped with a broadband white-light source and the following steps are applied during the measurement procedure.

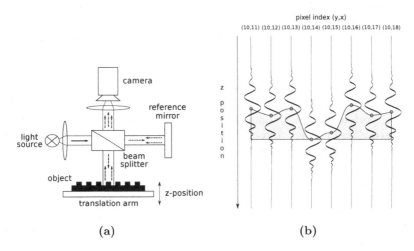

Fig. 1. (a) Michelson Interferometer. (b) Correlogram Signals.

The light source emitts a beam into the beam splitter, where it is split into two separated beams. One of these created beams follows a constant optical path to the reference mirror and the other one is passed to the object to be measured. Both beams are reflected from corresponding surfaces and directed through the beam splitter to a CCD-camera (cf. Fig. 1a). On the way back the reflected beams are superimposed. If the optical path difference of both beams is within the coherence length of used light, then interferences will occur. When the path difference is an even number of half-wavelengths, both waves are in phase leading to constructive interferences and thus to rising amplitude values in the interference signal. In case that the optical path difference is an odd number of half-wavelengths, the superimposed waves are 180° out of phase, destructive interferences arise and amplitude of the signal decreases [2]. Using a mechanical translation stage the object sample to be scanned is moved vertically, either closer to the scanning device or further away, resulting in a change of interference intensities recorded by the pixels of the CCD sensor at discrete time steps. The captured images can be interpreted as slices of the object at different heights [3]. When the optical path difference is zero for a particular x,y-position on object's surface, then the interference intensity for this position is maximal. The z-position of the translation arm at which the maximum interference was measured is considered as the height for that specific x,y-location.

For a correct height estimation of a scanned sample it is important that for each surface position the significant interference range (the region where the optical path length difference of reflected beams is less than the coherence length of the used light) is covered [4]. Fig. 1b shows such a two dimensional signal series, called correlogram or interferogram, for multiple pixels. The circles in Fig. 1b show the height positions of the translation arm, where the optical path length difference of reflected beams is zero for corresponding pixel positions and thus the constructive interferences are maximal. In case that the object is not a flat plane, each pixel has its interference maximum at different time step during the measurement. The 3D map of scanned sample is than assessed from these positions of the translation arm where maximum interferences were observed. Thus, the aim of the white-light interferometry analysis process is to find the corresponding maximum interference value for each pixel of the CCD sensor.

In conventional white-light interferometry surface analysis, hundreds or thousands of interference images may be necessary for a single surface area covered by the used camera sensor. Depending on used color depth, the resolution of the CCD sensor and the number of required scan positions to cover all necessary interferograms, the resulting data amount an be enormous. E.g., using a CCD sensor with a resolution of 4096×4096 pixels and a color depth of 16 bit, each captured image will be 32 MiB in size. In case that the complete scanning procedure requires 1000 scan positions, the resulting data volume would be 32 GiB. This is a huge data amount considering that the current scanning devices cover areas as small as 1 mm^2 at a time [5].

An application area for white-light interferometry surface scanning is the industrial production of e.g. optical lenses or integrated circuits (IC). Here the computationally intensive data processing must be carried out as fast as possible, to obtain correct results. Invalid results can occur due to time-dependent environmental influences like temperature drift, air turbulence and mechanical vibration. GPGPUs, consisting of hundreds of processing cores, provide such computational power to fulfill the timing constraints [6–8]. However, there are also small industrial companies and research laboratories, which do not yet use GPGPUs or have limited computational resources and thus cannot achieve high evaluation rates of collected measurement data. In these cases, it is necessary to use all available compute units (e.g. CPUs, older GPUs) in the best possible fashion. To achieve a good utilization of available resources the application developer of WLI data evaluation algorithms has to have deep knowledge of the specific hardware features of each compute device, which is crucial to get the maximum performance out of the hardware. E.g., today even the simplest

CPU has at least two cores, that are capable of parallel and independent instruction execution. Further, these CPU cores provide single-instruction multiple-data (SIMD) ALUs, that allow a simultaneous processing of multiple data using the same operation. To exploit the available cores and SIMD execution units, the software developer must use specific instructions, e.g. instructions used to create and destroy on CPU cores running threads or SIMD-intrinsics to use vector arithmetics. These steps have nothing to do with the actual task - e.g. elaboration of WLI data, and they blow up the code, making it harder to read and maintain. Additionally, minimizing data accesses to main memory is also important, as these can impair the overall performance. To achieve this, CPU cache hierarchy and the organization of available caches must be taken into consideration in the algorithm inception and implementation phases. In case of GPUs and other architectures, beside the already described programming tasks, the developer must learn and apply new programming paradigmas and even new programming languages (e.g. CUDA, OpenCL). However, often the WLI application developer is not a computer scientist, what makes this effort difficult for him. Thus, to make the software development more convenient a transparent usage of available heterogeneous resources is necessary.

The conventional approach for elaboration of WLI data to assess the 3D surface profile for a measured sample can be subdivided into two major parts: (i) Preprocessing, determines an approximate height position for each pixel. As shown in [7] or described in [9] there exist different preprocessing algorithms. In (ii) postprocessing, the envelope for each interferogram is computed and used for an iterative fitting approach to obtain a more precise height value for each pixel. There are also different fitting approaches, which vary in complexity and precision [10, 11]. Although, the existing algorithms achieve good accuracy and have acceptable runtime complexity, exploration of new algorithms and evaluation methods is going on. An important characteristic of these algorithms is the fact, that each pixel's correlogram is elaborated independently from those of the other pixels.

A last but not least important matter is the interoperability of the WLI application. Current proprietary WLI software solutions support the most common operating systems. Thus, to increase the usability of a new WLI application interoperability should be one of the features implemented.

In order to help the application developer with these tedious tasks we propose a generic WLI framework, which is capable of (i) detecting the available computational resources, (ii) determining the computational performance of each component and (iii) employing user specified C++-template functor objects, which implement the elaboration steps for a single pixel required in preprocessing and postprocessing phase, on selected computational resources, applying the provided algorithms to all pixels. Furthermore, avoiding operating system specific instructions the proposed framework supports the major operating systems available. In this full research paper the computational module of this framework is presented.

This paper is structured as follows: Section 2 describes former attempts to develop frameworks for managing computational resources. Additionally, optimization efforts targeting multi- and many-core architectures for WLI algorithms will be presented. Section 3 introduces the concepts behind the framework. Section 4 contains a detailed description of the computational module. Section 5 shows achieved performance results. Section 6 concludes the paper and discusses future work.

2 Related Work

Parallel computing has been a niche for scientific research in academia for decades. However, as common industrial applications become more and more performance demanding and raising the clock frequency of conventional single-core systems is hardly an option due to reaching technological limitations, efficient use of multi- or even many-core systems has become imperative. White-light interferometry is one of such high-performance demanding applications.

The first multi-threaded attempt in surface metrology applications was done, to the best of our knowledge, by Purde et al. in 2004 [12]. Using the High Level Shading Language (HLSL) they implemented an analysis algorithm for the so called electronic speckle pattern interferometry on GPUs.

M. Sylwestrzak et al. [13] used GPUs for the Spectral Optical Coherence Tomography (SOCT). In SOCT, a technique which is based on the white-light interferometry, the data acquisition is carried out by scanning a single vertical line (A-scan) at a time. The light reflected is brought to interference with a light beam that was previously propagated to a reference mirror, as in the original white-light interference setup. To obtain the 3D profile of the test object, it is scanned line by line and the collected data is processed, essentially using a Fourier transformation [13].

In 2011, we conducted a comparison study of achievable performance for the preprocessing task of WLI data analysis process on Intel's Xeon X5650 Hexcore CPUs, IBM's PowerXCell 8i processor and NVIDIA's Tesla C2050 GPUs [7]. Although, for the simplest algorithm used, CPUs were faster than the other architectures, which is due to required additional data transfers to GPU memory, GPUs exceeded the performance attained by other architectures, for most of compared algorithms.

Pacholik et al. [6] compared the achievable performance for the complete WLI elaboration process on FPGAs and GPUs. The attained results show that due to massive parallel processing capabilities exploitation of GPUs lead to highest data analysis throughput. The presented work shows also, that although the implementation effort for CPUs and GPUs is marginal compared to that of FPGAs, the power consumption during data analysis process of the FPGAs is clearly below that of CPUs and GPUs, allowing development of embedded hardware for WLI data elaboration process.

In 2012 You et al. [8] accomplished parallel processing of fringe data for low-coherence interferograms using GPUs. They developed a recursive, weighted algorithm, that allows the detection of height positions without postprocessing of computed values.

To the best of out knowledge beside the auto-tuning framework developed by T. Heller et al. [14], which targets stencil code applications, there exist no other framework for non-contact, non-destructive testing applications. However, there exist many generic frameworks for high-performance computing applications. Following, we give a brief overview of solutions proposed over the past few years, that have some features required for the WLI process. But none of the approaches fulfills completely the requirements in our case, which is why we preferred the development of a proprietary framework.

LibGeoDecomp developed by A. Schäfer [15] is an auto-parallelizing library for stencil codes. It provides state of the art features such as dynamic load balancing, exchangeable domain decomposition techniques, heterogeneous computation on grid systems etc. However, as it targets stencil code applications, where elements or cells depending only on cells within a fixed neighborhood radius are updated iteratively, it is not suitable for the elaboration process required in WLI.

G. Bosilca et al. [16] developed DAGuE, a generic framework for architecture aware scheduling and management of micro-tasks on heterogeneous cluster-systems. This framework takes over the responsibility of distributing computational tasks to compute devices available in cluster systems. To achieve this, the computational tasks must be encapsulated into sequential kernels using DAGuE's internal representation of Directed Acyclic Graphs, called JDF. Additionally, using JDF data dependencies between these kernels must be specified, to achieve correct and thus desired program execution. In [17] G. Bosilca et al. presented the GPU subsystem of DAGuE. Using Cholesky factorization they demonstrated that with the DAGuE framework only minimal efforts are required to enable GPU acceleration. However, to achieve GPU acceleration the application developer must implement and provide the CUDA code to DAGuE's scheduling system. Considering the case that the target platform for an application is a heterogeneous system, including CPUs and GPUs, multiple versions of the same computational problem are required. This increases the implementational efforts for the application developer.

Hardware Locality (hwloc) [18] is a generic framework, which gathers informations about available resources (CPUs, caches, memory nodes, GPUs) and make them accessible for other applications. It uses operating system specific strategies to get the desired informations. DAGuE framework is one known application that exploits hwloc functionality to assess the required hardware informations for computational and memory units. Although, hwloc enables the detection of available resources, to use the computational devices found, OpenMP, MPI, CUDA, OpenCL or other frameworks are required.

3 Framework

In our framework we use a twofold approach for exploitation of available heterogeneous computation devices. On one side we provide native support for x86-compatible CPUs and for CUDA-capable GPUs using C/C++ language library and NVIDIAs CUDA Runtime API. On the other side we employ OpenCL for those systems that have the necessary OpenCL drivers and libraries, allowing usage of x86-compatible CPUs, NVIDIAs and AMDs GPUs, Intels Xeon Phi Mic-architecture and others.

To simplify the development process of new WLI algorithms, tasks that have nothing to do with the actual WLI data analysis process, such as searching for computational devices and selecting the best one regarding the provided performance, memory management and so on are integrated within our framework. Furthermore, tasks that are common to the WLI analysis process, independently from the computational core of each major processing step, are also incorporated in our framework. The application developer must only provide code fragments, that describe the required processing steps within the preprocessing and postprocessing stages for a single pixel, which is sufficient because each pixel is processed in same way. For that, the conventional C++ programming scheme is used in usual manner, without any hardware or domain specific language statements, as proposed in [14].

The proposed framework is currenty available as an early prototype and consists out of two major modules: (i) resource finder and (ii) computational module. The resource finder module is responsible for information gathering of available computational resources (e.g. CUDA-capable GPUs, CPU-Cores) and the achievable performance of these devices for WLI data processing. The resource finder is work in progress because currently we select manually the best device available and use this device for all processing steps. However, a major objective behind the development of the resource finder is the automatic selection of resources using auto-tuning techniques. This should make it feasible for the framework user to achieve best performance attainable, without having the burden of analyzing the available resources and mapping the computational process to selected devices manually. The computational module is described in next section.

4   Computational Module

The computational module implements the pre- and postprocessing stages of the WLI analysis process. Fig. 2 shows the processing pipeline, with necessary inputs and generated outputs of each step. The following subsection explaines the processing steps briefly.

4.1   WLI Processing Pipeline

Preprocessing. The preprocessing stage takes the interference images captured by the interferometer as input and determines for each pixel the scan position where the significant interference pattern begins. For that two steps are necessary: (i) the image index corresponding to the scan position of the maximum interference value is determined and (ii) the image index is adjusted to the left by subtracting a user-defined offset. This shifted value defines the beginning of the

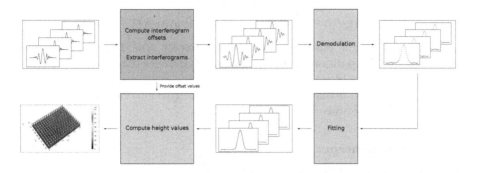

Fig. 2. The processing pipeline for WLI analysis process

significant interference pattern for an interferogram. As described in [7, 9] there exist a range of different approaches to compute the maximum interference values. Which algorithm is suitable for a specific measurement process depends on the surface characteristics and the signal-to-noise ratio of the environment [19].

Demodulation. The correlogram signal of each pixel, obtained during the scanning procedure consists of a carrier wave modulated on the envelope of the interference signal. To calculate the height data from interferograms the envelope maximum has to be detected [9]. In this step the preprocessed interference patterns are demodulated, separating the carrier wave from the envelope. In order to separate the two signals, same methods as in the preprocessing step can be used.

Fitting. The measurement and processing technology, which is used in optical metrology, is limited by the axial sampling period. Thus it allows only a discretized analysis of object properties. However, the envelope signals obtained in previous step possess a shape similar to a Gaussian function. This is due to the spectrum of the light source used in white-light interferometry. Exploiting this characteristic it is possible to get a more accurate approximation of the height values. For that, e.g. a Gaussian function [20] is used in a nonlinear regression process, in which the parameters for the Gaussian are iteratively adjusted, resulting in a curve that approximates the corresponding envelope in best-possible way. The mean value parameters of fitted curves are then used in final step for the computation of the actual height values.

Compute Height Values. In this step the results computed in preprocessing and fitting steps are agglumerated to get the surface profile of the analyzed object. Equation 1 shows how the height value $h(x, y)$ for pixel with coordinates x and y is computed.

$$h(x,y) = h_s + (I_{sip}(x,y) + \mu(x,y)) * SD \ . \tag{1}$$

h_s is the start height position of the interferometer arm. I_{sip} represents the image index to the position of the significant interference pattern within pixel's correlogram signal, which was obtained in the preprocessing stage. $\mu(x,y)$ is the mean value of the Gaussian function fitted to the envelope of this pixel. SD is the scan distance between two successive frames.

4.2 Module Concepts

Parameterisation of the Evaluation Process

Interferometer systems are frequently offered with varying characteristics, that have to be considered by our framework in a generic way. One important attribute is the employed camera system. Camera properties, such as the recording speed, camera resolution, supported transmission modes, the color depth and so on, affect in a non-negligible fashion the scanning procedure. E.g., using a high-resolution camera allows a more accurate investigation of object's surface profile, compared to a low-resolution setup. This is due to the fact that the lateral resolution, defining the distance between measurement points on object's surface, is obtained directly from the resolution of the camera. Thus, using a higher resolution, the surface area with same physical dimensions, is represented in more detail. However, these attributes have not only an influence on the measurement process itself, but also on the coding fashion required to cope with varying system characteristics. In case of parameters such as camera resolution or transmission mode, that do not require specific data types, it is sufficient if these parameters are provided to the application at runtime. On the other hand the color depth changes the data type of variables and memory entities which are used for measured interference values. In order to ensure a correct access to the pixel data, this fact must be considered at compile time.

Template metaprogramming techniques enable the generation of functions with the same behaviour for any type of data, by attaching a template classifier with a template parameter to the corresponding function. Then the compiler generates a data type specific specialisation from the general template using either template argument deduction provided by C++-compiler or the template argument explicitly defined by the programmer at the function call [21]. Thus, each function that access interference values must be declared only once. Our framework utilize the template functionality extensively to support the different camera setups and an another requirement, namely the ability of data processing in single and double precision.

As can be seen in Fig. 2 the WLI data analysis pipeline is fixed. This applies to the processing stage order as well as to the processing fashion, namely that the interferogram signal of a single pixel is elaborated independently from those of the other pixels using same computational steps. However, the algorithmic scheme used within a stage may differ between surface topology analysis processes of two different objects. E.g., for a plain object simple preprocessing and demodulation approaches may be used, like using single interference values to find the maximum intensity position, while for objects with rough and non-planar surface more complex and error-robust methods, like the sliding average algorithm (SLA) shown in 2, are necessary, due to e.g. light scattering effects.

$$contrast_z(x,y) = |I_z(x,y) - I_{z+1}(x,y)|$$
$$sum_k(x,y) = \sum_{j=k}^{k+C-1} contrast_j(x,y) \ . \tag{2}$$

I_z, with $z \in [0, N-2]$, and N as the number of scanning positions, is the interference intensity for pixel with coordinates x and y from zth scan position. In this approach the maximum value $max(sum_k(x,y))$, with $k \in [0, N-C-1]$, for the interferogram of a pixel is obtained using the sum of C absolute differences of successive sampling points. This approach exploits the fact that the gradient of interferogram oscillations is maximal near the actual maximum interference position.

```
template < typename DATA_T >
class SLAProcess {
  public:
    inline KERNEL_CONFIG DATA_T operator(
      PAR_DECL(VALUE_COUNT,DATA_T& rFrame)) {
      DATA_T ctr0,ctr1,ctr2;

      ctr0 = fabs(rFrame0_p - rFrame1_p);
      ctr1 = fabs(rFrame1_p - rFrame2_p);
      ctr2 = fabs(rFrame2_p - rFrame3_p);

      return ctr0 + ctr1 + ctr2;
}};
```

Listing 1. Class type functor implementing the SLA-Algorithm with C = 3

Finding a suitable algorithm leading to correct results for a processing stage in WLI is not a task that is easy to deal with, because not only the surface profile has an impact on the quality of the interference images, but also the environmental influences like temperature, vibration or ambient light. Thus, the computational module of our framework, provides the fixed operation chain for the WLI analysis process, maintaining at the same time the freedom of adjusting the processing manner within the process stages. To make this degree of freedom feasible, we use another C++-specific language construct, the so-called class type function objects (functors) [21]. This concept allows to encapsulate the functionality needed for a processing task within a class type functor and using this class type functor as a template parameter to insert the functionality in other function or class entities like in our case in framework class components which represents the pipeline stages shown in Fig. 2.

Listing 1 shows a class type functor, that implements the algorithmic scheme of the sliding average method shown in 2. The template parameter $DATA\_T$ is used to specify the data type used for intermediate values, e.g. *float* or *double*. $KERNEL\_CONFIG$ is employed to declare the function as a function of the host or the device if CUDA is used. In case that the framework is compiled for x86-compatible CPUs or the OpenCL-processing fashion is enabled, then $KERNEL\_CONFIG$ is defined as an empty makro and thus not used at all. Which configuration, native C/C++, CUDA or OpenCL, is used must be defined during the build process setup, where the user selects one of the available modes.

```
template < typename DATA_T >
class SLAProcess {
  public:
    inline DATA_T operator()(DATA_T& rFrame0_p, DATA_T& rFrame1_p,
                             DATA_T& rFrame2_p, DATA_T& rFrame3_p)
    ...
```

Listing 2. By C/C++-preprocessor generated intermediate version of the class type functor defined in previous listing

Because the algorithms utilized in preprocessing or demodulation stages require varying count of intensities for a single computation, e.g. in listings 1 four intensities are used to compute three contrast values agglumerated in a single return value, a compile-time facility to configure the count of intensity values used is necessary. For that we exploit the preprocessor metaprogramming functionality of the well established boost library [22]. Employing macros, provided by the boost library, it is possible to instruct the C/C++–preprocessor to expand a code statement, i.e. using the user-defined macro $VALUE\_COUNT$ the C/C++-preprocessor generates from the defined operator function an intermediate version with $VALUE\_COUNT$ many parameters as shown in listing 2.

```
template < class PREPROCESS_OP, class DEMODULATION_OP,
           class FITTING_OP, typename MEAS_DATA_T,
           typename PROCESS_DATA_T >
class WLIInterface { ... };

void ProcessData(...) {
  WLIInterface< SLAProcess< double >,
          SLAProcess< double >,
          FittingProcess< double >,
          uint8_t, double > wliProcess(...);
}
```

Listing 3. Inserting the sliding average method into framework for the preprocessing and demodulation stages

To provide the sliding average functor to the e.g. preprocessing and demodulation stages of the processing pipeline, the software interface, implemented as a template class entity can be used as shown in listing 3. The first three template parameters of the interface class represents the operations applied in pipeline stage preprocessing, demodulation and fitting to the correlogram data of each pixel. As already stated for preprocessing and demodulation steps the same algorithm and thus the same functor object can be used. Template parameter $MEAS\_DATA\_T$ is employed to define the data type of the interference values, whilst the parameter $PROCESS\_DATA\_T$ defines the computational precision.

Support for Heterogeneous Architectures

The white-light interferometry data analysis process is an application that puts high demands on the required computing power. Currently, to fulfill this requirement either homogeneous multi-core CPU systems, heterogeneous architectures such as CUDA-capable GPGPUs or even configurable hardware devices (FPGAs) are used. However, none of currently available WLI application software provides support for heterogeneous processing where different architectures can be used for same task at same time. Thus, one major aim for our framework was to agglumerate the support capabilities for different architectures in a single application, and to enable the framework user to exploit these heterogeneous resources through the same interface, without major adjustments to the code. To achieve this, the functionality required by a single computational step of the pipeline shown in Fig. 2 is implemented in a class hierarchy, where for all architectures common steps like memory allocation is embedded within a base class and all hardware and programming technique specific steps like function calls, data transfers and so on is contained within from that base class derived classes. Thus, an application developer can use any of supported architectures without dealing with hardware specific programming. However, the user must specify which execution scheme should be used in case that CUDA and/or OpenCL are available.

5   Results

In this section we present the performance results achieved. The example WLI data set consisted out of 1000 frames with a resolution of 1024×768 pixels each. A color depth of eight bit was used and a single precision processing manner was applyed. The width of the significant interference pattern was set to 64 frames. For the preprocessing and demodulation stage the sliding average method was provided through a function object. The fitting procedure was conducted using the Gaussian fitting method presented in [6]. To investigate the support capability for heterogeneous architectures we used the native C/C++– and CUDA–based framework classes. The OpenCL-based approach will be presented in a future paper. The architecture we used to investige the scalability on

Table 1. CPU and GPU processing steps performance (in sec)

Compute Unit	Processing step			
	Preprocessing	Demodulation	Fitting	Height compute
GeForce GTX 580	0.326	0.003	0.298	0.0004
Tesla C2050	0.229	0.005	0.519	0.0005
GeForce GTX 670	0.653	0.008	0.851	0.0007
Keppler K20	0.487	0.004	0.350	0.0004
Intel(R) Xeon(R) E5-2670	1.998	0.020	0.806	0.0116

multi-core CPUs is a dual-socket system equipped with Intel(R) Xeon(R) E5-2670 CPUs with 2.6 GHz clock speed and 64 GB DDR3 RAM. Further, we used different CUDA-capable GPU-architectures to survey performance changes due to architecture differences. The used GPU devices are an NVIDIA Tesla C2050, an NVIDIA K20, an NVIDIA GTX 580 and an NVIDIA GTX 670. Currently our framework does not have any auto-tuning facilities, such that the CUDA-thread count required to achieve the best possible performance on corresponding device could be determined previously to the actual WLI data analysis. Thus, we used a constant settings of 256×256 threads across all architectures. In Table 1 the achieved timings for NVIDIA's GPUs and Intels Xeon CPUs are shown. To get the best performance on CPUs we executed the WLI data analysis process using upto 32 threads supported natively by the described dual-socket system. Table 1 contains only the best measured values for the CPU version. In case of the preprocessing step using 12 threads the minimum runtime was achieved. This is due to the large data amount resulting from the correlogram reduction procedure resulting in memory bandwidth competition between the running threads. The demodulation processing step scaled upto 32 threads. This is explainable by the main characteristic of the underlying processing manner. Each pixel is elaborated independently from other ones and because no additional memory transfers were required, like in the reduction step of the preprocessing stage. The high operational intensity, measured in floating point operations per accessed byte, of the fitting procedure, where the derivatives of the Gaussian functions must be computed several times, allows a high parallelisation degree, such that for this stage also using 32 threads yielded the best performance. In the last processing step, where the actual height values are computed, the operational intensity is relatively low, thus this stage does not allow an optimal performance scaling behaviour, and in our case scaled only upto 14 threads. As can be seen, the Fermi-generation commercial GPU GeForce GTX 580 is always faster than the Keppler-counterpart GeForce GTX 670. This is due to major changes in the architecture of NVIDIAs GPUs. The GTX 670 has seven multiprocessors each with 192 scalar processors which executes the CUDA-threads, while GTX 580 has 16 multiprocessors with 32 scalar processors each. Thus, although 192 processors running in parallel are better, but because the L1-cache, that is inherent to multiprocessors, is also shared by the scalar processors and due to the fact

Table 2. GPU processing steps speed-up resulting from comparison to the fastest CPU version

GPU	Processing step			
	Preprocessing	Demodulation	Fitting	Height compute
GeForce GTX 580	6.13	6.78	2.70	28.94
Tesla C2050	8.73	4.07	1.55	23.15
GeForce GTX 670	3.06	2.54	0.95	16.54
Keppler K20	4.10	5.08	2.30	28.94

that it is same in size as in Fermi-GPUs, more scalar processors running at the same time compete for L1-accesses. Furthermore in Keppler generation NVIDIA changed the policy of using L1-cache for load-store spills resulting from memory accesses to data residing in global memory. Thus it is used only for thread local memory allocated within the kernel. This new architecture characteristics in Keppler GPUs induce for WLI data processing higher L1-cache-trashing and higher access rates to L2-cache hierarchy or even main memory impairing the available bandwidth. Same applies for the performance achieved by the Tesla C2050 in preprocessing stage compared to the Keppler K20. However, in fitting process the K20 is 1.48 times faster than the Tesla C2050. Because the Keppler-generation GPUs have a larger register set, the large number of intermediate results required in fitting procedure can be stored more efficiently reducing spills to L2-cache and main memory and leading to the better performance.

Table 2 contains the performance speed-up values resulting from comparison between the respective GPU performance and the maximal performance achieved by the CPU implementation. Here the benefit of our generic framework solution can be seen. Using same code fragments specified by the user, just by selecting a GPU device for computation a large speed-up compared to fastest CPU version can be achieved.

6   Conclusion

WLI data analysis is a high-performance demanding application. To fulfill the performance requirements, it is important to choose the right hardware platform and to fit the application on the underlying architecture. Further, to obtain accurate results the environmental influences, but also the material characteristics of the scanned sample must be considered by choosing the appropriate algorithms to be used in the WLI processing pipeline.

In this paper we presented the computational module of a new generic framework for the WLI elaboration process. Using sophisticated programming methods, C++–templates and function objects, this module enables the algorithm developer to concentrate on the task of finding or developing a suitable algorithmic solution, without major implementational efforts. Additionally, using native programming paradigmas and the functionality provided by CUDA- and

OpenCL, within our framework, the user can exploit a broad variety of heterogeneous resources without knowing all the details of these architectures and without the required know-how to program these devices.

The presented results show that using multi-core CPUs our framework is able to achieve the expected scalability for the embarrassingly parallel processing steps of the WLI process, even though the computational core is defined by the user outside of our framework. Further, we have shown that the framework module is also capable to apply the user-defined code fragment not only to CPUs but also to GPUs, achieving a tremendous speed-up compared to the CPU performance.

In future work we will integrate SIMD-support within the framework, such that the WLI applications may achieve even better performance on conventional CPUs, as it is possible just with multi-threading. We want also to look more into the High Performance ParalleX (HPX) runtime system as an alternative to currently used OpenMP API. Because HPX allows to reuse threads from a thread pool, thread instantiation and destruction overhead may be avoided. But more important is the *future*-feature provided by HPX. It allows to suspend a thread waiting for results or data transfers and to change to another waiting thread without considerable overhead.

Acknowledgments. This work was supported by the Federal Ministry of Economics and Technology which is funding the project "Kompakter Messkopf zur schnellen parameterfreien 3D-Oberflächenmessung für den produktionsnahen Einsatz (AiF KF202 7108AB2)"

References

1. Michelson, A.A.: On the application of interference methods to astronomical measurements. Philosophical Magazine Series 5 30(182), 1–21 (1890)
2. Leach, R.: Optical Measurement of Surface Topography. Springer (2011)
3. Shilling, K.M.: Mesoscale Edge Characterization. Dissertation, Mechanical Engineering. Georgia Institute of Technology (March 2006)
4. Larkin, K.G.: Topics in Multi-dimensional Signal Demodulation. PhD thesis, The Faculty of Science in the University of Sydney (2000)
5. Kapusi, D., Machleidt, T., Franke, K.H., Jahn, R.: White light interferometry in combination with a nanopositioning and nanomeasuring machine (NPMM), vol. 6616, pp. 661607-1–661607-10 (2007)
6. Pacholik, A., Muller, M., Fengler, W., Machleidt, T., Franke, K.H.: GPU vs FPGA: Example Application on White Light Interferometry. In: 2011 International Conference on Reconfigurable Computing and FPGAs (ReConFig), pp. 481–486 (2011)
7. Schneider, M., Fey, D., Kapusi, D., Machleidt, T.: Performance comparison of designated preprocessing white light interferometry algorithms on emerging multi- and many-core architectures. Procedia Computer Science 4, 2037–2046 (2011)
8. You, J., Kim, Y.J., Kim, S.W.: GPU-accelerated white-light scanning interferometer for large-area, high-speed surface profile measurements. International Journal of Nanomanufacturing 8(1), 31–39 (2012)

9. Hissmann, M.: Bayesian Estimation for White Light Interferometry. PhD thesis, Combined Faculties for the Natural Sciences and for Mathematics of the Ruperto-Carola University of Heidelberg, Germany (2005)
10. Chen, D.J., Chiang, F.P., Tan, Y.S., Don, H.S.: Digital speckle-displacement measurement using a complex spectrum method. Appl. Opt. 32(11), 1839–1849 (1993)
11. Kim, S.W., Kim, G.H.: Thickness-Profile Measurement of Transparent Thin-Film Layers by White-Light Scanning Interferometry. Appl. Opt. 38(28), 5968–5973 (1999)
12. Purde, A., Meixner, A., Schweizer, H., Zeh, T., Koch, A.: Pixel shader based real-time image processing for surface metrology. In: Proceedings of the 21st IEEE Instrumentation and Measurement Technology Conference, IMTC 2004, vol. 2, pp. 1116–1119 (May 2004)
13. Sylwestrzak, M., Szkulmowski, M., Szlag, D., Targowski, P.: Real-time imaging for Spectral Optical Coherence Tomography with massively parallel data processing. Photonics Letters of Poland 2(3) (2010)
14. Heller, T., Fey, D., Rehak, M.: An auto-tuning approach for optimizing base operators for non-destructive testing applications on heterogeneous multi-core architectures. In: SORT (2013) (invited paper)
15. Schäfer, A., Fey, D.: LibGeoDecomp: A Grid-Enabled Library for Geometric Decomposition Codes. In: Lastovetsky, A., Kechadi, T., Dongarra, J. (eds.) EuroPVM/MPI 2008. LNCS, vol. 5205, pp. 285–294. Springer, Heidelberg (2008)
16. Bosilca, G., Bouteiller, A., Danalis, A., Herault, T., Lemarinier, P., Dongarra, J.: DAGuE: A generic distributed DAG engine for High Performance Computing. Parallel Computing 38(1- 2), 37–51 (2012); Extensions for Next-Generation Parallel Programming Models.
17. Bosilca, G., Bouteiller, A., Herault, T., Lemarinier, P., Saengpatsa, N., Tomov, S., Dongarra, J.: Performance Portability of a GPU Enabled Factorization with the DAGuE Framework. In: 2011 IEEE International Conference on Cluster Computing (CLUSTER), pp. 395–402 (2011)
18. Broquedis, F., Clet-Ortega, J., Moreaud, S., Furmento, N., Goglin, B., Mercier, G., Thibault, S., Namyst, R.: hwloc: A Generic Framework for Managing Hardware Affinities in HPC Applications. In: 2010 18th Euromicro International Conference on Parallel, Distributed and Network-Based Processing (PDP), pp. 180–186 (2010)
19. Robinson, D.W.: Interferogram Analysis: Digital Fringe Pattern Measurement Techniques. Institute of Physics Publishing (1993)
20. Larkin, K.G.: Efficient nonlinear algorithm for envelope detection in white light interferometry. J. Opt. Soc. Am. A, 832–843 (1996)
21. Vandevoorde, D., Josuttis, N.M.: C++ Templates: The Complete Guide, 1st edn. Addison-Wesley Professional (November 2002)
22. Abrahams, D., Gurtovoy, A.: C++ Template Metaprogramming: Concepts, Tools, and Techniques from Boost and Beyond (C++ in Depth Series). Addison-Wesley Professional (2004)

Effect of Wind Field Parallelization on Forest Fire Spread Prediction*

Gemma Sanjuan, Carlos Brun, Ana Cortés, and Tomàs Margalef

Computer Architecture and Operating Systems department,
Universitat Autònoma de Barcelona, Spain
{gemma.sanjuan,carlos.brun,ana.cortes,tomas.margalef}@uab.es,
http://grupsderecerca.uab.cat/hpca4se/

Abstract. Forest fire spread prediction is a crucial issue to mitigate forest fire effects. Forest fire propagation models require several input parameters describing the conditions where the fire is taking place. However, some parameters, such as wind, present a different value on each point of the terrain due to topography. So, it is necessary to couple a wind field model that evaluates the wind on each terrain point. However, calculating the wind for each point on large maps is a time consuming task that can make the prediction unfeasible. So, it is necessary to parallelize the wind field computation. One approach is to apply a map partitioning technique, so that the wind field is calculated for each map part. The wind field obtained is lightly different from the one obtained with a single global map, and it is necessary to evaluate the effect of such difference on forest fire spread prediction.

Keywords: Wind field model, Forest fire spread prediction, Coupling models, Parallelization.

1 Introduction

It is well known that wind speed and direction are the parameters that most significantly affect forest fire propagation. Therefore, an accurate acknowledge of such values is mandatory to successfully estimate forest fire spread beforehand. Wind depends on meteorological conditions, but the meteorological wind is modified by the topography of the terrain so that the wind speed and direction are different on each point of the terrain under consideration. It means that the wind is not represented by a single value of speed and direction for the whole map, but there is a complete wind field with a value for wind speed and direction for each point of the terrain. However, it is not possible to measure the wind on each point of the terrain in a real emergency. Moreover, when predicting forest fire spread it is necessary to use meteorological values provided by weather forecast models, such as WRF [9], that provide values with low resolution (approximately 2.5 Km). Such resolution does not take into account

---

* This work was supported by Ministerio de Ciencia e Innovación (MICINN - Spain) under contract TIN2011-28689-C02-01.

B. Murgante et al. (Eds.): ICCSA 2014, Part IV, LNCS 8582, pp. 538–549, 2014.

the effect of terrain topography on wind variables. So it is necessary to use a wind field model that takes the meteorological wind or the values measured on certain meteorological stations and generates a wind field with high resolution (30 or even 10 meters depending on the digital elevation map available).

In this way, the meteorological wind data are introduced to a wind field model that evaluates the wind field at high resolution and the obtained wind field is introduced to a forest fire spread simulator to provide an accurate prediction of forest fire propagation [8]. In this work the wind field simulator used is Wind-Ninja [5][6] and the forest fire simulator is FARSITE [4]. The complete prediction scheme coupling wind field model and forest fire propagation model is shown in figure 1. FARSITE is one of the most widely used forest fire spread simulators in the forestry community. It has been designed to accept a wind field map as input data. WindNinja is a wind field model that was originally developed to be directly coupled to FARSITE.

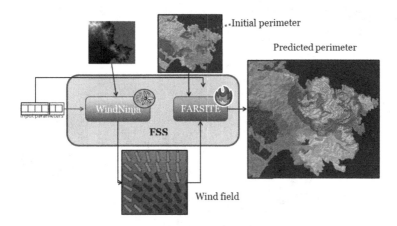

Fig. 1. Coupling wind field and forest fire spread models

This scheme is very promising and the accuracy in forest fire spread prediction is significantly increased, but the computation time to reach such a prediction is also significantly increased, specially when terrain map is large (30x30 Km) and the resolution is high (30x30 m). However, computation time is not the only constraint, but also data structures size is a significant constraint due to amount of memory available on a single node. To overcome these constraints it is necessary to apply some parallelization method that reduces execution time and distributes data structures along parallel system memory.

This work focuses on WindNinja Parallelization and studies the effect of the parallelization method selected on forest fire spread prediction, considering execution time and prediction accuracy. So, Section 2 describes WindNinja wind field simulator and determines the main constraints of such simulator. Section 3 shows the parallelization method proposed based on map partitioning considering overlapping. Section 4 presents the experimental results considering execution time and prediction accuracy. Finally, section 5 summarizes the main conclusions of this work.

2   WindNinja Wind Field Simulator

WindNinja is a wind field simulator that calculates the effect of topography on wind. WindNinja does not predict wind fields for future times, but computes the spatially varying wind field on the surface for one instant time. WindNinja requires as basic input parameters the elevation map of the underlying terrain (Digital Elevation Model -DEM- file), the global meteorological wind speed and wind direction and the required output resolution. As output, WindNinja delivers wind speed and wind direction at the specified output resolution. Usually, the resolution delivered in the output wind field is set to be the same resolution as the input elevation map. Using this one-to-one relationship, each map cell will have its own wind components. Figure 2 shows the internal structure of WindNinja and the steps that it carries out to calculate the wind field.

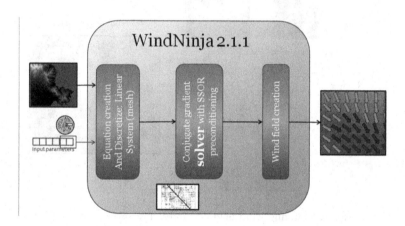

Fig. 2. WindNinja System

The main problem with WindNinja is that as the number of cells increases, the execution time increases significantly. The number of cells of the input map depends on the map size in absolute terms and on the resolution of the map. If the map size or the resolution are increased the execution time is increased. Moreover, the amount of memory required to solve the wind field increases linearly with the number of cells of the map making unaffordable to be solved a map with a large number of cells in a single node. Therefore, it is necessary to apply some parallelization technique to reduce execution time and memory requirements. It means that calculating the wind field of a 1500x1500 cells map on a single node with 4GB of main memory fails and no output is delivered. According to WindNinja documentation the maximum amount of memory required to execute the simulator can be expressed by equation 1. This equation shows that the amount of memory directly depends on the number of columns of the map (NColumn) and the number of rows of the map (Nrows).

$$M \text{ (Bytes)} = 20480 + 15360*\text{NRows} + 15360*\text{NCols} + 11520*\text{NRows}*\text{NCols} \quad (1)$$

Execution time is another issue to be considered, because forest fire prediction time must be much faster than real time in order to be operational. The execution time of WindNinja depends on the number of equations that form the system. So, it is directly proportional to the number of cells of the map. WindNinja has been executed on a DELL cluster based on Poweredge C6145 with a total of 8 CPUs with 16 cores and 128 GB of memory considering different map size and it has been determined that the relationship between execution time and the number of cells can be approximated to a straight applying linear regression. Figure 3 show the execution time depending on the number of cells of the map.

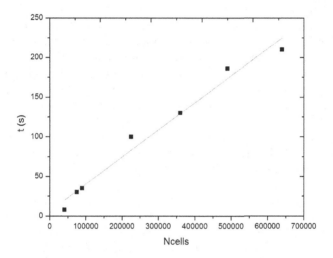

Fig. 3. Execution time depending on the number of cells

The expression obtained from linear regression is the following one:

$$t = 3.42 * 10^{-4} * \text{NCells} + 6.11 \quad (2)$$

Since it has been determined that the maximum time to evaluate the wind field should be around 60 seconds, from expression (2), it can be deduced that the maximum number of cells that can have a map is 160000. The results of the wind direction and speed of a cell depend on the cells that are around. Therefore, the squarer the part the fewer exposed cells were found in the partition, since the square is the geometric figure that has the minimum perimeter. So, the map should be of a maximum size of 400x400. For this map size the amount of memory required according to equation (1) is 1.7 GBytes.

The execution time and amount of memory required are reasonable, but this map size is very small and in most real cases maps will be larger. So it is necessary to apply some parallelization technique that reduces execution time and memory requirements.

3   Map Partitioning WindNinja Parallelization

The initial approach that has been considered to parallelize WindNinja is map partitioning. However, far from being an easy approach, this map partitioning scheme involves new issues that must be tackled. WindNinja is based on the equations that describe air flow variation in the atmosphere. Specifically, it is based on mass conservation and delimited boundary conditions. This implies that terrain slope variation generates wind changes and, due to boundary conditions, the obtained results in regions close to the borders of the map will not be correct since the system needs some cells to stabilized. Consequently, many external map cells have a non reliable value and, therefore, a set of cells around the evaluated map must be dismissed as a final result. When the global map is partitioned into parts, this problem is extrapolated to all parts and a direct combination of output wind fields results in an aggregation of boundary errors introducing additional uncertainty in wind values. So, if the map is partitioned and the wind field is calculated for each part of the map without considering the neighbour parts of the original map, the wind field resulting can be significantly different from the original one in the cells close to part boundaries. The inclusion of an overlapping is necessary to reduce the variation and uncertainty in the wind field near the part boundaries. Including a certain degree of overlapping among map parts increases execution time but reduce the degree of difference among the wind field calculated using a global map and the partitioned one.

To overcome this problem, a certain degree of overlapping among parts must be considered for all adjacent parts to soften border errors. An example of this partitioning and overlapping approach can be seen in figure 4 where the result of applying overlapping in AxB parts partitioning is shown. So, we propose a map partitioning with overlapping scheme for wind field evaluation as follows:

1. partition the input DEM map into X parts with a given overlapping,
2. run in parallel as many executions of the wind field model as parts have been generated at the partitioning process and,
3. combine the outputs of the X parts discarding the overlapping cells to obtain the global map.

Finally, the resulting wind field map, once the map partitioning scheme has been applied, has the same dimensions as the original one. It must be taken into account that even when some overlapping is introduced the system of equations to be solved is not exactly the same and the numerical solution obtained from the global map and the one obtained from the partitioning map may not be exactly the same. This map partitioning approach has been implemented in a Master/Worker MPI [7][3][1] application where the Master creates the map parts

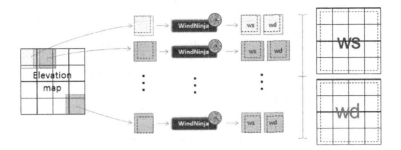

Fig. 4. AxB partitioning with overlapping

and distributes them to the workers and the workers calculate the wind field for each part and return the results to the Master that aggregate the wind fields in a complete wind field.

As it has been stated above a map part of 400x400 cells has reasonable execution time and memory requirements. But this map part size must include overlapping. An overlapping of 50 cells per side is reasonable, so, each part should contain around 300x300 cells of the global map.

The wind fields obtained from the global map and the partitioned one has been compared. To carry out this comparison the measures has been used to estimate the difference. These measures are:

1. The RMSE (Root Mean Square Error) that is a statistical measure of the difference in the wind speed (or direction) among the wind obtained in both cases for each cell. The main problem of such value is that it does not include information about the deviation of the differences.
2. The number of cells with an error higher than a particular values (1 mph for wind speed). This measure shows the points of the map where the differences are larger than this reference value.
3. The maximum difference value along the whole map.

Table 1 summarizes the differences obtained for a particular map of 1500x1500 cells considering different meteorological wind speed and different partitioning methods.

Figure 5 shows in red the points of the terrain map where the difference in wind speed in both fields is larger than 1mph. It can be observed that this points are concentrated on the map zones with abrupt slope change. But, this points are not so much. The point with a wind direction difference larger than 5° are approximately the same.

WindNinja parallelization by map partitioning is a feasible approach that significantly reduce execution time and memory requirements. But, as it has been shown there is a difference among the wind field obtained from a global map and a partitioned one. So, the next point to analyze is the influence of these wind field differences on the forest fire spread prediction. Therefore, an experimental study

Table 1. Similarity indexes for different partitioning methods

Speed (mph)	Partitioning	$RSME_{sp}$ (mph)	Speed ≥ 1 (mph)	Max_{sp} (mph)
5	5x5	0.187	7818	3.76
5	15x15	0.250	16369	4.16
10	5x5	0.373	49565	7.50
10	15x15	0.499	102412	8.33
15	5x5	0.559	140826	11.30
15	15x15	0.749	258006	10.20

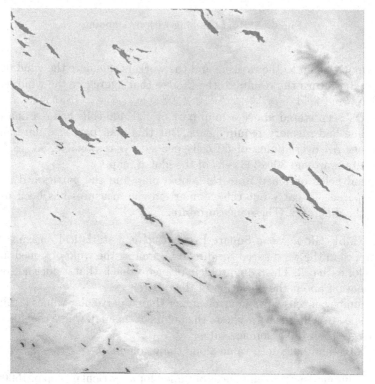

Fig. 5. Points with wind speed difference higher than 1mph

has been carried out to determine such effect using different terrain maps, different ignition points and different meteorological wind conditions. The experiments and the results obtained are described and analysed in the following section.

4   Effect of Map Partitioning on Forest Fire Spread Prediction

As it has been stated in previous section map partitioning does not generate extreme differences in wind fields, but it necessary to analyse the influence of such differences in forest fire spread prediction. To carry out such analysis it is necessary to execute a lot of propagation simulations considering maps with different topography terrains, different wind conditions, different vegetation types, different canopy covert and different fire positions. So, different terrain maps corresponding different areas of Spain has been selected. The raster maps used are composed by 1500 rows and 1500 columns with 30m resolution per cell. That means that the map has a dimensions of 45km x 45km.

The experiments have been executed in a DELL cluster based on Poweredge C6145 with a total of 8 CPUs with 16 cores per node and 128 GB of memory.

For comparison purpose, the propagation predicted by FARSITE has been obtained applying three different WindNinja configurations:

1. Global map WindNinja wind field: In this case the wind field is calculated from a global map. No partitioning is applied.
2. 5x5 Map partitioning wind field. The global map is partitioned in 25 parts. Each part is composed of 400x400 cells with an overlapping of 50 cell for each side.
3. 15x15 Map partitioning wind field. The global map is partitioned in 75 parts. Each part is composed of 200x200 cells with an overlapping of 50 cell for each side.

To generate the set of experiments the 4 more common vegetation types (brush, grass, conifer and rough), 3 wind directions (45, 180 and 270 degrees), 3 wind speeds (5, 10 and 15 mph) and 3 ignition point positions (over, near, far) have been considered to cover a wide range of combinations.

The difference among the predicted burned area using different WindNinja configurations is estimated by applying a fitness function expressed by equation (3). This equation calculates the difference in the number of cells burnt between the predicted area by 2 different WindNinja configurations. In this case, the area predicted using a global map wind field (GMWF) is used as reference propagation. Formally, this equation corresponds to the symmetric difference between the global map wind field area (GMWF) and the partitioned map (5x5WF or 15x15WF) divided by the GMWF area, so as to express a proportion. $\cup$(GMCell,PCell) is the union of the number of cells burned in the GMWF and the cells burned in the partitioned map, $\cap$(GMCell,PCell) is the intersection between the number of cells burned in the GMWF propagation and in the partitioned map wind field, and GMCell is the number of cells burned using Global map wind field.

$$D = \frac{\cup(\text{GMCell,PCell}) - \cap(\text{GMCell,PCell})}{\text{GMCell}} \tag{3}$$

Tables 2, 3 and 4 show the difference on burned areas for different types of vegetation and different map partitioning considering a meteorological wind of 15mph and a direction of 45°. This meteorological wind has been considered because it is the wind that generate larger differences in wind field.For each configuration, the evolution of such difference according to simulation time is

Table 2. Difference for ignition points over different wind zones

t	Over error 5x5					Over error 15x15				
	Brush	Conifer	Conifer	Grass	Rough	Brush	Conifer	Conifer	Grass	Rough
(h)	c10	c10	c50	c10	c10	c10	c10	c50	c10	c10
6	0.134	0.173	0.174	0.129	0.111	0.166	0.210	0.205	0.162	0.141
8	0.130	0.152	0.144	0.112	0.089	0.162	0.189	0.175	0.143	0.117
12	0.114	0.128	0.139	0.112	0.067	0.144	0.162	0.167	0.144	0.092
24	0.080	0.213	0.103	0.360	0.240	0.114	0.353	0.128	0.402	0.420

Table 3. Difference for ignition points near different wind zones

t	Near error 5x5					Near error 15x15				
	Brush	Conifer	Conifer	Grass	Rough	Brush	Conifer	Conifer	Grass	Rough
(h)	c10	c10	c50	c10	c10	c10	c10	c50	c10	c10
6	0.047	0.043	0.023	0.082	0.050	0.068	0.061	0.023	0.111	0.068
8	0.051	0.041	0.040	0.070	0.051	0.072	0.057	0.048	0.096	0.069
12	0.052	0.049	0.029	0.069	0.051	0.076	0.052	0.032	0,095	0.067
24	0.063	0.035	0.022	0.072	0.051	0.073	0.047	0.029	0.096	0.069

Table 4. Difference for ignition points far from different wind zones

t	Far error 5x5					Far error 15x15				
	Brush	Conifer	Conifer	Grass	Rough	Brush	Conifer	Conifer	Grass	Rough
(h)	c10	c10	c50	c10	c10	c10	c10	c50	c10	c10
6	0.035	0.028	0.040	0.032	0.032	0.037	0.030	0.040	0.033	0.034
8	0.037	0.040	0.000	0.028	0.028	0.038	0.038	0.001	0.028	0.029
12	0.034	0.032	0.028	0.029	0.023	0.034	0.033	0.029	0.029	0.026
24	0.037	0.028	0.043	0.025	0.013	0.038	0.030	0.040	0.026	0.014

shown. In particular, table 2 shows the results considering fire ignition points over terrain zones with a wind speed difference larger than 1mph, table 3 shows the results considering fire ignition points near the terrain point with differences larger than 1mph (it means that at some point of the propagation the fire front crosses that zones) and table 4 shows the results considering fire ignition points far from those different wind speed zones (it means that the fire front does not cross those zones).

From the experiments carried out it can be observed that as the number of parts is increased, the error increases proportionally to that number of parts. This is due to the fact that the wind field generated when the parts are very small has a larger difference from the global map wind field and this larger differences provoke larger differences in fire spread predicted area.

On the other hand, it can be observed, as it was expected, that the position of the fire is very significant. If the fire does not cross points with significant wind speed difference the spread area difference is negligible. When the fire ignition point is on large wind speed difference zones the difference in burned area is larger, but not extremely different. This results are also presented in figures 6,

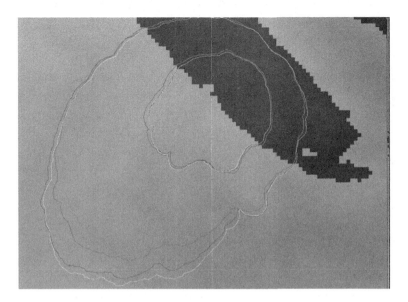

Fig. 6. Fire perimeters with ignition point over wind difference zones

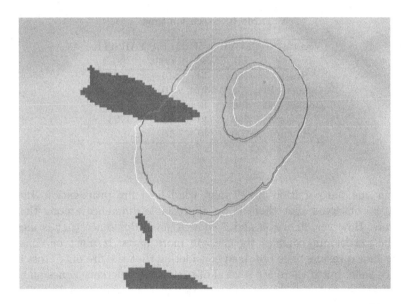

Fig. 7. Fire perimeters with ignition point near wind difference zones

7 and 8 that present an example of each one of the ignition point situation. In these figures the blue dots represent the zones with large wind speed difference, the yellow perimeter represents the Global Map Wind Field fire propagation, the red one represents the partitioning 5x5 map wind field fire propagation and

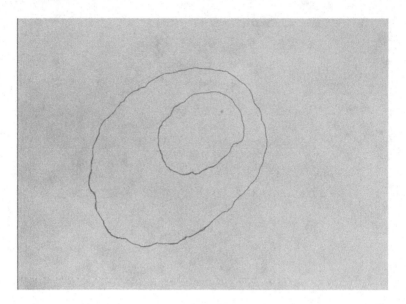

Fig. 8. Fire perimeters with ignition point far wind difference zones

Table 5. Execution time

Vegetation	GMWF (s)	5x5P (s)	15x15P (s)	FARSITE (s)
Brush	750	40	10	50
Conifer c10	750	40	10	51
Conifer c50	750	40	10	26
Grass	750	40	10	1725
Rough	750	40	10	163

the green one the partition 15x15 map wind field fire propagation. In figure 8 it can be observed that there is no appreciable difference among the three perimeters. However, figures 6 and 7 show a small difference that is increased when the partitioning divides the map in more parts. It must be considered that the shown results were obtained considering that the fire only cross a large difference zone, but if there are several of such large difference zones on the fire area, differences in fire propagation prediction will be larger since the effects are accumulative.

Another important measure is the execution time of the wind field calculation and the fire spread simulation. The average execution obtained from our experiments are shown is table 5. It can be observed that the execution time of a global map wind field takes 750 seconds for all the terrains and the map partitioning is a very effective measure since it reduces WindNinja execution time to 40 (5x5) or 10 seconds (15x15). On the other hand it can be observed that FARSITE execution time is very dependable from each particular scenario conditions [2].

5 Conclusions

Wind is a parameter that significantly affects fire propagation. However, meteorological wind is not a feasible estimation since wind is significantly modified by terrain topography. Therefore, it is necessary to introduce wind field models that calculate the wind at each point of the terrain given a meteorological wind and a terrain map. These wind field models are time consuming models when terrains maps are large and the time incurred can make unfeasible the use of such simulators. Therefore, a map partitioning approach has been introduced to generate wind fields faster. However, the wind fields obtained from map partitioning are lightly different from those obtained from a global map. So, it is necessary to study the influence of such differences on forest fire spread propagation. The experiments carried out show that the difference in fire spread prediction is not very significant and only when the fire crosses zones with a larger wind value difference the fire spread prediction is lightly different, but in any case it is not relevant to determine fire tendency.

References

1. Artés, T., Cencerrado, A., Cortés, A., Margalef, T.: Relieving the effects of uncertainty in forest fire spread prediction by hybrid mpi-openmp parallel strategies. Procedia Computer Science 18, 2278–2287 (2013)
2. Cencerrado, A., Cortés, A., Margalef, T.: Response time assessment in forest fire spread simulation: An integrated methodology for efficient exploitation of available prediction time. Environmental Modelling & Software 54, 153–164 (2014)
3. Chapman, B., Jost, G., Van Der Pas, R.: Using OpenMP: portable shared memory parallel programming, vol. 10. The MIT Press (2007)
4. Finney, M.A.: FARSITE, Fire Area Simulator–model development and evaluation. Res. Pap. RMRS-RP-4, Ogden, UT: U.S. Department of Agriculture, Forest Service, Rocky Mountain Research Station (1998)
5. Forthofer, J.M., Shannon, K., Butler, B.W.: Simulating diurnally driven slope winds with windninja. In: 8th Symposium on Fire and Forest Meteorological Society (2009)
6. Forthofer, J.M., Shannon, K., Butler, B.W.: Initialization of high resolution surface wind simulations using nws gridded data. In: Proceedings of 3rd Fire Behavior and Fuels Conference, October 25-29 (2010)
7. Gropp, W.: MPICH2: A new start for MPI implementations. In: Kranzlmüller, D., Kacsuk, P., Dongarra, J., Volkert, J. (eds.) PVM/MPI 2002. LNCS, vol. 2474, pp. 7–42. Springer, Heidelberg (2002)
8. Lopes, A.M.G., Cruz, M.G., Viegas, D.X.: Firestation–an integrated software system for the numerical simulation of fire spread on complex topography. Environmental Modelling & Software 17(3), 269–285 (2002)
9. William, C., Skamarock, J.B.: Klemp, Jimy Dudhia, David O Gill, Dale M Barker, Wei Wang, and Jordan G Powers. A description of the advanced research wrf version 2. Technical report, DTIC Document (2005)

Experiments with GPU-Acceleration for Solving a Radiative Transfer Problem

Paulo B. Vasconcelos[1,*] and Osni Marques[2]

[1] University of Porto and CMUP, Portugal
[2] Lawrence Berkeley National Laboratory, USA

Abstract. High performance computing systems are increasingly incorporating the computational power provided by accelerators, especially GPUs. With the programmability of GPUs greatly facilitated by OpenCL or NVIDIA's CUDA, with support for full double precision on GPUs, many challenging problems are benefiting from these processing units. It is well-known that memory latency is the speed limiting factor on GPUs. To hide memory latency, kernel instances must be executed in parallel on the same core, making sparse data more difficult to deal with than dense data. In this work we examine the numerical solution of a radiative transfer problem. We show that integral problem formulations relying on sparse linear algebra computations can benefit from the computing power of such devices, achieving an average speedup of 50× when compared to a representative CPU implementation.

Keywords: GPU implementation, integral operators, sparse linear iterative methods.

1   Introduction

The computing paradigm on emerging systems is being redefined because of the devices they are built upon, consequently requiring a recasting of many numerical algorithms that are necessary to efficiently solve scientific problems. Current supercomputers and emerging high-end nodes make use of accelerators, such as general purpose graphical processing units (GPGPUs), to perform computations that are usually done on CPUs. Going beyond nowadays levels of computing power is expected to be disruptive to the High Performance Computing (HPC) community, although it will present great opportunities for developers of applications and algorithms that may lead to new scientific discoveries. Numerical algorithms must be re-written and redefined in order to keep up with these changes. In fact, even for nowadays computer technology, hybrid programming models often need to be employed [8].

Basically, two approaches can be adopted for programming GPUs: CUDA (Compute Unified Device Architecture), a parallel computing architecture developed by NVIDIA [10]; and the OpenCL (Open computing Language), a standard

---

* Part of this work was done while the author was a Fulbright visiting scholar to Lawrence Berkeley National Laboratory, Computational Research Division, USA.

B. Murgante et al. (Eds.): ICCSA 2014, Part IV, LNCS 8582, pp. 550–559, 2014.

for program execution across heterogeneous platforms [16]. However, the effort required to use these languages is high, requiring a significant understanding of the underlying hardware. Recently, high-level libraries have been released that attenuate that effort. Among these libraries, in particular the ones that are relevant for sparse matrix computations, programmers can benefit from NVIDIA's Thrust [5], a flexible high-level interface for GPU programming, cuSPARSE [10], a collection of sparse basic linear algebra subroutines, and CULA Sparse [9], a GPU-accelerated library for linear algebra. Other important libraries with parallel algorithms for sparse matrix computations are Cusp [4], a collection of generic parallel algorithms for sparse linear algebra and graph computations on CUDA architecture, and ViennaCL [12] (The Vienna Computing Library) based on OpenCL. While Cusp is an open source library, CULA is delivered in three versions, each offering different levels of functionality, features, and distinct policy costs associated.

It is worth mentioning that mapping sparse linear algebra on GPUs is challenging. It is well-known that memory latency is the speed limiting factor on GPUs, and to hide memory latency kernel instances must be executed in parallel on the same core. This makes sparse data more difficult to tackle than dense data.

This work aims at evaluating the performance of a GPU implementation of a radiative transfer problem modeled through an integral equation, exploring sparse matrix computations. To harness the computational power of such processors, algorithms need to be decomposed into a large number of separate and fine-grained threads of execution. This is a demanding task for code developers and particularly for practitioners that are focused on the problems and not implementation intricacies. By implementing the integral problem in combination with high-level libraries, we provide a case study in which performance and efficiency can be achieved with the necessary level of abstraction; robustness is guaranteed.

We note that [18] addressed a parallel implementation of the radiative transfer problem considered here, and [19] studied a companion eigenvalue computation. The latter was extended to GPUs in [11]. For a comprehensive discussion of the problem formulation we refer to [2].

The rest of the paper is structured as follows. In section 2 we briefly describe the problem, and in section 3 some implementation details and the libraries used are presented. This is followed by numerical results in section 4. In section 5, a few concluding remarks and open questions are discussed.

2   The Problem

Consider the following integral equation

$$\frac{\varpi}{2} \int_0^{\tau^*} E_1 \left(|\tau - \tau'| \right) \varphi \left(\tau' \right) d\tau' - z\varphi(\tau) = f(\tau), \quad \tau \in [0, \tau^*] \tag{1}$$

modeling the emission of photons in stellar atmospheres [2,14], where τ^* is the optical depth of the stellar atmosphere, $\varpi \in \left]0,1\right[$ is the albedo[1], z is a parameter in the resolvent set of the integral operator $\mathcal{T}$, f is the source term, and E_1 is the first function of the family of exponential integrals [1]

$$E_\nu(\tau) = \int_1^\infty \frac{\exp(-\tau\mu)}{\mu^\nu} d\mu, \quad \nu \geq 1. \tag{2}$$

That is,

$$(\mathcal{T} - zI)\varphi = f \tag{3}$$

for $(\mathcal{T} - zI)$ invertible. It is noteworthy that $\mathcal{T}$ is a weakly singular operator since near 0 the function E_1 has a logarithmic singularity.

Approximations φ_m for the solution of (1) can be obtained by solving

$$(T_m - zI)\varphi_m = f \tag{4}$$

where (T_m) is a sequence of finite rank operators converging to $\mathcal{T}$. By evaluating the projected problem on a specific basis function, (4) is reduced to a linear system

$$(A - zI)x = b \tag{5}$$

for a finite matrix A, with coefficients given by

$$A(i,j) = \begin{cases} \frac{\varpi}{2(\tau_{n,i} - \tau_{n,i-1})}(-E_3(|\tau_{n,i} - \tau_{n,j}|) + E_3(|\tau_{n,i-1} - \tau_{n,j}|) + \\ \quad + E_3(|\tau_{n,i} - \tau_{n,j-1}|) - E_3(|\tau_{n,i-1} - \tau_{n,j-1}|)), & i \neq j \\ \\ \varpi[1 + \frac{1}{\tau_{n,i} - \tau_{n,i-1}}(E_3(\tau_{n,i} - \tau_{n,i-1}) - \frac{1}{2})], & i = j \end{cases} \tag{6}$$

Details on the discretization process can be obtained from [2].

To compute a good approximation to φ, the dimension of the matrix A, $m \times m$, must be high. Furthermore, for this particular kernel dependent of the E_1 function, A shows a sharp decrease in the magnitude of the entries from the diagonal, not necessarily within a structured form. By discarding small magnitude values, under a high threshold, matrix A becomes a large sparse matrix. Moreover, A is usually nonsymmetric since the discretization process is supported by a nonuniform mesh to better adapt the source term.

For the interested reader on a detailed presentation of the problem we recommend [2]. Fundamental information on radiative transfer and on scattering theory can be found, respectively, in [7] and [13].

3   Implementation Details and Hardware

To maintain the implementation at a reasonable level of abstraction while ensuring efficiency and robustness, we used CUDA [10], version v5.5 and explored

---

[1] The albedo is the ratio of reflected radiation from the surface to incident radiation upon it. In this work it is assumed constant.

the numerical libraries Thrust [5], version v1.7, and Cusp [4], version v0.3.1. Cusp was used instead of CULA since the former is an open source library while the latter, depending on the version, may have restrictions. These libraries implement Krylov subspace methods, which provide an attractive iterative procedure framework for the solution of large sparse linear systems by eliminating the need for the expensive factorization operations that are typical of direct methods. In particular, two nonsymmetric linear solvers, GMRES (Generalized Minimum RESidual) and Bi-Conjugate Gradient stabilized (BiCGstab) on Cusp were explored. Reference [15] provides a description of the main developments in iterative methods for solving linear systems as well as a comprehensible bibliography.

Depending on the characteristics of the coefficient matrix A in a system $Ax = b$, preconditioners may be a requisite. Our experiments focused on no preconditioner (NONE), diagonal preconditioner (DIAG) and algebraic multigrid preconditoner with smoothed aggregation (SA-AMG). For a complete survey on preconditioners we refer to [6], and for implementation details of the algebraic multigrid preconditoner for parallel architectures we refer to [17] and on GPUs to [3].

The impact of sparse matrix storage formats on the performance of integral problems on the GPU is also analysed. Numerical results for compressed sparse row, diagonal, coordinate format, ELLPACK and hybrid (combination of the last two) are assessed by comparing execution time.

The diagonal format (DIA) is formed by two arrays, storing the nonzero values and the offset of each diagonal from the main diagonal. The benefits of this format are that the row and column indices of each nonzero entry are implicitly defined and all memory access is contiguous.

The ELLPACK format stores the nonzero values of an $m \times n$ matrix with a maximum of k nonzeros per row in a dense $m \times k$ structure along with their column indices. This format was introduced as a way to compress a sparse matrix with the purpose of solving large sparse linear systems on vector computers.

The coordinate (COO) stores the matrix in three arrays: row indices, column indices and nonzeros. Unlike DIA and ELL, both row and column indices are stored explicitly.

The compressed sparse row (CSR) format is a general-purpose sparse matrix representation, where column indices and nonzero values are explicitly stored. A third array of $m + 1$ row pointers completes this representation, where the last entry stores the total number of nonzeros.

The hybrid format combines COO with ELLPACK. It stores the a typical number of nonzeros per row in the ELL data structure and the remaining entries in the COO format.

In terms of storage requirements, DIA and ELL can be more appropriate for structured meshes, and HYB and COO for unstructured meshes. The usual CSR, generally used on CPUs, can be inappropriate for GPUs if the aforementioned distinction, in terms of structured and unstructured meshes, is clear and markedly different. However CSR can be adequate for structured meshes with tight neighborhood relationships, as will be highlighted in this work.

The machine used in our experiments was Dirac, a 50 GPU node cluster connected with QDR IB, located at the National Energy Research Scientific Computing Center (NERSC) in Berkeley, California. This GPU cluster has 44 nodes with one NVIDIA Tesla C2050 Fermi (3GB of memory and 448 parallel CUDA processor cores), 4 nodes with one C1060 NVIDIA Tesla GPU (4GB of memory and 240 parallel CUDA processor cores), one node with four NVIDIA Tesla C2050 Fermi (each with 3GB of memory and 448 parallel CUDA processor cores), and one node with four C1060 Nvidia Tesla (each with 4GB of memory and 240 parallel CUDA processor cores). Dirac is a sub-cluster of Carver, an IBM iDataPlex system with 1202 compute nodes (each node contains two Intel Nehalem quad-core processors, summing 9,984 processor cores), capable of 106.5 Teraflops/sec theoretical peak performance.

4   Numerical Experiments

For the numerical tests presented here we set the tolerance to determine the convergence of the iterative linear solvers to 10^{-12} and the maximum number of iterations allowed was set to 1000. For GMRES a fixed restart parameter of 30 was chosen. Times are reported in milliseconds averaged over 5 operations, on the device (GPU) NVIDIA Tesla C2050 and on the host (CPU) Intel Nehalem quad-core processor.

One of the main constraints for full exploitation of GPUs is that all data fits in the memory of the device. This means that we could not solve problems of order 64000 (and above); the tests were performed using order matrices ranging from 8000 to 32000 (Table 1).

Table 1. Matrix dimension, number of unknowns and nonzeros

matrix order (m)	degrees of freedom	nonzeros
8000	6.40×10^7	1713655
16000	2.56×10^8	4507978
32000	1.02×10^9	26460424

Table 2 reports timings for the sparse linear solver on 6 CPUs (24 cores) and on one GPU, for the 16000×16000 problem, using BiCGstab and GMRES solvers preconditioned by diagonal and algebraic multigrid with smoothed aggregation preconditioners, as well as no preconditioning. The number of iterations for convergence is also reported. Due to the characteristics of the problem, namely the relatively small condition number and eigenvalues far from zero, preconditioning is not as crucial as in other problems. The number of iterations is greatly reduced with the use of SA-AMG but the time per iteration is very high for both iterative methods. The use of DIAG preconditioner is not sufficient to improve performance with respect to NONE. This behavior is persistent for higher values of m but since GMRES is more demanding in terms of memory requirements, for larger dimension problems BiCGstab shows to be a more favorable alternative.

Table 2. Comparison for Krylov type solvers on the GPU for several preconditioners ($m = 16000$)

preconditioner	BiCGstab	iterations	GMRES	iterations
NONE	36.97	14	36.00	22
DIAG	36.84	14	35.91	22
AMG	36.93	7	41.50	14

Table 3 shows the timings for BiCGstab on the host and on the device for three values of m. It should be mentioned that we are executing the same code both on the device and on the host. On the host, 24 cores (6 CPU's × with 4 cores each) are used. This means that we are not comparing the best version for CPU with an implementation for the GPU; rather, we are testing the same code implemented on both the CPU and the GPU. The relative gains of the GPU range from 50× up-to 75× compared to 24 CPU cores for the larger dimensional problem, being always superior to 30× irrespectively of the dimension.

Table 3. Timings for BiCGstab on the CPU and GPU for several values of m

matrix order	precond.	CPU solver	iter.	GPU solver	iter.	speedup
8000	NONE	686.96	14	16.45	14	42
	DIAG	621.24	14	14.78	14	42
	AMG	731.05	7	26.09	7	28
16000	NONE	1592.43	14	36.97	14	43
	DIAG	1598.68	14	36.84	14	43
	AMG	1681.31	7	36.93	7	46
32000	NONE	10744.71	15	138.97	15	77
	DIAG	10757.58	15	138.97	15	77
	AMG	12436.25	8	236.72	8	53

The above results use the CSR format on both the CPU and GPU. Iterative methods for large dimensional problems rely largely on sparse matrix-vector products. Due to sparsity, few arithmetic operations are computed per memory access, so this is a memory bound problem. In addition, the sparsity structure impacts the locality of memory accesses and the amount of work per thread.

We also evaluate the impact on the performance of the matrix formats provided by Cusp, namely, Coordinate (COO), Compressed Sparse Row (CSR), Diagonal (DIA), ELLPACK/ITPACK (ELL) and Hybrid ELL/COO structure (HYB). In Figures 1 and 2 we show the timings for several sparse matrix formats for values of m ranging from 8000 to 32000, respectively, using BiGGstab and GMRES solvers. Both solvers exhibit similar performances; BiCGstab showing to be superior to GMRES. The timings are comparable, being CSR the best overall followed by the ELL format. The worst approach was found to be the HYB

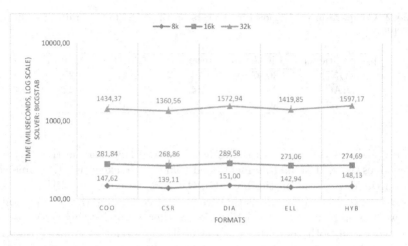

Fig. 1. Impact on the performance of sparse matrix compression formats (and associated data locality). BiCGstab used as linear solver.

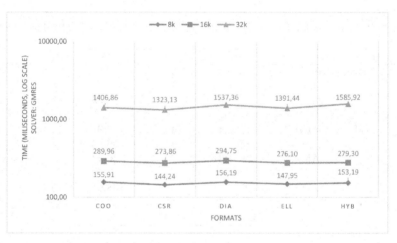

Fig. 2. Impact on the performance of sparse matrix compression formats (and associated data locality). GMRES used as linear solver.

format, usually one of the favorites for use on the GPU. From this study, and according to the matrix structure, we show that the CSR structure is adequate for exploring the GPU architecture in structured meshes with tight neighborhood relationships and nonuniform grid evaluations. This is noteworthy, since the CSR (or its version by columns) is the most frequently data structure used for CPUs.

5   Conclusions

Current GPGPU technology offers great computing power, at a very moderate cost, that can be explored to solve complex and time consuming scientific problems. To explore these devices, problems must be solved with numerical methods that can be cast into a large number of threads. This is crucial for the GPGPU to overcome the waiting time for memory calls.

Although the programmability of GPUs has been simplified by the introduction of CUDA (and OpenCL), more complex applications still require a reasonable effort to be solved on these devices. Linear algebra libraries, sparse in particular, are easing that effort and allowing significant performance increases but avoiding host-device communication is key for the highest performance. The difficulty is that it may not be possible to completely cast the problem into the device. This was the case for the larger problem tested, 64000. This suggests that hybrid algorithms, to exploit computations on both devices, is an alternative that needs to be continually investigated.

Memory bandwidth is the limiting factor for performance; problems that involve sparse matrices fit in this well-understood problem. However, excellent speedups were obtained for the type of problem tested. Speedups of about $50\times$ are noteworthy, even considering that the code was not tuned to the CPU architecture.

Our tests showed that the set of storage formats used led to comparable performances. The DIA and ELL formats are well-suited to matrices obtained from structured grids, whereas HYB is generally the fastest format for unstructured matrices. However, CSR was the most efficient for the radiative transfer problem, since the matrices involved were obtained with structured meshes but making use of nonuniform grid evaluation. Its efficiency also results from the balanced distribution of nonzeros per row is. This has favorable implications, as CSR is perhaps the most popular general-purpose sparse matrix format used for CPU computations. Overall, our results suggest that libraries are dealing well with memory coalescing, avoiding low bandwidth and inherent poor performance.

With respect to linear solvers, for the problem at hands, BiCGstab was slightly better than GMRES. BiCGSTAB can outperform GMRES and be the method of choice when solving a system of linear equations on GPUs.

In the present paper we show experiments that illustrate how GPUs can be used by non-experts in computer science, making use of specialized open source libraries tuned for these devices. The solution process for the problem, in this case a radiative transfer problem, is decomposed into building blocks. Tuning the computation of these blocks leads to an efficient computation of the overall problem and the effort to write optimized codes for the GPU is minimized by resorting to specialized numerical libraries. The paper also highlights the impressive gains achieved for sparse matrix data structures, even though GPUs will generally perform better on large dense matrices.

Acknowledgments. This research used resources of the National Energy Research Scientific Computing Center, which is supported by the Office of Science of the U.S. Department of Energy under Contract No. DE-AC02-05CH11231. This research received financial support from Fulbright visiting scholar program and logistic support from Lawrence Berkeley National Laboratory, Computational Research Division, USA.

References

1. Abramowitz, M., Stegun, I.A.: Handbook of Mathematical Functions. Dover, New York (1960)
2. Ahues, M., d'Almeida, F.D., Largillier, A., Titaud, O., Vasconcelos, P.: An L^1 refined projection approximate solution of the radiation transfer equation in stellar atmospheres. Journal of Computational and Applied Mathematics 140(1-2), 13–26 (2002)
3. Bell, N., Dalton, S., Olson, L.N.: Exposing fine-grained parallelism in algebraic multigrid methods. SIAM Journal on Scientific Computing 34(4), C123 – C152 (2012)
4. Bell, N., Garland, M.: Cusp: Generic parallel algorithms for sparse matrix and graph computations (2012), http://cusp-library.googlecode.com, version 0.3.0
5. Bell, N., Hoberock, J.: Thrust: A productivity-oriented library for CUDA. GPU Computing Gems 7 (2011)
6. Benzi, M.: Preconditioning techniques for large linear systems: a survey. Journal of Computational Physics 182(2), 418–477 (2002)
7. Busbridge, I.: The mathematics of radiative transfer, vol. 369. Cambridge University Press (1960)
8. Dongarra, J., Beckman, P., Moore, T., Aerts, P., Aloisio, G., Andre, J., Barkai, D., Berthou, J., Boku, T., Braunschweig, B., et al.: The international exascale software project roadmap. International Journal of High Performance Computing Applications 25(1), 3–60 (2011)
9. NVIDIA Corporation: CULA Sparse Reference Manual (2013), http://www.culatools.com/cula_sparse_programmers_guide/
10. NVIDIA Corporation: NVIDIA CUDA C Programming Guide (February 2014), http://docs.nvidia.com/cuda/
11. Roman, J.E., Vasconcelos, P.B.: Harnessing GPU power from high-level libraries: Eigenvalues of integral operators with SLEPc. Procedia Computer Science 18, 2591–2594 (2013)
12. Rupp, K., Rudolf, F., Weinbub, J.: ViennaCL - A High Level Linear Algebra Library for GPUs and Multi-Core CPUs. In: Intl. Workshop on GPUs and Scientific Applications, pp. 51–56 (2010)
13. Rutily, B.: Multiple scattering theory and integral equations. In: Constanda, C., Ahues, M., Largillier, A. (eds.) Integral Methods in Science and Engineering, pp. 211–232. Birkhäuser (2004)
14. Rutily, B., Chevallier, L.: The finite Laplace transform for solving a weakly singular integral equation occurring in transfer theory. Journal of Integral Equations and Applications 16(4), 389–409 (2004)
15. Saad, Y., van der Vorst, H.A.: Iterative solution of linear systems in the 20th century. Journal of Computational and Applied Mathematics 123, 1–33 (2000)
16. Stone, J.E., Gohara, D., Shi, G.: OpenCL: A parallel programming standard for heterogeneous computing systems. IEEE Design & Test 12(3), 66–73 (2010), http://dx.doi.org/10.1109/MCSE.2010.69

17. Tuminaro, R.S.: Parallel smoothed aggregation multigrid: aggregation strategies on massively parallel machines. In: Proceedings of the 2000 ACM/IEEE conference on Supercomputing (CDROM). p. 5. IEEE Computer Society (2000)
18. Vasconcelos, P.B., d'Almeida, F.: Performance evaluation of a parallel algorithm for a radiative transfer problem. In: Dongarra, J., Madsen, K., Waśniewski, J. (eds.) PARA 2004. LNCS, vol. 3732, pp. 864–871. Springer, Heidelberg (2006)
19. Vasconcelos, P.B., Marques, O., Roman, J.E.: Parallel eigensolvers for a discretized radiative transfer problem. In: Palma, J.M.L.M., Amestoy, P.R., Daydé, M., Mattoso, M., Lopes, J.C. (eds.) VECPAR 2008. LNCS, vol. 5336, pp. 336–348. Springer, Heidelberg (2008)

Optimizing Memory Usage and Accesses on CUDA-Based Recurrent Pattern Matching Image Compression

Patricio Domingues[1], João Silva[1], Tiago Ribeiro[1], Nuno M.M. Rodrigues[1,2],
Murilo B. De Carvalho[3], and Sérgio M.M. De Faria[1,2]

[1] School of Management and Technology,
Polytechnic Institute of Leiria,
Leiria, Portugal
[2] Instituto de Telecomunicações, Portugal
[3] TET/CTC, Universidade Federal Fluminense,
Niterói - RJ, Brazil

Abstract. This paper reports the adaptation of the Multidimensional Multiscale Parser (MMP) algorithm to CUDA. Specifically, we focus on memory optimization issues, such as the layout of data structures in memory, the type of GPU memory – shared, constant and global – and on achieving coalesced accesses. MMP is a demanding lossy compression algorithm for images. For example, MMP requires nearly 9000 seconds to encode the 512×512 Lenna image on a 2013's Intel Xeon. One of the main challenges to adapt MMP to manycore is related to the dependency over a pattern codebook which is built during the execution. This forces the input image to be processed sequentially. Nonetheless, CUDA-MMP achieves a 12× speedup over the sequential version when ran on an NVIDIA GTX 680. By further optimizing memory operations, the speedup is pushed to 17.1×.

Keywords: CUDA, memory optimization, manycore computing, image compression.

1 Introduction

Creation and usage of multimedia content is growing exponentially [1]. On one hand, the exploding growth of cheap commodity digital capture devices such as recorders, cameras, and more recently smartphones, tablets and other mobile gadgets enable not only professionals but also individuals to record and produce massive amounts of multimedia material. On the other hand, multimedia devices and standards are evolving toward delivering higher quality, namely higher definition. Indeed, some of the newest smartphones are able to capture video in so-called Full HD (1080p), with some high end models even delivering so-called 4K video [2]. These applications impose high demands on storage and bandwidth. For instance, streaming bandwidth for video content is 8 Mb/s for the standard television format (SD), 19.3 Mb/s for HD and 195 Mb/s for Ultra-HDTV (4K),

B. Murgante et al. (Eds.): ICCSA 2014, Part IV, LNCS 8582, pp. 560–575, 2014.

this last one considering the H.265 format [3]. All this contributes to a massive increase in the size of multimedia content, stressing not only storage but also networks, used to transfer content. Compression mechanisms targeting multimedia content are thus essential to reduce storage and bandwidth demands to a manageable level. Examples include lossless mechanisms such as the Portable Network Graphic (PNG) and JPEG XR formats for images, Windows Media Lossless (WMA Lossless) and Free Lossless Audio Codec (FLAC) for sounds, just to name a few. Lossy methodologies, which sacrifice part of the digital data to achieve higher compression ratios, have an important contribute to reduce the demands of multimedia content. Examples includes MP3 and Ogg Vorbis for sound, JPEG for images and MPEG-4 for video. Even so, current networks struggle to deliver HD content, let alone 4K content [3].

The Multidimensional Multiscale Parser (MMP) is a lossy pattern-matching-based algorithm for the compression of multimedia content, namely images [4,5]. It relies on a dynamically generated set of codebook patterns, which are used to approximate each block of the input image. It uses a multiscale approach, which enables the approximation of image blocks with different sizes.

MMP achieves high compression/quality ratio as measured through peak signal to noise ratio (PSNR), surpassing standards such as H.264 and JPEG2000 [6]. However, the compression/quality ratio achieved by MMP requires massive single precision floating point computation to deal with the pattern matching process. Indeed, MMP is computationally very complex, with the sequential CPU version requiring lengthy execution times when compressing images. For instance, the 512×512 Lenna 8-bit gray image requires more than two hours of wall clock time on a state of the art 2013's Intel Xeon machine.

MMP resorts to a dynamic codebook of blocks that are used to replace the original blocks of the image to compress. In fact, for each original block of the input image, many operations are performed in order to identify and if needed, to create, the replacement block(s) that yields the lowest distortion. Moreover, the best fitted blocks are added to the codebook after the encoding operation of each input block. The use of an adaptive codebook constitutes an important challenge to the parallelization of MMP and henceforth to successfully port the algorithm to parallel accelerators such as Graphical Processing Units (GPUs). In this work, we analyze the techniques used to achieve meaningful speedups over commodity GPUs resorting to the CUDA framework. We essentially focus on the memory optimization issues.

GPUs have revolutionized high performance computing, with affordable commodity hardware hosting thousands of execution cores, yielding massive speedups for certain types of applications. For instance, Lee et. al [7] report that a commodity GPU is, on average, 14 times faster than a state of the art 6-core CPU over a set of CPU- and GPU-optimized kernels. However, important adaptations need to be made to the CPU-based applications to achieve proper speedup on GPUs. The popularity of GPUs within the gamer community brings large scale volumes that not only help to reduce prices but also push performance forward by way of intense competition among manufacturers of GPUs, namely AMD and NVIDIA [8].

Paramount to the success of GPU within the high performance community, is the availability of programming platforms such as the CUDA framework [9] and the OpenCL standard [10], which fitted with the appropriate developing tools, bring simplicity and efficiency to programmers.

In this paper, we depict the adaptation of the MMP algorithm to CUDA, focusing on the main challenges and reporting the main performance results. We believe that the main contributions of this paper lie in i) in the way that a pattern matching-based algorithm, with a major sequential dependency over its codebook, can be successfully ported to CUDA manycore computing and ii) the memory optimization strategies that pay off in CUDA.

The remainder of this paper is organized as follows. Section 2 focuses on related work. Section 3 briefly presents CUDA. Section 4 describes the MMP algorithm. Section 5 presents the main memory adaptations that were performed to MMP-CUDA. In section 6, we assess the performance of the MMP-CUDA version. Finally, section 7 concludes the paper and outlines some venues for future work.

2   Related Work

We review related work. We first focus on i) image encoding algorithms adapted to GPUs and then on ii) studies regarding optimal data layout and placement for GPUs.

Victor Lee et al. analyze the performance of several applications comparing optimized versions of CPU vs. GPU, reporting a speedup of $14\times$, on average, for the GPU versions [7]. However, the execution of the JPEG2000 image encoding algorithm in GPU – NVIDIA GeForce 9800 and NVIDIA GTX 280 – is slower than the CPU versions ran over an INTEL Quad Core Q9450.

Ozsoy et al. present several optimizations, including memory-based ones, for the CUDA-based version of the Lempel-Ziv-Storer-Szymanski (LZSS) lossless data compression algorithm [11]. Regarding memory, the usage of texture memory brought, on average, a 16% speedup increase. With all optimization combined, the performance increase was 5.63 times relatively to a previous CUDA-based version. The optimized CUDA-LZSS is $34\times$ and $2.21\times$ faster than the serial and the parallel (non-CUDA) versions, respectively.

Sodsong et al. introduce a JPEG decoding scheme for systems comprising a multicore CPU and an OpenCL-programmable GPU optimized [12]. Under their approach, the non-parallel tasks of the JPEG decoder are kept in the CPU, while the parallel tasks are split between the CPU and the GPU. The GPU tasks are additionaly optimized for OpenCL memory hierarchies. This approach yields a speedup of $8.5\times$ and $4.2\times$ over the sequential decoder and the SIMD-version of libjpeg-turbo [12], respectively.

Sung et al. present a formulation and language extension that enables automatic data layout transformation for CUDA code [13]. This approach applies a transformation tool, providing a data layout optimized for the underlying CUDA memory hierarchy. The authors report a speedup comparable to hand-tuned code. Nevertheless, this approach is targeted solely for structured grid applications, where each output point is computed as a function of itself and its nearest neighbors [13]. Additionally, the code needs to be adapted to the tool, namely using FORTRAN-style subscripted array access. Likewise, Jaeger et al. also focus on optimizing layouts for stencil-oriented multithreaded programs over CPU and GPUs [14].

Sung, Lui and Hwu propose a runtime OpenCL DL library for optimal data layout placement on heterogeneous architectures [15]. DL implements Array of Structure of Tiled Arrays (ASTA). Additionally, DL supports the so called in-place data marshalling, in order to counteract the size bloat that frequently occurs when data are laid out for performance reasons. For certain problem types, DL yields speedups higher than so called discrete arrays layout – programmers break the structure by hand – but only when the overhead from the runtime marshalling is not taken into account.

Mei and Tian study the impact on CUDA performance of several data layouts, namely the Structure of Arrays (SoA) and the Array of Structures (AoS) approaches, for the Inverse Distance Weighting [16]. The former approach is based on a structure which holds all the needed data within its arrays, with the existence of one array per type of data. The AoS approach is the commonly used in sequential programming. It comprises a structure which holds the basic data, which is replicated in an array. They conclude that, due to the CUDA dependence over coalesced memory accesses, SoA-based layouts deliver the highest performance. Our approach confirms this dependency and reinforces the need to properly layout data on memory to maximize coalescing accesses and consequently performance.

3   CUDA

Modern GPUs from NVIDIA and AMD have thousands of cores, bringing affordable parallelism to the desktop. However hardware without proper APIs and tools to harness the massive parallelism is of limited use and thus of reduced attractiveness. NVIDIA's Computer Unified Device Architecture (CUDA) is a framework that offers a comprehensive software stack to facilitate the use and programmability of manycore GPUs. The CUDA software stack includes compilers, profilers, libraries and a vast set of examples and samples.

CUDA extends the C++ programming language adding a few keywords, identifiers, modifiers and functions. For instance, functions such as cudaMalloc and cudaMemcpy have names that identify its functionalities with the well known C malloc and memcpy functions. This allows C++ and C programmers to rapidly get acquainted with CUDA, once the parallel paradigm is mastered and contributes to the popularity of the framework. Indeed, although CUDA is proprietary, running solely in NVIDIAs GPUs, Stamatopoulos et al. report that, so far, CUDA is the only manycore technology which has achieved wide adoption and usage [17]. In this paper, we focus solely on the CUDA runtime API, which is simpler and hence attracts more users than the more verbose, but low level, CUDA driver level API [9].

At the hardware level, a CUDA GPU is comprised of multiple multiprocessors (MP), each one holding a given number of CUDA cores. The number of CUDA cores per multiprocessors is dependent on the so called *compute capability* of the device. For instance, CUDA devices with compute capability 3.0 have 192 CUDA cores per multiprocessor. For example, the NVIDIA GTX 680 GPU is comprised of 8 multiprocessors, holding a total of 1536 CUDA cores.

At the software level, a CUDA program is made up of code to run on the CPU and code to explicitly run on the GPU. The CPU code controls the main functions such as allocating memory on the GPU, controlling data transfers between the host main memory and the diverse types of memory of the GPU (global, constant, shared, etc.), and launching GPU code. The GPU code is organized in so called kernels and in device functions. A kernel represents an entry point into the GPU with its code being executed by the CUDA cores of the GPU. Specifically, the CUDA framework exposes the cores to the programmer through a so called execution grid which comprehends blocks organized in a layout that can have up to three dimensions. Furthermore, a block holds a given number of threads, with threads being the active executive entities. Like blocks, threads within a block are organized in a layout that can have up to three dimensions. The number of blocks at each dimension is specified by the programmer when calling a kernel. Within the same call, the programmer also specifies the dimensionality and number of threads per block. Within the GPU code, special CUDA identifiers such as *threadIdx* (available in the x, y and z dimensions) and *blockIdx* allow the programmer to localize and differentiate the code in order to distribute work among threads and blocks. Although programming in CUDA is rather trivial, extracting good performance is much more difficult, since many performance details are tied to the underlying architecture [18,19].

Examples of factors that decisively influence the performance of CUDA programs include adherence to the underlying memory model, namely coalesced accesses to memory [20,21], resorting to shared memory in order to take advantage of its lower memory latency [19] and using constant memory to offload read-only data in order to speedup the simultaneous access of warp of threads to these data. We study the effectiveness of these techniques in this paper.

4   The MMP Algorithm

The MMP algorithm is based on multiscale pattern matching with exhaustive search through an adaptive multiscale codebook, also referred to as dictionary. Specifically, MMP parses the input image into blocks, which are compared with every element of the codebook. Unlike conventional matching pattern algorithms, which use fixed-sized blocks, MMP segments the input block to search for a better representation of the signal as a concatenation of different sized patterns. The use of different block sizes, or scales, account for the important multiscale feature of the algorithm. The codebook is comprised by elements of all possible scales. The adaptive nature of the codebook means that while the algorithm is coding a signal, MMP uses information gathered in the parts already coded to create new codebook entries (code-vectors), thus yielding a dynamic codebook. These new code-vectors become available to approximate the blocks of the input signal not yet encoded.

Although the MMP algorithm is suited for input signals with any dimension, in this work we address standard grayscale images, *i.e.*, two dimensional arrays of bytes, which represent intensity values that range from 0 to 255. Furthermore, the elementary input signal (block) corresponds to a 16×16 pixel block, with the algorithm processing the input blocks of the image from left to right and top to bottom.

To find the best element of the codebook that should be used to represent the current block of the input signal, the MMP algorithm resorts to the Lagrangian cost J, given by the equation $J = D + \lambda.R$, where $\mathbf{D}$ measures the distortion and $\mathbf{R}$ represents the number of bits (rate) needed to encode the dictionary element. The numerical parameter λ is given by the user and remains constant throughout the encoding process. It allows to tune the balance between quality (small λ, e.g. $\lambda = 5$) and bitrate (large λ, e.g. $\lambda = 300$). Note that a small λ drives quality at the cost of computational complexity and a larger codebook. The used version of the MMP algorithm combine the adaptive pattern-matching with a predictive coding step. MMP uses an intra-frame prediction algorithm to determine prediction blocks, by using the values of the previously encoded neighboring pixels [5]. In this case, MMP does not deal directly with the input image. Instead, it operates on a residue block, which is the difference between each pixel of the prediction block and the corresponding pixel of the input block. The rationale to encode the residue instead of the input signal lies from the fact that the residue's values have a much lesser variability and hence can be more efficiently compressed [5]. MMP currently employs 10 prediction modes. This means, that for each input block, MMP has to analyze all the candidate residue blocks that arise from the 10 prediction modes.

A major part of the computational complexity of MMP is associated with determining the distortion between each residue block and every available code-vector. The distortion, $\mathbf{D}$, between an input block, $\mathbf{X}$ and a given code-vector $\mathbf{C}$ is given by the mean-squared error (MSE), defined by:

$$D = \sum_{i=1}^{M} \sum_{j=1}^{N} (X(i,j) - C(i,j))^2, \tag{1}$$

where $M \times N$ represents the size of $\mathbf{X}$ and $\mathbf{C}$.

For each residue block, the MMP algorithm computes the distortion for every code-vector available in the codebook. This procedure allows MMP to determine the code-vector that will be used to represent the input block. As was previously mentioned, MMP does not stop its analysis at the current block scale. It also computes the distortion for all the subscales of the block. For this purpose, the block is first segmented in half, either vertically into two blocks of $M/2 \times N$ pixels (vertical segmentation) or into two blocks of $M \times N/2$ pixels (horizontal segmentation). This process goes on recursively until scale 0 (1×1 blocks) is reached. A 16×16 pixels block can yield 961 segments: two 8×16, two 16×8, four 8×8 and so on, down to the 256 single-pixel blocks (scale 0). Thus a tree of blocks and subblocks is built, with the distortion being computed for all the blocks that correspond to the tree nodes. This requires a considerable amount of computation time to fully encode an input image. As we shall see later on, the MMP encoding of the 512×512 well known Lenna image requires slightly less than 9000 seconds on an 2013 high end machine, fitted with an Intel Xeon CPU.

In a first stage, MMP expands the entire tree, determining the cost for every possible node (sub-block). After this, a recursive pruning strategy compares the sum of the distortion of the children blocks against the distortion of their parent. If the distortion sum of the two children blocks is lower than the distortion of their parent block, the children blocks are kept, otherwise they are pruned. This results in a pruned tree, which represents the optimum segmentation of the original input block. After encoding the elements of the optimum tree, new blocks, generated by applying scale transformation over the concatenation of every pair of child nodes, are added to the codebook. The whole process – building the tree and its subsequent prune operation – is repeated for every block of the input image.

The codebook updating procedure, which takes place at the end of the encoding operation of each block of the input image, creates a sequential dependency. This impedes the traditional approach for parallelization of image-related algorithms, where several input blocks are processed in parallel. Thus parallelism needs to be exploited within the operations that are performed to encode an individual input block. For an individual block, parallelization opportunities arise from the usage of multiple prediction modes – each prediction mode forms a task – and for the computationally costly operations of computing the distortion, as well as finding the best suited block and updating the codebook.

5   Porting MMP to CUDA

The port of the MMP algorithm was preceded with the profiling of the sequential version to identify the performance hot spots, whose adaptation to CUDA could provide performance gain. The profiling pointed out that the most time consuming operations of the sequential version are i) the computation of the distortion of the current input block with all the blocks of the codebook and ii) the update of the adaptive codebook which is performed at the end of the encoding operation of every input block.

5.1   Naïve Approach

Two CUDA kernels were created – *kernelDistortion* and *kernelReduction* – in order to offload the computational hotspots to the GPU. The kernel *kernelDistortion* computes not only the distortion set of a given input block against all existing blocks in the codebook, but also the Lagrangian, while *kernelReduction* locates the block that has the lowest Lagrangian, effectively performing a reduction operation, hence its name [18]. The kernels are chained together, with the second one called right after the first one has computed the whole set of distortions. Note that the final part of the reduction is performed by the CPU as it is usually the case in GPU-based reduction [22]. This chained pair of calls – *KernelDistortion* and then *kernelReduction* – occur many times for every block of the input image due to the multiple paths that are exploited by MMP by way of prediction modes and multiscale.

Regarding data, the set of blocks that forms the adaptive MMP codebook is kept in the global memory, along with the distortion set, allowing the kernel *kernelReduction* to use this data, thus avoiding costly copies CPU-GPU. Nonetheless, in the naïve approach, the whole codebook is laid out in global memory in a similar way that the one adopted in the CPU version. This yields non-efficient accesses, especially non-coalesced accesses.

The poor performance results of the naïve approach confirms that although porting a program to CUDA is rather simple, achieving good performance requires great efforts to accommodate the code to the CUDA architecture [9]. Indeed, the MMP-CUDA naïve version performed as much as 10 times slower than the CPU sequential version.

5.2   Optimizations

Non-memory Related. The performance of the naïve approach was analyzed with the NVVP profiling tool to pinpoint the performance bottlenecks and to correct them. Then, we applied several non-memory related optimizations. For instance, the reduction kernel was optimized following the practice led out by Mark Harris [23]. Additionally, both the *kernelDistortion* and *kernelReduction* were revised with the goal of reducing the usage of registers in order to increase the occupancy of the GPU.

Specifically, the updated *kernelDistortion* has each CUDA thread computing the distortion between one partition of the residue block and one element of the codebook of the same level/scale. This way, a single call of the kernel computes all the distortions values for every partition of the given block sweeping across all the elements of the codebook. For instance, when dealing with a block of scale 24 (block with 16×16 pixels), the *kernelDistortion* computes the distortion between the block and all the block of the codebook, repeating the process for every partition of scale 24 down to scale 0 (individual pixels). The final output of *kernelDistortion* is an array holding the distortion of the input block and its partition for every element of the codebook. Regarding the CUDA execution grid, the updated *kernelDistortion* has no limitations or restrictions, and can be executed under any grid.

The kernel *kernelReduction*, which aims to select the best codebook's entry for each possible partition of the block being coded, was also heavily changed. It now returns, for each partition, the block that yields the lowest Lagrangian cost. For instance, for a coding block scale 24 (16×16), the kernel returns an array with 705 elements, that is, one per partition excluding the 256-single pixel partition. Indeed, a 16×16 block has 961 possible partitions, 256 of them corresponds to a single-pixel partition. To achieve its purpose, the kernel acts in two main steps.

First, each thread loads the previously computed distortion and computes the Lagrangian cost, looping over a set of values (e.g., thread 0 handles the values from index 0, index 128, etc.). Having finished its loop, each thread saves the lowest Lagrangian cost into shared memory, yielding an array which holds the lowest costs. The second step actually reduces the array built in the previous step, through a parallel GPU reduction [22]. Note that each block of CUDA threads deals with a given partition, hence there are as many blocks of CUDA threads as there are partitions.

A change that affected both kernels was the Lagrangian computation which was moved from the *kernelDistortion* to the *kernelReduction*. The change was performed to save variables and hence registers, thus allowing for a larger execution grid. The revision also allowed us to diminish the number of calls for each kernel, thus reducing overhead. As we shall see later on, these non-memory related optimizations yielded a meaningful $11.96\times$ speedup relatively to MMP-CPU.

We then focused on memory-related optimizations to pursue further performance improvement. Indeed, an important area of CUDA performance is the proper usage of the hierarchy of memory of CUDA [22,21]. CUDA presents several levels of memory: the relatively slow global memory, the fast but block-limited and size-limited shared memory and the specific-purpose constant memory [20]. Next, we report the efforts and results achieved through the optimization of memory related aspects of MMP-CUDA.

Using the Constant Memory. CUDA constant memory is a read-only memory from the point of view of the GPU, with values being loaded through CPU code. It has a limited size of 64KB. Although the constant memory is not the fastest CUDA memory – it comes third after the register file (fastest) and the shared memory – it is cached [21]. More importantly, the constant memory is specially adapted for broadcasting. In fact, it excels when all the threads of the warp simultaneously access the same value held in constant memory, yielding performance comparable to accessing the register file [21]. This broadcast scenario is exploited by MMP through the *kernelDistortion* and it is the main reason for the layout adopted to represent the codebook in GPU memory as described above. Indeed, as explained earlier, all the 32 threads of a warp running the *kernelDistortion* compute the distortion of the same pixel of the input block. Therefore, by storing the residue of the input block on the constant memory, the value of the same pixel is broadcast to the warp that also benefits from coalesced accesses to the blocks of the codebook. Since the value of a pixel is represented by an integer, the 64 KB of the constant memory are more than enough to hold the needed values. Likewise, some control parameters were moved to the constant memory to benefit from the broadcast effect.

Non-pageable Memory. Usage of non-pageable memory is a simple optimization technique since it only involves minor changes, namely switching some API calls [21]. The non-pageable memory is a set of memory pages from the host machine that cannot be swapped out by the operating system. In CUDA, usage of non-pageable memory speeds up the memory transfer between the CPU and the GPU, allowing the CUDA memory engine to copy data through Direct Memory Address (DMA) between the CPU and the GPU without the need to resort to extra internal copies as it is necessary when dealing with pageable memory. One of the problems of using non-pageable memory is the strain that it can cause on the memory management system, since it reduces the available quantity of pageable memory. However, since the codebook for each scale is limited to 50000 elements, the maximum size of the whole codebook is 183 MB, an amount that should cause no problem, since 4 GB or more of RAM are common nowadays.

Coalescing Memory Accesses to the Codebook. An important issue regarding performance on CUDA is properly accessing the global memory in order to reduce the number of transactions with the memory controller. The recommendation is to access memory in a so-called coalesced form, with neighbor threads of a warp accessing neighbor elements. For instance, thread 0 of a given CUDA block of threads accesses index 0 of the array, thread 1 accesses index 1 of the array and so forth. This way the access of the 32 threads that comprises a warp of threads only incurs the cost of a single memory transaction. This assumes that the CUDA array starts on properly aligned memory, which is the case whenever the CUDA memory allocation functions are used. On the other hand, non-coalesced memory accesses results in multiple memory transactions that might significantly slow down computation [21]. Note also that memory accesses and the effects of non-properly coalesced accesses depend on the CUDA capability of the GPU. For instance, for devices with compute capability 2.x, and as stated in [21], the concurrent accesses of the threads of a warp will coalesce into a number of transactions equal to the number of cache lines necessary to service all the of the threads of the warp, with a cache line holding 128 bytes.

The MMP-CPU version organizes the codebook in memory as a two-dimensional array, as follows. The first index represents the scale, with its values ranging from 0 to 24 which corresponds to blocks 1×1 (scale 0) to 16×16 (scale 24). The second index holds the offset of the pixel where the entry starts. For instance, for the scale 24, each block is represented by 16×16 integers, meaning that the first block starts at offset 0, the second at offset 256 and so forth.

The naïve CUDA implementation followed the MMP-CPU memory layout organization for the codebook. However, as easily seen, this yields non-coalesced memory accesses, since within the kernel, consecutive threads work on consecutive entries of the codebook. For instance, thread 0 accesses the first codebook entry for the scale it is dealing with, thread 1 performs the same for the second codebook entry and so forth. This way, serving the memory request for a warp requires multiple memory transactions, since neighbor accesses are 256 bytes apart for scale 24. To reduce the number of memory transactions, the memory layout used to keep the GPU's copy of the codebook was deeply changed. First, a single large block of GPU memory is used to hold the entire codebook. The size of the GPU's memory block corresponds to the maximum size that the codebook can attain. This maximum size is easily computed, since the maximum number of elements per scale that a codebook can hold is a parameter of the MMP algorithm, fixed at startup. In our experimental setup, the codebook was dimensioned for a maximum of 50000 elements per scale, corresponding roughly to 183 MB as reported earlier. Therefore the codebook fits easily in CUDA devices, since CUDA only run on devices that have at least 512 MB of memory. Moreover, modern GPUs all have 1 GB or more of memory.

The major changes with the codebook are related to its internal organization. Specifically, the codebook is laid out as follows: the first value corresponds to the value of the first pixel of codebook's block 0, the second value corresponds to the value of the first pixel of codebook's block 1, and so forth up to the 32th value which corresponds to the value of the first pixel of block 31. This means that the first 32 values correspond to the first pixel of the first 32 blocks of the codebooks. The second set of 32 values (from the 33th to the 64th value) holds the second pixel for the first 32 blocks. This goes on until all the pixels of all the first 32 blocks of the codebook are laid out. Then, the process is repeated for the second set of blocks (from 33 to 64). To insure that the codebook is properly aligned in memory, padding is used when needed to read 128-byte boundaries. It is important to note that the codebook comprises the blocks for all the 25 supported scales, ranging from scale 0 to scale 24. For each of them, the above layout is adopted.

The rationale for contiguously aligning 32 values per set stems from the fact that it allows all the 32 threads of a warp to access, within a single memory transaction, the value of the same order pixel. This way, thread 0 accesses the ith order pixel for block 0, thread 1 accesses the ith order pixel for block 1 and so forth. Under this memory layout, the codebook can be accessed efficiently not only when the distortion is being computed by kernel *kernelDistortion*, but also when the codebook is searched to check whether blocks to be inserted already exist in the codebook, as we shall see in the next section.

Reengineering the Codebook Update. At the end of the encoding operation of an input block, the codebook requires an update if new encoding block(s) has been used. However, first MMP needs to determine if these blocks are effectively new. This means that a search is needed over the codebook. For this purpose, a third CUDA kernel – *kernelSearchCodebook* – was devised for the memory optimized version of MMP. In reality, MMP does not use an exact comparison criterion to judge whether two blocks are equivalent or not. Instead, it resorts to the MSE metric computed between the two blocks: if the distance is less than a fixed threshold, then the blocks are considered similar, meaning that the codebook is not updated with the block. Therefore, the lookup of the codebook for update purposes requires computing the MSE between the new candidate to be inserted block and all the blocks of the codebook that belong to the same scale. This computation can be performed efficiently in parallel, precisely through the *kernelSearchCodebook*.

6   Performance Evaluation

6.1   Experimental Environment

To assess the performance of the various versions of MMP-CUDA, we resorted to a desktop machine fitted with two Intel Dual-Xeon E5-2630 CPUs, 64 GB of DDR3 RAM memory and a NVIDIA GTX 680 connected through the PCI Express 3.0. The GTX 680 is powered by the GK104 Kepler architecture chip, delivering compute capability 3.0. It has 2 GB of DDR5 memory, 1536 CUDA cores evenly distributed over 8 CUDA multiprocessors (192 CUDA cores per multiprocessor). The bandwidth as measured by the bandwidth benchmark provided in NVIDIA SDK is 11.081 GB/s between host to device, 10.957 GB/s between device to host and 145.865 GB/s between the device itself. We note that the last value regarding the device-device bandwidth is considerably less than the theoretical 192 GB/s. Additionally, although the GTX 680 has a double precision floating point performance which is only 1/24 of its single precision floating-point performance, this does not affect our tests, since MMP relies solely on single floating-point. The machine uses a 64-bit version of the Linux operating system with kernel 3.2.0. Finally, the GPU software stack comprises CUDA version 5.5 and version 304.88 of the Linux NVIDIA GPU driver.

6.2   Main Performance Results

The main results for the various CUDA versions of MMP are shown in table 1. All tests were run five times. The λ parameter was set to 10 which corresponds to a quality-oriented setup which is computationally demanding. The input was the well-known 512 × 512 pixel Lenna image with 8-bit gray intensity.

The kernels were run with the CUDA execution grids that experimentally yielded the best execution times. Specifically, the kernels *kernelDistortion* and *kernelSearchCodebook* were both set with a one-dimension execution grid, with 64 CUDA blocks, each block holding 256 threads. For the kernel *kernelReduction* a bi-dimensional grid of blocks was selected. Specifically, the number of CUDA blocks for this kernel corresponds on its x-dimension to the number of available partitions and on the y-dimension to the number of available prediction modes. Again, each CUDA block was set with 256 threads organized in a one-dimension layout.

The column *Execution time* displays the wall clock time in seconds, which corresponds to the average value of the five executions. The column *stdev* reports the standard deviation of the runs. The column *Speedup vs. CPU* holds the speedup of the current version relatively to the CPU version of MMP. Finally, the column *Speedup vs. reference* shows the speedup of the current version relatively to the *MMP-CUDA reference* version, that is, the MMP-CUDA version that holds all non-memory related optimizations.

It is important to note that the current CPU version is single-threaded, and thus does not take advantage of manycore CPUs. We aim to address this issue in the future. The low performing results of the naïve version confirm that attaining

Table 1. Main performance results

Version	Execution time (seconds)	stdev	Speedup vs. CPU	Speedup vs. reference
MMP-CPU	8990.584	14.797	1	–
MMP-CUDA Naive	23250.678	36.660	0.387	–
MMP-CUDA reference	752.012	11.107	11.955	1
ConstMem	732.440	5.590	12.275	1.027
ConstMemNP	716.570	1.178	12.547	1.049
ConstMemNPCoal	665.438	2.448	13.511	1.13
ConstMemNPCoalSearch	529.142	0.573	16.991	1.421
ConstMemNPCoalSearchConst	522.974	1.921	17.191	1.438
ConstMemNPCoalSearchShared	527.578	1.156	17.041	1.425

proper performance with CUDA is seldom feasible by just porting the sequential code to CUDA. Indeed, the first set of optimizations and reengineering yielded a nearly 12× performance improvement relatively to MMP-CPU as reported by the *MMP-CUDA reference* version. The problems with the naïve version were manifold. Firstly, the kernels were consuming many registers, thus reducing the available parallelism. Secondly, the code was performing a very large number of kernel calls and associated CPU/GPU memory copies, both situations contributing for a significant overhead. Finally, the *kernelReduction* was poorly balanced, with one thread assigned per partition of MMP.

We now focus on the results related with memory optimization. The *ConstMem* version makes uses of the constant memory features for the distortion computation, achieving a mere 2.7% improvement. A similar improvement is achieved when non-pageable memory is used on top of the constant memory optimization (version *ConstMemNP*). Interestingly, the use of non-pageable memory seems to reduce the standard deviation of the tests. We aim to assess this hypothesis in future work.

The coalesced version – *ConstMemNPCoal* also contributes with a small improvement to performance, around 8%. All together, the combined usage of constant memory, non-pageable and coalesced accesses deliver a 13% performance improvement.

A meaningful improvement was achieved through the creation of the kernel *kernelSearchCodebook*. As stated earlier, this kernel filters out candidate blocks that already exist in the codebook, thus preventing duplicate insertions and eliminating costly operations from the CPU. The performance benefits are shown by the *ConstMemNPCoalSearch* version that attains a 42.1% improvement relatively to *MMP-CUDA reference* and nearly 17× for the *MMP-CPU* version. Finally, the two last reported versions – *ConstMemNPCoalSearchConst* and *ConstMemNPCoalSearchShared* – are variant of the *ConstMemNPCoalSearch* version. In these versions, the kernel *kernelSearchCodebook* uses, respectively, constant and shared memory to hold the candidate blocks that are to be looked up over the codebook to assess whether they already exist or not. However, the performance improvements for both versions are marginal.

7 Conclusion and Future Work

We study the performance improvements achieved by resorting to memory-related optimizations of a CUDA-based version of the MMP algorithm. Although a final 17× speedup was attained relatively to the CPU sequential version, the contribution of memory-based optimizations only amounts for a relatively small part of the whole performance improvement. In fact, the main performance improvement derives from adapting costly operations of the CPU version to the GPU through the creation of appropriately crafted kernels. Our work also exemplifies that merely porting a relatively complex algorithm to CUDA is not enough. Indeed, deep adaptation needs to be made, otherwise merely porting the code to CUDA most probably generates negative speedups as happened with our *MMP-CUDA Naïve* version.

Taking into consideration the sequential restrictions of the MMP algorithm, namely the causality that exists between the encoding of successive input blocks – the encoding operation of the nth input block can only start after the completion of the encoding of the $n - 1^{th}$ input block – achieving a 17× performance improvement is an important testimony of the performance that can be attained with GPUs and CUDA.

As future work, we plan to assess the optimized version of MMP-CUDA in newer GPU hardware. Additionally, we plan to study its adaptation to OpenCL GPU-based platforms and to analyze the suitability of the sequential code to be adapted to multicore CPUs. Finally, we aim to further revise the MMP algorithm to identify optimizations at the algorithm level that can boost performance both for the CPU and the GPU versions.

Acknowledgments. Financial support provided by FCT (Fundação para a Ciência e Tecnologia, Portugal), under the grant PTDC/EIAEIA/122774/2010.

References

1. Seltzer, M.L., Zhang, L.: The data deluge: Challenges and opportunities of unlimited data in statistical signal processing. In: IEEE International Conference on Acoustics, Speech and Signal Processing, ICASSP 2009, pp. 3701–3704. IEEE (2009)
2. Murakami, T.: The development and standardization of ultra high definition video technology. In: Mrak, M., Grgic, M., Kunt, M. (eds.) High-Quality Visual Experience. Signals and Communication Technology, pp. 81–135. Springer, Heidelberg (2010)
3. Coughlin, T.: Evolving Storage Technology in Consumer Electronic Products (The Art of Storage). IEEE Consumer Electronics Magazine 2(2), 59–63 (2013)

4. De Carvalho, M.B., Da Silva, E.A., Finamore, W.A.: Multidimensional signal compression using multiscale recurrent patterns. Signal Processing 82(11), 1559–1580 (2002)
5. Rodrigues, N.M., da Silva, E.A., de Carvalho, M.B., de Faria, S.M., da Silva, V.M.M.: On dictionary adaptation for recurrent pattern image coding. IEEE Transactions on Image Processing 17(9), 1640–1653 (2008)
6. De Simone, F., Ouaret, M., Dufaux, F., Tescher, A.G., Ebrahimi, T.: A comparative study of JPEG2000, AVC/H.264, and HD photo, vol. 6696, pp. 669602–669602-12 (2007)
7. Lee, V.W., Kim, C., Chhugani, J., Deisher, M., Kim, D., Nguyen, A.D., Satish, N., Smelyanskiy, M., Chennupaty, S., Hammarlund, P., Singhal, R., Dubey, P.: Debunking the 100X GPU vs. CPU myth: an evaluation of throughput computing on CPU and GPU. SIGARCH Comput. Archit. News 38, 451–460 (2010)
8. de Verdiére, G.C.: Introduction to GPGPU, a hardware and software background. Comptes Rendus Mécanique 339(23), 78–89 (2011); High Performance Computing Le Calcul Intensif
9. Farber, R.: CUDA Application Design and Development. Morgan Kaufmann (2011)
10. Stone, J.E., Gohara, D., Shi, G.: Opencl: A parallel programming standard for heterogeneous computing systems. Computing in Science and Engineering 12(3), 66–73 (2010)
11. Ozsoy, A., Swany, M., Chauhan, A.: Optimizing LZSS compression on GPGPUs. Future Generation Computer Systems 30, 170–178 (2014); Special Issue on Extreme Scale Parallel Architectures and Systems. In: Cryptography in Cloud Computing and Recent Advances in Parallel and Distributed Systems, ICPADS 2012 Selected Papers
12. Sodsong, W., Hong, J., Chung, S., Lim, Y., Kim, S.D., Burgstaller, B.: Dynamic partitioning-based jpeg decompression on heterogeneous multicore architectures. In: Proceedings of Programming Models and Applications on Multicores and Manycores, PMAM 2014, pp. 80:80–80:91. ACM, New York (2007)
13. Sung, I.J., Stratton, J.A., Hwu, W.M.W.: Data layout transformation exploiting memory-level parallelism in structured grid many-core applications. In: Proceedings of the 19th International Conference on Parallel Architectures and Compilation Techniques, PACT 2010, pp. 513–522. ACM, New York (2010)
14. Jaeger, J., Barthou, D.: et al.: Automatic efficient data layout for multithreaded stencil codes on CPUs and GPUs. In: IEEE Proceedings of High Performance Computing Conference, pp. 1–10 (2012)
15. Sung, I., Liu, G., Hwu, W.: DL: A data layout transformation system for heterogeneous computing. In: Innovative Parallel Computing (InPar), pp. 1–11. IEEE (2012)
16. Mei, G., Tian, H.: Performance Impact of Data Layout on the GPU-accelerated IDW Interpolation. ArXiv e-prints (February 2014)
17. Stamatopoulos, C., Chuang, T.Y., Fraser, C.S., Lu, Y.Y.: Fully automated image orientation in the absence of targets. ISPRS - International Archives of the Photogrammetry, Remote Sensing and Spatial Information Sciences XXXIX-B5, 303–308 (2012)
18. Nickolls, J., Buck, I., Garland, M., Skadron, K.: Scalable Parallel Programming with CUDA. Queue 6, 40–53 (2008)
19. Nvidia, C.: NVIDIA CUDA Programming Guide (version 5.5). NVIDIA Corporation (2013)
20. Wilt, N.: The CUDA Handbook: A Comprehensive Guide to GPU Programming. Pearson Education (2013)

21. Nvidia, C.: NVIDIA CUDA C Best Practices Guide - CUDA Toolkit v5.5. NVIDIA Corporation (2013)
22. Kirk, D.B., Wen-mei, W.H.: Programming Massively Parallel Processors: a Hands-on Approach, 2nd edn. Newnes (2012)
23. Harris, M., et al.: Optimizing parallel reduction in CUDA. NVIDIA Developer Technology 2, 45 (2007)

Removing Inefficiencies from Scientific Code: The Study of the Higgs Boson Couplings to Top Quarks

André Pereira[1,2], António Onofre[1,2], and Alberto Proença[1]

[1] Universidade do Minho, Portugal
[2] LIP-Minho, Portugal
{ampereira,aproenca}@di.uminho.pt, onofre@fisica.uminho.pt

Abstract. This paper presents a set of methods and techniques to remove inefficiencies in a data analysis application used in searches by the ATLAS Experiment at the Large Hadron Collider. Profiling scientific code helped to pinpoint design and runtime inefficiencies, the former due to coding and data structure design. The data analysis code used by groups doing searches in the ATLAS Experiment contributed to clearly identify some of these inefficiencies and to give suggestions on how to prevent and overcome those common situations in scientific code to improve the efficient use of available computational resources in a parallel homogeneous platform.

Keywords: Scientific Computing, High Performance Computing, Code Efficiency, ATLAS Experiment.

1 Introduction

At the European Organization for Nuclear Research (CERN), the fundamental structure of the universe is studied using the most complex scientific instruments built by physicists and engineers up to now. CERN was founded in 1954 by 12 members states but has grown to the size of a world lab with 21 member states and more then 30 states to which were given the status of "observer" states. The instrumentation used in nuclear and particle physics research is essentially formed by particle accelerators and detectors. The Large Hadron Collider (LHC) speeds up groups of particles close to the speed of light, in opposite directions, inducing a controlled collision of protons at the detectors core. The detectors record various characteristics of the resultant particles of each collision (an event), such as energy and momentum, which originate from complex decay chains of particles produced in the interaction of the partons inside the colliding protons. The purpose of these experiments is to test models and predictions in High Energy Physics (HEP), such as the Standard Model, by confirming or discovering new particles and interactions.

The ATLAS Experiment [1] is one of the seven particle detectors at the LHC. ATLAS goals are to study the properties of the recently discovered Higgs boson [2], the search for new particles predicted by models of physics beyond the

B. Murgante et al. (Eds.): ICCSA 2014, Part IV, LNCS 8582, pp. 576–591, 2014.

Standard Model like Susy, searches for new heavy gauge bosons and precision measurements where the top quark is of utmost importance. Approximately 600 million collisions occur every second at the LHC. Particles produced in head-on proton collisions interact with the detectors, generating massive amounts of raw data. It is estimated that all the combined detectors produce 25 petabytes of data per year, and it is expected to grow after the ongoing LHC upgrade [3]. This data then passes a set of processing and reconstruction stages until it is ready to be used by specific analysis codes developed to search for interesting events predicted by several HEP models that may be present in data. Several research groups work in event reconstruction in the same experiment, enforcing positive competition to produce quality results in a fast and consistent way.

These factors enforce the need to process more data, more accurately, in less time, which often leads to investments on larger computing clusters to improve the quality of the research results. However, most scientific code was not designed and/or developed for an efficient use of the available computational resources. If these applications were adequately designed (or tuned), the event analysis throughput could be massively increased. An efficient parallel application can significantly improve its performance at a much lower cost [4].

This paper addresses inefficiencies in two stages of the data analysis application: the code development and application runtime. In the former, inefficiencies in the algorithm coding and data structuring are pinpointed and several solutions are suggested, based on a quantitative analysis of the bottlenecks. The latter identifies inefficiencies in threads accessing remote shared memory, and gives hints to overcome these limitations.

This paper is organized as follows: section 2 briefly presents the top quark and Higgs boson decay process and introduces a short characterization of the data analysis application used as case study; in section 3 the code inefficiencies are identified, analysed, and removed, with a final shared memory parallelization proposal; in section 4, runtime inefficiencies of the parallelization are identified and possible alternatives suggested, concluding with an assessment of the core affinity impact; finally, section 5 concludes the paper with suggestions for future work.

2   Top Quark and Higgs Boson Decay

At the LHC, two proton beams are accelerated close to the speed of light in opposite directions, set to collide inside a specific particle detector. This head-on collision triggers a chain reaction of decaying particles, and most of the final particles interact with the detector, allowing to record relevant data. One of the searches being conducted at the ATLAS Experiment relates to the study of the top quark and Higgs boson couplings. Figure 1 represents the final state topology of the associated production of two top quarks and one Higgs boson (that decays to two b-quarks), labelled from now on as $t\bar{t}H$ production.

The ATLAS detector can record the characteristics of the bottom quarks, detected as a jet of particles, and leptons (muon and electron). Neutrinos do not

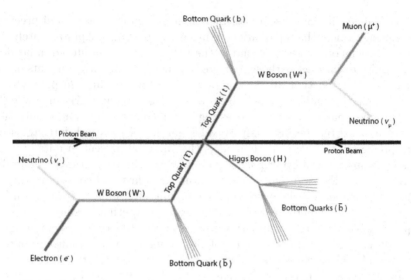

Fig. 1. Schematic representation of the $t\bar{t}$ system and Higgs boson decay

interact with the detector, so, their characteristics are not recorded. Since the top quark reconstruction requires the neutrinos, their characteristics are analytically determined with the known information of the system, through a kinematical reconstruction. However, the $t\bar{t}$ system may not have a possible reconstruction: the reconstruction has an intrinsic uncertainty associated which determines its accuracy.

The amount of jets from bottom quarks and leptons present in the events may vary according to the decay channel of the W bosons produced in the top quark decays. As shown in figure 1, four jets and two leptons are required to be present in the events. Two of the jets, together with two leptons are required to reconstruct the $t\bar{t}$ system, and the remaining two jets are used for the Higgs boson reconstruction. For the kinematical reconstruction, every possible combination of jets and leptons must be evaluated and only the most accurate reconstruction of each event is considered. In a first step, the $t\bar{t}$ system reconstruction is tried. If it has a possible solution, the Higgs boson is reconstructed from the jets of the two remaining bottom quarks. The Higgs reconstruction does not use the jets which were associated to the best $t\bar{t}$ system reconstruction. The overall quality of the event processing depends on the quality of both reconstructions.

For the global event reconstruction, several solutions can be tested if we assume that the ATLAS detector has an experimental energy-momentum resolution of $\pm 1\%$, by varying these quantities within their uncertainty. This uncertainty is propagated into the $t\bar{t}$ system and Higgs reconstructions, affecting their accuracy. To improve the quality of the reconstructions several random variations are applied to the measured values, within a maximum range of $|1\%|$ next to the measured values. The quality of the reconstructions and the application execution time is directly proportional to the amount of variations performed per

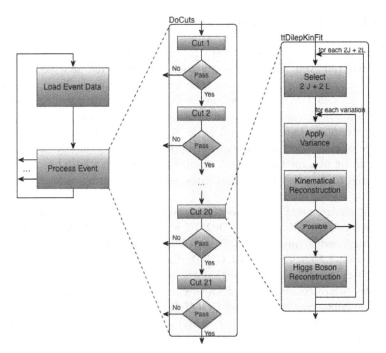

Fig. 2. Schematic representation for the ttH_dilep application flow

combination. The goal is to do as many variations as possible within a reasonable time frame.

To reconstruct the $t\bar{t}H$ system a data analysis application was developed, the ttH_dilep. The application flow is presented in figure 2. Each event data on an input file is individually loaded into a single global state, shared between the data analysis code and the LipMiniAnalysis toolbox[1], and it is overwritten every time a new event is loaded. The event is then submitted to a series of cuts, which filters events that are not suited for reconstruction. When an event reaches the cut 20, the $t\bar{t}$ system and Higgs boson are reconstructed in the function ttDilepKinFit, which is expected to be the most computing demanding. If the $t\bar{t}$ system reconstruction fails, the current combination is discarded and the next is processed. If an event has a possible reconstruction it passes the final cut and its final information is stored.

The application also depends on the ROOT framework[2] for part of the functionalities used in the reconstructions and for result output and visualisation. The code from both ROOT and the LipMiniAnalysis toolbox cannot be modified as many data analysis applications depend on them.

---

[1] The LipMiniAnalysis toolbox provides a skeleton to several data analysis applications under study in the Portuguese LIP institution, a CERN partner in the ATLAS Experiment.

[2] ROOT [5] is a C++ framework produced by CERN to help the development of particle data analysis code, by implementing specific features.

3   Coding Inefficiencies

Inefficiency removal is a two stage iterative process, where bottlenecks are identified and later removed. First, the application is profiled and analysed to identify the critical sections of the code that take longer to compute. Then, the critical section is optimized by modifying the code, algorithm, or parallelization. The identification of critical sections can be automated by using third party tools, such as gprof [6], Callgrind [7], or VTune [8], which produce reports listing the percentage of time spent in each of the application functions. A more detailed analysis can be obtained using tools similar to PAPI [9], where hardware counters are used to quantify cache miss rates, executed floating point instructions, and other low level information.

The test environment used in both this section and section 4 is a dual-socket system with two Intel Xeon E5-2670v2 [10] with 10 cores, with hardware support for 20 simultaneous threads, at 2.5 GHz each, 256 KB L2 cache per core and 25 MB shared L3 cache, with 64 GB DDR3 RAM. The K-Best measurement heuristic[3] was adopted to ensure that the only the best, but consistent, time measurements are considered. Software wise, the GNU Compiler version 4.8.2 with *O3* optimizations enabled and ROOT 5.34/17 were used. A 5% interval was used for a k of 4, with a minimum of 12 and maximum of 24 time measurements.

Profiling the data analysis code using Callgrind, the `ttDilepKinFit` was identified as the most time consuming function, taking 99% of the execution time for 1024 variations. `ttH_dilep` execution with this amount of variations was considered reasonable for all efficiency measurements unless stated otherwise, without compromising the application execution time.

A preliminary computational analysis concluded that the application is compute bound on the testbed system, where accesses to the system RAM memory are not a limiting factor with a ratio of 7 instructions per fetched byte for 1024 variations.

An analysis of the code showed two major inefficiencies restricting the performance: (i) the pseudo-random number generation is consuming a large part of the `ttDilepKinFit` execution time, (ii) the way data is structured in the LipMiniAnalysis prevents processing in parallel events from the same input file. These two issues are further detailed in the next subsections.

3.1   Pseudo-Random Number Generation Inefficiencies

Pseudo-random number generators (PRNGs) are common in many Monte Carlo simulation and reconstruction applications. A good PRNG deterministically generates uniform numbers with a long period, its produced values pass a set of randomness tests and, in HPC, it must be efficient and scalable. Repeatability is ensured by providing a seed to the PRNG prior to number generation, due to their deterministic execution.

---

[3] For a detailed explanation of the K-Best Measurement Scheme, see Chapter 9.4.3 of Computer Systems: A Programmer's Perspective (CS:APP), Randal E. Bryant and David R. O'Hallaron, Prentice Hall, 2003.

The reconstruction of $t\bar{t}H$ events depends crucially in the intrinsic energy-momentum resolution of the ATLAS Experiment. In order to increase the probability of finding the correct solution for the particular configuration of the $t\bar{t}H$ event under study, pseudo experiments are performed. For each pseudo experiment, the event kinematics is varied by applying an offset which changes the energy-momentum four vectors of final state particles, within detector resolutions. As a case study, the maximum offset has been set to ±1% of the original value and is computed with the help of PRNG. An analysis of the callgraph produced for 256 variations (higher variations made the Callgrind execution time infeasible) showed that 63% of ttH_dilep execution time was spent on the PRNG. However, 23% of the time was spent defining a new seed for the PRNG. Figure 3 presents the callgraph for the ttDilepKinFit function of ttH_dilep.

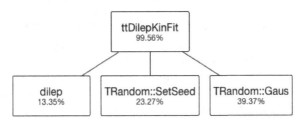

Fig. 3. Callgraph subset of the ttDilepKinFit most time consuming functions for 256 variations per combination

An analysis of the code showed that the application uses a PRNG available in ROOT, which uses the Mersenne Twister algorithm [11], resetting the seed for every parameter variation. The Mersenne Twister period is approximately $4.3 * 10^{6001}$, while the maximum amount of pseudo random numbers generated by the application, for the input file used and 1024 variations, is $3 * 10^9$, making the seed reset unnecessary. The removal of this inefficiency granted a 71% performance improvement.

3.2 Data Structure Inefficiencies

Once removed the PRNG seed reset inefficiency, the ttDilepKinFit still remained the critical region in the application, with no apparent code inefficiency. The most obvious solution is to process in parallel several events from the same input file. However, the function in LipMiniAnalysis that loads events from a file into memory assigns a single global space. This data structure contains information that is modified during the event reconstruction process, and it is overwritten for every event loaded. Changing the data structure to support multiple events in memory simultaneously, and loading all events in the input file at the beginning of the data analysis, would allow the parallel processing of events with low overhead. However, as mentioned in section 2 many data analysis applications depend on LipMiniAnalysis preventing any modifications to its structure, so alternative solutions were explored.

3.3   Alternative Parallel Approaches

Next step to improve the code execution time is to parallelize `ttDilepKinFit`. Note that it is not possible to parallelize the whole event processing since only one is loaded at a time and part of its information is stored in LipMiniAnalysis toolbox. Besides not allowing this parallelization, reading events individually is more inefficient than reading all events at once, where in the former slower random reads are made on the hard drive and in the latter the fast sequential reads are used.

Parallelizing `ttDilepKinFit` implies modifying its flow. Currently, for each different combination of jets and leptons from an event, the processed data of each variation of the detector measurements is overwritten. A new data structure is required to hold all combinations of each event. Picking a lepton/jet combination depends on all previous chosen combinations, which serializes the construction of the data structure. Each parallel task (indivisible work segment) selects a combination with variations still to compute, then varies the particles parameters, performs the kinematical reconstruction, and attempts to reconstruct the Higgs boson. A parallel merge is performed after all combinations are computed to get the most accurate reconstruction for the event. Figure 4 presents the sequential and parallel workflow for `ttDilepKinFit`.

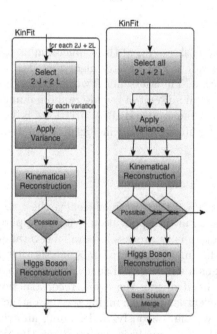

Fig. 4. Schematic representation of the `ttDilepKinFit` workflows: sequential (*left*) and parallel (*right*)

A shared memory parallelization using OpenMP [12] was devised, as it is the best approach for single shared memory systems. The parallel tasks are grouped into threads, which holds the best reconstruction to minimize the complexity of the merge by reducing through all the threads instead of tasks. The amount of tasks for each thread is balanced dynamically by the OpenMP scheduler, as the workload is irregular since the Higgs boson reconstruction execution is not always computed. Each thread has a private PRNG initialized with different seeds to avoid correlation between the numbers generated.

Figure 5 presents the speedups for different number of parallel threads. The purpose of the 1 thread test is to evaluate the parallelization overhead. The best efficient implementation occurs when using 2 and 4 threads, where the application is using almost all resources at each used core. The best overall performance occurs for 40 threads, but it only offers a speedup of 8.8, underusing the available 20 physical cores. Note that there is no significant overhead due to NUMA[4] accesses, as seen by the constant increase in performance from 10 to 16 threads. For more than 20 threads all available resources on both CPU devices.

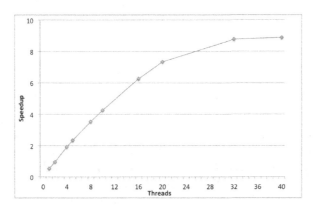

Fig. 5. Speedup for the ttH_dilep original parallel version of the application

The lack of scalability beyond a low number of parallel threads suggests that inefficiencies may still affect the application, probably caused by the paralleliza- tion overhead. Intel's VTune was used to search for hotspots (bottlenecks) on the parallel ttH_dilep, since this tool is best suited for profiling parallel appli- cations while providing a user friendly graphic interface. A preliminary analysis showed that the application was spending 20% of the execution time building the combination data structure for 256 variations.

An analysis of the coded data structure showed that inefficiencies were affect- ing the performance in specific situations. Data that is read-only on the parallel

---

[4] NUMA, Non-Unified Memory Access, since each Xeon device has its own memory controller with attached RAM: RAM access time for each core differs as the RAM is connected to the same device or the neighbour Xeon.

section is being replicated in each element of the data structure. If the elements were to share a pointer to such data, the overhead of constructing the data structure would be reduced. However, this could lead to worse cache management, due to cache line invalidations, since the application is accessing data on memory more frequently, and the data structuring did not efficiently separate read-only data from read/write data. This is particularly critical in NUMA environments, where communication costs are higher. This was implemented and tested (addressed as *pointer version*), with its speedups plotted in figure 6. The reference value for the speedup computation is still the same sequential version.

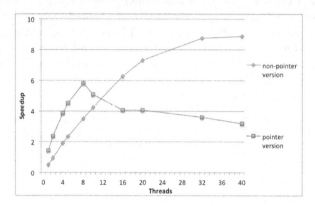

Fig. 6. Speedup for `ttH_dilep` parallel pointer and non-pointer implementations

As expected, the best speedup occurs when using only one CPU device. The performance degradation from 8 to 10 threads (on the same device) may be explained by the increase of concurrent accesses to the shared L3 cache. However, this implementation is more efficient than the non-pointer implementation when using only one device. This is a classical case where partitioning the data structure allows it to fit on the L2 cache, providing superscalar speedups for certain number of threads. The data structure has an average size of 30 KB, but can have up to 867 KB with the input used. Considering the 5 thread superscalarity, partitioning the data structure among 5 cores reduces its size per core to 173 KB, fitting their private L2 cache and avoiding the slower accesses to the L3 cache.

4   Runtime Inefficiencies

When submitting a job or application for execution on a given computing system, most users trust the default configurations of the submission environment. However, if the user needs to improve the efficiency of the code execution, he/she must be aware of the environment variables that can be controlled and how those can impair the performance. Two cases will be addressed here: (i) how to spread the code parallelism, between processes and threads, and (ii) how to allocate the available cores on each device to threads and processes.

4.1 Multithreading Inefficiencies

Without the sensibility provided by the tests in section 3, a scientist would incur in the pitfall of using all available cores on the system (and even all hardware threads, if each core supports hardware multithreading), hoping that it would provide the best performance. While it may be true for the non-pointer implementation, the system computational resources would be inefficiently used, and using the single device highly efficient pointer implementation would induce a even greater waste.

A closer look to the pointer based implementation shows in fact that it is the most efficient one. As seen in section 3.2, the scalability of the parallelization is limited by the NUMA organization on modern multiple CPU device systems. If the threads on cpu_1 do not share information with the threads on cpu_2, the NUMA bottleneck is removed by using multithreaded processes. However, parallelization at the process level, where each process performs a data analysis on a separate event, is not possible with the current implementation of LipMiniAnalysis, where a single global state is allocated to store data from each event processing.

Data analysis applications are individually executed for each file (around 1GB in size) in a very large set of files, at a terabyte scale, received weekly from CERN. An alternative approach to the process parallelization over a single input data file is to balance the execution of different ttH_dilep processes in the system on a set of distinct input files. This reduces the complexity of the implementation, with no changes needed for ttH_dilep, and avoids communication between processes. A simple scheduler was devised, which takes a set of input files and spawns a given amount of ttH_dilep processes. The scheduler dispatches the files to the different processes in a queue-like approach, and monitor their execution as shown in figure 7. A set of 20 input files was considered for testing and evaluation purposes, with different configurations of processes and threads per process.

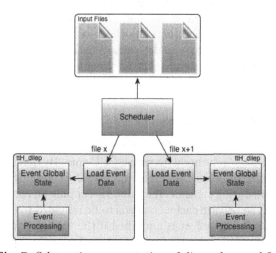

Fig. 7. Schematic representation of dispatcher workflow

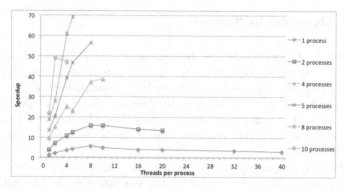

Fig. 8. Speedups for the scheduler with the pointer based implementation for several combinations of #processes and #threads per process

Figure 8 presents the speedups using 2, 4, 5, 8, and 10 processes for various thread configurations, with maximum number of threads limited to 40. A higher amount of processes was not tested as the efficiency decayed from 8 to 10 processes. The best speedups occur for 8 processes with 5 threads each, with a peak of 69.3, 7.8 and 11.7 times better than the best non-pointer and pointer implementations, respectively. A small number of threads such as this allows for a small overhead in the ttH_dilep parallelization, namely on load balancing and final best reconstruction merge for each event. For 10 processes the load on the system due to the lack of shared memory and I/O operations affects the performance, decreasing the speedups relatively to using 8 processes. A common behaviour is that when using the CPU devices hardware multithreading the speedups tend to stabilize, or even drop for 2, 4, and 5 processes. Overall, the best speedups occur when using all available cores on the system, with multithreading.

Since this implementation uses the parallel pointer version, which has superlinear speedups as presented at the end of subsection 3.3, the performance improvements tend to be higher than the theoretical maximum. Here, the superscalarity is enhanced because the data structure is less partitioned than with the pointer version for the same number of threads, as threads from the same process are paired in each core to allow sharing common data on the L2 cache.

4.2   Core Affinity Inefficiencies

One of the key issues in runtime efficiency is the thread affinity [13], namely to control the allocation of each thread to which CPU core. By default, OpenMP lets the operating system to manage the thread affinity; as a consequence, threads may migrate among cores during runtime. If a thread is running on core c_1 and moves to core c_2, all data on the private cache l_{c_1} needs to be reloaded to cache l_{c_2}, causing unnecessary overhead. This effect is amplified if the threads are moved between adjacent CPU devices. When multiple different, and (possibly) parallel processes are running on the same system, which is common in production environments, such scheduling occurrences happen more frequently. This subsection presents a preliminary study on thread affinity for data analysis code.

Defining the thread affinity of an application may provide a more predictable, or in some cases better, performance. In theory, an optimum thread affinity scheme allocates the threads to contiguous physical cores of one CPU device, uses the cores of the second CPU device only after the first is filled, and finally uses the multithreading capability after filling all physical cores. Note that using multithreading before the second CPU device is fully occupied may provide better performance in memory bound applications. This type of affinity must be defined prior to the application execution and depends on the system used. In this compute bound data analysis case, the affinity was specifically tuned to the 20-core testbed system for all threads or process/threads configurations for the scheduler.

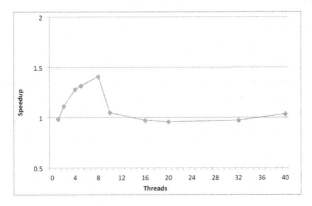

Fig. 9. Speedup of the ttH_dilep parallel pointer implementation with core affinity

By analysing the speedups of the pointer implementation of ttH_dilep with thread affinity, in figure 9, the specification of the affinity provides speedups for the previous most efficient number of threads, i.e., up to 8 threads. For 8 threads the performance increases by 41%, relative to its no affinity counterpart. With this number of threads, and the amount of shared data, moving threads between cores at runtime causes more cache warm ups to occur, significantly affecting the performance. When using more than 10 threads the application is roughly 4% slower as the operating system uses some multithreaded cores rather than using all available physical cores and it does a better job at managing the multithreading.

The same affinity study was performed on the scheduler, with the speedups presented in figure 10. The performance is increased for some specific configurations, with the exception of 5 processes that is always worse, providing improvements up to 52%, 90%, 8%, and 25% for 2, 4, 8, and 10 processes respectively.

The performance with core affinity is less susceptible to oscillations, as with no affinity it is sometimes affected by OS thread reallocations. It is when many

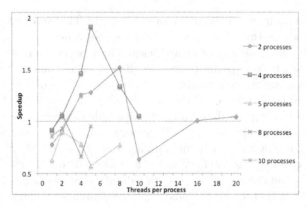

Fig. 10. Speedups of the scheduler with the pointer based implementation for various threads per process with core affinity

reallocations may occur that setting the core affinity provides the best performance. Hard setting the affinity may not allow for proper multithreading to hide the memory accesses latency, affecting the performance. It is not possible to use a theoretical affinity scheme to always improve the performance on every system, as it is highly dependent on:

- the algorithm, memory bound, suffers more from core reallocation due to losing all data on cache, where fixing their position on a specific core avoids unnecessary accesses to the RAM;
- the application execution time, as the impact from thread reallocations is higher in applications with low execution times;
- the operating system, as OpenMP, by default, lets it manage the thread allocation and it is susceptible to the overall system load, causing fluctuations in consecutive applications execution time.

With all optimizations considered, the best overall performance in a dual 10-core Xeon system is obtained using the scheduler combined with the pointer implementation, with 8 multithreaded processes (with 5 threads each), reaching a speedup of 112 over the original sequential application.

5   Conclusion

This paper presents a study of the inefficiencies in scientific code, using a particle reconstruction analysis application as a case study. Top quark and Higgs boson studies require reconstructing from measurements of a very large number of particle collisions, performed weekly by the ttH_dilep application on terabytes of data. A faster and more accurate analysis of the data allows to better reconstruct $t\bar{t}H$ events and improve the quality of the research results.

Execution inefficiencies of software applications may occur due to several factors, from algorithm and data structure design to numerical approaches, choice of library functions or compile tuning for code vectorisation, among others. Our focus in this work was on the identification and removal of inefficiencies at only two stages of the application - the code design and its submission for execution - since efficient numerical libraries were already used [14], as well as the Intel guidelines to develop efficient code [15,16,17]. The code inefficiencies we identified and corrected had a significant negative impact on performance: the removal of the unnecessary seed generation for the pseudo random number generation led to 71% performance improvement. Two parallelisation alternatives were proposed to overcome data structure inefficiencies, one that scales with two CPU devices and other much more efficient but only scales with one CPU device.

At application runtime, a multiprocess approach using the more efficient parallel implementation tackled its inefficiencies on NUMA systems, providing a superscalar speedup of 69.3 with 40 threads. This superscalarity was achieved due to the data partition, which led to a better cache usage. An efficient control on the thread affinity of this implementation provided a performance improvement close to 2x. However, the fluctuation in performance and the dependencies on many system characteristics prevented the definition of a generalized heuristic to aid to control the best affinity for the application, for any computing system.

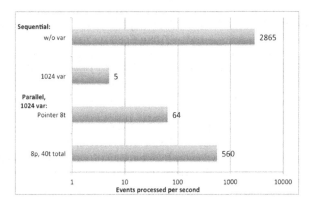

Fig. 11. Throughput of events processed for the original sequential ttH_dilep, with no and 1024 variations (*var*), and for the parallel pointer and multiprocess implementations (with p processes and t threads)

Figure 11 plots the number of events processed per second in four different situations using a dual 10-core Xeon computing system: (i) assuming the particle properties as given by the ATLAS detector with 100% confidence, (ii) the same code but assuming an error of $\pm 1\%$ on the detector measurements and considering 1024 variations on those values improving the analysis accuracy, (iii) a parallel single process implementation, with multiple threads, and (iv) the best parallel combination of processes and threads. By removing the code inefficiencies and developing efficient parallelisation strategies, the overall result of this work increased the event throughput by a factor of 112, from 5 to 560 events per second.

These promising results on code execution efficiency leave yet some room for further enhancements. The scheduler could be improved to automatically predict the best process/thread configuration for each system by analysing a set of micro-benchmarks or the application itself on a small input, and ultimately identify the best core affinity scheme. Also, the application efficiency could be improved using hardware accelerators, balancing the workload among accelerators and CPU devices in heterogeneous systems. The use of development frameworks for heterogeneous systems, such as StarPU [18], may further improve productivity and efficiency through aids in code parallelization and transparent workload distribution among multi-core devices and computing accelerators.

Acknowledgments. This work is funded by National Funds through the FCT - Fundação para a Ciência e a Tecnologia (Portuguese Foundation for Science and Technology) within project PEst-OE/EEI/UI0752/2014, by LIP (Laboratório de Instrumentação e Física Experimental de Partículas), and the SeARCH cluster (REEQ/443/EEI/2005).

References

1. The ATLAS Collaboration: The ATLAS Experiment at the CERN Large Hadron Collider. Journal of Instrumentation 3(08), S08003 (2008)
2. Aad, G.: et al.: Observation of a new particle in the search for the Standard Model Higgs boson with the ATLAS detector at the LHC. Phys. Lett. B716, 1–29 (2012)
3. Oliveira, V., Pina, A., Castro, N., Veloso, F., Onofre, A.: Even Bigger Data: Preparing for the LHC/ATLAS Upgrade. In: 6th Iberian Grid Infrastructure Conference (2012)
4. Pereira, A.: Efficient Processing of ATLAS Events Analysis in Homogeneous and Heterogeneous Platforms. Master's thesis, University of Minho (September 2013)
5. Rademakers, F., Canal, P., Bellenot, B., Couet, O., Naumann, A., Ganis, G., Moneta, L., Vasilev, V., Gheata, A., Russo, P., Brun, R.: ROOT (November 2012)
6. Graham, S.L., Kessler, P.B., Mckusick, M.K.: Gprof: A Call Graph Execution Profiler. SIGPLAN Not. 17(6), 120–126 (1982)
7. Developers, V.: Callgrind: a call-graph generating cache and branch prediction profiler (January 2013)
8. Intel: Profiling Runtime Generated and Interpreted Code with Intel VTune Amplifier. Technical report (January 2013)
9. Browne, S., Deane, C., Ho, G., Muccima, P.: PAPI: A Portable Interface to Hardware Performance Counters. In: Proceedings of Department of Defense HPCMP Users Group Conference (June 1999)
10. Intel: Intel Xeon Processor E5 v2 Family: Datasheet. Technical report (September 2013)
11. Matsumoto, M., Saito, M.: Mersenne Twister: A 623-dimensionally equidistributed uniform pseudorandom number generator. ACM Transactions on Modeling and Computer Simulations: Special Issue on Uniform Random Number Generation (1998)

12. Board, O.A.R.: OpenMP Application Program Interface. Technical report (July 2013)
13. Dow, E.: Take charge of processor affinity. IBM developerWorks (September 2005)
14. Blackford, L.S., Demmel, J., Dongarra, J., Duff, I., Hammarling, S., Henry, G., Heroux, M., Kaufman, L., Lumsdaine, A., Petitet, A., Pozo, R., Remington, K., Whaley, R.C.: An Updated Set of Basic Linear Algebra Subprograms (BLAS). ACM Trans. Math. Soft. 28(2) (2002)
15. Corporation, I.: Intel 64 and IA-32 Architectures Optimization Reference Manual. Technical report, Intel Corporation (2013)
16. Corporation, I.: Intel 64 and IA-32 Architectures Software Developers Manual. Technical report, Intel Corporation (February 2014)
17. Ott, D.: Optimizing Applications for NUMA. Technical report (February 2011)
18. Augonnet, C., Thibault, S., Namyst, R., Wacrenier, P.A.: Starpu: A unified platform for task scheduling on heterogeneous multicore architectures. Concurr. Comput.: Pract. Exper. 23(2), 187–198 (2011)

Distributed Prime Sieve in Heterogeneous Computer Clusters

Carlos M. Costa, Altino M. Sampaio, and Jorge G. Barbosa

Faculdade de Engenharia da Universidade do Porto
Rua Dr. Roberto Frias, s/n 4200-465 Porto, Portugal
{carlos.costa,pro09002,jbarbosa}@fe.up.pt

Abstract. Prime numbers play a pivotal role in current encryption algorithms and given the rise of cloud computing, the need for larger primes has never been so high. This increase in available computation power can be used to either try to break the encryption or to strength it by finding larger prime numbers. With this in mind, this paper provides an analysis of different sieve implementations that can be used to generate primes to near 2^{64}. It starts by analyzing cache friendly sequential sieves with wheel factorization, then expands to multi-core architectures and ends with a cache friendly segmented hybrid implementation of a distributed prime sieve, designed to efficiently use all the available computation resources of heterogeneous computer clusters with variable workload and to scale very well in both the shared and distributed memory versions.

Keywords: Sieve of Eratosthenes, wheel factorization, shared memory, distributed memory.

1 Introduction

Prime numbers have been a topic of wide research given their important role in securing online transactions using asymmetric public key algorithms such as RSA [1]. In this context prime numbers are used to create the public key that is used by the sender to encrypt the message contents. The reason to use a product of two prime numbers to create the public key is because in the current computer architecture, it is computation infeasible to factor large prime numbers, and thus find the original primes used to create the public key, and used then to decrypt the message (with the progress made in quantum computers, this may not hold for very long). Besides encryption, prime numbers can also be used in hash functions to reduce the risk of collision and in pseudo-random number generators.

Although nowadays is more common to use primality testing algorithms [2] to find large primes, sieves have been a known method to generate primes up to a given number. One of the most efficient prime number sieve was discovered by Eratosthenes [3] in the ancient Greece and can generate prime numbers up to a given number n with $O(n \log \log n)$ operations. Other prime sieves were discovered since then, such as the sieve of Atkin [4] or the sieve of Sundaram [5],

B. Murgante et al. (Eds.): ICCSA 2014, Part IV, LNCS 8582, pp. 592–606, 2014.

but a modified sieve of Eratosthenes is still considered to be the most efficient algorithm to use in the current computer architecture, and thus it was the one chosen to be used.

This paper contributes with an extensive analysis of modified sieve of Eratosthenes algorithm. For that, we have implemented several variants of the algorithm, ranging from sequential form, passing to parallel on multi-core architectures and ending in distributed computer clusters. Each of these 3 main implementations have several algorithm variations, to determine which strategy is more suitable for each usage. As such, it was developed algorithms variations that focus on using the minimum amount of memory, while others focus on computing time, and some make a trade-off between the two. Finally it was implemented several segmented versions of both parallel and distributed algorithms to allow the computation of primes to near the maximum number represented by the current computer architecture (2^{64}). Special attention was devoted to the distributed algorithms since they provide the most suitable implementation that can be used to calculate primes near 2^{64} in reasonable time. The restriction to limit the range to near 2^{64} aims to avoid overflow of numbers when sieving. To achieve this, the maximum limit is lowered to $2^{64} - 2\sqrt{2^{64}} - 1$. This guarantees that the maximum number that can be used as a seed will not cause overflow. This version is designed to scale well to large clusters and is implemented to perform dynamic allocation of segments to nodes in order to efficiently use all the computation resources of clusters with heterogeneous hardware and with variable workload. The implemented algorithms are assessed in terms of speedup, efficiency and scalability.

The remainder of this paper is organized as follows. Section II discusses related work. Section III first introduces some background information that are the starting point of the proposed algorithms, and then the several implementation versions are described. Section IV analyzes the results obtained and Section VI concludes the paper.

2   Related Work

The generation of prime numbers have been a topic of wide research essentially due to their need for the creation of key pairs or as a computation stage during various cryptographic setups. Singleton et al. [13] published in 1969 the first segmented version of the sieve of Eratosthenes. Based in the sieve of Eratosthenes, Shahid et al. [19] implemented a parallel version on the Flex/32 shared memory multiprocessor at NASA Langley Research Center. The author have studied the impact of several parameters in its performance, and concluded that there is no advantage in using more than four or five processors unless dynamic load balancing is employed. Anderson et al. [20] implemented an Ada program which uses a multitasking solution for the Sieve of Eratosthenes. Authors illustrate the power and flexibility of Ada in achieving concurrent solutions, and the capacity of the program to be used as a benchmark program for evaluating multiple CPU systems. Shapiro et al. [21] described a modification to the Sieve of Eratosthenes implementing a concurrent solution algorithm for multi-processor computers. Sorenson et al. [18] have presented two parallel prime number sieves. They proposed two algorithms and show that the first sieve runs in $O(\log n)$ time using $O(n/(\log n \log \log n))$ processors, while the second sieve runs in $O(\sqrt{n})$ time using $O(\sqrt{n})$ processors. Authors clarify that the second algorithm is more efficient when communication latency is high.

From all the publicly implementations analyzed, special interest was devoted to the Prime Sieve [10] developed by Kim Walisch, which is considered to be the fastest multi-core implementation publicly available at the present date. Other papers that influenced the strategies developed included [12] that details a very efficient use of the cache memory system for very large prime numbers. Additionally, [14] [15] explain how to use wheel factorization to considerably speed up the sieving process, and [16] [17] provide insights on how to implement the simple MPI [22] version. However, as far as we know, there is no available implementation that can compute prime numbers in a distributed memory architecture, in order to allow the computation of prime numbers up to 2^{64} in reasonable time.

3   Background

In this section we present a brief explanation of the main concepts used in our proposals.

3.1   Sieve of Eratosthenes

The sieve of Eratosthenes [3] was invented in ancient Greece by Eratosthenes around the 3rd century B.C., and describes a method for calculating primes up to a given number n in $O(n \log \log n)$ operations. The Algorithm 1 describes the pseudo code of the sieve of Eratosthenes:

Algorithm 1. Sieve of Eratosthenes

1: **Input:** an integer $n > 1$
2: ▷ Let P be an array of Boolean values, indexed by integers 2 to n, initially all set to true

3: **for** $i \in \{2, 3, 4, ..., n\}$ **do**
4:     **if** $P[i] == true$ **then**
5:         **for** $j \in \{2i, 3i, 4i, ..., n\}$ **do**
6:             $P[j] = false$

7: ▷ Now all i, such that P[i] is true, are prime numbers

This algorithm finds all the prime numbers less than or equal to a given integer n. For that, it creates a list of integers from 2 to n (line 3), and then, starting in 2, the first prime number, it marks all items in the list that are integers multiple of 2 (line 6). It then repeats the process for the next item in the list that is not marked (which is the next prime), until there is no such item.

3.2   Wheel Factorization

Wheel factorization [9] is a known optimization used to cross off multiples of several sieving primes. It can be used to considerable speed up the sieving process if used efficiently.

A modulo 30 wheel, which has sieving primes 2, 3 and 5, can cross off as much as 66% of composites. A modulo 210 wheel, which besides sieving primes 2, 3 and 5 also has prime 7, can increase the percentage of removed composites to 77%. Larger wheels may not have a reasonable improvement in % of composites crossed off, because the table required for the wheel factorization [14] may not fit in the cache memory. The creation of the wheel sieve table is described by Algorithm 2:

Algorithm 2. Wheel Factorization

1: **Input:** a P list of known prime numbers
2: ▷ Let m be the product of the known prime numbers in the P list
3: ▷ Let C be an array of Boolean values with numbers in [1, m], initially all set to false

4: **for** $i \in P$ **do**
5:     **for** $j \in \{2i, 3i, 4i, ..., m\}$ **do**
6:         $C[j] = true$

7: ▷ Now for all x numbers such as k=(x modulus m) in [1, m], if C[k] is false, then x is a probable prime number, and needs to be checked by other means to confirm if it is a prime number or not. If C[k] is true, then x is a composite number of one of the primes in list P

The purpose of wheel factorization is to skip multiples of small primes. In this regarding, if the kth wheel is added to the sieve of Eratosthenes, then any multiples that are divisible by any of the first k primes are skipped. For example, the 1st wheel considers only odd numbers, the 2nd wheel (modulo 6) skips multiples of 2 and 3, the 3rd wheel (modulo 30) skips multiples of 2, 3, 5 and so on [10].

By applying the wheel factorization method, the running time of the sieve of Eratosthenes can be reduced by a factor of $\log \log n$ if the wheel size is $\sqrt{n}$ [11].

4   Proposed Solutions

In the following sections we present the main concepts behind each prime sieve algorithm implemented. The order used denotes the development of each variation in relation to the previous ones and aims to facilitate the identification of the optimizations that were introduced in each development iteration.

In the simple algorithms the pseudo code is provided to facilitate the explanation of the algorithm. However, given the length of the algorithms, for the more optimized versions, it is only explained the main concepts. The implementation source code is available in [7].

Sequential Prime Sieve Using Trial Division (v.1): Algorithm 3 is one of the simplest methods to compute prime numbers. It works by crossing off all odd j numbers that are composite of a previous prime number i. This can be achieved by checking if a j number is multiple of a previous prime number (i) by using the modulus operator. If the remainder of the integer division between j and i is 0, then j is multiple of i and is marked as composite.

One optimization applied that improves upon the original Eratosthenes algorithm is that the sieving can start at i^2 instead of $2i$ in order to avoid marking a composite number multiple times. For example 21 is multiple of 3 and 7, so there is no need to mark it as composite when sieving with prime 7 because it was already sieved with the prime 3. For the same reason, it is only necessary to check i until $\sqrt{n}$, because when $i > \sqrt{n}$, the j will exceed n and as a result, the code in line 7 will never be executed.

Another optimization is to completely exclude even numbers, and this way reduce the computations and memory to half. This will require an adjustment in the way the numbers are mapped to memory positions ($[j - 3/2]$ instead of $[j]$) and will increase the increment necessary to transition to the next number ($+2$ instead of $+1$ in line 5).

Algorithm 3. Sequential Prime Sieve Using Trial Division (v.1)

1: **Input:** an integer $n > 1$
2: ▷ Let C be an array of Boolean values representing the odd numbers > 1, initially all set to false

3: **for** i $\in \{3, 5, 7, ..., \sqrt{n}\}$ **do**
4:     **if** $C[(i - 3)/2] == false$ **then**
5:         **for** $j \in \{i^2, i^2 + 2, i^2 + 4, ..., n\}$ **do**
6:             **if** $j \% i == 0$ **then**
7:                 $C[(j - 3)/2] = true$

8: ▷ Now all the positions in array C still marked as false, represent prime numbers

---

Sequential Prime Sieve Using Fast Marking (v.2): Algorithm 3 is very inefficient because it uses modulus operations to check for composites, which is a very computational intensive task, especially if the maximum range to sieve is a large number.

An alternative is to use additions instead of modulus operations, because they are much faster to compute. The idea is to cross off all j numbers that we already know that are composites, because $i^2 + k \times i$ is guaranteed to be a composite of number i.

As shown in Algorithm 4, another optimization is that the increment to mark the next composite can be $2i$ since adding i to an odd multiple of i will result in an even number (even numbers are not prime).

Algorithm 4. Sequential Prime Sieve Using Fast Marking (v.2)

1: **Input:** an integer $n > 1$
2: ▷ Let C be an array of Boolean values representing the odd numbers > 1, initially all set to false

3: **for** i $\in \{3, 5, 7, ..., \sqrt{n}\}$ **do**
4:     **if** $C[(i - 3)/2] == false$ **then**
5:         **for** $j \in \{i^2, i^2 + 2i, i^2 + 4i, ..., n\}$ **do**
6:             $C[(j - 3)/2] = true$

7: ▷ Now all the positions in array C still marked as false, represent prime numbers

Sequential Prime Sieve Using Fast Marking with Block Decomposition Optimized for Space (v.3): Although Algorithm 4 is considerable faster than Algorithm 3, it suffers from performance degradation by not using the cache memory system efficiently. This happens because the same areas of array C are being loaded to cache memory more times than necessary, since the algorithm is sieving the composites of each prime number i to the end of array C. This is extremely inefficiently, because each cache miss forces the CPU to wait hundreds of clock cycles to load data from the main memory.

To solve this problem, Algorithm 5 implements a sieve with block decomposition (splits the sieving range in blocks that fit in the cache), in order to load the values of the array C only one time (to the cache), and then sieve all primes numbers up to $\sqrt{n}$ in that block, before moving to the next.

Sequential Prime Sieve Using Fast Marking with Block Decomposition Optimized for Time (v.4): Algorithm 5 has a section of code (line 20) that is constantly repeating computations. That section is the calculation of the closest prime number in relation to the beginning of the block. In order to prevent this repetition, this version keeps in memory the last prime multiple associated with each prime (that was computed in the previous block). And since it is only necessary to store the prime multiples associated to primes up to $\sqrt{n}$, this represents a very reasonable trade-off between space and computation time.

This is the first variation in which a segmented sieve was implemented, in order to try to save some memory by keeping the bitset of booleans only for the current block. But since all the primes found are kept in memory instead of being compacted in the bitset, it ends up consuming much more memory because each prime has 64 bits of storage.

Sequential Prime Sieve Using Fast Marking with Block Decomposition Optimized for Space and Time (v.5): This variation is implemented to reduce even more the repetition of computation that occur in the previous algorithm and to revert to the previous memory scheme.

The improvement is to avoid the repetition of the computation of the double of a prime to use as an increment in the *removePrimesFromPreviousBlocks* function

Algorithm 5. Sequential Prime Sieve Using Fast Marking With Block Decomposition Optimized For Space (v.3)

1: **Input:** an integer $n > 1$
2: ▷ Let C be an array of Boolean values representing the odd numbers > 1, initially all set to false
3: ▷ Let bs be the block size in number of elements
4: ▷ Let nb = n / bs be the number of blocks to use in sieving

5: CALCULATEPRIMESINBLOCK(3, 3 + bs)   ▷ First block
6: **for** b $\in \{1, 2, 3, ..., nb\}$ **do**   ▷ Remaining blocks
7:     $a = b \times bs$
8:     $b = a + bs$ ▷ Last block should have an upper limit of n + 1, because they are sieved in range [a, b[
9:     REMOVEPRIMESFROMPREVIOUSBLOCKS(a, b)
10:    CALCULATEPRIMESINBLOCK(a, b)
11: ▷ Now all the positions in array C still marked as false, represent prime numbers

12: **function** CALCULATEPRIMESINBLOCK(a, b) ▷ Sieves in range [a, b[
13:    **for** i $\in \{a, a + 2, a + 4, ..., \sqrt{n}\}$ **do**
14:        **if** $C[i] == false$ **then**
15:            **for** j $\in \{i^2, i^2 + 2i, i^2 + 4i, i^2 + 6i, ..., b\}$ **do**   ▷ Not including b
16:                $C[j] = true$

17: **function** REMOVEPRIMESFROMPREVIOUSBLOCKS(a, b) ▷ Sieves in range [a, b[
18:    **for** i $\in \{0, 1, 2, ..., k\}$ **do** ▷ k not exceeding the position in C associated with $\sqrt{n}$
19:        **if** $C[i] == false$ **then**
20:            p = closest prime multiple of number associated with position i in relation to a
21:            **for** j $\in \{p, p + 2p, p + 4p, ..., b\}$ **do**   ▷ Not including b
22:                $C[j] = true$

---

in Algorithm 5 (line 21). As such, once the prime double is calculated for the first time in *calculatePrimesInBlock* function in the same algorithm, it is associated to the prime multiple. Since there are very few primes to sieve in comparison to the maximum range of sieving, saving a pair of the current prime multiple and its double to avoid constant repetition of computations is very reasonable.

Sequential Prime Sieve Using Fast Marking with Block Decomposition Optimized for Space With Modulo 30 or 210 Wheel Factorization (v.6 and v.7): This alternative introduces the wheel factorization to speed up the sieving process. Instead of using the wheel factorization to pre-sieve numbers and update the bitset of composite numbers, this implementation uses the wheel sieve to determine the next possible prime given a number. This way the bitset is only updated in the positions that represent possible primes. This variation of use of the wheel sieve is more efficient for sieving, but also implies that to extract the primes from bitset only the positions associated with possible primes must be evaluated, since the others are not considered in the sieving process. With this insight, the access to the bitset is changed in order to store only bits associated with possible primes, and the memory consumption is reduced.

A version with a 30 wheel and 210 wheel was implemented to see the impact of different wheel sizes.

Sequential Prime Sieve Using Fast Marking with Block Decomposition Optimized for Space and Time and With Modulo 30 or 210 Wheel Factorization (v.8 and v.9): Since the use of the wheel to store only bits of possible primes introduces a lot of overhead (in the computation of the index to store the result in the bitset), this version reverts to the previous scheme of storing all odd numbers, and uses the wheel to skip the sieving of the composites of the prime numbers 2, 3 and 5 (and also the number 7 in the 210 wheel version). This way, it is expect to obtain a significant performance boost in comparison with the algorithms that did not use wheels and to mitigate the overhead associated with the computation of the bitset indexes when storing only numbers in the wheel.

Sequential Prime Sieve Using Fast Marking with Block Decomposition Optimized for Time and with Modulo 30 or 210 Wheel Factorization (v.10 and v.11): Since memory access is a bottleneck in all sieving algorithms, these versions try to determine if using direct memory access without any need to offset calculations would improve performance. The only difference between these variations and the previous ones is that it reserves memory for all numbers, so that the number itself is the position within the bitset where the sieve analysis is going to be stored.

Parallel Prime Sieve Using Fast Marking with Block Decomposition Optimized for Space or Time and With Modulo 210 Wheel Factorization (v.12 and v.13): With current computer architectures, any computer intensive algorithm should be designed to take full advantage of the parallel computation resources available in multicore systems. Given that the previous algorithms are already implemented using block decomposition, and that the multicore systems use shared memory, porting the best sequential algorithm (v.9) to a parallel architecture using OpenMP [8] only requires minor changes to the source code. This approach results in assigning groups of blocks to different threads, and optimizing the allocation of these blocks in order to avoid load unbalancing between the different threads. Implementation 13 is a port of the version 11 to a parallel architecture.

Segmented Parallel Prime Sieve Using Fast Marking with Block Decomposition Optimized for Space and Time and with Modulo 210 Wheel Factorization (v.14): Since the parallel algorithms were very fast to sieve primes up to 2^{32} (about a second), it was developed a segmented variant to allow the computation of primes to near 2^{64} using only the amount of memory that is specified in the command line arguments. This way, the algorithm can adapt to different hardware and still compute all primes requested with very little overhead (about 3%) compared with the fastest OpenMP version (v.13). This overhead is associated with the management of the segments (groups of blocks) and reseting the bitset values (when moving to a new segment).

Distributed Prime Sieve Using Fast Marking with Block Decomposition Optimized for Space and Time and with Modulo 210 Wheel Factorization (v.15): This is the first implementation of a distributed prime sieve algorithm prepared to run in homogeneous computer clusters. It uses the best sequential algorithm (v.9), and splits evenly the workload among the processes (that can be in different computers). In order to keep communication between processes to a minimum, each process computes the primes up to $\sqrt{n}$ and then uses them to sieve their share of the workload. The share of the workload [startNumber, endNumber[is computed as presented in equation 1:

$$startNumber = \frac{processID \times maxRange}{numberProcessesInWorkgroup},$$
$$endNumber = \frac{(processID + 1) \times maxRange}{numberProcessesInWorkgroup}, \qquad (1)$$
$$processID \in [0, numberProcessesInWorkgroup - 1].$$

Distributed Prime Sieve Using Fast Marking with Block Decomposition Optimized for Time and with Modulo 210 Wheel Factorization (v.16): This is the same algorithm as above but using one less operation (shift left 1 for dividing the number by 2) per memory access. Is similar to previous algorithms optimized for time, but since the algorithm is now segmented, direct memory access can't be done.

Hybrid Distributed Prime Sieve Using Fast Marking with Block Decomposition Optimized for Space and / or Time and with Modulo 210 Wheel Factorization (v.17 and v.18): The previous implementation completely disregards the fact that each node in the workgroup may be a multicore system, and as such, a hybrid version would take better advantage of the shared memory architecture to avoid the replication of the computation of the sieving primes up to $\sqrt{n}$ in the same node. This way, OpenMPI can be used to coordinate the distribution of the workload between different nodes in the cluster and then inside each node can be used an OpenMP variant optimized to take full advantage of the shared memory architecture.

Implementation 18 is a modification of version 17 to use one less operation per memory access (similar to the memory access method introduced in version 16).

Hybrid Distributed Prime Sieve Using Fast Marking with Block Decomposition Optimized for Space and / or Time, with Modulo 210 Wheel Factorization and with Dynamic Scheduling (v.19 and v.20): The previous distributed hybrid algorithms didn't take into account that the cluster in which they may be used can have heterogeneous hardware nodes with different processing capabilities, and even if the cluster is homogeneous, it may very well have variable workload. These external factors may have a damaging effect on the performance of the previous algorithms because some nodes may finish much sooner than others and then became idle, while others are still processing at full load. In order to avoid this, it was implemented a dynamic scheduling algorithm based in the implementation 18, but with a control process that creates a given number of segments (specified by the user), and then distributes these segments dynamically according to the requests of the nodes that finish their segments. This way, if there is any node that finishes sooner than others, it can continue contributing to the computation by requesting a new segment from the control process. With this strategy, the algorithm is ready to adapt to heterogeneous clusters with variable workloads and can be fine-tuned to a given network topology by specifying how many segments should be created, how many processes should be started and in which nodes, and how many threads should each process use.

Implementation 20 is a modification of version 19 to use one less operation per memory access (similar to the memory access method introduced in version 16).

5   Results and Discussion

This section presents the obtained results and the analysis of the algorithms according to different performance metrics. Detailed explanation of the reasons behind the different performances in each algorithm is presented in Section 5.2.

These results were collected using Ubuntu 13.04 64 bits, and the source code (available in [7]) was compiled with -O3 -s flags using mpic++ with g++ 4.7.3 and OpenMPI 1.4.5.

The sequential and parallel versions were tested on a laptop (Clevo P370EM) with an Intel i7-3630QM (quad core processor with Hyper-Threading - 2400 MHz clock rate) and 16 GB of RAM DDR3 1600 MHz.

For the distributed algorithms, it was added a second laptop (Asus G51J) with an Intel i7-720QM (quad core processor with Hyper-Threading - 1600 MHz clock rate) and 4 GB of RAM DDR3 1066 MHz. The router for connecting the two laptops was a Technicolor TG582n with 100 Mbps Ethernet connection.

5.1   Global Performance Analysis

In Figure 1 is given a global overview of the performance of some of the implementations in ranges from 2^{25} to 2^{32}. The chart shows that most algorithms need a range of about 2^{28} to reach their maximum performance. The only exceptions are the distributed algorithms, because the computation capacity was increased roughly 50% by adding a second node to the workgroup. As such, the performance reaches its maximum in ranges larger than 2^{32}. The reason for this is related to network latency and initializations overhead of processes and threads taking a considerable percentage of total computation time when the useful computations are relatively small (less than a second).

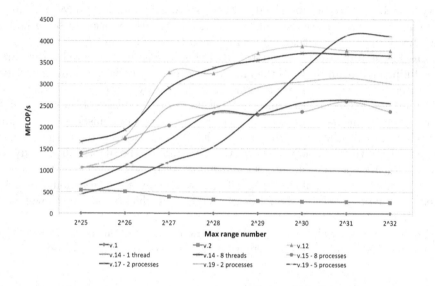

Fig. 1. Global performance comparison (Mflops)

Additionally, we notice that the sequential prime sieves using 30 or 210 wheel factorization achieved a significant performance boost in comparison with the algorithms that did not use wheels.

Since memory access has a significant impact on the runtime of the sieving algorithms, several access and store methods were tested. The fastest way to access a memory position is using the number itself as an index. But results shown that although this is true for small ranges, that doesn't hold for larger ranges (bigger than 2^{25}). The possible reason may be related to the fact that with this strategy, only half of the numbers can be stored in cache (comparing to implementations that only store even numbers). And as such, although the overhead to access memory is reduced (by not having to calculate the associated index to a given number), the net loss in cache misses starts to outweigh the improved memory access when the range is increased.

Moreover, given the results, implementation 14 is the recommend algorithm to use in single node multicore systems, when it is required to calculate very large primes (bigger than 2^{35}), using efficiently the memory available, because it is a segmented sieve optimized for good cache usage.

For very large ranges (bigger than 2^{48}), implementation 19 is the appropriate algorithm to use in a multi node, multi core system, since is can dynamically distribute the computation across several nodes with varying number of cores, by employing a dynamic scheduling / load balancing architecture.

5.2 Performance Analysis

In this subsection we analyze the obtained results according to different metrics.

Real Speedup Metric: Figure 2 shows the obtained speedup when comparing execution times of the recommended parallel implementation v.14 and distributed implementation v.19, with the best sequential implementation v.9.

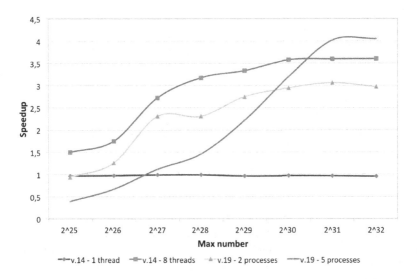

Fig. 2. Real speedup due to different strategies implemented

The results show that the parallel version in implementation v.14 runs best when using 8 threads on a quad core system with hyper-threading. But as expected, the result is not proportional to the number of threads but proportional to the number of real available cores. The speedup is not 4 because there is some sequential sections in the algorithm that cannot be parallelized. It is also clear that the distributed algorithm should only be used for large ranges, since the overhead and latency of MPI calls can damage performance if the total computations are small. The speedup of the distributed version isn't higher because the second node has 50% less processing capability and uses a 50% slower memory when compared with the first node. This was done on purpose to determine the adaptability of the algorithm to heterogeneous nodes.

Efficiency: Figure 3 shows the obtained efficiency when comparing execution times of the recommended parallel implementation v.14 and distributed implementation v.19, with the best sequential implementation v.9. It was calculated based on the real speedup presented earlier.

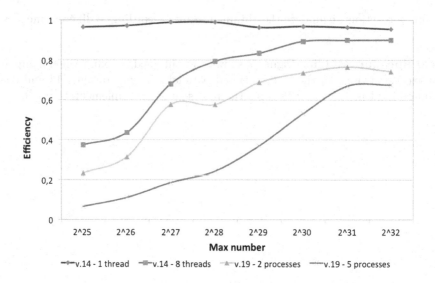

Fig. 3. Efficiency for two versions of the parallel algorithm

As expected, the best efficiency was obtained with the shared memory algorithm, since it keeps computation repetitions to a minimum (the distributed implementations need to calculate the sieving primes in each process) and doesn't suffer from overhead of MPI calls and latency associated with network communications. The reason for having 5 processes in implementation v.19 when running in 2 nodes is related to the fact that 1 process is only to control the dynamic allocation of segments to the other 4 slave processes, and because on the current testing system, using only one process per node was resulting in lower performance, due to latency in the MPI calls when requesting a new segment from the control process. As such, the performance improves with 2 processes per node, because this way, when 1 process is waiting for a new segment allocation, the other one can still be using the node computation resources. With this strategy, idle periods are reduced when network latency is high.

Scalability: The shared memory version maintains high levels of efficiency when more resources are used, showing good scalability.

The distributed version has less efficiency than the shared memory version due to network latency and load balancing overhead, but when a new node was introduced, the efficiency decreased only slightly (between 4% and 5%). On the other hand, given the current hardware architecture, this is the implementation that can be scaled more easily.

We can conclude that the algorithms scale relatively well, since their efficiency is not significantly reduced when more computing resources are added, and the computation time is reduced consistently.

6   Conclusions

In this paper we presented several algorithms and optimizations that can be used to efficiently compute prime numbers up to a specified number. It was discussed several variants of algorithms to maximize the efficiency in different computer architectures and cluster topologies. The final implementation shows very reasonable efficiency and scalability and it can be used to compute prime numbers up to 2^{64} in a distributed multicore computer architecture. For maximum efficiency was developed an OpenMP version to be used in common multi-core processors, and a hybrid OpenMP and MPI version with dynamic scheduling to be used in heterogeneous computer cluster, that may have computation nodes with different hardware capabilities and variable workload. To improve the current implementation, in the future, we intend to use a bucket sort algorithm [12] to increase the cache hit rate for very large ranges and port the best algorithms to use GPUs.

References

1. Standard, RSA Cryptography: RSA Public Key Cryptography Standard# 1 v. 2.2. RSA Laboratories (2012)
2. Lenstra Jr., H.W.: Primality testing algorithms (after Adleman, Rumely and Williams). In: Sminaire Bourbaki, vol. 1980/81 Expos 561578, pp. 243–257. Springer, Heidelberg (1981)
3. O'Neill, M.E.: The genuine sieve of Eratosthenes. Journal of Functional Programming 19(1), 95 (2009)
4. Atkin, A., Bernstein, D.: Prime sieves using binary quadratic forms. Mathematics of Computation 73(246), 1023–1030 (2004)
5. Aiyar, V.R.: Sundaram's Sieve for Prime Numbers. The Mathematics Student 2(2), 73 (1934)
6. Joye, M., Paillier, P., Vaudenay, S.: Efficient generation of prime numbers. In: Paar, C., Koç, Ç.K. (eds.) CHES 2000. LNCS, vol. 1965, pp. 340–354. Springer, Heidelberg (2000)
7. Distributed prime sieve C++ implementation (git repository), https://github.com/carlosmccosta/Distributed-Prime-Sieve
8. Dagum, L., Menon, R.: OpenMP: an industry standard API for shared-memory programming. Computational Science & Engineering 5(1), 46–55 (1998)
9. Pritchard, P.: Explaining the wheel sieve. Acta Informatica 17(4), 477–485 (1982)
10. Primesieve, http://primesieve.org/
11. Pritchard, P.: Fast compact prime number sieves (among others). Journal of Algorithms 4(4), 332–344 (1983)

12. Járai, A., Vatai, E.: Cache optimized linear sieve. arXiv preprint arXiv:1111.3297 (2011)
13. Singleton, R.C.: An efficient prime number generator. Communications of the ACM 12, 563–564 (1969)
14. Paillard, G. A. L.: A Fully Distributed Prime Numbers Generation using the Wheel Sieve. In: Parallel and Distributed Computing and Networks, pp. 651–656 (2005)
15. Sorenson, J.: An analysis of two prime number sieves. In: Computer Sciences Department. University of Wisconsin-Madison (1991)
16. David, J.W.: Parallel Prime Sieve: Finding Prime Numbers. In.: Parallel Computing Seminar Report, Institute of Information & Mathematical Sciences, Massey University at Albany, Auckland, New Zealand (2009)
17. Cordeiro, M.: Parallelization of the Sieve of Eratosthenes. In: Parallel Programming, Doctoral Program in Informatics Engineering, Engineering Faculty. University of Porto (2012)
18. Sorenson, J., Parberry, I.: Two Fast Parallel Prime Number Sieves. Information and Computation 114(1), 115–130 (1994)
19. Bokhari, S.H.: Multiprocessing the Sieve of Eratosthenes. Computer Journal 20(4), 50–58 (1987)
20. Anderson, G.: An Ada Multitasking Solution for the Sieve of Eratosthenes. Ada Lett. 8(5), 71–74 (1988)
21. Shapiro, E.: The family of concurrent logic programming languages. ACM Computing Surveys (CSUR) 21(3), 413–510 (1989)
22. Gropp, W., Lusk, E., Skjellum, A.: Using MPI: portable parallel programming with the message-passing interface, vol. 1. MIT press (1999)

Recent Memory and Performance Improvements in OCTOPUS Code

Joseba Alberdi-Rodriguez[1,2,3], Micael J. T. Oliveira[3,4], Pablo García-Risueño[5],
Fernando Nogueira[3], Javier Muguerza[1], Agustin Arruabarrena[1],
and Angel Rubio[2,6,7]

[1] Dept. of Computer Architecture and Technology,
University of the Basque Country UPV/EHU, M. Lardizabal, 1,
20018 Donostia-San Sebastián, Spain
[2] Nano-Bio Spectroscopy Group and European Theoretical Spectroscopy Facility,
Spanish node, University of the Basque Country UPV/EHU, Edif. Joxe Mari Korta,
Av. Tolosa 72, 20018 Donostia-San Sebastián, Spain
[3] Center for Computational Physics, University of Coimbra, Rua Larga,
3004-516 Coimbra, Portugal
[4] Unité Nanomat, Université de Liège, allée du 6 août, 17, B-4000 Liège, Belgium
[5] Institut für Physik und IRIS Adlershof, Humboldt Universität zu Berlin,
Zum Grossen Windkanal 6, 12489 Berlin, Germany
[6] Centro de Física de Materiales, University of the Basque Country UPV/EHU,
20018 Donostia-San Sebastián, Spain
[7] Fritz-Haber Institut der Max-Planck Gesellschaft, Faradayweg 4-6,
D-14195 Berlin-Dahlem, Germany

Abstract. In this work we present the improvements made to the
OCTOPUS code in order to reduce the memory requirements and to optimise parallel data distribution. Both topics are central for efficiency and
feasibility of calculations when the system must be run in a large HPC
environment. These modifications were mainly made in the real-space
mesh partitioning and mapping algorithms, and are thus transferable to
other codes using this type of real-space representation of data. The code
became much more efficient, and we present several scalability results
showing that it is now possible to address *ab-initio* quantum-mechanical
simulations of the interaction of light with big biomolecules, paving the
way for a better understanding of phenomena such as energy conversion
in plants.

Keywords: DFT, TDDFT, OCTOPUS software, HPC, memory optimisations.

1 Introduction

The understanding of phenomena at nanoscopic scale is a major goal of science
since the very moment of the appearance of quantum physics. Software simulation is an invaluable tool to study these types of phenomena. Currently, high
performance computing (HPC)[1] enables the simulation of atomic and molecular systems according to the fundamental equations of quantum mechanics.

B. Murgante et al. (Eds.): ICCSA 2014, Part IV, LNCS 8582, pp. 607–622, 2014.
© Springer International Publishing Switzerland 2014

OCTOPUS [2–4] is a popular software for quantum electronic structure calculations. It is based on Density Functional Theory (DFT)[5, 6] and Time-dependent Density Functional Theory (TDDFT) [7, 8], which are the most widely used approaches in the field of quantum *ab-initio* simulation.

In this work we will present recent memory improvements in the OCTOPUS code and we will show its extreme performance. Previous to this work, OCTOPUS was prepared to run with relatively small atomic systems (hundreds of atoms at most), in machines with hundreds of processor cores. In the aim to reach bigger systems with thousands of atoms, it is necessary to use also thousands of processors, and this is the scenario where new problems appear. Altough the results to be presented are limited to OCTOPUS, lessons learned in this work can be readily applied to any other real-space mesh based code. It is to be stressed that our data transfer improvements correspond to the calculation of the Hartree potential, which is customarily calculated using real-space meshes also for codes using basis sets.

This paper is structured as follows. In section 2 we give some comments on how the OCTOPUS code works. In section 3, we explain the strategies we followed to increase the efficiency and optimise the memory management. Some measurements on the current performance of the code are presented in section 4. Finally, in section 5 we outline the main conclusions of our work.

2   Introduction to the OCTOPUS Code

The simulation at a quantum level of systems consisting of thousands of atoms makes it possible to understand a wide variety of physical, chemical and biological phenomena. Despite the remarkable recent improvements in scientific codes and HPC infrastructures, the size of the systems that can be simulated using TDDFT is still very limited, and performing calculations with thousands of atoms is still a significant challenge.

OCTOPUS is a very efficient scientific software package used to study by first principles the properties of the excited states of large biological molecules, complex nanostructures, and solids. The code is mostly developed for density functional theory (DFT) and time-dependent density functional theory (TDDFT) calculations, which are convenient quantum-mechanic approaches to study the electronic structure of molecular systems and its time evolution behaviour. It has been proven that their capacity to provide accurate results on the description of a big variety of phenomena at a relatively cheap computational cost; specially, TDDFT is being used to accurately predict, *ab-initio*, the optical absorption spectra of biological systems. When properly validated, TDDFT calculations can be quite reliable, and they are increasingly used by non-experts to support and interpret experimental results.

In TDDFT the main quantities to be represented are three-dimensional functions: the density and the single particle orbitals (Kohn-Sham states). For big systems these are the most memory demanding variables. OCTOPUS uses MPI and OpenMP for parallelisation. For several tasks the code relies on external libraries. For example, linear algebra operations are handled using the BLAS and LAPACK libraries, and the Poisson equation is solved using very efficient massively parallel libraries (either interpolating scaling functions (ISF) [9] or parallel fast Fourier transforms (PFFT) [10]). Also the Laplacian of the states has to be evaluated for every mesh point. Octopus calculates it by finite differences, usually using a *star-stencil* with 24 neighbours. Support for BLACS and SCALAPACK is also available, but it has not been optimised yet.

Octopus is released under the GPL license, so it is freely available to the whole scientific community for use, study and modification. The code has been developed extensively in the last years to study systems up to hundreds of atoms, and it is a present goal to make it suitable for systems of thousands of atoms. Over the past years, Octopus has evolved into a fairly complex and complete tool, and it is now being used by dozens of research groups around the world.

2.1   Data Structures

As mentioned above, in TDDFT the main quantities to be represented are three-dimensional functions: the electronic density and the single particle orbitals (states). The single particle orbitals are evolved following the time-dependent Kohn-Sham equations [6] taking as initial condition in most cases the solution of the ground-state density functional theory problem, also obtained by Octopus. In the code the functions are represented in a real-space mesh, and differential operators are approximated by high-order finite differences. Octopus is able to combine multi-level parallelism by using MPI for coarse grain parallelism and OpenMP for a finer level. It can also take advantage of GPU architectures. The MPI parallelisation relies on a tree-based data parallelism approach. The main piece of data to be divided among processes are the single particle states, an object that depends on two main indices: the state index and the space coordinate. Each one of these indices is associated with a data parallelization level, where each process is assigned a section of the total range of the index. There is also a third index, which defines the k-point and spin values. This multi-level parallelization scheme ensures a very good scaling of real-time TDDFT.

As the size of the system grows, two factors affect the computing time: first, the space region to simulate is larger, and second, there are more electrons to simulate (which is directly related to the number of electronic states). By dividing each of these degrees of freedom among processors, multi-level parallelization ensures that the total parallel efficiency remains constant as we increase the system size and the number of processors.

Mesh. The mesh is the data structure used to represent the space. The real-space has to be bounded and discretised, and it can be done in one, two or three dimensions. On top of this mesh are represented such things as the potential energy (ν), the electronic density of the system (ρ), the system wavefunction, etc.

The shape of the mesh data structure can be *adaptive*, parallelepiped or spherical (Figure 1). In the former case (which is the default) the mesh is made by an union of spheres centred in the atoms of the simulated system; in the other two cases the shape is regular: a parallelepiped and a sphere, respectively. Two parameters define the mesh: the radius (from each atom in the adaptive case; the length of each edge in the parallelepiped case; and from the centre in the case of the sphere) and the spacing, which is the distance between two consecutive mesh points. All the mesh points are usually taken to be equally spaced.

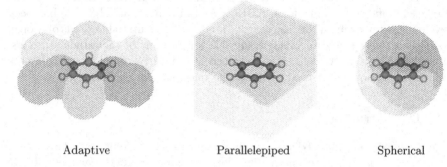

| Adaptive | Parallelepiped | Spherical |

Fig. 1. Different mesh partitions for the benzene molecule. The mesh is partitioned in 6 domains using adaptive, parallelepiped and spherical shapes.

State. A state is the representation of an electronic orbital. Each one of the electronic states under study is defined by a concrete data structure, which is represented over the mesh.

A lower bound of the memory requirements of the simulation can be estimated from the number of simulated states × number of mesh points.

Cube. The cube is an auxiliary data structure used to solve Fourier transforms (FFTs), which are needed to calculate the electrostatic Hartree potential. The evaluation of FFTs in 3D demands to represent the data in parallelepiped meshes. In addition, the cutoff technique [11] used in OCTOPUS to reduce the effect of the undesired periodic images of the inverse FFT also requires cubic meshes.

The edge of the cube is by default twice the length of the largest axis of the mesh; thus, it is at least 8 times bigger than the corresponding mesh. The cube is filled with the mesh points (explained in the next section), and all the extra points are padded with zeros.

2.2   Execution Modes

OCTOPUS simulates physical systems basically in two *phases*: firstly, the electronic ground-state is calculated, and secondly, the converged Kohn-Sham wavefunctions are propagated in time under the effect of an external perturbation. The ground-state (GS) is obtained applying density functional theory (DFT), whereas the TDDFT theory is used to obtain the time-dependent (TD) solution.

Memory needs increase (roughly) quadratically with the system size under simulation. Generally, the GS requires the use of real numbers for quantum states, while the TD runs require the use of complex numbers. Thus, the amount of memory in a time-dependent run usually doubles that of the corresponding ground-state. On the other hand, total computation amount (FLOP) does not increase in the same way for the two modes: the ground-state increases roughly as $S^{5/2}$, while the time-dependent increases as $S^{3/2}$ (S is the system size, i.e., the number of atoms).

These two run modes scale differently with respect to the number of processes, and pose different problems. More precisely, the time-dependent run mode scales better than the ground-state calculation, as we will see in the next section. The different scalabilities are due to the specific parallelization schemes available in each run mode.

3   Improvements in the Memory Usage

OCTOPUS has been used successfully during the last years to analyse complex nanostructures, but it had some limitations when a high number of processes must be used to simulate systems with thousand of atoms. Most of the limitations are related to the memory requirements, which is a limited resource in any computer. So, we have analysed how OCTOPUS uses the computer memory, to optimise its usage.

3.1   Mesh Partitioning

The largest data structures, namely the functions represented on the mesh, have to be distributed between the available computer nodes. An important issue in this domain parallelisation is selecting which points of the mesh are assigned to each processor. This task, known as mesh partitioning, is not trivial for meshes of adaptive shape. Not only the number of points must be balanced between processors but also the number of points in the boundary regions must be minimised. This issue is crucial, because communication costs are directly related to the number of boundary points. An example of a mesh partitioning is shown in Figure 1 (each colour represents a domain). OCTOPUS relies on external libraries for this task: METIS [12] and PARMETIS [13]. These libraries implement several algorithms and the quality of the partition will depend on the selected one.

Serial Partitioning. Previous to this work, METIS was the default library in OCTOPUS to partition the mesh. It is a serial library, so the process that calls it does need all the mesh data, which have to be gathered from all the processes. Moreover, to reduce communication needs, all the processes called the library and, consequently, all of them needed to store the whole data structure. This is a valid approach if small-medium size meshes are used, but it is unfeasible with million of mesh points, because the amount of memory per process is limited.

Parallel Partitioning. With the aim of avoiding the mentioned memory problems with the mesh partitioning, we have adapted OCTOPUS to use a new a highly parallel library called PARMETIS. PARMETIS is a parallel library implemented in MPI and built on top of the METIS library (version 5.1). As METIS, this library makes partitions of graphs, so a transformation of the OCTOPUS mesh structure into a graph is required previously. We use a graph data structure to store the neighbour information of all points. Therefore, each mesh point is represented by a vertex in the graph, and neighbours points are connected through an edge. The information of which the neighbours of a point are is necessary e.g. to calculate the discretised Laplacian operator, which is done using a *stencil* (a given function of the values of a function in a given point and in its neighbours). PARMETIS works in parallel and, so, all data structures are distributed. Each process works with a contiguous chunk of the graph, with local matrices of vertexes and edges, and all the processes know how the graph is initially distributed.

In order to use the new library, we have developed a new distributed version of the mesh partitioning for OCTOPUS. At the beginning, each process obtains an arbitrary (not optimised) portion of the mesh (called "division"), whose size is roughly N/P (being N the number of mesh points and P the number of MPI processes involved in the partitioning). Then, using this initial division, each process calls PARMETIS to obtain the actual domain partition of the graph. Finally, local results are informed to the corresponding processes (the real owners of the mesh points).

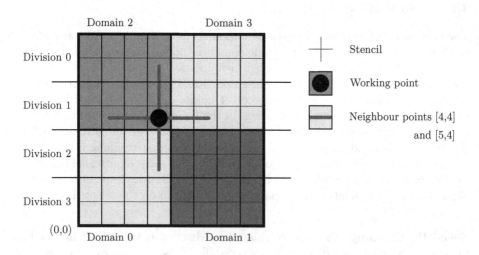

Fig. 2. A simplified example of the OCTOPUS mesh partition. The initial (non-optimised) division schedules chunks of 2×8 points (division i is assigned to process i). Afterwards, each process uses PARMETIS locally to create the actual domain partitions (4x4 points), aiming to minimize the boundary points.

So, the final domain partition of the mesh is saved in a distributed way. Consequently, when a process works with a neighbour point that does not belong to its partition, it must identify the owner of the point. This information can be obtained from the initial owner of the points, that processed it using PARMETIS.

Figure 2 shows an example of this procedure. Let us assume that we are working with the blue mesh point with the circle inside, point [3,4], belonging to domain 2 (process 2) and labelled as *working point*. Using a stencil of length 2, it wants to obtain the two neighbours on the right (points [4,4] and [5,4]). According to the initial division of the mesh, it knows that process 1 (division 1) calculated the final owner of those points. So, it will ask process 1 for this information (both points are assigned to process 3) before doing the actual communication.

This new distribution strategy, based in PARMETIS, allows us to use only local data to do the final partition/decomposition, reducing greatly the use of memory in each process.

3.2   Data Transfers between Different Partition Types

In the context of scientific simulation it is common to use external libraries to do standard operations in an highly optimised way. These libraries could use different data distributions. One example of this situation are the different mesh partitions used in OCTOPUS and in its Poisson solvers [14] (the interpolating scaling functions (ISF) [9] and the parallel fast Fourier transforms (PFFT) [10] libraries).

Both Poisson solvers are based on FFT methods, where the spatial functions are represented in a cubic mesh. The ISF library splits the mesh into domains which are parallel plane-like parallelepipeds (Figure 3C), while PFFT makes a two-dimensional split of the mesh into column-like domains (Figure 3B). However, OCTOPUS divides the mesh in a three-dimensional manner into more compact domains (Figure 3A). Therefore, the data that a process deals with in the OCTOPUS main program are not the same as the PFFT and ISF libraries use inside, and hence a data transfer is necessary, independently of the shape of the mesh.

The simple way to carry this data transfer out is to gather all the data (density ρ or potential v) in all the processes before distributing them accordingly to the new mesh partition. In fact, this was the option used in the previous version of OCTOPUS. This solution works if the number of processes is not very high, but efficiency decays when the number of processes increases, because of the global MPI communication. Therefore, a new data transfer strategy had to be implemented to be used in massively parallel machines.

We have overcome this problem in a highly efficient way. At the initialisation stage a mapping between the OCTOPUS mesh partition and the FFT mesh partition is established and saved. This mapping is used when running the actual solver to efficiently communicate only the strictly necessary data between processes. Each of the different domain decomposition (for instance, OCTOPUS and FFT partitions) is represented by a MPI group/communicator, that might differ in the number of processes (OCTOPUS might use processes for different

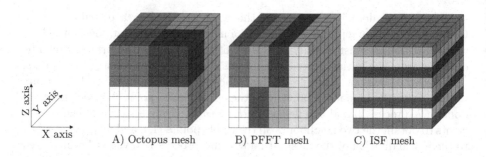

Fig. 3. Simplified domain decomposition of the simulation mesh. Each little cube represents a mesh point (8^3 points in total) and each colour represents a partition (8 domains). A) OCTOPUS mesh with a 3D domain decomposition; B) PFFT mesh with a 2D decomposition; C) ISF mesh with a 1D decomposition.

parallelisation levels), but includes all the mesh points. At a given time, data points have to be sent from one group to the other. Unfortunately, MPI does not allow to send information between different groups unless they are disjoint, which is not, by definition, the current case. This means that communication will have to be done inside one of the groups. This is not a problem, because we can determine the rank of the receiver processes through the `MPI_COMM_WORLD` global communicator. Therefore, all the data transfers between processes can be done in a unique `MPI_Alltoallv` function. It is important to note that, using this improvement, each process only transfers each point once. Therefore, the total amount of information that must be sent between all the processes is equal to the number of points in the mesh, and it is independent of the number of processes. The data transfer process has been encapsulated in an specific Fortran module in OCTOPUS,[1] achieving almost perfectly linear parallel scaling.

3.3 Other Improvements

Apart from the memory optimisations explained above, the older version of the code used as temporary arrays three matrices of size $P \times P$, being P the number of running processes. Those matrices are small with few processes, but they become problematic when we need to use a high number of processes > 2048. Assuming 8 byte integers, with 2048 processes the size of each matrix is of 32 MiB, yet not very large, but with 8192 processes the size increases to 0.5 GiB, and with 64K processors the size is already prohibitive (32 GiB) for current machines. After implementing a new logic, we got rid of them, because they were only used in the initialisation process.

---

[1] See the source code of our implementation in the file
`src/grid/mesh_cube_parallel_map.F90` at public Subversion repository:
`http://www.tddft.org/svn/octopus` and the documentation in the wiki:
`http://www.tddft.org/programs/octopus`).

Other inefficiencies related to initialisation processes (serial evaluation of the local and boundary points, for example) also had impact in the execution time when simulating big systems, and have been removed.

4   Results

Together with the optimisation of the data flow, the main goal of this work was to address the huge memory requirements of Octopus when running in a large HPC environment. The changes made in the code allow it to run in a much more efficient manner, opening the door for the simulation of the interaction of light with big biomolecules. In this section we will first present the improvements in the memory usage, and next will proceed to demonstrate the extreme scalability of the code in its current form. No scalability tests for older versions of the code will be shown, as they were not able to handle the largest systems discussed below.

Tests have been done for different chunks of the light harvesting complex molecule, of 180, 650, 1365 and 2676 atoms (Figure 4). Those systems have from 452,878 mesh points and 250 electronic states to 4,106,680 points and 3,656 states.

Tests were run using three different computers: a Blue Gene/P, a Blue Gene/Q and a small cluster (Corvo).

- The Blue Gene/P (BG/P) [15] system uses low power processors (IBM PowerPC 450, running at 850 MHz) connected with highly efficient communication networks. Each node has one chip of 4 cores, with only 2 GiB of RAM. In its largest installation it had 294,912 processor cores.
- The Blue Gene/Q (BG/Q) [16] is the next generation of the BG systems. Each compute card (which we call a *compute node*) of a BG/Q features an IBM PowerA2 chip with 16 cores working at a frequency of 1.6 GHz, 16 GiB of RAM and the network connections. A total of 32 compute nodes are plugged into a so-called *node card*. Then 16 node cards are assembled in one midplane, which is combined with another midplane and two I/O drawers to give a rack with a total of $32 \times 32 \times 16 = 16,386$ cores for each rack.
- The cluster Corvo represents a common parallel machine of a research group. Each node of Corvo has two Intel Xeon with 6 cores each and 48 GiB of RAM. There are 960 cores in total, connected with an Infiniband network.

We want to remark that the amount of memory per processor core is relatively low and it is not increasing significantly. For instance, in the computers that we have used, the quantity of memory per core goes from 0.5 GiB in the BG/P to 4 GiB in Corvo, if all the cores are used to calculate.

4.1   Memory Measurement

Octopus provides a tool to measure the memory usage and the timing of the most important functions. A deep profiling of the memory usage has been

Fig. 4. Different chunks of the chlorophyll molecule, from 180 atoms to 2676 atoms

done mainly using this internal profiler and validated with VALGRIND'S MAS-SIF tool[17]. For example, when running the system of 180 atoms in 32 MPI processes, MASSIF estimates memory usage up to 96 MiB, whereas OCTOPUS estimates 88.5 MiB. This difference is owing to the fact that OCTOPUS does not take into account the memory of linked libraries such as MPI, ISF, etc. Despite this small difference, we have used the internal profiler because its execution overhead is much lower than that of MASSIF tool.

Figure 5 shows the memory usage per process of the system of 180 atoms, before and after the optimisations we have done. Memory usage per process is decreasing as more processes are used, but unfortunately the oldest version (4.0) of OCTOPUS tends to increase the use of memory after a relatively small number of processes (256) and, so, makes impossible the use of thousands of processes. On the contrary, the last version has eliminated this increasing tendency. In fact, the memory need has been decreased by a factor of four in the studied range, with a clear tendency to be more efficient when using more processes. To realise the importance of this restriction it has to be considered that the amount of memory per CPU core hardly goes beyond 4 GiB.

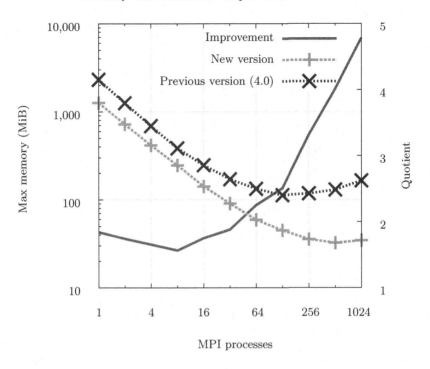

Fig. 5. Ground-state maximum memory usage (from OCTOPUS profiler) per MPI process for the system of 180 atoms

4.2 Data Transfer Time Measurement

The new implementation of the mesh data transfer between different partitions (explained in section 3.2) shows really good performance as can be observed in the Figure 6. The time spent transferring data is really low, compared with the execution time of the library (PFFT in this case). It is shown a particular case of a cubic mesh of edge 31.6 and spacing of 0.2, which makes a total amount of 4,019,679 mesh points and needs $325^3 = 34,328,125$ cube points for the PFFT. The data transfer accounts from 0.5% to 55% of the total Poisson execution time.

4.3 Execution Times

Improvements made in the OCTOPUS code allowed us to simulate bigger systems than ever. Next paragraphs show the results obtained with both executions modes of OCTOPUS: ground-state and time-dependent.

Ground-State. Figure 7 shows the execution time of the ground-state calculation of systems of 180, 441 and 650 atoms in Blue Gene/P and Corvo.

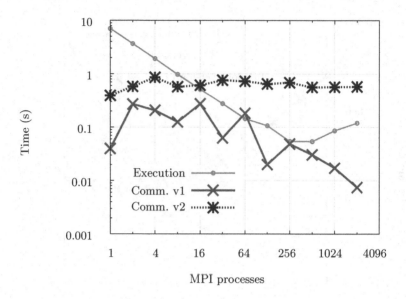

Fig. 6. Time for the new parallel communication pattern (*Comm. v1*), compared with the old serial communication (*Comm. v2*), for a system of 4,019,679 mesh points in the Blue Gene/P machine. While the old version does not scale and becomes a bottleneck, the improvement of the new approach is huge and it has a much better scalability. Total execution time of the Poisson solver is the sum of the *Execution* and *Comm.* times.

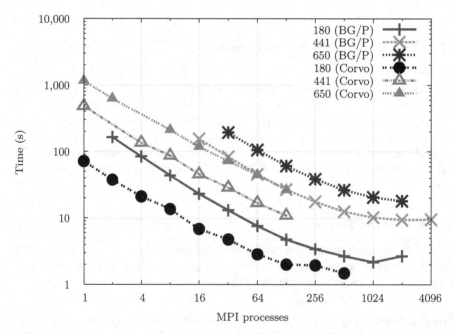

Fig. 7. Ground-state iteration execution time in Blue Gene/P and Corvo (x86-64)

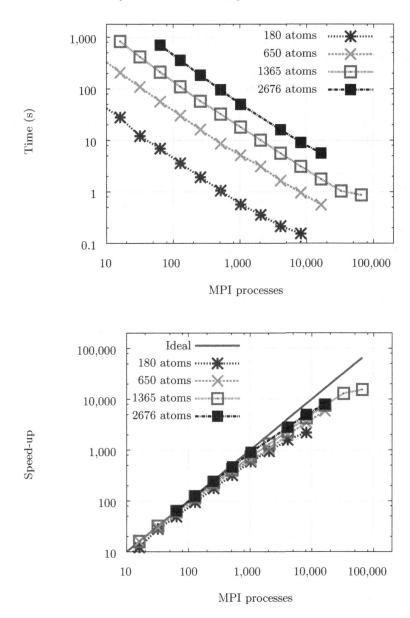

Fig. 8. 180, 650, 1365 and 2676 atoms system running in the Blue Gene/Q up to 65,536 MPI processes (1 MPI process per core)

The trend for all the atomic systems and for both machines is equivalent, being Corvo 3 times faster. The scalability is more than acceptable; for instance, the 180 atoms system scales well up to 256 processes, while with the 650 atoms system the scalability is almost perfect until 256 processes, showing improvements up to 2048 MPI processes.

Time-Dependent. The execution times and relative speed-ups of the time-dependent simulations are show in the Figure 8. Our improvements in the OCTOPUS code have allowed us to go further in the scalability and also to be able to run bigger systems. Executions correspond to the time-dependent in the BG/Q machine and to systems of 180, 650, 1365 and 2676 atoms. Runs are made using one MPI process per core, thus 16 processes per node. Extremely good scalability is shown for the smallest system, that, because of the reduction of the memory needs, can be now run in only 4 MPI processes and it is still efficient with 8k processes. For the system of 650 atoms, there is no sign of saturation up to 16k processes. The system of 1365 atoms is highly parallel up to 32k processors and we have been able to run it in 64k processors, improving by far previous results. Finally, almost perfect scalability has been reached with the system of the 2676 atoms up to 16k processors (we were limited to the tests shown here, because we had a limited CPU quota for our project).

Scaling tests also were done in the BG/P machine, showing a very similar behaviour. In this case, runs were made using one MPI process per node and four OpenMP threads (one per node core). Specially remarkable is the performance we have obtained with the 650 atoms system in this machine, which does not saturate until the maximum available CPU cores (128k). Deserves to mention as well the system of 180 atoms, which is really efficient from 4 up to 64k MPI processes. The new efficient memory usage allowed us to run in this machine a bigger system of 5879 atoms in 32k processors.

5   Conclusions

OCTOPUS is a scientific software package based on TDDFT theory, which is being used successfully by dozens of research groups around the world. To date, it was mostly used to analyse medium-sized complex nanostructures. Modifications were necessary to enable the code to work with bigger systems (thousands of atoms and beyond).

In this work we have analysed and solved some of the limitations that OCTOPUS had in this new scenario, mainly the amount of memory and the transfers of the main data structures in some phases of the computation. Both problems have been solved in a very efficient manner. On the one hand, we have used PARMETIS to do the partitioning of the mesh using only local data, avoiding the use of the whole data mesh in each process and, consequently, reducing greatly memory needs. On the other, we have optimised data transfer between processors when different domain decompositions must be used.

The benefits of these improvements are significant. OCTOPUS uses now a much more limited amount of memory per processor, and it is ready to simulate larger systems. Indeed, we have shown excellent scalability up to 64k and 128k cores, almost independent of the system size (number of atoms) beyond a given size. Although we have shown the executions of different chunks of the light harvesting complex, the real aim is to be able to run the entire light harvesting complex with more than 16,000 atoms, a work which is in progress. The recent improvements make possible to obtain scientifically relevant results for these large systems at a reasonable time.

Acknowledgements. This work was funded by the University of the Basque Country UPV/EHU (Aldapa, GIU10/02) and by the Department of Education, Universities and Research of the Basque Government (IT395-10 research group grant). We acknowledge financial support from the European Research Council Advanced Grant DYNamo (ERC-2010- AdG-267374), Spanish Grant (FIS2010-21282-C02-01), Grupos Consolidados UPV/EHU del Gobierno Vasco (IT578-13), Ikerbasque and the European Commission projects CRONOS (Grant number 280879-2). We acknowledge PRACE for awarding us access to resources Jugene and Juqueen based in Germany at JSC. J. Alberdi-Rodriguez acknowledges the scholarship of the University of the Basque Country UPV/EHU and the the support of project PRACE-3IP. P. García-Risueño is funded by the Humboldt Universität zu Berlin. M. J. T Oliveira acknowledges financial support from Belgian FNRS-FRFC project "Control of Attosecond Dynamics" number 2.4545.12.

References

1. García-Risueño, P., Ibáñez, P.E.: A review of high performance computing foundations for scientists. International Journal of Modern Physics C (IJMPC), 1230001 (2012)
2. Andrade, X., Alberdi-Rodriguez, J., Strubbe, D., Oliveira, M., Nogueira, F., Castro, A., Muguerza, J., Arruabarrena, A., Louie, S., Aspuru-Guzik, A., et al.: Time-dependent density-functional theory in massively parallel computer architectures: the octopus project. Journal of Physics: Condensed Matter 24(23), 233202 (2012)
3. Castro, A., Appel, H., Oliveira, M., Rozzi, C.A., Andrade, X., Lorenzen, F., Marques, M.A.L., Gross, E.K.U., Rubio, A.: Octopus: a tool for the application of time-dependent density functional theory. Phys. Status Solidi (b) 243(11), 2465–2488 (2006)
4. Marques, M.A.L., Castro, A., Bertsch, G.F., Rubio, A.: Octopus: a first-principles tool for excited electron–ion dynamics. Comput. Phys. Commun. 151(1), 60–78 (2003)
5. Hohenberg, P., Kohn, W.: Inhomogeneous Electron Gas. Phys. Rev. 136(3B), B864–B871 (1964)
6. Kohn, W., Sham, L.J.: Self-Consistent Equations Including Exchange and Correlation Effects. Phys. Rev. 140(4A), A1133–A1138 (1965)
7. Fiolhais, C., Nogueira, F., Marques, M.A.L. (eds.): A Primer in Density Functional Theory, 1st edn. Lecture Notes in Physics, vol. 620. Springer (2003)
8. Marques, M.A., Maitra, N.T., Nogueira, F.M.: Fundamentals of Time-Dependent Density Functional Theory, vol. 837. Springer, Heidelberg (2012)
9. Genovese, L., Deutsch, T., Neelov, A., Goedecker, S., Beylkin, G.: Efficient solution of Poisson's equation with free boundary conditions. J. Chem. Phys. 125(7), 074105 (2006)
10. Pippig, M.: PFFT - An extension of FFTW to massively parallel architectures. SIAM J. Sci. Comput. 35, C213 – C236 (2013)

11. Rozzi, C.A., Varsano, D., Marini, A., Gross, E.K.U., Rubio, A.: Exact coulomb cutoff technique for supercell calculations. Phys. Rev. B 73, 205119 (2006)
12. Karypis, G., Kumar, V.: METIS - Unstructured Graph Partitioning and Sparse Matrix Ordering System, Version 2.0. Technical report (1995)
13. Karypis, G., Kumar, V.: Parallel multilevel k-way partitioning scheme for irregular graphs. In: Proceedings of the 1996 ACM/IEEE Conference on Supercomputing (CDROM), Supercomputing 1996. IEEE Computer Society, Washington, DC (1996)
14. García-Risueño, P., Alberdi-Rodriguez, J., Oliveira, M.J.T., Andrade, X., Pippig, M., Muguerza, J., Arruabarrena, A., Rubio, A.: A survey of the performance of classical potential solvers for charge distributions. Journal of Computational Chemistry 35(6), 427–444 (2014)
15. Sosa, C., Knudson, B.: IBM system Blue Gene solution: Blue Gene/P application development. IBM International Technical Support Organization (2008)
16. Gilge, M.: et al.: IBM System Blue Gene Solution Blue Gene/Q Application Development. IBM Redbooks (2013)
17. Nethercote, N., Walsh, R., Fitzhardinge, J.: Building workload characterization tools with valgrind. In: 2006 IEEE International Symposium on Workload Characterization, p. 2 (2006)

Two High-Performance Alternatives to ZLIB Scientific-Data Compression

Samuel Almeida[1], Vitor Oliveira[2], António Pina[2],
and Manuel Melle-Franco[3]

[1] Departamento de Informática, Universidade do Minho,
Campus de Gualtar, Braga, Portugal
[2] Laboratório de Instrumentação e Física Experimental
de Partículas, Campus de Gualtar, Braga, Portugal
[3] Departamento de Informática, Centro de Ciências e
Tecnologias da Computação, Universidade do Minho,
Campus de Gualtar, Braga, Portugal
{sam.alm321,manuelmelle}@gmail.com, {vspo,amp}@di.uminho.pt

Abstract. ZLIB is used in diverse frameworks by the scientific community, both to reduce disk storage and to alleviate pressure on I/O. As it becomes a bottleneck on multi-core systems, higher throughput alternatives must be considered, exploring parallelism and/or more effective compression schemes. This work provides a comparative study of the ZLIB, LZ4 and FPC compressors (serial and parallel implementations), focusing on CR, bandwidth and speedup. LZ4 provides very high throughput (decompressing over 1GB/s versus 120MB/s for ZLIB) but its CR suffers a degradation of 5-10%. FPC also provides higher throughputs than ZLIB, but the CR varies a lot with the data. ZLIB and LZ4 can achieve almost linear speedups for some datasets, while current implementation of parallel FPC provides little if any performance gain. For the ROOT dataset, LZ4 was found to provide higher CR, scalability and lower memory consumption than FPC, thus emerging as a better alternative to ZLIB.

Keywords: Data Compression, Scientific Data, ROOT, Parallel Compression, ZLIB, LZ4, FPC.

1 Introduction

Technological developments have lead to increasingly more powerful data processing systems, higher-resolution sensors and higher bandwidth communication networks. Yet systems are getting increasingly farther from Amdahl's balanced computing system [1], as the abilities to create, process, distribute and store data have not grown equally. In particular, the availability of multi-core systems has greatly amplified the ratio between the available computational power and the available input/output bandwidth [2]. As the potential of "big data" repositories in our society is explored [3,4], this balance between the system's ability to analyse, to transmit and to store data becomes even more important.

B. Murgante et al. (Eds.): ICCSA 2014, Part IV, LNCS 8582, pp. 623–638, 2014.

The same applies to the techniques that can affect it, such as caching, prefetching and compressing the required data.

Data compression, in particular, has been used for many years to trade computing power for storage capacity as well as for network and storage bandwidth whenever the first more abundant than the later ones. It has been used to improve disk space utilization with compressed packages or file-systems, to speedup WAN traffic over slow links and to increase write bandwidth in some solid-state disks. It also presents a large potential for specialized libraries such as NetCDF, HDF5 and ROOT to store scientific data in more compact forms. Taking into account the 2:1 file size reduction allowed by the LZ77 compression, very significant savings in terms of network bandwidth and disk space are achieved in datasets such those from the LHC's ATLAS (A Toroidal LHC Apparatus) experiment [5], which start in the multi-petabyte range, and that after several pre-processing stages, the data reaches the analysis applications still in the tens of terabytes range.

In [6] it was identified that in the analysed application the processing time associated with compression was indeed significant, where the ZLIB library took 18% of the total CPU-time to expand data prior to analysis. Some of that overhead can be offset by the reduced time associated with writing and reading less data to disk, as long as compression/decompression bandwidth exceeds storage bandwidth. But modern storage systems with well behaved usage patterns can provide much more bandwidth than ZLIB decompression can handle, both because a single modern disk is already capable of sequentially reading faster than the 120 MB/s decompression bandwidth achieved by ZLIB in a single processor core (Intel Xeon E5620) and because ZLIB does not support multiple threads.

As ZLIB compression and decompression becomes a bottleneck, techniques must be devised to improve performance while maintaining the advantages of compression. This work focuses both alternative compression methods, that could provide higher bandwidth, and implementations with a level of parallelism, that could properly explore multi-core systems. We present a comparative performance study of six compressors, in terms of bandwidth, compression ratio (CR) and scalability, using a group of different numeric scientific datasets. This approach permits assessing the compressor behaviour in the context of ROOT files and a few other application areas where compression can be beneficial.

This work focuses on the study of the serial compressors gzip, LZ4 and FPC and their respective multi-threaded counterparts pigz, lz4mt and pFPC.

2   Background Information

Scientific data compression has been used extensively, and *in situ* techniques are also on the rise [7,8,9]. For instance, in the ROOT toolkit [10], compression is used to deal with large volumes of information such as that generated by the Large Hadron Collider's (LHC) experiments. The approach there is to compress objects with the general purpose ZLIB or LZMA compressors prior to writing the data nodes to disk.

ZLIB and LZMA compression is asymmetric, the decompression time can be very small compared to the compression time, which makes it fitting for data that has to be read multiple times. Although they achieve very good CR on general purpose data, it is at the cost of both being time-consuming compressors. Several high-bandwidth compression algorithms have been developed in recent years, sacrificing some CR for faster execution times. LZ4 is a good representative of the modern high-performance compressors, and a potential alternative to ZLIB for ROOT, as it is about an order of magnitude faster than ZLIB when compressing, and around five times faster at decompression [6].

Recent research has focused on the development of compression techniques that explore domain knowledge to achieve higher CR or increased throughput, including techniques with large potential such as compressive sensing, but different application domains may require different techniques and yield very different results, so, that potential is hard to explore. The FPC lossless compressor is one example that uses context information to effectively compress and decompress 64-bit double precision floating-point data. Another one is the compression scheme based on entropy coding that is used in ALICE (A Large Ion Collider Experiment) hosted at CERN, where the differences of the times in two consecutive bunches (group of adjacent samples coming from the sensor pad), produced by ALICE's Time Projection Chamber detector, are coded [11], reducing the entropy of the source by exploiting the time correlation present in the data.

2.1 Compressor Algorithms

The gzip compressor implements the DEFLATE algorithm, which is an evolution of original LZ77 [12]. It is a well-known general-purpose compression system, providing high CR at the cost of performance, due to the use of a form of entropy encoding (Huffman coding). The algorithm is very asymmetric, with compression taking between 2.5 and 10 times the decompression speed, depending on the compression level selected and the input data.

LZ4 by Collet [13] is also a lossless compressor based on LZ77 algorithm, but on current multi-core systems it can reach throughputs of more than 400 MB/s per core when compressing, while during decompression it can achieve more than 1.8 GB/s[1], bound only by RAM bandwidth. The algorithm works by finding matching sequences and then saving them in a LZ4 sequence using a token, that stores the literals length (uncompressed bytes) and the match length, followed by the literals themselves and the offset to the position of the match to be copied from (i.e. a repetition). There are optional fields for literals and match length if necessary, and the offset can refer up to 64 KB. With the offset and the length of the match the decoder is able to proceed, and copy the repetitive data from the already decoded bytes. The simplicity of the algorithm together with the fact that entropy coding is not used makes LZ4 decompression very fast.

FPC [14,15] is a lossless compression algorithm for linear streams of 64-bit floating-point data. This compressor is based on a fast compression algorithm,

---

[1] `http://code.google.com/p/lz4/` Accessed March 6th.

tailored for scientific floating-point data compression on high-performance environments, where low latencies and high throughput are essential. The FPC algorithm can be implemented entirely with fast integer operations, resulting in a compression and decompression time one to two orders of magnitude faster than other more generic algorithms. It starts by predicting each value in the sequence and performing an exclusive-or operation (xor) with the actual value. A good prediction results in a substantial number of leading-zeroes in the calculated difference, which are then encoded by simply using a fixed-width count. After each prediction the predictor tables are updated with the actual double value to ensure that the sequence of predictions are the same during both compression and decompression. The remaining uncompressed bits are output in the end, after the count of leading-zeroes.

2.2   Block-Oriented Parallel Compression

When a stream can be divided in blocks, and in order to increase throughput, the processing can be parallelized as each block is delivered to a different thread and handled independently. But this approach carries some restrictions, as resulting compressed blocks have to be joined into the final output and that may or may not be a simple operation, depending on how the stream is concatenated. While LZ4's output stream is byte aligned, in ZLIB the intermediate output streams are bit aligned, so concatenating them implies a time-consuming data shift. To bypass the shift, intermediate blocks have to be terminated explicitly, which consumes more space (5-6 bytes per block) on the final output stream.

The compressor can also take longer to reach an effective dictionary, lowering the CR, as the contextual information is less rich. Due to this, it should be noted that block parallelism is effective only when processed in blocks of a significant size, and, produces less compressed outputs than a single block implementation.

The use of large data blocks may also not be straightforward. ROOT files, for example, have a tree-like internal structure in which only data nodes are compressed, with many being less than 10 K bytes. Multiple threads can still be used in this case, each writing a different data node, but it requires that multiple threads access ROOT's internal structures concurrently, which is only possible after ROOT version 6.

2.3   Multi-pass Compression

When data streams must be compressed immediately as they reach the compressor (such as low-memory or low-latency communication systems), or when forced to use small data blocks, improving performance by block parallelization becomes more difficult. Multi-pass compression is an alternative approach that, in very particular circumstances that depend on the type of input data, may yield significant compression rates.

Multi-pass compression consists of performing multiple passes of the compressor, with a high-bandwidth compressor on a first stage and a high-compression compressor in a second stage. The *rationale* is that a high-performance first stage compressor can reduce the stream significantly, leaving a more thorough analysis to the second compressor that has less data to deal with. To use this technique the higher level redundancy must remain visible to the second compressor, which happens with LZ4 but not so much with the bit-oriented output streams generated by ZLIB. It also requires that not all redundancy is seen by the first level compressor, otherwise the second level will see no compression improvement. And while a more capable first stage compressor could be aware of the higher-level redundancies and eliminate them, a real advantage of this approach is that it may achieve similar results in less time.

In very redundant log data the results are surprising enough to be worth mentioning[2], but we found no evidence of this being the case in our scientific-data streams. In particular, we found that a 3.5 GB binary log file from a profiling application could be compressed with ZLIB to 54 MB on a first pass and to 17 MB in a second. With LZ4 on both stages, the first pass left the file at 56 MB and the second at just over 9 MB. With high-compression LZ4 (LZ4HC) on both stages, the first pass got a 44 MB file, the second pass a 2 MB, and after a few more rounds the file was left with a 750 KB size. In terms of throughput, a single step from that application using ZLIB took 265ms, while the same step with LZ4 and two additional passes of LZ4HC took only 16ms. It also proved effective with some text files, where one pass with LZ4 and one with LZ4HC ended up producing a similar size file as a single LZ4HC, in one fifth of the time.

3   Evaluating Compressor Performance

In this section the main considerations regarding testing the compression performance are presented, including characterization of the datasets and the testing procedures.

3.1   Evaluation Datasets

The compressors were tested with several numerical datasets from different backgrounds and sources (see Table 1). The six different disciplines covered by the 33 datasets, mostly from simulation programs, go from molecular and electronic structure modelling, through message, numeric and observational data to particle collision simulation data. After an initial evaluation of the 33 datafiles, a selection of five was made consisting of only one datafile per datagroup, based on its properties and characteristics being representative of each group (Table 2). The entire dataset can be consulted on Table3.

3.2   Dataset Preparation

The datafiles used came in many different source formats (some in binary and others in text format). FPC compresses only double-precision floating-point

---

[2] A public discussion on this topic is available at
https://groups.google.com/forum/#!msg/lz4c/DcN5SgFywwk/AVMOPri0O3gJ.

Table 1. Characteristics of the datagroups

Datagroup	#Files	Research Area	Software	Data Type
waterglobe	6	water nanodroplet	TINKER	text
engraph	3	graphene flake	TINKER	text
gauss09	4	graphene nanoribbon	Gaussian 09	text
sci-files	13	message, numeric, observational	*diverse sources*	doubles
NTUPs	7	particle collision simulation	LIP code	ROOT files

data, so in order to compare the compressors properly the input files had to undergo some transformation in order to be stripped of additional information and converted to binary, 64bit floating-point numbers.

3.3   Dataset Information Content

The information contents of the datasets are presented in Table 2. Equation (1) describes the percentage of uniques in a dataset, where V is the original vector consisting of all values, and V_{Unique} is the vector with duplicates removed.

$$\text{Uniqueness} = \frac{|V_{Unique}|}{|V|} \times 100\% . \tag{1}$$

$$H(V) = -\sum_{i=1}^{N} p(x_i) \times log_2 p(x_i) . \tag{2}$$

$$\text{Randomness} = \frac{H(V)}{H(\text{Random}_{unique}(|V|))} \times 100\% . \tag{3}$$

Equation (2) represents the Shannon entropy $H(V)$, where N is the number of distinct elements x_i, and $p(x_i)$ the probability of those elements, i.e., the number of x_i occurrences divided by the total number of elements in the file. An element of a dataset depends on the datatype that composes it (8bits ASCII, 32bits single, 64bits double). The randomness is closely related with the entropy as described in (3). Its value reflects how close the Shannon entropy of the datafile is to that of a true 100% unique random datafile with the same number of elements.

These datasets have high degrees of random entropy, in average 81.43%, which indicate that entropy coding will not be very effective and low compression ratios should be expected. The molecular mechanics data correspond to Cartesian coordinates and velocities which by definition have very low content in zeros. Interestingly, the waterglobe.vel has a lower value of randomness than the Cartesian coordinates as the atomic velocities on a molecular dynamics simulation of an equilibrated system are by definition periodical. The electronic density matrix (gauss09_density) was chosen as it has a very low zero content and, expected, high randomness. It is relevant to point out that engraph and the gauss09_density are calculations on graphene systems which are highly regular.

The uniqueness varies more and reaches an average of 44.66%, whilst some files barely contain unique values, others are almost entirely composed of them.

Table 2. Information details from the selected five datasets. R-ness is the Randomness.

Datafiles	Size(MB)	#elements	Unique%	Zeros%	Entropy	R-ness%
waterglobe.vel	1318	172800000	3.30%	0.00%	21.466	78.44%
engraph1_100	650	85190400	72.88%	0.00%	25.664	97.42%
gauss09_density	128	16753366	39.78%	0.05%	22.535	93.90%
msg_sp	277	36263232	98.95%	0.00%	25.032	99.68%
NTUP2_floats	1433	375644746	28.70%	38.77%	15.116	53.07%
AVG (all)	860	163954134	44.66%	10.47%	NA	81.43%

What is interesting is that even the files with low uniqueness are highly random (high randomness%). With this early but quite insightful statistical characterization we can already predict that NTUP datagroup (only one shown) should have the best CR with entropy coders (gzip/pigz). No prediction can be made for the FPC compressor, as that would require knowledge about the smoothness, or data continuity of the datasets, which was not analysed.

The metrics calculated here agree with the values from the sci-files datagroup as described in [15,7]. The overall dataset seems well balanced, with datafiles that cover many possible combinations.

3.4   Measuring Parallel Scalability

In order to assess the scalability potential of the algorithms two metrics were considered: speedup ratio and parallel efficiency. Speedup compares the performance of the parallel version to that of the serial version. It is the ratio between the execution time of a compression cycle using the serial compressor and the execution time of the same compression using the multi-threaded compressor, as shown in (4):

$$\text{Speedup} = \frac{\text{exec time}_{serial}}{\text{exec time}_{parallel}} \implies Sp_t = \frac{T_s}{T_t}. \tag{4}$$

where T_s is the execution time of the serial version and T_t is the execution time of the parallel version, executed with t threads.

Efficiency is the ratio of the achieved speedup with the expected maximum gain, as defined in (5):

$$\text{Efficiency} = \frac{\text{Speedup}_{parallel}}{\#\text{threads used}} \implies Ef_t = \frac{Sp_t}{t}. \tag{5}$$

where Sp_t is the measured speedup for that number t of threads used.

3.5   Parallel Decompression

In pFPC the decompression is symmetric to compression, so in both cases data is chunked and assigned to threads as needed. Due to its internal mechanics, pFPC performs slower on decompression, but with an approximately linear behaviour where higher compression levels come with increasing compression and decompression times.

In ZLIB and LZ4 the decompression is simpler and much faster than compression. In fact, a single LZ4 decompression stream reaches a large portion of the available memory bandwidth. The decompression does not depend on the selected compression level (asymmetric), and there is very limited opportunity to explore parallelism. In fact, in lz4mt and pigz, decompression in parallel is not really implemented. Although pigz uses a single main thread for decompression, it creates three other threads for reading, writing, and checksum calculations, which can speed up decompression under some circumstances.

3.6   Test-Bench Characterization

The results presented in the following section were performed in cluster nodes containing two six-core Intel Xeon X5650 @ 2.66 GHz CPUs and at least 12 GB of RAM. Hyper-Threading technology was enabled, given that integer performance benefits from it, so a maximum of 24 threads per node were tested.

All timing measurements are represented by the walltime reported by the routine omp_get_wtime() from the OpenMP API. To obtain consistent results, for each file there are *nRuns* executions of compression and decompression per compression level (dependent on the compressor), written to the local hard drive disk and to /dev/null (i.e. data is discarded), which are then transferred compressed (but not measured), through the local network, as evaluated in [16].

The parallel scalability tests ran the same sequential tests, and the number of concurrent threads used (*nthreads*) is passed as a parameter. In each compression/decompression loop, the call for the *compress(file)* or *decompress(file)* receives *nthreads*. The amount of *nthreads* used correspond to the sequence (1,2,4,6..24), i.e. one, two, four, six... until twenty four, hence thirteen different tests in total.

4   Results and Discussion

The following sections present the CR, serial and parallel throughput, speedup ratio and memory requirements of the aforementioned compressors, as described in the previous section.

4.1   Compression Ratio and Serial Throughput

Figure 1 depicts the throughput of the three serial compressors, for both compression and decompression, and the CR yield for the datafile NTUP2. Interval bars are used for the three metrics, representing the minimum and maximum values measured from the five selected datafiles. Three compression levels are used for each compressor, with the exception of LZ4 that only has two modes. The levels are the minimum compression allowed by the algorithm, a selected

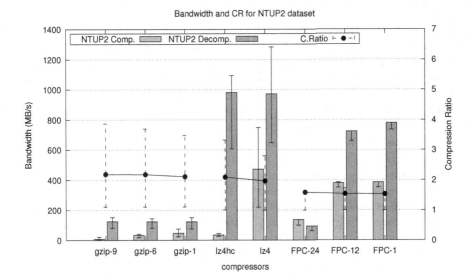

Fig. 1. Bandwidth of compression and decompression from the three different algorithms using NTUP2. The attained compression ratio for each compressor is read on the right y axis.

intermediate level and the maximum (FPC compression has no theoretical maximum, but is bound by the amount of RAM).

The bars in Fig.1 demonstrate immediately the performance advantage of both LZ4 and FPC compared to gzip's lower compression and decompression bandwidth. The compression bandwidth is largely dependent of the compression level selected. In case of gzip (ZLIB), while the highest level of compression is barely readable (8 MB/s), the lowest level is able to reach 48 MB/s. In the case of LZ4 the difference is dramatic, as the highest level of compression reaches 34 MB/s while the lowest reaches 472 MB/s. Regarding FPC, it ranged from 136 MB/s to almost 400 MB/s.

Comparing to the 120 MB/s of ZLIB, LZ4 showed a decompression bandwidth of 1.3 GB/s for one of the five datafiles. For NTUP2, LZ4 reached almost 1 GB/s, while FPC decompression bandwidth ranged from almost 800 MB/s and down to 63 MB/s (depending on the compression level).

Nevertheless, ZLIB provides the highest CR of 2.19 (with a maximum of 3.87), followed by lz4hc with 2.10 (with a maximum of 3.33), lz4 with 1.97 (with a maximum of 2.80), and finally FPC with 1.59. For some other datasets FPC reached the highest CR by a huge difference, e.g. datafile num_plasma attains a CR of 15 (see Table 3), but for the selected five the maximum CR of FPC is only 1.72 (reached when working with the gauss09_alpha dataset) and compression level 12. The CR varies a lot because the advantages of domain knowledge and the heuristics used depend strongly on the data.

4.2 pFPC Irregular Compression Ratio

When it comes to the variability of CR there are some unexpected events with pFPC, as higher compression levels may lead to lower compression. pFPC assigns each thread with chunks, representing a quantity number of double-precision floats to compress (8192 was the elected chunk size), and as more threads are used they will only compress certain parts of the data for the input datafile. Depending on the dimensionality (e.g. number of variables) of the data, the threads receive a chunk and can end up getting the values from the same dimension (variable) as they process the file. Therefore, this will affect the predictions and CR for the best if the same dimension ends up with same thread, or for worse if the threads get chunks from different dimensions.

Figure 2 presents the CR obtained with pFPC for the five datasets varying the compression level. The pattern that appears with gauss09_alpha repeats itself, much more subtly, with engraph1_100. Both datasets show a decrease in CR which then starts to recover with higher levels. With waterglobe.vel.bin the CR line starts to decline after compression level 17 (not visible on current scale), while with NTUP2 it starts to increase after level 14. These events depend on the file itself and the compression level of FPC/pFPC, as these algorithms are based on predictors. The predicted values change with each compression level, thus giving a chance to expose these behaviours. Summing up, higher compression levels in FPC/pFPC do not always yields higher compression ratios.

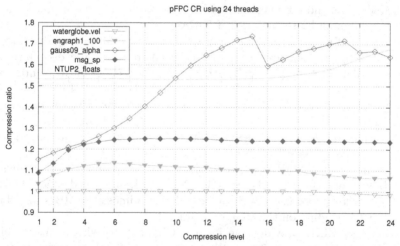

Fig. 2. Compression ratio versus compression level in pFPC

4.3 Parallel Speedup Ratio

Figure 3 presents comparative plots for the speedups of the parallel compressors versus the serial compressors. To compare the speedup ratio from different compression levels, three levels were used, low, medium, and high compression, with

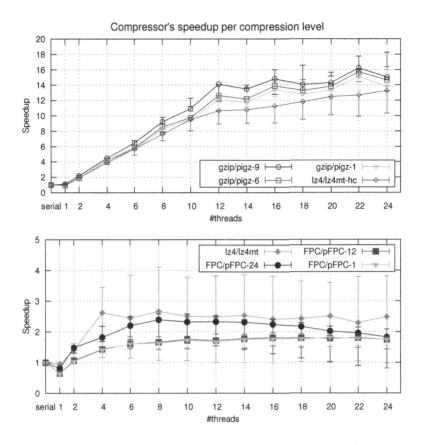

Fig. 3. Speedup versus nthreads used for the NTUP2 datafile. pigz and lz4mthc (*top plot*) with higher speedup, and lz4mt and pFPC (*bottom plot*) with lower speedup.

the exception of lz4/lz4mt that only have low and high compression levels. The NTUP2 dataset is used as the baseline, the remaining four datasets are represented as interval bars (minimum and maximum). The value 'one' corresponding to $x=0$ represents the baseline for the speedup, and the values on $x=1$ represent one-threaded version of each compressor/parameter.

Figure 3 shows that speedup varies significantly, either with the datafiles, the compression level used or the compressors themselves. It is split in high and low speedups in order to improve readability. The best performing compressor is pigz when used with maximum compression. After 12 threads, when the maximum number of cores run is reached, the speedup increases more modestly, which demonstrates the benefits of hyper-threading in this kind of workload.

Speedups are larger with more demanding compression levels, as higher compression levels usually mean more computations, thus, longer execution times and a better chance to explore parallelism. This is indeed observable in almost every case for the initial *nthreads*, most notably on pigz with compression level

9, that yields the highest speedups of this study. On the opposite, lz4mt and all pFPC levels show very poor speedups. Using LZ4 in the fast mode is so fast that using multiple threads can actually reduce performance (when the datasets are small the execution times are really low). However, when datasets are bigger and/or the compression level is increased it leads to longer execution times, which in turn yields higher compression speedup (lz4mthc achieves a speedup close to 11 using 12 threads).

Super-linear speedups, or ratios above one, appear mostly with higher compression levels on pigz, and less frequently lz4mt, depending on the datafile. This arises from the fact that the parallel implementation is at times faster than the simpler serial algorithm, which was unexpected. This is particularly noticeable when 12 threads are used.

pFPC shows the worst scaling, presenting the lowest speedup values since the first *nthreads*. The overhead of the multi-threaded version is specially negative for pFPC.

Depicted in Fig.4 is the speedup versus the compression level, using 12 and 24 threads, in order to assess the scalability of the algorithm when the level of compression is increased. The top left plot (pigz 12 threads) shows once again that pigz was faster with one thread than gzip, as the speedups consistently surpass the theoretic limit of 12. The same does not happen with pFPC, that shows a peak followed by a drop with higher compression levels. The behaviour is the same with 12 or 24 threads, with the nuance that the speedups of two files

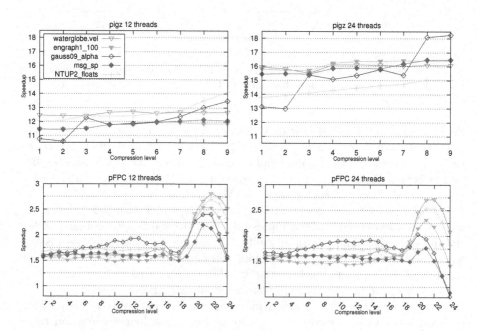

Fig. 4. Speedup versus compression level for pigz (*top plots*) and pFPC (*bottom plots*)

(gauss09_alpha and msg_sp) drop below one with 24 threads when maximum compression settings are used.

4.4 Memory Requirements

The memory required for ZLIB and LZ4 serial compression and decompression is insignificant in a modern system. Nevertheless, the respective parallel compressors, pigz and lz4mt, do use more memory than their serial counterparts. The values observed for pigz were around 10 MB with 12 threads and 18.5 MB with 24 threads, while lz4mt reserves about 100 MB and 196 MB with 12 and 24 threads respectively. For decompression the memory requirements are very similar, and only 100-400 KB lower for both the gzip/pigz and LZ4/lz4mt compressors.

When it comes to FPC memory usage, and in particular to the parallel pFPC, the requirements can be much higher. Serial FPC allocates a table with 2^{n+4} bytes of memory, while pFPC allocates 2^{n+4} bytes for each thread, with n being the compression level selected. This means that the amount of memory used

Table 3. Highest CR and speedup values for each datafile. The third and sixth columns contain, enclosed in square brackets, the algorithm, number of threads and compression level that originated these values.

Datafiles	CR	Largest CR Conditions	MB/s	Sp_t	Largest Sp_t Conditions	MB/s	Ef_t
waterglobe.arc.txt	2.14	[gzip 1 8]	7.3	17.94	[pigz 24 9]	131.6	0.75
waterglobe.1col.arc.txt	2.20	[gzip 1 8]	6.2	18.17	[pigz 24 8]	111.9	0.76
waterglobe.vel.txt	2.19	[gzip 1 8]	6.7	18.10	[pigz 24 8]	122.1	0.75
waterglobe.1col.vel.txt	2.27	[gzip 1 8]	5.3	18.41	[pigz 24 9]	97.1	0.77
waterglobe.arc.bin	1.20	[gzip 1 3]	20.0	16.35	[pigz 24 8]	286.0	0.68
waterglobe.vel.bin	1.48	[gzip 1 5]	19.1	16.12	[pigz 24 5]	307.6	0.67
engraph1_100.txt	2.33	[gzip 1 9]	5.7	18.60	[pigz 24 8]	106.5	0.78
engraph1_100.1col.txt	2.44	[gzip 1 9]	5.0	19.18	[pigz 24 9]	95.5	0.80
engraph1_100.bin	1.22	[gzip 1 3]	20.3	16.42	[pigz 24 7]	276.4	0.68
gauss09_alpha.txt	4.37	[gzip 1 9]	1.9	20.85	[pigz 24 9]	39.4	0.87
gauss09_density.txt	2.36	[gzip 1 9]	4.1	19.28	[pigz 24 9]	78.6	0.80
gauss09_alpha.bin	3.87	[gzip 1 9]	4.6	18.25	[pigz 24 9]	84.7	0.76
gauss09_density.bin	1.09	[FPC 1 24]	82.8	14.83	[pigz 24 9]	294.5	0.62
msg_bt	1.29	[FPC 1 24]	82.4	16.06	[pigz 24 9]	263.5	0.67
msg_lu	1.17	[FPC 1 20]	173.7	15.92	[pigz 24 7]	279.4	0.66
msg_sp	1.26	[FPC 1 24]	116.7	16.47	[pigz 24 9]	217.9	0.69
msg_sppm	7.43	[gzip 1 9]	314.8	15.84	[pigz 24 8]	360.0	0.66
msg_sweep3d	3.09	[FPC 1 24]	166.3	15.40	[pigz 24 7]	272.4	0.64
num_brain	1.16	[FPC 1 24]	96.0	15.68	[pigz 24 5]	263.2	0.65
num_comet	1.16	[gzip 1 9]	88.7	15.76	[pigz 24 9]	236.3	0.66
num_control	1.16	[gzip 1 9]	18.0	15.53	[pigz 24 7]	280.0	0.65
num_plasma	15.00	[FPC 1 24]	127.3	13.05	[pigz 24 5]	322.3	0.54
obs_error	3.54	[FPC 1 24]	91.4	15.94	[pigz 24 8]	193.2	0.66
obs_info	2.27	[FPC 1 24]	65.7	12.30	[pigz 20 7]	232.8	0.62
obs_spitzer	1.23	[gzip 1 3]	18.0	16.45	[pigz 24 9]	203.7	0.69
obs_temp	1.04	[gzip 1 4]	18.3	13.56	[pigz 24 6]	247.7	0.56
NTUP1_floats.bin	2.19	[pigz 1to24 9]	107.8	16.19	[pigz 22 9]	116.2	0.74
NTUP2_floats.bin	2.19	[pigz 1to24 9]	107.9	16.20	[pigz 22 9]	116.3	0.74
NTUP3_floats.bin	2.19	[pigz 1to24 9]	107.6	14.47	[pigz 22 8]	257.2	0.66
NTUP4_floats.bin	2.19	[pigz 1to24 9]	107.8	14.44	[pigz 22 8]	257.1	0.66
NTUP5_floats.bin	2.19	[pigz 1to24 9]	107.7	14.49	[pigz 22 8]	257.6	0.66
NTUP1to5_doubles.bin	4.27	[pigz 1to24 9]	94.9	14.76	[pigz 22 3]	884.7	0.67
NTUP1to5_floats.bin	2.19	[pigz 1to24 9]	114.4	14.84	[pigz 24 8]	263.8	0.62

grows exponentially with the compression level selected. For decompression both FPC and pFPC require about the same memory, because it is needed to refill prediction tables upon the decompression process.

In [15] FPC was tested with n=25, so it took $2^{25+4} = 512M$ bytes of memory, but in order to use the same compression level with 24 threads the tables would occupy 12 GB of memory, which was unavailable in the testing nodes. To stay within the limit n was set to $n = 24$, which allows for the use of all available threads $24 \times 2^{24+4} = 6\text{GB}$ of memory.

4.5   Compression Ratio and Speedup over the Full Dataset

The highest CR and highest speedup measured are presented on Table 3, with CR summarized on the left three columns and speedup on the rightmost four columns. The speedup values have an extra fourth column that presents the associated parallel compression efficiency of the best attained speedup. As one can verify, the serial algorithms have the best CR, with the exception of NTUPs that are best compressed with pigz, which is also the compressor that achieve highest speedups. The best compression ratios come mostly from higher compression levels, which is expected. pigz has the best speedup for every dataset, with mostly 24 threads, and an overall efficiency above 66%.

5   Conclusions

Selecting an effective compressor for large volumes of scientific data is not an easy task. Firstly, scientific data can be generated and processed with many different usage patterns, which means that the balance between compression and decompression effort, and storage and communication requirements, are not easy to find. Secondly, data can be highly heterogeneous, and the quality of compression depend strongly on the very data. This is relevant, as for large data volumes a poor compressor choice will needlessly increase the CPU time.

We studied high-performance alternatives to ZLIB for high energy physics data and a combination of representative scientific datasets, the general purpose ZLIB/gzip against LZ4 (another general LZ dictionary with increased performance) and the floating-point data compressor FPC. Five datasets with an average random entropy (randomness) of 84.5% were selected, out of 33 datasets, for a more thorough analysis.

In terms of performance, LZ4 is the fastest decompressor, and the faster compressor but only on the lowest compression level. For parallel speedup, pigz yields the best values, but is the slowest compressor when it comes to absolute serial execution times. The most efficient parallel compressor is pigz, but closely followed by lz4mt, presenting efficiency around one when 12 threads are used on a 12 core machine.

As expected, FPC achieved in some cases CR that were unattainable by general purpose compressors [15], arising from the domain-knowledge in the algorithm. We tested if FPC could be a potential alternative to the LZ4 compression system being implemented in the ROOT toolkit [17]. For 9 datasets FPC had the highest CR of all compressors, while the gzip/pigz pair achieved the highest CR of the remaining 24 datasets. In the five datasets analysed in more detail, FPC shows the highest CR in two cases, but the maximum CR in those cases was 1.29 on data that was difficult to compress by other means. In terms of performance, FPC topped 20% below the performance of LZ4 low compression, which is still much higher than ZLIB. However, parallel scalability was poor, as the multi-threaded implementation seems not to be fully mature yet.

In the context of the high energy physics datasets studied, nevertheless, the behaviour of FPC does not seem competitive. On the ROOT dataset the FPC CR, less than 1.8, was much lower that both LZ4's CR of 2 and ZLIB's CR of 2.1. Also, on practical systems the memory requirements for the pFPC compressor will become critical. In fact, when high compression levels are used together with many threads in pFPC several gigabytes of RAM memory are necessary. On the other hand LZ4 performance is very good, CR degradation is limited and even the parallel implementations use memory sparingly, not becoming an obstacle for other application's activities.

Acknowledgments. This work is funded by National Funds through the FCT - Fundação para a Ciência e a Tecnologia (Portuguese Foundation for Science and Technology) within project PEst-OE/EEI/UI0752/2014, UT Austin — Portugal FCT grant SFRH/BD/47840/2008, and the resources from the project SeARCH funded under contract CONC-REEQ/443/2005. We would also like to thank Nuno Castro and Rafael Silva for their contributions.

References

1. Bell, G., Gray, J., Szalay, A.: Petascale computational systems. Computer 39(1), 110–112 (2006)
2. Hilbert, M., López, P.: The worlds technological capacity to store, communicate, and compute information. Science 332(6025), 60–65 (2011)
3. Staff, S.: Challenges and opportunities. Science 331(6018), 692–693 (2011)
4. Lohr, S.: The age of big data. The New York Times (February 11, 2012)
5. Search for pair production of heavy top-like quarks decaying to a high-p_T W boson and a b quark in the lepton plus jets final state at $\sqrt{s}$=7 TeV with the ATLAS detector
6. Oliveira, V., Pina, A., N.C.F.V.A.O.: Even bigger data: Preparing for the LHC/atlas upgrade. Ibergrid 2012 submission (November 2012)
7. Schendel, E., Jin, Y., Shah, N., Chen, J., Chang, C., Ku, S.H., Ethier, S., Klasky, S., Latham, R., Ross, R., Samatova, N.: ISObar preconditioner for effective and high-throughput lossless data compression. In: 2012 IEEE 28th International Conference on Data Engineering (ICDE), pp. 138–149 (April 2012)
8. Schendel, E.R., Pendse, S.V., Jenkins, J., Boyuka II, D.A., Gong, Z., Lakshminarasimhan, S., Liu, Q., Kolla, H., Chen, J., Klasky, S., Ross, R., Samatova, N.F.: ISObar hybrid compression-I/O interleaving for large-scale parallel I/O optimization. In: Proceedings of the 21st International Symposium on High-Performance Parallel and Distributed Computing, HPDC 2012, pp. 61–72. ACM, New York (2012)

9. Lakshminarasimhan, S., Shah, N., Ethier, S., Ku, S.H., Chang, C.S., Klasky, S., Latham, R., Ross, R., Samatova, N.F.: Isabela for effective in situ compression of scientific data. Concurrency and Computation: Practice and Experience 25(4), 524–540 (2013)

10. Brun, R., Rademakers, F.: Root - an object oriented data analysis framework. Nuclear Instruments and Methods in Physics Research Section A: Accelerators, Spectrometers, Detectors and Associated Equipment 389(1-2), 81–86 (1997), New Computing Techniques in Physics Research V

11. Nicolaucig, A., Mattavelli, M., Carrato, S.: Compression of tpc data in the alice experiment. Nuclear Instruments and Methods in Physics Research Section A: Accelerators, Spectrometers, Detectors and Associated Equipment 487(3), 542–556 (2002)

12. Ziv, J., Lempel, A.: A universal algorithm for sequential data compression. IEEE Transactions on Information Theory 23(3), 337–343 (1977)

13. Collet, Y.: Development blog on compression algorithms, http://fastcompression.blogspot.in/2011/05/lz4-explained.html

14. Burtscher, M., Ratanaworabhan, P.: High throughput compression of double-precision floating-point data. In: Data Compression Conference, DCC 2007, pp. 293–302 (March 2007)

15. Burtscher, M., Ratanaworabhan, P.: FPC: A high-speed compressor for double-precision floating-point data. IEEE Transactions on Computers 58(1), 18–31 (2009)

16. Welton, B., Kimpe, D., Cope, J., Patrick, C., Iskra, K., Ross, R.: Improving I/O forwarding throughput with data compression. In: 2011 IEEE International Conference on Cluster Computing (CLUSTER), pp. 438–445 (September 2011)

17. Peters, A.J.: Lz4hc compression for root and io baseline evaluation. In: ROOT IO Workshop (December 2013)

Design of a Relocation Staff Assignment Scheme for Clustered Electric Vehicle Sharing Systems *

Junghoon Lee and Gyung-Leen Park

Dept. of Computer Science and Statistics,
Jeju National University, Republic of Korea
{jhlee,glpark}@jejunu.ac.kr

Abstract. This paper presents a design and evaluates the performance of a relocation staff allocation scheme for electric vehicle sharing systems, aiming at overcoming the stock imbalance problem and thus improving the service ratio. Basically, the relocation procedure moves vehicles from overflow stations to underflow stations according to the future demand estimation. For a given target distribution and the relocation pairs, the number of staff members for each cluster is decided to reduce relocation distance and time. The proposed scheme preliminarily runs the unit scheduler with minimal staff allocation to build an empirical distance estimation model. It repeats estimating the relocation cost for each cluster and assigning a staff member to the cluster having the worst relocation distance one by one. The performance measurement results show that the proposed scheme can reduce the relocation distance by up to 31.7 % compared with the even allocation scheme. It invokes the unit scheduler just twice, but achieves the performance comparable to the long loop scheme which runs the unit scheduler as many times as the number of staff members.

Keywords: Electric vehicle sharing system, vehicle relocation, staff allocation, two-phase allocation, relocation distance.

1 Introduction

Electric vehicles, or EVs, are expected to replace gasoline-powered vehicles thanks to their eco-friendliness, particularly in low carbon emissions and better energy efficiency [1]. However, due to high price and not so easy maintenance, EVs are not yet affordable for personal ownership. Hence, EV sharing is a promising and reasonable business model for the time being, and many countries are building their own city-wide EV sharing systems [2]. The examples include *Green Move* in Milan and *Move About* in Oslo [3]. In addition, Seoul city has also begun its EV sharing service with 18 stations. They are trying to accelerate the penetration of EVs into

---

* Prof. Gyung-Leen Park is the corresponding author.

This research was financially supported by the Ministry of Trade, Industry and Energy (MOTIE), Korea Institute for Advancement of Technology (KIAT) through the Inter-ER Cooperation Projects.

B. Murgante et al. (Eds.): ICCSA 2014, Part IV, LNCS 8582, pp. 639–651, 2014.

our daily lives, providing a convenient reservation mechanism and alleviating the burden of maintenance. It is also important to make the sharing system work with other transportation methods by means of intermodal vehicle scheduling [4].

In sharing systems consisting of multiple sharing stations, a user rents out an EV at a station and returns to a different station, leading to the stock imbalance problem [5]. Here, some underflow stations cannot serve sharing requests. To overcome this problem and enhance the system-wide service ratio, EVs must be explicitly relocated from overflow stations to underflow stations [6]. While a partial relocation procedure can be initiated any time on necessary basis, the system-wide relocation takes place during non-operation hours. In the latter case, if the size of sharing systems gets larger and the number of sharing stations increases, it's not an easy problem to decide how to relocate EVs. The difficulty lies in that unlike bike relocation systems, the number of simultaneously relocatable EVs is very small [7]. As both service staff assignment and operation scheduling are very complex problems, it is necessary to exploit sophisticated computer algorithms. Here, objects in the physical world are modeled to data structures and cost functions in the cyber world [8].

According to a national enterprise named *Jeju Smart Grid Testbed City*, Jeju city is now hosting a variety of developments and tests for leading-edge technologies in smart grid. This city currently possesses about 500 EVs, and the number of EVs will keep increasing. Necessarily, EV charging infrastructure is being facilitated over the city. Hence, a city-wide EV sharing system is now considered for the fast penetration of EVs, and thus efficient operation and management mechanisms are essential. The relocation procedure combined with service staff allocation is one of the most important utilities for EV sharing systems. In this regard, this paper designs a service staff allocation scheme for clustered sharing stations to reduce global relocation distance and time, taking advantage of well-defined computational intelligence, especially, genetic algorithms. Here, clustering divides a large problem into a set of subproblems, namely, local relocation plans and global coordination [9]. As an extended version of [10], this paper includes more detailed explanation on the system development and experiment results.

2   Related Work

There are several sharing systems currently in operation and having their own relocation strategies. To begin with, as an open-ended reservation system working in Genoa, Italy, PICAV (Personal Intelligent City Accessible Vehicles) develops a fully user-based relocation scheme [2]. The system operator assigns a station to return EVs to renters by giving incentives. Next, [6] has designed a three-phase decision-making system for intelligent relocation while testing it by means of operational data obtained from the Honda carsharing system in Singapore. Based on the formulation of a mixed integer linear programming model, the problem is solved using branch-and-bound techniques. It consists of optimizer, trend filter, and simulator phases. In addition, UCR (University of California at Riverside) Intellishare provides both reservation-based and on-demand sharing mechanisms, while relocation can be performed through towing and ridesharing [4]. Here, a passenger team can take separate vehicles if the destination lacks EVs. On the contrary, multiple trips can be merged into a single trip.

Demand pattern analysis and forecast are important prerequisites for relocation planning. [11] has developed a relocation scheme consisting of an offline demand pattern capture module and an online optimization module. Continuously monitoring and comparing optimum and current states, it decides whether to trigger the relocation procedure. [12] has developed a focus-forecasting model for sharing services and estimates the future vehicle distribution. It exploits popular time-series techniques to calculate the selective moving average, finding an efficient relocation path using a microscopic simulator [13]. Next, Green Move figures out the dynamics in the user demand in sharing systems based on real-time feedback and rating mechanisms. Being tested in Milan area, this service gives benefits to those trips bound for lacking stations, providing a different fare policy. In addition, [14] has proposed an evolutionary neural network for forecasting net flow of a car sharing system. It is built on top of both a mixed optimization with genetic algorithms as well as back propagation training for neural networks.

It is true that the locations of sharing station are usually restricted by geographic factors such as accessibility from many users, charging facility availability, and the like. However, the stock imbalance problem can be alleviated by selecting optimal station locations. In the design of a vehicle sharing system developed under the framework of SUCCESS (Smaller Urban Communities in CIVITAS for Environmentally Sustainable Solutions), the preferences of habitants are analyzed to find the criteria capable of evaluating the station locations [15]. Then, tens of fuzzy logic rules infer the potential attitude of each subscriber. Next, [5] has created a trip matrix collected by a geo-coded survey in Lisbon and developed mathematical optimization models. Its objective is to maximize the profits of a carsharing organization in terms of revenues and costs. In addition, the demand pattern model will be very useful to large scale regional employment and renewable energy integration into the power grid [16].

3   Relocation Procedure Overview

Figure 1 overviews our relocation strategy. When the operation center decides to initiate the relocation procedure, the current distribution must be obtained first. Each station reports the number of EVs currently parked in it. The current distribution is denoted by an integer vector, in which each element represents the number of EVs in the associated station. The number of elements in this vector is equal to the number of stations over the EV sharing system. A target distribution accounts for the number of EVs after relocation for each station. Several intuitive strategies have been considered in [17]. First, the even distribution makes equal the number of EVs for all stations. Second, the number of EVs is made to be proportional to the number of daily rent-out requests. Necessarily, other strategies can be developed and integrated into our relocation procedure.

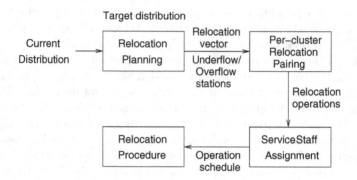

Fig. 1. Overview of the relocation procedure

Anyway, after this step, we know which station has more EVs and which has less. Let $T = \{T_1, T_2, ..., T_n\}$ be the target distribution and $C = \{C_1, C_2, ..., C_n\}$ the current distribution, respectively, where n is the number of stations. Then, the relocation vector R will be $\{C_1 - T_1, C_2 - T_2, ..., C_n - T_n\}$. In the relocation vector, a station having a positive (negative) number is an overflow (underflow) station.

Next, it is necessary to decide the set of relocation pairs, each of which consists of an overflow station and an underflow station. It indicates that an EV will be moved from the first to the second. With m relocation teams, the stations are partitioned into m clusters, mainly by geographical factors, and each cluster is assigned to a relocation team. This assumption is practical and reduces the overall relocation distance, as most EVs are relocated within a cluster. Here, sometimes, it is necessary to move EVs across different clusters. This inter-cluster relocation can be accomplished through intermediary stations which belong to multiple clusters as shown in Figure 2. Namely, the relocation team in an overflow cluster moves EVs to an intermediary station, as if it is an underflow station. Now, the intermediary station will be an overflow station in the underflow cluster. As the relocation procedure is carried out in each cluster, the two stations in a relocation pair are belonging to a same cluster. After all, per-cluster sets of relocation pairs are created. In Figure 2, Cluster 5 has 4 intermediary stations and works as a relay cluster. The downtown area can easily host such an intermediary cluster.

For each cluster, the service staff operation is scheduled to be able to trigger the relocation procedure. It decides an execution order for the set of relocation pairs. Different execution orders lead to different relocation distance and time. In addition, [18] finds a suboptimal schedule for staff operations using genetic algorithms. In this work, only a two-man team is assumed to be assigned to each cluster. Basically, two service men move to an overflow station driving a single service vehicle. One drives an EV to the underflow station while the other follows by the service vehicle. Then, two head for the next overflow station together in the service vehicle [17]. However, if more staff members are employed, more than

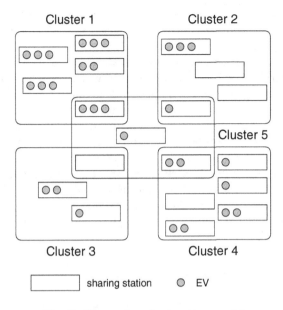

Fig. 2. Cluster-based relocation model

one EV can be relocated simultaneously. Hence, it is necessary to recognize those clusters having higher relocation load and assign more members to them. Here, how to assign human resources is a very complex problem researched in many areas [19][20], and the complexity gets worse when the effect of a staff addition cannot be accurately estimated.

4 Relocation Scheme

4.1 Unit Scheduler

Once staff members are assigned to each cluster, the staff operation must be decided for each cluster. We define the relocation distance as the distance a relocation vehicle takes during a relocation staff. Here, each cluster will have different relocation distance and time. The maximum of them is the global relocation distance (or time). This can be cut down when the relocation load is distributed evenly. We define the cluster-level staff operation scheduler as the unit scheduler, and it is developed based on genetic algorithms [21]. Within a cluster, the relocation distance is calculated in the cost function, and it is also affected by the number of staff members, as more than one EV can be moved simultaneously. If EVs are to be moved from a common overflow station to a common underflow station, m service men move to an overflow stations while one drives the service vehicle. Then, $(m\text{-}1)$ members drive $(m\text{-}1)$ EVs to an underflow station, respectively, while the driver of the service vehicle follows them. The merge of multiple pairs reduces the relocation distance by up to $(m\text{-}1)$ times.

The genetic algorithm is one of the most commonly used suboptimal techniques inspired by the process of national selection [22]. To find an efficient relocation team plan by genetic algorithms, it is necessary to encode an operation schedule to the corresponding integer-valued vector called a chromosome. To represent a service staff operation schedule by an integer-valued vector, overflow stations are listed sequentially, appearing as many times as the number of overflow EVs, creating *Overflow* list. Underflow stations are also listed in the same way, creating *Underflow* list. Then, in an integer vector, each element represents a single relocation pair. For the i-th element having the value of j, i is the index to *Overflow* list, while j to *Underflow* list. Those two elements indexed by i and j are one of overflow stations and one of underflow stations, respectively. Now, the relocation problem is to find the best sequence having the minimal cost, just like the well-known traveling salesman problem. This encoding scheme makes it possible to run standard genetic operators. For further details, refer to [21].

4.2   Relocation Staff Allocation

We can expect the reduction of relocation distance with more than 2 staff members, only when a relocation plan includes equivalent relocation pairs. If they are all different, the relocation distance cannot be reduced even if more service men are assigned to the cluster, as all EVs must be moved one at a time. With station clustering, the number of stations in a cluster gets smaller. It enhances the probability of creating the same relocation pairs in a relocation plan. Moreover, the inter-cluster relocation is performed through the intermediary stations which belong to more than one cluster. This relocation creates many equivalent relocation pairs, as destinations of many relocation pairs in an overflow cluster will be the intermediary stations, even if their final destinations are different in respective underflow clusters. The genetic iteration gives precedence to those relocation plans having many equivalent pairs.

For the different number of relocation staff members, the cost function for a cluster must take into account the number of staff members. After deciding the set of relocation pairs for each cluster, how to assign staff members is what this paper is to solve. Basically, each cluster is assigned one relocation team consisting of two members, making it possible for them to move just one EV at a time. The rest of them are extra members. The problem lies in that the number of service men does not linearly improve the relocation performance. The relocation distance can be known only when the unit scheduler, working for a cluster, finishes creating its schedule, not before scheduling. After all, to get the optimal allocation, it is necessary to compute every feasible allocation. If there are c clusters and e extra members, the number of feasible allocations will be e^c. It must be mentioned that even if the best one out of e^c allocations is selected, the subsequent schedule is not an optimal schedule, provided that the per-cluster schedule is just suboptimal.

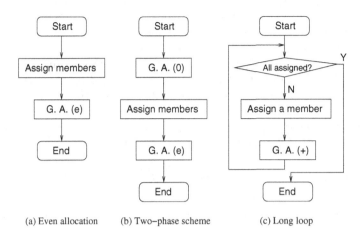

(a) Even allocation   (b) Two–phase scheme   (c) Long loop

Fig. 3. Feasible staff assignment strategies

The simplest form of the staff allocation is to decide the number of staff members in each cluster first and create a local schedule for each cluster, as shown in Figure 3(a). As it is very hard to estimate how much the relocation distance will be improved with more members, this scheme just evenly assigns service men to each cluster. Named as *Even allocation*, it runs the unit scheduler just once with all staff members fully allocated. In this figure, G.A. (e) means that the unit scheduler based on the genetic algorithm runs with e extra members assigned. In the other extreme shown in Figure 3(c), the staff members are assigned to each cluster one by one. For each assignment, relocation schedules for respective clusters are created to find the one having the largest relocation distance. The unit scheduler is invoked e times. Even though this scheme, named *Long loop* scheme, can reduce the global relocation distance incrementally, the unit scheduler runs e times. G.A. (+) means that the unit scheduler is invoked with 1, 2, ..., e extra members assigned incrementally.

If we investigate all of feasible allocations, the unit scheduler is executed e^c times. In the mean time, it is possible to compromise both schemes for better computation speed and reasonable schedule quality, as depicted in Figure 3(b). The proposed scheme runs the unit scheduler at first to approximate how much the relocation distance will be reduced by a specific staff allocation scheme. Based on the empirical observation that the relocation distance decreases roughly by the ratio of $\frac{2}{e+1}$ from the initial estimation, we can incrementally assign staff members with simple arithmetic adaptation not with the execution of unit scheduler for each assignment. Actually, we have tested diverse EV distributions and found out that this estimation works quite well when e is not so large and there are sufficiently many mergeable relocation pairs. Moreover, as the locations of sharing stations hardly change, we can refine the estimation model for a specific station distribution.

For formal description, let D_i be the relocation distance of a cluster i for the current allocation and E_i be the number of allocated extra staff members. After the initial scheduling step, $\{D_i\}$ is the relocation distance without extra staff allocation, while all E_i's will be set to 0. In addition, the number of maximum members assigned to a cluster is E_{max}, which is limited by the service vehicle capacity. Then, the following steps iterate until all extra service men are assigned.

(Step 1) Find the cluster, say i, for which D_i is largest and E_i is less than E_{max}.
(Step 2) Update D_i as in Eq. (1) and increase E_i by one.

$$D_i = D_i \times \frac{E_i + 2}{E_i + 3} \tag{1}$$

, where multiplying $\frac{E_i+2}{E_i+3}$ sequentially makes the series of $\frac{2}{3}$, $\frac{2}{4}$, and $\frac{2}{5}$ each time E_i increases by 1.

5   Performance Measurement

We have implemented a prototype of the proposed relocation scheme to assess its performance. The main performance metric is the global relocation distance, which is the maximum of all D_i's. The performance parameters considered in the experiment include the number of staff members, the number of clusters, the number of stations per cluster, the number of moves, and population size. Each cluster has the same number of stations not explicitly taking into account geographic factors, to concentrate on the relocation distance. With a smaller relocation distance, we cannot only minimize the out-of-service interval for sharing stations but also reduce the employment cost. The experiment measures the effect of each parameter by changing one while the other four fixed. Default parameter setting assumes 5 clusters, 5 stations for each cluster, 20 moves, 5 extra staff members, and the population of 32 chromosomes. Next, the distance between two stations exponentially distributes with the average of 3.0 km. The number of EVs per cluster is 64. For each parameter setting, 10 sets are generated and their results are averaged.

The first experiment measures the effect of the number of staff members to the relocation distance. As the number of clusters is set to 5, 10 staff members are basically employed to organize 5 two-man teams, each of which will be assigned to a cluster. The experiment changes the number of extra members from 0 to 10, and the measured relocation distance is plotted in Figure 4. If there is no extra service man, each cluster performs the relocation procedure just with a single two-man team. The 3 curves for the even allocation, proposed two-phase, and long loop schemes begin with the same point which corresponds to the case of no extra staff member. Apart from this point, the even allocation shows the longest relocation distance, but the other two show the almost same performance. Both of them can improve the performance by up to 31.7 % when there are 4 extra service men, or 14 members in total. The maximum difference between the proposed and long loop schemes is observed to be below 1.0 %. From the point of 10 extra staff members, all three schemes the same performance, as an addition of staff members does not result in the further reduction of the relocation distance.

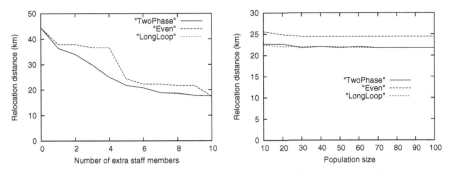

Fig. 4. Effect of the number of staff members **Fig. 5.** Effect of the population size

Next, Figure 5 plots the effect of the population size, one of the most representative genetic algorithm-specific parameters. With large population, we can expect better diversity in finding a feasible schedule. But, it can make longer the execution time, as the genetic loop necessarily involves sorting of all chromosomes in population. Here, the number of extra service men is fixed to 8, while the population size ranges from 10 to 100. Similarly to Figure 4, the proposed and long loop schemes show the same performance, the difference limited by 2.6 %. This difference takes places when the population size is 20, namely, for relatively small population. However, from the population size of 30, we can hardly find a difference. Anyway, Figure 5 shows that the proposed scheme can stably reduce the relocation distance by 21.7 ∼ 22.6 %, compared with the even allocation. All three schemes show slow reduction in the relocation distance, namely, 4.4 % at maximum, according to the increase in the population size. This result indicates that we can select population size compromising the execution time and the system performance.

Next experiment measures the effect of the number of moves to the relocation distance, and the result is shown in Figure 6. We can tune the number of EVs in overflow stations in our simulator, and it is equal to the number of moves for a relocation process. The per-pair relocation distance distributes randomly, as overflow and underflow stations are chosen randomly in a cluster. The larger the number of moves, the more duplicated relocation pairs will be generated. Hence, the efficient extra staff allocation better contributes to the reduction of relocation distance. When the number of moves is less than 15, the performance behavior is similar to the previous experiment. Namely, both the proposed and long loop schemes show the same performance, outperforming the even allocation by up to 21.5 %. However, for 30 moves, the performance gap to the long loop scheme reaches 6.7 %, while that from the even allocation scheme is 2.9 %. This result indicates that our approximation shown in Eq. (1) cannot work so efficiently for the large number of relocation pairs.

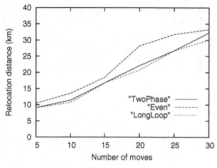 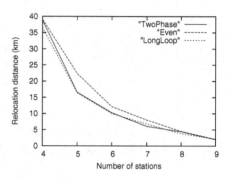

Fig. 6. Effect of the number of moves   **Fig. 7.** Effect of the number of stations

Figure 7 plots the relocation distance when we change the number of stations from 4 to 9. The number of moves for each cluster is fixed to 20. So, if there are more stations, it is possible to find a relocation pair in which the distance between two pairs is shorter. Hence, the relocation distance decreases according to the increase in the number of stations. Just like the previous experiment, the proposed scheme shows a relocation distance comparable to the long loop scheme for the whole experiment range except the case of 4 stations. In this case, the per-pair relocation distance is long, so even a small difference in the allocation vector will increase the difference in the total relocation distance. After all, the proposed scheme shows the relocation distance longer than the long loop scheme by 5.8 % for 4 stations, while outperforming the even allocation scheme by up to 24.9 % for the given parameter values.

Finally, Figure 8 shows the relocation distance according to the number of clusters. Here, the number of stations for each cluster is fixed to 5, while the number of clusters ranges from 3 to 8. As the number of stations linearly increases according to the increase in the number of clusters, the relocation distance also largely gets longer. Actually, the proposed and long loop schemes are not optimal, but just highly likely to approach the optimal solution. The even allocation scheme might also work equally or better than them. For 4 clusters, all the three have the almost same relocation distance. Except that point, the proposed scheme outperforms the even allocation by up to 39.7 %. However, this experiment discovers not a little performance gap to the long loop scheme for 6 and 7 clusters, namely, 6.7 % and 7.0 %, respectively. In these cases, the number of system-wide moves increases, making small estimation error lead to significant performance degradation. Additionally, for the case of 8 clusters, three schemes can reduce the relocation distance by merging more relocation pairs.

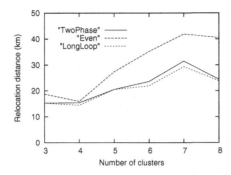

Fig. 8. Effect of the number of clusters

6 Conclusions

Even though EVs have many benefits in environmental aspects, they are still expensive and not easy for maintenance. Accordingly, vehicle sharing systems are now appearing as a promising business model for EVs ahead of their complete personal ownership. However, in one-way rentals, the stock imbalance problem, stemmed from uneven rent-out and return patterns, must be overcome by explicit relocation to improve service ratio, as sharing requests at underflow stations cannot be served. How to relocate EVs for multiple stations over multiple clusters is very important to employment cost and service time. It is a very complex problem and can be solved by intelligent computer algorithms. In our design, the relocation procedure consists of relocation planning, per-cluster station pairing, staff allocation, and actual relocation procedure.

Among these, this paper has designed a two-phase staff allocation scheme capable of reducing relocation distance with just two invocations of the unit scheduler, which is also an instance of the genetic algorithm-based implementation. It takes into account not only the relocation distance for the basic service staff allocation but also the empirical observation that the relocation distance is cut down by the ratio of $\frac{2}{e+1}$, where e is the number of added extra staff members. Staff members are iteratively assigned to the cluster having the worst relocation distance one by one. Here, the bottleneck cluster keeps changing each time a staff member is assigned. The performance measurement results, obtained by a prototype implementation, shows that the proposed scheme improves the relocation efficiency by up to 31.7 % compared with even allocation scheme, just with overhead much less than the long loop scheme.

As future work, we are planning to develop a decision making scheme for the selection of sharing station locations on the target city, namely, Jeju city, Republic of Korea. For this purpose, the travel pattern in the city will be analyzed. In addition, the sharing system must cooperate with public transportation systems. It will be a time-dependent vehicle scheduling problem combined with efficient reservation mechanisms. This system will interact with the public transport information service and the charging facility tracking systems.

References

1. Ipakchi, A., Albuyeh, F.: Grid of the Future. IEEE Power & Energy Magazine, 52–62 (2009)
2. Cepolina, E., Farina, A.: A New Shared Vehicle System for Urban Areas. Transportation Research Part C, 230–243 (2012)
3. Lue, A., Colorni, A., Nocerino, R., Paruscio, V.: Green Move: An Innovative Electric Vehicle-Sharing System. Procedia-Social and Behavioral Sciences 48, 2978–2987 (2012)
4. Barth, M., Todd, M., Xue, L.: User-based Vehicle Relocation Techniques for Multiple-Station Shared-Use Vehicle Systems. Transportation Research Record 1887, 137–144 (2004)
5. Correia, G., Antunes, A.: Optimization Approach to Depot Location and Trip Selection in One-Way Carsharing Systems. Transportation Research Part E, 233–247 (2012)
6. Kek, A., Cheu, R., Meng, Q., Fung, C.: A Decision Support System for Vehicle Relocation Operations in Carsharing Systems. Transportation Research Part E, 149–158 (2009)
7. Caggiani, L., Ottomanelli, M.: A Modular Soft Computing based Method for Vehicles Repositioning in Bike-Sharing Systems. In: International Scientific Conference on Energy Efficient Transportation Networks (2012)
8. Kim, J., Kim, H., Lakshmanan, K., Rajkumar, R.: Parallel Scheduling for Cyber-Physical Systems: Analysis and Case Study on a Self-Driving Car. In: International Conference on Cyber-Physical Systems, pp. 31–40 (2013)
9. Lian, L., Castelain, E.: A Decomposition Approach to Solve a General Delivery Problem. Engineering Letters 18(1) (2010)
10. Lee, J., Park, G.: Per-Cluster Allocation of Relocation Staff on Electric Vehicle Sharing Systems. In: ACM Symposium on Applied Computing (to appear, 2014)
11. Weikl, S., Bogenberger, K.: Relocation Strategies and Algorithms for Free-Floating Car Sharing Systems. In: International Conference on Intelligent Transportation Systems, pp. 355–360 (2012)
12. Wang, H., Cheu, R., Lee, D.: Logical Inventory Approach in Forecasting and Relocating Share-Use Vehicles. In: International Conference on Advanced Computer Control, pp. 314–318 (2010)
13. Wang, H., Cheu, R., Lee, D.: Dynamic Relocating Vehicle Resources Using a Microscopic Traffic Simulation Model for Carsharing Services. In: International Joint Conference on Computational Science and Optimizations, pp. 108–111 (2010)
14. Xu, J., Lim, J.: A New Evolutionary Neural Network for Forecasting Net Flow of a Car Sharing System. In: IEEE Congress on Evolutionary Computation, pp. 1670–1676 (2007)
15. Ion, L., Cucu, T., Boussier, J., Teng, F., Breuil, D.: Site Selection for Electric Cars of a Car-Sharing Service. World Electric Vehicle Journal (2009)
16. Sagosen, O., Molinas, M.: Large Scale Regional Adoption of Electric Vehicles in Norway and the Potential for Using Wind Power as Source. In: International Conference on Clean Electric Power, pp. 189–196 (2013)
17. Lee, J., Kim, H.-J., Park, G.-L.: Relocation Action Planning in Electric Vehicle Sharing Systems. In: Sombattheera, C., Loi, N.K., Wankar, R., Quan, T. (eds.) MIWAI 2012. LNCS, vol. 7694, pp. 47–56. Springer, Heidelberg (2012)

18. Lee, J., Park, G.-L.: Planning of Relocation Staff Operations in Electric Vehicle Sharing Systems. In: Selamat, A., Nguyen, N.T., Haron, H. (eds.) ACIIDS 2013, Part II. LNCS, vol. 7803, pp. 256–265. Springer, Heidelberg (2013)

19. Wen, F., Lin, C.: Multistage Human Resource Allocation for Software Development by MultiObjective Genetic Algorithm. The Open Applied Mathematics Journal 2, 95–103 (2008)

20. Murakami, K., Tasan, O., Gen, M., Oyabu, T.: A Solution of Human Resource Allocation Problem in a Case of Hotel Management. In: 40th International Conference on Computers and Industrial Engineering (2010)

21. Lee, J., Park, G.-L.: Design of a Team-Based Relocation Scheme in Electric Vehicle Sharing Systems. In: Murgante, B., Misra, S., Carlini, M., Torre, C.M., Nguyen, H.-Q., Taniar, D., Apduhan, B.O., Gervasi, O. (eds.) ICCSA 2013, Part III. LNCS, vol. 7973, pp. 368–377. Springer, Heidelberg (2013)

22. Sivanandam, S., Deepa, S.: Introduction to Genetic Algorithms. Springer, Berlin (2008)

The Relationship between Human and Smart TVs Based on Emotion Recognition in HCI

Jong- Sik Lee and Dong-Hee Shin

Department of Interaction Science, Sungkyunkwan University, Seoul, South Korea
jongsic@hotmail.com, dshin@skku.edu

Abstract. This study focuses on the relation between humans and smart TVs with emotion recognition function. The study hypothesizes that the emotion-based user interface is most effective in human interaction in comparison to other user-centered devices with passive user interface. Forty participants were given three types of user interfaces such as the remote controller, gesture recognition system, voice recognition system, and emotional recognition system to be used for watching the smart TV. They were given interesting contents and sad contents to choose from within a specific time limit. We use the Fraunhofer IIS SHORE™ demo software to automatically detect the facial expressions of the participants they exhibited in response to the contents. A study on the preference of contents according to the individual emotional responses of the users was done simultaneously. Additionally, it figured out the relation between four types of emotion as emotion recognition UI on SMART TV screen, the contents, and the study concentrated on the satisfaction and usability of four different user interfaces. As a result of studying the relations of emotions and content preferences from this research, it is recognized that comedy programs are much more preferred based on the degree of happy emotion detected, and that men appear to have higher preference than women. When they were in sad emotion, the valid result value was not achieved though, which shows slight preference for exciting contents. It is noted that women have a little higher preference than men. These results were similar to the standard mood based management of Zilinman's theory. Another result is that out of the four user interfaces, the emotion recognition-based user interface which selects the contents on the screens of the smart TV showed a higher satisfaction than other passive user interfaces.

Keywords: Cognitive Smart TV, Emotion recognition based Smart TV, Usability in Smart TV, User interface in Smart TV, Contents preference.

1 Introduction

From the time early forms of Television appeared at the end of the 19th century, display method or design was continuously developed but it was only at the end of the 20th century that television was firmly considered as the most popular information delivery device. But no matter how definition has improved over the years and despite the numerous TV channels that have increasingly become available, there is no sign

B. Murgante et al. (Eds.): ICCSA 2014, Part IV, LNCS 8582, pp. 652–667, 2014.
© Springer International Publishing Switzerland 2014

of change in the fundamental format that viewers only accept one-way information which is unilaterally delivered by broadcasting. Such limitation of TV has been evident more clearly as [1]PC and Internet have gained popularity since 1990s. As cable TV, [2]IPTV, etc. emerged; TV was evolving into direction of possible interactive service, though partially, such as the [3]VOD service. The emergence of Smart TVs can be traced from this. Smart TV is multifunctional TV which can do things like combining Internet connection to TV, web surfing and VOD watching by setting up all sorts of application, [4]SNS, game and so forth. It was once called as hybrid TV (Hybrid: combination) because it adds the function of a PC to the existing TV. Matching Smart TV's capabilities, Smart Phones have prevailed remarkably since around 2010.   They are aptly called Smart TVs as well. It is without a doubt that the continued evolution of the Smart TV, the rapid development of technology, and the smart technology are greatly beneficial to the users. However, these technology and services should prioritize user-based designs that reflect sufficient user experiences.

Today, the people using SMART TV experience numerous entertainment contents on screens. Moreover, SMART TVs are designed to enlarge content to increase the user's satisfaction and benefits of service providers. Various contents correspond to user's requests, but there is a limitation with regard to easily accessing contents, depending on the kind and the amount of contents. This research is undertaken to address this limitation and design an effective UI. The availability of existing user interface might be inconsistent with the selection of the various contents on SMART TV screen. Therefore this research evaluates the usability, accessibility and satisfaction of the three types of UI (remote, gesture, voice) and   UI with the newly developed emotion recognition functionality. With regard to the user experience in each interface, the key fundamental design principles that will allow successful user experience are explained as user interfaces trends to look forward to in the future; they focus on quick and easy user task, flexibility to allow users to have a seamless visual experience as they switch between different devices, effortlessness by keeping the look simple, clean and consistent, and finally, emotional engagement [40]. The reasons for studying the application of the emotion recognition   are summarized into five items in this paper.

Fig. 1. Usability problem in Smart TV screen

Usability Problem for Smart TV in HCI

First, the current Smart TV has a problem on usability, particularly in choosing contents on the screen as with problems of other IPTV, and similar devices. This study evaluates the usability of 3 typical types of user interfaces (remote controller UI, voice recognition UI, and gesture recognition UI) of Smart TV, centering on the user who needs to choose contents to view, and evaluates the relation between the Smart TV and the user. In a previous research by Chorianopoulos, Konstantinos, and Diomidis Spinellis.[5], it is noted that a diverse user population employs interactive TV (ITV) applications in a leisure context for entertainment purposes. The reality is, in using Smart TV, service providers also consider user's preference for entertainment so various contents are for entertainment.

User Interface Problem for Smart TV in HCI

Second, it's not common that user interface on Smart TV communicates enough with user while choosing or adjusting contents on the screen. In the current HCI field, the problem in user interface is that only one of the parts or features is highlighted; the researchers think that the key to success depends on how a smart device could communicate well with people. Such concept is the very basis for developing user interface of HCI .

According to this concept, the user interface of Smart TV is marketed as having voice recognition method and gesture recognition method in addition to the common remote controller method. What would be the method of the next generation user interface after this? In this study, the researchers suggest emotional recognition-based user interface based on the concept for which patent application has already been completed. The satisfaction on user interface of each user might be varied as expected. The researchers evaluate the satisfaction of each of the users and the usability of all user interfaces.

User Centered in Media Service in Smart TV

Third, it is not very evident that user interface of current Smart TV is designed and mostly structured enough for users. The current Smart TV user interfaces in electronic products are passive and do not properly fit in users' needs. For a user-centered device, the researchers propose that the emotion based interactive television is the most effective in interaction compared to other passive input products.

User Experiment in Media Service in Smart TV

Fourth, the study aims to check whether the method of user interface of current Smart TV is maintained or not and whether the situation needs to change or not. The researchers are able to recognize these through checking the trend of current user interfaces. The multimedia has brought lots of changes so far and anyone has a desire to acquire a multimedia product easily and more conveniently. The existing user interface is getting intellectualized gradually by applying this part to TV. Consumers wish certain device reads their minds and operates conveniently. In this sense, the user interface is changing. Therefore, the researchers evaluated the satisfaction and usability of all user interfaces (remote controller method, voice recognition method, gesture recognition method) which are used for Smart TV in addition to emotion recognition user interface. The researcher also studied the preference relation of contents among users through experiment.

User Needs & Trend of User Interface in Smart TV

Fifth, as demands for user interface are changing, the researchers focus on the relationship between humans and reactive television when emotion recognition through facial expression mechanism is used. The study was done based on these three types of interfaces and a new emotion recognition based user interface applied to the Smart TV. Recently, as researches are conducted and technologies are enhanced, it seems that things around us will slowly be replaced by non-touchable interaction. Interaction will change from written or spoken languages to gestures and facial or conversational emotions, enabling the user to respond accordingly.

1.1   Structure of Emotion Recognition Detection

The experiment was carried out following procedure as shown above. The camera receives a fixed image of a participant's face which is automatically sent to the software. Then, the software detects the facial expression of the participant. The image is then used to recognize emotion expression. According to the categorization of data, four emotions are displayed [25].

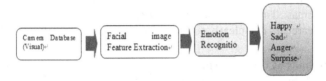

Fig. 2. Architecture of emotion recognition as face expression

Different emotions can be viewed as structures that differ in one or more of these components. [37] Experiment was conducted with the idea that different facial expressions would be affected by different depths of emotion. As we all know, the level of emotion being felt is not the same for everybody at any time. Thus, the personality of the user must accurately detect the emotion process from facial expression in real time.

2   Theoretical Backgrounds

Most of people who watch TV consume the contents according to their emotions and characters. Thus, the more complex the user-centered interface is, the higher the demand for user-centered smart devices in HCI [41] is. This study investigates the level of user satisfaction and impact of the interface on individuals, especially of four major kinds of user interfaces used in smart TVs today, and proposes the most appropriate user interfaces to individuals. Although each user interface type is different, the theoretical background for the most effective user interface will be discussed. This paper aims to discuss that most of today's user interfaces in electronic products are passive and do not properly fit into users' needs. So, in terms of user-centered device, we propose that emotion based interactive television is the most effective way to interact compared to other passive input products. In this respect, we focus on the effect of interaction between humans and reactive television when emotion recognition through facial expression mechanism is used.

The interaction is an episode or its series of physical actions and reactions embodied in human and between the worlds [17], including the environment, objects and beings in the world. The user-centered design is a widespread practice in the domain of user interface design [1]. People should not have to change radically to "fit in with the system", the system should be designed to match their requirements [14][19].

Emotionally interactive Smart TV can detect emotion from facial expressions of people within a fixed time, and then if happy mode is detected then the programs of TV would be shifted to funny or interesting shows and if sad mode is detected then it would be changed to sad dramas, comedies and game shows contents [21]. Through the test on user interface and emotion recognition, the researchers figured out the emotional status of participants; in addition, the researchers found the preference of contents with the results on emotion. The universal purpose of watching TV is to get information or to get entertained [16]. The latter part is especially important in watching TV. Based on these reasoning, in order to attain the representativeness and universality of contents in different genre, the pilot tests were taken and programs and genres were set based on the results. The researchers looked for the most appropriate user interface that work best within a limited period of time, and also put forward appraisal and proposal on three different contents (Game Show: Gangnam Style (Psy)/ Sad Dramas: Titanic/ Comedies: Gag Concert) [21]. Through the following theory, the researchers can clearly see that emotion plays a significant role in selecting contents.

Different Methods of User Interface
1) Remote controller   2) voice recognition
3) Emotional recognition 4) Gesture recognition systems

Remote Controller Unit (RCU) Based Interface in Smart TV
The effect of the early television remote control is that it allowed audiences, for the first time, to interact with their TV without touching it. They no longer watched programs just because they did not want to get up to change the channel. They could also channel surf during commercials, or turn the sound off [29].

Voice Recognition Based Interface in Smart TV
Voice recognition and control actually work in two ways. Firstly, the user's voice is transformed into workable text which is speech to text, and then the device understands what the user has said. The second one is Artificial Intelligence (AI) in the HCI domain which brings it back to life after so many years. Apple is pioneering model in this field with the introduction of Siri in the new iPhone 4S. The experiment has been carried out as well.

Emotion-Based Facial Expression Recognition Interface
1) Emotion
People usually use their facial expressions, gestures, verbal pitches, and postures to express their emotional state [42]. Emotions play an important role in human-to-human communication and interaction. The emotion-expression relationship is greatly clarified by the componential approach to emotion [11][26]. According to that approach, emotions are structures of moderately correlated components. Affect, appraisal, action disposition, and physiological response are the major components. Emotional feelings are considered as one's awareness of one or more, if components exist. Different emotions can be viewed as structures that differ in one or more of these components [26].

2) Facial Expression

Until the late 1960s, it was widely assumed that facial expression was a noisy, unreliable system with little reliable communicative value. Authors cited myriad examples; individuals smiling at the decapitation of a rat [27] that challenged notions of one-to-one correspondence between facial expression and the experience of emotion [7]. Facial expression was assumed to be like the phonemes of a language. The units of communication were considered to be attached to specific events and experiences in a specific way as a part of the cultural construction of emotion. More recently, it has been claimed that facial expressions of emotion do not relate to the experience of emotion [12], but are instead determined by context-specific social motives [13]. Although attempts to document relations between facial expression and other markers of emotion face numerous difficulties related to the elicitation, timing, and measurement of emotion [39], several relevant studies now exist. These studies have documented consistent and even substantial links between facial expression and other markers of emotion [9].

3) Content Choice (Mood-Based Media Use)

In the mass communication field, research on mood-based media use as a mechanism to help cope with psychological difficulties and negative feelings has been examined using two different theoretical approaches: uses and gratifications (Rubin, 2002). Affect-dependent media stimulus arrangement is known as "mood management" [20][35]. The interaction between emotions and moods is also important. Moods tend to be biased against the experienced emotions, and lowered the activation thresholds for mood-related emotions. When assessing user response to an interface, it is important to consider the biasing effects of user mood. Users entering a usability or experimental study in a good mood, for instance, are more likely to experience positive emotion during an interaction than users in a bad mood [02].

According to Zillmann 1988, [46] Zillmann & Bryant 1985 [47] affect-dependent media stimulus arrangement theory directly examines mood itself as a predictor of media choices. This theory suggests that individuals who experienced negative affective states are likely to select media content they see as fulfilling the goal of maximizing pleasure and minimizing pain. Uses and gratifications research categorizes different types of motivation in regard to media use, and one of them is a motivation to regulate mood. Research has shown that when individuals are in negative moods, they are more likely to opt for media content including hedonically pleasing values such as comedies [32] or energetic and joyful music [22]. In terms of arousal regulation, [03] it has been reported that stressed participants were likely to watch calm and relaxing programming (e.g., documentaries featuring natural scenery), whereas bored participants were likely to select exciting and arousing programming. Zillmann (2000) presents theoretical propositions based on hedonic motivations to explain the formation of preferences between these characteristics of mood-impacting media content and individuals' affective states. Based on the existing theory, this experiment evaluated the preference of contents according to the emotion of user by using emotional recognition mechanism through facial expression to know the emotion and preference of contents in this study.

Gesture Recognition Based Interface in Smart TV

Today's smart TV has a gesture recognition program but it is not perfect. This is because gesture recognition itself is very complex and it is fraught with errors according to the environment. The size of gesture affects the device's recognition capability. Despite those limits, some devices are being experimented to make a great use of gesture recognition. The following discusses theories on gesture recognition.

Recognizing gestures is a complex task which involves many aspects such as motion modeling, motion analysis, pattern recognition, machine learning, and even psycholinguistic studies. There are already several survey papers in human motion analysis [15][45] and interpretation [36]. Since gesture recognition is receiving more and more attention in recent research, a comprehensive review on various gesture recognition techniques developed in recent years is needed [45].

3   Literature Review

Concept of User Interface

As we see the user interface with theoretical views on the TV side from last related research, interactivity can be explained as follows.

1) "Interactivity means two-way communication between source and receiver, or, more broadly multidirectional communication between any number of sources and receivers" [36] (Pavlik et al 1998) two-way communication

2) "An interaction is an episode or series of episodes of physical actions and reactions of an embodied human with the world, including the environment and objects and beings in the world." [17] Action and reaction

3) "The essence of interactivity is exchange" [19] Exchange

4) "Interactivity should be defined in terms of the extent to which the communicator and the audience respond to, or are willing to facilitate, each other's communication needs" [18] Responsiveness

5) "Interactivity is the extent to which users can participate in modifying the form and content of a mediated environment in real time" [43] Real-time participation

6)   "Interactive speed is a construct that contributes to flow and is based on measures such as waiting time, loading time, and degree to which interacting with the web is "slow and tedious" [33] Time required for interaction.

7) Individuals rated interactivity of sites on the basis of their perceptions of two-way communication, level of control, user activity, sense of place, and time sensitivity [30] Perception of two-way communication, control, activity, sense of place, and time sensitivity

8) Interactivity has been defined using multiple processes, functions, and perceptions. However, three elements appear frequently in the interactivity literature: direction of communication, user control, and time [31].

9) Direction of communication: Researchers who examine ways that computers facilitate human interaction often focus on the importance of enabling two-way communication [38](Rafaeli & Sudweeks 1997).

10) Another time element important to interactivity is the ability of users to navigate through a wealth of information quickly and easily find what they are seeking [28]

(Mahood, Kalyanaraman, and Sunder 2000; Nielsen 2000; Wu 1999).

Previous Research of User Interface

In last user interface research which is a study about effects of voice vs. remote control interface [44], there was a significant interaction. Japanese participants completed tasks more easily and thought the interface was better with a remote control. Conversely, United States participants completed tasks more easily and thought the interface was better with voice control. The participants from both cultures liked content more and felt more uncomfortable when using voice control [44]. Facial expression and gesture are the two most common ways to manifest emotion in screen-based characters [6][23]. Usability is defined as ease of use, and it is associated with the efficient use of an interactive product [34].

Also concerning the user interface, USA Patent application publication (Philips Electronics 2003) which is genre recommendation system has been existed as a patent since long ago [14].

In this research, the emotional recognition user interface through facial expression, which has been applied as Korea Patent-pending # 10-2013-0009568 (Jong Sik Lee 2013)[24], was applied to Smart TV. It was compared to other user interface in terms of satisfaction level by applying the method of this patented emotional recognition user interface. The user preference side of contents was also checked according to the emotion and tendency of the person specifically.

Emotion Based Facial Expression Recognition Interface

People usually use their facial expressions, gestures, verbal pitches, and postures to express their emotional state.[7] Emotions can be divided into two layers: momentary emotion and mood. Momentary emotions are the behaviors that we display briefly when interacting to events (e.g., angry, happy, or sad). Ekman's six emotional expressions [8] show happiness, sadness, anger, surprise, fear, and disgust/ contempt. Emotions and moods are displayed in our bodily movement more apparently than in personality. The attitude towards other people displays the interpersonal relationship.

There are different emotions of sadness contents in gender. Males and females understand, interpret, and regulate the emotion of sadness differently in accordance with gender-specific social norms. Sadness is stereotypically perceived as a Feminine emotion associated with weakness and lack of control [4]

4 Research Questions & Hypotheses

Research Question

RQ1) Which user interface was more effective to males and females?

RQ2) What is the most satisfactory user interface on Smart TV?

RQ3) Is the user satisfied with the customized contents based on the recognition of the happy & sad emotion through individualized facial expression?

Hypotheses

H1) The emotion recognition interface from facial expression on Smart TV is the most interactive interface of all the interfaces.

H2) Individually optimized interface based on emotion is the most effective interface of all the interfaces.

H3) The person who has happy emotions will prefer the contents of music and comedy genre more in Smart TV.

H4) The person who has sad emotions will prefer the contents of comedy genre more in Smart TV.

H5) The person who has sad emotions will prefer the contents of Sad genre more in Smart TV.

5   Method of Experiment

Participants
Sample size: 40 people (20 males and 20 females)
Age: Between 20 and 30

Procedure
Before performing this experiment, this research carried out the availability sample test of contents to increase credibility of the experiment more, and the researchers decided on the genre of the contents that target one hundred people besides the experiment participants.

40 people will randomly experience 4 types of user interfaces, which are emotion recognition interface, remote control, voice recognition interface and gesture recognition interface on Smart TV. With each interface, the participants will use Smart TV for 20 seconds. With emotion recognition interface, their emotion will be detected automatically through their facial expressions for 3 seconds, and then specific contents according to the detected emotion will be shown for 17 seconds. If the emotion is detected as happy or surprising, the programs of TV will shift to funny or interesting shows (Gangnam style) whereas if sad or angry mode is detected, it will change to moving or touching shows (The Titanic). To show the preference between emotion and contents, after analyzing the user's emotion by seven point scale, the participants were made to experience each of the three contents for five minutes each and then were evaluated by seven point scale each according to the preference of contents. After 10 minutes, the feeling response from the user is defined.

Fig. 3. Test on user interface of four types in Smart TV

Measurement (Evaluation of User Interface)
IV: Type of User Interface (Remote Controller, Voice Recognition, Emotional Recognition, Gesture recognition systems)
CV: Contents (Sad Dramas Comedies Game Shows)
DV: Satisfaction

Measurement (User Preference between Emotion and Contents)
IV: Emotion Type (Angry, Happy, Sadness, Surprise)
CV: Contents (Sad Dramas Comedies Game Shows)
DV: Preference

Emotion Recognition Interface through Facial Expressions in Smart TV
Fraunhofer IIS SHORE™ [10] Emotion recognition software based on the tested for
cognitive TV models as in the following.

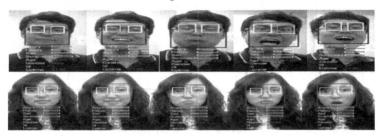

Fig. 4. Recognition of participants' emotion through the software

From Ekman's six basic categories of emotion (i.e. happy, sad, angry, fearful, dis-
gusted, and surprised) [8], user interfaces trends to watch in the future focus on helping
the user perform tasks quickly and easily, flexibility & smart contents in Smart TV.

6   Results

Remote Controller Based User Interface in Smart TV
Although the results may not be significant, expectation (M_male = 5.25, SE = .31)
was the highest for men when experiment with remote control was conducted and
women had the highest perceived level (M_female = 5.35, SE = .24). Men enjoyed
(M_male = 3.78, SE = .28) using the remote control at least and QUIS(Questionnaire
for User Interaction Satisfaction) (M_female = 4.51, SE = .24) was at the lowest for
women.

Voice Recognition Based User Interface in Smart TV
Although the results may not be significant, perceived level (M_male = 5.77, SE =
.30) was the highest for men when experiment with voice recognition was conducted,
and women also had the highest perceived level (M_female = 5.93, SE = .26). Men
had the lowest satisfaction level (M_male = 4.48, SE = .34) on using the remote control and QUIS satisfaction (M_female = 5.12, SE = .24) was at the lowest for women.

Gesture Recognition Based User Interface in Smart TV
Results for expectation is significant (T(38)= -3.78,P <0.05). In addition, the level of
perceived (M_male = 5.45, SE = .35) was the highest for men when experiment with
gesture recognition was conducted, and women had the highest expectation
(M_female = 5.95, SE = .26). Men had the lowest satisfaction level (M_male = 4.27,
SE = .41) using the remote control and usefulness (M_female = 4.37, SE = .35) was at
the lowest for women.

Emotional Recognition Based User Interface in Smart TV
Results for evaluation (T(38)= -2.134,P <0.05) and satisfaction is significant (T(38)=
-3.388,P <0.05) In addition, perceived level (M_male = 5.25, SE = .29) was the highest for men when experiment with gesture recognition was conducted, and women

had the highest satisfaction (M_female = 5.85, SE = .28). Men had the lowest usefulness (M_male = 3.88, SE = .32) using the remote control and usefulness (M_female = 4.53, SE = .26) was also at the lowest for women.

Table 1. Descriptive Statistics

	N	Minimum	Maximum	Mean	Std. Deviation
E[5]_satisfaction	40	1	7	4.9668	1.85838
V[6]_satisfaction	40	1	6.67	4.8512	1.34382
R[7]_satisfaction	40	1.67	7	4.8255	1.23768
G[8]_satisfaction	40	1	7	4.6248	1.75485
Valid N (list wise)	40				

Overall, the satisfaction level was the highest on emotion based recognition interface (Mean– 4.96) and next came voice recognition (Mean = 4.85). The next was the remote control (Mean= 4.82) and the next was gesture recognition (Mean = 4.62). In particular, women (Mean=5.85) were very satisfied with emotion recognition interface compared to men (Mean=4.08). Thus, the following Hypotheses are applicable: *H1) The emotion recognition interface from facial expression on Smart TV is the most interactive interface of all the interfaces. H2) Individually optimized interface based on emotion is the most effective interface of all the interfaces.* It is recognizable that H1, H2 is adopted and especially user interface of emotional recognition on the Smart TV screen is considered as better user interface to women when it comes to choosing the contents.

When correlation analysis was conducted on the four interfaces, people who liked the remote control also liked voice recognition interface, those who liked voice recognition also liked gesture recognition, those who liked gesture recognition liked voice recognition and those who liked emotion recognition liked both voice and gesture recognition. In this part, it is known that the person who likes gesture recognition also prefers voice recognition the most. On the other hand, the person who likes remote control method showed tendency which did not like gesture recognition.

In terms of satisfaction, those who were highly satisfied with remote control also had high level of satisfaction of voice recognition, and those who were highly satisfied with voice recognition also had high levels of remote control and gesture recognition. Those who were satisfied with gesture recognition were also satisfied with voice recognition and those who were highly satisfied with emotion recognition interface were also highly satisfied with gesture recognition.

Analysis of Emotion Based on Content Preference
The results in Table 4 above indicate preference of 3 types (Comedy, Sad Movie, Music Show) of contents according to the four types of emotion of experiment participants, especially the result value of correlation under 0.03 of significant level about

---

[5] Emotion recognition based on user interface.
[6] Voice recognition based on user interface.
[7] Remote controller based on user interface.
[8] Gesture recognition based on user interface.

comedy when they are happy. On other items, the level is also not significant but person correlation value is + value when emotion is angry and the user has a tendency to prefer Gangnam Style of music show, and even in the sad mood the user still prefers Gangnam Style of music show as person correlation value is + value.

Table 2. Correlation Analysis on all user interfaces in Smart TV

		R_total	V_total	G_total	E_total
Remote controller _total	Pearson Correlation	1	.549[**]	.350[*]	.365[*]
	Sig. (2-tailed)		0	0.027	0.02
	N	40	40	40	40
Voice recognition _total	Pearson Correlation	.549[**]	1	.589[**]	.474[**]
	Sig. (2-tailed)	0		0	0.002
	N	40	40	40	40
Gesture recognition _total	Pearson Correlation	.350[*]	.589[**]	1	.474[**]
	Sig. (2-tailed)	0.027	0		0.002
	N	40	40	40	40
Emotion recognition _total	Pearson Correlation	.365[*]	.474[**]	.474[**]	1
	Sig. (2-tailed)	0.02	0.002	0.002	
	N	40	40	40	40

**. Correlation is significant at the 0.01 level (2-tailed).
*. Correlation is significant at the 0.05 level (2-tailed).

Table 3. Analysis on Satisfaction of all user interfaces in Smart TV

		R_satisfaction	V_satisfaction	G_satisfaction	E_satisfaction
Remote controller _satisfaction	Pearson Correlation	1	.499[**]	.023	.080
	Sig. (2-tailed)		.001	.888	.622
	N	40	40	40	40
Voice recognition _satisfaction	Pearson Correlation	.499[**]	1	.498[**]	.411[**]
	Sig. (2-tailed)	.001		.001	.008
	N	40	40	40	40
Gesture recognition _satisfaction	Pearson Correlation	.023	.498[**]	1	.519[**]
	Sig. (2-tailed)	.888	.001		.001
	N	40	40	40	40
Emotion recognition _satisfaction	Pearson Correlation	.080	.411[**]	.519[**]	1
	Sig. (2-tailed)	.622	.008	.001	
	N	40	40	40	40

**.Correlation is significant at the 0.01 level (2-tailed).

Therefore, the following hypothesis is selected: *H3) The person who has will prefer the contents of happy emotions will prefer the contents of music and comedy genre more in Smart TV. H4) The person who has sad emotions will prefer the contents of comedy genre more in Smart TV.* The fifth hypothesis is dismissed: *H5) The person who has sad emotions will prefer the contents of Sad genre more in Smart TV.*

After checking the result on the preference of each content, the Gag contents is $F(3,36)=1.607, p>.05$ on the Table 20 above which didn't make significant result value and even Titanic didn't make a significant result value as $F(3,36)=0.554, p>.05$ on the preference side. And finally, Gangnam Style of music show didn't make valid result value as $F(3,36)=0.827, p>.05$ on the preference side. However, the preference has a tendency to appear in the order of Gag Comedy, next, the Music Show Gangnam Style, and lastly, Titanic, with standard of F value.

Table 4. Correlation Analysis on User Preference between Emotion and Content

		Angry	Happy	Sad	Sur-prise	Gag Satisfaction	Titanic Satisfaction	Gangnam Satisfaction
Angry	Pearson Correlation	1	-.134	.407**	.133	.020	-.150	.118
	Sig. (2-tailed)		.410	.009	.415	.901	.357	.469
	N	40	40	40	40	40	40	40
Happy	Pearson Correlation	-.134	1	-.519**	.433**	.451**	.027	-.152
	Sig. (2-tailed)	.410		.001	.005	.003	.867	.349
	N	40	40	40	40	40	40	40
Sad	Pearson Correlation	.407**	-.519**	1	-.049	-.110	-.159	.070
	Sig. (2-tailed)	.009	.001		.766	.499	.327	.668
	N	40	40	40	40	40	40	40
Surprise	Pearson Correlation	.133	.433**	-.049	1	.260	.053	-.110
	Sig. (2-tailed)	.415	.005	.766		.106	.744	.498
	N	40	40	40	40	40	40	40
Gag_ Satisfac-tion	Pearson Correlation	.020	.451**	-.110	.260	1	-.132	-.282
	Sig. (2-tailed)	.901	.003	.499	.106		.418	.078
	N	40	40	40	40	40	40	40
Titanic_ Satisfac-tion	Pearson Correlation	-.150	.027	-.159	.053	-.132	1	-.077
	Sig. (2-tailed)	.357	.867	.327	.744	.418		.637
	N	40	40	40	40	40	40	40
Gangnam _Satisfact ion	Pearson Correlation	.118	-.152	.070	-.110	-.282	-.077	1
	Sig. (2-tailed)	.469	.349	.668	.498	.078	.637	
	N	40	40	40	40	40	40	40

**.Correlation is significant at the 0.01 level (2-tailed)

Table 5. ANOVA Analysis on Content Preference in Smart TV

		Sum of Squares	df	Mean Square	F	Sig.
Gag	Between Groups	10.285	3	3.428	1.607	.205
	Within Groups	76.815	36	2.134		
	Total	87.100	39			
Titanic	Between Groups	4.285	3	1.428	.554	.649
	Within Groups	92.815	36	2.578		
	Total	97.100	39			
Gangnam	Between Groups	4.255	3	1.418	.827	.488
	Within Groups	61.720	36	1.714		
	Total	65.975	39			

7   Conclusions

In this study, in order to find out the most satisfactory user interface, the researchers experimented with 20 men and 20 women of ages between 19 and 31, average age of 25 and they spend one hour in average, watching smart TV.

Among the four user interfaces, Gesture recognition interface had the highest level of expectation. Especially, the emotion recognition interface had the highest evaluation and satisfaction level when compared to other user interfaces. Overall, the satisfaction level was the highest on emotion based recognition interface and next came voice recognition. The next was the remote control and the next was gesture recognition. Particularly, women were very satisfied with emotion recognition interface. Men's satisfactory level on emotion recognition interface was.

With the experiment on emotion recognition user interface, the researchers also found out that in terms of preferences of contents, those who had negative feelings preferred watching entertaining contents ([9]Gag & [10]Music show) in term of correlation number is + value (Sad, Angry & Gag, Music show) and correlation number is - value ([11]Sad contents) but it is not significant at the 0.01 level (2-tailed)). Those who had positive feelings preferred watching Gag contents. (Correlation number = 0.451** (Happy & Gag). Correlation is significant at the 0.01 level (2-tailed)). Since women tend to be more sensitive to emotion when watching the contents, ways to approach women in a different way is needed.

To settle the emotional recognition interface based on such emotion and tendency as a user interface, firstly and fundamentally, analyzing the emotion of user exactly at first is far more important. Thus, it is important to provide right contents according to the exact feeling of user and such mechanism will affect satisfaction of user. Of course this emotional recognition user interface cannot be applied to all smart devices universally yet, but the researchers think it will be applied very effectively according to the feelings of the user for choosing the contents which is increasing continuously on Smart TV screen. Therefore, when the usability and satisfaction of user interface of Smart TV were viewed, the emotional recognition user interface which is based on tendency and emotion of the user and recommendation of the proper contents related to emotion and the tendency of user will present a future direction of innovative user interface.

References

1. Brooke, J., Bevan, N., Brigham, F., Harker, S., Youmans, D.: Usability statements and standardisation: Work in progress in ISO. In: Proceedings of the IFIP TC13 Third International Conference on Human-Computer Interaction, pp. 357–361. North-Holland Publishing Co. Ltd. (August 1990)
2. Brave, S., Nass, C.: Emotion in human-computer interaction. In: The Human-Computer Interaction Handbook: Fundamentals, Evolving Technologies and Emerging Applications, 81–96 (2002)
3. Bryant, J., Zillmann, D.: Using television to alleviate boredom and stress: Selective exposure as a function of induced excitational states. Journal of Broadcasting & Electronic Media 28(1), 1–20 (1984)
4. Brody, L.R.: Gender differences in emotional development: A review of theories and research. Journal of Personality 53(2), 102–149 (1985)
5. Chorianopoulos, K., Spinellis, D.: User interface evaluation of interactive TV: a media studies perspective. Universal Access in the Information Society 5(2), 209–218 (2006)
6. Cassell, J.: Embodied conversational interface agents. Communications of the ACM 43(4), 70–78 (2000)
7. Ekman, P.: Cross-cultural studies of facial expression. Darwin and Facial Expression: A Century of Research in Review, pp. 169–222 (1973)

---

[9] Gag concert-Gag contents (Korea Gag genre)

[10] Gangnam style- music contents (Korea Psy-music video genre)

[11] Titanic movie- the sinking of the Titanic – Sad contents (USA movie)

8. Ekman, P.: An argument for basic emotions. Cognition & Emotion 6(3-4), 169–200 (1992)
9. Ekman, P.: Facial expression of emotion, 2nd edn. Handbook of Emotions, ch. 15. Guilford Publications, Inc., New York (2000)
10. Fraunhofer IIS SHORE™,
    http://www.iis.fraunhofer.de/bf/bsy/produkte/shore/
11. Frijda, N.H.: The emotions. Cambridge University Press (1986)
12. Fernández-Dols, J.M., Ruiz-Belda, M.A.: Spontaneous facial behavior during intense emotional episodes: Artistic truth and optical truth. In: The Psychology of Facial Expression, p. 255 (1997)
13. Fridlund, A.J.: The behavioral ecology and sociality of human faces (1992)
14. Lekakos, G., Chorianopoulos, K., Spinelis, D.: Information Systems in the living room: A case study of personalized interactive TV design. In: Global Co-Operation in the New Millennium The 9th European Conference on Information Systems, Bled, Slovenia, June 27-29 (2001)
15. Gavrila, D.M.: The visual analysis of human movement: A survey. Computer Vision and Image Understanding 73(1), 82–98 (1999)
16. Haridakis, P.M., Rubin, A.M.: Motivation for watching television violence and viewer aggression. Mass Communication and Society 6(1), 29–56 (2003)
17. Heeter, C.: Interactivity in the context of designed experiences. Journal of Interactive Advertising 1(1), 75–89 (2000)
18. Ha, L., Lincoln James, E.: Interactivity reexamined: A baseline analysis of early business web sites. Journal of Broadcasting & Electronic Media 42(4), 461 (1998)
19. Haeckel, S.H.: About the nature and future of interactive marketing. Journal of Interactive Marketing 12(1), p63 (1998)
20. Knobloch-Westerwick, S.: Mood management: Theory, evidence, and advancements (2006)
21. Kim, J.: Do we improve, disrupt, or embrace sadness? Exploring sadness-based media choice and its anticipated effects on coping (Doctoral dissertation, The Pennsylvania State University) (2007)
22. Knobloch, S., Zillmann, D.: Mood management via the digital jukebox. Journal of Communication 52(2), 351–366 (2002)
23. Kurlander, D., Skelly, T., Salesin, D.: Comic chat. In: Proceedings of the 23rd Annual Conference on Computer Graphics and Interactive Techniques, pp. 225–236. ACM (August 1996)
24. Korea Patent-pending # 10-2013-0009568 (Jong Sik Lee 2013)
25. Lee, J.S., Shin, D.-H.: A Study on the Interaction between Human and Smart Devices Based on Emotion Recognition. In: Stephanidis, C. (ed.) HCII 2013, Part I. CCIS, vol. 373, pp. 352–356. Springer, Heidelberg (2013)
26. Lang, P.J.: The emotion probe: Studies of motivation and attention. American Psychologist 50(5), 372 (1995)
27. Landis, C.: Studies of Emotional Reactions. II. General Behavior and Facial Expression. Journal of Comparative Psychology 4(5), 447 (1924)
28. Mahood, C., Kalyanaraman, S., Sundar, S.S.: The effects of erotica and dehumanizing pornography in an online interactive environment. In: Annual Conference of the Association for Education in Journalism and Mass Communication, Phoenix, AZ (2000)
29. Bellis, M.: History of the Television Remote Control / Remote control technology was first developed for military use (1997),
    http://inventors.about.com/od/rstartinventions/a/remote_control.htm

30. McMillan, S.J.: Interactivity is in the eye of the beholder: function, perception, involvement, and attitude toward the web site. In: Proceedings of the Conference-American Academy of Advertising, pp. 71–78 (2000)
31. McMillan, S.J., Hwang, J.-S.: Measures of perceived interactivity: An exploration of the role of direction of communication, user control, and time in shaping perceptions of interactivity. Journal of Advertising 31(3), 29–42 (2002)
32. Meadowcroft, J.M., Zillmann, D.: Women's comedy preferences during the menstrual cycle. Communication Research 14(2), 204–218 (1987)
33. Novak, T.P., Hoffman, D.L., Yung, Y.-F.: Measuring the customer experience in online environments: A structural modeling approach. Marketing Science 19(1), p29 (2000)
34. Nielsen, J.: Guerrilla HCI: Using discount usability engineering to penetrate the intimidation barrier. In: Cost-Justifying Usability, pp. 245–272 (1994)
35. Oliver, M.B.: Mood management and selective exposure. In: Communication and Emotion: Essays in Honor of Dolf Zillmann, pp. 85–106 (2003)
36. Pavlovic, V.I., Sharma, R., Huang, T.S.: Visual interpretation of hand gestures for human-computer interaction: A review. IEEE Transactions on Pattern Analysis and Machine Intelligence 19(7), 677–695 (1997)
37. Russell, J.A., Bachorowski, J.-A., Fernández-Dols, J.-M.: The psychology of facial expression. Cambridge University Press, Cambridge (1997)
38. Rosenberg, E.L., Ekman, P.: Coherence between expressive and experiential systems in emotion. Cognition & Emotion 8(3), 201–229 (1994)
39. Rafaeli, S., Sudweeks, F.: Networked interactivity. Journal of Computer Mediated Communication 2(4) (1997)
40. Asimakopoulos, S.: New Human Computer Interaction Trends Focus on User Experience p6 p3 (2012)
41. Stephanidis, C.: User interfaces for all: New perspectives into human-computer interaction. User Interfaces for All-Concepts, Methods, and Tools 1, 3–17 (2001)
42. Su, W.-P., Pham, B., Wardhani, A.: Personality and emotion-based high-level control of affective story characters. IEEE Transactions on Visualization and Computer Graphics 13(2), 281–293 (2007)
43. Steuer, J.: Defining virtual reality: Dimensions determining telepresence. Journal of Communication 42(4), p84 (1992)
44. Steuer, J.: Defining virtual reality: Dimensions determining telepresence. Journal of Communication 42(4), p84 (1992)
45. Tan, G., Brave, S., Nass, C., Takechi, M.: Effects of voice vs. remote on US and Japanese user satisfaction with interactive HDTV systems. In: CHI 2003 Extended Abstracts on Human Factors in Computing Systems, pp. 714–715. ACM (April 2003)
46. Wu, Y., Huang, T.S.: Vision-based gesture recognition: A review. Urbana 51, 61801 (1999a)
47. Zillmann, D.: Mood management: Using entertainment to full advantage. In: Donohew, L., Sypher, H.E., Higgins, E.T. (eds.) Communication, Social Cognition, and Affect. Lawrence Erlbaum, Hillsdale (1988b)
48. Zillmann, D., Bryant, J.: Affect, mood, and emotion as determinants of selective exposure. In: Selective Exposure to Communication, pp. 157–190 (1985)

An Intelligent Broadcasting Algorithm for Early Warning Message Dissemination in VANETs

Ihn-Han Bae

School of IT Engineering, Catholic University of Daegu, Gyeongsan 712-702, Korea
ihbae@cu.ac.kr

Abstract. Vehicular Ad-hoc network (VANET) has gaining much attention recently to improve road safety, reduce traffic congestion, and to enable efficient traffic management because of its many important applications in transportation. In this paper, we propose an early warning intelligence broadcasting algorithm, EW-ICAST to disseminate a safety message for VANETs. The proposed EW-ICAST uses not only the early warning system on the basis of time to collision (TTC) but also the intelligent broadcasting algorithm on the basis of fuzzy logic. Thus, the EW-ICAST resolves effectively broadcast storm problem and meets time-critical requirement. The performance of EW-ICAST is evaluated through simulation and compared with that of other alert message dissemination algorithms. From the simulation results, we know that EW-ICAST is superior to *Simple*, *P-persistence* and *EDB* algorithms.

Keywords: Broadcast, collision warning, fuzzy logic, routing, VANET.

1   Introduction

VANETs have been considered as an important communication infrastructure for the Intelligent Transportation Systems (ITS). VANETs have been considered as an important communication infrastructure for the Intelligent Transportation Systems (ITS). In IEEE 802.11p, the Dedicated Short Range Communication (DSRC) is a core function and it is a US government project for vehicular network communication for the enhancement of driving safety and comfort of automotive drivers. DSRC-based communication devices are expected to be installed in future [1].

VANETs raise new challenges to the design of data communication protocols due to the high dynamicity of the underlying topology, the intermittent connectivity, and fast changing density. Broadcasting is the message delivery task from a source node to all other nodes in a network to enhance the safety of drivers and provide the comfortable driving environment. Many important VANET services, ranging from safety applications to location-based advertisement, rely on the reliability and efficiency of underlying broadcast protocols. Applications have different requirements on broadcast protocol design. Location-based advertisement emphasizes reliability in order to achieve higher coverage of vehicles, while warning delivery, which broadcasts emergent information to approaching vehicles, requires both low propagation delay and reliability. Because of the shared wireless medium, blindly broadcasting packets may lead to frequent contention and collisions among transmitting neighboring nodes.

B. Murgante et al. (Eds.): ICCSA 2014, Part IV, LNCS 8582, pp. 668–681, 2014.

This problem is sometimes referred to as the broadcast storm problem. While multiple solutions exist to alleviate the effects of the broadcast storm problem in MANET, only a few solutions have been proposed to resolve this issue in VANET [2, 3].

The main contribution of this paper is to present EW-ICAST, an Early Warning Intelligence broadCASTing algorithm for safety message dissemination in VANET. The proposed EW-ICAST uses not only the early warning system on the basis of TTC but also the intelligent broadcasting algorithm on the basis of fuzzy logic. In EW-ICAST, when a driving vehicle recognizes that the brake light of a vehicle right ahead is on or a vehicle of lateral lane is lane changing through vehicle-to vehicle (V2V) communication, the vehicle computes TTC. If the TTC was less than or equals to a threshold of TTC, the vehicle broadcasts an alert message to following vehicles. Then a vehicle receives an alert message for the first time, the vehicle determines rebroadcast degree using fuzzy logic rules, where the rebroadcast degree depends on the current traffic density of road and the distance between previous-hop vehicle and current receiving vehicle. The probability of rebroadcasting the message, as well as the rebroadcast delay dependent on the computed rebroadcast degree. If the vehicle didn't receive the rebroadcasted alert message from another vehicle until a time-out delay expires, the vehicle rebroadcasts the alert message with the rebroadcast probability.

The remainder of this paper is organized as follows. Section 2 reviews related work. Section 3 describes the proposed EW-ICAST algorithm. Section 4 offers a performance evaluation of EW-ICAST through simulation. Finally, Section 5 concludes the paper and discusses directions for future investigations.

2   Related Work

2.1   VANET Routing Mechanism

VANET routing mechanism is classified into four broad categories; unicast, multicast, geocast and broadcast approaches [4, 5]. Unicast routing is a fundamental operation for vehicle to construct a source-to-destination routing in a VANET Multicast is defined by delivering multicast packets from a single source vehicle to all multicast members by multi-hop communication. Geocast routing is to deliver a geocast packet to a specific geographic region. Vehicles located in this specific geographic region should receive and forward the geocast packet; otherwise, the packet is dropped. Broadcast protocol is utilized for a source vehicle sends broadcast message to all other vehicles in the network as shown.

2.2   Broadcast Protocols for VANET

The primary goal for safety alert application is to deliver the alert message to all vehicles approaching the incident site, so that drivers may be alerted prior to their natural visual reaction. So end-to-end delay for the alert message has to be minimized. The previous systems for alert message broadcast dissemination in VANETs are divided two groups, one where vehicles are not equipped with GPS and the other where they are not equipped with GPS.

(1)   GPS Not Equipped

Simple broadcast [6, 7] is the simplest protocol used in V2V safety alert application for VANET in the literal sense of the words. When there is an accident, safety alert application will send alert messages to all vehicles approaching towards accident site. When a vehicle receives a broadcast message for the first time, it retransmits the message. The vehicle ignores all subsequent broadcast messages it receives, from other vehicles rebroadcasting the same message.

P-persistence [7, 8] tries to reduce the broadcast storm problem by using a stochastic selection strategy to decide the vehicles that will rebroadcast the alert message. When a vehicle receives a broadcast message for the first time, the vehicle will rebroadcast the alert message with a random probability P. This method will help reduce the number of re-broadcasting vehicles and alleviate the effects of the broadcast storm. However failures to extend the alert message decide not to, which will cause the loss of alert message.

(2)   GPS Equipped

Unlike the P *persistence* or gossip-based scheme, *weighted P-persistence* [3] assigns higher probability to nodes that are located farther away from the broadcaster given that GPS information is available and accessible from the packet header.

Upon receiving a packet from node i, node j checks the packet ID and rebroadcasts with probability P_{ij} if it received the packet for the first time; otherwise, it discards the packet. Denoting the relative distance between nodes i and j by D_{ij} and the average transmission range by R, the forwarding probability, P_{ij}, can be calculated on a per packet basis using the following simple equation:

$$P_{ij} = \frac{D_{ij}}{R}. \tag{1}$$

Li [9] proposed a novel broadcast protocol called Efficient Directional Broadcast (EDB) for urban VANET using directional antennas. Due to the topology of VANET changed rapidly, EDB makes receiver-based decisions to forward the packet with the help of the GPS information. The receiver only needs to forward the packet in the opposite direction where the packet arrives. After a vehicle receives a packet successfully, it waits for a time before taking a decision whether to forward the packet or not. During this time, the vehicle listens to other relay of the same packet. The waiting time can be calculated using the following equation:

$$WaitingTime = \left(1 - \frac{D}{R}\right) \times \max WT. \tag{2}$$

where D is the distance from the sender which can be obtained using the sender's location information added in the packet and its own, and R is the transmission range. The *maxWT* is a configurable parameter which can be adjusted according to the density of the vehicle.

2.3   Collision Warning Systems

This section describes the models proposed for the collision warning system based on sensor information and for driver evasive action in response to collision warnings issued by the system. In multi-lane road environment, vehicles move in both longitudinal and lateral directions. Accordingly, vehicles can be expected to cause both longitudinal collisions and lateral collisions. Therefore, the collision warning system model is conceived of as encompassing two sub-systems: a forward vehicle collision warning system to address longitudinal collisions and a side collision warning system to address lateral collisions as shown in Fig. 1 [10].

Collision Warning System

Collision Warning Sub-System	Collision Warning Sub-System
Forward Vehicle Collision Warning System (Longitudinal)	Side Collision Warning System (Lateral)

Fig. 1. Structure of a collision warning system

(1)   Forward Vehicle Collision Warning

Fig. 2 illustrates the variables used in forward vehicle collision warning. The current time is set to zero. The position, velocity, and acceleration of the preceding vehicle at that time are defined as $x_{p_0}, v_{p_0}, a_{p_0}$, and the position, velocity, and acceleration of the following vehicle are defined as $x_{f_0}, v_{f_0}, a_{f_0}$. Furthermore, the relative position, relative velocity, and relative acceleration are defined as $x_{r_0} = x_{f_0} - x_{p_0}$, $v_{r_0} = v_{f_0} - v_{p_0}$, $a_{r_0} = a_{f_0} - a_{p_0}$.

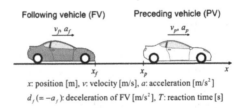

x: position [m], v: velocity [m/s], a: acceleration [m/s^2]

$d_f (= -a_f)$: deceleration of FV [m/s^2], T: reaction time [s]

Fig. 2. Definition of variables

Assume that the preceding vehicle suddenly decelerates and that the following vehicle then decelerates after the reaction time T, the inter-vehicular distance d when the two vehicles stop can be described as equation (3) [11]:

$$d = -x_r - x_f \cdot T + \left(\frac{v_f^2}{2a_f} - \frac{v_p^2}{2a_p} \right). \tag{3}$$

The condition of forward collision warning is satisfied when the following distance obtained by the sensors D becomes smaller than the calculated inter-vehicular distance d, called the stopping.

(2)   Side Collision Warning

Sensors are capable of detecting both distance to and speed of (both laterally and longitudinally for each) all vehicles with radius R [m] and viewing angle $\pm\phi$ [°].

Based on the accumulated data every refresh cycle, the system decides to issue a warning when warning assessment criteria Equation (4) are satisfied both laterally and longitudinally [10].

$$x_r \geq v_r \times T. \tag{4}$$

(3) Time to Collision

One of the most representative indices for assessing the warning provision timing of the forward obstacle collision warning system (FOCWS) is TTC [11]. The TTC is defined as follows:

$$TTC = -\frac{x_r}{v_r} = \frac{x_f - x_p}{v_f - v_p}. \tag{5}$$

The TTC represents the predicted time to collision on the assumption that the current relative velocity is maintained.

In the situation which the adjacent vehicle avoids the collision by applying the brakes when the subject vehicle changes lanes, the following conditions are required to prevent the adjacent vehicle from colliding with the preceding vehicle in the lane change, which means that the headway distance before lane changing must be less than the necessary distance for the following vehicle's deceleration [12].

$$v_r \cdot TTC > v_r \cdot T + \frac{v_r^2}{2\alpha},$$
$$TTC > T + \frac{v_r}{2\alpha}, \tag{6}$$

where α represents the following vehicle's deceleration.

When it is assumed that $T = 1$ second, $v_r = 30$ Km/h and $\alpha = 4$ m/s^2, TTC required to avoid the collision is calculated to be over 2.04 seconds. Braking alone will not avoid the collision when TTC is less than 2 seconds.

3   EW-ICAST Design

In this paper, we present EW-ICAST to improve the propagation of traffic safety application in VANET. In the design of EW-ICAST, we assume the following:

- Before transmitting an alert message, the on-board GPS is used to calculate the distance between the previous-hop vehicle and the current receiving vehicle;
- Further, the GPS is used to calculate the speed of the current receiving vehicle;
- All vehicles are equipped with multiples directional antennas that are the antennas which radiate greater power in one or more directions allowing for increased performance on transmit and receive and reduced interference from unwanted sources.

The proposed EW-ICAST uses not only the early warning system on the basis of TTC but also the intelligent broadcasting algorithm on the basis of fuzzy logic. In EW-ICAST, when a driving vehicle recognizes that the brake light of a vehicle right ahead is on or a vehicle of lateral lane is lane changing through V2V communication, the vehicle computes TTC. If the TTC was less than or equals to a threshold value of TTC, the vehicle broadcasts an alert message to following vehicles. When a vehicle receives an alert message for the first time, if the current speed of the vehicle was higher than a threshold value of vehicle speed, HI-CAST (hybrid intelligent broadcast) is performed. Otherwise, I-CAST (intelligent broadcast) is performed. The structure of EW-ICAST algorithm is shown in Fig. 3.

In I-CAST, the receiving vehicle determines rebroadcast degree using fuzzy logic rules, where the rebroadcast degree depends on the current traffic density of road and the distance between previous-hop vehicle and current receiving vehicle. The probability of rebroadcasting the message, as well as the rebroadcast delay dependent on the computed rebroadcast degree. If the vehicle didn't receive the rebroadcasted alert message from another vehicle until a time-out delay expires, the vehicle rebroadcasts the alert message with the rebroadcast probability. HI-CAST uses I-CAST in conjunction with alert token protocols, RPB-TOKEN, where RPB (Relative Position Based)-TOKEN sends an alert token to the neighboring vehicle in opposite direction.

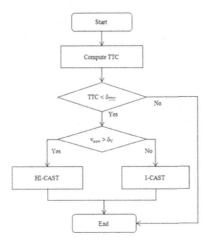

Fig. 3. Structure of EW-ICAST algorithm

We map the speed of the current receiving vehicle (v) to five basic fuzzy sets: VF (very fast), F (fast), M (medium), S (slow), VS (very slow) using the fuzzy function as shown in Fig. 4. The membership function of v represents fuzzy set of v. The membership function which represents a fuzzy set of v is usually denoted by, where V represents the maximum speed of vehicles.

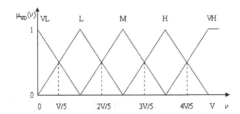

Fig. 4. Membership function for the current speed

Figure 5 shows a few examples of EW-ICAST, where $S0$, $S1$, $S2$, $S3$ and $S4$ represents the segments that divide the transmission range into the equal-size blocks, respectively. $S0$ and $S4$ represent the nearest and the farthest segments from a collision warning point, respectively. Firstly, consider the I-CAST scenario in Figure 5(a) where vehicles exist in the transmission range. Vehicle A which detects the collision warning broadcasts an alert message to all vehicles in its transmission range.

Vehicle D travelling in $S4$ has very short waiting time, but vehicle B which is in $S2$ has long waiting time. If the current speed of vehicle D was medium, the vehicle D has high rebroadcast probability. Vehicle D rebroadcasts with high probability if the vehicle D received the alert message for the first time and has not received any duplicates before its waiting time; otherwise, it discards the alert message. Secondly, consider the HI-CAST scenario depicted in Figure 5(b) where no vehicles exist within transmission range. If the current traffic density of the road was very low or the current speed of receiving vehicle is very fast, the RPB-TOKEN sends an alert token to the neighboring vehicle in opposite direction. Vehicle B receives the alert token from the vehicle A which detects collision warning, then sends the alert token to the vehicle C travelling straight head. The vehicle C sends the alert token to the vehicle D travelling in opposite direction. The vehicle D broadcasts the alert message to all vehicles in transmission range, and the vehicle sends the alert token to the neighboring vehicle E in opposite direction.

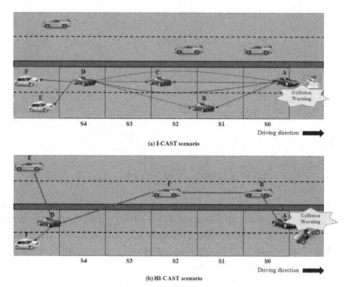

Fig. 5. Illustrating the working of EW-ICAST

For I-CAST, the control rules of rebroadcast degree which consider the current speed of receiving vehicle and the distance between previous-hop vehicle and receiving vehicle are shown in Table 1.

Upon receiving an alert message from vehicle i, vehicle j calculates $Rebroadcast_{Prob}(i, j)$ and $segWT(i, j)$ through equation (7) and equation (8), where $Rebroadcast_{Prob}(i, j)$ is the non-fuzzy control output and $defuzzifier$ is the defuzzification operator [13]. The vehicle j rebroadcasts with $Rebroadcast_{Prob}(i, j)$ if the vehicle j received the alert message for the first time and has not received any duplicates before $segWT(i, j)$; otherwise, it discards the alert message.

$$Rebroadcast_{Prob}(i,j) = defuzzifier \left(\begin{array}{c} a\ lingustic\ weighted \\ factor\ for\ rebroadcasting \end{array} \right), \qquad (7)$$

where $0 \leq Rebroadcast_{Prob}(i, j) \leq 1$,

$$defuzzifier\left(\left\{\begin{matrix}VH\\H\\M\\L\\VL\end{matrix}\right\}\right)=\left\{\begin{matrix}1.0\\0.8\\0.6\\0.4\\0.2\end{matrix}\right\}.$$

$$segWT(i,j)=\left(1-\frac{SN(j)}{N}\right)\times maxsegWT, \tag{8}$$

where $0 \le segWT(i,j) \le maxsegWT$.

And $SN(j)$ represents segment number which the current receiving vehicle j is travelling, N is the largest number of segments and $maxsegWT$ represents the maximum segment waiting time which is determined by considering the number of segments and the transmission delay of a VANET.

Table 1. The control rules for rebroadcast degree

Segment	VD				
	VS	S	M	F	VF
S0	VL	VL	L	L	M
S1	VL	L	L	M	M
S2	L	L	M	M	H
S3	L	M	M	H	VH
S4	M	M	H	VH	VH

Notes: (input variables) VD: VF (very fast), F (fast), M (medium), S (slow),
VS (very slow)
(output variables) rebroadcast degree: VH (very high), H (high), M (medium),
L (low), VL (very low)

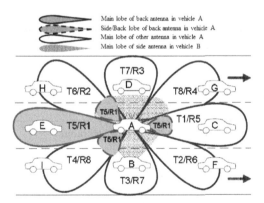

Fig. 6. Directional antenna with dedicated channel pair

We also propose an alert token passing algorithm on the basis of RPB-MACn [14] that is called RPB-TOKEN. Fig. 6 depicts such a directional antenna enabled MAC design based on the run-time static relative position property in VANETs. Here, all vehicles are equipped with 8 statically configured directional antennas, each dedicated to one relative position vicinity, operating over the single wireless channel to communicate with its 1-hop neighbors, and Tn indicates the transmission over wireless channel n, while Rn indicates the reception over channel n.

When a traffic accident has occurred, if the current traffic density of road was low, the vehicle that detects the accident initiates RPB-TOKEN. RPB-TOKEN sends an alert token with initial hop count value to a neighboring vehicle in the opposite direction. Upon receiving the token the vehicle sends the alert token to the vehicle ahead of it on the road. The next receiving vehicle sends the alert token to the vehicle traveling in opposite direction. If the vehicle next after received the alert message for the first time, the receiving vehicle broadcasts the alert message to all vehicles in transmission range. And the receiving vehicle sends the alert token with initial hop count value to the neighboring vehicle in opposite direction. Figure 7 shows the structure of RPB-TOKEN algorithm in case the vehicle which initializes RPB-TOKEN is in the right lane.

```
procedure RPB-TOKEN
  var ch, hops;
  event Initialize_TOKEN
  begin
    hops = 2;
    ch = TR6/R2 and T7/R3;
    send TOKEN(alert, hops, ch) to the neighboring node in opposite
      direction;
  end
  event Receive_TOKEN
  begin
    if (the hops of received token == 2) then
      hops = hops-1;
      ch = T1/R5 and (T8/R4 or T2/R6);
      send TOKEN(alert, hops, ch) to the straight ahead traveling
        vehicle;
    else if (the hops of received token == 1) then
      hops = hops-1;
      ch = T2/R6;
      send TOKEN(alert, hops, ch) to the neighboring vehicle in
        opposite direction;
    else
      ch = T5/R1 and (T6/R2 or T4/R8);
      if (receiving the alert message for the first time == yes)
      then broadcast MESSAGE(alert, ch) to all vehicles in
          transmission range;
    end
    hops = 2;
    ch = T6/R2 and T7/R3;
    send TOKEN(alert, hops, ch) to the neighboring node in
      opposite direction;
  end
end
```

Fig. 7. RPB-TOKEN algorithm

4   Performance Evaluation

The primary objective of EW-ICAST is to improve the success rate of safety message dissemination, that is, the percentage of vehicles that receive the safety alert message. We also aimed to mitigate the effect of the broadcast storm problem that afflicts most of the VANET's safety alert protocols. We use three metrics to evaluate different protocols.

- Collision: The number of alert message collisions that occur during the period of simulation;
- Success rate: Percentage of vehicles that received alert message;
- Time: Time delay from accident occurred until last vehicle received alert message.

The parameters and values of the performance evaluation for EW-ICAST are shown in Table 2, where the alert region represents the circular area within which the message is transmitted and PHY/MAC layers are compliant to IEEE 802.11p draft standard [15]. . The current speed of vehicles depends on the traffic density of the road. Thus, the higher the traffic density, the lower the vehicle speed; similarly, the lower the traffic density, the faster the vehicle speed. Accordingly, the current speed of a vehicle is computed from equation (9).

$$v_{now} = v_{max} \times \left(1 - \frac{\rho_{now}}{\rho_{max}}\right), \tag{9}$$

where v_{max} represents the maximum allowable speed of the road, ρ_{max} represents the traffic density that the vehicle speed is zero when a traffic jam occurred, and ρ_{now} represents the current traffic density of the road.

We evaluate the performance of EW-ICAST in the MATLAB 7.0 [16], where the dataset of traffic density follows Gaussian distribution, and the evaluation results are derived from the simulation program running of 5 times a case.

Table 2. Simulation parameters

Parameter	Value
Radius of alert region	2~10 Km
Transmission range (R)	500 m
The length of a segment	100 m
Deceleration (α)	4.0 m/s^2
Reaction time (T)	1 sec
Threshold value of TTC (δ_{TTC})	2 sec
Threshold value of vehicle speed (δ_V)	75 Km/h
Traffic density (TD)	20~140 vehicles/Km
Maximum traffic density ($maxTD$)	160 vehicles/Km
Traffic deviation	($1.5*maxTD$)/TD
Maximum speed of vehicles	100 Km/h
Initial hop count value	2
Number of lanes	4
The broadcast probability in p-Persistence	0.5
Transmission delay	20 ms/hop
Maximum waiting time	120 ms
Maximum segment waiting time	110 ms

Fig. 8 shows average number of alert messages that are transmitted by RPB-TOKEN accordingly to traffic densities in case the radius of alert region is 10 Km. The threshold value of traffic density is the traffic density that may have no vehicles in transmission range. Thus, the threshold value of traffic density is determined through the simulation result of Fig. 8 and the equation (9).

Fig. 9(a) shows the number of alert message collisions that is occurred accordingly to the radius of alert region in case traffic density is low. While the number of collisions of EW-ICAST is smaller than Simple and p-Persistence algorithms, the number of collisions of EW-ICAST is approximately equal to EDB algorithm. Fig. 9(b) shows respectively numbers of alert message collisions for EW-CAST and EDB that are occurred accordingly to the radius of alert region in case traffic density is high, where the performance of EW-ICAST is extremely better than that of EDB because EW-ICAST uses the fuzzy control rules for rebroadcast degree that consider the speed of a receiving vehicle and the distance segment between the previous-hop vehicle and the receiving vehicle.

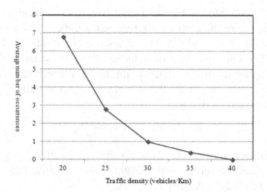

Fig. 8. Average number of transmitted alert message with RPB-TOKEN

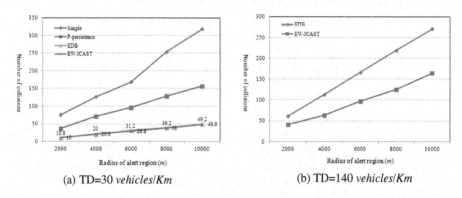

(a) TD=30 *vehicles/Km*   (b) TD=140 *vehicles/Km*

Fig. 9. Number of collisions with alert region radius

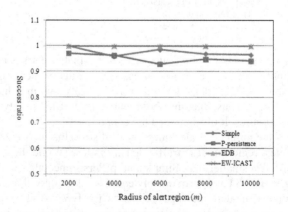

Fig. 10. Success rate with alert region radius

The most important result, the success rate for different algorithms in case traffic density is 30 vehicles per kilometer is shown in Fig. 10. The loss of alert message causes low success rate. The success rate of EW-ICAST is higher than that of Simple and p-Persistence algorithms, and the success rate of EW-ICAST is equal to that of EDB algorithm which achieves perfect success rate through broadcasting an alert message every *10 maxWT* until a next hop neighbor appears.

Message dissemination delay in case traffic density is 30 vehicles per kilometer is shown in Fig. 11, where we only consider the network transmission time for alert message, but the delay time of PHY/MAC layers due to alert messages congestion and collision doesn't considered. EW-ICAST uses TTC based early collision warning system. The vehicle that expects a collision broadcasts an alert message to all following vehicles in advance before the vehicle collision occurs. The delay time of EW-ICAST algorithm is very shorter than other algorithms. While, the delay time EDB has the worst delay time because that multiple *maxWT* delays are continued until a next hop neighbor appears.

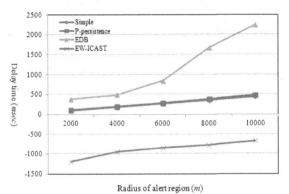

Fig. 11. Delay time with alert region radius

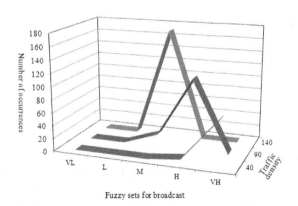

Fig. 12. Number of occurrences for fuzzy sets in EW-ICAST

Fig. 12 shows the number of occurrences of the fuzzy sets for rebroadcast in EW-ICAST in three cases that the current traffic density is 40, 90 and 140 vehicles per kilometer respectively, where the radius of alert region is 10 *Km*. The fuzzy set which rebroadcast probability is very high (VH) is occurred more frequently in case traffic density is low (TD=40). But the fuzzy set which rebroadcast probability is medium (M) is only occurred in case traffic density is high (TD=140).

5   Conclusion

Most VANET applications favor broadcast transmission that addresses the broadcast storm problem to avoid unnecessary loss of information during dissemination. Emergency warning for public safety is one of the many applications that are highly time-critical and require more intelligent broadcast mechanism than just blind flooding. In this paper, we propose EW-ICAST which uses not only the early warning system on the basis of TTC but also the intelligent broadcasting algorithm on the basis of fuzzy logic. The performance of EW-ICAST is evaluated through simulation and compared with that of other alert message dissemination algorithms. From the simulation results, we know that our Hi-CAST is superior to *Simple*, *p-Persistence* and EDB algorithms. Therefore, EW-ICAST Therefore, the EW-ICAST resolves effectively broadcast storm problem and meets time-critical requirement. Our future work includes studying an adaptive alert message dissemination algorithm which considers road conditions and shapes.

References

1. Chitra, M., Sathya, S.S.: Efficient Broadcasting Mechanisms for Data Dissemination in Vehicular Ad Hoc Networks. International Journal of Mobile Network Communications & Telematics 3, 47–63 (2013)
2. Liu, J., Yang, Z., Stojmenovic, I.: Receiver Consensus: On-Time Warning Delivery for Vehicular Ad-Hoc Networks. IEEE Transaction on Emerging Topics in Computing 1, 57–68 (2013)
3. Wisitpongphan, N., Tonguz, O.K., Parikh, J.S., Mudalige, P., Bai, F., Sadekar, V.: Broadcast storm mitigation techniques in vehicular ad hoc networks. IEEE Wireless Communications 14, 89–94 (2007)
4. Schoch, E., Kargl, F., Weber, M.: Communication Patterns in VANETs. IEEE Communication Magazine 46, 119–125 (2008)
5. Lin, Y.-W., Chen, Y.-S., Lee, S.-L.: Routing Protocols in Vehicular Ad Hoc Networks: A Survey and Future Perspective. Journal on Information Science and Engineering 26, 913–932 (2010)
6. Tonguz, O.K., Wisitpongphan, N., Bait, F., Mudaliget, P., Sadekart, V.: Broadcasting in VANET. In: Proceedings of Mobile Networking for Vehicular Environments, pp. 7–12. IEEE Press, New York (2007)
7. Suriyapaibonwattana, K., Pomavalai, C.: An Effective Safety Alert Broadcast Algorithm for VANET. In: Proceedings of International Symposium on Communications and Information Technologies, pp. 247–250. IEEE Press, New York (2008)
8. Suriyapaibonwattana, K., Pornavalai, C., Chakraborty, G.: An adaptive alert message dissemination protocol for VANET to improve road safety. In: IEEE International Conference on Fuzzy Systems, pp. 1639–1644. IEEE Press, New York (2009)

9. Li, D., Huang, H., Li, X., Li, M., Tang, F.: A Distance-Based Directional Broadcast Protocol for Urban Vehicular Ad Hoc Network. In: Proceedings of International Conference on Wireless Communications, Networking and Mobile Computing, pp. 1520–1523. IEEE Press, New York (2007)
10. Takatori, Y., Hasegawa, T.: Stand-alone Collision Warning Systems based on Information from On-Board Sensors. IATSS Research 30, 39–47 (2006)
11. Hiraoka, T., Tanaka, M., Kumamoto, H., Isumi, T., Hatanaka, K.: Collision Risk Evaluation Index Based on Deceleration for Collision Avoidance (First Report). Review of Automotive Engineering 30, 429–437 (2009)
12. Wakasugi, T.: A Study on Warning Timing for Lane Decision Aid Systems Based on Driver's Lane Change Maneuver. In: International Technical Conference on the Enhanced Safety of Vehicles, pp. 1–7. National Highway Traffic Safety Administration, Washington (2005)
13. Lee, C.C.: Fuzzy Logic in Control Systems: Fuzzy Logic Controller-Part I. IEEE Transactions on Systems, Man, and Cybernetics 20, 404–418 (1990)
14. Chigan, C., Oberoi, V., Li, J.: RPB-MACn: A Relative Position Based Collision-free MAC Nucleus for Vehicular Ad Hoc Networks. In: IEEE Global Telecommunications Conference. IEEE Press, New York (2006)
15. Toor, Y., Muhlethaler, P., Laouiti, A., Fortelle, A.D.L.: Vehicle Ad Hoc Networks: Applications and Related Technical Issues. IEEE Communication Surveys & Tutorials 10, 74–88 (2008)
16. Kay, M.G.: Basic Concepts in Matlab (2010),
    http://www.ise.ncsu.edu/kay/Basic_Concepts_in_Matlab.pdf

Enhancing the *Status Message Question Asking* Process on Facebook

Cleyton Souza[1], Jonathas Magalhães[1], Evandro Costa[2],
Joseana Fechine[1], and Ruan Reis[1]

[1] Laboratory of Artificial Intelligence - LIA,
Federal University of Campina Grande - UFCG,
Campina Grande-PB, Brazil
[2] Group of Intelligent, Personalized and Social Technologies - TIPS,
Federal University of Alagoas - UFAL,
Maceió-AL, Brazil

Abstract. People have been using Social Networks to search for help by broadcasting messages that reflect their information needs. However, several factors, usually not considered by the user, influence the outcome of receiving or not an answer. In this work, we aim to increase the users' chances of finding someone who could help them. For this purpose, we propose a mobile app called *Social Query*, which guides the users through some steps before they share the problem with their friends. As far as we know, this is the first work to merge these three aspects of the social search: Question Rephrasing, Expert Search Filtering and Expertise Finding. To evaluate our proposal, we ran a questionnaire in which users considered Useful most functions of the app.

Keywords: Social Search, Social Query, Social Network, Status Message Question Asking.

1 Introduction

Currently, Social Networks (SNs) are the most popular service on the Web, surpassing even E-mail [23]. In this scenario, Facebook stands out as the most popular worldwide SN, with more than one billion users [16]. These SN sites were first designed to allow remote interaction among geographically dispersed people [6]. One of the goals of interacting with others is the exchange of knowledge [4]. Thus, versions of knowledge exchange emerged from these virtual spaces.

One of the ways that knowledge exchange may occur is by means of *Status Message Question Asking* (SMQA), that is, members of SN sites making use of status messages to express information needs to friends and contacts [15]. It is an attempt to transform social relationships in practical knowledge [6]. This strategy is particularly useful when we are dealing with high-contextualized problems or when we are looking for personalized information [13].

Broadcasting has become a popular way to share questions on SNs. However, it is not the best, especially, in the context of Facebook. First, on Twitter, the

B. Murgante et al. (Eds.): ICCSA 2014, Part IV, LNCS 8582, pp. 682–695, 2014.

status would appear in all your follower's timeline, but there is no guarantee that people able to help will see your status while it stays on the top, and the probability of being seen decreases as it falls down [2]. However, on Facebook, the feed shown to each user is based on a personalized algorithm [11]; therefore, when someone broadcasts a question, there is no guarantee that it will be seen at any moment by people able to help. To ensure that the question will be at least visualized, some works defend that directing questions is more effective than broadcasting [14], but knowing who the question should be directed to is not always easy and, if we choose poorly, the person could just ignore the question or even give a wrong answer [20]. In addition, the way the question is phrased could be decisive to receive an answer.

Teevan et al. [21] found that characteristics of the question itself predicted the quality, number, and speed of responses. Thus, we notice that making use of SNs to find information is not always simple, because too many factors must be taken into account.

In this work, we aim to improve the SMQA process. To help the user, we propose an app called *Social Query*, which enhances the probability of receiving an answer, guiding the user through some steps before the disclosure of the question on Facebook. We propose a tool to help people looking for help. It is not only an Expertise Finding System (EFS), but also a tool to assist users to phrase their problems and restrict the social search to a certain demographic group. As far as we know, this is the first work to merge these three aspects of the social search (question rephrasing, expert search filtering and expertise finding). Through this app, users inform their questions and receive suggestions to increase the probability of receiving answers. The suggestions range from tips to rephrase the question to indications about who probably knows the answer (a person or group), so the user could direct the question to specific contacts, ensuring that it will be visualized by someone who could help.

We used a questionnaire to get feedback about our proposal. Through the questionnaire, people could express their opinions about the functions available in the Social Query app. The results were excellent; people considered useful most of the available functions, but we highlight the acceptance of the Expertise Finding engine and the Filtering engine.

The remainder of this paper is organized as follows. Section 2 presents a brief literature review. Section 3 describes the internal work of our proposal, and Section 4 presents the Social Query app. In Section 5, we present the questionnaire results and, finally, Section 6 ends the work with Conclusions and Future Work.

2   Related Work

The literature review is presented in the next sections. In the first part, we discuss about the practice of sharing questions on SNs; next, we list some EFS and, later, we discuss about the main differentials of our work to previous research.

2.1   About Status Message Question Asking

The habit of sharing questions on the web born on Community Questions and Answering (CQA) sites and was extended to SNs [19]. Asking a question on SN is an explicit action performed by users in order to convert the social relationships maintained on the site into actionable information and other social capital outcomes [6]. SMQA serve many purposes, including creating social awareness, encouraging the asker to reflect on a current information need, building social ties, and, of course, finding answers [21]. The motivations to answer vary but are mainly Altruism, Feel like an Expert, and Interestingness [13].

Morris et al. [13] presented statistics confirming SMQA as a viable method to find on-line answers. In their case study, 93.5% of users had their questions answered after sharing them and these responses, in 90.1% of the cases, were provided within one day. The main motivations pointed by the users who practice the SMQA were (1) their trust on their contacts and (2) the hope of a personalized answer [13]. These motivations highlight the advantages of posing questions on SNs compared to more generic CQAs sites; in addition, some patterns, identified by the studies on information seeking, suggest that certain information needs, such as those revolving around quotidian occurrences, are more commonly solved by individuals one already knows (e.g., "Was there a test in English class today? I overslept.").

Teevan et al. [21] found that characteristics of the question itself predicted the quality, number, and speed of responses: requests that were stated as a question, posed as a single sentence, and explicitly scoped received better responses. In addition, Nichols and Kang [14] confirmed that directing questions significantly improve the response rate, while the quality of the answer depends on who the question will be directed to [18]. In this sense, EFSs play an important role: if we identify an expert on the topic of the question and direct it to that expert, the answer would come faster and with higher quality [18]. Next, we will detail some interesting systems that could be used to this end and are related to ours.

2.2   About Query Routing Systems

The process of directing questions to appropriate helpers is called in literature as Query Routing (QR) and it is a well explored theme, specially, in the context of CQAs.

Tomiyasu et al. [22] propose a mobile CQA with the QR feature. Basically, the users describe their expertise using keywords and, when someone has a question, keywords are used to describe the subject and the question will be broadcasted to people who have these same keywords in their profiles. This probably is the simplest way to implement a QR policy. In [12], a similar app is presented, but with the differential that questions are only directed to the most suitable people connected to the questioner and they have the option to forward the question to people connected to them. In addition, the system maintains a history of previous answered questions to keep forwarding questions that fit with user's current interests. Aardvark [7] was a CQA that belonged to Google in which people ask and answer questions directed to them according to their expertise. The expertise was informed by the user and his/her friends, and learned through their question and answering history and the information captured through other channels (e.g., Gmail, Orkut, Google search history). In addition, users were allowed to ask questions specifically to their private SN of friends who also were in the system. Unfortunately, the project was closed in 2011 [8].

Regarding QR on the SN context, [5] proposes a topic EFS to Twitter named Cognos, which infers expertise by crowd wisdom captured by user's List information. Ghosh's work focuses on finding global experts on Twitter: the more times a user appears in lists related with "computing", greater the evidence that he is a computer expert; each appearance is taken as an implicit vote. Lin et al. [9] present SmallBlue, a corporative SN with an EFS. The expertise profiles are built based on corporative information (e.g., email, personal webpage, chat history). Moreover, when recommending experts, SmallBlue can prioritize people close to the requester and there is a filter to restrict the search using predefined criteria like department or position. Davitz et al. [3] proposed iLink; a tool for social search and message routing on SN. Davitz's work included an entity named *supernode*, which monitors the SN and decides to whom the questions will be routed. In some cases, the supernode is also able to offers answers based on Frequently Asked Questions (FAQ). However, due the computational cost to monitor the entire SN, they only evaluate in small forums.

2.3   The Differential of our Research

The works presented so far are all EFS. Expertise Finding (EF) usually involves a global context of candidates, for instance [5]. However, we aim at the detection of specialists in the set of the questioner's friends (local context). Thus, the "amount" of expertise necessary to characterize someone as "an expert" changes according the set that we are looking into.

Moreover, EF often considers only the expertise about some topic like [22] and [12]; what we are proposing is taking into account several factors to improve the probability of finding relevant information through the help of friends. In [20], we had proposed a QR system that routed questions to followers on Twitter based on three criteria: knowledge, trust and activity. However, in this work, we are not proposing a system that just recommends experts.

Here, we propose a tool that also assists the user in the process of sharing his problem like [7] and [3]. (I) The Social Query app will analyze the question and suggest modifications; (II) it will suggest restricting the search for help to a certain demographic group (as people with your age or the same profession as you); and (III) it will suggest people based on their bounds, availability and expertise. As far as we know, this is the first work to merge these three aspects of the social search (question rephrasing, expert search filtering and expertise finding). In addition, we propose this to the context of the most popular SN nowadays, instead a private access context, as done in our previous studies.

3   Project and Architecture

In this Section, we explain the internal organization of the Social Query app. Figure 1 presents the architecture of the proposal. The white arrows represent input, the black arrows represent output, and the gray arrows have multiple meanings (e.g., HTTP request, SQL execution, HTTP response, etc.).

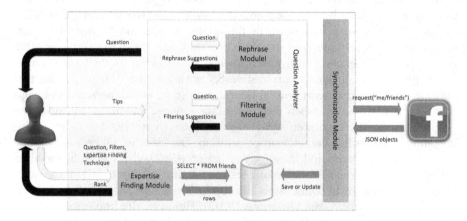

Fig. 1. Social Query app architecture

As can be seen, this app has four main modules; each module represents an independent function, namely: 'Expertise Finding Module', 'Synchronization Module', 'Rephrase Module' and 'Filtering Module'. Together, these last two form the Question Analyzer, which could be also considered as a module, where the input is the question and the output is the set of suggestions to rephrase the question (established by the 'Rephrase Module') and to filter the expert search to a certain demographic group (established by the 'Filtering Module') .

The 'Synchronization Module' is responsible for crawling the information from Facebook and updating the local database. This information consists of data about the users and their contacts. Figure 2 illustrates the entity 'User' in our model.

The attributes of the entity 'User' are self-explanatory. In upcoming releases, we are planning to allow users to update their own profiles through the Social Query app, but in the current version all information is retrieved from their Facebook profiles. The 'Synchronization Module' retrieves the information when users connect the app with their Facebook accounts and every time they use the re-sync option in the app's menu. This information is essential to the 'Expertise Finding Module'. The inputs of this Module are the EF technique, the active Filters, and the Question. The output is a rank of the questioner's friends ordered by their fitness with the Question and the Filters.

Fig. 2. Entity 'User' and its attributes

4   Social Query on Facebook: Mobile App

The Social Query app was developed for Android. It helps users to use the potential of their social capital in order to turn social connections into practical knowledge. In the next sections, we will detail how our app works and the ideas behind its views.

4.1   First View

The First View of the app is the Login page, shown in Figure 3 (left), where users must inform their Facebook credentials (right).

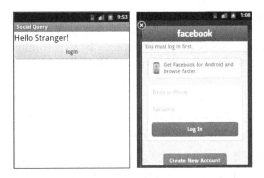

Fig. 3. First view of the app (left) and Facebook's mobile login dialog (right)

After logon, the user must give us permission to access their Facebook account information and to publish content in their feeds, as presented in Figure 4 (left). After that, they are directed to the Main Page, as shown in Figure 4 (right).

Fig. 4. Permission's dialog (left) and main view (right)

The options in the Main page are: Logout (a); go to Settings (c); Synchronize again with the Facebook account (d); and Go make a question (b). The Logout option directs the user to the Login page again. The Settings option allows users to choose what EF model to use (currently, there are three available ones) and to define Filters to the EF search. The Synchronize option is an opportunity for users to update the app information about them (catch more recent information about them, their contacts and new connections); it will start the same thread initiated after the Login. The Go button guides the user to the main functionality of the app. Next, we detail the Settings option and will later talk about how our proposal fits the Q&A process.

4.2   Settings

Currently, the Settings option is limited to choosing the EF model and active Filters to the Expert Search. The EF model is the technique that will be used to represent the contact's expertise. The Filters to Expert Search restrict the recommendation to a certain group of contacts. Both are inputs of the 'Expertise Finding Module'. Next, we will discuss about the EF Models and Filters available in Social Query app.

Expertise Finding Models. In the current version of the app, there are three EF models available to the users, namely:

– **Voting Model:** Proposed by Macdonald and Ounis [10], it considers the task of ranking experts as a voting problem. The profile of each expert candidate is associated to a set of documents that represent their expertise. The request for an expert is assumed as a query in a search engine that retrieves some of these documents. Each retrieved document is associated to one or many users and counts as an implicit vote for them. The ranking of experts is based on the total of votes of each candidate. Several strategies could be used for retrieving the documents, associating the document to the users or weighting the votes.

- **Vector Space Model:** A classical approach from Information Retrieval (IR), proposed originally in [17]. The idea behind the model is to represent content in multidimensional vectors. In our context, the vector represents the content associated with each user, the coordinates represent the words, and the coordinate values are calculated using TD-IDF. The expertise score is the similarity between the expertise profile and the question vector using cosine similarity.
- **PageRank:** A classical algorithm that measures the importance of a node counting the number and quality of nodes pointing to it [1]. If we consider that the scenario where "a user X, author of question Q, receives an answer A, from user Y" represents a graph like $X \to Q \to A \to Y$, that could be simplified to $X \to Y$. One of the goals of PageRank is to estimate the probability of randomly getting into a node; the higher this probability, the greater the odds of the node being a good recommendation.

Filters. The Filters are used to restrict the social search to a certain social group. Currently, there are five Filters implemented. Restrict by: age, gender, profession, formation and location. In the current version of the app, each filter restricts the expert search to people with the same characteristics of the user. For instance, if the user is a man and he checks the "Filter by gender", only men will be recommended; if he lives in Paris and he also checks the "Filter by location", only men who live or lived in Paris will be recommended as well.

However, for the upcoming releases of the app, we are planning improvements to the filter engine. One of the improvements will allow users to choose the filter value (e.g., the hometown that they want to use to restrict the search). Another improvement will be the automatic prediction of the ideal filter value (e.g., find what would be the most indicated hometown). In the literature, there is already some research in this direction, e.g., [7]. In addition, we are constantly thinking about new Filters.

4.3   Q&A Process

The main goal of the app is to help users before they share their questions on Facebook. For this purpose, we planned to split the Q&A process into three main screens: one to phrase the question; one to receive suggestions from the Question Analyzer; and, one for users to tag people in text of the question. Figure 5 presents these three screens.

The First screen (Part 'A' of Figure 5) is where the users phrase their problems. After that, they are directed to the screen where the Question Analyzer's suggestions are displayed (Part 'B' of Figure 5). Then, they must choose what suggestions they will accept. Finally, they see the rank of their friends and check those who they want to tag in text of the question (Part 'C' of Figure 5).

In Part 'A' of Figure 5, the user 'Ted Mosby' has a question about places to visit in Paris. After typing the question, the user will click on the *Ask* button, being directed to the Tips View. The Part 'B' of Figure 5 illustrates some of

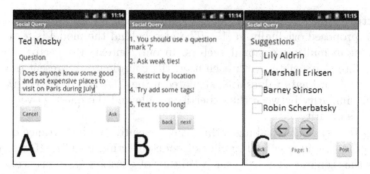

Fig. 5. The three screens of the Q&A process. The user enters a question (Part 'A'); Question Analyzer's suggestions (Part 'B') and; Friends recommendation (Part 'C').

the tips that could be given to the user. The decision about which tips will be presented is determined by our Question Analyzer, based on the characteristics of the question. The Question Analyzer processes the question and extracts its characteristics. Briefly, the Rephrase Module and the Filtering Module look into the text of the question, searching for specific information (e.g., terms or mentions to place or people). They associate these characteristics to pre-established tips, which were decided based on the literature review and on the interviews that we conducted. The chosen tips are displayed to the users, who will have the option to accept them or not. If they decide to accept any tip, they must click the *Back* button (to edit the question text) or Settings Menu (to turn on some filters). After that, they can click the *Next* button to be directed to the Recommendation List View, where they chooses who they want to forward the question to. This view is Part 'C' of Figure 5. Questioner's friends are ordered according to their utility score, calculated by the EF model chosen on Settings. The users check the people and click the *Post* button. Then, the Social Query app posts on the Facebook user feed the question tagging the friends that they checked.

5   Results

To validate our tool, we shared a questionnaire with Facebook groups. This questionnaire was answered by 250 volunteers. To know about our volunteers, the first part of questionnaire asked them about their experience with SMQA; the second part request them to value the main function of the Social Query app. We decide by the questionnaire instead a case study because it allows us to capture more impressions. The next Subsection summarizes our results.

5.1   Who Answered the Questionnaire?

We shared the questionnaire in Facebook groups. It was answered by 250 volunteers. However, only 159 confirmed that had already shared questions through an SN. In this section, we will briefly describe the experience and habits of these volunteers regarding the social query.

Regarding their habits before sharing the question, most volunteers search for the answer by themselves before turning to friends for help, only 5% of them admitted that they go straight to SNs. In addition, most people (84%) often think carefully about how to phrase the problem. It is known that a short period and a well-defined audience are associated with better answers [21]. However, only 1/3 thinks about people they know who probably can help. Moreover, 1/3 of volunteers also make the "mistake" of being thorough.

Regarding their opinion about how easy to it is find help through SMQA, 130 (81%) consider it easy while 29 (19%) consider it hard, but 94% said that they usually do not need share their problem multiple times to receive an answer. The most common sharing strategy is to publish the question in a group with the same topic (70%), followed by asking someone directly through chat (62%) and sharing the question through feed (53%). The same strategies, but combined with tagging, were not so popular: sharing in group tagging a member obtained 10% and sharing in the feed tagging some friend (s), 20%.

This describes some of the volunteers who answered our questionnaire.

5.2   What Did They Think about the Mobile App?

The volunteers evaluated the aspects of the application described in previous sections. Initially, we asked volunteers to value the main functions of our proposal. There was a template question like "How useful would be a tool with this [function]?" followed by one of the functions of the Social Query app. The answer options were "Don't know", "Somewhat Useful", "Useful" and "Very Useful". The results are summarized in Table 1.

As can be seen in Table 1, most functions were labeled as "Useful". We used a one-tailed binomial test to statistically compare the percentages of each row. The results are also displayed on Table 1, the 'Useful' label occurrence was

Table 1. The results of the user questionnaire. The symbol (*) means significance at the level $p < 0.05$ and (**) means level $p < 0.01$).

Question	Don't Know	Somewhat Useful	Useful	Very Useful
How useful would be a tool that indicates friends who probably can answer your question?	4%	12%	47%**	37%
How useful would be a tool that suggests changes to your question in order to increase the likelihood of getting an answer?	4%	33%	40%*	23%
How useful would be a tool that indicates what group of your friends is able to help you?	2%	12%	43%**	42%**

statically higher in all cases than the second place label, except in the third row (Filtering Module) where both the "Very Useful" label and the "Useful" label have statistically more occurrences than "Somewhat Useful" (third place). The most useful functions, according the answers, were the Expertise Finding engine and the Filtering engine.

Regarding the Expertise Finding, we asked volunteers about what they look for in answers from their Facebook friends. Figure 6 presents a summary of these answers.

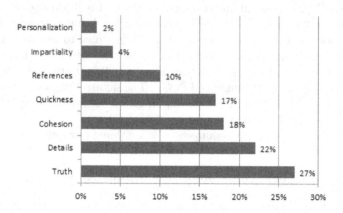

Fig. 6. The results about what users expect from answers from their Facebook friends

"Truth" (27%) was the most desired characteristic followed by "Detail" (21%), which means that the utility score function should prioritize the Expertise over other subjective criteria like friendship, trust, bond, availability, distance, etc. We obtained an unexpected result, because "Personalization" was the less desirable characteristic (2%), while in [13] many appreciated that their private SN was familiar with their additional context, such as knowledge of their location, family situation, or other preferences. Impartiality did not receive many votes either (4%).

Later, we asked if they believed that certain questions were implicitly directed to people in a demographic group. We used a template question like "Do you agree that some questions can only be answered by a certain [characteristic]?" followed by each Filter option. This question aimed to evaluate the practical utility of the Filtering engine and their results are summarized in Figure 7.

In general, all the Filters were considered useful by most of volunteers, except the gender filter, which was a polemic subject. We believe that this rejection was due to the sexist aspect of our question. Unfortunately, we do not have information about gender from our volunteers; however, based on the email address, we think that 15% of them are women, 24% are men, and 61% are unknown. Thus, 38% of "women" (by our standards) considered the Gender Filter useless, while 45% of "man" thought it was useful. When we observe the difference between

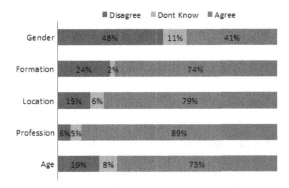

Fig. 7. The results about the Filtering engine utility

the filter acceptance percentage by men and women, we realize that the Gender filter has the greatest difference (14%, while others did not exceed 5%). This may be absurd, but men and women may have understood that they were not able to answer questions made by the other gender and rejected the filter by this reason. But this is just a guess; we could not confirm this without individually interviewing each respondent. The fact is that the Gender filter was not well received by our audience.

6   Conclusion and Future Work

In this work, we presented the Social Query app to assist users to search for information on SNs. While most part of previous work focused on the Expertise Finding engine, we propose a tool to help the users through several steps of the social search process. First, our solution helps the users to rephrase the question, enhancing its probability of being answered. Second, the app offers three different approaches to finding experts. Last, there is an option to filter the expert finding search to a certain group with the same demographic characteristics as the users (age or gender, for instance).

To evaluate our proposal, we ran a questionnaire, which was answered by 250 Facebook users. Through the questionnaire, these users could give their impressions about the functional aspect of the Social Query app. The results were excellent. The main functions (Expertise Finding mechanism, Filtering engine and Rephrase engine) of the app, in average, were considered at least useful by more than 40% of users. In addition, we obtained great feedback that allows us to think in improvements to our proposal.

As Future Work, we are planning the following improvements: (1) use of other Expertise Finding models, including those which consider semantics; (2) improve the Question Analyzer, besides suggesting changes in problem specification, automatically applying some or all of these changes; (3) improve the Filtering use to specify the input; (4) allow users to maintain a list of contacts; (5) allow users to maintain lists of friends; (6) considering additionally the users' reputation, based on previous; and (7) make friends of friends available as expert candidates.

Acknowledgments. We want to thank the people who answered our questionnaire.

References

[1] Brin, S., Page, L.: The anatomy of a large-scale hypertextual web search engine. Comput. Netw. ISDN Syst. 30(1-7), 107–117 (1998)

[2] Comarela, G., Crovella, M., Almeida, V., Benevenuto, F.: Understanding factors that affect response rates in twitter. In: Proceedings of the 23rd ACM Conference on Hypertext and Social Media, HT 2012, pp. 123–132. ACM, New York (2012)

[3] Davitz, J., Yu, J., Basu, S., Gutelius, D., Harris, A.: ilink: Search and routing in social networks. In: Proceedings of the 13th ACM SIGKDD International Conference on Knowledge Discovery and Data Mining, KDD 2007, pp. 931–940. ACM, New York (2007)

[4] Furlan, B., Nikolic, B., Milutinovic, V.: A survey of intelligent question routing systems. In: 2012 6th IEEE International Conference on Intelligent Systems (IS), pp. 014–020 (September 2012)

[5] Ghosh, S., Sharma, N., Benevenuto, F., Ganguly, N., Gummadi, K.: Cognos: Crowdsourcing search for topic experts in microblogs. In: Proceedings of the 35th International ACM SIGIR Conference on Research and Development in Information Retrieval, SIGIR 2012, pp. 575–590. ACM, New York (2012)

[6] Gray, R., Ellison, N.B., Vitak, J., Lampe, C.: Who wants to know?: Question-asking and answering practices among facebook users. In: Proceedings of the 2013 Conference on Computer Supported Cooperative Work, CSCW 2013, pp. 1213–1224. ACM, New York (2013)

[7] Horowitz, D., Kamvar, S.D.: The anatomy of a large-scale social search engine. In: Proceedings of the 19th International Conference on World Wide Web, WWW 2010, pp. 431–440. ACM, New York (2010)

[8] Karch, M.: Google labs aardvark, about.com guide (2010), http://google.about.com/od/experiment_graveyard/g/Google-Labs-Aardvark.htm (retrieved February 2014)

[9] Lin, C.-Y., Cao, N., Liu, S.X., Papadimitriou, S., Sun, J., Yan, X.: Smallblue: Social network analysis for expertise search and collective intelligence. In: IEEE 25th International Conference on Data Engineering, ICDE 2009, pp. 1483–1486 (March 2009)

[10] Macdonald, C., Ounis, I.: Searching for expertise: Experiments with the voting model. The Computer Journal 52(7), 729–748 (2009)

[11] Magid, L.: Facebook tweaks news feed algorithm again (January 2014), http://www.forbes.com/sites/larrymagid/2014/01/21/facebook-tweaks-news-feed-algorithm-again/ (retrieved February 2014)

[12] Manoranjitham, G., Veeraselvi, S.: Mobile question and answer system based on social network. International Journal of Advanced Research in Computer and Communication Engineering 2, 3620–3624 (2013)

[13] Morris, M.R., Teevan, J., Panovich, K.: What do people ask their social networks, and why?: A survey study of status message q&a behavior. In: Proceedings of the SIGCHI Conference on Human Factors in Computing Systems, CHI 2010, pp. 1739–1748. ACM, New York (2010)

[14] Nichols, J., Kang, J.-H.: Asking questions of targeted strangers on social networks. In: Proceedings of the ACM 2012 Conference on Computer Supported Cooperative Work, CSCW 2012, pp. 999–1002. ACM, New York (2012)

[15] Oeldorf-Hirsch, A., Hecht, B., Morris, M.R., Teevan, J., Gergle, D.: To search or to ask: The routing of information needs between traditional search engines and social networks. In: Proceedings of the 17th ACM Conference on Computer Supported Cooperative Work & Social Computing, CSCW 2014, pp. 16–27. ACM, New York (2014)

[16] Ross, M.: Facebook turns 10: the world's largest social network in numbers (February 2014), http://www.abc.net.au/news/2014-02-04/facebook-turns-10-the-social-network-in-numbers/5237128 (retrieved February 2014)

[17] Salton, G., Wong, A., Yang, C.-S.: A vector space model for automatic indexing. Commun. ACM 18(11), 613–620 (1975)

[18] de Souza, C.C., de Magalhães, J.J., de Costa, E.B., Fechine, J.M.: Predicting potential responders in twitter: A query routing algorithm. In: Murgante, B., Gervasi, O., Misra, S., Nedjah, N., Rocha, A.M.A.C., Taniar, D., Apduhan, B.O. (eds.) ICCSA 2012, Part III. LNCS, vol. 7335, pp. 714–729. Springer, Heidelberg (2012)

[19] Souza, C., Magalhães, J., Costa, E., Fechine, J.: Routing questions in twitter: An effective way to qualify peer helpers. In: Web Intelligence, pp. 109–114 (2013)

[20] Souza, C., Magalhães, J., Costa, E., Fechine, J.: Social query: a query routing system for twitter. In: Proc. 8th International Conference on Internet and Web Applications and Services (ICIW), pp. 147–153. IARIA Press (2013)

[21] Teevan, J., Morris, M.R., Panovich, K.: Factors affecting response quantity, quality, and speed for questions asked via social network status messages. In: ICWSM (2011)

[22] Tomiyasu, H., Maekawa, T., Hara, T., Nishio, S.: Profile-based query routing in a mobile social network. In: Proceedings of the 7th International Conference on Mobile Data Management, MDM 2006, pp. 105–109. IEEE Computer Society, Washington, DC (2006)

[23] Wayne, T.: Social networks eclipse e-mail (May 2009), http://www.nytimes.com/2009/05/18/technology/internet/18drill.html (retrieved February 2014)

Medium Access Scheme with Power Control to Improve Performance of FlashLinQ*

Jun Suk Kim, Jaheon Gu, Hee-Woong Yoon, and Min Young Chung**

College of Information and Communication Engineering,
Sungkyunkwan University,
300, Chunchun-dong, Jangan-gu, Suwon, Gyeonggi-do, 440-746, Republic of Korea
{jsk7016,mageboy,hiwoong,mychung}@skku.edu

Abstract. Device-to-Device (D2D) communications has been considered to reduce heavy traffic load on Base Station (BS) since it enables devices to directly communicate with each other without a relay of BS. Recently, Qualcomm Inc. introduced FlashLinQ (FLQ) for D2D communications, which has a radio frame based on Orthogonal Frequency Division Multiplex (OFDM). In FLQ, D2D User Equipments (DUEs) distributively access to medium based on Signal to Interference Ratio (SIR). When DUEs operate on a real-time service such as Voice over Internet Protocol (VoIP) or File Transfer Protocol (FTP), DUE with low link quality may cause excessive yielding for medium access of other links. Thus, the service requirements of the other links cannot be guaranteed. In order to solve this problem, we propose a medium access scheme including transmission power control for DUEs in FLQ. In the proposed scheme, DUEs determine their transmission power based on the number of neighbor DUEs, and try accessing to medium with the controlled transmission power. Through simulations, we evaluate performance of the proposed scheme by comparing those of the conventional FLQ and Probabilistic Medium Access Scheme (PMAS). We show that the proposed scheme can improve performance of FLQ by simulation results.

Keywords: FlashLinQ, Medium Access Scheme, Device to Device, Connection Scheduling, Yielding, Power Control.

1 Introduction

As use of smart devices has been expanded, mobile traffic has rapidly increased [1]. In infra-based wireless network such as cellular network, Cellular User Equipment (CUE) can exchange data traffic with each other only through a BS [2]. The increasing of data traffic causes heavy loads on BS. As one of technologies

---

* This research was supported by Basic Science Research Program funded by MOE (NRF-2010-0020210) and supported by Next-Generation Information Computing Development Program through the National Research Foundation of Korea (NRF) funded by the Ministry of Science, ICT & Future Planning (2010-0020727).
** Corresponding author.

B. Murgante et al. (Eds.): ICCSA 2014, Part IV, LNCS 8582, pp. 696–708, 2014.

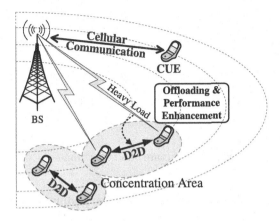

Fig. 1. Mobile communication network with D2D communications

offloading the burden of BS, D2D communications through physically direct communication link between DUEs [3]. The example of mobile communication network with D2D communications is shown in Figure 1. Since DUEs exchange data with their pair DUEs without relaying of BS, D2D communications can reduce traffic loads on BS.

As one of technologies supporting D2D communications, Qualcomm introduced a medium access technology, called FLQ [4][5]. In FLQ, DUE can establish a D2D communications link with their pair DUE through their own Connection Identification (CID) which is locally unique and used for determining priority of the D2D communications link. Each D2D link distributively tries accessing to medium by exchanging single-tone signals through the CID. The signals are separated into time and frequency phase based on OFDM structure.

For DUEs to access the wireless medium, FLQ defines a procedure called connection scheduling in which DUEs reserve medium for D2D communication. In connection scheduling period of FLQ, according to the priority of D2D link, a DUE calculates SIR between itself and the other DUEs with higher priorities. By using the SIR, DUEs determine whether they conduct medium access or not. For example, D2D link with low priority and SIR may not access to medium, because it expects that its D2D communications is not successful through the D2D link due to its low SIR.

D2D links with low link quality cannot be guaranteed sufficient performance in D2D communications even though they obtain opportunities of medium access. In addition, if a DUE with low link quality participates in connection scheduling, it may excessively cause yielding of the other DUEs for medium access. As a result, DUE having high link quality can give up their transmissions because they receive interference from a DUE with higher priority but low link quality. This may cause degradation of system performance. Especially, if DUEs use real-time services, a DUE with high link quality cannot be guaranteed its service

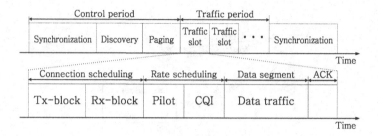

Fig. 2. Structure of a super frame

requirement, because it may excessive yield its medium access due to a DUE with low link quality.

In order to reduce the excessive yielding of a DUE in real environment, a medium accessing scheme is proposed in this paper, which includes a power control scheme of DUEs. In the proposed scheme, each DUE determines its transmission power depending on the number of neighbor DUEs. The rest of this paper is organized as follows. In Section 2, we briefly introduce frame structure of FLQ and connection scheduling procedure of terminals for their medium access. The proposed scheme is introduced in Section 3. In Section 4, we analyze performance of the proposed scheme. Finally, Section 5 gives a conclusion of this paper.

2   Related Works

For D2D communications, FLQ defines a periodic super-frame with length of 1 sec [4][5]. The super-frame is composed of control period and several traffic period as shown in Figure 2. The control period is located at the start of the frame and the traffic slots with length of 2.08 msec follow after the control period. The control period is used for controlling basic functionality of DUEs such as synchronization, discovery and D2D link establishment. In the traffic period, DUEs distributively perform scheduling for accessing to medium and they transmit data traffic to their pair DUEs based on the result of the scheduling.

The control period consists of synchronization, discovery, and paging period. In synchronization period, DUEs basically synchronize time and frequency with neighbor DUEs through geographic information system such as Global Positioning System (GPS). In discovery period, DUEs broadcast beacon signals including their information and detect the existence of neighbor DUEs through received beacon signals from them.

In paging period, DUEs establish D2D link with their pair DUEs to conduct D2D communications. For this, a D2D link should acquire a locally unique CID which is used for identifying its link among neighboring links. In order to find available CIDs, DUEs inquire their pair DUEs by exchanging available CIDs list between them. If a DUE finds available CIDs in its neighbor, it forms its own available CIDs list. After forming the list, a DUE exchanges the list with its pair DUE by signaling based on OFDMA. By exchanging the list of available CIDs and selecting a CID in the list, they can obtain a locally unique CID. Then, DUE and its pair DUE can establish D2D link with the selected CID.

Traffic period consists of several traffic slots in which DUEs exchange their data traffic with the pair DUEs through the established D2D links. A traffic slot is composed of connection scheduling, rate scheduling, data segment, and ACK as illustrated in Figure 2. For D2D communications, the D2D links perform connection scheduling in order to determine whether to access medium or not with consideration of their SIR and priorities which are changed every traffic slot. At rate scheduling, the scheduled D2D link in the connection scheduling exchange a wide-band pilot signal and Channel Quality Indicator (CQI) to determine its own code rate and modulation scheme for data segment based on the D2D link quality. At data segment, D2D transmitter transmits data traffic to its pair D2D receiver with the determined code rate and modulation scheme. In response to successful reception of data traffic, D2D receiver transmits ACK to its pair D2D transmitter.

Connection scheduling consists of two blocks, namely Rx block and Tx block that have an OFDM based structure. Each block is composed of several resource elements, and a DUE and its pair DUE exchange single-tone signals for decision whether to access medium or not through a resource element. By exchanging the single-tone signals, a DUE can obtain information such as its D2D link quality and interference from/to neighboring D2D links. By using the information, each DUE distributively decides whether it communicate with its pair DUE or not.

In Rx block, D2D transmitter transmits a single-tone signal, called Direct Power Signal (DPS), with a predefined power to its pair DUE on the location corresponding to its priority. D2D receiver receives DPSs not only from its pair D2D transmitter but also from the other pairs. Based on the strength of the signals it decides whether to try accessing medium or not. When the D2D receiver decides to try accessing medium, it transmits a sing-tone signal, called Inverse Power Echo (IPE), at a power which is inverse to the power of received signal from its pair D2D transmitter. In Tx block, a D2D transmitter receives IPEs from other D2D receivers and distributively make decision whether to yield their medium access or not based on the strength of the received signals.

Figure 3 shows an example of the connection scheduling between two D2D links. D2D transmitter A and D2D receiver B have established a D2D link with first priority, and D2D transmitter C and D2D receiver D have established a D2D link with second priority. $|h_{XY}|^2$ represents the channel gain between transmitter X and receiver Y. The connection scheduling is composed of two steps; Rx-yielding decision and Tx-yielding decision. For Rx-yielding decision, A and C transmit DPS to their pair DUEs with power P_A and P_C, respectively. D receives two DPSs from A and C with power $P_A|h_{AD}|^2$ and $P_C|h_{CD}|^2$. D regards the DPS from A as interference signal, and the DPS from C as oriented signal. D measures

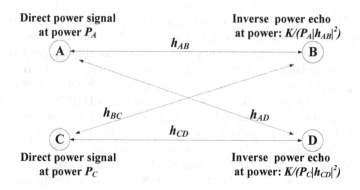

Fig. 3. Scenario of a connection scheduling

SIR of the received signals based on the strength of two DPSs. D compares the measured SIR with a predefined Rx-yielding threshold (γ_{Rx}) as follows,

$$\frac{P_C \times |h_{CD}|^2}{P_A \times |h_{AD}|^2} > \gamma_{Rx} \qquad (1)$$

where γ_{Rx} is a minimum SIR required for successful medium access of a receiver. If the measured SIR is higher than γ_{Rx}, D makes decision that it can successfully receive data traffic from C, nevertheless the interference from A. On the other hand, if the measured SIR is lower than γ_{Rx}, D decides that it will successively not be able to receive data traffic from C since it will severely be interfered by the transmission from A to B. Then D immediately ceases the connection scheduling procedure, and does not participate in following Tx-yielding decision procedure (Rx-yielding decision).

If B and D decide to be able to receive data traffic from their pair DUEs, B and D transmit IPE signals to their pair as response to the received DPS at power $K/(P_A|h_{AB}|^2)$ and $K/(P_C|h_{CD}|^2)$, respectively. K is a constant and can be changeable to adapt to systems. C receives the IPE signal from B with power $K|h_{AC}|^2/(P_A|h_{AB}|^2)$. As multiplying the strength of received IPE by P_C/K and inverting it, C can calculates SIR of B, when it transmits data traffic to D as follows,

$$\frac{P_A \times |h_{AB}|^2}{P_C \times |h_{BC}|^2} > \gamma_{Tx} \qquad (2)$$

where γ_{Tx} is a minimum SIR required for successful data traffic transmission of a transmitter to its pair receiver by considering interference of D2D receiver having higher priority. If the calculated SIR is more than γ_{Tx}, C decides that it can transmit data traffic to its pair. On the other hand, if the calculated SIR is less than γ_{Tx}, C decides that its transmission causes much interference to B. Thus, it decides not to participate in following rate-scheduling (Tx-yielding decision).

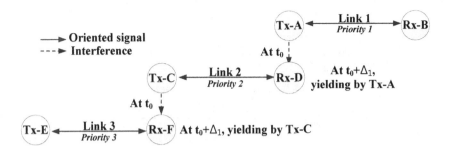

Fig. 4. A example for an excessive yielding problem

In connection scheduling of FLQ, DUEs distributively decide whether to access to medium based on their SIRs and priorities. However, FLQ has a problem that a D2D link can contiguously cause yielding of other links for accessing to medium. when the D2D link has low link quality. It means that a D2D link cannot perform D2D communications because of the D2D link with low link quality. We refer to the problem as the excessive yielding problem of DUE.

For a example, we consider a case of that transmitter A, C and E have established a D2D link 1, 2 and 3 with receiver B, D and F, respectively, as illustrated in Figure 4. Link index represents a priority of the link. Link 1 and 3 have high link quality while link 2 has low link quality. D and F have received stronger interference from A and C than oriented signal strength from its pair C and E. In this case, link 1 can access to medium regardless of existence of link 2 and 3, since it has the highest priority. On the other hand, link 2 and 3 distributively decide whether to access to medium with consideration of interference from/to link(s) with higher priority. D with low SIR may give up medium access to link 1 at $t_0 + \Delta_1$ due to interference from A. Simultaneously, F receives much interference from A and C. F may also yield medium access for link 1 and 2 at $t_0 + \Delta_1$, even though link 3 has a high link quality. Despite link 1 and 3 can simultaneously conduct medium access, link 3 yields the medium access because of much interference from link 2. Even if link 2 does not yield the medium access, it may cause Tx-yielding of link 3. Link 2 cannot efficiently conduct D2D communications than that of link 3 in terms of data transmission rate, since link 2 has lower link quality than link 3. In this case, it is better that link 2 does not try accessing to medium or control its transmission power in order to mitigate interference from itself to other links.

In real-environment where each D2D link uses different service, the requirements of the services are also different from each other. Some D2D links use service which requires high data transmission rate while has tolerance about transmission delay. On the other hand, some D2D links use other services which require short transmission delay instead of high data transmission rate. In connection scheduling of FLQ, some D2D links with low link quality may excessively cause the yielding of other D2D links, and FLQ cannot guarantee the service requirement of the other link.

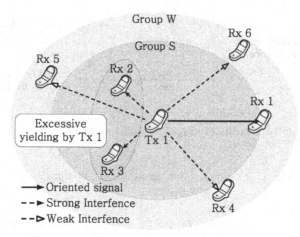

Fig. 5. Scenario of D2D signals and interference in the proposed scheme

3   Medium Access Scheme with Power Control for FLQ

In order to solve a problem that D2D links with low link qualities cause excessive yieldings of other D2D links in connection scheduling of FLQ, a medium access scheme with transmission power control of DUE is proposed in this paper. In the proposed scheme, a D2D transmitter tries accessing to medium by controlling its transmission power based on the link quality and density of DUEs in the proximity of itself. The density of DUEs represents how many other DUEs exist in proximity of a DUE. It can be estimated by exchanging the CID lists between a DUE and its pair DUE in paging period prior to the connection scheduling. The proposed scheme aims to reduce excessive Rx- or Tx-yielding of DUEs by controlling the transmission power of DUE.

The method of estimating the density is as follows. In order to explain the method, a scenario is considered as shown in Figure 5. Tx and Rx refer to D2D transmitter and D2D receiver, respectively. Tx 1 and Rx 1 try to establish a D2D link, and 5 other D2D Rxs are deployed around Tx 1 and the Rxs respectively try to establish a D2D link with its pair Txs. For establishing a D2D link, each Rx exchanges its own CID list with its pair Tx by transmitting a single-tone signal in paging period. Tx 1 receives the signals from the Rxs and can measure link gain between Rxs and itself by using strength of the received signals. Based on the measured link gains, Tx 1 can estimate the amount of interference from itself to each Rx. With consideration of the amount of interference from Tx 1 to each other Rxs, Tx 1 divides its neighbor Rxs into two groups: S and W. S is the group of Rxs expected to be severely interfered by Tx 1. Rxs which are expected to be relatively less interfered by Tx 1 are grouped in W.

In order to determine whether a Rx is severely interfered by a Tx or not, we define a criterion as strength of the signal received at Tx 1 from its pair D2D receiver, Rx 1. If the strength of interference from a Tx to a neighbor Rx is stronger than the criterion, the Tx groups the Rx into group S. Otherwise, Tx groups the Rx into group W. By grouping the Rxs, Tx 1 can know how many Rxs receive strong interference from itself. Tx 1 calculates α that is the ratio of severely interfered Rx to total Rxs as follows,

$$\alpha = \frac{N_S}{N} \tag{3}$$

where N_S and N denote the number of Rx in S, the total number of Rxs around Tx 1, respectively. α represents the proportion of Rxs which are severely interfered by Tx among total Rxs.

In order to reduce the excessive yieldings, Tx 1 controls its transmission power by considering the calculated ratio α. Based on the α, Tx 1 determines transmission power when trying accessing to medium. We consider two scenarios according to priority of D2D link belonging Tx 1. In case that Tx 1 has a highest priority, since it does not yield its medium access to other D2D links, it tries accessing to medium by transmitting DPS with a predefined transmission power without power controlling. This guarantees an opportunity for the D2D link's medium access, even though it has low link quality.

On the other hand, if the priority of Tx 1 is not highest, since it may yield its medium access to a D2D link with higher priority than its own, it controls its transmission power and transmits DPS with the controlled transmission power for accessing medium. The controlled transmission power of Tx 1 is determined as follows,

$$P_{DPS} = P_{pre} \times (1 - \alpha) \tag{4}$$

where P_{DPS} and P_{pre} denote the controlled transmission power of DPS signal and a predefined transmission power for Tx 1, respectively. This means that D2D link quality of Tx 1 and the density of Tx 1 determines the transmission power of Tx 1. If Tx 1 has low link quality or there are many Rxs in proximity of Tx 1, the number of other Rxs in group S may be large since the criterion is low or there are many other Rxs severely interfered by Tx 1. Then it transmits DPS to its pair D2D receiver with low transmission power. Otherwise, it tries accessing to medium with high transmission power.

The effects of the proposed scheme are classified into two cases. In point of Rx-yielding, more Rxs can try accessing to medium than those of conventional FLQ, since interference from Tx 1 to other Rxs can be reduced. In point of Tx-yielding, Tx 1 does not perform excessive Tx-yielding by controlling its transmission power. Since the controlled transmission power of Tx 1 causes less interference to the D2D receivers with higher priorities than that of the Tx 1, the Tx 1 can perform D2D communications with its pair Rx.

4   Performance Evaluation

For evaluating the performance of the proposed scheme, simulations based on C programming are performed. In the simulations, an environment that there are only DUEs in network is considered. There are up to 110 D2D links and

Table 1. Simulation Parameters

Parameters	Value
Dimension	1 km^2
Carrier frequency	2.4 GHz
System bandwidth	5 MHz
Size of Tx/Rx blocks	28 tones × 4 symbols
Service types	FTP, VoIP
Threshold for Tx/Rx yielding	9 dB
Transmission power of DUE(P_{pre})	20 dBm
Path loss model	ITU-R P1411 Outdoor[6]
Traffic model	User Traffic Mix[7]
Spectral efficiency	MCS table[8]
Noise power spectral density	-174 dBm/Hz

they are uniformly distributed in a 1 km x 1 km rectangular area. A role of each DUE is predefined as a transmitter or a receiver at the start of simulations. In considered environment, DUEs distributively access to medium based on their measured SIR and priority without support of eNB.

In order to evaluate performance of real condition, we considered normal traffic environment where a part of D2D links has data to transmit and the other D2D links does not has data to transmit according to the service used by each D2D link. As the considered service, there are two service types, namely FTP and VoIP service, respectively. Typically, FTP service requires a relatively high transmission data rate and greater tolerance for delay. On the other hand, VoIP service requires a relatively low transmission data rate but a short transmission delay [9]. In considered simulation environment, 10 percentages of total D2D links use FTP service and the remainder D2D links use VoIP service.

As the traffic parameters of each service, we consider the traffic generation model in [7]. According to the considered traffic model, a FTP service packet is generated with truncated lognormal distribution. The average packet size, standard deviation and reading time are 2 Mbytes, 0.722 Mbytes and 180 seconds, respectively. The reading time represents that the interval between end time of download of previous file and time that the user requests for the next file. VoIP traffic consists of talk-spurts and silent periods, with relatively small packets transmitted quite rarely [10]. Voice packet is generated with interval of 20 msec in talk-spurts period. Silence Indicator Packet (SIP) is generated with interval of 160 msec in silence period. The size of each packet type is 40 bytes and 15 bytes, respectively. Other detailed simulation parameters are given Table 1.

In order to evaluate performance of the proposed scheme, we compared the proposed scheme with two schemes: conventional FLQ and PMAS [11]. In conventional FLQ, DUEs determines whether to access medium or not without consideration of the number of DUEs in their proximity and transmission power control. PMAS supports that DUEs probabilistically determine their medium access based on only the number of DUEs in their proximity without transmission power control.

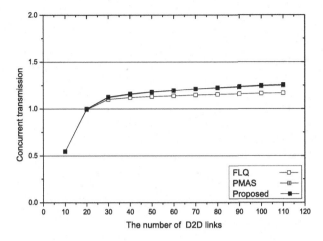

Fig. 6. The average number of concurrent transmission as varying the number of D2D links

The average number of concurrent transmission as increasing the number of D2D links is shown in Figure 6. The average number of concurrent transmission is defined as the average number of D2D links which access to medium at the same time. Since more D2D links try accessing to medium as the number of D2D links increases, more D2D links simultaneously conduct D2D communications at a traffic slot. The proposed scheme achieves a higher performance than FLQ in terms of the average number of concurrent transmission. In proposed scheme, a D2D Tx which has many D2D Rxs in its proximity tries accessing to medium with lower transmission power than the other D2D Txs which has few D2D Rxs in their proximity. Overall interference of network can be reduced by controlling the transmission power and D2D links can efficiently perform connection scheduling with low interference. As a result, the number of D2D links which excessively yield its medium access decreases and the average number of D2D links that can simultaneously access to medium at a data traffic slot increases. In addition, the performance of the proposed scheme is almost equal to that of PMAS. It means that the SIR of each DUE in the proposed scheme is similar to that in PMAS. In PMAS, both the strength of the oriented signal and interference from other D2D links increase since a DUE transmits data with a fixed transmission power. In the proposed scheme, both the strength of the oriented signal and interference from other D2D links are reduced since a DUE controls its transmission power based on the density of its proximity. Thus, the proposed scheme can achieve similar performance with the PMAS.

In real-time service, a delay of packet is a significant indicator to evaluate the performance of the service. In order to evaluate performance in terms of a delay, we analyze average delay of a generated VoIP packet depending on the schemes. Figure 7 shows average delay of a generated VoIP packet as the number of D2D links increases. Average delay of the packet increases as the number of D2D links increase. Since more D2D links try accessing to medium, a D2D link

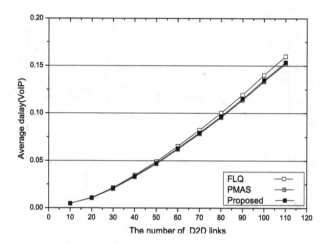

Fig. 7. Average delay of a VoIP packet as varying the number of D2D links

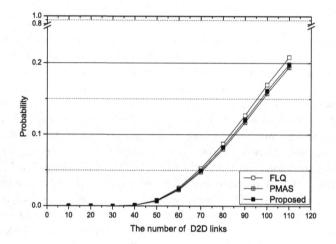

Fig. 8. Dissatisfying probability with VoIP service requirement

can obtain less opportunity for D2D communications and a packet generated by the D2D link cannot be transmitted during a certain period. And a period in which a packet cannot be transmitted becomes longer. Thus, the delay of a packet increases as the number of D2D links increases. The proposed scheme can slightly improve performance of FLQ and achieve similar performance to that of PMAS in terms of average delay of a VoIP packet. Since the average number of concurrent transmission of the proposed scheme and PMAS are larger than that of FLQ, the proposed scheme and PMAS make the period be shorter in which a packet cannot be transmitted.

Figure 8 shows a probability that delay of a VoIP packet is over than service requirement of VoIP. We considered 250 msec as the service requirement of VoIP [7]. When there are few of D2D links in the network, the probability is

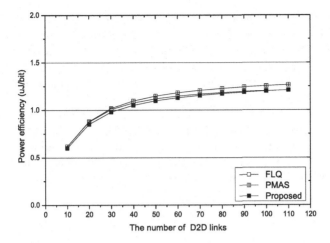

Fig. 9. Power consumption of DUE as varying the number of D2D links

close to zero. It means that the most of D2D links can access medium without Tx-/Rx-yielding at least once during 250 msec. As the number of D2D links increases, the probability rapidly rises. As shown in Figure 6, a part of D2D links among total D2D links can perform D2D communications at a specific data traffic duration. For this reason, a D2D link can obtain less opportunity to access medium at a specific data traffic slot as the number of D2D links increases. Since a D2D link more yields their medium access, the packet of the D2D link cannot be transmitted at contiguous data traffic slots. Thus, the dissatisfaction probability for service requirement of VoIP rapidly rises as the number of D2D links increases. The proposed scheme achieves slightly lower probability than conventional FLQ, since it makes more D2D links to simultaneously conduct D2D communications at a specific data traffic slot.

Figure 9 shows power efficiency of DUE as varying the number of D2D links. The power efficiency is defined as the consumed power of DUE's transmission for one bit. PMAS achieved the lowest power efficiency of DUEs. Since more number of D2D link conduct D2D communications with a fixed transmission power, the interference between D2D links increases, and it makes transmission power of a DUE increase to transmit for one bit. On the other hand, the proposed scheme achieved the highest power efficiency of DUEs. Since the proposed scheme achieved similar performance with PMAS in terms of concurrent transmission as shown in Figure 6, and interference between D2D links is reduced by controlling transmission power. It means that the proposed scheme enable DUEs to consume less transmission power to transmit for one bit.

5   Conclusion

In this paper, we propose a medium access scheme including transmission power control in order to improve performance of FLQ. In the proposed scheme, each

DUE distributively determines their transmission power based on the number of neighbor DUEs and tries accessing to medium for its D2D communications with the controlled transmission power. The proposed scheme can reduce the overall interference and the excessive yieldings of DUEs. DUEs may efficiently perform connection scheduling with low interference. In order to evaluate performance of the proposed scheme in real condition, we consider simulation environment where DUEs use real-time services, such as VoIP and FTP. Simulation results show that the proposed scheme improves performance of FLQ in terms of the number of concurrent data transmission, the packet delay and power consumption. Through simulation results, we conclude that the proposed scheme can enable DUEs to efficiently perform D2D communications.

References

1. Yeh, S., Talwar, S., Wu, G., Himayat, N., Johnsson, K.: Capacity and Coverage Enhancement in Heterogeneous Networks. IEEE Wireless Communications 18(3), 32–38 (2011)
2. Hakola, S., Chen, T., Lehtomaki, J., Koskela, T.: Device-to-Device (D2D) Communication in Cellular Network - Performance Analysis of Optimum and Practical Communication Mode Selection. In: IEEE Wireless Communication and Networking Conference (WCNC), pp. 1–6 (2010)
3. Doppler, K., Rinne, M., Wijting, C., Ribeiro, C., Hugl, K.: Device-to-Device Communication as an Underlay to LTE-Advanced Networks. IEEE Communications Magazine 47(12), 42–49 (2009)
4. Corson, M.S., Laroia, R., Junyi, L., Park, V., Richardson, T., Tsirtsis, G.: Toward Proximity-Aware Internetworking. IEEE Wireless Communications 17, 26–33 (2010)
5. Wu, X., Tavildar, S., Shakkottai, S., Richardson, T., Li, J., Laroia, R., Jovicic, A.: FlashLinQ: A Synchronous Distributed Scheduler for Peer-to-Peer Ad Hoc Networks. In: IEEE Allerton Conference, pp. 514–521 (2010)
6. Cichon, D.J., Kerner, T.: Propagation Prediction Models., COST 231. Final Rep. (1995)
7. Liu, G., Xiadong, S., Kramer, J., Abeta, S., Salzer, T., Jacks, E., Buldorini, A., Wannemacher, G.: Next Generation Modbile Networks Radio Access Performance Evaluation Methodology. A white Paper by the NGMN Alliance (2008)
8. 3GPP: Evolved Universal Terrestrial Radio Access (E-UTRA); Radio Frequency (RF) System Scenarios (Release 10). TR 36.942 V10.2.0 (2011)
9. Szymanski, T., Gilbert, D.: Provisioning mission-critical telerobotic control system over interent backbone networks with essentially-perfect QoS. IEEE Journal on Selected Areas in Communication 28(5), 630–643 (2010)
10. Puttonen, J., Kolehmainen, N., Henttoenen, T., Moisio, M.: Persistent packet scheduling performance for Voice-over-IP in evolved UTRAN downlink. In: IEEE PIMRC Conference, pp. 1–6 (2008)
11. Yoon, H.-W., Lee, J., Bae, S.J., Chung, M.Y.: A Probabilistic Medium Access Scheme for D2D Terminals to Improve Data Transmission Performance of FlashLinQ. In: Murgante, B., Misra, S., Carlini, M., Torre, C.M., Nguyen, H.-Q., Taniar, D., Apduhan, B.O., Gervasi, O. (eds.) ICCSA 2013, Part I. LNCS, vol. 7971, pp. 131–141. Springer, Heidelberg (2013)

Kinect-Based Monitoring System to Prevent Seniors Who Live Alone from Solitary Death

Jae-Gwang Lee, Jae-Pil Lee, Il-Kwon Lim, Young-Huyk Kim,
Hyun-Namgung, and Jae-Kwang Lee

Hannam Univ. Department of Computer Engineering,
Daejeon, Korea
{jglee,jplee,yhkim,iklim,ghnam,jklee}@netwk.hnu.kr

Abstract. In this study, a monitoring system was implemented to prevent solitary death of seniors who live alone. When monitoring activities of seniors who live alone, in order to protect their privacy, color (RGB) camera and depth (3D) sensor of Kinect are used to extract the skeleton and collect data. The collected data use WCF(Windows Communication Foundation) to record the images on a regular basis and send the images to DB with a designated interval. Sencha touch was used for monitoring the images saved in DB in mobile and PC.

Keywords: Kinect, Monitoring, Security Operation.

1 Introduction

Today, the South Korean society is under influence of various factors such as change of demographic environment, supply of medical resources, emergence of new diseases, development of medical technology, and healthcare system. The problem is that, with development of modern medicine and economy, the country is rapidly entering into aging society [1]. [Table 1] shows the composition of estimated population aged 65 or older (South Korea) by National Statistical Office. By the end of 2005, the senior population over the age of 65 shared 8.9% of the entire population, which already qualifies the country as an aging society [2]. Added by a low birthrate and increased average lifespan, from 2020, the population over 65 will take more than 15.6% of the entire population, making the country a population aging [3].

As the society is aging, various problems related to the seniors began to appear, and, particularly, because of nuclear families and changing perception and value system towards parents, the number of seniors who live with their children is decreasing while those who live alone are increasing. According to Ministry of Health and Welfare, as of 2010, the number of senior citizens aged 65 or older are 4,383,000 and 783,000 of them, 17.8%, live alone [4].

In Chungcheongnam-do (province), numbers of senior citizens who live alone are 82,349 compared with 215,373 of elder people aged 65 and older on the basis of 2013, so it is the level applicable to 27%. In addition, numbers of elder people living

B. Murgante et al. (Eds.): ICCSA 2014, Part IV, LNCS 8582, pp. 709–719, 2014.

along aged 65 and older were 540,000 in Korea in 2000, and it was 16% among about 3.4 million of all the older people. However, Ministry of Health and Welfare reports that the figure has more than doubled in 2013, with 1.19 million. The proportion of seniors who live alone has also risen, surpassing 20% of the entire senior population. In other words, one out of five seniors in South Korea lives alone [5].

Table 1. Proportion of population over 65 estimated by NSO (Korea)[2] (Number, %, Person/female 100)

	Male	Over 65	%	Female	Over 65	%
2000	23,666,769	1,299,786	5.5	23,341,342	2,095,110	9
2005	24,190,906	1,733,661	7.2	23,947,171	2,632,981	11
2010	24,540,316	2,189,996	8.9	24,334,223	3,166,857	13
2015	24,706,848	2,678,037	10.8	24,570,246	3,702,782	15.1
2020	24,679,762	3,303,494	13.4	24,645,927	4,397,631	17.8
2030	24,190,354	5,217,615	21.6	24,444,217	6,593,092	27
2040	22,854,325	6,647,579	29.1	23,488,692	8,393,328	35.7
2050	20,734,181	7,131,572	34.4	21,608,588	9,024,185	41.8

Seniors who live alone have difficulty interacting with other people or the society in general due to their housing and living environment, and this can lead to a serious societal issue, that is, 'solitary death.'

According to the statistics published by MHW, in South Korea, the number of the deceased without relatives is continuously increasing from 587 in 2009, 636 in 2010, and 727 in 2011 [6].

Solitary death refers to dying alone without any caretaker. This phase has been coined in Japan, which experienced aging society earlier than South Korea and where deceased seniors were often found after death, and now is used by South Korea media. Recently, reports on solitary death of seniors who live alone are increasing. This issue must be perceived as a social issue rather than an individual one that only concerns seniors or their families, and measures should be taken in order to prevent it[7].

As part of the solution, various services including Dolbomi Service [7], and Contact Service[2]. However, to provide these services efficiently requires a large amount of workforce and costs. Also, the CCTV monitoring service is causing many problems related to privacy of the seniors who live alone.

Developed by Microsoft, Kinect sensor is a low-price depth camera that provides real-time depth information as well as RGB video and joint tracking. The data provided by Kinect sensor (depth, RGB, joint position) eliminates the need for detecting person/body part and estimating a pose that are required for gesture recognition, and facilitates developing application relating to game or human-computer interaction [8,9,10].

Therefore, in this study, a observation and remote monitoring system for mobile and PC using Kinect color (RGB) camera and depth (3D) sensor and skeleton is researched in order to protect privacy of seniors.

2   Related Studies

2.1   Kinect

Kinect was launched by Microsoft in November 2010 as a new input device for XBOX-360. The name is a compound word of kinetic and connect, and the device recognizes human motion and inputs it in computer.

As various attempts were made to apply the Kinect sensor to different fields, Microsoft launched 'Kinect for Windows', which can be used for embedded devices rather than game, in February 2012, as well as 'Kinect for Windows SDK,' which allows Kinect sensors to be used for Windows-based embedded devices. As a result, it is being recognized as a next-generation interface for smart TV, medical equipment, and other state-of-art IT devices. As shown in [Figure 1], Kinect sensor is composed of color (RGB) camera, depth(3D) sensor, four multi-array mics, and motorized tilt, which allows moving the sensor vertically [11].

A key feature of Kinect sensor, the depth (3D) sensor is made up of a monochrome CMOS sensor and infrared laser beam projector. As shown in [Figure 2], when the infrared laser beam is projected, the CMOS sensor receives the reflected beam points and the distance from each pixel is measured. And then, the data is processed by the image processor in order to recognize the user in front of Kinect sensor. [Table 2] shows specification of Kinect [13].

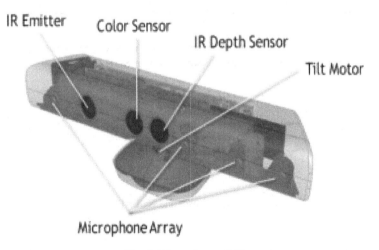

Fig. 1. Kinect Sensor[12]

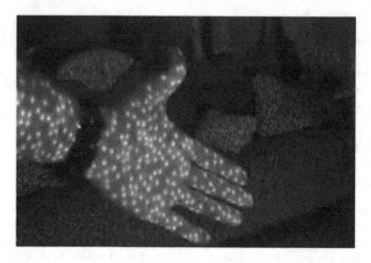

Fig. 2. Kinect's Infrared ray laser beam[13]

Table 2. Kinect specification[13]

Kinect sensor	Description
Sensor item	Describes the range
Viewing angle	Field of view, vertically 43 degrees, horizontally 57 degrees
Mechanized tilt range	Up 27 degrees, down 27 degrees
Frame rate	30 FPS(Frames per second)
Resolution, depth stream	VGA (640 X 480)
Resolution, color stream	VGA (640 X 480)
Audio format	16-kHz, PCM(pulse code modulation)
Audio input characteristics	24-bit analog

Software and library related to Kinect sensor interacts with the application as shown in [Figure 3]. [Figure 4] shows components of SDK [14].

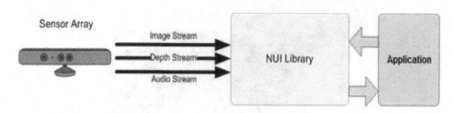

Fig. 3. Kinect Sensor and NUI API[14]

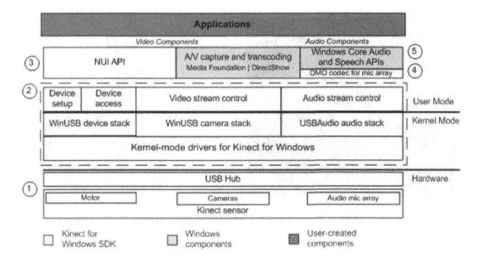

Fig. 4. Structure of Kinect for Windows SDK [14]

2.2　Monitoring Program Using Kinect

In previous research using color image, depth image data, and skeleton recognition of API [15], in order to record the image, byte-arrayed image data is received from the video stream of Kinect sensor and saved in the buffer space. Array data is copied as a factor in the function for image output or recording, and converted into a BGR(Blue Green Red) bitmap object as a result of function. Image recording uses VideoWriter class of the image processing library emguCV and Xvid for compression codec. And the image is recorded at 30 frames per second, and in 640*480, the maximum resolution that can be outputted by the sensor. Also, the motion can be detected for recording and, the most commonly used method of determining motion uses the difference between two frames and, when the difference is above a particular level, it

Fig. 5. Image recorded bright by the program[15]

Fig. 6. Image recorded dark by the program[15]

is determined as a motion and image saved. The skeleton-based method can track a person even at night, and can monitor motion by recording both color image and depth image during the night. [Figure 5], and [Figure 6] are recorded images of a program in previous research. The previous study [15] saved the image as a video, which increases the size and can be only monitored in PC. Also, recording a video can violate privacy. Therefore, the mobile monitoring system in this study for PC and mobile environment takes photographs in order to reduce the size and extracts skeleton and collects image data.

3   Design and Implementation

[Figure 7] is the overall structure of the monitoring system to prevent solitary death of seniors who live alone. When recognizing a human body, it uses Kinect sensor to extract skeleton and record the image. Also, images were designed to be recorded every 30 seconds, and sent every 30 seconds by using WCF (Windows Communication Foundation), which allows an application program to communicate through multiple computers connected to a computer or network. Since WCF does not have a separate UI(user interface), it requires a service application program. Kinect can expand and use WCF.

[Figure 7] is the overall structure of the monitoring system to prevent solitary death of seniors who live alone. People are shot by using Kinect's Depth (3D) sensor and infrared laser beam. As image information at this time, image data created transmitted by extracting skeleton information are transmitted rather than receiving RGB data. When transmitting data, the subsystem of communication making applications through one computer or multiple computers connected to the network transmits it to DB at 30-second intervals by using WCF(Windows Communication Foundation). Not having separate UI (User Interface), WCF needs service applications to be used through connection and transmission can be done more easily because Kinect can be used by extending the function of WCF.

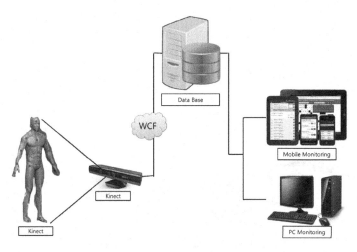

Fig. 7. Monitoring System Design

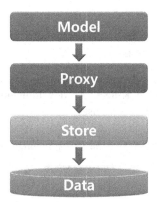

Fig. 8. Sencha Touch Service Structure

For monitoring image data transmitted to DB, Sencha Touch 2.0, the program to develop hybrid app is used. When passing data for monitoring through Sencha Touch 2.0 at the server, image data in DB is transmitted by using Ajax communication. In addition, Sencha Touch 2.0 can process data as XML or JSON. While XML is an easy language, the tags in the beginning and end increase the overall packet size, and it is necessary to reduce the size in order to provide service both in PC and mobile environment. As a solution to this issue, JSON format is used for processing data. [Figure8] shows the service structure of Sencha Touch 2.0.

Model defines the data structure while proxy creates direct connection to the server. Store processes received data for the local storage. The processed data, after the stage 3, is sent to mobile or PC for monitoring. [Figure 9] shows transmission for monitoring.

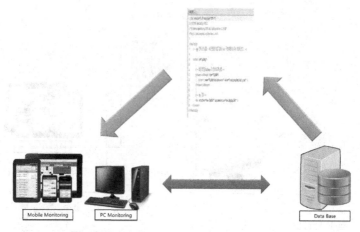

Fig. 9. Transmission Method for Monitoring

4   Result and Analysis

4.1   Implementation Environment

[Table 3] shows the implementation environment. Sencha Touch ver.2.0 was used and C# was used for implementing Kinect. As development environment, Windows 7 64bit OS, 4G RAM, and 1TB HD were used.

Table 3. Implementation Environment

Item	Contents
Language	C#
Mobile framework	Sencha Touch 2.0
OS	Windows 7 64bit
RAM	DDR 4GB
HD	1TB

[Figure 10] shows the monitoring screen using the Android-based Galaxy Note 10.1 and iOS-based iPad mini, and [Figure 11] using a computer screen.

4.2   Image of Service Implementation

[Figure 12] is part of the source to create image data. Skeleton is extracted to create image data. WCF is possible to use from adjusting a name space and service name space of Kinect library as a class to be used. Kinect is initialized by InitializeNui function and executed through nui_ColorFrameReady as an event handler whenever a frame comes in. Through m-nCount, make regular time intervals without performing every time. When saving, it is transmitted after saving it into as a file form of '.png', and it is designated as yymmdd_hhmmss for its data name.

The image data is sent at an interval of 30 seconds to DB by using WCF. [Figure 13] is part of the source implementing WCF. However, there was 1 to 2 seconds of error due to delay in computer performance and program operation.

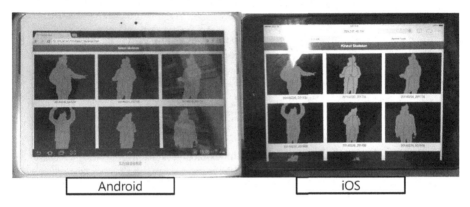

Fig. 10. Mobile monitoring

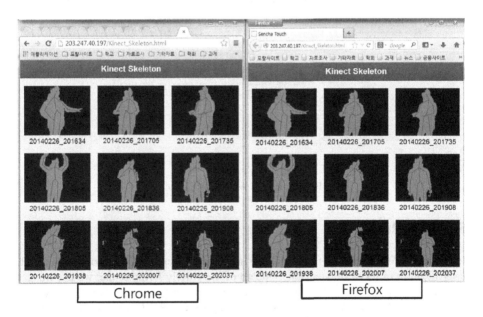

Fig. 11. PC monitoring

```
KinectSensor nui = null;
void InitializeNui()
{
    nui = KinectSensor.KinectSensors[0];

    nui.DepthStream.Enable(DepthImageFormat.Resolution320x240Fps30);
    nui.DepthFrameReady += new EventHandler<DepthImageFrameReadyEventArgs>(nui_D

    nui.DepthStream.Enable();
    nui.SkeletonStream.Enable();
    nui.AllFramesReady += new EventHandler<AllFramesReadyEventArgs>(nui_AllFrame

    nui.Start();
}

void SetRGB(byte[] nPlayers, int nPos, byte r, byte g, byte b)
{
    nPlayers[nPos + 2] = r;
    nPlayers[nPos + 1] = g;
    nPlayers[nPos + 0] = b;
}

private byte[] Players(DepthImageFrame PImage, short[] depthFrame, DepthImageStre
{
    byte[] nPlayers = new byte[PImage.Width * PImage.Height * 4];

    for (int i16 = 0, i32 = 0; i16 < depthFrame.Length && i32 < nPlayers.Length;
    {
        int player = depthFrame[i16] & DepthImageFrame.PlayerIndexBitmask;

        SetRGB(nPlayers, i32, 0, 0, 0);
        if (player == 1) SetRGB(nPlayers, i32, 0xFF, 0x00, 0x00); // 붉은색
        if (player == 2) SetRGB(nPlayers, i32, 0xFF, 0x7F, 0x7F); // 연붉은색
        if (player == 3) SetRGB(nPlayers, i32, 0x00, 0xFF, 0x00); // 녹색
        if (player == 4) SetRGB(nPlayers, i32, 0x7F, 0xFF, 0x7F); // 연두색
        if (player == 5) SetRGB(nPlayers, i32, 0x00, 0x00, 0xFF); // 파란색
```

Fig. 12. Part of source for extracting skeleton

```
<system.web>
    <compilation debug="true" targetFramework="4.0" />
</system.web>
<system.serviceModel>
    <behaviors>
        <serviceBehaviors>

            <behavior name="Service1Behavior">

                <!-- 메타데이터 정보를 공개하지 않으려면 배포하기 전에 아래의 값을 false로 설정
                <serviceMetadata httpGetEnabled="true"/>
                <!-- 디버깅 목적으로 오류에서 예외 정보를 받으려면 아래의 값을 true로 설정하십
                <serviceDebug includeExceptionDetailInFaults="false"/>
            </behavior>
        </serviceBehaviors>
    </behaviors>
    <serviceHostingEnvironment multipleSiteBindingsEnabled="true" />

    <services>
        <service name="KinectSkeletonService.Service1" behaviorConfiguration="Service1Beha
            <endpoint binding="basicHttpBinding" contract="KinectSkeletonService.IService1"/
            <endpoint contract="IMetadataExchange" binding="mexHttpBinding" address="mex" na
        </service>
    </services>

</system.serviceModel>
<system.webServer>
    <modules runAllManagedModulesForAllRequests="true"/>
</system.webServer>

</configuration>
```

Fig. 13. Part of WCF source code

5   Conclusion

In this study, When shooting images by using Kinect's Depth (3D) sensor and infrared
laser beam in order to prevent lonely death of the elderly living alone, we use image
data extracting human skeleton information rather than using RGB image data like
existing CCTV or IP camera. This method does not use RGB method and therefore,
the subject's physical information and private information could be minimized. Image
data extracted like this is shot at 30-second intervals and transmitted to DB through
WCF. Image data saved in the server could implement monitoring available system in
PC environment as well as mobile environment through hybrid app developed by

using Sencha Touch 2.0. Also, data capacity could be minimized because image data such as photos not video method are processed and rapid transmission rate was shown due to small capacity when running on a mobile through hybrid app.

Also, it is expected to enable fast response to death of seniors who live alone as well as reducing workforce and costs for Dolbomi Service and Contact Service. In addition, by linking with 119 and police, it can prevent solitary death through rapid response. However, in this study, there was no security applied to skeleton extraction data and DB. Therefore, future research will need to address the issues like data encryption, DB security, and mobile authentication.

References

1. Lim, J.-H., Park, C.-Y., Park, S.-J.: Based on international standards for IEEE 11073/ISO TC215 U-health platform Technology. Information and Communications Magazine 27(9), 15–22 (2010)
2. Kim, Y.-S., Lee, C.-M., Namgung, S.-J., Kim, H.-G.: A Study on the Social Networks Effectiveness to Prevent the Lonely Death of the Elderly who Live Alone. Social Science Research 50(2), 143–169 (2011)
3. Do, S.-R.: Medical Utilizations of the Aged: Issues and Policy Tasks. Health-welfare Policy Forum 157, 66–79 (2009)
4. Kim, Y.-H., Lim, I.-K., Lee, J.-P., Lee, J.-G., Lee, J.-K.: Development of Mobile Hybrid MedIntegraWeb App for Interoperation between u-RPMS and HIS. In: Murgante, B., Gervasi, O., Misra, S., Nedjah, N., Rocha, A.M.A.C., Taniar, D., Apduhan, B.O. (eds.) ICCSA 2012, Part III. LNCS, vol. 7335, pp. 248–258. Springer, Heidelberg (2012)
5. Kwon, H.-N.: Ethical Reflection on Lonely Death Problem of the Elderly in the Aging Society. Humanities Research Treatises 35, 245–277 (2013)
6. Kwon, H.-N.: Legal and Ethical Issues on Lonely Death - Personal choice or social problem? Institute for Humanities & Sciences 38, 463–479 (2013)
7. Kwon, M.-H., Kwon, Y.-E.: A Study on the Subjectivity of the Elderly who Live Alone Caregivers in Perception of Lonely Death. Korean J. Adult Nurs. 24(6), 647–658 (2012)
8. Oikonomidis, I., Kyriazis, N., Argyros, A.A.: Efficient model-based 3D tracking of hand articulations using Kinect. In: British Machine Vision Conference, pp. 101.1–101.11 (2011)
9. Bergh, M.V., Carton, D., Nijs, R.D., Mitsou, N., Landsiedel, C., Kuehnlenz, K., Wollherr, D., Gool, L.V., Buss, M.: Real-time 3D hand gesture interaction with a robot for understanding directions from humans. In: Symposium on Robot and Human Interactive Communication, pp. 357–362 (2011)
10. Choa, S.-Y., Byun, H.-R., Lee, H.-K., Cha, J.-H.: Hand Gesture Recognition from Kinect Sensor Data. JBE 17(3), 447–458 (2012)
11. http://blog.naver.com/PostView.nhn?blogId=jay_korea&logNo=30135162627
12. Park, K.-W., Chae, J.-G., Moon, S.-H., Park, C.-S.: A Landmark Based Localization System using a Kinect Sensor. The Transactions of the Korean Institute of Electrical Engineers 63(1), 99–107 (2014)
13. Go, J.-G.: Kinect Programming. Korea Electronics Association (February 2012)
14. Microsoft Kinect SDK vs PrimeSense OpenNI, http://deviak.com/category/Study
15. Sung, H.-G., Kim, J.-I., Choi, S.-W., Kim, G.-H.: The Development of Real-Time monitoring program using Kinect. In: Korea Information Communication Society, Conference Materials 2012 (2012)

The Analysis and Countermeasures
on Security Breach of Bitcoin

Il-Kwon Lim, Young-Hyuk Kim, Jae-Gwang Lee, Jae-Pil Lee,
Hyun Nam-Gung, and Jae-Kwang Lee

Hannam Univ., Department of Computer Engineering, Daejeon, Korea
{iklim,yhkim,leejk,jplee,ghnam}@netwk.hnu.kr, jklee@hnu.kr

Abstract. Due to the strength of bitcoin including the convenient payment and transfer, exchange into legal tender, low transfer fee, and others, its usage is increasing dramatically. Bitcoin is an e-money and virtual currency currently used as a means of payment in about 20,000 online companies and 1,000 offline stores as of Feb. 2014. Different from previous e-money, there is no issuing institution and it has trait of no regulation by countries or companies. However, as there is an increase in value of bitcoin followed by Cyprus financial crisis in past March, issues regarding the security breach on the rise. Since the exchange between legal tender and virtual currency is free in case of bitcoin, direct damage caused by security breach is very large. Also, as there is no governing institution, the user is responsible for all loss in case of damage occurrence. Therefore, the purpose of this study is to look into transaction method and security breach trend of bitcoin and its countermeasures.

Keywords: bitcoin, trends in bitcoin security breaches, bitcoin security countermeasures.

1 Introduction

With the rapid expansion and popularization of internet worldwide, the necessity for currency to be used in online was on the rise other than the form of traditional currency. As a means to fulfill such needs, e-money was introduced. At the early stage, e-money was mainly used to purchase digital contents of small sum including game items, movies, music, and others but it is developed and used in various formats including transportation payment system, PayPal, Amazon coin, and others [1, 2].

Recently, virtual currency such as Bitcoin which is digital bit like e-money but without full establishment of legal foundation was introduced. According to the classification of European Central Bank, both virtual currency and e-money are saved as digital bit but there is almost no legal regulation on virtual money whereas e-money is under strict legal regulation. As virtual currency is used as means to conduct money laundering, drug trafficking, prepare for devaluation of legal tender, and others, more attention is paid upon virtual currency [2].

Bitcoin which was first introduced in a study by Satoshi Nakamoto (alias) in 2009 is virtual currency in which most attention is paid to and it is a typical fiat money

B. Murgante et al. (Eds.): ICCSA 2014, Part IV, LNCS 8582, pp. 720–732, 2014.

which exists in encrypted code without cash value. Although it was first issued in Jan. 2009, the value of bitcoin increased after Cyprus financial crisis in Mar. 2013. With the opinion that FEC may receive bitcoin as contribution to federal election campaign on Nov. 2013, the value of bitcoin reached its peak. Below [Fig. 1] is a value change graph of bitcoin [2-5].

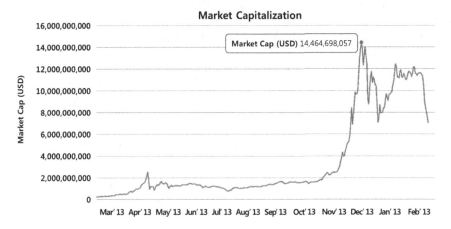

Fig. 1. Value change graph of bitcoin

With the strength of low fee and quick payment method, bitcoin is used as the means of payment in 20,000 online companies and 1,000 offline stores as of Feb. 2014. There is a continuous increase in Bitcoin affiliate members and a dramatic expansion in affiliated members is estimated unless it is regulated by the government of each country [2, 6].

Although bitcoin is known to be safe as it uses p2p method and the distributed processing of transactions conducted to all users, it may be susceptible to personal security as the bitcoin wallet is managed as private key and public key in personal PC or smart phone. Also, with the increase in value of bitcoin, various security breaches of bitcoin are reported including the hacking toward the exchange and bitcoin mining group, malicious mining through numerous anonymous PC with the distribution of malignant code, and others. Therefore, the transaction and production methods of bitcoin are examined in Chapter 2, bitcoin related security breaches are checked upon in Chapter 3, and safe bitcoin usage is looked into in Chapter 4 of this study. At last, the conclusion is made in Chapter 5.

2   Related Studies

2.1   Transaction Method of Bitcoin

In regards to the technology, the transaction of bitcoin is achieved through the sale between the e-wallet holders based on public key encryption and such key value is saved to distributed database in file format. In the transaction of bitcoin, the buyer sends his public key which plays a role of address to the bitcoin seller. Next, the bitcoin seller conducts e-signature with the use of his own secret key on the hash which contains all transactions history and public key of bitcoin buyer. Then, the transaction is completed when the bitcoin is transferred from the e-wallet of bitcoin seller to e-wallet of bitcoin buyer. All transaction history of transaction target bitcoin including recent transaction history are saved to the hash within bitcoin and all bitcoin transactions occurred at bitcoin network is recorded to the block chain and are opened to all bit coin users in peer to peer format thus the stability of system itself is secured [2,7].

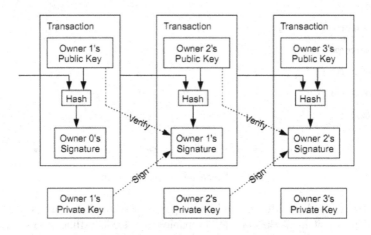

Fig. 2. Transaction procedure of bitcoin

2.2   Production Method of Bitcoin

Central institution or device that issue and manage the currency do not exist for bitcoin and the production is achieved by solving mathematical encryption necessary for security and acquiring certain amount of bitcoin. Such process is called a mining. The bitcoin is issued only through mining and individual or group who mines is called as a miner. In order for a miner to solve a single mathematical encryption, he needs to calculate 670 trillion worth hash. As the latest quad-core 2GHz central processing unit calculates 10 million hashes per second, it takes 670 million seconds (about 21 years) to mine one bitcoin. As considerable amount of time is consumed to mine bitcoin, a group which uses software for bitcoin mining to mine together and share bitcoin according to one's contribution appeared. As a representative mining group, there is BTC Guide and this group has been mining 25 BTC (currency unit of bitcoin) per 20 minutes as of 2013. Since it is difficult for an individual to mine bitcoin, the easiest way to exchange legal tender into bitcoin is to go through the exchange. Following [Table 1] illustrates the transaction available bitcoin exchanges [2].

Table 1. The transaction available bitcoin exchanges

Name	URL
Mt.Gox	www.mtgox.com
BitStamp	www.bitstamp.net
btc·e	btc-e.com
BTC China	btcchina.com
bitcoin.de	www.bitcoin.de
Kraken	www.kraken.com
Bitcurex	pln.bitcurex.com
Asia Nexgen	anxbtc.com
Canadian Vitual Exchang	www.cavirtex.com/home
LocalBitcoins	localbitcoins.com

3   Trend of Bitcoin Security Breaches

Although bitcoin is a virtual currency, it can be exchanged to legal tender different from other virtual currencies. Therefore, there are unceasing reports of security breaches including the theft of bitcoin in the exchange by hacking, distribution of malignant code for mining, and others.

3.1   Mining through Distribution of Malignant Code

Although bitcoin can be produced only by mining, there is a limitation with the use of personal PC as it was illustrated above. Therefore, the mining application is being operated with the account of malignant code creator by distributing malignant bitcoin mining application. Those incidents of security breaches are as following.

Last April, it was discovered that Trojan horse virus was distributed through Skype which is a free internet call service program of Microsoft. Starting with India, it spread quickly in Europe including Italia, Russia, Poland, Costa Rica, Spain, Germany, Ukraine, and others with weak security. Russian security firm Kaspersky first discovered the malignant code (Trojan.Win32.Jorik.IRCbot.xkt). When a user clicks the address linked to Skype message, malignant program makes user PC as a bitcoin miner and conceals the infection with the use of fake currency other than bitcoin so that the user would not find out about it. In addition, the amount of bitcoin intercepted is in proportion to the CPU usage and it brings about the decrease in CPU performance as it clicks 2,000 times per hour and 11,917 times per 24 hours. Then, the bitcoin is given to the distributor of malignant code[8, 9].

From last Dec. 31st to Jan. 3rd, a malignant code was distributed through virus-laden Ad of Yahoo. Some Ads which do not conform to the edition guideline of Yahoo were introduced and it attacked by taking advantage of the weakness of Java. As well as the bitcoin mining malware, other software installed includes ZeuS, which attempts to steal banking information; Andromeda, which turns the computer into part of a "botnet" for use by third parties, and "adjacking" malware which hijacks the user's browser to click on adverts, thus channeling income to corrupt site owners. Schematically the exploit looks like [Fig. 5].

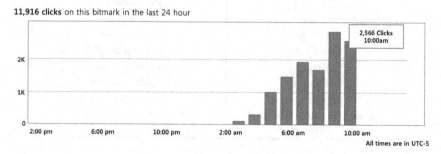

Fig. 3. Number of clicks for each time zone of infected PC

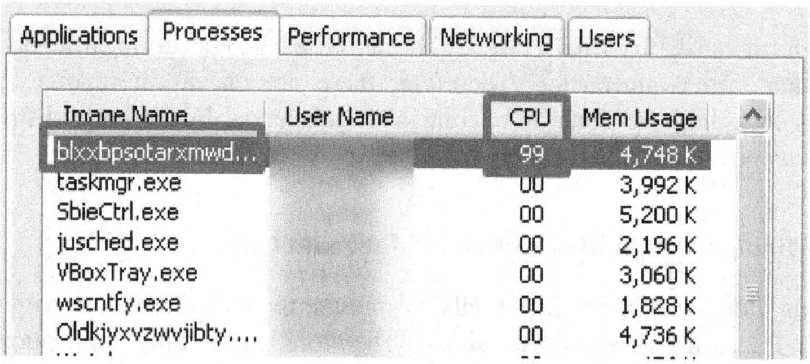

Fig. 4. In this case, the threat Trojan maxes out the computer's CPU

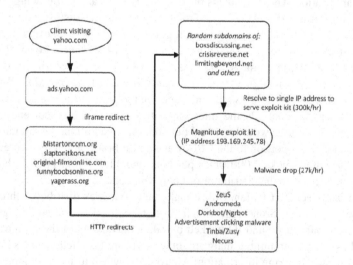

Fig. 5. Schematically the exploit flow

Fox IT, the Dutch cyber security firm which first disclosed the vulnerability to the public estimated that there were around 27,000 infections every hour the malware was live on the site. If the malware was being served consistently for the three days, it may be the case that almost 2 million computers were infected [10, 11].

3.2   Money Extortion through Bitcoin Account Extortion

According to Inca internet correspondence team of Korea last December, it was discovered that partially mutated version of malignant file with the function of online game account extortion contains the account extortion function of bitcoin account. Numerous malignant codes for account extortion have been introduced for several years as the cash transaction of online game items was possible. As more attention is paid to bitcoin recently, it led to the introduction of mutated malignant code for account extortion of bitcoin exchange. Below [Fig. 6] presents packet analysis arranged with the items for account extortion of bitcoin exchange in Korea.

```
      0
'.bitup.co.kr',0,0,0,0,'.korbit.co.kr',0,0,0,'Pass',0,0,0,0,'PWD',0,'PASS',0,0,0,
'password',0,0,0,0,'pwd',0,'pass',0,0,0,0,'Pwd',0,'Password',0,0,0,0,'bitpay.com'
0,0,'coinbase.com',0,0,0,0,'multibit.org',0,0,0,0,'bitcoindomains.blogspot.k'
'r',0,0,'.bitcointalk.org',0,0,0,0,'.bitcoin.org',0,0,0,0,'.mtgox.com',0,0,'.x'
'bitcoinx.com',0,0,'.sellbitcoins.co.kr',0,'.buyanythingwithbitcoin.c'
'om',0,'.buybitcoin.co.kr',0,0,0,'blockchain.info',0,'weburl=',0,'lck=',0
0,0,0
```

Fig. 6. Packet analysis for bitcoin account extortion confirmation of Korea

Table 2. bitcoin malignant code distributed sites

http://www.daum.net/**001.jpg**
http://www.dombyshop.co.kr/bbs2/data/**001.jpg**
http://www.btdot.com/**001.jpg**
http://www.srsr.co.kr/bbs2/data/**001.jpg**
http://www.dompage.co.kr/bbs2/data/**001.jpg**
http://www.v3lite.com/**001.jpg**
http://www.nate.com/**001.jpg**
http://www.msn.com/**001.jpg**
http://www.hangame.com/**001.jpg**
http://www.cbs.co.kr/**001.jpg**
http://www.joinsmsn.com/**001.jpg**
http://m.ahnlab.com/**001.jpg**
http://m.ahnlab.com/0048253/**001.jpg**
http://www.tistory.com/start/**001.jpg**
http://www.nexon.com/**001.jpg**
http://user.nexon.com/**001.jpg**
http://www.netmarble.net/**001.jpg** ...etc.

The extortion targets of malignant file included total 14 websites including a representative bitcoin exchange of Korea korbit.co.kr. A malignant file related to distribution as illustrated in below [Table 2] also performed the function of downloading an image (001.jpg) file that does not exist in reality by accessing to popular websites of Korea. Also, it was revealed that it contained a function to sabotage the free security programs [12].

3.3   DDoS Attack against Bitcoin Exchange

DDoS (Denial of Service attack) refers to an attack which maliciously makes numerous access attempts to certain server as [Fig. 7] in order to block normal service use of other users or an attack which runs out TCP connection of server. As bitcoin exchanges provide exchange service to legal tender, there are unceasing DDoS attacks of malicious objective.

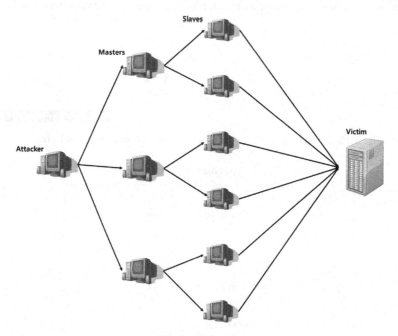

Fig. 7. DDoS Attack

A large scale DDoS attack was made against BTC China, no.3 bitcoin exchange worldwide, in last Sep. The traffic reached up to 100 gigabit and it lasted for 9 hours. According to Incapsula which corresponded to the DDoS attack at the moment, DDoS attack was made through SYN flood attack. The attackers used the method to send numerous SYN packets in small size and small amount of SYN packets in large size and others. For this attack, not a group of zombie PC but network made of numerous hacked servers were used. It was revealed that many Word Press servers susceptible to the security were the target. In addition, the bitcoin withdrawal was suspended in Bitstamp, a bitcoin exchange, in this Feb. due to DDoS attack and BTC-e also received a DDoS attack presenting the delay in transaction processing for two days [13, 14].

4   Security Countermeasures

Since there is no governing country or group for bitcoin, the users are responsible for the loss of bitcoin due to hacking or security breach. Since bitcoin can be exchanged into legal tender different from other virtual currencies, the scale of damage is much larger than normal hacking.

4.1   Safe Bitcoin Use for Individual User

As it was examined above, more attention needs to be paid upon the computer of personal user and personal information management since it can be used for personal account extortion or mining through a malignant code. Safe management methods for e-wallet of bitcoin are as following [15].

1. In regards to the client e-wallet in use, maintain the latest version. In case of bitcoin QT wallet, the version update is completed just with the installation of execution file.
2. Arrange preparatory measures for the failure or damage of electronic devices by performing frequent backup after installing e-wallet. In case of bitcoin QT, the backup of e-wallet refers to 'wallet.dat file' encrypted with password. More precisely, it refers to backup of private key in encrypted format and storage to electronic device.
3. The encryption of e-wallet shall be conducted with long encryption syntax that is not used in daily basis.
4. Conduct distributed management of e-wallet such as for offline access, for online transaction, watch-only, and others. In regards to the client e-wallet related to above, Armory in case of desktop computer and Mycelium in case of mobile application provide similar function.
5. Offline (or backup) copy shall not be carried and it shall be stored to a place that does not arouse the suspicion of insiders or outsiders and a place difficult to be reached.
6. In case of Hot Wallet in which e-wallet is used online, it may enable the access by unauthorized person followed by the virus infection thus the anti-malware system shall be constructed and update shall be conducted at all times. .
7. Large quantity of money shall be stored offline not online.
8. Additional hardware security device just like TREZOR or BitSafe as shown in [Fig. 8] shall be used[16, 17].

Fig. 8. bitcoin hardware wallet 'TREZOR' and 'BitSafe'

9.   Offline copy can be stored to analog medium with long hardware life cycle instead of electronic device such as USB. If a person can transcribe private key itself or QR code (partial) or perform printing and coating with the use of a printer and then store it to 'a secrete place in daily lives' of his own, the safety of non-electronics increases. However, with the exposure of this private key, all of bitcoin can be lost.

Also, the latest version of operating system and virus vaccine program shall be maintained and test on virus and malignant code shall be conducted on regular basis. Programs with unknown origin shall not be installed and one should refrain from installing illegally distributed commercial program. It is safer to use a program trusted and used by many users for a long period of time. The password of web service used in bitcoin web wallet shall be long and complex and special attention needs to be paid upon the storage of private key for bitcoin wallet

4.2   Security Measures for Safe Bitcoin Transaction

Bitcoin is not currently considered as an official currency in the legal perspective in many of the countries and hence is not established with legal protective policies. This seems to be how bitcoin is not regarded as an official currency according to the characteristics of mock currency or is considered as a mere speculation object along with rapid decrease of the value of it. However, bitcoin is currently serving as a real currency in diverse purposes in both on/off lines that currency exchange is being occurred at the bitcoin trading office almost everyday. Therefore, security policy is now specifically required. Therefore, a study is needed in each of the countries recognizing how realistic it is for bitcoin to serve as a role in the real e-financial transaction and preparing for measures. In addition, bitcoin trading office of the bitcoin shall fulfill responsibility as a financial institution beyond a mere e-currency trading organization.

User authorization is specifically required for safe transaction at the bitcoin trading office. However, as shown in the [Fig. 9], user authorization is proceed only with ID/Password as how it has been previously applied in the traditional computer system for bitcoin trading office.

Fig. 9. How to authorize user of the bitstamp

Password consisting of combination of particular alphabets or words that only a user knows is exposed to simple combination due to limitation of the word mixing matrix that a human can come up with as well as to fishing attack or key logger. The user authentication conducted by only one single factor like password is called Single-Factor Authentication. Therefore, there exists a security issue. Hereupon, e-financial transaction has been studied and improved with authorization method in various methods for the safe authorization of users. The authentication, mixing with various authentication techniques and using multiple authentication means more than two besides password that only user knows, is called Two-Factor Authentication or Multi-Factor Authentication, which has absolutely higher security for user authentication. The following [Fig. 10] is the classification of user authorization techniques designed for e-financial transaction [18-20].

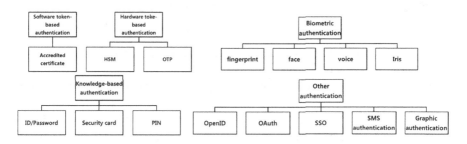

Fig. 10. Classification of user authorization techniques

Such authorization techniques are developed to improve safety in various means such as mock keypad or scramble pad for user authorization in the simple ID/Password system as well as other types of measures of enhancing safety of it via techniques such as hardware token methods including card-shaped token or OTP (One time password) and PKI (Public Key Infrastructure).

NIST (National Institute of Standards and Technology) classifies such authentication techniques into 4 electronic authentication guideline levels. The authentication means to satisfy those levels are cryptologic hardware device, one-time password, cryptologic software device, password, etc. Each authentication means has different security level and the authentication means that satisfy an upper level satisfy also all the lower levels as shown in [table 3].

Table 3. Authentication method applying security level

Authentication means	Level 1	Level 2	Level 3	Level 4
Cryptologic hardware device (HSM: Hardware security module)	○	○	○	○
One-time password (OTP)	○	○	○	
Cryptologic software device (Accredited certificate)	○	○	○	
Knowledge-based authentication means (ID/Password, PIN)	○	○		

Therefore, for user authentication in an online bitcoin exchange, Two-Factor Authentication should be conducted using authentication means such as HSM, OTP or accredited certificate together with ID/password. The characteristics of such authentication means are as follows;

Table 4. Features authentication method

Means	Characteristics
Accredited certificate	- Accredited certificate is based on ITU-T x.509, the Public Key Infrastructure Standard. - A kind of electronic identification made by adding the owner's information on a public key. - Guarantee the forgery prevention or non-repudiation of transactional information, the identification of trader, etc. - Apt to be leaked by malicious code and difficult to be recognized by user when leaked.
HSM	- Hardware-type security token that stores software-form authentication token such as accredited certificate. - As a processor and cryptographic operation device are embedded in device, it is possible to create electronic signature key or create and validate electronic signature. - Disadvantages of low convenience and higher price than other hardware tokens.

Table 4. (*Continued*)

OTP	- It uses one-time password created randomly. - High security because impossible to analogize the password used for authentication. - Possible to use in various financial transactions with one device throu gh combining with other easy authentication techniques. - Relatively high convenience because it doesn't need any additional sec urity software. - Disadvantages of higher cost than other authentication means, inconve nient portability, high risk in case of losing OTP token and inconveni ence of exchanging token battery.

If the above authentication means are used, the 3rd level of the NIST's guideline levels can be secured and safer transactions can be possible. For bitcoin transactions, such authentication techniques should be introduced so that any potential security accidents such as account extortion may be prevented through safe user authentication.

5   Conclusion

With the introduction of its service in 2009, the value of bitcoin skyrocketed to 1,203 dollars at its peak but now its value has been lowered as of Feb. 2014 due to various security threats, transaction suspension crisis of Mt. Gox and others. Since bitcoin can be exchanged to legal tender, it is most exposed to various security threats. There were countless incidents of security breaches during 2013 when the value of bitcoin started to increase and it was revealed that all types of security breaches previously occurred such as DDoS attack, private account hacking, and others set the bitcoin exchange websites and personal user as their target. Since anonymous possession and transaction of bitcoin is available, it is true that there are dark sides to bitcoin including as it can be used as a means for money laundering, drug trafficking, and speculation.

However, with continuous increase in number of internet users worldwide and internet available devices with the development of mobile technology, the use of e-money is estimated to increase. Also, as bitcoin is decentralized currency without issuing institution, the exchange is more convenient in comparison with previous virtual currencies and it use at online and offline stores are gradually increasing with the strength of low fee and quick payment. Accordingly, currencies similar to bitcoin are under continuous production including Litecoin, Peercoin, and others.

Accordingly, the trend of security breaches and their countermeasures were examined in this study. However, the reality is that there is no legal foundation or systematic safety net for virtual currency but for personal users to pay special attention to it. Therefore, for the transactions in the bitcoin exchanges, security authentication means applicable to electronic financial services should be applied, and minimum two-factor authentication should be used utilizing authentication technologies such as HSM, OTP, accredited certificate, etc. Also, many studies shall be conducted for not only security technology of virtual currency but also systematic and legal foundation for safe and convenient use of bitcoin.

References

1. Korea Internet & Security Agency, Internet & Security Weekly, fifth week of May (May 30, 2013)
2. Kim, T.-O.: The using status and implications of virtual currency: Focusing on Bitcoin and Linden Dollar. The Payment Settlement and Information Technology 53 (July 2013)
3. Jeon, J.-Y.: Korea information society development institute, The understanding and implications of bitcoin. KISDI Premium report (October 31, 2013)
4. An, S.-U.: bloter.net, Bitcoin, from 10,000 won, Until 1 million won (February 1, 2013), http://www.bloter.net/archives/171449
5. Blockchain, Bitcoin market capitalization, http://blockchain.info/en/charts/market-cap
6. Sisainlive.com, I want to put BitCoin into my computer, http://www.sisainlive.com/news/articleView.html?idxno=19225
7. Nakamoto, S.: Bitcoin: A Peer-to-Peer Electronic Cash System, http://www.bitcoin.org
8. Korea Internet & Security Agency, Internet & Security Weekly, Second week of April (April 11, 2013)
9. Neowin.net, New Bitcoin mining malware spreading on Skype at 2,000 clicks per hour (April 7, 2013)
10. The guardian, Yahoo malware turned European computers into bitcoin slaves, http://www.theguardian.com/technology/2014/jan/08/yahoo-malware-turned-europeans-computers-into-bitcoin-slaves
11. FOX IT, Malicious advertisements served via Yahoo, http://blog.fox-it.com/2014/01/03/malicious-advertisements-served-via-yahoo/
12. nProtect Response Team Official Blog, Appeared malicious files aimed at domestic Bitcoin exchange users, http://erteam.nprotect.com/456
13. DailySecu, 3 of the World BitCoin Exchange, 100 gigabit DDoS attacks, http://dailysecu.com/news_view.php?article_id=5468
14. ITWord, BitCoin industry added insult to injury... DDoS attacks after SW bug, http://www.itworld.co.kr/t/61022/%EB%94%94%EC%A7%80%ED%84%B8%20%EB%A7%88%EC%BC%80%ED%8C%85/86010
15. xbitcoinx.com, BitCoin Safety keeping method, http://xbitcoinx.com/Bitcoin_Wallet/66907
16. Trezor, bitcoin hardware wallet, http://www.bitcointrezor.com/#about-trezor
17. BitSafe, bitcoin hardware wallet, http://www.butterflylabs.com/bitcoin-hardware-wallet/
18. Lim, H.-J., Shim, H.-W., Seo, S.-H., Kang, W.-J.: The certified technology trend analysis of electronic financial transaction environment. Journal of The Korea Institute of Information Security & Cryptology 18(5) (October 2008)
19. Bitstamp.net, Bitstamp login web page, https://www.bitstamp.net/
20. Yeom, H.-Y., Jo, H.-J., Lee, D.-H., Jeong, Y.-G., Jang, G.-H., Lee, S.-R.: Research on security criteria for extension to electronic authentication method usage-based. Final Research Report, Korea internet & security agency (December 7, 2011)

Battery Consumption Modeling for Electric Vehicles Based on Artificial Neural Networks*

Junghoon Lee[1], Min-Jae Kang[2], and Gyung-Leen Park[1],**

[1] Dept. of Computer Science and Statistics
[2] Dept. of Electronic Engineering,
Jeju National University, Republic of Korea
{jhlee,minjk,glpark}@jejunu.ac.kr

Abstract. This paper presents how to develop a battery consumption model taking advantage of state-of-charge streams acquired from real-life electric vehicles. From the record consisting of timestamp, longitude, latitude, and battery remaining, learning patterns are generated to build a neural network for each of 4 major roads, essentially taken by long-distance trips in Jeju city. Our 3-layer neural network model is made up of an input node, 10 hidden nodes, and an output node. The input variable takes the approximated distance while the output variable represents the battery consumption from the start point of a road. Neural networks, being able to efficiently tracing non-linear data streams, accurately keep track of battery consumption irrespective of road shapes and elevation changes. The assessment result shows that the average errors for each road range from 0.22 to 0.33 km, indicating that this model can estimate battery demand for a given route for navigation applications.

Keywords: Electric vehicle, battery consumption, state-of-charge stream, neural network, trace model.

1 Introduction

Smart grid is aiming at enhancing energy efficiency in many areas, embracing great diversity of power entities, not just restricted to power transmission facilities [1]. EVs (Electric Vehicles) are one of its key elements, allowing even transport systems to cooperate with the power network. EVs get energy from rechargeable batteries, not burning fossil fuels as in gasoline-powered vehicles. The battery obtains electricity from many types of energy sources such as nuclear power, water, steam, and even renewable energies including wind or sunlight. Hence, EVs cannot just achieve energy efficiency but also reduce greenhouse gas emissions. Accordingly, many countries are encouraging their large deployment by implementing relevant business models such as shared vehicles, taxis, and rent-a-cars [2][3][4].

---

* This work was supported by the research grant from the Chuongbong Academic Research Fund of Jeju National University in 2013.
** Corresponding author.

B. Murgante et al. (Eds.): ICCSA 2014, Part IV, LNCS 8582, pp. 733–742, 2014.

However, there are still many obstacles for EVs to penetrate into our daily lives promptly. First, they are too expensive for personal ownership. Next, due to the limitation in battery capacity, battery charging time is quite long and driving distance is short. It takes about 30 minutes to fully charge a single EV with fast chargers and 6 ~ 7 hours with slow chargers [5]. A fully charged EV can drive just about 100 *km*, bringing so called *range anxiety* especially to long-distance drivers. Above mentioned problems can be alleviated by computational intelligence developed in information technologies, until a significant innovation appears in the battery industry [6]. It can create energy-efficient routes possibly combined with charging plans, reserve chargers, manage EVs in sharing systems, and so on.

One of the most fundamental concerns in route selection is whether an EV can reach the destination with its current battery SoC (State of Charge) [7]. If not possible, drivers want to decide where to charge somewhere on the route. Such questions can be answered only if the battery consumption model along a specific route is known in advance. However, battery consumption depends on a variety of factors such as basic battery discharge dynamics, terrain effects, drivers' driving style, and the like. Moreover, SoC readings from the BMS (Battery Management System) inevitably include estimation errors. Hence, it is reasonable to trace the trend of SoC levels rather than build component-by-component mathematical models. The can be done by an analysis of spatio-temporal SoC record streams along frequently used roads.

In this regard, this paper develops SoC trace models for 4 major roads in Jeju city, Republic of Korea. Those roads are taken by commuters and tourists. As a smart grid model city, Jeju area hosts many EV projects, pursuing an ambitious plan of replacing all vehicles with EVs by 2030. SoC records are now being accumulated here and need to be analyzed. Specifically, we take advantage of artificial neural networks, which are widely used for nonlinear behavior tracing. At this stage, we begin with rather a straightforward trace model, taking the distance from the predefined start point as an input parameter. In addition, the SoC difference will be the out variable. This model can integrate more input variables such as weather conditions, road elevation, and in-vehicle appliance operations in the future.

This paper is organized as follows: After issuing the problem in Section 1, Section 2 introduces related works on EV battery consumption modeling. Section 3 details our neural network-based trace scheme after introducing the road layout of the target city. Section 4 demonstrates the tracing results and calculates the estimation error. Finally, Section 5 summarizes and concludes the paper with a brief description of future work.

2   Related Works

To begin with, in the course of scheduling battery charge, discharge, and rest operations based on the cyber-physical system paradigm, [8] models Lithium-ion battery-specific features in terms of both discharge efficiency and recovery efficiency. It is observed that at a high discharge rate, battery cell efficiency sharply drops down. The authors investigate the voltage drop curves for different discharge rates and build a reference-based estimation scheme. In addition, the recovery efficiency accounts for the battery behavior that a cell recovers by a rest operation to a certain level after temporary voltage drops brought by a high discharge rate. How much the battery recovers depends on the discharge rate, the discharge time, and the rest time. Particularly, an efficiency factor is defined and obtained by a multivariate linear regression method.

Second, [9] presents general SoC estimation models based on Kalman filtering and a neural network. It asserts that battery SoC estimation is affected by so many electrical factors such as current, temperature, battery capacitance, internal resistance, and so on. Hence, it is necessary not only to obtain accurate information from inaccurate data but also to filter out unwanted noise. Kalman filtering is known to be appropriate for SoC determination as it can keep track of the inner states of a dynamic process, while neural networks can learn by themselves and adaptively create trace models according to the given input pattern. However, EVs show a different behavior from general batteries, as they keep moving just along the road network and this road effect is critical. Moreover, the EV model has a set of in-vehicle status parameters.

Interestingly, battery power requirements for EVs can be predicted based on previous power consumption, speed, acceleration, and road networks [10]. With a prediction model integrated into a BMS, it is possible to efficiently schedule the allocation of battery cells in real time. The PRP (Power Requirement Predictor) deals with many physical elements including traffic conditions to make them descriptively captured by cyber elements [11]. The authors classify involved parameters into stable, dynamic but easy to predict, and dynamic and difficult to predict groups, respectively. The last group includes acceleration and speed. Here, the acceleration is estimated by the user's driving pattern, traffic flaws, and traffic regulations. It can work with an advanced BMS design equipped with vehicle activity and battery status monitoring.

3   Battery Consumption Model

3.1   Preliminaries

Figure 1 depicts the road layout of Jeju city, in which two major downtown areas are located at the north and south parts of the island. 4 main roads connect two areas while most long-distance trips include at least part of them. Their names are 1100 road, 516 road, West road, and East road, respectively. Officially, the last two have their own Korean names, but West and East roads are better to recognize. In our research and development project, a probe EV drives along each road, collecting SoC readings. An SoC value indicates the distance the EV can reach with current battery remaining. Many EVs display current SoC in such a form on their dash boards. Here, the research team has developed a DCC (Driving Condition Collector) module which creates an SoC record by integrating a coordinate from a GPS receiver and an SoC reading from the underlying BMS [12].

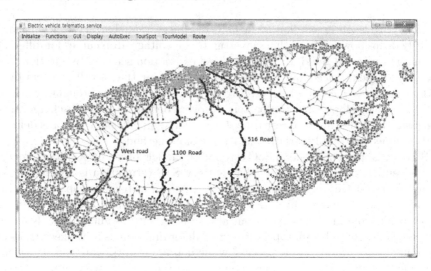

Fig. 1. Road layout

After validating the basic SoC dynamics model, more information fields will be included one by one.

Each road has its own features. For example, going through high mountain area, 1100 and 516 roads are quite winding. Those roads provide beautiful sceneries, attracting many tourists. On the contrary, laid on plain area, West and East roads are largely straight. With longitude and latitude fields in a record, the altitude can be retrieved from Google Map. We invoke the interface for each pair of longitude and latitude via the Wininet API, and the result is plotted in Figure 2. In this figure, curve segments having neither + nor × marks take places when Google Map does not find a value matched to the inquiry. However, by Figure 2, we can infer the elevation change of 4 roads. If all records get the altitude from Google Map, it may be possible to integrate elevation field. Actually, 1100 road is named according to the highest altitude it goes up. Soon, the road elevation will be available from our own DCC module.

3.2   Data Analysis

A series of SoC records comprise an SoC stream, while each record consists of time stamp (t), longitude (x), latitude (y), and SoC value (s), as illustrated in Figure 3. Sampling period is 10 seconds, so the difference between two temporally adjacent records is 10 seconds. Our modeling procedure employs FANN (Fast ANN), which provides abundant ANN library functions and a comprehensive interface specification [13]. Its interface defines a text file format for a set of learning patterns, generated from SoC records. By calling library functions in a C language program, we can build a neural network that traces the SoC stream. FANN also provides API functions to retrieve estimated outputs for any input parameter value. After all, we can compare the estimated output and actual

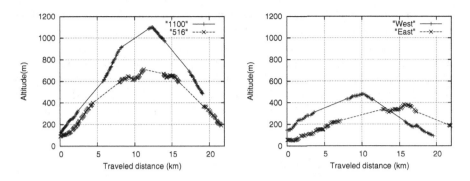

Fig. 2. Road elevation

SoC stream

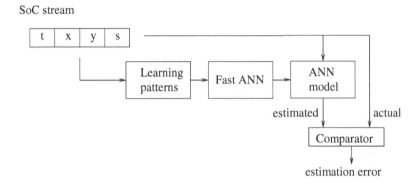

Fig. 3. Neural network modeling

SoC values, calculating the estimation error and thus assess the accuracy and feasibility of neural network models based on diverse statistical analysis methods.

Our model wants to verify the validity of neural network modeling for EV battery consumption. To this end, it is necessary to define input and output variables and select the number of hidden nodes. Trial-and-error finds the case of 10 hidden nodes works best, having the smallest mean square error. Then, the battery consumption model has just a single variable, namely, the approximated distance from the start point of a road, which will be denoted by d. The output variable, represented by c, is the amount of battery consumption. It is also represented by a distance reachable with the current SoC to be consistent with SoC readings. This formulation leads to a straightforward relation function, F. Namely,

$$F(d) = c \qquad (1)$$

Figure 4 illustrates how to generate learning patterns from SoC records, each of which is denoted by (x_i, y_i, s_i) for each valid record i. The SoC records will be created only on the road segment if there is no GPS error. Even though the network distance is more accurate, we take the Euclidean distance between two records for simple calculation. This error is not as large as the inherent reading error provided by the current BMS. Next, output values, namely, the amount of battery consumption from (x_0, y_0), is $s_i - s_0$ for each valid record i. Here, it must be mentioned that two or more records may have the same fields, namely, $x_i = x_j$, $y_i = y_j$, and $s_i = s_j$, for different i and j, if the vehicle does not move waiting for a traffic signal change. We remove such duplicated records from the learning pattern set. After all, a SoC stream is converted to a series of learning patterns as follows:

$$F(\sum_{j=1}^{i} D_j) = s_i - s_0, for \ \ 0 \leq i < n \tag{2}$$

, where n is the number of valid records. In addition, D_i is calculated as shown in Eq. (3).

$$D_i = \sqrt{(x_i - x_{i-1})^2 + (y_i - y_{i-1})^2} \tag{3}$$

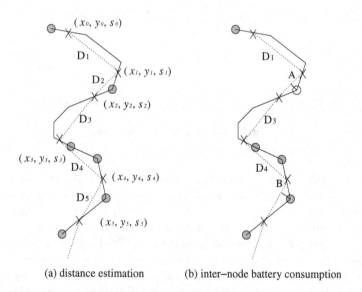

(a) distance estimation   (b) inter–node battery consumption

Fig. 4. Learning pattern generation

In Figure 4, the actual road shape is plotted by solid lines while virtual trajectory of the probe vehicle by dotted lines. Each dotted line segment corresponds to the Euclidean distance between two consecutive sensing points. Taking the Euclidean distance makes the distance estimation much simpler, bringing a small approximation error just in winding roads. If the sensing period gets smaller, this error term will decrease. In addition, Figure 4(b) explains how to obtain the SoC consumption between two nodes, or intersections. For two nodes A and

B, we can find the closest line segments for each one. Then, after calculating the orthogonal points on those two segments, we can estimated the offsets from (x_0, y_0), say, $F(D_A)$ and $F(D_B)$, respectively. Finally, the battery consumption between A and B is $F(D_B) - F(D_A)$.

4   Tracing Results

Figure 5 plots the traced and actual SoC values for each of 4 roads. As can be seen from the figure, the roads are about $20 \sim 25$ km long. 1100 road, which goes up highest, shows the largest discrepancy between two values around the area having the highest altitude. When an EV goes down slopes, the SoC gets increased due to the regenerative braking system. Hence, the SoC consumption is cut down even though the EV drives not a little distance. For the roads on plain area, modeling error is not significant as shown in Figure 5(c) and Figure 5(d). For all 4 roads, our neural network model traces SoC very accurately for the upslope area. Anyway, Figure 5 discovers that the neural network builds an efficient SoC tracing model even with a simple input parameter selection, regardless of road characteristics.

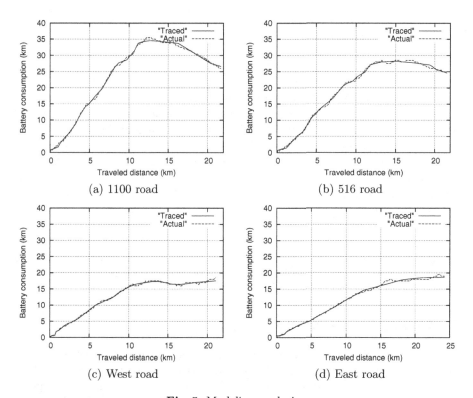

(a) 1100 road   (b) 516 road

(c) West road   (d) East road

Fig. 5. Modeling analysis

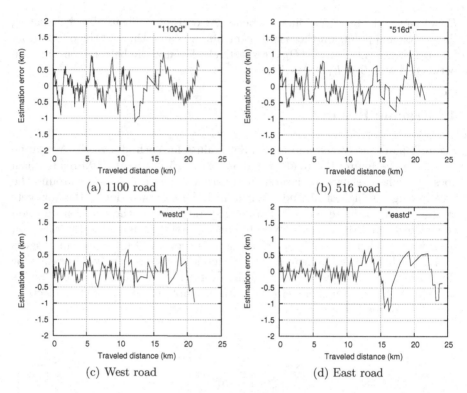

Fig. 6. Modeling error

In addition, Figure 6 demonstrates the estimation error for each road. The tracing error is obtained by subtracting the estimated value from the actual value for each point fed into the neural network. The sign changes according to whether the error comes from overestimation or underestimation. The error lies within the range from -1.0 to 1.0 km except for 6 out of 227 records for 1100 road, 1 out of 185 for 516 road, 0 out of 138 for West road, and 5 out of 154 for East road. The average error is 0.33 km, 0.32 km, 0.22 km, and 0.31 km for each road in the same order. If we calculate the relative error by calculating the ratio of this error to the maximum SoC value, the maximum errors are 3.1 %, 3.8 %, 5.1 %, and 6.0 % for each road, respectively. It's reasonably small compared with the inherent SoC reading error in the BMS, which is known to be about 10 %. There is no bias between underestimation and overestimation errors, underestimated cases occupying 53 %, 51 %, 51 %, and 44 %.

5   Concluding Remarks

EVs, powered by rechargeable batteries, make the transport system be a part of power networks. To overcome their drawbacks in long charging time, short driving range, and high cost, sophisticated information services are indispensable. Those services can be provided either by in-vehicle computer devices or telematics-style framework built upon modern information and communication technologies. According to the performance improvement of mobile devices and wireless communication technologies, more computing intensive applications can be developed for EVs.

In this paper, we have developed a battery consumption model with real-life SoC streams collected in Jeju city, Rep. of Korea. Our neural network model takes the approximated distance from the start point as input and the amount of consumed battery as output. Valid SoC stream records are filtered and converted into learning patterns and fed to the neural network. Assessment results have found that even with a simple variable selection, it is possible to trace the SoC change along specific roads with reasonable accuracy. The average error falls in the range from 0.22 to 0.33 km for the given roads. This model can be embedded in an EV navigation application to provide an energy-efficient route selection, which is one of the most important function blocks in most EV information services. In addition, our research team is expanding information fields collected from EVs. They will help us to refine our model and enhance the model accuracy for various driving environments.

References

1. Ipakchi, A., Albuyeh, F.: Grid of the Future. IEEE Power & Energy Magazine, 52–62 (2009)
2. Cepolina, E., Farina, A.: A New Shared Vehicle System for Urban Areas. Transportation Research Part C, 230–243 (2012)
3. Lue, A., Colorni, A., Nocerino, R., Paruscio, V.: Green Move: An Innovative Electric Vehicle-Sharing System. Procedia-Social and Behavioral Sciences 48, 2978–2987 (2012)
4. Lee, J., Park, C.J., Park, G.-L.: Design of a Performance Analyzer for Electric Vehicle Taxi Systems. In: Nguyen, N.T., Attachoo, B., Trawiński, B., Somboonviwat, K. (eds.) ACIIDS 2014, Part II. LNCS (LNAI), vol. 8398, pp. 237–244. Springer, Heidelberg (2014)
5. Botsford, C., Szczepanek, A.: Fast Charging vs. Slow Charging: Pros and Cons for the New Age of Electric Vehicles. In: International Battery Hybrid Fuel Cell Electric Vehicle Symposium (2009)
6. Ramchurn, S., Vytelingum, P., Rogers, A., Jennings, R.: Putting the 'Smarts' Into the Smart Grid: A Grand Challenge for Artificial Intelligence. Communications of the ACM 55(4), 86–97 (2012)
7. Rahman, M., Dua, Q., Al-Shaer, E.: Energy Efficient Navigation Management for Hybrid Electric Vehicles on Highways. In: International Conference on Cyber-Physical Systems, pp. 21–30 (2013)
8. Kim, H., Shin, K.: Scheduling of Battery Charge, Discharge, and Rest. In: 30th IEEE Real-time Systems Symposium, pp. 13–22 (2009)
9. Chen, Z., Qiu, S., Mansur, M., Murphey, Y.: Battery State of Charge Estimation Based on a Combined Model of Extended Kalman Filter and Neural Networks. In: International Joint Conference on Neural Networks, pp. 2156–2162 (2011)
10. Kim, E., Lee, J., Shin, K.: Real-Time Prediction of Battery Power Requirements for Electric Vehicles. In: International Conference on Cyber-Physical Systems, pp. 11–20 (2013)

11. Kim, J., Kim, H., Lakshmanan, K., Rajkumar, R.: Parallel Scheduling for Cyber-Physical Systems: Analysis and Case Study on a Self-Driving Car. In: International Conference on Cyber-Physical Systems, pp. 31–40 (2013)
12. Lee, J., Park, G.-L., Lee, B.-J., Han, J., Kang, J.K., Kim, B., Kim, J.: Design and Development of a Driving Condition Collector for Electric Vehicles. In: Jeong, Y.-S., Park, Y.-H., Hsu, C.-H.(R.), Park, J.J.(J.H.) (eds.) Ubiquitous Information Technologies and Applications. LNEE, vol. 280, pp. 1–6. Springer, Heidelberg (2014)
13. Nissen, S.: Neural Network Made Simple. Software 2.0 (2005)

The Multi-level Security for the Android OS

Ji-Soo Oh[1], Min-Woo Park[1], and Tai-Myoung Chung[2,*]

[1] Department of Electrical and Computer Engineering,
Sungkyunkwan University, Suwon, Korea
{jsoh,mwpark}@imtl.skku.ac.kr
[2] College of Information and Communication Engineering,
Sungkyunkwan University, Suwon, Korea
tmchung@ece.skku.ac.kr

Abstract. Recently, the Android smartphone has become a frequent target of attackers. The smartphone provides personalized services such as finance and healthcare application, so attackers may invade the user's privacy by compromising the smartphone. The Android platform has permission based security model, but there are several vulnerabilities. First, when components communicate through message passing mechanism, attackers can eavesdrop or intercept the message. It may lead to leakage of user's sensitive information. Second, a non-privileged caller can access more privileged callee by exploiting vulnerable interfaces. In this paper, we propose multi-level security framework to protect against the attacks described above. Our security framework assigns security level to all applications, and it regulates communication between components at runtime.

Keywords: Android, Mobile Phone Security, Android Permission Mechanism.

1 Introduction

Mobile phones have exponentially evolved both in hardware and software over the past decades. Nowadays they play prominent roles in our daily life as one of the important tools of communication. Because of personalized service by means of third-party application, smartphones has become increasingly popular unlike feature phones. Gartner says worldwide smartphone sales accounted for 55 percent of overall mobile phone sales in third quarter of 2013, and Google Android surpassed 80 percent of the smartphone operating system market [1]. Unfortunately, Android has also attracted considerable interest from attackers due to its popularity. New mobile threat families and variants rose by 49 percent in first quarter of 2013 from last quarter from 100 to 149 according to mobile threat report by F-Secure [2].

There are two main types of attacks in Android, which are Intent-based attack and privilege escalation attack. Firstly, the Intent is a message object that

---

* Corresponding author.

B. Murgante et al. (Eds.): ICCSA 2014, Part IV, LNCS 8582, pp. 743–754, 2014.

is used for Android's message passing mechanism. The Intent mechanism is a facility for late run-time binding between components in the same or different applications, and it encourages inter-application collaboration. However, numerous vulnerabilities of Intent messaging allow malicious applications to eavesdrop or intercept Intents, and these attacks are called Intent-based attack. These attacks may invade smartphone user's privacy. Secondly, Android does not ensure that a non-privileged caller is restricted to access more privileged callee. Important resources like Internet access or user contacts are protected via Android's permission based security model. However, malicious applications can gain unauthorized access to protected resources by exploiting vulnerable interfaces of another privileged application.

In this paper, we propose multi-level security framework which is inserted into Android platform by modifying middleware layer of Android. Our security framework assigns security level to all of installed applications, and it controls Inter-Component communication (ICC) depending on the security levels. The Level Manager in the framework is responsible for assigning security level according to the system's security policies. Platform developers and security administrators in enterprises or organizations can adopt out security framework for reflecting their different security requirements. Therefore when an application is installed, the Level Manager analyzes the manifest and decides a security level in accordance with the result of analysis. After that, all the ICC establishments are regulated by the security levels. Our security framework can prevent from Intent-based attack and privilege escalation attack through multi-level security model.

The rest of the paper is organized as follows. In Section 2, we describe the Android architecture and its security mechanism. We then present our adversary model in Section 3. Section 4 shows our multi-level security framework. Section 5 presents sample policies of assigning security levels by way of illustration. Finally, we conclude the paper in Section 6.

2   Android

Before presenting our work, we simply describe the Android architecture and its security mechanism.

2.1   Android Architecture

Android is an open source software stack consists of applications at the top, middleware framework in the middle, and a Linux kernel with various drivers at the bottom. Fig. 1 shows the layered architecture and possible communication channels. Android application layer includes third party and core applications, and all applications are written in the Java. Applications consist of several modular components. There are four types of components, which are Activities, Services, Broadcast Receivers, and Content Providers. The middleware layer includes an application framework, Dalvik Virtual Machine (DVM), and Java and native libraries. The application framework, which is preinstalled on Android devices, provides high-level services for supporting applications. For example, Activity Manager and Package Manager are both included in the application framework. Activity Manager controls an application's life cycle and Package Manager lets an application learn about other application packages. The DVM is an optimized version of a Java Virtual Machine, which allows every Android application to run in its own process. The Linux kernel provides core system functionalities to upper layers, such as threading, process management, a network stack, and a driver model[3,4].

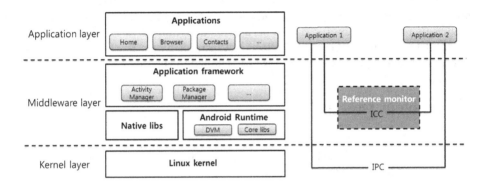

Fig. 1. Android architecture and application communication channel

Although applications run in isolated environment for security and privacy, the Android platform provides some channels for communication among applications. Communication at kernel layer can be established via Linux's standard Inter-Process Communication (IPC) mechanisms. At middleware layer, applications communicate with each other through Inter-Component Communication (ICC) channels. To establish an ICC channel, component send an Intent which is a message object that describes operations to accomplish. The Android middleware defines three types of ICC like Fig. 2 [5]. Fig. 2a depicts that a component starts an Activity and expects a return value. Communicating between a component and a Service described in Fig. 2b is also one type of ICC. Lastly, there is a broadcast Intent from Activity, Service, and system to related Broadcast receiver in the types of ICC like Fig. 2c.

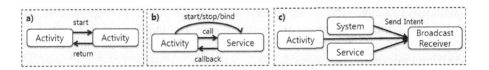

Fig. 2. Communication between application components

2.2   Security Mechanism

There is a reference monitor which enforces mandatory access control (MAC) on ICC calls in middleware layer. Android protects sensitive resources like Internet access or user contacts database through the security enforcement of a reference monitor. The application developer defines own security policy via assigning permission labels, and the permission labels are included in a XML manifest file. When an application is installed, the application asks for the necessary permissions from the user. The user decides whether or not to grant all the requested permissions. Once granted, application permissions cannot be changed until the application is reinstalled because Android's security enforcement is mandatory. Android's reference monitor checks permission assignments at runtime to mediate all ICC establishments, and it allows ICC establishment to proceed or denies ICC calls if the required permission is not included in the caller's manifest file[6].

3   Problem Description

Android developers specify own security policy and easily upload applications on the market store. Developers can also distribute the application to user through unauthorized link. Moreover, Android users often install the application without any concern about unnecessary permissions. It means malicious applications can be distributed to user without difficulty. Once installed, the application can have the privileges correspond to its permissions. Because reference monitor only restricts access with permissions and does not provide information flow guarantees, there are several vulnerabilities in Androids permission model. In this section, we will describe Intent-based attacks which use unjustified implicit Intents. In addition, we explain some problem of privilege escalation attacks that exploit vulnerable interfaces of benign applications.

3.1   Intent-Based Attack

The Intent is a message object that is passed between components in the same or different applications for component interaction. The Intents can be divided into an explicit Intent and an implicit Intent. An explicit Intent specifies the target component by its name and an implicit Intent does not designate a target's name. An implicit Intent is delivered to a target component only if the Intent matches the component's Intent filter. Broadcast Intents are a kind of an implicit Intent which is delivered to all interested receivers. When a component sends an implicit Intent, the Intent can be received by the unintended recipient. For this reason, an attacker could launch the Intent-based attack by unnecessarily receiving implicit Intents.

First of all, a malicious application can eavesdrop on some broadcast Intent by declaring an Intent filter. The eavesdropper can read contents of the broadcast, so that smartphone user's privacy can be invaded if the Intent includes sensitive data. A malicious application can also launch denial of service attack on ordered broadcasts. An ordered broadcast Intent is delivered to a single receiver one by one in order of priority. The receivers choose to propagate a result to a next receiver or to abort the broadcast. An attacker receives the ordered broadcast Intent first with abnormally high priority and impedes delivery to other benign receivers by aborting it. For example, when a SMS message is sent to an Android smartphone, an ordered broadcast about the message is delivered to a related Broadcast Receiver with the highest priority. If the receiver is malicious, it may abort the broadcast, and then other receivers never know anything about the SMS message. These attacks take place in a type of ICC described in the Fig. 2c. In addition to the above-mentioned, a malicious component can launch Activity and Service hijacking attacks. An attacker intercepts a request for Activity or Service and starts its own component in place of the intended one. The hijacking attack occurs in cases depicted in Fig. 2a and Fig. 2b [7].

3.2   Privilege Escalation Attack

Android's security model does not guarantee that a non-privileged caller is restricted to access more privileged callee. Malicious applications can escalate their privileges to gain unauthorized access to protected resources. They exploit defenseless interfaces of a privileged application. There are potential risk of privilege escalation attack in both third party applications and Android core applications. Fig. 4 describes a privilege escalation attack at the application level. App A, B and C run on Android, and they are isolated in its own sandbox. App A has no granted permissions, but component C1 can access to components protected by permission P1. Component C2 is not protected by any permission label, and it can access C3 because the App B is granted P1 permission. Therefore, component C1 is able to access C3 via C2 of App B.

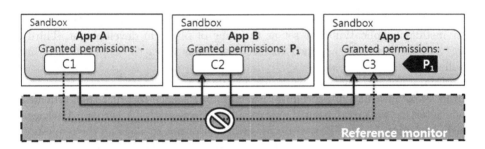

Fig. 3. Communication between application components [8]

4   Multi-level Security

Android's reference monitor mediates ICC calls with mandatory access control (MAC) enforcement. Classical MAC policies regulate access on the basis of classification of users and objects which are assigned security levels. However, the MAC enforcement in Android does not provide information flow with the security level. Therefore, we introduce multi-level security to Android's security model for reducing its vulnerabilities. Our security framework is adopted when an enterprise needs to provide secure platform to employees or when platform developer takes on the work for handling sensitive information. In this section, we explain overall architecture and process of our security framework.

4.1 Overview

Our security framework controls access between components using security level. The framework assigns proper security level to applications according to the pre-defined policy and regulates ICC establishment depending on the result of checking security levels. The ICC restriction prevents Intent-based attack and privilege escalation attack and controls information flow between applications. Whole architecture is described in Fig. 4. Level Manager and Application Levels database are newly added into the platform, and Package Manager, Package Installer, and reference monitor are modified. At first, Level Manager is responsible for assigning security level. Security administrators have to configure a programmed policy with their security requirements and input this policy into the Level Manager. The policies should include the number of security levels, rules for determining application's security level, and so on. Level Manager decides the security level, and then it stores them in the Application Levels database. After this assignment process, reference monitor mediates ICC using the security levels of caller and callee in accordance with the multi-level security rules which embedded in the system. If a security administrator needs sophisticated control, the rules can be re-established.

4.2 Application Installation

Android applications are divided pre-installed and third-party. Security levels of pre-installed applications are set at the time of platform development. Third-party applications are assigned the security levels when the application is installed.

When a smartphone user wants to download a third-party application, Android starts installation procedure. First of all, Package Installer that provides user interface for installation shows required permissions list through PackageInstallerActivity API. The user decides whether or not to allow all the requested permissions, and if the user grants the permissions, Package Manager is called for installing APK file via InstallAppProgress API. Package Manager extracts Android permissions from the Manifest and stores them in the Permission database. After that, user returns to InstallAppProgress, and it shows user that the installation terminates. We add assignment of security level into the installation procedure. We insert getLevelManager() function which calls Level Manager API into Package Manager like Fig. 5. When the Level Manager is called, it checks the application in the middle of installation using pre-defined policy which is established by security administrator. The checking process may involve analysis

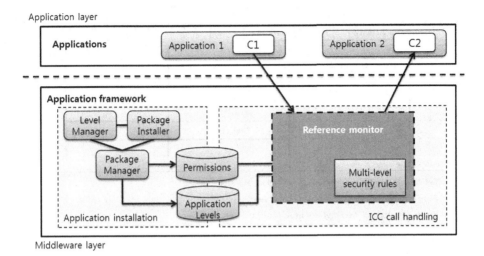

Fig. 4. Framework Architecture

of parsed manifest data. When Level Manger determines proper security level of the application, it subsequently stores the security level in Application Levels database. The database contains security levels of pre-installed applications entered by security administrators in advance. Pseudo code of Level Manager is showed in Fig. 5.

4.3 ICC Call Handling

Whenever an ICC occurs, reference monitor checks permission assignments. We modified reference monitor to handle ICC call depending on the security levels. Our security framework applies multi-level security rules to three types of ICC described in Fig. 2, and implicit Intent and explicit Intent are controlled differently. In case of implicit Intent, high level component's messages cannot flow to low level component, while explicit Intent from low level component restrict to be delivered to high level component. Setting information flow in the opposite direction prevents Intent-based attack and alleviates privilege escalation attack. Additionally, a security administrator who needs sophisticated controls can modify these rules himself.

Implicit Intent. Communication between high level caller and low level callee using implicit Intent is restricted by security level. We can prevent Intent-based attacks. In case of one types of ICC depicted in Fig. 2a, a component can start Activity which is included in high level or same level application. In other words, if high level component send an implicit Intent which starts low level Activity, reference monitor have to intercept the Intent and deny it. As a result, important information which is dealt with by high level application is protected from

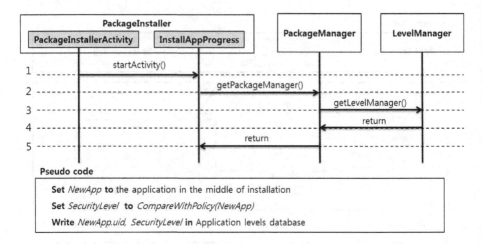

Fig. 5. Level Manager

leaks to low level application. Security administrator can block Activity hijacking attacks occurs in case of Fig. 2a by setting security level assignment policy appropriately. Besides, security administrator can also relax multi-level security rules for using the security level as a priority when the system requires flexibility. Communication between a component and a Service described in Fig. 2b is also controlled by the same multi-level security rules as mentioned above, and it can block Service hijacking attacks. Finally, in case of Fig. 2c, our security framework obstructs that low level Broadcast receiver receives a broadcast Intent from high level Activity or Service. Broadcast Intents are divided into normal broadcasts and ordered broadcasts. If the broadcast Intents are sent by the system, we handle two types of broadcast Intents differently. Normal broadcasts are delivered to all interested receivers without any restriction, while ordered broadcasts are delivered to receivers who are assigned higher security level than the system or the same level with the system. We modify existing platform which delivers broadcast Intent indiscreetly, so that our security framework can prevent eavesdropping, hijacking, and denial of service attacks on ordered broadcast Intent.

Explicit Intent. In contrast with the case of previously mentioned, communication between low level caller and high level callee using explicit Intent is restricted for alleviation of privilege escalation attack. If a component requests a Activity or a Service using explicit Intent, reference monitor decides whether or not to deliver the Intent according to the security levels. Therefore, applications which have many permissions for sensitive resources must be assigned high level, and untrusted or malicious applications must be assigned low level. Security administrators should note that during the configuration of level assignment policy. Consequently, if a component of malicious application requests ICC establishment with more privileged component, our security framework denies the request through multi-level security rule. However, there is a limitation that the framework hard to prevent from privilege escalation attack which use implicit Intent.

In middleware layer of Android, reference monitor is a part of Activity Manager which manages the life cycle of applications. Therefore, if a component sends an Intent, Activity Manager is responsible for delivery the Intent. For example, when a component starts an Activity using startActivity() method, system performs a execution sequence described in Fig. 6, and ActivityManagerService API is last performed. Instrumentation API executes startActivity call made by the application, and then ActivityManagerProxy provides binding with ActivityManagerService.

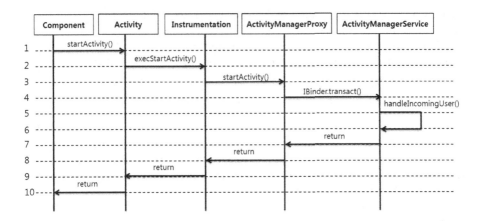

Fig. 6. startActivity()

Pseudo code

```
Case callee's component is
    When Activity or Service =>
        If explicit Intent && (caller's security level >= callee's security level) then
            connection established
        If implicit Intent && (caller's security level <= callee's security level) then
            connection established
    When Broadcast Receiver =>
        If Caller is a system && the broadcast Intent is normal broadcast then
            connection established
        If caller's security level <= callee's security level then
            connection established
reject the connection
```

Fig. 7. Checking security level

We analyzed the Android framework in depth, and we found out that handleIncomingUser method of ActivityManagerService class checks all Intents include request for Activity. The method confirms that the caller of the Intent has

INTERACT_ACROSS_USERS and INTERACT_ACROSS_FULL permissions. INTERACT_ACROSS_USERS permission removes restrictions on where broadcasts can be sent and allows other types of interactions. INTERACT_ACROSS_USERS_FULL permission allows an application to open windows that are for use by parts of the system user interface, and it is not used for third-party applications. Accordingly, we modified handleIncomingUser method to call security level check process of our security framework at the end of function. The pseudo code of level check is described in Fig. 7.

5   Basic Security Policy

Security administrators divide the security levels according to the system properties. More fine grained security level can satisfy more security requirements. But, it involves complex issues of definition and management. In this section, we explain samples of policy which are two level security and three level security.

5.1   Two Level Security

The most basic way to apply multi-level security to Android platform is the adoption of two level security. Our example of two level security shows that the framework protects trusted applications from the others. Applications which must be protected or reliable are assigned high level, and the rest applications are assigned low level.

First, pre-installed applications are allocated high level at the time of platform development. Because these applications are installed by the manufacturer and they do not have potential risk of trying to Intent-based or privilege attack. The security levels of pre-installed applications are stored in Application Levels database as high level in advance. Third-party applications are assigned security level at install time. Because they need to be diagnosed potential harmfulness, the security framework performs various security analysis. In this paper, we inspect third-party applications using following simple checklist.

1. Download the application from official market store like Google Play
2. Require permissions for sensitive resources
3. The number of permissions that the application requires

All applications are assigned high or low security level through this check process, and our security framework mediates ICC establishment using these security levels.

5.2   Three Level Security

Several companies produce Android device exclusively for finance or health-care, so they installed specialized applications in advance. In this case, pre-installed applications handle sensitive information and require high security to protect it. Therefore, we show an example of three level security that provide high security for pre-installed applications. Because three level security needs more complex policy, security administrator has to pay special attention to making the policy.

In this example, we divide all applications into pre-installed application, trusted third-party application, and untrusted third-party application, and they are assigned high, medium, and low security level respectively. Pre-installed applications are assigned high level first. Third-party applications are analyzed at install time and assigned medium or low security level according to the result of analysis. Applications which have medium applications are restricted to request ICC call to pre-installed applications even though they are reliable. This restriction prevents from leakage of sensitive information. Moreover, pre-installed applications have some permission for sensitive resources, and the restriction protects these permissions from risk of privilege escalation attack.

There are various ways to divide third-party applications into trusted and untrusted applications. We can find an application which has dangerous permissions or intent filter combinations as suggested in [5], and this kind of application are assigned low level. We can also differentiate untrusted application by using the method proposed in [9]. [9] extracts specific information from manifest, and compare it with the key word list for calculating the malignancy score. After that, it compares the score with the threshold values to detect Android malware. If you focus on the privilege escalation attack, you may check vulnerability of applications by analyzing manifest proposed in [10]. In this case, vulnerable applications which must be protected are assigned medium level.

6   Conclusion

In this paper, we describe the Android platform and its two main types of attacks which are Intent-based attack and privilege escalation attack. We propose multi-level security model for Android platform. Our security framework allocates proper security level to application according to the pre-defined policy and controls ICC establishment using the security level. Security administrators may adopt our security framework to reflect their different security requirements. For our future work, we plan to extend our security framework to prevent a variety of attacks. As the Android is becoming more and more popular, the number of new attacks is increasing. Therefore, we aim to prepare for new kinds of attacks.

Acknowledgments. This research was funded by the MSIP(Ministry of Science, ICT & Future Planning), Korea in the ICT R&D Program 2014[2014044072 003, Development of Cyber Quarantine System using SDN Techniques].

References

1. Gupta, A.: Market Share Analysis: Mobile Phones, Worldwide, 3Q13. Garter (November 2013)
2. F-Seure, Mobile Threat Report January-March 2013 (April 2013)

3. Android Developers Google, http://developer.android.com/
4. Bygiel, S.: Towards Taming Privilege-Escalation Attakcs on Android. In: Proceedings of the 19th Annual Symposium on Network and Distributed System Security (February 2012)
5. Enck, W.: On Lightweight Mobile Phone Application Certification. In: Proceedings of the 16th ACM Conference on Computer and Communications Security (CCS) (November 2009)
6. Enck, W.: Understanding Android Security. IEEE Security & Privacy Magazine (January 2009)
7. Chin, E.: Analyzing Inter-Application Communication in Android. In: Proceedings of the 9th International Conference on Mobile System, Applications, and Services, MobiSys 2011 (June 2011)
8. Davi, L.: Privilege Escalation Attacks on Android. In: Proceedings of the 3rd Information Security Conference (October 2010)
9. Sato, R.: Detecting Android Malware by Analyzing Manifest Files. In: Proceedings of the Asia-Pacific Advanced Network, APAN (2013)
10. Chan, P.P.: A Privilege Escalation Vulnerability Checking System for Android Applications. In: 13th IEEE International Conference on Communication Technologies (ICCT) (September 2011)

A Study of Reducing Resource Waste for Mobile Grid with Software Defined Network

Jun Kwon Jung[1], Ji Hoon Hong[1], and Tai-Myoung Chung[2]

[1] Department of Electrical and Computer Engineering,
Sungkyunkwan University, Suwon, Korea
{jkjung,jhhong88}@imtl.skku.ac.kr
[2] College of Information and Communication Engineering,
Sungkyunkwan University, Suwon, Korea
tmchung@ece.skku.ac.kr

Abstract. This paper introduces a model that provides reduced resource waste. A resource waste is main issue of performance of mobile grid. To solve this issue, we utilize new type network system. That is Software Defined Network(SDN). In traditional network environment, there are various network technologies on Internet. People enjoy a high speed Internet service by advanced network technologies now. However, this advanced network becomes a hurdle that the new network technology is hard to be applied. Many researchers try to overcome this problem. They introduce a new network concept that is called SDN. SDN divides a switch into a control plane and a data plane. SDN provides network administrators with more comfortable management. In addition, SDN is a variety of other benefits. These benefits are used to the mobile grid. Mobile grid has an issue of management that resources of a node are freely controlled due to unstable mobile network environment. SDN overcomes this problem using additional tag of a switch and programmable controller of own networks. In this paper, we introduce a model of mobile grid with SDN. Also, we assert benefits of mobile grid with SDN.

Keywords: Software Defined Network, Mobile Grid, Optimization, Resource Waste.

1 Introduction

Network environment is evolving and people's lives become even more affluent. Over the years, Internet speed is increasing faster. For Internet speed faster, network devices are evolved fast more and more. Vendors apply very advanced hardware / software technologies to their devices. While each network device is upgrading, network administrator's works are very particular to preserve compatibility. Due to this environment, some new technologies are difficult to be inserted in existing network environment. This is called ossified network. While the network is ossified, new network technology is difficult to be instituted and network development is slowing. To overcome this challenge, world researchers try new network technologies on special

B. Murgante et al. (Eds.): ICCSA 2014, Part IV, LNCS 8582, pp. 755–765, 2014.

network for researching. NSF(National Science Foundation) provides GENI that is built to try a next generation network technology. In GENI network, various research institutes test network technologies. Also, vendors of network devices establish an organization for new network technology which replaces existing ossified network. In these research topics, the most popular one is Software Defined Network(SDN).

Mobile devices along with development of smartphone permeate into people's life. As years go by, the Internet of things(IoT) technology which provides people with enrich life is realized. Various mobile devices have common limits that are processing resource leaks. This limitation is difficult to serve users to powerful computing services. Grid computing is one of the solutions for this problem. To use grid computing, large computing problem is divided to many tiny tasks and these tasks scatter many grid nodes. Grid architecture shows it virtualized super computer. In grid computing, one of main issues is addition/deletion of grid node. Grid with mobile environment is particular about this issue. Because each mobile node has a mobility characteristic, grid controller must deal with node connection management often. In mobile grid environment, disconnection and reconnection are main issues. Therefore, mobile grid administrator wants to solve problems by these issues. SDN is next generation network to easily manage existing networks that are complex. If mobile grid utilizes SDN technology, the mobile grid network is more effective and controllable. In this paper, we introduce a model of mobile grid with SDN for one performance issue of mobile grid.

SDN, the new type network technology, will be applied to various industries such as cloud system, ubiquitous system. SDN is also helps mobile grid to work more effectively. In this paper, we introduce mobile grid with SDN and how it is more useful using SDN. The one of main issues of mobile grid is efficient task scheduling. If the network is more efficient applying SDN, each node is more easily controllable. Thus, grid scheduler can reduce unnecessary resource disuse. In this paper, we introduce a model of mobile grid with SDN. Also, we suggest more effective resource use scenario.

2 Related Work

2.1 Mobile Grid

Grid computing is a system that shares computing resources of computers in networks and computes complex tasks. Grid computing with utility computing and other computing concepts becomes cloud computing. Grid computing is started by making virtual supercomputer. To provide supercomputing, grid controller divides a huge task into many tiny tasks and distributes each task. To make a grid, each node in grid network runs a handshake with grid scheduler. A scheduler receives node's available resource information and splits a large process from a grid user. A scheduler sends divided tasks to each node. Each node processes the received task and returns the result to sender. The scheduler assembles received results from each node and sends a complete result to a querying user. Grid computing is used for huge computing such as SETI@home project, cloud computing, and etc. Mobile grid is a type of grid that is added mobile nodes. Mobile nodes are portable machines with computing resources such as a notebook, a smartphone, a PDA, a tablet PC and etc. These mobile nodes become users that request a job or task nodes that provide resources. Especially, a node which is a component of mobile grid can connect a wired grid network and it needs a mobile node to provide a quick notification and a task processing. Each mobile node may lose a connection often than wired grid node. This challenge is a main issue of mobile grid computing.

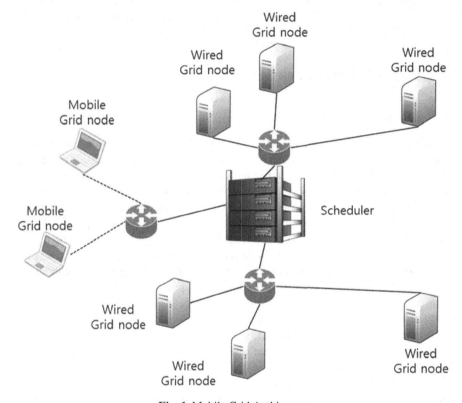

Fig. 1. Mobile Grid Architecture

2.2   Performance Issue of Mobile Grid

Mobile grid node has a characteristic of mobility. This means each mobile node can freely move itself and a mobile node may disconnect mobile grid network anytime. Although its position is in mobile grid area, it may occur this disconnection event because an obstacle can harm a wireless network connection. Disconnected node cannot send a task result to scheduler, so a scheduler discards a task that is from disconnected node when a scheduler knows a disconnection of the node. A scheduler resends the task that is discarded to another node due to complete grid process. This problem causes considerable performance trouble on mobile grid.

There are two aspects of overall performance in mobile grid. They are Response time(RT) and Resource waste(RW). The RT is a time of communication between a scheduler and a node. The RT is related with network performance rather than node process performance. The RW is a task failure when a node cannot process and send a task result to a controller due to network disconnection. If a scheduler throws away the task once, the task cannot be used and is thrown away. If the RT is shorter and the RW is fewer, the mobile grid performance is better.

For better service quality, a scheduler considers a countermeasure about abandoned tasks. However, in traditional network environment, it is hard to solve performance problems of mobile grid. In this paper, we focus on Resource waste problem. We introduce SDN and propose utilization of SDN for mobile grid.

3   Mobile Grid Based on SDN

3.1   SDN(Software Defined Network)

The network technology advances, various devices have been provided to the user. Many networks are connected to each other and these networks are faster and faster. This internet technology grows higher and faster. This environment provides users with convenient Internet services, but network researchers are difficult to study and apply new network technologies. Existing network telecommunication model is a 'client-server' model. Network architecture is hierarchical and main packet flow is north-south flow. However, the latest packet flow is more parallel. East-west flow is more than 75% of total packet flow on datacenter. Therefore, today's data flow is not suitable for existing network architecture. In addition, existing network technologies are mix installed, so a new network technology must provide compatibility for various existing network technologies. Existing network technologies become a hurdle of institution to new network technologies. This situation is called a network is ossified. To overcome the ossified network, many researchers make an effort to provide ossified network with flexible and replace existing networks to new flexible networks. Main problem of existing networks is that various network vendors provide own network interface and narrow compatibility. Each vendor is reluctant to open their inside architecture. Also, they provide different I/O interface. Thus, existing inter-networking environment can only communicate with each other restrictively. Network administrators are difficult to construct and maintain own network. SDN is made by this background.

SDN divides a network switch into two planes – a control plane and a data plane. A data plane provides a switch with packet forwarding using a flow table. If a data plane receives a packet from ingress port, a data plane does packet matching process using a flow table. A flow table consists of a match field, an action field, counter and so on. Actions in a flow table send targeted packet to next switch by output port or modify header data of targeted packet. A data plane just manages packet flow. A data plane is located in switches. Each switch has several hosts. Many data plane switches

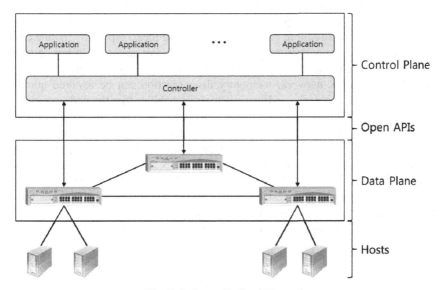

Fig. 2. Software Defined Network

connects to one or several controllers. A control plane consists of a controller and network applications. It can make, modify, and remove a flow table. Also a control plane notifies a next switch of packet flow from a switch. A control plane creates virtual network topology and provides comfortable user control by programmable control interface. The network administrator uses own control application in a control plane. This application is based on controller APIs. The network administrator just considers control API for own management. Various switches provide common API for SDN. The most notable SDN project is OpenFlow. The OpenFlow project is founded by Open Networking Foundation(ONF). ONF develops OpenFlow specifications and makes public deployment version 1.0.0. OpenFlow operates a data plane just in a switch that supports OpenFlow. OpenFlow works a bridge with a switch and a control plane. For a role of control plane, NOX is developed. SDN provides integrated management of the entire network through a combination of user, controller, and programmable switches. A picture shows a simple architecture of SDN with NOX and OpenFlow.

SDN has many advantages due to programmable network control. Network administrator just needs to understand and use SDN API. Then the controller and SDN switches are operating administrator's needs. Thus, network management also becomes easy. Furthermore, network administrator divides own network virtually. This allows the network division of the user or the packet according to the security level.

3.2   Model of Mobile Grid with SDN

Mobile grid has some performance issues. In these issues, resource waste problem is solved by SDN. A node of mobile grid may disconnect grid network, so a task of the node may be discarded. This resource waste is a cause of performance decline of mobile grid. Using SDN, the problem of RW is able to be solved. There are some ways of disconnecting reason in grid networks. Among them, especially mobile environment, node may disconnect the network when moving across buildings or move passed by obstacle. These disconnections are temporary. In large public mobile networks such as campus network, temporary disconnection can be restored quickly. Because this network is dense and there is high probability that the node position is inside the network. Thus, temporary disconnected node have not discard immediately received task. If a task discard is reduced, RW is also reduced. However, scheduler can classify whether or not the node is inside the network. The scheduler cannot solve this issue alone, but SDN is able to solve it. In SDN environment, controller uses additional information for switch control. Switch's location information is additional fields to determine connection lost point. If a node is disconnected at campus boundary, we think that it is a high probability that a node may get out of campus network. Figure n shows a simple model of mobile grid with SDN.

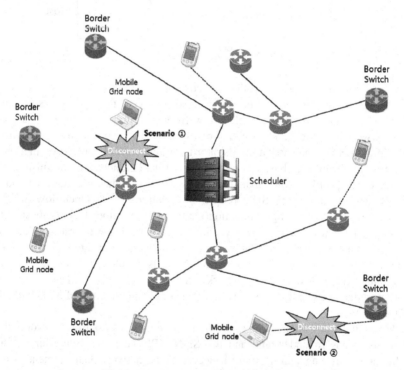

Fig. 3. Disconnect scenario in mobile grid

First scenario describes situation when a mobile node on a switch inside the network when it is disconnected in the network. A node of mobile grid connects to a grid scheduler in advance. If a node is connected, a scheduler sends connected information to the node. This scheduler includes SDN controller. Thus, the scheduler can use a location tag of switches. A location tag determines each switch's position whether inside network or boundary network. A tag is attached with a control message.

A control message is similar to OpenFlow channel message. This message is usually used to switch control or determining undetermined packet flow. In this model, grid message is sent with SDN control message. To determine grid message, SDN switch checks destination address. If a destination address is scheduler's address, switch knows this packet is for scheduler. If a scheduler receives disconnected message from inside network switch, a scheduler modifies a tag of task scheduling table entry to 'reserved'. This means this task would be inside own grid network with high probability and this task is just undetermined location in grid network. When a grid node with reserved task is reconnected in grid network, grid node sends a result of reserved task to a scheduler. A scheduler reassembles received results from all grid nodes.

Second scenario describes situation when a mobile node on a switch of the grid network boundary when it is disconnected in the network. When a switch knows a disconnection, this switch sends a scheduling message to a scheduler with SDN controller. If a scheduler receives this message, this scheduler removes task information of disconnected task and reallocates another grid node because a boundary node has high probability of escaping grid network. Thus, a scheduler does not reserve this task for whole task processing.

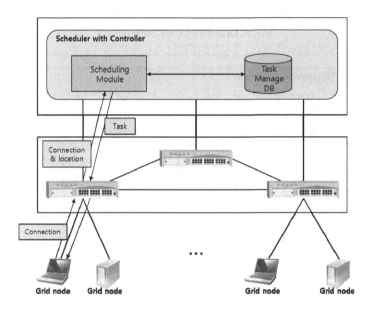

Fig. 4. First connection in inside switch

If a mobile grid node connects to inside switch, the node sends connection packet to a scheduler that is combined SDN controller. Connected SDN switch that is inside grid network sends connection packet with own location information to a scheduler. When a scheduler receives a connection packet, it compares packet's location data with own data in scheduling DB. If a scheduler determines that a connected node is new grid node, it allocates new task to the grid node.

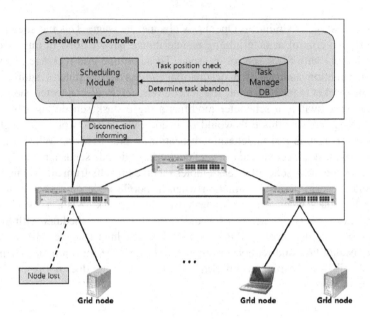

Fig. 5. Disconnect node event control

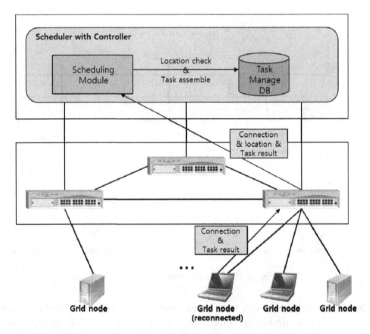

Fig. 6. Reconnection of grid node event

When a mobile grid node is disconnected, a switch informs node disconnection to a scheduler. A scheduler requests the position of lost grid node to own manage DB. If lost grid node is located in inside network, a scheduler does not discard the task that is allocated to lost grid node and this task is renamed to 'reserved' task. If lost grid node is located in border network, however, a scheduler discards this task and reallocates the task to the other grid node.

When the grid node that is disconnected before is reconnected, the grid node sends a connection packet to a scheduler. If the grid node completes received task, it sends the result also. A SDN switch relays this packet to a scheduler with own location information. A scheduler checks received packet from connected node. If this node is reserved task's node, this task is renamed to 'allocated'. If reconnected node sends a task result also, a scheduler receives the result and assembles results.

To do this, a scheduler needs combining SDN controller and a scheduler uses specific task scheduling table. This table consists of task type tag, task ID and node ID. Task type tag value is unprocessed, reserved or complete. This tag informs how each task is processed. Reserved tag is sent to designated node but that node is disconnected temporary. Thus, a scheduler distributes unprocessed tasks first, reserved task is sent when all unprocessed task is run out. The task of mobile grid can reassemble complete tasks. In a typical mobile grid, a task on disconnected is discarded but proposed system provides reducing resource waste.

3.3   Performance of Mobile Grid with SDN

To determinate of disconnection point for grid scheduler, the scheduler should interact the SDN controller or have the role of SDN controller. SDN controller can determine restoring connection probability using switch position information. Controller assigns the tag that is defined switch's position and grasps switch's approximate location. Grid scheduler can determine a position of each grid node by connected switch's location information. Grid scheduler reserves tasks that have high probabilities of node reconnection. Then, grid scheduler reduces resource waste. Thus, grid scheduler with SDN controller makes mobile grid performance improvement.

In mobile grid, a resource waste is important performance issue. In mobile environment, each node has high probability of disconnection. If a node disconnection is occurred, a task in that node is corrupted typically. Thus, a scheduler should reschedule corrupted task to other live node. A node which has corrupted task can complete its task and reconnect often. This corrupted task is a duplicated task. Thus, this corrupted task occurs resource waste. In sungkyunkwan university, we test reconnecting university own network. We make an app that check disconnection and reconnection of mobile AP. This app measures frequent reconnection of certain AP. Frequent reconnection means it may have many RW from disconnection. We probe AP's reconnection 07:00~24:00. In this test, we check about 360 minutes of connection time and average 30 reconnections in one day. The shortest reconnection interval is 1 minute and average reconnection interval is 16minutes. Also, reconnection counts in 5minutes are 15. If each task is processed for 5minutes, maximum 75 minutes are wasted. It shows about 20% time waste in basic mobile grid network.

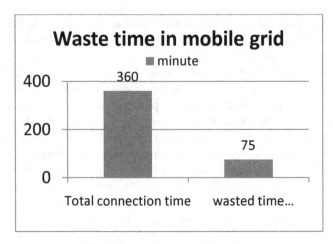

Fig. 7. Waste time in mobile grid

4   Conclusion

Network technology's development gives more comfortable and fast Internet services for people. However, this advance makes new network technology to be hard to apply. To overcome this challenge, many researchers work on. In these efforts, SDN is the most realized network technology model. In campus network, a mobile grid is interesting system to use many mobile nodes. This mobile grid has performance issue as a necessity. In this paper, we propose new model which reduces resource waste rate in mobile grid. To archive this, we use SDN. Grid scheduler is combined with SDN controller and combined scheduler uses information of switches location and additional task scheduling tag. Inside nodes except nodes connected to boundary switch can save grid resource in node disconnection due to scheduler's additional control information. Thus, considerable disconnected task will be not corrupted. However, mobile grid has many issues also. Mobile node has limitation of power. Especially, smartphone can be an element of grid resource pool, but power limitation of smartphone is hard to be solved. This paper does not concern power limitation of mobile node. In addition, this paper does not introduce accurate mathematical evaluation of performance advantage. Thus, we study mathematics for evaluation next time and other performance issue in mobile grid.

Acknowledgement. This research was funded by the MSIP(Ministry of Science, ICT & Future Planning), Korea in the ICT R&D Program 2014[2014044072003, Development of Cyber Quarantine System using SDN Techniques].

References

1. Park, S.-M., Ko, Y.-B., Kim, J.-H.: Disconnected Operation Service in Mobile Grid Computing. In: Orlowska, M.E., Weerawarana, S., Papazoglou, M.P., Yang, J. (eds.) ICSOC 2003. LNCS, vol. 2910, pp. 499–513. Springer, Heidelberg (2003)

2. Sim, K.M.: Grid Resource Negotiation:Survey and New Directions. IEEE Transactions on Systems, Man, and Cybernetics, Part C: Applications and Reviews 40(3), 245–257
3. Hendersona, T., et al.: The changing usage of a mature campus-wide wireless network. Computer Networks 52(14), 2690–2712
4. Lin, L.: QGrid: an Adaptive Trust Aware ResourceManagement Framework. IEEE Systems Journal 3(1), 78–90
5. Ding, Z., et al.: VJM - A Deadlock Free Resource Co-allocation Model for Cross Domain Parallel Jobs. In: High Performance Computing Asia 2007, Korea, pp. 9–12 (September 2007)
6. Katsaros, K.: Optimizing Operation of a Hierarchical Campus-wide Mobile Grid for Intermittent Wireless Connectivity. In: LANMAN 2007, pp. 111–116 (2007)
7. Katsaros, K., Polyzos, G.C.: Towards the realization of a mobile grid. In: CoNEXT 2007, vol. (31), pp. 1–2 (2007)
8. Sezer, S., et al.: Are we ready for SDN? Implementation challenges for software-defined networks. IEEE Communications Magazine 51(7), 36–43 (2013)
9. Lara, A., Kolasani, A., Ramamurthy, B.: Network Innovation using OpenFlow: A Survey. IEEE Communications Surveys & Tutorials 16(1), 493–512 (2014)
10. McKeown, N., et al.: OpenFlow: enabling innovation in campus networks. ACM SIGCOMM Computer Communication Review 38(2), 69–74 (2008)

On Cost-Reduced Handoff Signaling for Secure ITS Communication Mechanism

Haenam Jeon[1], Jae-Young Choi[2], and Jongpil Jeong[2,*]

[1] Solution Biz Division, Mobile Security Development Team
Infosec Co., Ltd.
1008-4, Daechi-dong, Kangnam-gu, Seoul 135-851, Korea
karas1237@gmail.com
[2] College of Information and Communications
Sungkyunkwan University
2066 Seobu-ro Jangan-gu, Suwon, Kyunggi-do, Korea
{jaeychoi,jpjeong}@skku.edu

Abstract. Globally, intelligent vehicles and telematics R&D through the integration of IT technologies in the vehicle are significant increasing. Real-time data communication for intelligent transportation system (ITS) is very important. It collects real-time data from the vehicle and provides the information collected from ITS center. We propose an effective and secure communication scheme for these communication procedures. In particular, our proposed SIP-based MVPN reduces signaling cost and has many advantages in security aspects. In addition, our proposed scheme performs the mobility management applying Network Mobility (NEMO) for the communication between the vehicles. In other words, we propose an ITS communication mechanism of SIP-based mobile VPN and V2V NEMO. Finally, our performance analysis show that the ITS of SIP-based MVPN is significantly reducing the handoff signaling cost.

Keywords: Intelligent Transportation System, Vehicle to Vehicle, Network Mobility, Session Initiation Protocol, Mobile Virtual Private Network.

1   Introduction

An Intelligent Transportation System (ITS) is expected to be the high-tech transportation system of the future. The rapid development of IT technologies and a variety of services to ITS implementation have being realized, since those concepts were introduced in the mid-1980s. It uses advanced technologies, such as electronic communication traffic control, and maximizes the usage efficiency to collect, manage/provide real-time traffic information services to the existing transportation facilities. In this paper, we study the communication technologies of vehicle-to-vehicle and vehicle-to-server, to implement an efficient and secure system of ITS. We then propose a cost-efficient scheme [1].

Session Initiation Protocol (SIP) supports user mobility and terminal mobility. Terminal mobility is done by sending a new INVITE (RE-INVITE) command to the

---

* Corresponding author.

B. Murgante et al. (Eds.): ICCSA 2014, Part IV, LNCS 8582, pp. 766–778, 2014.

Correspondent Node (CN) using the same Call-ID as that of the first session. In this paper, we propose to use a combination Secure Real-time Transport Protocol (SRTP) and Compressed RTP (cRTP) to not only protect the payload data, corresponding to the protection of the header [2-4].

In addition, we use the Multimedia Internet KEYing (MIKEY) for key management on RTP and SRTP. MIKEY was designed for small or Peer-to-Peer (P2P) group. NEMO VPN gateway is a possible representative of all MNs for an address registration request of an MN [5]. Thus it is possible to reduce the signaling cost of the address registration request [6]. This study's design is based on the architecture of [2], and reduces handoff signaling costs by applying better performance of ITS.

The organization of this paper is as follows. Related work is discussed in Section 2. Section 3 describes the mechanism proposed in this paper. We evaluate the performance of the proposed SIP-based MVPN in Section 4. Section 5 outputs the results of the numerical analysis. Section 6 concluded the paper.

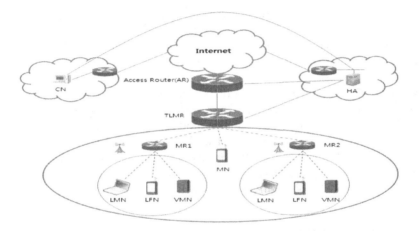

Fig. 1. NEMO System Architecture

2   Related Work

Network Mobility (NEMO) [5] is a technique to provide a seamless service network connected to the Mobile Network Nodes (MNN). The NEMO Basic Support Protocol (BSP) is depicted in Fig. 1, with a plurality of MR for MNN. MNN is classified into three types. Local Fixed Node (LFN), which means that you do not move in relation to mobile networks. LFN is present in the mobile network in general, you will not be able to move to a different network from the current network. Visiting Mobile Network (VMN) belongs to another network, but it is dependent on the MR of the mobile network. When it is registered to a network belonging to MR, MR is able to know the presence of the Home Agent (HA). We also define the CN to communicate with MNN. Each station belonging to MNN has related behavior mobility, instead of running. NEMO a MR is a router on behalf of the MNN to the Mobile Route (MR). It is

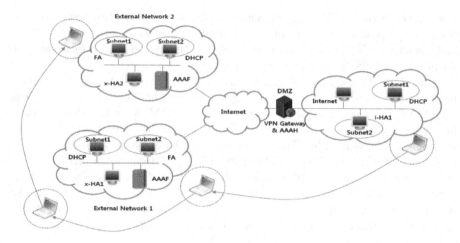

Fig. 2. MVPN architecture and protocol

possible to have a more efficient use of the network, because the MR is executed instead a signal, existing terminals is executed directly.

IETF has previously defined the architecture and protocol for MVPN [7,8]; it is shown in Fig. 2. Here, the internal HA (i-HA) and external HA (x-HA) are present in the intranet and Internet and the two HAs. A new care-of address (CoA) is first obtained from the dynamic host configuration protocol (DHCP) server or foreign agent (FA) when the MN moves out of the intranet. This CoA is registered in x-HA. Then, MN creates a VPN gateway and IPSec tunnel using its external home address (x-HoA). An IPSec tunnel is created by using internet key exchange (IKE) [9]. Fig. 2 shows the three tunnels (x-MIP, IPSec, and i-MIP).

3   Cost-Reduced Handoff Signaling for Secure ITS Communication

3.1   System Architecture

In this section, we describe our proposed the mechanism applying the SIP-based MVPN. The mechanism has a similar communication structure, such as the existing u-TSN. In our proposed scheme, MN is the target network entity of all traffic and mechanism is a networking technology in which it is possible to configure the network automatically and transmit/receive the state information and location information between nodes.

The ITS communication mechanism is configured with the ubiquitous transportation center (UTC), ubiquitous infra-structure sensor (UIS), and ubiquitous vehicle sensor (UVS), as shown in Fig. 3. First, we describe about signaling, secure transport, key management, Authentication, Authorization, and Accounting (AAA), and Application Level Gateway (ALG). SIP is a signaling protocol of the application layer.

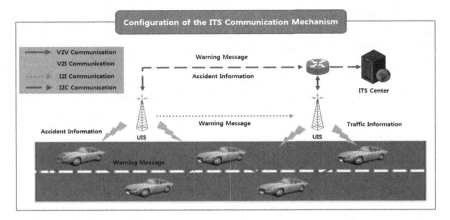

Fig. 3. Configuration of the ITS Communication Mechanism

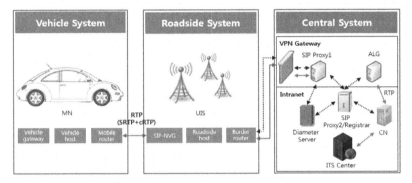

Fig. 4. System architecture of proposed mechanism

The SIP-based MVPN proposed in this paper is used for user identification and authentication of mobile users.

In addition, the SIP supports user mobility and terminal mobility [3]. Terminal mobility uses the first session of the mobility device with the same Call ID to CN by sending a new INVITE (RE-INVITE) command. The new INVITE command includes the MN address of the new connection acquired at the new location. After receiving the RE-INVITE command, CN redirects to a new location of the MN from the next traffic. SRTP is defined as the framework for the integrity and encryption of messages RTCP and RTP stream [4]. Encapsulation of the Security Payload of IPsec provides the data origin authentication and confidentiality of all packets. In contrast, SRTP supports only the protection of the data payload of the RTP packet. However, in this paper, we combine SRTP and cRTP for header protection. cRTP is used to compress all the headers of IP / UDP / RTP. After that, the results are included in the extension header of the RTP. SRTP protects the RTP data that contains the extension header of RTP. By sending the SRTP packet, it decrypts the RTP data, and releases the recompressed IP / UDP / RTP Headers for original cRTP data, as shown in Fig. 4.

Diameter SIP-based applications are based on diameter-based protocol [10], and authenticates/identifies the clients on the SIP server from the Diameter server. In the proposed architecture, when the authentication of the SIP message is required, the SIP server adds the MIKEY initiator message received from the AAA server for TGK transfer to the SDP of 200 OK, when the INVITE request is successful. The SIP proxy server provides a Data SA that TGK is added for the protection of the SRTP to ALG.

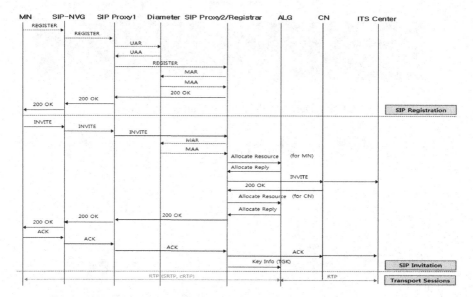

Fig. 5. Roaming from intranet to internet

3.2 Operational Procedure

Fig. 5 shows that the communication with the Center in the intranet and via vehicles in the external networks. The SIP NEMO VPN Gateway (SIP-NVG) in the external network is the gateway of the mobile network. It is required when moving to another network. It flows the SIP standard and maintains seamless session roaming to another subnet, SIP-NVG provides data in a secure manner. Moreover, the SIP-NVG performs traffic management for the mobile network. The VPN gateway consists of ALG and SIP Proxy 1, which is a SIP proxy server. If the MN roams to the intranet or accesses the intranet, the SIP Proxy 1 authenticates the incoming SIP message through the Diameter server and routes the message to SIP Proxy 2. Then, ALG monitors all data traffic. MN must register its new location with the SIP Registrar when roaming outside the intranet and roaming back to the intranet. It is based on the Diameter SIP Application. MN sends a REGISTER first Registrar. If the authentication of signaling messages is required, the Registrar generates the Multimedia Auth Request (MAR) and sends it to the Diameter Server.

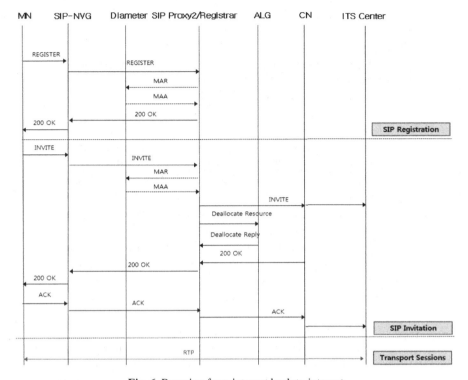

Fig. 6. Roaming from internet back to intranet

Fig. 6 is a procedure when the MN returns to the intranet from the Internet. When the MN is located on the intranet, the message does not need to go through the SIP Proxy 1. First, the MN registers its new address with the SIP Proxy 2/Registrar and sends a RE-INVITE message. After the SIP Proxy 2/Registrar receives the RE-INVITE message, the SIP Proxy 2/Registrar will de-allocate all resources in advance. Then, without going through the ALG, the MN directly communicates with the CN, as shown in "Transport Sessions" in Fig. 6.

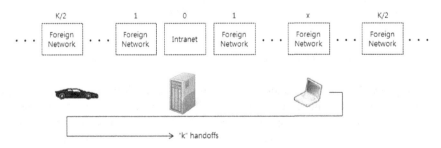

Fig. 7. Network topology for performance analysis

Table 1. Parameter for performance analysis

Parameter	Description
K	The number of networks an MN visits before it goes back to the intranet.
x	The network in which an MN currently resides, $0 \leq x \leq K$, When $x=0$, the MN is inside the intranet.
β_x	The probability distribution of x. We consider uniform, linear, and exponential distributions.
$C_{t,i}$	The handoff signaling cost that consists of registration cost and RE-INVITE cost when MN has i sessions and the handoff type is t.
D_t	The handoff signaling cost that consists of x-MIP registration cost, IPsec tunneling cost, and i-MIP registration cost when MN has handoff type t.
S_ϕ	The handoff signaling cost per unit time for SIP-based MVPN, where ϕ can be a uniform, linear, or exponential distribution.
$S'_{i,\phi}$	The average handoff signaling cost of an MN incurred between two consecutive events, where i is the number of sessions and ϕ can be a uniform, linear, or exponential distribution.
R_h / R_f	Average handoff cost when a mobile network (sent by SIP-NVG) when the mobile network enters its home or foreign network.
$V_{hf} / V_{ff} / V_{fh}$	Average cost for the first part of re-INVITE when a mobile network moves from its home network to a foreign network or from a foreign network to another foreign network or from a foreign network to its home network.
$I_{hf} / I_{ff} / I_{fh}$	Average cost for the second part of re-INVITE of a session when a mobile network moves from its home network to a foreign network or from a foreign network to another foreign network or from a foreign network to its home network.
$T_{l,m} / U_{l,m}$	The transmission cost of SIP registration or RE-INVITE between Node l and node m.
$X_{l,m}$	The transmission cost of User-Authorization-Request (UAR)/User-Authorization-Answer (UAA) or MAR/MAA between Node l and Node m.
Y_l	The processing cost of SIP message in Node l.
Z_l	The processing cost of UAR/UAA or MAR/MAA in Node l.

4   Performance Analysis

Our proposed ITS communication mechanism applies the SIP-based MVPN [2] in V2V-NEMO environments and attempt to reduce the handoff that occurs frequently in the ITS. In this section, we analyze the performance of the in proposed mechanism.

Similar to [11-14], the cost function of the signal is configured in the processing cost and transmission costs. The transmission fee is proportional to the distance of the network between two nodes. The processing cost of network nodes, authentication and processing cost of messages, such as re-encapsulation and de-encapsulation of packets that are tunneling, is included. The signal flow is almost the same, because it is based on the architecture proposed in the IETF [15-17]. It is assumed on the basis of the network topology, such as in Fig. 7, that CN is present on the intranet, MN moves between intranets and the Internet. The parameters in Table 1 is a parameter defined for the performance analysis.

Considering that an MN starts moving from the intranet and goes back to the intranet after it visits K foreign networks $K=0,...,\infty$. In the SIP-based MVPN, as in the aforementioned discussion, an MN needs to register with the SIP Registrar to update its location and send a RE-INVITE to the CNs that have ongoing sessions(s) with the MN. We assume that session arrival rate, denoted as λ, of an MN, and follows the Poisson process. The session serving time is exponentially distributed with rate μ. Each MN has a maximum number of sessions denoted as c. The session arrival at MN, therefore, are regarded as an M/M/c/c queuing system. By the deviations for M/M/c/c queuing system [11,12], we obtain the probability π_i that MN have i sessions for $i=0,...,c$.

$$\pi_i = \frac{\lambda^i}{i!\mu^i}\left(\sum_{x=0}^{c}\frac{\lambda^x}{x!\mu^x}\right)^{-1}$$

As in [14], in this paper, we derive the probability $\alpha_i(k)$ that MN moves across k networks between two consecutive events. An event can be a new session arrival or an ongoing session departure. Suppose that $t_{e,i}$ is the time interval between two consecutive events and MN has i sessions during $t_{e,i}$.

$$E[t_{e,i}] = \frac{1}{\phi} = \begin{cases} \dfrac{1}{\lambda+i\mu}, & 0 \leq i < c \\ \dfrac{1}{c\mu}, & i = c \end{cases}$$

$\alpha_i(k)$ can be derived as follows:

$$\alpha_i(k) = \begin{cases} 1 - \dfrac{\lambda_n\left[1-f_n^*(\phi_i)\right]}{\phi_i}, & k = 0 \\ \dfrac{\lambda_n\left[1-f_n^*(\phi_i)\right]^2 f_n^*(\phi_i)^{k-1}}{\phi_i}, & k > 0 \end{cases}$$

It is assumed that the network residence time follows a Gamma distribution with rate λ_n. Thus $f_n^*(s)$ is derived as follows:

$$f_n^*(\phi_i) = \left(\frac{\lambda_n\gamma}{\phi_i+\lambda_n\gamma}\right)^\gamma$$

In particular, when $\gamma=1$, the network residence time follows the exponential distribution. To consider different handoff types and K, we analyze the handoff signaling cost for $(1 \le x < K)$ and $(x=0)$. First, if an MN currently resides in a network X $(1 \le x < K)$ in the internet and performs handoff k times between two events, it hands off from the foreign network to the intranet $\lfloor (x+k)/K \rfloor$ times, from the intranet to the foreign network $\lfloor (x+k-1)/K \rfloor$ times, and from the foreign network to another foreign network in the remainder $k - \lfloor (x+k)/K \rfloor - \lfloor (x+k-1)/K \rfloor$ times. Second, if an MN currently resides in the intranet $(x=0)$ and performs handoff k times between two events, it hands off from the foreign network $\lfloor k/K \rfloor$ times, from the intranet to the foreign network $\lfloor k/K \rfloor$ times, and from the foreign network to another foreign network in the reset $k - \lfloor k/K \rfloor - \lceil k/K \rceil$ times. Therefore, assuming an MN has i sessions, the average handoff signaling cost functions for $1 \le x < K$ and $x=0$ can be derived as follows:

For $1 \le x < K$,

$$\sum_{k=0}^{\infty} \left\{ \left\lfloor \frac{x+k}{K} \right\rfloor \cdot C_{fh,i} + \left\lfloor \frac{x+k-1}{K} \right\rfloor \cdot C_{hf,i} + \left(k - \left\lfloor \frac{x+k}{K} \right\rfloor - \left\lfloor \frac{x+k-1}{K} \right\rfloor \right) \cdot C_{ff,i} \right\} \cdot \alpha_i(k)$$

For $x=0$,

$$\sum_{k=0}^{\infty} \left\{ \lfloor \frac{k}{K} \rfloor \cdot C_{fh,i} + \lceil \frac{k}{K} \rceil \cdot C_{hf,i} + \left(k - \lfloor \frac{k}{K} \rfloor - \lceil \frac{k}{K} \rceil \right) \cdot C_{ff,i} \right\} \cdot \alpha_i(k)$$

Referring to [12], let $g_i = f_n^*(\phi_i)$ and $\rho_i = \phi_i / \lambda_n$, for $i=0,...,c$. As $k=pK+q, p=0,1,...,\infty$, $\alpha_i(0)$ is not considered. As an MN can move across several networks, let $k=pK+q, p=0,1,...,\infty$ and $0 \le q < K$. Then

$$\alpha_i(pK+q) = \frac{(1-g_i)^2}{\rho_i g_i} (g_i^K)^p g_i^q = y_i z_i^p w_i^q$$

As a result, the average handoff signaling cost per unit time can be derived as follows:

$$S_\phi(K) = \sum_{i=0}^{c} S'_{i,\phi}(K,\rho_i) \cdot \phi_i \cdot \pi_i$$

The subscripts l and m can be mn, nvg, pro, dia, reg, cn, alg and its, which refer to MN, SIP Proxy l, Diameter server, SIP Proxy 2/Registrar, CN, and ALG, respectively, Referring to Figs. 4, 5, the registration cost and the RE-INVITE cost can be determined as follows:

$$R_h = Y_{mn} + Y_{nvg} + 2Y_{reg} + 2Z_{reg} + 2Z_{dia} + 2T_{mn,nvg} + 4T_{nvg,reg}$$
$$+ 4X_{reg,dia}$$

$$R_f = Y_{mn} + Y_{nvg} + 4Y_{pro} + 2Y_{reg} + 2Z_{pro} + 2Z_{reg} + 4Z_{dia} + 2T_{mn,nvg}$$
$$+ 4T_{nvg,pro} + 4T_{pro,reg} + 4X_{pro,dia} + 4X_{reg,dia}$$

$$V_{hf} = 2Y_{mn} + 2Y_{nvg} + 5Y_{pro} + 6Y_{reg} + 2Y_{alg} + Y_{cn} + Y_{its} + 2Z_{reg} + 2Z_{dia}$$
$$+ 5U_{mn,nvg} + 5U_{nvg,pro} + 5U_{pro,reg} + 3U_{reg,alg} + 3X_{reg,cn} + 4X_{reg,dia}$$

$$V_{ff} = 2Y_{mn} + 2Y_{nvg} + 5Y_{pro} + 4Y_{reg} + Y_{alg} + 2Z_{reg} + 2Z_{dia} + 5U_{mn,nvg} + 5U_{nvg,pro}$$
$$+ 5U_{pro,reg} + 3U_{reg,alg} + 4X_{reg,cn}$$

$$V_{fh} = 2Y_{mn} + 2Y_{nvg} + 5Y_{reg} + Y_{alg} + Y_{cn} + 2Z_{reg} + 2Z_{dia} + 5U_{mn,nvg}$$
$$+ 5U_{nvg,reg} + 2U_{reg,alg} + 3U_{reg,cn} + 2U_{cn,its} + 4X_{reg,dia}$$

$$I_{hf} = 2Y_{nvg} + 5Y_{pro} + 6Y_{reg} + 2Y_{alg} + Y_{cn} + Y_{its} + 2Z_{reg} + 2Z_{dia} + 5U_{nvg,pro}$$
$$+ 5U_{pro,reg} + 3U_{reg,alg} + 3X_{reg,cn} + 4X_{reg,dia}$$

$$I_{ff} = 2Y_{nvg} + 5Y_{pro} + 4Y_{reg} + Y_{alg} + 2Z_{reg} + 2Z_{dia} + 5U_{nvg,pro} + 5U_{pro,reg}$$
$$+ 3U_{reg,alg} + 4X_{reg,cn}$$

$$I_{fh} = 2Y_{nvg} + 5Y_{reg} + Y_{alg} + 2Z_{reg} + 2Z_{dia} + 5U_{mn,nvg}$$
$$+ 5U_{nvg,reg} + 2U_{reg,alg} + 3U_{reg,cn} + 2U_{cn,its} + 4X_{reg,dia}$$

The handoff signaling cost of the SIP-based MVPN can be expressed as:

$$C_{hf,i} = R_f + V_{hf} + i \cdot I_{hf}, \quad C_{ff,i} = R_f + V_{ff} + i \cdot I_{ff}, \quad C_{fh,i} = R_h + V_{fh} + i \cdot I_{fh}$$

Table 2. Parameter values for performance analysis

Parameter	Value	Parameter	Value
Y_{mn}	5.0	Y_{pro}/Z_{pro}	10.0
Y_{reg}/Z_{reg}	30.0	Y_{alg}	50.0
Y_{cn}	5.0	Y_{reg}	30.0
Y_{nvg}	25.0	Z_{dia}	10.0
$T_{mn,reg}/U_{mn,reg}$	1.0	$T_{mn,pro}/U_{mn,pro}$	3.0
$T_{mn,nvg}/U_{mn,nvg}$	1.0	$T_{nvg,reg}/U_{nvg,reg}$	5.0
$T_{nvg,pro}/U_{nvg,pro}$	5.0	$U_{cn,its}$	0.5
$T_{pro,reg}/U_{pro,reg}$	0.1	$U_{reg,alg}$	0.1
$U_{reg,cn}$	1.0	$X_{reg,dia}$	0.1
$X_{pro,dia}$	0.1	R_h	149.4
R_f	238.2	V_{ff}	326.2
V_{hf}	444.4	V_{fh}	303.6

5   Numerical Results

This section provides the numerical results for the analysis presented in Section 4. We assume that $\lambda_m = 0.1$, $c=5$, $K=5$. Also, x was randomly selected between 0 and K. Therefore, various handoff scenarios have been considered. Table 2 lists the parameter values used in the performance analysis.

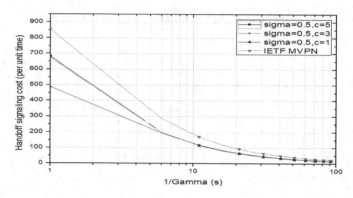

Fig. 8. Handoff signaling cost when MN moves from the intranet to a foreign network

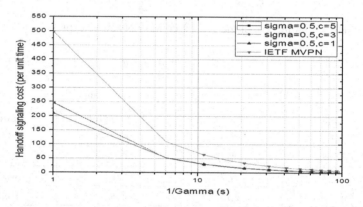

Fig. 9. Handoff signaling cost when MN moves from the foreign network to another foreign network

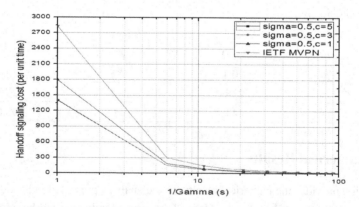

Fig. 10. Handoff signaling cost when MN moves from the foreign network to the intranet

Fig. 8 shows the signaling cost for the handoff from the Internet to the intranet. We assume that sigma=5, and vary c. Remember that c is the number of sessions in the MN. In proposed mechanism, an MN needs to perform SIP registration and SIP RE-INVITE for each session. Fig. 9 shows the signaling cost for the handoff from the intranet to the Internet. Fig. 10 depicts the results for the handoff from the internet back to the intranet. The discussion is similar to that in Fig. 9. Fig. 11 simply adds up the results in Fig. 8-10 and shows the average handoff cost with different handoff scenarios.

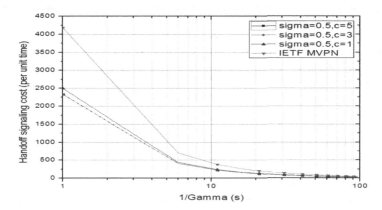

Fig.11. Handoff signaling cost for summation of Figs. 8-10

6   Conclusion

In this paper, we propose the ITS communication mechanism applying the method V2V NEMO and SIP-based MVPN for effective communication in ITS environments. Our proposed SIP-based MVPN reduces signaling cost and has many advantages in security aspects. In addition, our proposed scheme performs the mobility management applying Network Mobility (NEMO) for the communication between the vehicles. Finally, our performance analysis show that the ITS of SIP-based MVPN is significantly reducing the handoff signaling cost.

Acknowledgments. This research was supported by Next-Generation Information Computing Development Program through the National Research Foundation of Korea (NRF) funded by the Ministry of Science, ICT & Future Planning (No.2010-0020737) and Basic Science Research Program through the National Research Foundation of Korea (NRF-2010-0024695). Also, this research was supported by the Ministry of Trade, Industry and Energy (MOTIE), KOREA, and Korea National Industrial Convergence Center (KNICC) through the Special Education program for Industrial Convergence.

References

1. He, J., Zeng, Z., Li, Z.: Benefit Evaluation Framework of Intelligent Transportation Systems, vol. 10, pp. 81–87 (2010)
2. Liu, Z.-H., Chen, J.-C., Chen, T.-C.: Design and Analysis of SIP-Based Mobile VPN for Real-Time Applications. IEEE Transactions on Wireless Communications 8(11) (November 2009)
3. Rosenberg, J., Schulzrinne, H., Camarillo, G., Johnston, A., Peterson, J., Sparks, R., Handley, M., Schooler, E.: SIP: session initiation protocol, IETF RFC3261 (June 2002)
4. Schulzrinne, H., et al.: RTP: A Transport Protocol for Real-Time Applications. IETF RFC 3550, Standard 64 (July 2003)
5. Devarapalli, V., Wakikawa, R., Petrescu, A., Thubert, P.: Network Mobility (NEMO) Basic Support Protocol. IETF RFC 3963 (January 2005)
6. Lin, Y.-B.: Reducing location update cost in a PCS network. IEEE/ACM Trans. Networking 5(1), 25–33 (1997)
7. Vaarala, S., Klovning, E.: Mobile IPv4 Traversal Across IPsec-Based VPN Gateways. IETF RFC 5265 (June 2008)
8. Liu, Y.-W.: dynamic external Home Agent Assignment in Mobile VPN. In: Vehicular Technology Conference (2004)
9. Harkins, D., Carrel, D.: The Internet Key Exchange (IKE). IETF RFC 2409 (November 1998)
10. Calhoun, P.: Diamter Mobile IPv4 Application, RFC4004 (2005)
11. Xie, J., Akyildiz, I.F.: A novel distributed dynamic location management scheme for minimizing signaling costs in Mobile IP. IEEE Trans. Mobile Comput. 1(3), 163–175 (2002)
12. Ma, W., Fang, Y.: Dynamic hierarchical mobility management strategy for mobile IP networks. IEEE J. Select. Areas Commun. 22(4), 664–676 (2004)
13. Rummler, R., Chung, Y.W., Aghvami, A.H.: Modeling and analysis of an efficient multicast mechanism for UMTS. IEEE Trans. Veh. Technol. 54(1), 35–365 (2005)
14. Fu, S., Atiquzzaman, M., Ma, L., Lee, Y.-J.: Signaling cost and performance of SIGMA: a seamless handover scheme for data networks. Wireless Commun. and Mobile Computing 5(7), 825–845 (2005)
15. Chen, J.-C., Liu, Y.-W., Lin, L.-W.: Mobile virtual private networks with dynamic MIP home agent assignment. Wireless Commun. and Mobile Computing 6(5), 601–616 (2006)
16. Chen, J.-C., Liang, J.-C., Wang, S.-T., Pan, S.-Y., Chen, Y.-S., Chen, Y.-Y.: Fast handoff in mobile virtual private networks. In: Proc. IEEE International Symposium on a World of Wireless Mobile and Multimedia Networks (WoWMoM 2006), Buffalo, NY, pp. 548–552 (June 2006)
17. Dutta, A., Zhang, T., Madhani, S., Taniuchi, K., Fujimoto, K., Katsube, Y., Ohba, Y., Schulzrinne, H.: Secure universal mobility for wireless internet. ACM Mobile Computing and Commun. Review 9(3), 45–57 (2005)

The Study of the Controlling Resonant Wireless Power Transfer Using Bluetooth Communication

Yongju Park, Younghan Kim, Hyunseok Ahn, and Yongseok Lim

Wireless Convergence Platform Research Center
Korea Electronics Technology Institute
11. World cup buk-ro 54-gil, Mapo-gu, Seoul,
Republic of Korea
121-835
{suede8247,ekmyph,hsahn,busytom}@keti.re.kr

Abstract. Wireless power transfer technology has become a recent big issue for mobile phones industry, which can be classified into two types: inductive type and resonant type. Inductive type usually has higher efficiency but requires short distance between the transmitter and receiver. And resonant type has much better freedom from distance.

Wireless power transfer system combines with several major blocks including the transmitter circuits, receiver circuits, and communication block of both sides. Among the numerous resonant wireless power transfer mechanisms, the one Resonant Power Tranfer Consortium announce using 6.78MHz or 13.56MHz band for wireless power transfer.

The efficiency of the communication algorithm is also important since there are a lot of informations using bluetooth, NFC communication for wirelss power transfer. In this paper, we suggest a new communication algorithm ande circuit for wireless power transfer with simple structure and high efficiency.

Keywords: A4WP, Wireless power Transfer, Resonant transfer, Control communication algorithm.

1 Introduction of Wireless Power Transfer

In the early 20th century, Nikola Tesla suggested scientific concept for wireless power transfer technology using schemes to transport power wirelessly. However, this typical conceptual wireless power transfer coils involved undesirably large electric fields. [1], [2].

Wireless power transfer can be classified into two types: inductive type and resonant type. Inductive type usually has higher efficiency but requires short distance and precise alignment between the transmitter and receiver. From the viewpoint of convenience, resonant type has much better for the point of freedom from distance and alignment under a handicap of somewhat less efficiency. [3], [4].

The technology has developed rapidly in recent decade, and we use the enormous electronic devices such as laptops, cellular phones, robots, sensors, etc. As a consequence,

B. Murgante et al. (Eds.): ICCSA 2014, Part IV, LNCS 8582, pp. 779–789, 2014.
© Springer International Publishing Switzerland 2014

interest in distance wireless power technology has risen for the high technology. Radiate transfer is suitable perfectly for transferring power and information together by using wireless power transfer applications.

And this advance in wireless communication and power transfer technology has enabled a wide variety of portable consumer electronic, medical, and industrial devices. However, as portable devices shrink, power and communication line connectors become a larger fraction of system size. Wireless power should offer the possibilities of connection free electronic devices. Thus, there are many desires for people to use wireless power technology to eliminate the remaining any wired connection including power line and communication LAN line. [1], [2], [7].

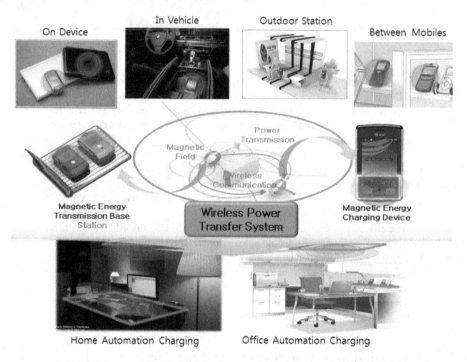

Fig. 1. Conceptual pictures of resonant type wireless power transfer system

2   General Techniques for Wireless Power Transmission

Radio frequency technology is already used as the principle of electromagnetic induction electric motor or transformer. As radio frequency technology developed, electromagnetic radiation such as radio waves or laser and a method for transmitting electrical energy was also tried. [1], [5].

We have already used electric toothbrush and some cordless shaver is charged using the principle of electromagnetic induction method. But electromagnetic induction method has a handicap of charging distance and alignment issue. Whereas the magnetic resonant wireless power transferring method is relatively free from a distance and alignment issue. Using antenna radiating method, a relatively small output is applied on but can achieve a distance of 10m or more, so it seems that satisfy desires of the application with free from distance and place anywhere and anytime.

Magnetic resonance wireless power transfer method is now on spotlight by power transferring for mobile devices because it makes people free from distance and alignment problems in wireless charging technology, whereas inductive power transfer suffer from. Wireless power transmission system is largely classified into 3 major blocks: self-resonant transmitting (TX) and receiving (RX) unit, the communication control unit, an antenna matching unit. Here, in this section, we introduce three kinds of units in detail first. [8], [9], [10]

2.1   Magnetic Resonance TX-RX System in Wireless Power Transmission

In brief, wireless power transfer system has alternated AC power from power source, and used the TX radiation antenna emitting AC power from the transmitter. And there is receiving antenna directly to the power supply DC power to convert a valid receiver.

In wireless power transmitter, first we change AC power supply into DC power for changing the frequency what we want for the effective irradiative power transmission. After converting around 1V DC current, we use Oscillator for making AC signal with a frequency that we want to convert to. In oscillator, we changes spec or properties of circuit, passive device including inductors and capacitors for achieving desired frequency that we want to make. Converted AC signal with appropriate frequency and around 1V passes through the power amp and signal is converted into a large signal suitable for transmission. Converted signal is radiated while passing through the TX antenna and TX-RX resonance antenna pair should be matched for achieving high quality Q factor.

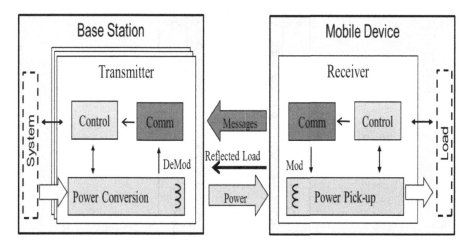

Fig. 2. Block diagram of Magnetic resonance wireless power transmission system Implementation proposed by WPC (Wireless Power Consortium) [6]

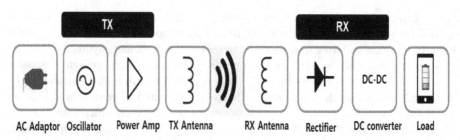

Fig. 3. Block diagram of components between wireless power transmission and receiving system

In resonance status, High-frequency AC electric power transmission behaves like conduction current particularly followed Ohm's characteristics like line current. A current flowing between the transmitter and the receiver appears in the same direction.

In wireless power receiver, ac signal which is received via the antenna is converted to a dc signal in rectifier circuit. The converted signal while passing through DC DC converter is converted appropriate size of DC signal suitable for the load stage. Device efficiency of DC-DC converter is outstandingly important technology for the wireless power transfer system.

2.2   Wireless Power Transmission Control Communication Unit

A wireless power transmission control communication can be divided into four steps. First is a Selection step, and we detect and decide devices for power transfer by using particular way of several communication protocols. After selection, in the Ping-Pong

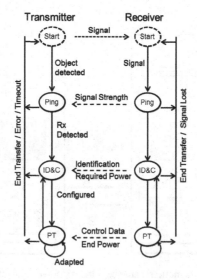

Fig. 4. Communication control algorithm be proposed in WPC (Wireless Power Consortium) [6]

step, TX and RX side send their information packet such as their load current, voltage, impedance to notify each other. And in third Notifying step as the, TX and RX transmit and receive a unique ID and the extension information to adjust wireless power transfer configure including dynamic current, voltage and power information. Finally, the power transmission jumped into step of Power Transfer stage and TX-RX system should be continuously transmitted and received wireless power and dynamic phase information such as their current, voltage and power level.

RX unit transmit the load modulated control information toward TX continuously by sending control instructions of wireless power transfer level. TX unit receive control information from reflected load of demodulation via the message from the RX unit and control supplying the power requested by the load.

By calculating the difference between the transferred power from TX side and the requested power of the RX side, and TX changes power transfer level by adopting differences calculated. And TX unit calculates and generates the newly current applied to compensate for differences by measuring the current to be applied. An adaptive control algorithm, the control parameters would determine the operating point of wireless power transfer system.

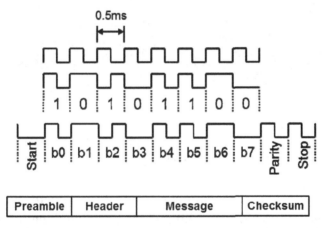

Fig. 5. Communication packet structure of the wireless power transfer system be proposed in WPC (Wireless Power Consortium) [6]

TX and RX terminal emit the communication signal continuously each other with according of an appointment to a packet of ID information and control information. For the battery management in wireless power transfer, communication between the TX-RX units becomes very important. So, I suggest the communication packet and architecture of the packet largely consists of a preamble, a header, the message and a checksum.

First, the preamble is disposed at the front of the packet indicates the start of the communication signal. And the header contained communication signal in the packet header indicates the category, contents about wireless power transfer etc. Message contains real control communication information and the length and category of contents are defined in the header already. Finally, the checksum a tool that can be determined an error of the current command.

2.3   Antenna and Matching Theory

In air space between the transmitter and the receiver, oscillating electromagnetic field behaves like virtual conduction properties shown in Fig6 (b). Resonant antenna produce amplified magnetic field H which is greater than the electromagnetic field, all the electric field vector is started at a point in three-dimensional space to spread out symmetrically spherical in the ideal case.

In wireless power transmission system, transmitting primary coil current is induced by the magnetic field, and the magnetic field induced in the secondary receiver make a wireless power transfer between TX and RX.

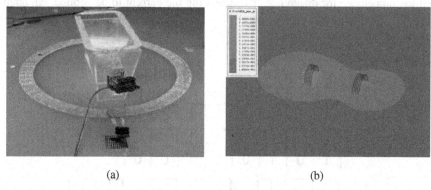

(a)   (b)

Fig. 6. (a) Resonant coupling TX-RX antennas and (b) HFSS 3D magnetic calculation simulation results between TX-RX antennas with resonant coupling

3   Wireless Power Transmission Control Implementation between Transmitter - Mobile Using Bluetooth communication

So far, we introduce briefly about components of the wireless power transmission circuit including TX and RX unit, communication unit and the antenna which is described. In this chapter, we implement algorithm proposed communication to control wireless power transfer between wireless power transmission control unit (TX) and mobile phone (RX) using the Bluetooth communication.

3.1   Overview of Bluetooth communication

Bluetooth communication in the 2.4GHz region as a single method of communication can be built smaller and lighter than other communication method, so it is easily adapted in cellular phones which is featured by low prices, low power consumption by appropriate communication method for under 100m distance.

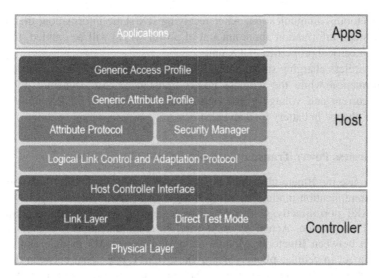

Fig. 7. Bluetooth architecture of communication layer

Time (us)	Master Tx	Radio Active (us)	Slave Tx
0		176	ADV_DIRECT_IND
326	CONNECT_REQ	352	
1928	Empty Packet	80	
2158		144	Attribute Protocol Handle Value Indication
2452	Empty Packet (Acknowledgement)	80	
2682		96	LL_TERMINATE_IND
2928	Empty Packet (Acknowledgement)	80	

Fig. 8. Timing diagram of Bluetooth communication

The client (TX unit) generate advertisement being continuously for detecting other side device (RX unit) using the Bluetooth communication. If there is a response from RX device (load fluctuation, or replying communication format), and then client transmit the amount power can wake up Bluetooth chip to make a connection between TX-RX for starting the wireless power transfer.

We propose communication protocol and information packet for controlling wireless power transfer. After waking RX up by big signal from TX, RX sends its own static information including load impedance, frequency of RX antenna and so on by using Bluetooth communication. After receiving RX static information, then the TX sends its static information about amount of power which will be emitted, frequency and other electrical properties of TX unit.

After wireless charging starts, both TX-RX units send the dynamic electrical property information while the wireless power transferring last. Dynamic information includes current and voltage at the impedance load of both TX-RX units and the amount of change in battery load and etc.

3.2   Wireless Power Transfer Control Circuit Configuration

In order to use the Bluetooth communication, we connect Cortex M3 board and Bluetooth communication module by using UART serial communication port with Bluetooth UART port directly. Cortex M3 module send a control communication signal via UART connection. Actual physical wireless communication is implemented by connection between Bluetooth Module connected Cortex M3 module and the Bluetooth chip set in cellular phone. Again, via a UART RX signal between devices is provided in the return value, and Cortex M3 module board show configuration information in the LCD of UI format.

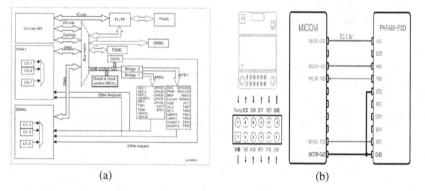

(a)   (b)

Fig. 9. (a) Cortex M3 structure for controlling Bluetooth and (b) Bluetooth chip pin definition and connection between FPGA and Bluetooth module

During transmitting and receiving information between TX-RX Bluetooth modules, we use communication protocol and packet based on the algorithm which we suggested. Initially Cortex M3 module board acting as the TX-end continues to send the advertisement signal nearby Bluetooth devices. When TX notice change in load impedance of RX, wireless power transmission is found to be relatively encapsulated large power sent to wake up Bluetooth chipset in cellular phone, if RX device support wireless power transfer.

Waking RX up by the TX by advertisement signal of Bluetooth chipset of wireless charging process, cellular phone sends its static information to the TX about availability permit of wireless power transfer and the possibility of connection between TX and RX. After receiving information from the cellular phone, TX confirms the correct connection between TX and RX and provides information of TX electrical properties to the RX device.

After RX finally connected to the wireless charging system and both TX and RX continuously exchange dynamic changing information about load of current, voltage, charging power requested in RX device, frequency change etc. And TX-RX control the amount transferring power and frequency based on control information by Bluetooth communication.

3.3   Implementation of Wireless Power Transfer Control Communication

We proposed communication protocol to control magnetic resonance wireless power transmission power this paper. First, TX looks for RX device which supports wireless power transfer. And transmitting and receiving a communication signals based on pre-determined protocol and packet and wireless power starts between TX and RX.

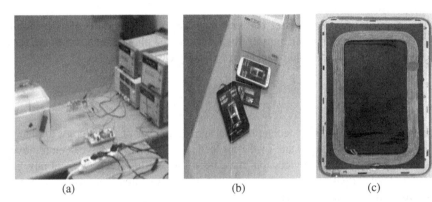

(a)	(b)	(c)

Fig. 10. (a) The picture of wireless power charging for cellular phone using 13.56MHz frequency, (b) power charging state and (c) 13.56MHz frequency receiver antenna

We implemented the wireless power transfer system by using TX control module and RX cellular phone together. The power transfer module is composed of a DC-DC converter, DC-AC inverter, primary coil, series resonant capacitor which is controlled by the voltage and frequency. A TX side consists of a power control circuit via V/I sensing output power and an algorithm which controls antenna matching system.

The RX side composed of a high efficiency rectifier circuit, DC-DC conversion circuit. We implemented 13.56MHz matched antenna and regulator module at the RX side rectifying the voltage suited for portable IT equipment in a constant voltage level. After detecting the acquisition module of the presence or absence of the load, TX-RX exchange control information about the power consumption of the battery, electrical properties and etc.

Cortex M3 MCU module connected the TX and RX Bluetooth module for achieving TX-RX state and properties during the wireless power transfer. Current sensing circuitry sense the level of battery charge, and continuous charging operation is required until charging the battery fully. The fig 11 shows a wireless communication

(a)   (b)   (c)

Fig. 11. (a) Photograph of the proposed set of control device and cellular phone for wireless power transfer, (b) received RX information from Cellular phone and (c) receiver TX information from Cortex M3 control circuit

system controlling power transfer between the transmission module and the mobile device. During wireless power transfer, the LCD and the phone UI shown in the form of power transmission, it could be identified visually easily to the user.

4   Conclusion4

Wireless power transmission technology as the big impact of technology to change the paradigm of a very large state-of-the-art technology. All wires have been removed in all areas of electronics devices that have been set position freely without any line and cable. If electrical energy is delivered and used in the area, you can move freely in everyday life so far, there will be other changes.

If wireless power transmission technology expanded in the industry or office area, people can use or move electric device freely at a little more effort without any power or communication line. It leads economic and industrial changes in terms of assuming a new revolutionary would be expected. If wireless power transmission technology expanded all over the Korea, I believe IT industry can create new industries and technologies evolution so far.

Acknowledgments. This work was supported by the IT R&D Program of MKE/KEIT (10045400, HF-Band Magnetic Resonant Wireless Charging SoC for Mobile Device).

References

1. Brown, W.: The history of power transmission by radio waves. IEEE Trans. Microw. Theory Tech. MTT-32(9), 1230–1242 (1984)
2. Sample, A., Yeager, D., Powledge, P., Mamishev, A., Smith, J.: Design of an RFID-based battery-free programmable sensing platform. IEEE Trans. Instrum. Meas. 57(11), 2608–2615 (2008)

3. Karalis, A., Joannopoulos, J., Soljacic, M.: Efficient wireless non-radiative mid-range energy transfer. Ann. Phys. 323(1), 34–48 (2008), January Special Issue 2008
4. Kurs, A., Karalis, A., Moffatt, R., Joannopoulos, J.D., Fisher, P., Soljacic, M.: Wireless power transfer via strongly coupled magnetic resonances. Science 317(5834), 83–86 (2007), Michalewicz, Z.: Genetic Algorithms + Data Structures = Evolution Programs, 3rd edn. Springer, Heidelberg (1996)
5. Sekitani, T., Takamiya, M., Noguchi, Y., Nakano, S., Kato, Y., Hizu, K., Kawaguchi, H., Sakurai, T., Someya, T.: A large-area wireless power-transmission sheet using printedorganic transistors and plastic MEMS switches. Nature Materials, 413–417 (2007)
6. Wireless Power Consortium, http://www.wirelesspowerconsortium.com
7. Karalis, A., Joannopoulos, J., Soljacic, M.: Efficient wireless non-radiative mid-range energy transfer. Ann. Phys. 323(1), 34–48 (2008), January Special Issue 2008
8. McSpadden, J., Mankins, J.: Space solar power programs and microwave wireless power transmission technology. IEEE Microw. Mag. 3(4), 46–57 (2002)
9. Cannon, B., Hoburg, J., Stancil, D., Goldstein, S.: Magnetic resonant coupling as a potential means for wireless power transfer to multiple small receivers. IEEE Trans. Power Electron. 24(7), 1819–1825 (2009)
10. Kim, Y.-H., Kang, S.-Y., Lee, M.-L., Yu, B.-G., Zyung, T.: Optimization of wireless power transmission through resonant coupling. In: Proc. CPE, pp. 426–431 (May 2009)

Wireless Sensor Networks in Next Generation Communication Infrastructure: Vision and Challenges

Saad Bin Qaisar[1], Salman Ali[1,*], and Emad A. Felemban[2]

[1]School of Electrical Engineering and Computer Science,
National University of Sciences and Technology, Islamabad, Pakistan
{saad.qaisar,salman.ali}@seecs.edu.pk
[2]College of Computer and Information System,
Umm-ul-Qura University, Mecca, Kingdom of Saudi Arabia
eafelemban@uqu.edu.sa

Abstract. Last decade saw the development of Wireless Sensor Networks with multitude of applications built around the sensors. Though most of the issues at protocol and device level remain solved for Wireless Sensor Networks, there is a growing trend in integration of sensors and sensor based systems with Cyber Physical Systems, Machine-to-Machine and Device-to-Device communication both in infrastructure and ad-hoc modes. The emergence of Cloud Computing, highly intelligent 'Smart' devices and efforts towards 5th Generation telecom systems is expected to lead the revolution of a networked world with increased demand for situational awareness leading proliferation of sensors at the edge of physical world. Mounting intelligence built in 'smart' devices, scarcity of frequency spectrum, disruptive technologies and limitations on further reduction in base station size may render base station centric-architecture of today's wireless networks old-fashioned. Due to low data rate bursty nature of sensor data, massive-MIMO antennas, device-centric architectures, smarter devices and native support for Machine-to-Machine and Device-to-Device communication, sensor data originating from these systems and disruptive technologies is expected to grow manifold through passing years. Communication and network engineers face an emerging challenge of designing communication and network protocols that can efficiently get integrated with the emerging wireless cellular and computing paradigms. In this paper, we list down these communication and networking challenges and provide our vision for Wireless Sensor Networks related platforms and enabling technologies of next decade.

Keywords: Wireless sensor network, cyber physical system, cloud computing, next generation network, seamless integration.

1 Introduction

Over the past two decades, wireless communication and networking has seen rapid advancement with introduction of new standards and protocol that provision short

---

* Corresponding author.

B. Murgante et al. (Eds.): ICCSA 2014, Part IV, LNCS 8582, pp. 790–803, 2014.
© Springer International Publishing Switzerland 2014

range personal communications to large geographical scaled monitoring applications. Significant contributions have been made to protocols related to Wireless Sensor Network (WSN) for applications involving real time as well as buffered data. WSNs provide low-cost, low power, multifunctional miniature devices, with multiple sensory interfaces that gather environment related information (Fig 1) and communicate it in an untethered manner over short distances [1]. Various WSN standards have evolved over the years according to the changes and requirements for sensory environments at different times [2]. More recently, with maturity in WSN and related wireless standards, researchers have started looking into leveraging WSN functionality with more control, interaction and reliability. In terms of interaction between real and virtual environments, Cyber Physical Systems (CPS) and Cloud Computing concept and method has been greatly discussed and included. CPS provides an intuitive mechanism for human-to-human, human-to-object and object-to-object interactions. Quite recently, WSN and similar wireless applications have started to investigate cloud aspects of the system for expanded network connectivity and finer integration. By definition, CPS and related futuristic network technologies provide a system of collaborating and interacting elements with physical input and output intended to provide full remote control giving a feeling of virtual environment instead of a sole standalone system [3] [10].

Currently, CPS and Cloud Computing for WSN have been discussed separately for their features, requirements and challenges while only a vague distinction exists to identify the overlapping areas where both can be seamlessly merged. With this perspective, there is a need to highlight, and emphasize with clear distinction as to where the communication layers for WSN and other network related platforms merge and how the challenges in making the widely distinct operation and functionality be made compatible. The possibility for integration of CPS and Cloud with WSN has also been made somewhat possible with accelerated maturity in wireless technology and embedded computing with applications like micro sensing MEMS, inertial motion detection, bio-signal sensing, environment parameter sensing, location and vehicular movement detection. While a common platform is being sought for merging and connecting common parameters, the main technological distinctions need to be made clear. WSN has been designed and implemented majorly with the perspective of communicating sensing related data with coordination over some limited geographical and a characteristic environment. CPS, on the other hand, utilizes a broader definition and dimension of sensing data over multiple networks with a Cloud specific link to the internet with the aim of providing elevated controlled intelligence [4].

A vast amalgamation of basic sensors has now been connected to a network that has global perspectives. This has thus created a phenomenon where information revolution and an explosion in the expertise to create, store and provide data mining to digital information from these sensors has come about. Wireless Sensor Networks has therefore moved from an early research topic to the pinnacle of sensor application development and deployment. Cloud Computing is the main principle to provide global perspective to the sensor data and is designed and promoted with the sense of being data centric with efficient interaction with the outside world. Though WSNs are designed to collect data in the real world, yet, issues still persist as to provide an automated method to discard or reduce the actually required information that can be analyzed at a later stage. An efficient aggregation method needs to be accompanied

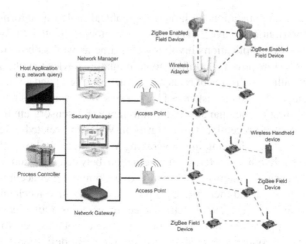

Fig. 1. A typical WSN architecture with network elements

with the Cloud oriented facility to cater this aspect [25]. Keeping this as the main focal area of work, we link real world requirements and industrial standards to provide a likely picture of where the WSN and next generation networks merger would be seen in near future and directions as to how it would be met.

2   Disruptive Network Technologies and Wireless Sensor Network

The use of wireless communication technologies, particularly environment monitoring applications has become ubiquitous due to the freedom of distributed capabilities, and cost savings the technology offers [3]. WSNs have therefore emerged as the next wave of penetrating wireless technology, enabling greatly distributed efficient measurements methods across vast physical systems and environments. With WSNs, one can effectively analyze and examine everything from infrastructure health and forests to the health and safety of living beings. The ability of WSN to allow use of spatially distributed measurement devices that utilize sensors to monitor physical or environmental conditions has provided an open platform wherein rapid applications and implementations are emerging [26]. In addition to many wireless measurement nodes, the WSN system may include several other nodes like gateway, relay and aggregators that collects data and provides connectivity back to a remotely hosted application on a Cloud or some monitoring station or even an automated embedded controller [27].

2.1   Wireless Sensor Networks with OpenFlow Technology

To enable implementation of vast protocols and emerging applications on the basic sensor platform, the OpenFlow technology concept has created another intense research dimension. The OpenFlow platform can basically provision splitting the traffic path into a data packet that is maintained by an underlying router or switch and the other being the control packet that would be maintained by a controller or control server itself. The controller would turn the physical sensor device into a much simpler one instead of a complicated mode since major complex intelligence related to basic programs and system level codes are removed. So, the platform becomes easier for the researchers to investigate and examine under different experimental traffic, various protocols and overall network would be simplified as well as being more advanced in terms of application support. The use of open platforms with WSN is the current need and it is much expected that in near future these research will lead to a momentous achievement in sensor network modernization [1]. The OpenFlow method can be very useful in this regard since from attainable throughput and feasible result viewpoint for sensor network and other platform integration, the amount of traffic and header definition rate is reasonably low and almost insignificant [11]. A FlowVisor device has been defined in the architecture that is basically a software-defined networking controller that enables network virtualization by slicing a physical network into multiple logical networks (Fig 2).

Fig. 2.   A typical WSN implementation with OpenFlow technology

OpenFlow for sensor networks is one of the most debated topics now a day since none has ever tried this protocol extensively on wireless sensor networks till date. OpenFlow based sensor network, is generally claimed to be much more reliable and flexible in comparison to typical sensor node since data packets, control packets, data route and even the sensor nodes themselves can be easily experiential, synchronized and routed whenever required. The system can be considered a fully centralized system from physical layer viewpoint but a distribution of services related to the sensing information are still maintained. A central system can be used to monitor and control all sorts of sensor traffics. This can help to achieve a better data rate, bandwidth efficiency, reliability, robust data routing. Ultimately this will lead to a better QoS provision for the monitored environment. But presently the sensor devices are mainly vendor oriented and individual software applications need to be maintained that present a lack of service coordination [11] [12]. By use of

OpenFlow inside sensor network domain allow can start and stop of traffic whenever required through a central coordination system.

2.2   Consistent WSN Application Development

Consistency of WSN application is one of the most requisite factors and difficult as well to achieve for these sensor nodes due to their short resources and communicational abilities. In the current innovations, the management and maintenance of the unstructured sensor devices with connection to the undetermined and uncertain number of wireless sensor nodes and their establishment in harsh and tough terrains and areas has provided a major implementation challenge. The sensor network can be dynamically self-structured and self-constructed and the nodes within the span of the network must possess the ability to deliver and maintain mesh connectivity at all times [13] [14]. As seen from a message transfer view point, hop to hop, node to node or end to end reliability issues and challenges are still unsolved. Several innovations have tried to work with reliability limits or to obtain partial reliability. It is noted that there is always a compromise with the reliability and total reliability has been proven to be almost impossible for all major wireless networks. So, it's beneficial to use various platforms for transporting sensor flows where traffic can be remotely controlled and sensor nodes can be monitored remotely. Currently different protocol changes and applications for sensor network are being experimented and supported by several major switch/router vendors that provide a set of functions which may be common at some point but support all sort of OSI layer headers. These technologies including simulators for senor network can provide a single platform for typical TCP/IP implementation and cross layer experiments. It will also be able to allow integration of the circuit and packet switching related technology and applications. Further these applications can be treated separately too. The core sensor network also gains noteworthy benefits due to control, management related policies and cost benefits including much needed energy effectiveness and overall better network performances and data gathering and aggregating capability.

3   Architectures for Future Sensor Network Deployment

The structured WSN deployment simplifies the routing protocols and increases efficiency and cost-effectiveness of the network. A hierarchical structure of monitoring case based on WSN allows the basic functionalities to be distributed among the nodes [22]. The nodes at the lowest level are the sensing nodes that sense the parameters from the environment Relay nodes closer to sensing nodes collect sensed data and pass it on to data dissemination nodes which finally transmit the data over long haul communication link to the control center (Fig 3). The advantage of such a topology is that it adds redundancy to the whole architecture while reducing the range of each node that ultimately induces low energy consumption. The functionalities of nodes on each level can be made distinctively diverse so as to make the data collecting system more intelligent. The CPS architecture resembles traditional embedded systems that aim to integrate abstract computations with physical processes. Contrary to traditional embedded systems, CPS provides an interconnected interactive with output and input that pertain physical existence and are standalone devices. The main layers of CPS are the virtual layer and physical layer. For the physical layer, an intelligently deployed network of actuators and sensors collect information and actually control the physical world. By converting the analog information into a digital format, the information is sent to a virtual layer input which serves as the decision making setup. This information is further used to calculate abstract computations that feed into the real world actuation system to drive and control physical world output.

In contrast to CPS, the WSN architecture focuses more on node design and inter-node communication and networking. In WSN, the node converts measuring metrics from various environment monitoring sensors related to physical, biomass and chemical parameters into digital information to be read and inferred by a remote monitoring facility. By application perspective, we can classify important hardware resources in WSN node as the Sensing Unit, Processing Element, Transceiver Device and Power Manager. The communication and networking for WSN needs to handle joining and leaving of hundreds of nodes while providing scalability. Current research involving data transportation adds re-configurability to nodes and manipulating the hardware during run-time and designing nodes such that they consume extremely low power.

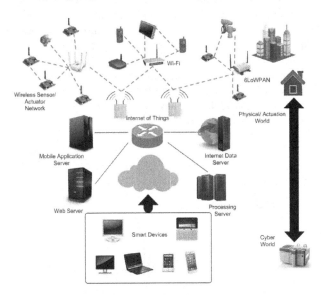

Fig. 3. WSN architecture integration with different platforms

3.1   A Virtual Sensor Infrastructure

Cloud computing services for wireless networks can provide virtual servers. Users can use the remote servers with no concern as to where the server is located and how much resource specifications it can provide [15]. The cloud computing infrastructure for sensor networks can provision virtualization for multiple physical sensors as 'virtual sensors' (Fig 4). For examples if there are multiple temperature sensors for building monitoring application, a set of sensors could be defined as virtual sensors for each floor providing a direct map for the monitoring area division. In a sensor cloud application, the users or nodes should be able to control virtual sensors with standard functions. Dynamically grouped virtual sensors can be used to provision automatic responses to different requests. The user of the sensor application should also be able to destroy the virtual sensor setting once it is no longer required, hence it should be provisioned with flexible formation and removal.

Monitoring virtual sensors would elevate the quality of service for wireless monitoring applications. The sensor cloud infrastructure needs to have a user interface either at the site or through some remote setup whereby the adding or deleting of physical sensor the cloud can be performed. Other functions for virtual sensors should include request, controlling, monitoring and registering different allowed functions for the specific sensors or sensor nodes.

For designing the virtual sensor system, a platform is required wherein the functionality of the virtualization is defined encompassing relationship between sensor groups, virtual sensors and physical sensors. Different types of sensors would have different specifications, so in order to create a virtual environment with different sensors, standardization method would be required. This also accounts for the automation process for the sensor network relating to hardware as well as software operation. A continuous monitoring system should be deployed that would confirm the division between the virtual and physical systems. Grouping sensors can cause several challenges, for example when setting the frequency of reporting from a group of virtual sensors, different sensor nodes would have different capabilities and ma not respond evenly, hence a mechanism for controlling the commands and its manipulation in a group is required. Defining a service model for the sensor cloud can provide different sensors as a service for a particular application. Finally the division between sensor owners and cloud administrator is required.

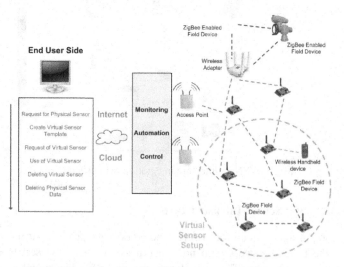

Fig. 4. A virtual sensor cloud infrastructure example

4   Future Wireless Sensor Network Applications

For monitoring remote WSN applications over the IP framework, cloud computing can provide a middleware cost effective solution to both domains that provisions a rich interactive communication platform. Since network communication costs a lot of bandwidth overhead for linking VMs in data intensive environments, a decentralized approach, where migration of VM services is provided with monitoring of traffic fingerprints can relieve the wasted overhead. Also, in particular cases, faults can occur in the middle of a query from distributed databases. This can be fixed by dividing queries into subqueries and mapping them in an intelligent way such that the results return on different nodes. In CPS, location for different data generating and terminating points serve as a first class knowledge for many applications. As compared to outdoor location detection through GPS and similar approaches, indoor location estimation or localization proves more challenging [5]. Proposed systems relate to smaller scale environments while to cope up with newer demands of extended scaled up systems, pattern matching of data can be applied as a useful solution.

CPS provides a bridge to link the cyber world with communication, intelligence and information components and the physical world counterpart providing sensing and actuation capabilities [6]. The CPS platform may be broadly classified as an integration of intelligent control design system with the mobile or static sensor or actuator system [23]. When considering individual WSN networks, issues like network formation, security, mobility and power management remain almost the same on a broader perspective [16]. However, major advanced technical differences from the WSN approach include the use of heterogeneous information flow, multi dimensional sensor cooperation and high level of intelligence and algorithm behind the actuation and decision framework [24].

From the applications point of view, CPS holds a wide range of useful features that can be used to provide elevated services to users with a wide range of implementations. For example, CPS can be used in cooperation with WSN to assist in management of greenhouse sensing information at large geographical distances. More complex system would include multiple sensors and actuators that can be used for applications such as environment related climate control settings with humidity, heating, carbon dioxide generation, fertilizing and watering system features.

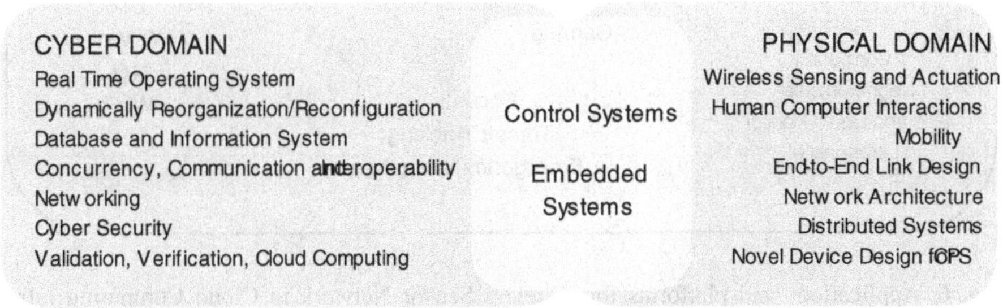

CYBER DOMAIN		PHYSICAL DOMAIN
Real Time Operating System		Wireless Sensing and Actuation
Dynamically Reorganization/Reconfiguration	Control Systems	Human Computer Interactions
Database and Information System		Mobility
Concurrency, Communication and Interoperability	Embedded	End-to-End Link Design
Networking	Systems	Network Architecture
Cyber Security		Distributed Systems
Validation, Verification, Cloud Computing		Novel Device Design for CPS

Fig. 5. Overlapping areas for cyber and physical worlds in sensing platform

4.1   Device Specifics for Future Sensor Networks

Sensing devices in the WSN are basically organized and placed with a data processing unit and communicational abilities that are required to measure the specific parameters from the surroundings and convert those parameters into relevant analyzable form. These kinds of sensing devices can significantly raise the effectiveness of both the environment conditions and general information data base for inspection, safety, and failure management. Also it is useful where usual and normal data access attempt have proved to be very expensive and uncertain. Unstructured wireless sensors also have the capabilities to observe and collect a large amount of environmental conditions such as sound, pressure, motion and temperature etc. These sensor nodes consume resources such as a small bandwidth, processing power, communication range, memory and capability. The processing of signals and communication activities are felt to be the significant factors to improve the data relay capabilities and reliability turns better thereby [18].

For monitoring remote WSN applications over the IP framework, cloud computing can provide a middleware cost effective solution to both domains that provisions a rich interactive communication platform. Since network communication costs a lot of bandwidth overhead for linking VMs in data intensive environments, a decentralized approach, where migration of VM services is provided with monitoring of traffic fingerprints can relieve the wasted overhead. Also, in particular cases, faults can occur in the middle of a query from distributed databases. This can be fixed by dividing queries into subqueries and mapping them in an intelligent way such that the results return on different nodes. In CPS, location for different data generating and terminating points serve as a first class knowledge for many applications. As compared to outdoor

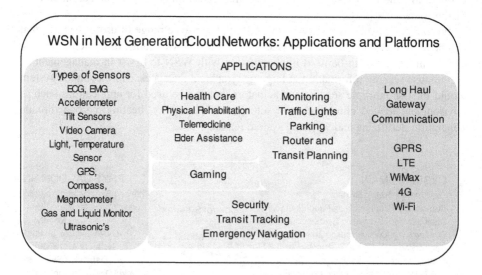

Fig. 6. Applications and platforms for Wireless Sensor Network in Cloud Computing infrastructure

location detection through GPS and similar approaches, indoor location estimation or localization proves more challenging. Proposed systems relate to smaller scale environments while to cope up with newer demands of extended scaled up systems, pattern matching of data can be applied as a useful solution. CPS for WSN provides a bridge to link the cyber world with communication, intelligence and information components and the physical world counterpart providing sensing and actuation capabilities. The CPS platform may be broadly classified as an integration of intelligent control design system with the mobile or static sensor or actuator system. When considering individual WSN networks, issues like network formation, security, mobility and power management remain almost the same on a broader perspective. However, major advanced technical differences from the WSN approach include the use of heterogeneous information flow, multi dimensional sensor cooperation and high level of intelligence and algorithm behind the actuation and decision framework [17] [19] [20].

5   Issues and Challenges

The massive deployment of WSNs in future platforms is expected to increase exponentially in the next few years, thus allowing millions of wireless devices to work autonomously for newer applications and platforms. These next-generation WSNs are also expected to interact with other devices such as Radio Frequency tags, home appliances, mobile equipment and automobiles [28]. When integrated with the cloud and similar platforms, these sensor networks could provide ubiquitous and pervasive services to the users by providing unlimited and powerful storage infrastructure. The current challenges for next-generation WSNs are scalability, resource scarcity, heterogeneity, decentralization and dynamicity. These challenges and requirements cannot be fulfilled by traditional WSNs and related approaches. In addition, the Network, Medium Access Control (MAC), and Physical (PHY) layer protocols developed for traditional WSN platforms [7] are not applicable to the next-generation WSNs, where thousands of battery-powered nodes need to be operated for longer durations. Said otherwise, the next-generation WSNs require the development of novel and innovative protocols at each layer that must be able to extend the network lifetime from months to several years. These protocols may allow seamless integration of next-generation WSNs and similar platforms with other networks and platforms including internet of things and cloud computing [29].

5.1   Data Reliability

Data consistency and reliability becomes much complex when WSN is viewed from a CPS perspective [21]. Major reliability design requirements for implementing a well planned CPS include 1) Service oriented architecture (SOA) 2) QoS aware communication and networking 3) intelligent resource management and 4) QoS aware power management at different networking levels. Service oriented architecture plays an important role in reducing the complexity of the overall infrastructure by decomposing CPS functions into smaller distinguishable units each viewed as a separate service. This allows rapid, efficient and scalable development of a CPS application through reusable service units. Application level QoS however needs to be defined for intelligent cross layered communication while in future, linking WSN with CPS environments would greatly require dynamic system settings for unpredictable environment. Self management policies would be needed so that allocated resources like CPU time, bandwidth memory, energy profiles to be controlled intelligently in a high level QoS constrained setting. Major concern with high level reliability provisioning always pertains to the minimization of energy consumption for major network elements. In this regard, cloud computing provides a possible solution wherein two major enabling technologies for this setup are virtualization and ubiquitous connectivity. Virtualization related technological methods allow provision of dynamically changing and altering resources based upon service isolation thus enabling scaling and managing of resources in more controlled way. With the addition of different payment levels for service variations and on-the-go routine, the cloud computing services can be classified under (1) Infrastructure as a Service (IaaS) (2) Platform as a Service (PaaS) (3) Software as a Service (SaaS). IaaS generally provides isolation between lower layers generally termed 'real machines' for use in the user oriented cloud infrastructure.

5.2   Web Enablement for the Cloud

Use of web access approach enables the integration of different networks into the internet with provision of user friendliness and platform independence. Enablement of the web for sensor network applications can be supported by HTTP over either TCP or UDP. Despite having overhead related to congestion control and flow control, HTTP/TCP is more suitable sine it provides interoperability. For HTTP/UDP however, there would be a requirement of an additional component at the sensor network border or gateway that will play the role of a TCP-UDP translator. Though the gateway can perform the role of a translator, the protocol differences may cause problems such as inconsistency, redundant transmission and delays. The draft of the IETF CoRE WG proposes to use HTTP over UDP where connection related response and requests can be packed into a single UDP connection and packets may be delivered using multiple datagram's. The condition associated with use of multiple datagrams is the support for fragmentation and reassembly at the end connection side. Hence the system might not be suitable for sensor networks that provide rich content delivery and multimedia data. This approach will also cause unnecessary retransmissions because a renewal command is sent at the higher level application, the backend languages like CSS, HTML and Javascript need to perform compatible actions. Hence a design choice needs to be decided for enabling effective web enablement solution for sensor networks reporting to remote locations and how future networks would be integrated smoothly into the IPv6 system.

5.3   Quality of Network Service

Prediction of accurate output decisions and reliability of sensing information are considered critical for CPS systems. These factors also form the Quality of Service (QoS) basis for achieving a real time intelligent system for high stress and constrained environments like mining, healthcare and warfare [8]. The real challenge lies in maintaining QoS factors like seamless flows and timely delivery and when CPS is integrated with other technologies like semantic agents and hybrid states. Deployment of CPS architectural parts require placement of sensing and actuator devices at strategically critical points with intelligent algorithms for node localization and geo-location detection. The Medium access control (MAC) should cater that the negotiation between neighboring data collection devices and sensors must conserve resources like bandwidth, number of channels, buffer storage and transmission energy [9].

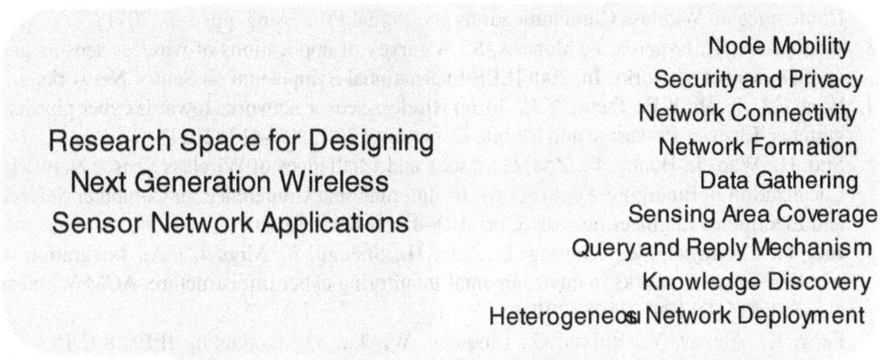

Research Space for Designing
Next Generation Wireless
Sensor Network Applications

Node Mobility
Security and Privacy
Network Connectivity
Network Formation
Data Gathering
Sensing Area Coverage
Query and Reply Mechanism
Knowledge Discovery
Heterogeneosi Network Deployment

Fig. 7.   Research Space for designing future Wireless Sensor Network applications

6   Conclusion

While the domain of WSN focuses more on the designs for sensing, data-retrieving, event-handling, communication, and coverage problems, CPS community focuses more on the development of cross-layered and cross domain intelligence from multiple WSNs and the interactions between the virtual world and the physical world. A CPS application may provide a bridge between multiple remote WSNs and invoke actuation based on inference from the sensed information. A lot of successful vehicle- and mobile phone-based CPS services have been developed over time. Data from such applications may be expected to be of continuous form at a very large volume, so storing, processing, and then intelligent interpreting of it in real-time manner is essential. Important factors to the success of CPS include management of cross-domain sensor related data, embedded and mobile sensing technologies and applications, elastic computing and storage related technologies with integrated privacy and security designs. We have also reviewed different platforms for environment monitoring, navigation and rescue services, ITS, social networking and gaming with related challenges in these systems. The specifics of future WSN platforms are expected to stimulate an interested reader with current CPS technological development and expected features of future enabled WSN platforms.

Acknowledgments. The authors would like to acknowledge research support by King Abdul Aziz City for Science and Technology (KACST) Saudi Arabia grants: NPST-11-INF1688-10 & NPST-10-ELE1238-10 and National ICTRDF Pakistan grant SAHSE-11.

References

1. Akyildiz, I.F., Su, W., Sankarasubramaniam, Y., Cayirci, E.: Wireless Sensor Networks: A Survey. IEEE Computer Journal 38(4), 393–422 (2002)
2. Shi, J., Wan, J., Yan, H., Suo, H.: A Survey of Cyber-Physical Systems. In: International Conference on Wireless Communications and Signal Processing, pp. 1–6 (2011)
3. Arampatzis, T., Lygeros, J., Manesis, S.: A survey of applications of wireless sensors and wireless sensor networks. In: 20th IEEE International Symposium on Sensor Networks
4. Wua, F.J., Koab, Y.F., Tseng, Y.C.: From wireless sensor networks towards cyber physical systems. Elsevier Pervasive and Mobile Computing 7(4), 397–413 (2011)
5. Suo, H., Wan, J., Huang, L., Zou, C.: Issues and Challenges of Wireless Sensor Networks Localization in Emerging Applications. In: International Conference on Computer Science and Electronics Engineering, vol. 3, pp. 447–451 (March 2012)
6. Jue, Y., Chengyang, Z., Xinrong, L., Yan, H., Shengli, F., Miguel, F.A.: Integration of wireless sensor networks in environmental monitoring cyber infrastructure. ACM Wireless Networks 16(4), 1091–1108 (2010)
7. Feng, X., Alexey, V., Ruixia, G., Linqiang, W., Tie, Q.: Evaluating IEEE 802.15.4 for Cyber-Physical Systems. EURASIP Journal on Wireless Communications and Networking 2011
8. Sangeeta, B., Abusayeed, S., Chenyang, L., Gruia-Catalin, R.: Multi-Application Deployment in Shared Sensor Networks Based on Quality of Monitoring. In: IEEE Real-Time and Embedded Technology and Applications Symposium, pp. 259–268 (2010)
9. Xia, F., Kong, X., Xu, Z.: Cyber-Physical Control Over Wireless Sensor and Actuator Networks with Packet Loss. In: Wireless Networking Based Control, pp. 85–102 (2011)
10. Ilic, M.D., Xie, L., Khan, U.A., Moura, J.M.: Modeling of Future Cyber–Physical Energy Systems for Distributed Sensing and Control. IEEE Transactions on Systems, Man and Cybernetics 40(4), 825 (2010)
11. Misra, P.K., Mottola, L., Raza, S., Duquennoy, S., Tsiftes, N., Hoglund, J., Voigt, T.: Supporting cyber-physical systems with wireless sensor networks: an outlook of software and services. Journal of the Indian Institute of Science 93(9) (2013)
12. Xia, F., Ma, L., Dong, J., Sun, Y.: Network QoS Management in Cyber-Physical Systems. In: International Conference on Embedded Software and Systems Symposia, vol. 307, p. 302 (July 2008)
13. Ganti, R.K., Tsai, Y.E., Abdelzaher, T.F.: SenseWorld: Towards Cyber Physical Social Network. In: IEEE International Conference on Information Processing in Sensor Networks, pp. 563–564 (2008)
14. Mottola, L., Picco, G.P.: Programming wireless sensor networks: fundamental concepts and state of the art. ACM Computing Surveys 43(3) (2011)
15. Elson, J., Estrin, D.: Sensor Networks: A bridge to the Physical World. In: Wireless Sensor Networks, pp. 3–20 (2004)
16. Li, M., Zhao, W.: Visiting Power Laws in Cyber-Physical Networking Systems. Mathematical Problems in Engineering 2012, article ID 302786, 13 pages (2012)

17. Sztipanovits, J., Koutsoukos, X., Karsai, G., Kottenstette, N., Antsaklis, P., Gupta, V., Goodwine, B., Baras, J., Wang, S.E.: Toward a Science of Cyber–Physical System Integration. Proceedings of the IEEE 100(1), 29–44 (2012)
18. Craciunas, S.S., Haas, A., Kirsch, C.M., Payer, H., Rock, H., Rottmann, A., Sokolova, A., Trummer, R., Love, J., Sengupta, R.: Information-acquisition-as-a-service for cyber-physical cloud computing. In: Proceedings of the 2nd USENIX Conference on Hot topics in Cloud Computing, no.14 (2010)
19. Kim, J.E., Mosse, D.: Generic framework for design, modeling and simulation of cyber physical systems. ACM Special Interest Group on Embedded Devices 5, Article 1 (January 2008)
20. Priyantha, N.B., Kansal, A., Goraczko, M., Zhao, F.: Tiny web services: design and implementation of interoperable and evolvable sensor networks. In: Proceedings of the 6th ACM Conference on Embedded Network Sensor Systems, pp. 253–266 (2008)
21. Kim, K.D., Kumar, P.R.: Cyber–Physical Systems: A Perspective at the Centennial. Proceedings of the IEEE 100(Special Centennial Issue), 1287–1308 (2012)
22. Stehr, M.-O., Kim, M., Talcott, C.: Toward Distributed Declarative Control of Networked Cyber-Physical Systems. In: Yu, Z., Liscano, R., Chen, G., Zhang, D., Zhou, X. (eds.) UIC 2010. LNCS, vol. 6406, pp. 397–413. Springer, Heidelberg (2010)
23. Xia, F., Tian, Y.C., Li, Y., Sung, Y.: Wireless Sensor/Actuator Network Design for Mobile Control Applications. IEEE Sensors 7(10), 2157–2173 (2007)
24. Kang, K.D., Son, S.H.: Real-Time Data Services for Cyber Physical Systems. In: 28th International Conference on Distributed Computing Systems Workshops, pp. 483–488 (June 2008)
25. Nakamura, E.F., Loureiro, A.F.A., Frery, A.C.: Information fusion for wireless sensor networks: Methods, models, and classifications. ACM Computer Survey 39(3), article 9 (2007)
26. Conti, M., Das, S.K., Bisdikian, C., Kumar, M., Ni, L.M., Passarella, A., Roussos, G., Troster, G., Tsudik, G., Zambonelli, F.: Looking ahead in pervasive computing: Challenges and opportunities in the era of cyber–physical convergence. Pervasive and Mobile Computing 8(1), 2–21 (2012)
27. Vuran, M.C., Akyildiz, I.F.: XLP: A Cross-Layer Protocol for Efficient Communication in Wireless Sensor Networks. IEEE Transactions on Mobile Computing 9(11), 1578–1591 (2010)
28. Bhattacharya, S., Saifullah, A., Chenyang, L., Roman, G.: Multi-Application Deployment in Shared Sensor Networks Based on Quality of Monitoring. In: 16th IEEE Real-Time and Embedded Technology and Applications Symposium, pp. 259–268 (April 2010)
29. Ilic, M.D., Xie, L., Khan, U.A., Moura, J.M.: Modeling future cyber-physical energy systems. In: IEEE Power and Energy Conference, pp. 1–9 (July 2008)

Author Index

Printed in the United States
By Bookmasters